T0350288

CAMBRIDGE STUDIES IN ADVANCED MATHEMATICS 168

PERIOD MAPPINGS AND PERIOD DOMAINS

This up-to-date introduction to Griffiths' theory of period maps and period domains focusses on algebraic, group-theoretic and differential geometric aspects. Starting with an explanation of Griffiths' basic theory, the authors go on to introduce spectral sequences and Koszul complexes that are used to derive results about cycles on higher-dimensional algebraic varieties such as the Noether–Lefschetz theorem and Nori's theorem. They explain differential geometric methods, leading up to proofs of Arakelov-type theorems, the theorem of the fixed part and the rigidity theorem. They also use Higgs bundles and harmonic maps to prove the striking result that not all compact quotients of period domains are Kähler.

This thoroughly revised second edition includes a new third part covering important recent developments, in which the group-theoretic approach to Hodge structures is explained, leading to Mumford–Tate groups and their associated domains, the Mumford–Tate varieties and generalizations of Shimura varieties. This viewpoint also leads to a factorization of the period map which has an arithmetic flavor. Higgs bundles reappear in connection with Shimura varieties.

James Carlson is Professor Emeritus at the University of Utah. From 2003 to 2012, he was president of the Clay Mathematics Institute. Most of Carlson's research is in the area of Hodge Theory.

Stefan Müller-Stach is Professor for number theory at Johannes Gutenberg Universität Mainz, Germany. He works in arithmetic and algebraic geometry, focussing on algebraic cycles and Hodge theory, and his recent research interests include period integrals and the history and foundations of mathematics. Recently, he has published monographs on number theory (with J. Piontkowski) and period numbers (with A. Huber), as well as an edition of some works of Richard Dedekind.

Chris Peters is a retired professor from the Université Grenoble Alpes, France and has a research position at the Technical University of Eindhoven. He is widely known for the monographs *Compact Complex Surfaces* (with W. Barth, K. Hulek and A. van de Ven), as well as *Mixed Hodge Structures*, (with J. Steenbrink). He has also written shorter treatises on the motivic aspects of Hodge theory, on motives (with J.P. Murre and J. Nagel), and on applications of Hodge theory in mirror symmetry (with Bertin).

CAMBRIDGE STUDIES IN ADVANCED MATHEMATICS

All the titles listed below can be obtained from good booksellers or from Cambridge University Press. For a complete series listing visit: www.cambridge.org/mathematics.

Already published

129 D. Goldfeld & J. Hundley *Automorphic representations and L-functions for the general linear group, I*
130 D. Goldfeld & J. Hundley *Automorphic representations and L-functions for the general linear group, II*
131 D. A. Craven *The theory of fusion systems*
132 J. Väänänen *Models and games*
133 G. Malle & D. Testerman *Linear algebraic groups and finite groups of Lie type*
134 P. Li *Geometric analysis*
135 F. Maggi *Sets of finite perimeter and geometric variational problems*
136 M. Brodmann & R. Y. Sharp *Local cohomology (2nd Edition)*
137 C. Muscalu & W. Schlag *Classical and multilinear harmonic analysis, I*
138 C. Muscalu & W. Schlag *Classical and multilinear harmonic analysis, II*
139 B. Helffer *Spectral theory and its applications*
140 R. Pemantle & M. C. Wilson *Analytic combinatorics in several variables*
141 B. Branner & N. Fagella *Quasiconformal surgery in holomorphic dynamics*
142 R. M. Dudley *Uniform central limit theorems (2nd Edition)*
143 T. Leinster *Basic category theory*
144 I. Arzhantsev, U. Derenthal, J. Hausen & A. Laface *Cox rings*
145 M. Viana *Lectures on Lyapunov exponents*
146 J.-H. Evertse & K. Győry *Unit equations in Diophantine number theory*
147 A. Prasad *Representation theory*
148 S. R. Garcia, J. Mashreghi & W. T. Ross *Introduction to model spaces and their operators*
149 C. Godsil & K. Meagher *Erdős–Ko–Rado theorems: Algebraic approaches*
150 P. Mattila *Fourier analysis and Hausdorff dimension*
151 M. Viana & K. Oliveira *Foundations of ergodic theory*
152 V. I. Paulsen & M. Raghupathi *An introduction to the theory of reproducing kernel Hilbert spaces*
153 R. Beals & R. Wong *Special functions and orthogonal polynomials*
154 V. Jurdjevic *Optimal control and geometry: Integrable systems*
155 G. Pisier *Martingales in Banach spaces*
156 C. T. C. Wall *Differential topology*
157 J. C. Robinson, J. L. Rodrigo & W. Sadowski *The three-dimensional Navier–Stokes equations*
158 D. Huybrechts *Lectures on K3 surfaces*
159 H. Matsumoto & S. Taniguchi *Stochastic analysis*
160 A. Borodin & G. Olshanski *Representations of the infinite symmetric group*
161 P. Webb *Finite group representations for the pure mathematician*
162 C. J. Bishop & Y. Peres *Fractals in probability and analysis*
163 A. Bovier *Gaussian processes on trees*
164 P. Schneider *Galois representations and* (φ, Γ)*-modules*
165 P. Gille & T. Szamuely *Central simple algebras and Galois cohomology (2nd Edition)*
166 D. Li & H. Queffelec *Introduction to Banach spaces, I*
167 D. Li & H. Queffelec *Introduction to Banach spaces, II*
168 J. Carlson, S. Müller-Stach & C. Peters *Period mappings and period domains (2nd Edition)*

Period Mappings and Period Domains

Second Edition

JAMES CARLSON
University of Utah

STEFAN MÜLLER-STACH
Johannes Gutenberg Universität
Mainz, Germany

CHRIS PETERS
Université Grenoble Alpes, France

CAMBRIDGE
UNIVERSITY PRESS

University Printing House, Cambridge CB2 8BS, United Kingdom

One Liberty Plaza, 20th Floor, New York, NY 10006, USA

477 Williamstown Road, Port Melbourne, VIC 3207, Australia

4843/24, 2nd Floor, Ansari Road, Daryaganj, Delhi – 110002, India

79 Anson Road, #06–04/06, Singapore 079906

Cambridge University Press is part of the University of Cambridge.

It furthers the University's mission by disseminating knowledge in the pursuit of education, learning, and research at the highest international levels of excellence.

www.cambridge.org
Information on this title: www.cambridge.org/9781108422628
DOI: 10.1017/9781316995846

First published 2003
Second edition 2017

A catalogue record for this publication is available from the British Library.

ISBN 978-1-108-42262-8 Hardback
ISBN 978-1-316-63956-6 Paperback

Contents

Preface to the Second Edition

In the fourteen years since the first edition appeared, ample experience with teaching to graduate students made us realize that a proper understanding of several of the core aspects of period domains needed a lot more explanation than offered in the first edition of this book, especially with regards to the Lie group aspects of period domains.

Consequently, we decided a thorough reworking of the book was called for. In particular Section 4.3, and Chapters 12 and 13 needed revision. The latter two chapters have been rearranged and now contain more, often completely rewritten sections. At the same time relevant newer developments have been inserted at appropriate places. Finally we added a new "Part Four" with additional, more recent topics. This also required an extra Appendix D about Lie groups and algebraic groups.

Let us be more specific about the added material. There is a new Section 5.4 on counterexamples to infinitesimal Torelli. In Chapter 6 the abstract and powerful formalism of derived functors has been added so that for instance the algebraic treatment of the Gauss–Manin connection could be given, as well as a proper treatment of the Leray spectral sequence. In Chapter 13 we have devoted more detail on Higgs bundles and their logarithmic variant. This made it possible to also include some geometric applications at the end of that chapter.

"Part Four" starts with a chapter explaining the by now standard group theoretic formulation of the concept of a Hodge structure. This naturally leads to Mumford–Tate groups and their associated domains. The chapter culminates with a result giving a factorization of the period map which stresses the role of the Mumford–Tate group of a given variation. In Chapter 16 Mumford–Tate domains and their quotients by certain discrete groups, the Mumford–Tate varieties, are considered from a more abstract, axiomatic point of view. In this chapter the relation with the classical Shimura varieties is also explained. In the next and final chapter we study various interesting subvarieties of Mumford–Tate varieties, especially of low dimension.

One word about the prerequisites. Of course, they remain the same (see page xi), but we should mention a couple of more recent books that may serve as a guide to the reader. There are now many introductory books to algebraic geometry and this is not the place to mention all of these. However, Donaldson's book (2011) starts as we do, from Riemann surfaces, and focusses on Hodge theory. So it serves as a particularly adequate introduction; moreover, its scope is broad and leads to some fascinating recent mathematics. Secondly, in the First Edition, we unfortunately failed to mention explicitly Chern's wonderful introduction to complex manifolds (see Chern, 1967), as well as Hartshorne (1987) although both figured in the bibliography. These are not needed to understand the text, but serve to complement it, Chern's book from the differential geometric side; Hartshorne's book from the algebraic side.

We acknowledge support from the University of Mainz, the French CNRS, the Technical University of Eindhoven, as well as the Deutsche Forschungsgemeinschaft (SFB, Transregio 45). Finally we thank Ana Brecan, Ariyan Javanpeykar, Daniel Huybrechts, Ben Moonen, Jan Nagel and Kang Zuo for their remarks on a preliminary version of this second edition.

Preface to the First Edition

What to expect of this book?

Our aim is to give an up to date exposition of the theory of period maps originally introduced by Griffiths. It is mainly intended as a text book for graduate students. However, it should also be of interest to any mathematician wishing to get introduced to those aspects of Hodge theory which are related to Griffiths' theory.

Prerequisites

We assume that the reader has encountered complex or complex algebraic manifolds before. We have in mind familiarity with the concepts from the first chapters of the book by Griffiths and Harris (1978) or from the first half of the book by Forster (1981).

A second prerequisite is some familiarity with algebraic topology. For the fundamental group the reader may consult Forster's book (loc. cit.). Homology and cohomology are at the base of Hodge theory and so the reader should know either simplicial or singular homology and cohomology. A good source for the latter is Greenberg (1967).

Next, some familiarity with basic concepts and ideas from differential geometry such as smooth manifolds, differential forms, connections and characteristic classes is required. Apart from the book by Griffiths and Harris (1978) the reader is invited to consult the monograph by Guillemin and Pollack (1974). To have an idea of what we actually use in the book, we refer to the appendices. We occasionally refer to these in the main body of the book. We particularly recommend the exercises which are meant to provide the techniques necessary to calculate all sorts of invariants for concrete examples in the main text.

Contents of the book

The concept of a period-integral goes back to the nineteenth century; it was introduced by Legendre and Weierstraß for integrals of certain elliptic functions over closed circuits in the dissected complex plane and of course is related to periodic functions like the Weierstraß $\wp$-function. In modern terminology we would say that these integrals describe exactly how the complex structure of an elliptic curve varies. From this point of view the analogous question for higher genus curves becomes apparent and leads to period matrices and Torelli's theorem for curves. We have treated this historical starting point in the first chapter.

Since we introduce the major concepts of the book by means of examples, this chapter can be viewed as a motivation for the rest of the book. Indeed, period mappings and period domains appear in it, as well as several other important notions and ideas such as monodromy of a family, algebraic cycles, the Hodge decomposition and the Hodge conjecture. This chapter is rather long since we also wanted to address several important aspects of the theory that we do not treat in later chapters but nevertheless motivate parts in it. Below we say more about this, but we pause here to point out that the nature of the first chapter makes it possible to use it entirely for a first course on period maps.

For instance, we already introduce mixed Hodge theory in this chapter and explain the geometry behind it, but of course only in the simplest situations. We look at the cohomology of a singular curve on the one hand, and on the other hand we consider the limit mixed Hodge structure on the cohomology for a degenerating family of curves. This second example leads to the *asymptotic* study and becomes technically complicated in higher dimensions and falls beyond the modest scope of our book. Nevertheless it motivates certain results in the rest of the book such as those concerning variations of Hodge structure over the punctured disk (especially the monodromy theorem) which are considered in detail in Chapter 13.

The beautiful topic of Picard–Fuchs equations, treated in relation to a family of elliptic curves, does not come back in later chapters. We certainly could have done this, for instance after our discussion of the periods for families of hypersurfaces in projective space (Section 3.2). Lack of time and space prevented us from doing this. We refer the interested reader to Bertin and Peters (2002) where some calculations are carried out which are significant for important examples occurring in mirror-symmetry and which can be understood after reading the material in the first part.

The remainder of the first part of the book is devoted to fleshing out the ideas presented in this first chapter. Cohomology being essentially the only available invariant, we explain in Chapter 2 how the Kähler assumption implies that one can pass from the type decomposition on the level of complex forms to the level of cohomology classes. This is the Hodge decomposition. We show how to compute the Hodge decomposition in a host of basic examples. In the next chapter we pave the way for the introduction of the period map by looking at invariants related to cohomology that behave holomorphically (although this is shown much later, in Chapter 6, when we have developed the necessary tools). Griffiths' intermediate Jacobians and the Hodge (p, p)-classes are central in this chapter; we also calculate the Hodge decomposition of the cohomology of projective hypersurfaces in purely algebraic terms. This will enable us on various occasions to use these as examples to illustrate the theory. For instance, infinitesimal Torelli is proved for them in Chapter 5, Noether–Lefschetz type theorems in Chapter 7 and variational Torelli theorems in Chapter 8.

In Chapter 4 the central concepts of this book finally can be defined after we have illustrated the role of the monodromy in the case of Lefschetz pencils. Abstract variations of Hodge structure then are introduced. In a subsequent chapter these are studied from an infinitesimal point of view.

In Part Two spectral sequences are treated and with these, previous loose ends can be tied up. Another central tool, to be developed in Chapter 7 is the theory of Koszul complexes. Through Donagi's symmetrizer lemma and its variants these turn out to be crucial for applications such as Noether–Lefschetz theorems and variational Torelli, treated in Chapter 7 and Chapter 8, respectively.

Then in Chapter 9 we turn to another important ingredient in the study of algebraic cycles, the normal functions. Their infinitesimal study leads to a proof of a by now classical theorem due to Voisin and Green stating that the image of the Abel-Jacobi map for "very general" odd-dimensional hypersurfaces of projective space is as small as it can be, at least if the degree is large enough.

We finish this part with a sophisticated chapter on Nori's theorem which has profound consequences for algebraic cycles, vastly generalizing pioneering results by Griffiths and Clemens.

In Part Three of the book we turn to purely differential geometric aspects of period domains. Our main goal here is to explain in Chapter 13 those curvature properties which are relevant for period maps. Previous to that chapter, in Chapters 11 and 12 we present several more or less well known notions and techniques from differential geometry, which go into the Lie theory needed for period domains.

Among the various important applications of these basic curvature properties we have chosen to prove in Chapter 13 the theorem of the fixed part, the rigidity theorem and the monodromy theorem. We also show that the period map extends as a proper map over the locus where the local monodromy is finite and give some important consequences. In the same chapter we introduce Higgs bundles and briefly explain how these come up in Simpson's work on nonabelian Hodge theory.

In Chapter 14 we broaden our point of view in that we look more generally at harmonic and pluriharmonic maps with target a locally symmetric space. Using the results of this study, we can, for instance, show that compact quotients of period domains of even weight are never homotopy equivalent to Kähler manifolds.

To facilitate reading, we start every chapter with a brief outline of its content. To encourage the reader to digest the considerable amount of concepts and techniques we have included many examples and problems. For the more difficult problems we have given hints or references to the literature. Finally, we end every chapter with some historical remarks.

It is our pleasure to thank various people and institutions for their help in the writing of this book.

We are first of all greatly indebted to Phillip Griffiths who inspired us either directly or indirectly over all the years we have been active as mathematicians; through this book we hope to promote some of the exciting ideas and results related to cycles initiated by him and pursued by others, like Herb Clemens, Mark Green, Madhav Nori and Claire Voisin.

Special thanks go to Domingo Toledo for tremendous assistance with the last part of the book and to Jan Nagel who let us present part of his work in Chapter 10. Moreover, he and several others critically read firsts drafts of this book: Daniel Huybrechts, James Lewis, Jacob Murre, Jens Piontkowski and Eckart Viehweg; we extend our gratitude to all of them.

We furthermore acknowledge the support we have received (on various occasions) from the University of Grenoble, the University of Utah at Salt-Lake City and the University of Essen.

Finally we want to thank our respective wives and companions Nicole, Siggi and Annie for their patience and endurance over all the years that it took us to prepare this book.

Part ONE

BASIC THEORY

1

Introductory Examples

The basic idea of Hodge theory is that the cohomology of an algebraic variety has more structure than one sees when viewing the same object as a "bare" topological space. This extra structure helps us understand the geometry of the underlying variety, and it is also an interesting object of study in its own right. Because of the technical complexity of the subject, in this chapter, we look at some motivating examples which illuminate and guide our study of the complete theory. We shall be able to understand, in terms of specific and historically important examples, the notions of Hodge structure, period map, and period domain. We begin with elliptic curves, which are the simplest interesting Riemann surfaces.

1.1 Elliptic Curves

The simplest algebraic variety is the Riemann sphere, the complex projective space $\mathbf{P}^1$. The next simplest examples are the branched double covers of the Riemann sphere, given in affine coordinates by the equation

$$y^2 = p(x),$$

where $p(x)$ is a polynomial of degree d. If the roots of p are distinct, which we assume they are for now, the double cover C is a one-dimensional complex manifold, or a Riemann surface. As a differentiable manifold it is characterized by its genus. To compute the genus, consider two cases. If d is even, all the branch points are in the complex plane, and if d is odd, there is one branch point at infinity. Thus the genus g of such a branched cover C is $d/2$ when d is even and $(d-1)/2$ when d is odd. These facts follow from Hurwitz's formula, which in turn follows from a computation of Euler characteristics (see Problem 1.1.2). Riemann surfaces of genus 0, 1, and 2 are illustrated in Fig. 1.1. Note that if $d = 1$ or $d = 2$, then C is topologically a sphere. It is not hard to prove that it is also isomorphic to the Riemann sphere as a complex manifold.

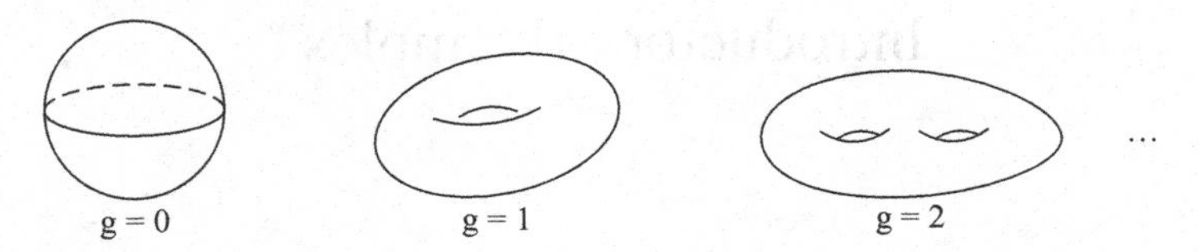

Figure 1.1 Riemann surfaces.

Now consider the case $d = 3$, so that the genus of C is 1. By a suitable change of variables, we may assume the three roots of $p(x)$ to be 0, 1, and λ, where $\lambda \neq 0, 1$:

$$y^2 = x(x-1)(x-\lambda). \tag{1.1}$$

We shall denote the Riemann surface defined by (1.1) by $\mathscr{E}_\lambda$, and we call the resulting family the *Legendre family*. As topological spaces, and even as differentiable manifolds, the various $\mathscr{E}_\lambda$ are all isomorphic, as long as $\lambda \neq 0, 1$, a condition which we assume to be now in force. However, we shall prove the following.

Theorem 1.1.1 *Suppose that $\lambda \neq 0, 1$. Then there is an $\epsilon > 0$ such that for all λ' within distance ϵ from λ, the Riemann surfaces $\mathscr{E}_\lambda$ and $\mathscr{E}_{\lambda'}$ are not isomorphic as complex manifolds.*

Our proof of this result, which guarantees an infinite supply of essentially distinct elliptic curves, will lead us directly to the notions of period map and period domain and to the main ideas of Hodge theory.

The first order of business is to recall some basic notions of Riemann surface theory so as to have a detailed understanding of the topology of $\mathscr{E}_\lambda$, which for now we write simply as $\mathscr{E}$. Consider the multiple-valued holomorphic function

$$y = \sqrt{x(x-1)(x-\lambda)}.$$

On any simply connected open set which does *not* contain the branch points $x = 0, 1, \lambda, \infty$, it has two single-valued determinations. Therefore, we cut the Riemann sphere from 0 to 1 and from λ to infinity, as in Fig. 1.2. Then analytic continuation of y in the complement of the cuts defines a single-valued function. We call its graph a "sheet" of the Riemann surface. Note that analytic continuation of y around δ returns y to its original determination, so δ lies in a single sheet of $\mathscr{E}$. We can view it as lying in the Riemann sphere itself. But when we analytically continue along γ, we pass from one sheet to the other as we pass the branch cut. That path is therefore made of two pieces, one in one sheet and one in the other sheet.

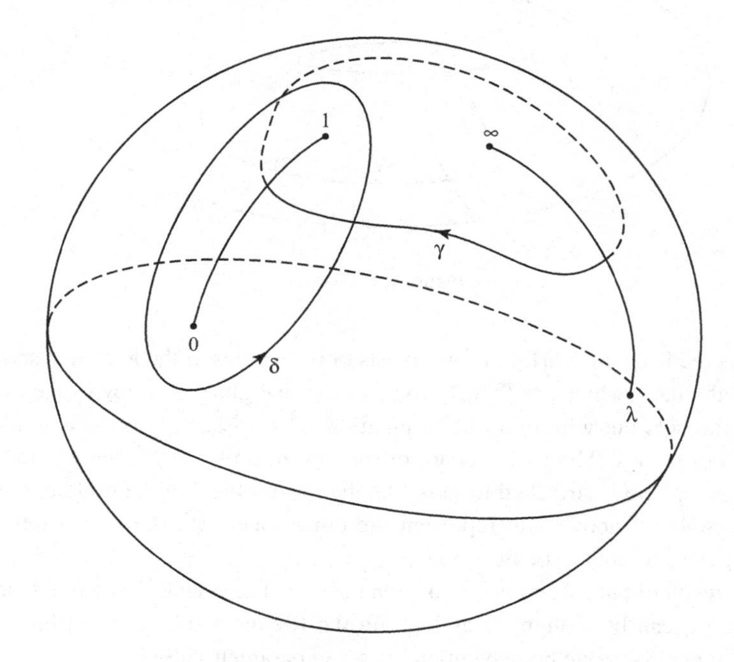

Figure 1.2 Cuts in the Riemann sphere.

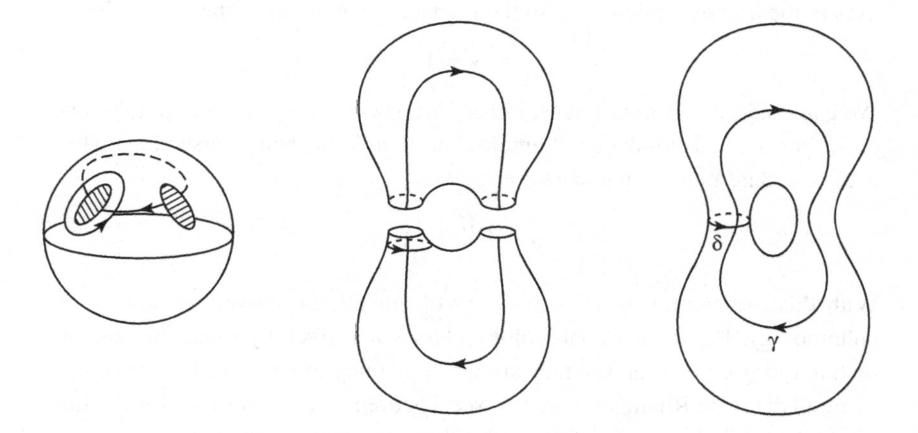

Figure 1.3 Assembling a Riemann surface.

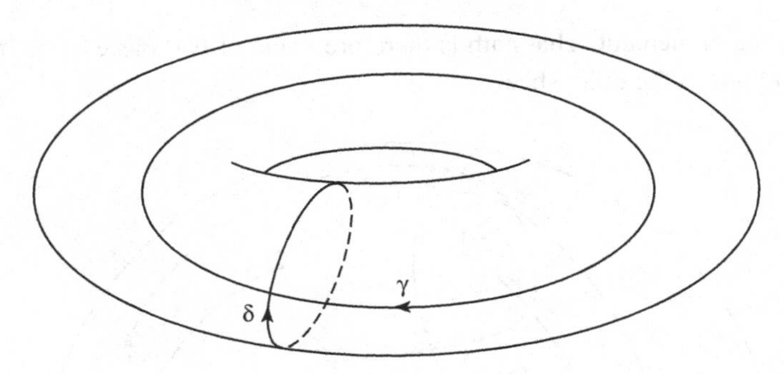

Figure 1.4 Torus.

Thus the Riemann surface of y consists of two copies of the Riemann sphere minus the cuts, which are then "cross-pasted": we glue one copy to the other along the cuts but with opposite orientations. This assembly process is illustrated in Fig. 1.3. The two cuts are opened up into two ovals, the opened-up Riemann sphere is stretched to look like the lower object in the middle, a second copy is set above it to represent the other sheet, and the two sheets are cross-pasted to obtain the final object.

The result of our assembly is shown in Fig. 1.4. The oriented path δ indicated in Fig. 1.4 can be thought of as lying in the Riemann sphere, as in Fig. 1.2, where it encircles one branch cut and is given parametrically by

$$\delta(\theta) = 1/2 + (1/2 + k)e^{i\theta}$$

for some small k. The two cycles δ and γ are oriented oppositely to the x and y axes in the complex plane, and so the intersection number of the two cycles is

$$\delta \cdot \gamma = 1.$$

We can read this information off either Fig. 1.3 or Fig. 1.2. Note that the two cycles form a basis for the first homology of $\mathscr{E}$ and that their intersection matrix is the standard unimodular skew form,

$$J = \begin{pmatrix} 0 & 1 \\ -1 & 0 \end{pmatrix}.$$

With this explanation of the homology of our elliptic curve, we turn to the cohomology. Recall that cohomology classes are given by linear functionals on homology classes, and so they are given by integration against a differential form. (This is de Rham's theorem – see Theorem 2.1.1). In order for the line integral to be independent of the path chosen to represent the homology class, the form must be closed. For the elliptic curve $\mathscr{E}$ there is a naturally given differential one-form that plays a central role in the story we are recounting. It is defined by

$$\omega = \frac{dx}{y} = \frac{dx}{\sqrt{x(x-1)(x-\lambda)}}. \tag{1.2}$$

As discussed in Problem 1.1.1, this form is holomorphic, that is, it can be written locally as

$$\omega = f(z)\, dz,$$

where z is a local coordinate and $f(z)$ is a holomorphic function. In fact, away from the branch points, x is a local coordinate, so this representation follows from the fact that $y(x)$ has single-valued holomorphic determinations. Because f is holomorphic, ω is closed (see Problem 1.1.7). Thus it has a well-defined cohomology class.

Now let δ^* and γ^* denote the basis for $H^1(\mathscr{E}; \mathbf{Z})$ which is dual to the given basis of $H_1(\mathscr{E}; \mathbf{Z})$. The cohomology class of ω can be written in terms of this basis as

$$[\omega] = \delta^* \int_\delta \omega + \gamma^* \int_\gamma \omega.$$

In other words, the coordinates of $[\omega]$ with respect to this basis are given by the indicated integrals. These are called the *periods* of ω. In the case at hand, they are sometimes denoted A and B, so that

$$[\omega] = A\delta^* + B\gamma^*. \tag{1.3}$$

The expression (A, B) is called the *period vector* of $\mathscr{E}$.

From the periods of ω we are going to construct an invariant that can detect changes in the complex structure of $\mathscr{E}$. In the best of all possible worlds this invariant would have different values for elliptic curves that have different complex structures. The first step toward constructing it is to prove the following.

Theorem 1.1.2 *Let $H^{1,0}$ be the subspace of $H^1(\mathscr{E}; \mathbf{C})$ spanned by ω, and let $H^{0,1}$ be the complex conjugate of this subspace. Then*

$$\boxed{H^1(\mathscr{E}; \mathbf{C}) = H^{1,0} \oplus H^{0,1}.}$$

The decomposition asserted by this theorem is the *Hodge decomposition* and it is fundamental to all that follows. Now there is no difficulty in defining the $(1,0)$ and $(0,1)$ subspaces of cohomology: indeed, we have alreaddy done this. The difficulty is in showing that the defined subspaces span the cohomology, and that (equivalently) their intersection is zero. In the case of elliptic curves, however, there is a quite elementary proof of this fact. Take the cup product of (1.3) with its conjugate to obtain

$$[\omega] \cup [\bar{\omega}] = (A\bar{B} - B\bar{A})\,\delta^* \cup \gamma^*.$$

Multiply the previous relation by $\mathrm{i} = \sqrt{-1}$ and use the fact that $\delta^* \cup \gamma^*$ is the fundamental class of $\mathscr{E}$ to rewrite the preceding equation as

$$\mathrm{i} \int_{\mathscr{E}} \omega \wedge \bar{\omega} = 2\,\mathrm{Im}(B\bar{A}).$$

Now consider the integral above. Because the form ω is given locally by $f\,\mathrm{d}z$, the integrand is locally given by

$$\mathrm{i}|f|^2\,\mathrm{d}z \wedge \mathrm{d}\bar{z} = 2|f|^2\,\mathrm{d}x \wedge \mathrm{d}y,$$

where $\mathrm{d}x \wedge \mathrm{d}y$ is the natural orientation defined by the holomorphic coordinate, that is, by the complex structure. Thus the integrand is locally a positive function times the volume element, and so the integral is positive. We conclude that

$$\mathrm{Im}(B\bar{A}) > 0.$$

We also conclude that neither A nor B can be 0 and, therefore, that the cohomology class of ω cannot be 0. Consequently the subspace $H^{1,0}(\mathscr{E})$ is nonzero.

Because neither A nor B can be 0 we can rescale ω and assume that $A = 1$. For such "normalized" differentials, we conclude that the imaginary part of the normalized B-period is positive:

$$\mathrm{Im}\,B > 0. \tag{1.4}$$

Now suppose that $H^{1,0}$ and $H^{0,1}$ do not give a direct sum decomposition of $H^1(\mathscr{E}; \mathbf{C})$. Then $H^{1,0} = H^{0,1}$, and so $[\bar{\omega}] = \lambda[\omega]$ for some complex number λ. Therefore

$$\delta^* + \bar{B}\gamma^* = \lambda(\delta^* + B\gamma^*).$$

Comparing coefficients, we find that $\lambda = 1$ and then that $B = \bar{B}$, in contradiction with the fact that B has a positive imaginary part. This completes the proof of the Hodge theorem for elliptic curves, Theorem 1.1.2.

An Invariant of Framed Elliptic Curves

Now suppose that $f : \mathscr{E}_\mu \longrightarrow \mathscr{E}_\lambda$ is an isomorphism of complex manifolds. Let ω_μ and ω_λ be the given holomorphic forms. Then we claim that

$$f^*\omega_\lambda = c\,\omega_\mu \tag{1.5}$$

for some nonzero complex number c. This equation is certainly true on the level of cohomology classes, although we do not yet know that c is nonzero. However, on the one hand,

$$\int_{[\mathscr{E}_\mu]} f^*\omega_\lambda \wedge f^*\bar{\omega}_\lambda = |c|^2 \int_{[\mathscr{E}_\mu]} \omega_\mu \wedge \bar{\omega}_\mu,$$

and on the other,

$$\int_{[\mathscr{E}_\mu]} f^*\omega_\lambda \wedge f^*\bar{\omega}_\lambda = \int_{f_*[\mathscr{E}_\mu]} \omega_\lambda \wedge \bar{\omega}_\lambda = \int_{[\mathscr{E}_\lambda]} \omega_\lambda \wedge \bar{\omega}_\lambda.$$

The last equality uses the fact that an isomorphism of complex manifolds is a degree-one map. Because $\mathrm{i}\omega_\lambda \wedge \bar{\omega}_\lambda$ is a positive multiple of the volume form, the integral is positive and therefore

$$c \neq 0. \tag{1.6}$$

We can now give a preliminary version of the invariant alluded to above. It is the ratio of periods B/A, which we write more formally as

$$\tau(\mathscr{E}, \delta, \gamma) = \frac{\int_\gamma \omega}{\int_\delta \omega}.$$

From Eq. (1.4) we know that τ has a positive imaginary part. From the just-proved proportionality results (1.5) and (1.6), we conclude the following.

Theorem 1.1.3 *If $f : \mathscr{E} \longrightarrow \mathscr{E}'$ is an isomorphism of complex manifolds, then $\tau(\mathscr{E}, \delta, \gamma) = \tau(\mathscr{E}', \delta', \gamma')$, where $\delta' = f_*\delta$ and $\gamma' = f_*\gamma$.*

To interpret this result, let us define a *framed elliptic curve* $(\mathscr{E}, \delta, \gamma)$ to consist of an elliptic curve and an integral basis for the first homology such that $\delta \cdot \gamma = 1$. Then we can say that "if framed elliptic curves are isomorphic, then their τ-invariants are the same."

Example 1.1.4 In the Legendre family, consider the fiber for $\lambda = -1$, the elliptic curve E. From its equation,

$$y^2 = x^3 - x,$$

we see that the map $(x, y) \mapsto (-x, \mathrm{i}y)$ is an automorphism of this curve of order 4. By Chowla and Selberg (1949), one can explicitly calculate its periods:

$$\int_\gamma \omega = \frac{\Gamma(1/4)^2}{\sqrt{2\pi}}, \quad \int_\delta \omega = -\mathrm{i} \int_\gamma \omega,$$

so that $\tau = \mathrm{i}$. In other words, $E = \mathbf{C}/\mathbf{Z} \oplus \mathbf{Z}\mathrm{i}$ and we see that the lattice defining E admits multiplication by i, an extra isomorphism of order 4. In

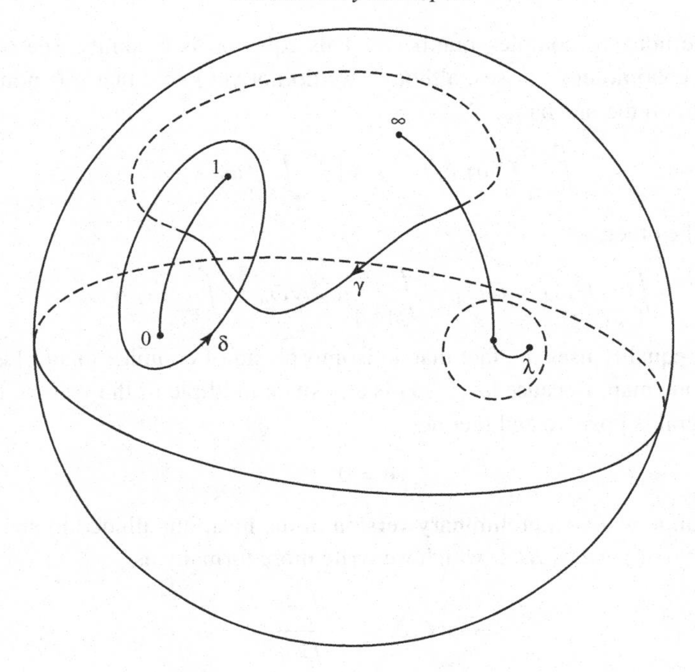

Figure 1.5 Modified cuts in the Riemann sphere.

fact, the lattice is stable under multiplication by the Gaussian integers $\mathbf{Z}[i] = \{m + in \mid n, m \in \mathbf{Z}\}$. We say that E admits *complex multiplication by* $\mathbf{Z}[i]$.

Holomorphicity of the Period Mapping

Consider once again the Legendre family (1.1) and choose a complex number $a \neq 0, 1$ and an $\epsilon > 0$ which is smaller than both the distance from a to 0 and the distance from a to 1. Then the Legendre family, restricted to λ in the disk of radius ϵ centered at a, is trivial as a family of differentiable manifolds. This means that it is possible to choose two families of integral homology cycles δ_λ and γ_λ on $\mathscr{E}_\lambda$ such that $\delta_\lambda \cdot \gamma_\lambda = 1$. We can "see" these cycles by modifying Fig. 1.2 as indicated in Fig. 1.5. A close look at Fig. 1.5 shows that we can move λ within a small disk Δ without changing either δ_λ or γ_λ. Thus we can view the integrals defining the periods A and B as having constant domains of integration but variable integrands.

Let us study these periods more closely, writing them as

$$A(\lambda) = \int_\delta \frac{dx}{\sqrt{x(x-1)(x-\lambda)}}, \quad B(\lambda) = \int_\gamma \frac{dx}{\sqrt{x(x-1)(x-\lambda)}}.$$

We have suppressed the subscript on the homology cycles in view of the remarks made at the end of the previous paragraph. The first observation is the following.

Proposition 1.1.5 *On any disk Δ in the complement of the set $\{0, 1, \infty\}$, the periods of the Legendre family are single-valued holomorphic functions of λ.*

The proof is straightforward. Since the domain of integration is constant, we can compute $\partial A/\partial\bar{\lambda}$ by differentiating under the integral sign. But the integrand is a holomorphic expression in λ, and so that derivative is 0. We conclude that the period function $A(\lambda)$ is holomorphic, and the same argument applies to $B(\lambda)$.

Notice that the definitions of the period functions A and B on a disk Δ depend on the choice of a symplectic homology basis $\{\delta, \gamma\}$. Each choice of basis gives a different determination of the periods. However, if δ' and γ' give a different basis, then

$$\delta' = a\delta + b\gamma,$$
$$\gamma' = c\delta + d\gamma,$$

where the matrix

$$T = \begin{pmatrix} a & b \\ c & d \end{pmatrix}$$

has determinant 1. The periods with respect to the new basis are related to those with respect to the old one as follows:

$$A' = aA + bB,$$
$$B' = cA + dB.$$

Thus the new period vector (A', B') is the product of the matrix T and the old period vector (A, B). The τ-invariants are related by the corresponding fractional linear transformation:

$$\tau' = \frac{d\tau + c}{b\tau + a}.$$

The ambiguity in the definition of the periods and of the τ-invariant is due to the ambiguity in the choice of a homology basis. Now consider a simply connected open set U of $\mathbf{P}^1 - \{0, 1, \infty\}$ and a point λ_0 and λ of U. The choice of homology basis for $\mathscr{E}_{\lambda_0}$ determines a choice of homology basis for all other fibers $\mathscr{E}_\lambda$.

Thus the periods $A(\lambda)$ and $B(\lambda)$ as well as the ratio $\tau(\lambda)$ are single-valued holomorphic functions on U. On the full domain $\mathbf{P}^1 - \{0, 1, \infty\}$, however, these functions are multivalued.

We can now state a weak form of Theorem 1.1.1.

Theorem 1.1.6 *The function τ is nonconstant.*

If τ, defined on a simply connected open set U, is a nonconstant holomorphic function then its derivative is not identically zero. Therefore its derivative has at most isolated zeroes. For a randomly chosen point, different from one of these zeroes, τ is a locally injective function.

There are at least two ways to prove that τ is nonconstant. One is to compute the derivative directly and to show that it is nonzero. The other is to show that τ tends to infinity as λ approaches infinity along a suitable ray in the complex plane. We give both arguments, beginning with an analysis of τ along a ray.

Asymptotics of the Period Map

Let us show that τ is a nonconstant function of λ by showing that τ approaches infinity along the ray $\lambda > 2$ of the real axis. Indeed, we will show that $\tau(\lambda)$ is asymptotically proportional to $\log \lambda$. To see this, assume $\lambda \gg 2$, and observe that

$$\int_\delta \frac{dx}{\sqrt{x(x-1)(x-\lambda)}} \sim \int_\delta \frac{dx}{x\sqrt{-\lambda}} = \frac{2\pi}{\sqrt{\lambda}}.$$

By deforming the path of integration, we find that

$$\int_\gamma \frac{dx}{\sqrt{x(x-1)(x-\lambda)}} = -2\int_1^\lambda \frac{dx}{\sqrt{x(x-1)(x-\lambda)}}.$$

The difference between the last integrand and $1/[x(\sqrt{x-\lambda})]$ is $1/[2x^2]$ + higher powers of $1/x$, an expression with asymptotically negligible integral $-\frac{1}{4} \cdot \lambda^{-1}$ + higher powers of λ^{-1}. The residual integral,

$$-2\int_1^\lambda \frac{dx}{x\sqrt{(x-\lambda)}},$$

can be computed exactly:

$$-2\int_1^\lambda \frac{dx}{x\sqrt{x-\lambda}} = \frac{4}{\sqrt{\lambda}} \arctan \frac{\sqrt{1-\lambda}}{\sqrt{\lambda}} \sim \frac{2i}{\sqrt{\lambda}} \log \lambda.$$

Thus one finds

$$\tau(\lambda) \sim \frac{i}{\pi} \log \lambda, \tag{1.7}$$

as claimed. Note also that $\tau(\lambda)$ has a positive imaginary part, as asserted in (1.4).

Derivative of the Period Map

We now prove the strong form of Theorem 1.1.1 by showing that $\tau'(\lambda) \neq 0$ for $\lambda \neq 0, 1$ for any determination of τ. To this end, we write the holomorphic differential ω_λ in terms of the dual cohomology basis $\{\delta^*, \gamma^*\}$:

$$\omega_\lambda = A(\lambda)\delta^* + B(\lambda)\gamma^*.$$

The periods are coefficients that express ω_λ in this basis, and the invariant $\tau(\lambda)$ is an invariant of the line spanned by the vector ω_λ. The expression

$$\omega'_\lambda = A'(\lambda)\delta^* + B'(\lambda)\gamma^*$$

is the derivative of the cohomology class ω_λ with respect to the "Gauss–Manin connection". This is by definition the connection on the bundle

$$\bigcup_{\lambda \in \mathbf{P}^1 - \{0,1,\infty\}} H^1(\mathscr{E}_\lambda)$$

of cohomology vector spaces with respect to which the classes δ^*, γ^* are (locally) constant. Then

$$[\omega_\lambda] \cup [\omega'_\lambda] = (AB' - A'B)\,\delta^* \cup \gamma^*.$$

However, $\tau' = 0$ if and only if $AB' - A'B = 0$. Thus, to establish that $\tau'(\lambda) \neq 0$, it suffices to establish that $[\omega_\lambda] \cup [\omega'_\lambda] \neq 0$.

Now the derivative of ω_λ is represented by the meromorphic form

$$\omega'_\lambda = \frac{1}{2}\frac{dx}{\sqrt{x(x-1)(x-\lambda)^3}}. \tag{1.8}$$

This form has a pole of multiplicity two at $p = (\lambda, 0)$. To see this, note that at the point p, the function y is a local coordinate. Therefore the relation $y^2 = x(x-1)(x-\lambda)$ can be written as $y^2 = u(y)\lambda(\lambda-1)(x-\lambda)$, where $u(y)$ is a holomorphic function of y satisfying $u(0) = 1$. Solving for x, we obtain $x = \lambda +$ terms of order ≥ 2 in y. Then setting $p(x) = x(x-1)(x-\lambda)$, we have

$$\omega_\lambda = \frac{dx}{y} = \frac{2\,dy}{p'(x)} \sim \frac{2\,dy}{\lambda(\lambda-1)},$$

where $a \sim b$ means that a and b agree up to lower-order terms in y. Using (1.8), the previous expression, and the expansion of x in terms of y, we find

$$\omega'_\lambda = \frac{1}{2}\frac{dx}{y(x-\lambda)} \sim \frac{dy}{\lambda(\lambda-1)(x-\lambda)} \sim \frac{dy}{y^2} + \text{a regular form.} \tag{1.9}$$

To explain why such a form represents a cohomology class on $\mathscr{E}$, not just on $\mathscr{E} - \{p\}$, we first note that its residue vanishes. Recall that the residue of ϕ is defined as

$$\frac{1}{2\pi \mathrm{i}} \int_{C_p} \phi = \operatorname{res}(\phi)(p),$$

where C_p is a small positively counterclockwise-oriented circle on S centered at p. Next, note that the residue map in fact is defined on the level of cohomology classes (just apply Stokes' theorem). In fact the resulting map "res" is the coboundary map from the exact sequence of the pair $(\mathscr{E}, \mathscr{E} - \{p\})$,

$$0 \longrightarrow H^1(\mathscr{E}) \longrightarrow H^1(\mathscr{E} - \{p\}) \xrightarrow{\text{res}} H^2(\mathscr{E}, \mathscr{E} - \{p\}),$$

provided we identify the third vector space with $\mathbf{C}$ using the isomorphism

$$H^2(\mathscr{E}, \mathscr{E} - \{p\}) \cong H^0(\{p\}) \cong \mathbf{C}.$$

See Problem 1.1.10, where this sequence is discussed in more detail.

Observe that the sequence above is the simplest instance of the so-called *Gysin sequence* for a smooth hypersurface (here just the point p) inside a smooth variety (here the curve $\mathscr{E}$). The Gysin sequence is at the heart of many calculations and is treated in detail in Section 3.2.

From the Gysin sequence we see that ω'_λ represents a cohomology class on $\mathscr{E}$, not just on $\mathscr{E} - \{p\}$. We now claim that

$$\int_{\mathscr{E}} [\omega_\lambda] \cup [\omega'_\lambda] = \frac{4\pi \mathrm{i}}{\lambda(\lambda - 1)}.$$

By establishing this formula we will complete the proof that $\tau'(\lambda) \neq 0$ for $\lambda \neq 0, 1$. To do this, first observe that the formula (1.9) implies that $\omega'_\lambda + \mathrm{d}(1/y)$ has no pole at p.

To globalize this computation, let U be a coordinate neighborhood of p on which $|y| < \epsilon$, and let $\rho(z)$ be a smooth function of $|z|$ alone which vanishes for $|z| > \epsilon/2$, which is identically one for $|z| < \epsilon/4$, and which decreases monotonically in $|z|$ on the region $\epsilon/4 < |z| < \epsilon/2$. Then the form

$$\vartheta_\lambda{}' = \omega'_\lambda + \mathrm{d}(\rho(y)/y)$$

lies in the same cohomology class on $\mathscr{E} - \{p\}$ as does ω'_λ. By construction, it extends to a form on $\mathscr{E}$ and represents the cohomology class of ω'_λ there. Because ω_λ and $\vartheta_\lambda{}'$ are both holomorphic one-forms on the complement of U,

$$\int_{\mathscr{E}} \omega_\lambda \wedge \vartheta_\lambda{}' = \int_U \omega_\lambda \wedge \mathrm{d}(\rho/y).$$

Because $\omega_\lambda \wedge d(\rho/y) = -d(\rho\omega_\lambda/y)$, Stokes' theorem yields

$$\int_U d\left(\frac{\rho\omega_\lambda}{y}\right) = -\int_{|y|=\frac{\epsilon}{4}} \frac{\rho\omega_\lambda}{y} = -\int_{|y|=\frac{\epsilon}{4}} \frac{\omega_\lambda}{y}.$$

A standard residue calculation of the line integral then yields

$$\int_{|y|=\epsilon/4} \frac{\omega_\lambda}{y} = \frac{4\pi i}{\lambda(\lambda-1)}.$$

This completes the proof.

Picard–Fuchs Equation

In computing the derivative of the period map, we proved that the form ω_λ and its derivative ω'_λ define linearly independent cohomology classes. Therefore the class of the second derivative must be expressible as a linear combination of the first two classes. Easing notation by dropping the subscripts λ, we thus find a relation

$$a(\lambda)\omega'' + b(\lambda)\omega' + c(\lambda)\omega = 0 \tag{1.10}$$

in cohomology. The coefficients are meromorphic functions of λ, and on the level of forms the assertion is that the left-hand side is exact on $\mathscr{E}_\lambda$. Let ξ be a one-cycle and set

$$\pi(\lambda) = \int_\xi \omega.$$

Then (1.10) can be read as a differential equation for the period function

$$a\pi'' + b\pi' + c\pi = 0.$$

One can determine the coefficients in this expression. The result is a differential equation with regular singular points at 0, 1, and ∞:

$$\lambda(\lambda-1)\pi'' + (2\lambda-1)\pi' + \frac{1}{4}\pi = 0. \tag{1.11}$$

Solutions are given by hypergeometric functions (see Clemens, 1980, section 2.11). To find the coefficients a, b, c above, we seek a rational function f on $\mathscr{E}_\lambda$ such that df is a linear combination of ω, ω', and ω'' whose coefficients are functions of λ. Now observe that

$$\begin{aligned}
\omega' &= \tfrac{1}{2}x^{-1/2}(x-1)^{-\frac{1}{2}}(x-\lambda)^{-\frac{3}{2}}\,dx,\\
\omega'' &= \tfrac{3}{4}x^{-\frac{1}{2}}(x-1)^{-\frac{1}{2}}(x-\lambda)^{-\frac{5}{2}}\,dx.
\end{aligned}$$

Thus it is reasonable to consider the function

$$f = x^{\frac{1}{2}}(x-1)^{\frac{1}{2}}(x-\lambda)^{-\frac{3}{2}}.$$

Indeed,

$$df = (x-1)\omega' + x\omega' - 2x(x-1)\omega'',$$

which is a relation between ω' and ω''. This is progress, but the coefficients are not functions of λ. Therefore consider the equivalent form

$$\begin{aligned} df = {} & [(x-\lambda)+(\lambda-1)]\omega' + [(x-\lambda)+\lambda]\omega' \\ & -2[(x-\lambda)+\lambda][(x-\lambda)+(\lambda-1)]\omega'' \end{aligned}$$

and use the relations

$$\begin{aligned} (x-\lambda)\omega' &= \frac{1}{2}\omega, \\ (x-\lambda)\omega'' &= \frac{3}{2}\omega' \end{aligned}$$

to obtain

$$-\frac{1}{2}\,df = \frac{1}{4}\omega + (2\lambda-1)\omega' + \lambda(\lambda-1)\omega''.$$

This completes the derivation.

One can find power series solutions of (1.11) which converge in a disk Δ_0 around any $\lambda_0 \neq 0, 1$. Analytic continuation of the resulting function produces a multivalued solution defined on $\mathbf{P}^1 - \{0, 1, \infty\}$. Now let π_1 and π_2 be two linearly independent solutions defined on Δ_0, and let γ be a loop in $\mathbf{P}^1 - \{0, 1, \infty\}$ based at λ_0. Let π_i' be the function on Δ_0 obtained by analytic continuation of π_i along γ. Because the π_i' are also solutions of the differential equation (1.11), they can be expressed as linear combinations of π_1 and π_2:

$$\begin{pmatrix} \pi_1' \\ \pi_2' \end{pmatrix} = \begin{pmatrix} a & b \\ c & d \end{pmatrix} \begin{pmatrix} \pi_1 \\ \pi_2 \end{pmatrix}.$$

The indicated matrix, which we shall write as $\rho(\gamma)$, depends only on the homotopy class of α, and is called the *monodromy matrix*. We determine this matrix in the next section using a geometric argument. For now we note that the map that sends α to $\rho(\alpha)$ defines a homomorphism

$$\rho : \pi_1(\mathbf{P}^1 - \{0, 1, \infty\}, \lambda_0) \longrightarrow \mathrm{GL}(2, \mathbf{C}).$$

It is called the *monodromy representation* and its image is called the *monodromy group*.

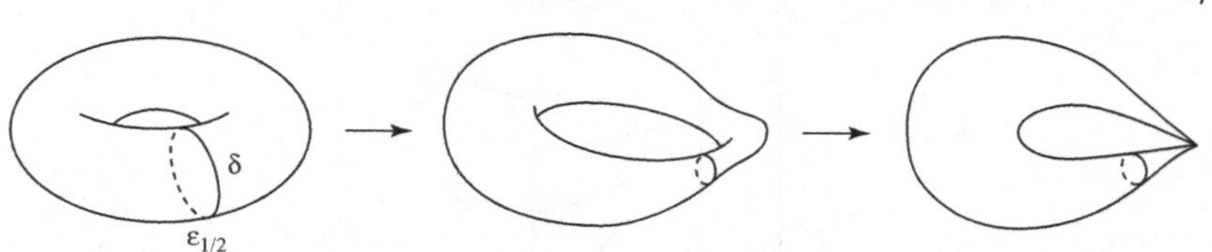

Figure 1.6 Degeneration of an elliptic curve.

The Local Monodromy Representation

To better understand the monodromy representation, consider the family of elliptic curves $\mathscr{E}_s$ defined by

$$y^2 = (x^2 - s)(x - 1).$$

The fiber $\mathscr{E}_0$, given by $y^2 = x^2(x-1)$ has a node at $p = (0,0)$. As s approaches 0, the fiber undergoes the changes pictured in Fig. 1.6. A copy of the loop δ is slowly contracted to a point, producing the double point at p. Note that in the limit of $s = 0$, the cycle δ is homologous to 0.

Now restrict this family to the circle $|s| = \epsilon$ and consider the vector field $\partial/\partial\theta$ in the s-plane. It lifts to a vector field ξ on the manifold

$$M = \left\{(x, y, s) \mid y^2 = (x^2 - s)(x - 1)\right\}$$

which fibers over the circle via $(x, y, s) \mapsto s$. By letting the flow which is tangent to ξ act for time ϕ, one defines a diffeomorphism g_ϕ of the fiber at $\theta = 0$ onto the fiber at $\theta = \phi$. This is illustrated in Fig. 1.7. (We think of a fluid flow transporting points of $\mathscr{E}_0$ to points of $\mathscr{E}_\phi$, with streamlines tangent to the vector field.)

Now consider the diffeomorphism $g_{2\pi}$. It carries the fiber at $\theta = 0$ to itself and therefore defines a map T on the homology of the fiber which depends only on the homotopy class of $g_{2\pi}$. This is the *Picard–Lefschetz transformation* of the degeneration $\mathscr{E}_\lambda$.

Through a careful study of the pictures in Fig. 1.8, we obtain the matrix of T in the "standard" basis $\{\delta, \gamma\}$. Because the matrix is not the identity, we conclude that the diffeomorphism is not homotopic to the identity map.

The left panel in Fig. 1.8 represents the cycle δ and γ on the fiber $\mathscr{E}_s$ for $s = r$ for some small r. The middle panel in Fig. 1.8 shows how the flow has mapped these cycles to $\mathscr{E}_s$ with $s = re^{\pi i}$. The right panel shows the result for $s = re^{2\pi i}$.

It is clear from the pictures in Fig. 1.8 that $T(\delta) = \delta$. To determine $T(\gamma) = \gamma'$, we observe that $\gamma' = a\delta + b\gamma$, and we compute intersection numbers as follows.

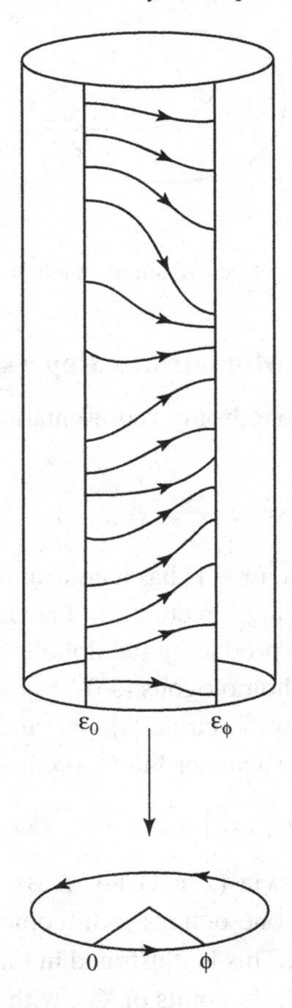

Figure 1.7 Diffeomorphism $g_\phi : \mathscr{E}_0 \longrightarrow \mathscr{E}_\phi$.

Using the sign convention as explained above in relation to Fig 1.4, we claim that $\gamma' \cdot \delta = -1$, $\gamma' \cdot \gamma = +1$. The former is clear from the right picture in Fig. 1.8, while the latter follows by superimposing the left and right pictures in Fig. 1.8. Thus $b = 1$ and $a = 1$, and so

$$T(\gamma) = \gamma + \delta.$$

The matrix of T relative to the basis $\{\delta, \gamma\}$ is

$$T = \begin{pmatrix} 1 & 1 \\ 0 & 1 \end{pmatrix}. \tag{1.12}$$

Equivalently, we have the *Picard–Lefschetz formula*

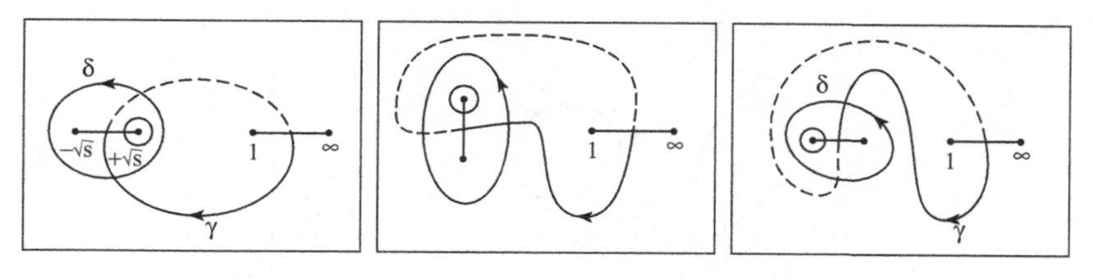

Figure 1.8 Picard–Lefschetz transformation.

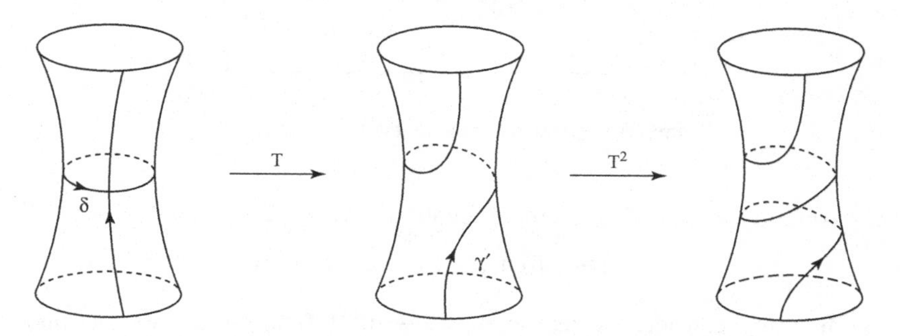

Figure 1.9 Dehn twist.

$$T(x) = x - (x \cdot \delta)\delta \tag{1.13}$$

for an arbitrary homology cycle x.

The Picard–Lefschetz formula is valid in great generality: it holds for any degeneration of Riemann surfaces acquiring a node where the local analytic equation of the degeneration is $y^2 = x^2 - s$. For such a degeneration the cycle δ is the one that is "pinched" to obtain the singular fiber, as in Fig. 1.6. This is the so-called *vanishing cycle*: under the inclusion of $\mathscr{E}_s$ into the total space of the degeneration, δ is homologous to 0. In a neighborhood of the vanishing cycle the Picard–Lefschetz diffeomorphism $g_{2\pi}$ acts as in Fig. 1.9: it is a so-called *Dehn twist*.

The Global Monodromy Representation

The Picard–Lefschetz transformation determines the *local monodromy representation* $\rho : \pi_1(\Delta^*, p) \longrightarrow \mathrm{GL}(2, \mathbf{C})$ for a family of Riemann surfaces defined on the *punctured* disk $0 < |s| < \epsilon$, where the fiber at $s = 0$ has a node. Let us now determine the *global monodromy transformation* for the *Legendre family*

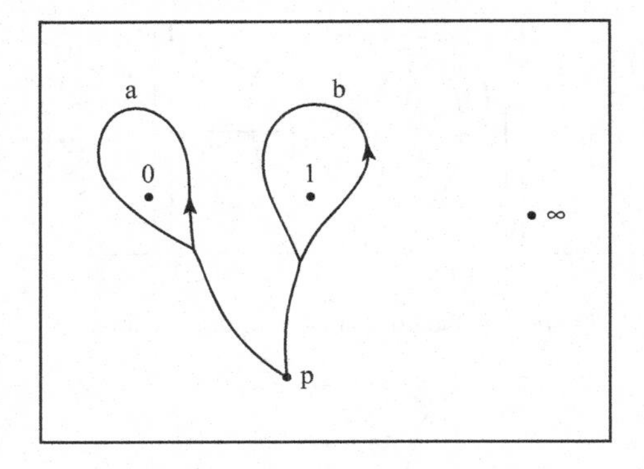

Figure 1.10 Parameter space for the Legendre family.

$y^2 = x(x-1)(x-\lambda)$. This is a representation

$$\rho : \pi_1(\mathbf{P}^1 - \{0, 1, \infty\}, p) \longrightarrow \mathrm{GL}(2, \mathbf{C}).$$

As a first step, consider the degeneration $\lambda \longrightarrow 0$. If λ moves clockwise around the circle of radius $r < 1$ centered at 0, then the slit connecting the branch cuts $x = \lambda$, and $x = 0$ turns through a full circle. Comparing with Fig. 1.8, where the slit makes a half turn, we see that the monodromy transformation for $\lambda \longrightarrow 0$ is the square of the matrix T in (1.12).

Now fix a base point p and choose generators a and b for the fundamental group of the parameter space as in Fig. 1.10.

Let $A = \rho(a)$ and $B = \rho(b)$ be the monodromy matrices relative to the basis indicated in the left picture of Fig. 1.8. From the discussion in the previous paragraph, we have

$$A = \begin{pmatrix} 1 & 2 \\ 0 & 1 \end{pmatrix}.$$

We claim that

$$B = \begin{pmatrix} 1 & 0 \\ -2 & 1 \end{pmatrix}.$$

To see that this is so, consider the degeneration $\lambda \longrightarrow 1$ and recall, as shown in Fig. 1.11, how the standard homology basis is defined relative to the standard branch cuts.

The set of branch cuts in Fig. 1.11 is ill adapted to computing the monodromy matrix of the degeneration $\lambda \longrightarrow 1$. Instead we consider the cuts in Fig. 1.12. The first frame gives the standard homology basis relative to this set of cuts. The second frame shows the result of rotating the branch slit connecting λ

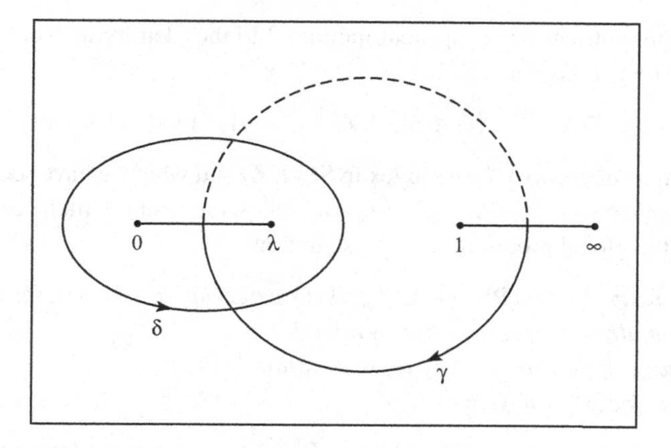

Figure 1.11 Standard homology basis.

and 1 through a half circle. The cycle δ' is obtained by dragging γ along with this rotation via the Picard–Lefschetz diffeomorphism. Computing intersection numbers, we find $\delta' = \delta - \gamma$. Thus the Picard–Lefschetz transformation is

$$S = \begin{pmatrix} 1 & 0 \\ -1 & 1 \end{pmatrix}.$$

Note that this matrix can also be computed using (1.13), taking note of the fact that the vanishing cycle is γ. The monodromy transformation $\rho(b)$ is therefore given by S^2, which is the indicated matrix B.

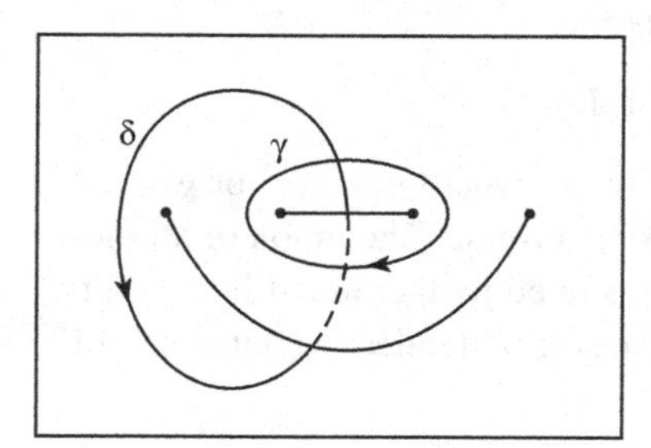

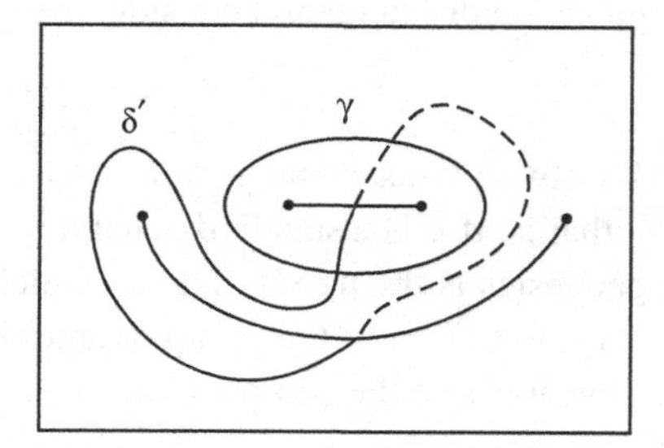

Figure 1.12 Monodromy for $\lambda \to 1$.

Now let $\Pi = \rho(\pi_1)$ denote the monodromy group. According to the preceding discussion, it is the group generated by the matrices A and B:

$$\Pi = \langle A, B \rangle .$$

The given matrices are congruent modulo 2 to the identity matrix, so every matrix in Π has this property. Let

$$\Gamma(N) \stackrel{\text{def}}{=} \{C \in \mathrm{SL}(2, \mathbf{Z}) \mid C \equiv \mathrm{id} \mod N\}.$$

It is a normal subgroup of finite index in $\mathrm{SL}(2, \mathbf{Z})$ and what we have seen is that Π is a subgroup of $\Gamma(2)$. We now assert the following result, which completely describes the global monodromy representation.

Theorem 1.1.7 (a) *$\pi_1(\mathbf{P}^1 - \{0, 1, \infty\})$ is a free group on two generators.*
(b) *The monodromy representation is injective.*
(c) *The image of the monodromy representation is* $\Gamma(2)$.
(d) *$\Gamma(2)$ has index six in $\Gamma = \mathrm{SL}(2, \mathbf{Z})$.*

Proof The proof of (a) is standard, since $\mathbf{P}^1 - \{0, 1, \infty\}$ is homotopy equivalent to the space obtained by joining two circles together at a point. The proof of (d) is an easy exercise. For the proof of (b), observe that monodromy matrices $\rho(\gamma)$ operate as linear fractional transformations on the part of the complex plane with a positive imaginary part. A fundamental domain for this action is given by the region indicated in Fig. 1.13. It is an ideal quadrilateral, with sides formed by two semicircles with endpoints on the real axis and two vertical rays with an endpoint on the real axis.

For our purposes it is better to look at this fundamental domain in the disk model of hyperbolic space. It is then the central quadrilateral in Fig. 1.14. The transformations A and B as well as their inverses act as reflections in the sides of the quadrilateral, and repeated applications of these transformations tile the disk by quadrilaterals congruent to the given one. Now take the center E of the given quadrilateral and consider the set of points

$$V = \{g(E) \mid g \in \Gamma\}.$$

Join two points x and y by a geodesic if $y = gx$ where $g = A^{\pm 1}$ or $g = B^{\pm 1}$ – that is, if g is a standard generator of Γ or its inverse. The union of all these geodesics is the tree $\mathcal{T}$, part of which is illustrated by the dotted lines in Fig. 1.13. Let $\Gamma(\mathcal{T})$ be the group of automorphisms of $\mathcal{T}$ defined by elements of Γ. Thus we have the composition

$$\pi_1 \xrightarrow{\rho} \Gamma \xrightarrow{\sigma} \Gamma(\mathcal{T}).$$

Consider now an element γ of π_1. It is a word spelled with the letters $a^{\pm 1}$, $b^{\pm 1}$. By considering the action of $\rho(\gamma)$ on $\mathcal{T}$ – indeed, by considering the position of $\rho(\gamma)(E)$ – one finds that $\rho(\gamma) \neq 1$ if $\gamma \neq 1$. See Problem 1.1.13 for more details.

For the proof of (c), the main idea is to compare the fundamental domain for

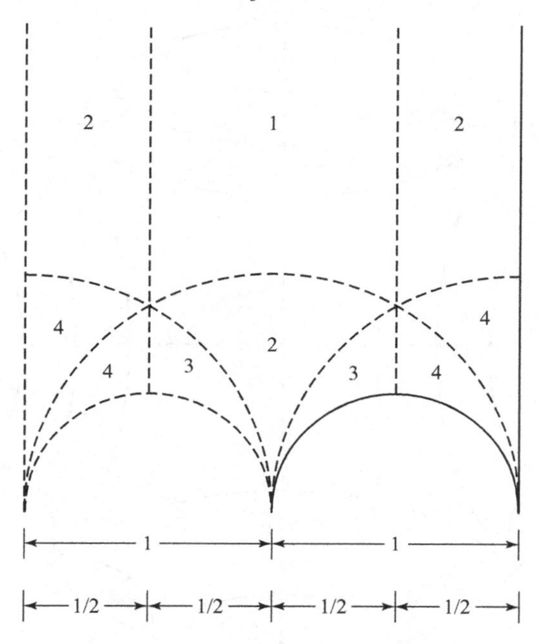

Figure 1.13 Fundamental domain for Γ and $\Gamma(2)$.

the action of $\Gamma = \mathrm{SL}(2, \mathbf{Z})$ with that of $\Gamma(2)$. A fundamental domain of the latter is made of six copies of the fundamental domain of the former, as indicated in Fig. 1.13. See Clemens (1980) for more details. □

Remark In the preceding example the kernel of the monodromy representation was trivial and its image was of finite index. The latter is typical (Beauville, 1986b), but the former is not. Thus, while the kernel of the monodromy representation for cubic curves is finite (see Libgober and Dolgachev, 1981), it is "large" for most families of hypersurfaces (see Carlson and Toledo, 1999). The notion "large" can be made precise; in particular, large groups are infinite and in fact contain a nonabelian free group.

Remark 1.1.8 Now that we have seen that $\Gamma(2)\backslash\mathfrak{h}$ is a parameter space for (isomorphisms classes of) framed elliptic curves, it follows that the quotient $\Gamma\backslash\mathfrak{h}$ parametrizes isomorphism classes of elliptic curves. Note that the first carries a tautological family, the Legendre family, while the second cannot carry such a family because of the isomorphisms you always have when a framing is absent. See also the later sections on moduli, especially Example 8.4.1. The invariant which establishes the isomorphism $\Gamma\backslash\mathfrak{h} \simeq \mathbf{C}$ is the j-invariant which is related

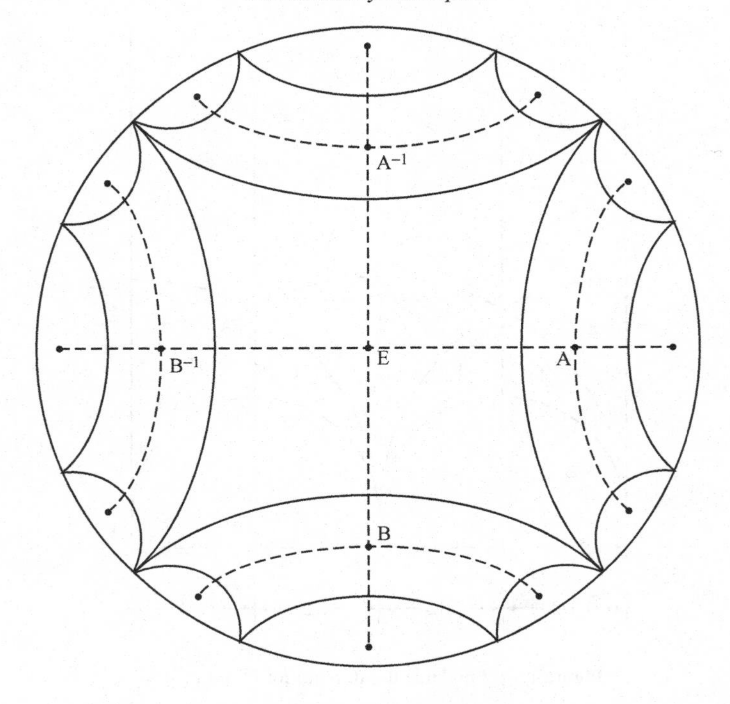

Figure 1.14 Fundamental domains for $\Gamma(2)$ – disk model.

to the λ-parameter as follows (see Silverman, 1992, Ch. III, Prop. 1.7):

$$j = j(\lambda) \stackrel{\text{def}}{=} 2^8 \frac{(\lambda^2 - \lambda + 1)^3}{\lambda^2(\lambda - 1)^2}.$$

In other words, there is a commutative diagram

$$\begin{array}{ccc} \mathfrak{h} & \searrow^{j} & \\ \downarrow & & \\ \Gamma(2)\backslash\mathfrak{h} & \xrightarrow[j(\lambda)]{6:1} & \Gamma\backslash\mathfrak{h}. \end{array}$$

From Serre (1979, Ch. VII.3) we quote the beautiful q-expansion with *integral* $c(n)$:

$$j(\tau) = \frac{1}{q} + 744 + \sum_{n=1}^{\infty} c(n)q^n, \quad q = \exp(2\pi i\tau),$$

$$c(1) = 196.884, \quad c(2) = 21.493.760.$$

Monodromy of the Picard–Fuchs Equation

Let us now return to the problem of understanding the monodromy of the period functions $\pi_i(\lambda)$ which solve the Picard–Fuchs equation (1.11). Analytic continuation of a solution

$$\pi(\lambda) = \int_{\xi} \omega$$

along a loop α transforms the solution into

$$\pi'(\lambda) = \int_{\rho(\xi)} \omega.$$

Thus the monodromy representation for solutions of the Picard–Fuchs equation is the same as the geometric monodromy representation. In particular, the representation must take values in the group of matrices with integer coefficients. In the case of the Legendre family, it is precisely the group $\Gamma(2)$ described in Theorem 1.1.7.

Problems

1.1.1 Show that the differential (1.2) is holomorphic on $\mathscr{E}$.

1.1.2 Show that the Euler characteristic of the Riemann sphere is two. Compute the Euler characteristic of the Riemann sphere with d points deleted. Let k be a divisor of d. Then there is a k-fold unbranched cover of the Riemann sphere defined by the equation $y^k = p(x)$, where p is a polynomial of degree d. Compute the Euler characteristic of this unbranched cover. Then compute the Euler characteristic of the corresponding k-fold branched cover, defined by the same equation. Finally, compute the genus of that branched cover.

1.1.3 Show that the only singular fibers of the Legendre family $y^2 = x(x-1)(x-\lambda)$ are at $\lambda = 0, 1$. Consider next the family of elliptic curves $x^3 + y^3 + z^3 + \lambda xyz$. What are its singular fibers?

1.1.4 Consider the family of elliptic curves $\mathscr{E}_{a,b,c}$ defined by $y^2 = (x-a)(x-b)(x-c)$. What is the locus in $\mathbf{C}^3 = \{(a, b, c)\}$ of the singular fibers (the discriminant locus)? More difficult: describe the monodromy representation.

1.1.5 Consider the family of elliptic curves $\mathscr{E}_{A,B,C}$ defined by $y^2 = x^3 + Ax^2 + Bx + C$. What is the locus in $\mathbf{C}^3 = \{(A, B, C)\}$ of the singular fibers? More difficult: describe the monodromy representation.

1.1.6 Consider the compact Riemann surface M with affine equation $y^2 = p(x)$ where p has degree two. Show that M is isomorphic as a complex manifold (or as an algebraic curve) to the Riemann sphere.

1.1.7 A holomorphic one-form on a Riemann surface is a differential one-form which is locally given by $f(z)\,\mathrm{d}z$, where z is a local holomorphic coordinate. Show that such a form is closed. Formulate and investigate the analogous assertion(s) for holomorphic forms on a complex manifold of complex dimension 2.

1.1.8 Prove the following identities: $\arccos u = \mathrm{i}\log(u+\sqrt{u^2-1})$, $\arccos u = \arctan(\sqrt{(1-u^2)}/u)$, $\arctan u = \arccos(1/\sqrt{1+u^2})$. Then show that all the integrals and estimates leading up to (1.7) hold as asserted.

1.1.9 Consider the family of elliptic curves defined by $y^2 = (x^2-s)(x-1)$. Find the asymptotic form of the period map $\tau(s)$ as s approaches 0. Comment on the relation between what you find and the asymptotic form in (1.7).

1.1.10 Let S be a Riemann surface, and let $A \subset S$ be a nonempty finite set. Show that there is an exact sequence

$$0 \longrightarrow H^1(S) \longrightarrow H^1(S-A) \xrightarrow{\text{res}} H^0(A) \longrightarrow H^0(S) \longrightarrow 0.$$

Note that elements of $H^0(A)$ are linear functionals on the vector space spanned by the points of A. They can be viewed as the pointwise residue as defined previously, and they can be combined to form the globally defined map "res". Your argument should show that the above sequence is defined on the level of integral cohomology.

1.1.11 Consider the degeneration of elliptic curves $\mathscr{E}_t$ defined by $y^2 = x^3 - t$. Find all values of t for which $\mathscr{E}_t$ is singular. By drawing a series of pictures of branch cuts, show that the monodromy transformation for $t = 0$ has order six, and find the corresponding matrix.

1.1.12 Let $\{\mathscr{E}_t\}$ be a family of elliptic curves with just two singular fibers, one at $t = 0$, the other at $t = \infty$. Show that the complex structure of $\mathscr{E}_t$ does not vary.

1.1.13 Let F be the free group with two generators a and b. Assign a graph $\mathscr{T}(F)$ to this group by letting the vertices be the elements of F, i.e., the words in a and b. We connect the vertices represented by w and wa by an edge, and we likewise draw an edge between w and wa^{-1}, between w and wb, and between w and wb^{-1}. No other vertices are connected. This defines also an action of F on $\mathscr{T}(F)$. Show that $\mathscr{T}(F)$ is a tree (compare the graph with the tree of Fig. 1.14). Show that the action of F on $\mathscr{T}(F)$ is free and faithful, i.e., if some word $w \in F$ fixes a vertex, then $w = 1$.

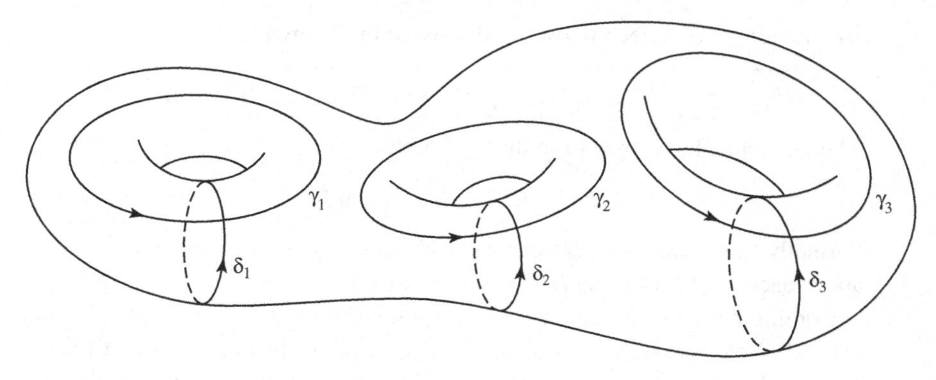

Figure 1.15 Riemann surface of genus 3.

1.2 Riemann Surfaces of Higher Genus

Let us now consider the Hodge theory and period mapping for Riemann surfaces of genus bigger than 1, as illustrated in Fig. 1.15. The cycles δ_i and γ_i form a "standard basis" for the first homology. For such a basis the intersection matrix has the form

$$J = \begin{pmatrix} 0 & \mathbf{1}_g \\ -\mathbf{1}_g & 0 \end{pmatrix},$$

where $\mathbf{1}_g$ is the $g \times g$ identity matrix. One can define a Riemann surface S of this kind by the equation

$$y^2 = (x - t_1)(x - t_2) \cdots (x - t_n), \tag{1.14}$$

where $n = 2g + 2$. However, for $g > 2$, there are Riemann surfaces which are not given by such equations (see Problem 1.2.1). Those which have an equation (1.14) are called *hyperelliptic Riemann surfaces*.

So far we have used a topological definition of the genus – the number of handles, which we can compute from the Euler characteristic. For another point of view, consider the vector space $\Omega^1(S)$ of holomorphic one-forms. These, which we have encountered already in the case of elliptic curves, are differentials which can be written locally as $f(z)\,\mathrm{d}z$ where $f(z)$ is a holomorphic function. Riemann's contribution to the Riemann–Roch theorem can be succinctly written as

$$\dim \Omega^1(S) = g(S). \tag{1.15}$$

This formula implies, as in the case of an elliptic curve, that the complex

1-cohomology of S decomposes as discussed in Theorem 1.1.2:

$$\boxed{H^1(S;\mathbf{C}) = H^{1,0}(S) \oplus H^{0,1}(S),\ H^{1,0}(S) = \Omega^1(S),\ H^{0,1}(S) = \overline{H^{1,0}(S)},}$$

where conjugation comes from the isomorphism

$$H^1(S;\mathbf{C}) = H^1(S,\mathbf{R}) \otimes_{\mathbf{R}} \mathbf{C}.$$

Formally this means that H^1 carries a *Hodge structure of weight* 1, and the above decomposition is the *Hodge decomposition*.

Coming back to the equality (1.15), we note that it is not at all obvious in general. What is easy (see Problem 1.2.1 (d)) is the inclusion $\Omega^1(S) \oplus \overline{\Omega^1(S)} \subset H^1_{\mathrm{DR}}(S;\mathbf{C})$, whence comes an inequality. However, for hyperelliptic Riemann surfaces the one-forms

$$\omega_i = \frac{x^i\,\mathrm{d}x}{y}, \qquad i = 0,\dots,g-1 \tag{1.16}$$

are independent. Indeed, given any polynomial $p(x) = v_1 + v_2x + v_3x^2 + \cdots + v_gx^{g-1}$ of degree $\leq (g-1)$, the one-form $p(x)\,\mathrm{d}x/y$ is 0 if and only if p is the zero polynomial. So, in view of the preceding, the forms (1.16) give a basis for $\Omega^1(S)$. In this case the Hodge decomposition for H^1 therefore follows.

Let us now consider the period map for the Riemann surfaces $S = S_t$ given by Eq. (1.14), where $t = (t_1,\dots,t_n)$. To this end we fix a standard basis, and we denote the elements of the dual basis by δ^i, γ^i. Thus $\delta^i(\delta_j) = \delta^i_j$, where the last symbol is Kronecker's δ, equal to 0 if $i \neq j$ and equal to 1 if $i = j$. This basis, or "marking," gives an isomorphism

$$H^1(S;\mathbf{Z}) \xrightarrow{m} \mathbf{Z}^{2g}$$

which extends to an isomorphism

$$H^1(S;\mathbf{C}) \xrightarrow{m} \mathbf{C}^{2g}.$$

Then the subspace

$$m(H^{1,0}(S)) \subset \mathbf{C}^{2g} \tag{1.17}$$

defines a point in the Grassmannian of g-planes in $2g$-space. It depends on both S and the marking.

Let

$$U = \{(t_1,\dots,t_n) \mid t_i \neq t_j\} = \mathbf{C}^n - \Delta$$

be the parameter space for the nonsingular Riemann surfaces S. The so-called *discriminant hypersurface* Δ is the union of hyperplanes $t_i = t_j$. The set of cohomology groups $H^1(S_u;\mathbf{Z})$ for u in U forms a local system of $\mathbf{Z}$-modules of rank $2g$ which we denote by H^1_U. Let $\widetilde{U}$ be the universal cover of U, and consider the pullback of the local system, which we denote by $H^1_{\widetilde{U}}$. Because $\widetilde{U}$ is simply connected, this system is trivial. Thus there is an isomorphism m from it to the trivial local system $\mathbf{Z}^{2g} \times \widetilde{U}$. Let

$$m(\tilde{u}) : H^1(S_{\tilde{u}}) \longrightarrow \mathbf{Z}^{2g} \times \{\tilde{u}\} \longrightarrow \mathbf{Z}^{2g}$$

denote the isomorphism of the fibers over $\tilde{u}$, composed with the projection to the first factor. Thus $m(\tilde{u})$ is an isomorphism of $H^1(S_{\tilde{u}}, \mathbf{Z})$ with $\mathbf{Z}^{2g}$. Since H^1_U is a local system, one has the relation

$$m(\gamma \cdot \tilde{u}) = \rho(\gamma) m(\tilde{u}),$$

where ρ is the monodromy representation and $\gamma \cdot \tilde{u}$ is the action of $\pi(U, u_0)$ on $\widetilde{U}$ by covering transformations. It follows that formula (1.17) defines a map

$$\widetilde{\mathcal{P}} : \widetilde{U} \longrightarrow \mathrm{Grass}(g, 2g)$$

which satisfies the equivariance condition

$$\widetilde{\mathcal{P}}(\gamma \cdot \tilde{u}) = \rho(\gamma) \widetilde{\mathcal{P}}(\tilde{u}). \tag{1.18}$$

Our aim is to understand this map in the same way that we understood the period map for elliptic curves. The first result is both important and easy to prove.

Theorem 1.2.1 *The period map is holomorphic.*

Proof Observe that $\widetilde{\mathcal{P}}(S, m)$ is the same as the row space of the $g \times 2g$ matrix (A, B), where

$$A_{ij} = \int_{\delta_i} \omega_j$$

and

$$B_{ij} = \int_{\gamma_i} \omega_j .$$

Thus $\widetilde{\mathcal{P}}$ will be holomorphic if the integrals A_{ij} and B_{ij} are holomorphic functions of $t = (t_1, \ldots, t_n)$. We leave the proof of this fact as an exercise. The reader should adapt the argument used for elliptic curves to the hyperelliptic case. □

Let us now consider the values of the period map. We claim first, in analogy with the proof of Theorem 1.1.2, that A, the first half of the period matrix, is nonsingular. To see that this is so, consider a vector v such that $vA = 0$. Let $\omega = v_1\omega_1 + \cdots + v_g\omega_g$, and observe that

$$\int_{\delta_j} \omega = \sum_i \int_{\delta_j} v_i\omega_i = \sum_i v_i A_{ij} = (vA)_j.$$

Thus, v is a null vector of A if and only if all the integrals of ω over the δ_j vanish. Consequently the cohomology class of ω is a linear combination of the γ^j if v is a null vector,

$$[\omega] = \sum_j w_j \gamma^j$$

for some coefficients w_j. Then

$$\overline{[\omega]} = \sum_j \bar{w}_j \gamma^j.$$

However,

$$\gamma^i \cup \gamma^j = 0 \text{ for all } i \text{ and } j,$$

and so $[\omega] \cup \overline{[\omega]} = 0$. As in the case of elliptic curves,

$$([\omega] \cup \overline{[\omega]})[S] = \mathrm{i} \int_S \omega \wedge \bar{\omega} \geq 0.$$

Moreover, the integral is identically 0 if and only if ω is 0 as a one-form. In the case of a hyperelliptic Riemann surface,

$$\omega = \frac{p(x)\,\mathrm{d}x}{y},$$

where $p(x) = v_1 + v_2 x + v_3 x^2 + \cdots + v_g x^{g-1}$. We have seen that such a holomorphic differential is 0 if and only if p is the zero polynomial, that is, if and only if v is the zero vector. Thus, if $vA = 0$, then $v = 0$, as required to prove the claim.

Now let

$$\vartheta_i = \sum_j A^{ij} \omega_j,$$

where A^{ij} is the ij component of the inverse of the matrix A. With the basis $\{\tilde{\omega}_i\}$ in place of the basis $\{\omega_i\}$, the matrix of A-periods is the identity. The nature of the matrix of B-periods is given by the following.

Theorem 1.2.2 *If the matrix of A-periods is the identity, then the matrix of B-periods is symmetric and has a positive definite imaginary part.*

Proof The positivity assertion of the theorem mirrors the corresponding result (1.4) for elliptic curves; the symmetry statement is a new phenomenon. To prove symmetry, note that the two-form $\vartheta_i \wedge \vartheta_j$ vanishes identically because it has the local form $f(z)\,\mathrm{d}z \wedge g(z)\,\mathrm{d}z$ for some holomorphic functions f and g. Therefore,

$$\int_S \vartheta_i \wedge \vartheta_j = 0$$

for all i and j. Now write ϑ_i in terms of the standard cohomology basis:

$$[\vartheta_i] = \delta_i + \sum_k B_{ik}\gamma^k.$$

Because of our integral formula above, the cup product $[\vartheta_i] \cup [\vartheta_j]$ vanishes. However,

$$[\vartheta_i] \cup [\vartheta_j] = \left(\delta_i + \sum_k B_{ik}\gamma^k\right) \cup \left(\delta_j + \sum_\ell B_{j\ell}\gamma^\ell\right) = (B_{ij} - B_{ji})[S].$$

It follows that B is symmetric. For the positivity assertion, consider the abelian differential $\vartheta = v_1\vartheta_1 + \cdots + v_g\vartheta_g$. Then

$$\int_S \vartheta \wedge \overline{\vartheta} = ([\vartheta] \cup \overline{[\vartheta]})[S].$$

On the one hand, the integral on the left is positive for nonzero ϑ. On the other hand, the expression on the right can be evaluated by evaluating the cup product

$$\mathrm{i}\left(\sum_i v_i\delta_i + \sum_{ik} v_i B_{ik}\gamma^k\right) \cup \left(\sum_j \bar{v}_j\delta_j + \sum_{j\ell} \bar{B}_{j\ell}\bar{v}_j\gamma^\ell\right).$$

One obtains the identity

$$2\sum_{ij} v_i\bar{v}_j \operatorname{Im}(B_{ij}) = 2v \operatorname{Im}(B)^{\mathsf{T}}\bar{v}[S].$$

Here v is a row vector and ${}^{\mathsf{T}}v$ denotes its transpose. Thus B has a positive definite imaginary part, as claimed. □

The *Siegel upper half-space* of genus g is given as follows.

$$\mathfrak{h}_g \stackrel{\text{def}}{=} \{Z \text{ a symmetric } g \times g \text{ matrix with } \operatorname{Im} Z > 0\}. \tag{1.19}$$

The period map takes values in $\mathfrak{h}_g$ viewed as a subset of the Grassmann manifold via the map

$$Z \mapsto \text{row space of } (\mathbf{1}_g, Z),$$

where $\mathbf{1}_g$ is the $g \times g$ identity matrix. The period map for the family (1.14) of hyperelliptic Riemann surfaces takes the form

$$\widetilde{\mathscr{P}} : \widetilde{U} \longrightarrow \mathfrak{h}_g.$$

Since a monodromy diffeomorphism associated with the path γ preserves the cup product form, the monodromy matrices $\rho(\gamma)$ preserve the skew-symmetric form J. Thus the monodromy representation takes values in the group

$$\boxed{\mathrm{Sp}(g, \mathbf{Z}) = \left\{ M \in \mathrm{GL}(2g, \mathbf{Z}) \;\middle|\; {}^{\mathsf{T}}MJM = J \right\}.}$$

This is the integer *symplectic group*. One may also consider $\mathrm{Sp}(g, \mathbf{R})$, the symplectic group with real coefficients. A matrix T in the symplectic group can be decomposed into $g \times g$ blocks as

$$T = \begin{pmatrix} K & L \\ M & N \end{pmatrix}.$$

Such matrices operate on $g \times 2g$ matrices (A, B) via multiplication on the right,

$$(A, B) \mapsto (A, B)T,$$

and there is a corresponding action on the row spaces. One checks (Problem 1.2.3) that a symplectic matrix preserves the set of row spaces corresponding to matrices $(\mathbf{1}, Z)$, where Z is symmetric with a positive definite imaginary part. Thus the symplectic group also acts on $\mathfrak{h}_g$, specifically, via

$$Z \mapsto T\langle Z \rangle \stackrel{\text{def}}{=} (K + ZM)^{-1}(L + ZN).$$

This is a kind of generalized fractional linear transformation which in the case of $g = 1$ reduces to the standard action of $\mathrm{SL}(2, \mathbf{R})$ on the upper half-plane. We want to establish some basic facts about this action. To begin, it acts transitively. This we can see by looking at the image of $\mathrm{i}\mathbf{1}_g$ under the map

$$T = \begin{pmatrix} \mathbf{1}_g & X \\ 0 & \mathbf{1}_g \end{pmatrix} \begin{pmatrix} {}^{\mathsf{T}}W & 0 \\ 0 & W^{-1} \end{pmatrix}. \tag{1.20}$$

We find $T\langle \mathrm{i}\mathbf{1}_g \rangle = X + \mathrm{i}\,{}^{\mathsf{T}}WW$. Since every positive definite Hermitian matrix Y can be written as $Y = {}^{\mathsf{T}}WW$, this shows that the action is indeed transitive.

Consider next the orbit map

$$\begin{aligned} \pi : \mathrm{Sp}(g, \mathbf{R}) &\to \mathfrak{h}_g, \\ T &\mapsto T\langle \mathrm{i}\mathbf{1}_g \rangle. \end{aligned}$$

We claim that this map is proper: any sequence $M_n \in \mathrm{Sp}(g)$ whose π-images $X_n + \mathrm{i}Y_n$ converge in $\mathfrak{h}_g$ has a convergent subsequence. From this it quite easy (see Problem 1.2.4) to show that $\mathrm{Sp}(g, \mathbf{Z})$ acts properly discontinuously on $\mathfrak{h}_g$, i.e., for any two compact sets $K_1, K_2 \subset \mathfrak{h}_g$, there are at most finitely many elements $\gamma \in \mathrm{Sp}(g, \mathbf{Z})$ such that $K_1 \cap \gamma K_2 \neq \emptyset$. Then a standard general result (see Cartan, 1957) asserts that the quotient of a complex manifold by a proper action action of a group is an *analytic space*: a (possibly singular) space on which the notion of holomorphic function is defined.

For the proof of the properness assertion, define T_n according to (1.20) so that $T_n = M_n U_n$ with $U_n \in \mathrm{U}(g) \cap \mathrm{Sp}(g)$, a compact group. Passing to a subsequence, we may assume that the U_n converge, and so it suffices to see that $\{T_n\}$ has a convergent subsequence. Since by assumption $X_n + \mathrm{i}Y_n$ converges to a point in $\mathfrak{h}_g$, the X_n converge and so it suffices to see that $\{{}^{\mathsf{T}}W_n\}$ and $\{{}^{\mathsf{T}}W_n^{-1}\}$ have convergent subsequences. The first is clear since ${}^{\mathsf{T}}W_n W_n = Y_n$ converges and so $\{W_n\}$ is a bounded set. Replacing $\{W_n\}$ by a converging subsequence, put $W = \lim_{n\to\infty} W_n$. Then ${}^{\mathsf{T}}WW > 0$ by assumption and so W is invertible; it follows that $W^{-1} = \lim_{n\to\infty} W_n^{-1}$, which completes the proof of our assertion.

The functional equation (1.18) asserts that the period map is equivariant as a map of $\widetilde{U}$ to $\mathfrak{h}_g$. Therefore, in the light of the previous two paragraphs, there is a quotient map

$$\mathcal{P} : U \longrightarrow \mathrm{Sp}(g, \mathbf{Z})\backslash\mathfrak{h}_g, \tag{1.21}$$

where, as noted, the right-hand side is an analytic space. It is definitely not a complex manifold because of the presence of fixed points of the action of $\mathrm{Sp}(g, \mathbf{Z})$ on $\mathfrak{h}_g$: see Problem 1.2.4. Indeed, the quotient has codimension 2 singularities, as in the model example of the quotient of $\mathbf{C}^2$ the group of transformations $(x, y) \mapsto (\pm x, \pm y)$. See Problem 1.2.5. Nonetheless, the notion of holomorphic function makes sense, and we have the following.

Theorem 1.2.3 *The period map* (1.21) *is a holomorphic map of analytic spaces.*

Degenerations

Consider now the family of genus 2 Riemann surfaces S_t given by

$$y^2 = (x - a_1) \cdots (x - a_5)(x - t).$$

The normalized period matrix of the fibers S_t has the form

$$Z = \begin{pmatrix} Z_{11} & Z_{12} \\ Z_{21} & Z_{22} \end{pmatrix},$$

where all entries are multivalued holomorphic functions of t. Let us suppose that $a_1 = 0$, and let us examine the behavior of the period matrix for t near $t = 0$.

If we use the standard bases illustrated in Fig. 1.16, then δ_1 is the vanishing cycle and the local monodromy transformation is given by

$$T(x) = x - 2(x \cdot \delta_1)\delta_1.$$

The factor of 2 comes from the fact that as t travels around a small circle centered at the origin, the branch slit connecting t to 0 makes a full turn: twice a half-turn, so twice the contribution of the vanishing cycle to monodromy. Following the same line of argument as used to establish (1.7), we find that

$$Z_{11}(t) = \frac{\mathrm{i}}{\pi} \log t + h(t),$$

where $h(t)$ is a holomorphic function, and where the remaining entries of the period matrix are holomorphic functions of t. Thus the multivaluedness of Z is of a very controlled sort.

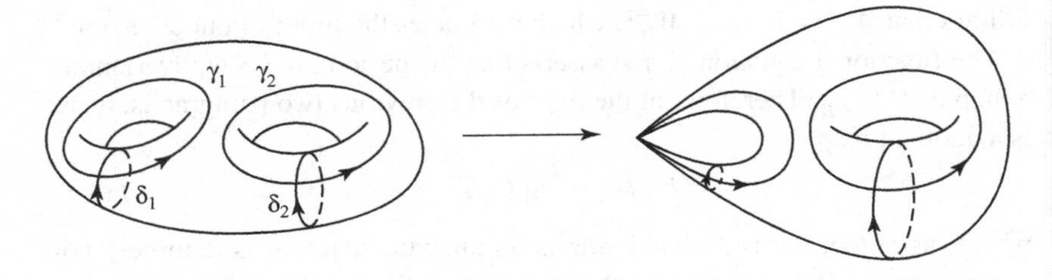

Figure 1.16 Degeneration.

Consider the linear transformation defined by $N(x) = (x \cdot \delta_1)\delta_1$. It is nilpotent and satisfies $T = 1 + 2N$. The results of the previous paragraph may be restated by saying that the matrix

$$\hat{Z}(t) = \exp\left(-\frac{\mathrm{i} \log t}{\pi} N\right) Z(t) \tag{1.22}$$

is single-valued and holomorphic in the punctured disk $0 < |t| < \epsilon$, and in fact is holomorphic in the disk $|t| < \epsilon$. Thus the period matrix itself can be written as

$$Z(t) = \exp\left(\frac{\mathrm{i} \log t}{\pi} N\right) \hat{Z}(t),$$

where $\hat{Z}(t)$ is holomorphic. This equation, which expresses the period matrix in terms of an exponential involving $\log t$, the logarithm of the Picard–Lefschetz transformation, and a holomorphic matrix, is a very special case of the *nilpotent orbit theorem* of Wilfried Schmid (1973). It is equivalent to the statement that in general the Picard–Fuchs equation has regular singular points.

Let us now inquire into the meaning of the entries of $\hat{Z}(0)$. Note first that if we replace the parameter t by ct, for some nonzero constant c, we replace $\hat{Z}_{11}(0)$ by $\hat{Z}_{11}(0) + \mathrm{i} \log c/\pi$. Thus the value of $Z_{11}(0)$ has no significance. The remaining entries, $Z_{ij}(0)$ for $(i, j) \neq (1, 1)$, however, are well defined and equal to $\hat{Z}_{ij}(0)$. To interpret them, write the elements of the basis for the space of abelian differentials as

$$\omega_1(t) = \frac{\mathrm{d}x}{\sqrt{x(x-t)(x-a)(x-b)(x-c)(x-d)}}$$

and

$$\omega_2(t) = \frac{x\,\mathrm{d}x}{\sqrt{x(x-t)(x-a)(x-b)(x-c)(x-d)}}.$$

For $t = 0$ these expressions become

$$\omega_1(0) = \frac{\mathrm{d}x}{x\sqrt{(x-a)(x-b)(x-c)(x-d)}}$$

and

$$\omega_2(0) = \frac{\mathrm{d}x}{\sqrt{(x-a)(x-b)(x-c)(x-d)}}.$$

Note that they make sense as possibly meromorphic differentials on the elliptic curve $\tilde{\mathscr{E}}$ defined by $y^2 = (x-a)(x-b)(x-c)(x-d)$. The Riemann surface $\tilde{\mathscr{E}}$ is the normalization of the algebraic curve $\mathscr{E} = S_0$ defined by $y^2 = x^2(x-a)(x-b)(x-c)(x-d)$ (see Problem 1.2.6).

The differential $\omega_1(0)$ is a differential of the third kind: it has simple poles at the two points corresponding to $x = 0$. Now consider the normalization of S_0, as illustrated in Fig. 1.17, and let ϑ be the normalized differential corresponding to $\omega_2(0)$: it is $\omega_2(0)$ divided by the integral of $\omega_2(0)$ over δ_2. We observe the following. First, the integral $Z_{22}(0)$ is a normalized period of $\omega_2(0)$:

$$Z_{22}(0) = \int_{\gamma_2} \vartheta.$$

Second, the integral $Z_{12}(0) = Z_{21}(0)$ is the normalized abelian integral for the divisor $p - q$:

$$Z_{12}(0) = \int_p^q \vartheta.$$

Each of these integrals, we emphasize, may be viewed as an integral on $\widetilde{\mathscr{E}}$. Thus the limiting values of entries of the period map can be interpreted as integrals on the normalization of the singular fiber of the degeneration, where that normalization has been marked in such a way that we "remember" what the

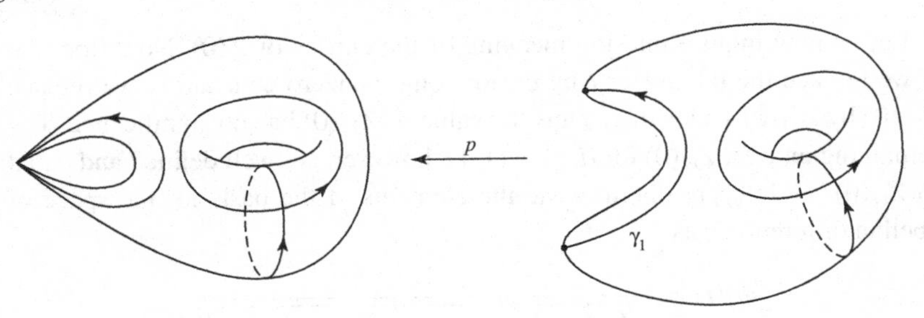

Figure 1.17 Normalization.

singular fiber was. Indeed, if we glue p and q in $\widetilde{\mathscr{E}}$, the result is biholomorphic to S_0.

The integral $Z_{12}(0)$ is just the Abel–Jacobi class associated with the divisor $p - q$. This class determines the location of p and q on $\widetilde{\mathscr{E}}$ up to translation, and so determines S_0, given $\widetilde{\mathscr{E}}$. Note also that $Z_{22}(0)$ is the normalized period of $\widetilde{\mathscr{E}}$, and so it determines $\widetilde{\mathscr{E}}$. Therefore the limit period matrix determines the singular fiber S_0.

Generalizing Hodge Theory

Our study of the period map for a degeneration of Riemann surfaces $\{S_t\}$ leads us to ask whether it makes sense to take a limit of the Hodge structure $H^1(S_t)$, and whether it is possible to define a (suitably generalized) Hodge structure for the singular variety S_0. The answer to both questions is "yes." We shall take up the question of generalizing Hodge theory to singular varieties first, and then consider limits of Hodge structures.

Let us begin with an easy observation. Since the cohomology of S_0 (see Fig. 1.17) has rank 3, it cannot carry a Hodge structure of weight 1: these have even dimension in view of the relation $\overline{H^{1,0}} = H^{0,1}$. Nonetheless, the cohomology of S_0 carries considerable structure, both topological and complex analytic. To understand the topology, consider the normalization map $p : \widetilde{S}_0 \longrightarrow S_0$ and its induced map on cohomology,

$$H^1(S_0) \xrightarrow{p^*} H^1(\widetilde{S}_0).$$

It is easy to see that p^* is surjective, for example by showing that the corresponding map p_* on homology is injective (simply look at Fig. 1.17). Thus there is an exact sequence

$$0 \longrightarrow K \longrightarrow H^1(S_0) \xrightarrow{p^*} H^1(\widetilde{S}_0) \longrightarrow 0,$$

for some kernel K. This kernel is the $\mathbf{Z}$-module generated by γ^1, which we think of as the natural cocycle associated with the node. Note that the quotient by K is isomorphic to $H^1(\widetilde{S_0})$, which carries a Hodge structure. However, there is more information to be extracted from the analytic structure of S_0. The key observation is the one already made above: both S_0 and $\widetilde{S_0}$ carry an abelian differential, namely $\omega_2(0)$. Let F^1 be its span, viewed as a subspace of either $H^1(S_0)$ or $H^1(\widetilde{S_0})$. Thus we now have the following data, which constitute a simple case of a *mixed Hodge structure*:

(a) a subspace $F^1 \subset H^1(S_0; \mathbf{C})$ defined by the complex structure of the central fiber,

(b) a subspace $K \subset H^1(S_0; \mathbf{Z})$ defined by the topology of the normalization map.

These filtrations have the following properties:

(c) the subspace which F^1 defines on $H^1(S_0)/K \cong H^1(\widetilde{S_0})$ is the natural subspace $F^1 H^1(\widetilde{S_0})$,

(d) $F^1 \cap K = 0$.

Property (c) asserts that the data (W, F^1) define a Hodge structure of weight 1 on $H^1(S_0)/K \cong H^1(\widetilde{S_0})$. It is the natural one with $H^{1,0} = F^1$, $H^{0,1} = \overline{F^1}$.

Property (d) is (in general) the statement that if the pullback of a holomorphic one-form is 0 as a cohomology class on $\tilde{S}_0$, then it is 0 as a cohomology class on S_0. Indeed, suppose that ω is a one-form whose cohomology class on $\widetilde{S_0}$ vanishes. Then

$$\int_{\tilde{S}_0} \mathrm{i}\omega \wedge \bar{\omega} = 0,$$

where ω is given locally as $f(z)\,\mathrm{d}z$, where $f(z)$ is holomorphic. Because the integrand is the nonnegative expression $\mathrm{i}|f(z)|^2\,\mathrm{d}z \wedge \mathrm{d}\bar{z}$, we conclude that $f(z)$ vanishes identically. It follows that ω, viewed as a form on S_0, vanishes. However, it then vanishes as a cohomology class on S_0 as well.

Formal Considerations: Hodge Structures and Polarizations

To interpret (d) in its proper general context, and to set the stage for work with higher-dimensional algebraic varieties, we must now make a series of formal definitions. These begin with the notion of a Hodge structure of arbitrary weight, modeled on the structure one finds on the cohomology of a projective algebraic manifold. This notion generalizes the idea we have become familiar with for Riemann surfaces.

Definition 1.2.4 (i) A *real Hodge structure* of weight k is a real vector space

H on the complexification of which there is a decomposition

$$\boxed{H_{\mathbf{C}} = \bigoplus_{p+q=k} H^{p,q}, \quad \overline{H^{p,q}} = H^{q,p}.}$$

The conjugation is relative to the real structure on $H_{\mathbf{C}} = H \otimes \mathbf{C}$. We call such an object a *Hodge structure* if in addition there is given a lattice $H_{\mathbf{Z}} \subset H$, i.e., a free abelian group such that $H_{\mathbf{Z}} \otimes \mathbf{R} \cong H$.

(ii) Suppose that a Hodge structure of any kind carries a bilinear form b satisfying the bilinear relations
   (a) $b(x, y) = 0$ if x is in $H^{p,q}$ and y is in $H^{r,s}$ for $(r, s) \neq (k-p, k-q)$;
   (b) $\mathrm{i}^{p-q} b(x, \bar{x}) > 0$ if x is a nonzero vector in $H^{p,q}$.

Then we say that the Hodge structure is *polarized*. When an integral lattice is part of the structure, one requires that b take integer values on it.

Remark 1.2.5 Note that p and q need not be positive and so, even if the weight is positive, p or q can be negative. If for a Hodge structure the non-zero Hodge numbers $h^{p,q}$ only occur when $p, q \geq 0$, we speak of an *effective Hodge structure*.

The Hodge decomposition on a compact Riemann surface, given by the holomorphic and anti-holomorphic one-forms, defines a Hodge structure of weight 1 which is polarized by the cup product form. As we see in Chapter 2, the k-th cohomology of a projective algebraic manifold carries a Hodge structure of weight k, where $H^{p,q}$ is represented by closed k-forms whose local expressions contain p $\mathrm{d}z$'s and q $\mathrm{d}\bar{z}$'s. Suppose further that the projective algebraic manifold in question, which we denote by X, is n-dimensional and that Y is a smooth hyperplane section. Define the *primitive cohomology* to be

$$\mathrm{Prim}^n(X) := \mathrm{Ker}[H^n(X) \longrightarrow H^n(Y)].$$

As we shall see (Theorem 2.3.3), this is a polarized structure of weight n and as a consequence (Corollary 2.3.5) also $H^n(X)$ carries a polarization.

Closely related to the Hodge decomposition is the *Hodge filtration*

$$\boxed{F^p = \bigoplus_{a \geq p} H^{a,b}} \tag{1.23}$$

which satisfies, for a weight-k structure, the relation

$$H_{\mathbf{C}} = F^p \oplus \overline{F^{k-p+1}}. \tag{1.24}$$

Just as a Hodge decomposition defines a Hodge filtration, a filtration satisfying (1.24) defines a decomposition

$$H^{p,q} = F^p \cap \overline{F^q}$$

which satisfies (1.23). Any filtration satisfying (1.24) is called a Hodge filtration. In the case of Riemann surfaces, the Hodge filtration is $[F^1 \subset F^0]$, where $F^1 = H^{1,0}$ and $F^0 = H^1_{\mathbf{C}}$.

The introduction of the Hodge filtration is not simply a matter of book keeping. As we see in Chapter 3, for a family of projective algebraic manifolds $\{S_t\}$, the filtration varies holomorphically, while the decomposition does not. We have verified this assertion for the special case of hyperelliptic Riemann surfaces. It then follows that $H^{0,1}(S_t)$, which is the complex conjugate of a holomorphically varying space, does not vary holomorphically.

The following examples are also useful.

Examples 1.2.6 (Tate structures) (i) The *trivial Hodge structure* $\mathbf{Z}$ is defined by the lattice $\mathbf{Z} \subset \mathbf{R}$ and the Hodge decomposition $\mathbf{C} = \mathbf{C}^{0,0}$.
(ii) The k-th *Tate twist* $\mathbf{Z}(k)$ is the Hodge structure of weight $-2k$ given by the lattice $(2\pi \mathrm{i})^k\mathbf{Z}$ on the real line $\mathrm{i}^k\mathbf{R}$ and by the decomposition $\mathbf{C} = \mathbf{C}^{-k,-k}$.

Let us link these Tate structures to geometry. One can think of $\mathbf{Z}$ as the Hodge structure on $H^0(\{x\})$, the cohomology of a point. Likewise, one can think of $\mathbf{Z}(-1)$ as the Hodge structure on $H^1(\mathbf{C}^*)$, where $\mathbf{C}^* = \mathbf{C} - \{0\}$. The motivation is that the cohomology is spanned by $\mathrm{d}z/z$ which has integral $2\pi\mathrm{i}$ on a loop that makes one counterclockwise turn around the origin. Note that $\mathrm{d}z/z$ is a holomorphic differential on $\mathbf{C}^*$ and so (counting $\mathrm{d}z$'s) has Hodge level 1, i.e., lies in F^1. Thus $\mathbf{Z}(-1)$ has type $(1, 1)$.

The set of Hodge structures forms the objects of a category where the morphisms preserve the lattice and the decomposition, i.e., $\phi : A \to B$ satisfies $\phi(A^{p,q}) \subset B^{p,q}$. One can extend the notion of morphism somewhat, allowing maps of type (p, p), i.e., ones satisfying

$$\phi(A^{r,s}) \subset B^{p+r,p+s}.$$

Hodge structures on cohomology behave functorially:

Lemma 1.2.7 *Let $f : X \longrightarrow Y$ be a holomorphic map of algebraic manifolds. Then $f^* : H^k(Y) \longrightarrow H^k(X)$ is a morphism (of type $(0, 0)$).*

It can be seen without much difficulty that the kernel, image, and cokernel of a morphism carry natural Hodge structures. For instance, the fact that a morphism ϕ preserves the real structure implies that the complex conjugate of $\phi(A^{p,q})$ is exactly $\phi(A^{q,p})$, and so the Hodge decomposition of B induces a true Hodge decomposition on the image $\phi(A)$. According to this discussion the primitive cohomology, defined above, is a sub-Hodge structure. The fact that it is polarized is more subtle. We come back to this in Chapter 2.

In the category of Hodge structures one can form direct sums, tensor products, and duals. Let us look at the details in the latter two cases. For the tensor product $A \otimes B$ of two Hodge structures $A = \bigoplus_{p+q=m} A^{p,q}$ of weight m and $B = \bigoplus_{r+s=n} B^{r,s}$ of weight n we set

$$(A \otimes B)^{i,j} := \bigoplus_{p,q} A^{p,q} \otimes B^{i-p,j-q}.$$

Then $i + j = (p + q) + (i - p + j - q) = m + n$, as expected for a structure of weight $m + n$. The complex conjugate of the right-hand side is the direct sum of $A^{q,p} \otimes B^{j-q,i-p}$, which by definition is $(A \otimes B)^{j,i}$. Thus the reality constraint is verified. Finally, every pure tensor $a \otimes b$ decomposes into Hodge types and so its components belong to the various Hodge components of $A \otimes B$.

If X and Y are smooth projective varieties, then $H^k(X \times Y)$ is, by the Künneth theorem, the direct sum of the tensor products $H^m(X) \otimes H^n(Y)$ where $m+n = k$. The Hodge structure on $H^k(X \times Y)$ imposed by the Künneth theorem agrees with the natural Hodge structure. As an example, consider the Hodge structure on $H^2(\mathscr{E} \times \mathscr{E})$, where $\mathscr{E}$ is an elliptic curve with abelian differential ω. Let $p_i : \mathscr{E} \times \mathscr{E} \longrightarrow \mathscr{E}$ be projection onto the i-th factor. From the Künneth theorem, one finds that $H^{2,0}(\mathscr{E} \times \mathscr{E})$ is one-dimensional and spanned by $p_1^* \omega \cup p_2^* \omega$. See Problem 1.2.9 for additional details.

For the dual $A^\vee$ of $A = \bigoplus_{p+q=n} A^{p,q}$, we set

$$(A^\vee)^{-p,-q} = \{ f : A \to \mathbf{C} \mid f|A^{r,s} = 0, (r,s) \neq (p,q) \}.$$

This is a Hodge structure of weight $-n$ (Problem 1.2.7). The standard example of a Hodge structure of negative weight is the homology $H_k(X; \mathbf{Z})$ modulo torsion, which is dual to the cohomology $H^k(X; \mathbf{Z})$ modulo torsion. As for functoriality we have the following.

Lemma 1.2.8 *If $f : X \longrightarrow Y$ is a morphism of projective varieties, the induced map in cohomology $f_* : H_k(X) \to H_k(Y)$ is a morphism of Hodge structures. However, taking Poincaré duals (see Appendix B), the induced* Gysin map *in cohomology $f : H^k(X) \longrightarrow H^{k+2c}(Y)$, $c = \dim Y - \dim X$, is a morphism of type (c, c).*

The proof, which involves study of the Hodge type of a fundamental class, is given in Problem 2.2.5 in Chapter 2.

Mixed Hodge Structures Again

We can now, as claimed above, interpret (d) on page 37 in its proper general context and, in so doing, arrive at the proper general notion of mixed Hodge structure. First, the Hodge filtration for $H^1(S_0)$ is given by the space of abelian differentials F^1 and by $F^0 = H^1(S_0)$. Thus the filtration has the form $F^1 \subset F^0$, just as in the case of a smooth compact Riemann surface. On the quotient $H^1(S_0)/K$, the filtration induces a filtration isomorphic to that defined naturally on $H^1(\widetilde{S}_0)$. On the subspace K, the filtration is

$$F^1 \cap K = 0, F^0 \cap K = K.$$

From this we deduce that K carries a Hodge structure of weight 0, where

$$K_{\mathbf{C}} = K^{0,0}.$$

Thus the data (K, F^1) define two Hodge structures, one of weight 0, the other of weight 1.

With the preceding example as a guide, we can make a general definition.

Definition 1.2.9 (Deligne 1971b, p. 30) A *mixed Hodge structure* consists of a triple $(H_{\mathbf{Z}}, F^\bullet, W_\bullet)$, where
(i) $H_{\mathbf{Z}}$ is a $\mathbf{Z}$-module of finite rank,
(ii) $F^\bullet$ is a finite decreasing filtration on $H_{\mathbf{C}} = H_{\mathbf{Z}} \otimes_{\mathbf{Z}} \mathbf{C}$, the *Hodge filtration*,
(iii) $W_\bullet$ is a finite increasing filtration on $H_{\mathbf{Q}} = H_{\mathbf{Z}} \otimes_{\mathbf{Z}} \mathbf{Q}$, the *weight filtration*,
satisfying in addition the requirement that the graded quotients for the weight filtration, $\mathrm{Gr}_k^W H = W_k/W_{k-1}$, together with the filtration induced by $F^\bullet$, form a (pure) Hodge structure of weight k.

In our example, the Hodge filtration is already defined and the weight filtration is given by $W_1 = H^1(S_0, \mathbf{Q})$, $W_0 = K$. By construction, $F^\bullet$ defines a Hodge structure of weight 1 on $W_1/W_0 \cong H^1(\widetilde{S}_0)$ and a Hodge structure of weight 0 on W_0.

In general, one may assume that $H_{\mathbf{Z}}$ is free (replacing, for instance, $H^*(X; \mathbf{Z})$ by its image in $H^*(X, \mathbf{R})$). For mixed Hodge structures, the notion of *Hodge number* still makes sense: $h^{p,q}$ is the dimension of the (p, q) component of the pure Hodge structure on the graded quotient Gr_{p+q}^W. In the case of a Riemann surface of genus 2 that has acquired a node, $h^{1,0} = h^{0,1} = 1$ and $h^{0,0} = 1$. There is an obvious notion of morphism of mixed Hodge structure. The kernel, image, and cokernel of a morphism carry a natural mixed Hodge structure. One can also form direct sums, tensor products, and duals, just as for Hodge structures.

The mixed Hodge structure on $H^1(S_0)$ carries more information than is present in the pure Hodge structures W_0 and W_1/W_0. This can be seen from the expression for the differential $\omega_2(0)$. As a cohomology class on S_0, it can be written as

$$\omega_2(0) = \delta^2 + Z_{21}(0)\gamma^1 + Z_{22}(0)\gamma^2,$$

and as a cohomology class on $\widetilde{S}_0$ it is written

$$\omega_2(0) = \delta^2 + Z_{22}(0)\gamma^2.$$

The coefficient $Z_{21}(0)$ can be viewed as the obstruction to an isomorphism of mixed Hodge structures

$$H^1(S_0) \cong W \oplus H^1(\widetilde{S}_0).$$

We have seen above that the period $Z_{21}(0)$ represents the divisor class $p - q$ associated with the inverse image of the node under normalization. Because the divisor class of $p - q$ is never 0 for $p \neq q$, the mixed Hodge structure of the Riemann surface $y^2 = x^2(x - a)(x - b)(x - c)(x - d)$ is never split. For a more formal treatment of these ideas, see Carlson (1980).

We now turn to the problem of whether it is possible to define the limits of Hodge structures. In some cases limits exist as Hodge structures. In general, however, the limit is a mixed Hodge structure. To see what the fundamental idea is, we turn again to degenerations of genus 2 curves and seek natural definitions of weight and Hodge filtrations. For the Hodge filtration, recall that as in (1.22), the period matrix of the degeneration can be written as a standard matrix exponential that carries the singularity of the mapping at $t = 0$, times a matrix that is holomorphic at $t = 0$. Thus,

$$Z(t) = \exp\left(\frac{\mathrm{i}\log t}{\pi}N\right)\hat{Z}(t),$$

where N is the logarithm of the monodromy transformation. Consequently, the period map is given asymptotically by the matrix-valued function

$$\exp\left(\frac{\mathrm{i}\log t}{\pi}N\right)\hat{Z}(0).$$

This asymptotic expression consists of two parts. The first is a part that is the same for all degenerations of the same topological type and is given by the exponential factor. The other is the constant matrix $\hat{Z}(0)$, which depends on the particular degeneration. We use it to define a Hodge filtration where

$$F^1 = \text{row space of } \hat{Z}(0), F^0 = \mathbf{C}^4.$$

The vector space $\mathbf{C}^4$ is viewed as having distinguished basis $\{\delta^1, \delta^2, \gamma^1, \gamma^2\}$ and so is identified with $H^1(S_t; \mathbf{C})$ for any t, with t small. Given this identification, we define a weight filtration by

$$W_0 = \{\gamma^1\}, W_1 = W_0 + \mathbf{Q}\{\delta^2, \gamma^2\}, W_2 = W_1 + \mathbf{Q}\{\delta^1\}.$$

In more intrinsic terms,

$$W_0 = \text{image of } N, W_1 = \text{kernel of } N.$$

Then one can check that $(\mathbf{Q}^4, W_\bullet, F^\bullet)$ is a mixed Hodge structure with weights 0, 1, and 2. Moreover,

$$W_1 \cong H^1(S_0)$$

is a mixed Hodge structure. Thus, in this case, the limit mixed Hodge structure determines the mixed Hodge structure of the central fiber.

What is the type of the Hodge structure W_2/W_1? It is spanned modulo W_1 by δ^1, hence it is also spanned modulo W_1 by

$$\hat{\omega}_1(0) = \delta^1 + \hat{Z}_{11}(0)\gamma^1 + \hat{Z}_{12}(0)\gamma^2, \tag{1.25}$$

that is, by the first row of the matrix $\hat{Z}(0)$. Thus, $W_2/W_1 = F^1 W_2/W_1$. The complex conjugate of the class (1.25) also spans W_2 modulo W_1, from which we conclude that $W_2/W_1 = \bar{F}^1 W_2/W_1$. Thus, referring to the induced filtration on the quotient, we have

$$W_2/W_1 = F^1 \cap \bar{F}^1.$$

Consequently, the quotient is a Hodge structure of type (1, 1).

We have now determined the types that occur in the limit mixed Hodge structure: (0, 0), (1, 0), (0, 1), and (1, 1). We have also determined the meaning of the mixed sub-Hodge structure W_1, which has types (0, 0), (1, 0), and (0, 1). What is the meaning of the quotient mixed Hodge structure W_2/W_0, which has types (1, 0), (0, 1), and (1, 1)? Part of the answer is clear: the sub-Hodge structure W_1/W_0 is isomorphic to $H^1(\widetilde{S}_0)$. For a complete answer, observe that

$$\hat{\omega}_1(0) \equiv \delta^1 + \hat{Z}_{12}(0)\gamma^2 \text{ modulo } W_0.$$

The coefficient $\hat{Z}_{12}(0)$ is the obstruction to an isomorphism of mixed Hodge structures

$$W_2/W_0 \cong H^1(\widetilde{S}_0) \oplus W_2/W_1.$$

We can interpret the coefficient $\hat{Z}_{12}(0)$ as follows. Let S' be the normalization of S_0 minus the two points p and q which correspond to the nodes, as illustrated in Fig. 1.18.

Writing the normalization of S_0 as $\widetilde{S}$, we have the following sequence deduced from the exact sequence of the pair $(\widetilde{S}, S')$:

$$0 \longrightarrow H^1(\widetilde{S}) \longrightarrow H^1(S') \longrightarrow Q \longrightarrow 0,$$

where

$$Q = \text{kernel } [H^1(S', \widetilde{S}) \longrightarrow H^1(\widetilde{S})].$$

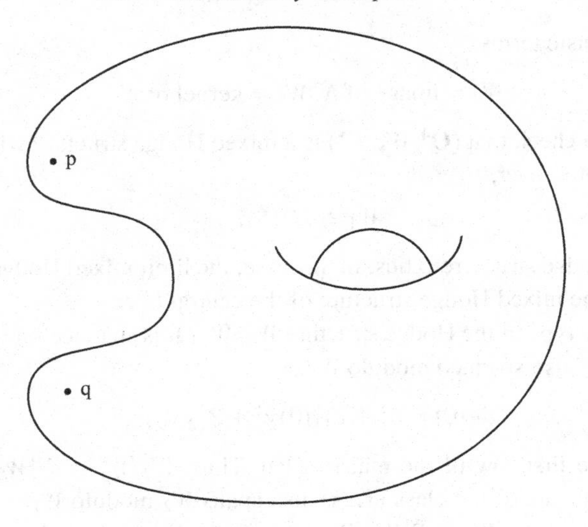

Figure 1.18 The punctured surface.

Define a weight filtration on $H^1(S')$ by setting $W_1 = H^1(\widetilde{S})$ and $W_2 = H^1(S')$. To define the Hodge filtration, let ω_1 be the unique differential of the third kind on $\widetilde{S}$ which has residue $+1$ at p and -1 at q and which has integral zero on δ_2. Let ω_2 be the unique abelian differential on $\widetilde{S}$ normalized to have integral 1 on δ_2. The last condition can be satisfied by adding a suitable multiple of ω_2 to a differential of the third kind with the prescribed residues. Define F^1 to be the span of ω_1 and ω_2, and define F^0 to be $H^1(S'; \mathbf{C})$. The result is a mixed Hodge structure such that

$$H^1(S') \cong W_2/W_0,$$

where the right-hand side refers to the limit mixed Hodge structure. The essential fact is that the periods of the differentials ω_1, ω_2 are given by the rows of the matrix $\hat{Z}(0)$.

As a footnote to this example, we claim that there is an exact sequence of mixed Hodge structures

$$0 \longrightarrow H^1(S_0) \longrightarrow H^1(S_t) \xrightarrow{N} H^1(S_t), \tag{1.26}$$

where $H^1(S_t)$ carries the limit mixed Hodge structure and where N has type $(-1, -1)$. To see this, note that we have shown above that the kernel of N is isomorphic to $H^1(S_0)$. This proves most of our claim already. We leave the rest as an exercise (Problem 1.2.11).

In general, one has the *Clemens–Schmid sequence* (Clemens, 1977):

$$\cdots \longrightarrow H^{k-2}(X, X^*) \longrightarrow H^k(X_0) \longrightarrow H^k(X_t) \xrightarrow{N} H^k(X_t) \longrightarrow \cdots .$$

It holds when $X \longrightarrow \Delta$ is a degeneration with smooth total space and with central fiber X_0, a normal crossing variety with components of multiplicity 1. As explained in Clemens (1977, section 2), a general result, which sometimes is called "semistable reduction", states that any degeneration can be put in this form by modifying X along the central fiber and by replacing the parameter $t \in \Delta$ by t^n for suitable n. The Clemens–Schmid sequence is an exact sequence of mixed Hodge structures where the maps in the exact sequence above are of type $(1, 1)$, $(0, 0)$, $(0, 0)$, and $(-1, -1)$, respectively. This sequence enforces a tight relationship between the geometry of the singular fiber and the linear algebraic properties of the monodromy transformation, $T = \exp(N)$.

The notion of mixed Hodge structure is due to Deligne (1971a). By Deligne (1971b, 1974) the k-th cohomology of any projective variety has natural mixed Hodge structure where $H^k = W_k$ and $W_{-1} = 0$. We say that H^k has weights 0 through k. Thus, as in our example, the weights of the first cohomology of an algebraic curve are 0 and 1. For the k-th cohomology of a smooth variety minus an arbitrary subvariety, the weights run from k to $2k$. We have seen this range of weights in our examples. For a singular Riemann surface the weights are 0 and 1, while for a smooth punctured Riemann surface the weights are 1 and 2. For an arbitrary quasi-projective variety, the range is from 0 to $2k$. This is the same range of weights that one sees for the limit mixed Hodge structure on $H^k(S_t)$, where $\{S_t\}$ is a degeneration.

Deligne's mixed Hodge structure is natural in the sense that if $f : X \longrightarrow Y$ is an algebraic map, then $f^* F^p \subset F^p$ and $f^* W_\ell \subset W_\ell$.

Problems

1.2.1 There are indeed nonhyperelliptic curves. One can see this by looking at smooth plane quartic curves.

(a) Assuming that a plane quartic curve C is also hyperelliptic, prove that the hyperelliptic involution extends to a projective involution I of the plane. Hint: you may use the fact that the canonical bundle K_C is induced by the hyperplane bundle (in Appendices A and C we review some basic facts from algebraic geometry that enable one to prove this).

(b) Show that in suitable homogeneous coordinates (X, Y, Z) we have $I(X, Y, Z) = (-X, Y, Z)$. Deduce that there are four fixed points.

(c) Show that the genus of a smooth plane quartic curve is 3 (use the fact that the degree of K_C is always $2g(C) - 2$) whereas there should be eight branch-points for the quotient $C \to \mathbf{P}^1$ by the hyperelliptic involution.

(d) Show that for any Riemann surface S we have a natural inclusion $\Omega^1(S) \oplus \overline{\Omega^1(S)} \subset H^1(S; \mathbf{C})$. Hint: if a function f satisfies $\partial\bar{\partial} f = 0$, it is harmonic and obeys a maximum principle.

1.2.2 Let $\{S_t\}$ be a family of Riemann surfaces with just two singular fibers, one at $t = 0$, the other at $t = \infty$. Show that the period map for this family is constant. Also, give an example of a family of Riemann surfaces of genus g which has just two singular fibers.

1.2.3 Prove that for any matrix $Z \in \mathfrak{h}_g$ and any symplectic matrix $\begin{pmatrix} K & L \\ M & N \end{pmatrix}$ the matrix $(K + LZ)^{-1}$ is invertible so that the expression for the action of the symplectic group on $\mathfrak{h}_g$ makes sense. Hint: prove that the symplectic action preserves the row space of the matrix $(\mathbf{1}_g, Z)$.

1.2.4 (a) Show that the upper half-plane $\mathfrak{h}_g$ can be written as $\mathrm{Sp}(g)/\mathrm{U}(g)$.

(b) Show that $\mathrm{Sp}(g, \mathbf{Z})$ acts properly discontinuous on $\mathfrak{h}_g$.

(c) Show that the isotropy group of any fixed points of the action of $\mathrm{Sp}(g, \mathbf{Z})$ on $\mathfrak{h}_g$ is a finite group.

(d) Let $g = 2$ and determine the isotropy group Γ of $i\mathbf{1}_g$.

1.2.5 Consider $\Delta^2 = \{(z, w) \mid |z| < 1, \ |w| < 1\}$ and let Γ be the group of transformations

$$(z, w) \mapsto (\pm z, \pm w).$$

Determine the ring of invariant functions and show that Δ^2/Γ has the structure of a quadratic cone.

1.2.6 Show that the algebraic curve $\tilde{y}^2 = f(x)$ is the normalization of the algebraic curve $y^2 = x^2 f(x)$, where $f(x)$ is a polynomial with distinct roots different from 0.

1.2.7 Show that if the module A has a Hodge structure of weight n and the module B has one of weight m, the module $\mathrm{Hom}(A, B)$ inherits a Hodge structure of weight $m - n$ by setting

$$\mathrm{Hom}(A, B)^{i,j} = \left\{ f : A \to B \;\middle|\; f(A^{p,q}) \subset B^{p+i,q+j} \right\}.$$

Show that this is in agreement with our definition for the Hodge structure on the dual of A, thereby showing that this indeed defines a Hodge structure of the asserted weight.

1.2.8 Show that the natural pairing of A and $A^\vee$ defines a morphism

$$A \otimes A^\vee \to \mathbf{Z}$$

of Hodge structures, where, as before, $\mathbf{Z}$ is the trivial Hodge structure.

1.2.9 Work out the full Hodge structure on $H^2(\mathscr{E} \times \mathscr{E})$, where $\mathscr{E}$ is an elliptic curve with holomorphic differential ω. Let $s : (x, y) = (y, x)$ be the symmetry which exchanges factors. Show that the Hodge structure splits into sub-Hodge structures corresponding to the eigenvalues of s^*, and describe these structures as explicitly as possible.

1.2.10 Explain the meaning of the symmetry of the period matrix $\hat{Z}$ (1.22) as a kind of duality between the mixed Hodge structures W_1 and W_2/W_0.

1.2.11 Show that (1.26) is an exact sequence of mixed Hodge structures.

1.3 Double Planes

Let us now consider the topology and Hodge theory of algebraic surfaces. The prototypical example of such an object is the surface S defined by

$$z^2 = f(x, y),$$

where $f(x, y)$ is a polynomial of even degree d. We consider various ways of looking at such surfaces, called double planes. In particular, we explicitly describe a compactification of S analogous to the compactifications discussed above for Riemann surfaces. This preliminary work done, we take up the study of the topology and Hodge theory of these algebraic manifolds.

There are various ways to think of the surface S. First, it is a subset of $\mathbf{C}^3$. If the algebraic curve $f(x, y) = 0$ is smooth, then so is the algebraic surface S. Consequently it is a two-dimensional complex manifold: using the implicit function theorem, one can parametrize open sets on S by open sets in $\mathbf{C}^2$, and the transition functions relating these parametrizations are holomorphic functions. It follows that S is also a four-dimensional real manifold.

Another way of looking at S is as a branched cover of $\mathbf{C}^2$. The map p that sends (x, y, z) to (x, y) sends points of S to points of $\mathbf{C}^2$. This map is surjective, and for each (x, y) in $\mathbf{C}^2$ there are in general two points of S which correspond to it under p. Only when (x, y) is in the branch curve

$$B = \{(x, y) \mid f(x, y) = 0\}$$

is there just one inverse image point. It is for this reason that we call S a *double plane*.

Compactification

We shall now construct an explicit compactification $\bar{S}$ of S which is modeled on the compactification of an affine hypersurface by the corresponding projective hypersurface. To this end, suppose that $f(x, y)$ has even degree d, and let

$$F(w, x, y) = w^d f(x/w, y/w)$$

be the corresponding homogeneous equation. Its zero set $\bar{B} \subset \mathbf{P}^2$ is a compact Riemann surface which compactifies the affine set $B \subset \mathbf{C}^2$ defined by $f(x, y) = 0$. The compactification $\bar{S}$ to be constructed will be endowed with a 2-to-1 projection $p : \bar{S} \longrightarrow \mathbf{P}^2$ whose branch locus is $\bar{B}$ and whose singularities correspond to singularities of $\bar{B}$. Thus, if the branch curve is smooth, then so is the double plane.

To construct $\bar{S}$, consider the affine algebraic set $S' \subset \mathbf{C}^4$ defined by

$$z^2 = F(w, x, y). \tag{1.27}$$

The set S' stands in the same relation to $\bar{S}$ as the affine algebraic set $B' = \{(x, y, z) \in \mathbf{C}^3 \mid F(x, y, z) = 0\}$ stands to the projective algebraic curve $\bar{B}$: the latter is the quotient of the former by an equivalence relation. In the case of $\bar{B}$, the points $(w, x, y) \neq 0$ and $(\lambda w, \lambda x, \lambda y)$ are equivalent for any nonzero scalar. In the present case we impose the equivalence relation

$$(w, x, y, z) \sim (\lambda w, \lambda x, \lambda y, \lambda^{d/2} z).$$

It has the property that a vector equivalent to a solution of (1.27) is also a solution of (1.27). Now consider the set of points in $S'_w \subset S'$ such that $w \neq 0$. The corresponding set in $\bar{S}$, which we denote by $\bar{S}_w$, is open. Moreover, each element of $\bar{S}_w$ has a unique representative in S' such that $w = 1$. Therefore, the open set $\bar{S}_w$ can be identified with the affine hypersurface

$$z^2 = F(1, x, y) = f(x, y).$$

Consequently, $\bar{S}_w$ can be identified with S.

The space $\bar{S}$ is covered by the open sets $\bar{S}_w$, $\bar{S}_x$, $\bar{S}_y$, and $\bar{S}_z$. The sets $\bar{S}_x$ and $\bar{S}_y$ can be treated just as $\bar{S}_w$ was treated. If $F(w, y, z) = 0$ is smooth, then so are $\bar{S}_w$, $\bar{S}_x$, and $\bar{S}_y$. The last set, $\bar{S}_z$, requires a slightly different argument. If $z \neq 0$, then we may impose the normalization $z = 1$. However, points of the form $(w, x, y, 1)$ are not unique in their equivalence class, because $(w, x, y, 1) \sim (\lambda w, \lambda x, \lambda y, 1)$ where $\lambda^{d/2} = 1$. Note, however, that the action by the $d/2$-th roots of unity has no fixed points on the set S'_z defined by the equation $F(w, x, y) = 1$. Thus, S'_z is an unramified $d/2$-fold cover of $\bar{S}_z$. We can view the map $S'_z \longrightarrow \bar{S}_z$ as a kind of generalized coordinate chart for the open set $\bar{S}_z$. For most purposes this chart is sufficient. For example, a function or a differential form on $\bar{S}_z$ is said to be holomorphic if its pullback to S'_z is holomorphic.

Notice that $\bar{S}$ is the disjoint union of $\bar{S}_w$, which we have already identified with S, and the complement of this set, which is the algebraic curve $\bar{C}$ defined by $z^2 = F(0, x, y)$. Thus,

$$\bar{S} = S \cup \bar{C},$$

where $\bar{C}$ is a two-sheeted cover $\mathbf{P}^1$ branched at the $2d$ roots of $F(0, x, y) = 0$. The curve C "at infinity" is thus a double branched cover of the line at infinity used to compactify $\mathbf{C}^2$. From this observation one can deduce that $\bar{S}$ is compact.

Euler Characteristic

Let us now try to understand the topology of a double plane $\bar{S}$ with branch locus $\bar{B}$. The first task is to compute the Euler characteristic. To this end, note that the branch curve can be viewed either as a subset of $\bar{S}$ or as a subset of $\mathbf{P}^2$. Viewed in the former way, we have a decomposition

$$\bar{S} = (\bar{S} - \bar{B}) \cup \bar{B}$$

into two disjoint sets. Using the second point of view, we see that $\bar{S} - \bar{B}$ is a two-sheeted unramified cover of $\mathbf{P}^2 - B$. Thus,

$$e(\bar{S}) = 2e(\mathbf{P}^2 - \bar{B}) + e(\bar{B}).$$

Since the Euler characteristic is additive over disjoint sets, this simplifies to

$$e(\bar{S}) = 2e(\mathbf{P}^2) - e(B).$$

The Euler characteristic of a plane curve of degree d is

$$e(\text{plane curve}, d) = -d^2 + 3d.$$

Thus, bearing in mind that d must be even, we have

$$e(\text{double plane}, d) = d^2 - 3d + 6.$$

Examples 1.3.1 (i) Suppose that the branch locus is a conic (degree 2). A typical affine equation is $z^2 = x^2 + y^2$, and so the surface is a smooth quadric, isomorphic to $\mathbf{P}^1 \times \mathbf{P}^1$. According to the formula, the Euler characteristic is 4, which agrees with $e(\mathbf{P}^1 \times \mathbf{P}^1) = e(\mathbf{P}^1) \times e(\mathbf{P}^1) = 2 \times 2 = 4$.
(ii) Suppose that the branch locus is a quartic (degree 4). Then the Euler characteristic is 10, and the surface is a so-called Del Pezzo surface. It can be obtained by blowing up seven points in $\mathbf{P}^2$.
(iii) Suppose that the branch locus is sextic (degree 6). The Euler characteristic

is 24, and the surface is a so-called $K3$ surface. By definition, a $K3$ *surface* is a surface with $b_1 = 0$ and trivial canonical bundle. That our example has these invariants follows from calculations below: the Betti numbers are calculated in the next subsection, and the holomorphic forms on double planes are investigated further below: see Eq. (1.30).

Betti Numbers

Let us now determine the Betti numbers of a double plane S. First, S is connected, and is, as a complex manifold, oriented. Thus $H^0(S, \mathbf{Z}) \cong H^4(S, \mathbf{Z}) \cong \mathbf{Z}$. By Zariski (1971, §8.1), for any smooth a smooth plane curve C of degree d one has

$$\pi_1(\mathbf{P}^2 - C) \cong \mathbf{Z}/d.$$

The fundamental group is generated by the oriented boundary of a small complex disk which meets the branch curve transversally. If $\bar{B}$ is a smooth curve of degree $2d$, then $\pi_1(\mathbf{P}^2 - \bar{B}) \cong \mathbf{Z}/2d$ and $\pi_1(\bar{S} - \bar{B}) \cong \mathbf{Z}/d$. The latter group is also generated by the oriented boundary of a small complex disk which meets the branch curve transversally. When $\bar{B}$ is attached to $\bar{S} - \bar{B}$ along its normal bundle, the homotopy class of the oriented disk boundary is killed, and so

$$\pi_1(\bar{S}) = \{1\}\,.$$

Thus $H_1(\bar{S})$, which is the abelianization of the fundamental group, is 0. By Poincaré duality, it follows that $H^3(\bar{S}; \mathbf{Z}) = 0$. Thus the middle Betti number is determined:

$$\dim\, H^2(\bar{S}, \mathbf{Q}) = e(S) - 2 = d^2 - 3d + 4.$$

Examples 1.3.2 (i) Suppose that the branch locus is a conic. Then the second Betti number is 2. Thus the rank of the second homology is also 2. Can we find a basis for the homology cycles? Since $\bar{S} \cong \mathbf{P}^1 \times \mathbf{P}^1$, one has cycles $\{a\} \times \mathbf{P}^1$ and $\mathbf{P}^1 \times \{b\}$. More directly, one has projective lines from each of the two rulings of the quadric surface $\bar{S}$. These lines intersect in a single point, say (a, b). The self-intersection of a line is 0, since $\{a\} \times \mathbf{P}^1$ is homologous to the disjoint cycle $\{a'\} \times \mathbf{P}^1$. Thus the intersection matrix of the given set of homology cycles is

$$\begin{pmatrix} 0 & 1 \\ 1 & 0 \end{pmatrix}.$$

It follows that the two rulings generate the homology.
(ii) Suppose that the branch locus is a quartic. Then the second Betti number is 8. An alternative model of $\bar{S}$ is the projective plane blown up at seven points in

general position. Let ℓ be a line which does not pass through the given points. Let E_i be the exceptional locus corresponding to the i-th blown up point p_i. Concretely, E_i is the set of lines through p_i. Thus it is a $\mathbf{P}^1$. The intersection matrix of the cycles $(\ell, E_1, \ldots, E_7)$ is the diagonal matrix $(1, -1, \ldots, -1)$, where there are eight (-1)s. Thus the given cycles are a free basis for the $\mathbf{Z}$-module $H_1(\bar{S}; \mathbf{Z})$.
(iii) Suppose that the branch locus is a sextic. Then the second Betti number is 22. The inverse image of a line in $\mathbf{P}^2$ is a cycle on $\bar{S}$ and one sees that it is nontrivial. Indeed, its self-intersection is 2. Can we find other cycles which generate the second homology? It turns out that this case is completely different from the others. In the previous cases the other homology cycles were *algebraic*, that is, defined by polynomial equations. In the present case, the inverse image of the line, up to multiplying by a rational number, is the *only* algebraic homology cycle, provided that $\bar{S}$ is "generic." Generic varieties, which should be opposed to "special" ones, have the property that they arise from a random choice of coefficients with probability 1. To make this assertion more precise (though not to prove it), observe that the space of polynomials in three homogeneous variables (the space of the $F(w, x, y)$ defining the branch locus) is a vector space of dimension 28. Excluding the zero polynomial and identifying F with λF, we see that the space of all possible branch loci is a 27-dimensional projective space. There is a countable union $\mathcal{F}$ of algebraic hypersurfaces with the property: if $\bar{B} \notin \mathcal{F}$ then all algebraic cycles are rational multiples of the obvious algebraic cycle. This fact is not easy to see and is a consequence of Hodge theory. Further on in this section we give a partial explanation of the assertion just made.

The Hodge Conjecture

In the previous paragraph we observed that for some double planes the middle homology is spanned by algebraic cycles and that for some it is not. When the branch locus has degree two or four, all of the cohomology is algebraic, but when the branch locus has degree six or higher, there exist nonalgebraic cycles. The explanation for this phenomenon is Hodge-theoretic. By de Rham's theorem, every cohomology class with complex coefficients can be represented by a closed differential two-form. If $(x_1, \ldots, x_4)$ are local coordinates, then such a differential form can be written as

$$\phi = \sum_{ij} a_{ij} \, \mathrm{d}x_i \wedge \mathrm{d}x_j,$$

where

$$d\phi = \sum_{ij} \frac{\partial a_{ij}}{\partial x_k} \, dx_k \wedge dx_i \wedge dx_j = 0.$$

Given a two-cycle, one may form the integral

$$\int_c \phi.$$

If $c' = c + \partial A$ is a cycle homologous to c, then

$$\int_{c'} \phi = \int_c \phi.$$

This is because of Stokes' theorem:

$$\int_{\partial A} \phi = \int_A d\phi = 0.$$

Thus ϕ defines a linear functional not just on two-chains, but on two-dimensional homology classes. Therefore it represents a cohomology class.

Because $\bar{S}$ carries a complex structure, we can distinguish several types of differential forms. Those of type $(2, 0)$ are those which can be written locally as $f(z, w)\, dz \wedge dw$, where z and w are local holomorphic coordinates. The coefficients of a $(2, 0)$-form that is closed are holomorphic. This is because

$$\begin{aligned} d(f \, dz \wedge dw) &= \frac{\partial f}{\partial z} \, dz \wedge dz \wedge dw + \frac{\partial f}{\partial w} \, dw \wedge dz \wedge dw \\ &\quad + \frac{\partial f}{\partial \bar{z}} \, d\bar{z} \wedge dz \wedge dw + \frac{\partial f}{\partial \bar{w}} \, d\bar{w} \wedge dz \wedge dw \\ &= \frac{\partial f}{\partial \bar{z}} \, d\bar{z} \wedge dz \wedge dw + \frac{\partial f}{\partial \bar{w}} \, d\bar{w} \wedge dz \wedge dw. \end{aligned}$$

Thus the Cauchy–Riemann equations hold for f if $f \, dz \wedge dw$ is closed:

$$\frac{\partial f}{\partial \bar{z}} = 0 \text{ and } \frac{\partial f}{\partial \bar{w}} = 0.$$

Consequently, closed $(2, 0)$-forms on an algebraic surface are, like closed $(1, 0)$-forms on a Riemann surface, holomorphic.

In addition to the conjugates of $(2, 0)$-forms, which have the local expression $g \, d\bar{z} \wedge d\bar{w}$, and which are said to "be of type $(0, 2)$," there are forms of mixed type, or "type $(1, 1)$." These have the local form

$$\phi = a \, dz \wedge d\bar{w} + b \, d\bar{w} \wedge dz.$$

We cannot say anything interesting in general about the coefficients of a closed $(1, 1)$-form, except that they are C^∞ functions.

As with Riemann surfaces, the closed forms of type (p, q) define a subspace of cohomology which we write as $H^{p,q}(S)$. What is more difficult to establish, but is nonetheless true, is that these subspaces give a direct sum decomposition:

$$H^2(S;\mathbf{C}) = H^{2,0}(S) \oplus H^{1,1}(S) \oplus H^{0,2}(S).$$

This is the Hodge decomposition for two-cohomology. It holds for all algebraic surfaces, but not for all complex surfaces, and it has the additional property that complex conjugation reverses type:

$$\overline{H^{p,q}} = H^{q,p}.$$

Let us reconsider the nature of algebraic cycles on an algebraic surface. Such an object is formally a sum of algebraic curves

$$Z = n_1 Z_1 + \cdots + n_k Z_k,$$

where the n_i are integers and the Z_i are algebraic curves, that is, Riemann surfaces with possibly a finite number of singularities. Given a closed two-form ϕ, one can form the integral

$$\int_Z \phi = n_1 \int_{Z_1} \phi + \cdots + n_k \int_{Z_k} \phi.$$

The component integrals have the form

$$\int_W \phi,$$

where W is an algebraic curve. Given a smooth point p of W, consider local holomorphic coordinates (z, w) where $w = 0$ defines W. Then the integrand $\phi = f(z, w)\, dz \wedge dw$ vanishes identically on W, and so we have

$$\int_W \phi = 0$$

for any closed form of type $(2, 0)$. We have established the following important principle.

Theorem 1.3.3 *Let ϕ represent a cohomology class of type $(2, 0)$, and let Z be an algebraic cycle. Then the equation*

$$\int_Z \phi = 0$$

is satisfied.

The preceding theorem gives a necessary condition for a homology class to be algebraic, that is, to be represented as an algebraic cycle. That this is also sufficient has been shown by Lefschetz and Hodge (see Hodge, 1959, pp.

214–216). The modern proof due to Kodaira and Spencer (1953) applies more generally to divisors, i.e., algebraic cycles of codimension 1 in any complex projective manifold.

Theorem 1.3.4 *Let S be an algebraic surface and let $c \in H_2(S, \mathbf{Q})$ be a nonzero rational homology class. Suppose that*

$$\int_c \phi = 0$$

for all $\phi \in H^{2,0}(S)$. Then there is an integer m such that mc is represented by an algebraic cycle.

One often states the preceding result in purely cohomological terms. A class $\alpha \in H^2(S, \mathbf{Q})$ has a decomposition

$$\alpha = \alpha^{2,0} + \alpha^{1,1} + \alpha^{0,2},$$

where $\alpha^{p,q} \in H^{p,q}$. Suppose that it is of type $(1,1)$ in the sense that $\alpha = \alpha^{1,1}$. Then Theorem 1.3.4 asserts that α is Poincaré dual to a rational multiple of an algebraic cycle.

One can attempt to generalize the previous result to projective algebraic manifolds of higher dimension. If $Z \subset M$ is an algebraic subvariety of complex dimension p, then one finds that

$$\int_Z \phi = 0$$

if ϕ is a cohomology class of type $(r, s) \neq (p, p)$. Thus an algebraic cycle of dimension p – a homology class given by a rational linear combination of subvarieties of dimension p – satisfies a certain set of equations given by the vanishing of certain integrals. The *Hodge conjecture* asserts that this set of equations is sufficient, up to rational multiples. The conjecture has been verified for $p = 1$ and $p = n - 1$, where n is the dimension of M. The latter case, as we saw above, is due to Kodaira and Spencer (1953), and the former follows by Lefschetz duality, as we see in Section 3.4.

Holomorphic Two-Forms

We have seen that the presence of forms of type $(2,0)$ on an algebraic surface controls the extent to which homology classes can be represented by algebraic cycles. We have also asserted double planes with branch curves of degree six or more carry such forms which are nonzero. To see why this is so, consider the expression

$$\phi = \frac{a(x, y)\ dx \wedge dy}{\sqrt{f(x, y)}}, \tag{1.28}$$

for a double plane $z^2 = f(x, y)$. This expression generalizes in a natural way the expression used to define holomorphic one-forms on branched double covers of the projective line. We will show that it defines a holomorphic two-form on $\bar{S}$ provided that

$$\text{degree } a \leq \frac{1}{2}\text{degree} f - 3. \tag{1.29}$$

Thus double planes with branch curves of degree six or more have nonzero holomorphic two-forms.

There are two issues to confront in establishing the previous result. The first is to show that ϕ is holomorphic at points of S. The argument here follows the same reasoning as with the corresponding expression in the case of Riemann surfaces. Because

$$\phi = \frac{a(x, y)\ dx \wedge dy}{z},$$

the given form is clearly holomorphic on the set $z \neq 0$. For the set $z = 0$, differentiate the relation $z^2 = f(x, y)$ to obtain $2z\ dz = f_x\ dx + f_y\ dy$. Multiply by dy to find $2z\ dz \wedge dy = f_x\ dx \wedge dy$. Then write

$$\phi = \frac{2a(x, y)\ dz \wedge dy}{f_x},$$

and note that there is a similar expression with f_y in the denominator. Since the equation $f(x, y) = 0$ is assumed to define a smooth algebraic curve, at least one of the partial derivatives is nonzero at each of its points. Thus, at least one of the possible expressions for ϕ is manifestly holomorphic, as required.

The second issue is to show that ϕ extends to a holomorphic differential on $\bar{S}$ at the points "at infinity," that is, at the points of $\bar{S} - S$. For this it is sufficient to show that there is a compatible set of holomorphic forms defined on the open sets $\bar{S}_w$, $\bar{S}_x$, $\bar{S}_y$, and $\bar{S}_z$. To this end, consider the form Φ deduced from

$$\phi = \frac{a(x, y)\ dx \wedge dy}{z}$$

by the substitutions $x \to x/w$, $y \to y/w$, $z \to z/w^{d/2}$. The result is

$$\Phi = \frac{w^{d/2-a-3}A(w, x, y)(w\ dx \wedge dy - x\ dw \wedge dy + y\ dw \wedge dx)}{z},$$

a form which is defined on $\mathbf{C}^4$ and which defines forms on $\bar{S}_w$, etc., upon the substitutions $w = 1$, etc. These substitutions define a compatible set of forms on the given open sets, and the form defined on $\bar{S}_w$ is ϕ. The argument of the previous paragraph shows that the forms on $\bar{S}_x$ and $\bar{S}_y$ are holomorphic if $F(w, x, y) = 0$ is smooth. The open set $\bar{S}_z$ remains to be considered. On this set the resulting form is

$$w^{\frac{d}{2}-a-3} A(w, x, y)(w\ dx \wedge dy - x\ dw \wedge dy + y\ dw \wedge dx).$$

It is holomorphic if and only if $a \leq d/2 - 3$, as claimed.

We have shown that the expressions (1.28) subject to condition (1.29) define holomorphic two-forms on the given double plane. One can show that all holomorphic two-forms are given by such expressions. Therefore, the Hodge number $h^{2,0}$ is the dimension of the space of polynomials of degree at most $d/2 - 3$ in two variables. This is the same as the dimension of the space of polynomials of degree exactly $d/2 - 3$ in three variables. Thus,

$$h^{2,0} = (d/2 - 1)(d/2 - 2)/2.$$

Recall that the degree of the branch curve is d.

From our computation of the middle Betti number and of $h^{2,0}$ we find that

$$h^{1,1} = \frac{3}{4}d^2 - \frac{3}{2}d + 2.$$

Thus, the Hodge numbers are determined. For a double plane with branch locus of degree two the vector of Hodge numbers $(h^{2,0}, h^{1,1}, h^{0,2})$ equals $(0, 2, 0)$. For a quartic branch locus, this vector is $(0, 8, 0)$, and for the sextic case it is $(1, 20, 1)$. If the branch locus has affine equation $f(x, y) = 0$, where f has degree six, then all holomorphic two-forms are proportional to

$$\phi = \frac{dx \wedge dy}{f(x, y)^{\frac{1}{2}}}, \tag{1.30}$$

quite like the case of elliptic curves. Since $K3$ surfaces play a recurring role later on, let us set apart what we found:

$$S \text{ a } K3 \text{ double plane} \implies h^{2,0}(S) = h^{0,2}(S) = 1, \quad h^{1,1}(S) = 20. \tag{1.31}$$

Periods

To describe the Hodge decomposition of an algebraic surface in somewhat more explicit terms, consider a basis $c_1, \ldots, c_n$ for the second homology and a basis $\alpha^1, \ldots, \alpha^k$ for $H^{2,0}$. The period matrix of the surface relative to these bases is the matrix with entries

$$P^i_j = \int_{c_j} \alpha^i.$$

Let c^i be the cohomology basis dual to the c_j in the sense that

$$c^i(c_j) = \delta_{ij}$$

is 1 if $i = j$ and is 0 otherwise. Then

$$\alpha^i = P^i_1 c^1 + P^i_2 c^2 + \cdots P^i_n c^n,$$

and so P determines $H^{2,0}$ as a subspace of $H^2(S; \mathbf{C})$. More precisely, the row space of P is the same as the subspace of $\mathbf{C}^n$ corresponding to $H^{2,0}$ via the identification $H^2(S; \mathbf{C}) \longrightarrow \mathbf{C}^n$ which sends a class α to the vector with coordinates $\alpha(c_i)$. In this way we can associate with a surface S and a choice of homology basis a k-dimensional subspace of $\mathbf{C}^n$.

Consider now a family of algebraic surfaces S_t, for example, the pencil of double planes with branch curve B_t given by $F(w, x, y) + tG(w, x, y)$. As with the family of Riemann surfaces considered above, it is possible to construct a "topologically trivial" family of homology bases $(c_i(t))$. Likewise, it is possible to construct a family of bases for the space of holomorphic two-forms. In the case of the family of double covers just considered, one can take

$$\alpha_i = \frac{a_i \, dx \wedge dy}{\sqrt{f_t}},$$

where $f_t = f(x, y) + tg(x, y)$ and $a_i(x, y)$ is a basis for the space of polynomials of degree at most $d/2 - 3$, and where d is the degree of f. One can then show that the map from the t-disk to the Grassmann variety of k-planes in n-space is holomorphic. This result, whose proof can be constructed along the lines of that presented for Riemann surfaces, gives a crude version of the period map for algebraic surfaces. In the case of a double plane with a sextic branch curve, the crude period matrix is a vector in $\mathbf{C}^{22}$ and its row space is the corresponding line. Thus the period matrix takes values in $\mathbf{P}^{21}$. We come back to this later.

A closer study of the period matrix reveals that it is not an arbitrary $k \times n$ matrix. This is because the subspace $H^{2,0}$ that it represents is not an arbitrary subspace of the vector space $H^2(S; \mathbf{C})$. To understand the kind of restrictions naturally imposed, consider the symmetric bilinear form

$$Q(\alpha, \beta) = \int_S \alpha \wedge \beta.$$

If α and β are elements of $H^{2,0}$, then the integrand has the local form

$$f(z, w) \, dz \wedge dw \wedge g(z, w) \, dz \wedge dw.$$

This expression vanishes since $dz \wedge dz = 0$. Thus $H^{2,0}$ is an isotropic space for the bilinear form given by the cup product: a space on which the cup product is identically 0. Note the analogy with the first bilinear relation for Riemann surfaces.

To put the analogy just noted in more concrete form, let $Q(x, y)$ denote the cup product and write $Q(c^i, c^j) = Q^{ij}$ and $\phi^i = \sum_j P^i_j c^j$, where $\{c^i\}$ is a basis for $H^2(S; \mathbf{Z})$ and $\{\phi^j\}$ is a basis for $H^{2,0}(S)$. Because $H^{2,0}$ is an isotropic subspace for Q, one has $Q(\phi^i, \phi^j) = 0$, which translates as

$$\sum Q^{mn} P^i_m P^j_n = 0,$$

where the summation is over repeated indices. Thus the period map takes values in a subvariety X of the Grassmannian defined by quadratic equations. In the case of a double plane with a sextic branch locus, the variety X is a 20-dimensional quadric in $\mathbf{P}^{21}$. However, the integral

$$\int_C \phi$$

vanishes for the curve C defined by $x = 0$, or for any of the homologous subvarieties defined by the vanishing of a linear function of the coordinates x, y. Equivalently, the period vector $(P_0, \ldots, P_{21})$ satisfies a fixed linear relation, which we can, by a suitable choice of homology basis, assume to be $P_{21} = 0$. Therefore the period map takes values in a fixed 19-dimensional quadric lying in a fixed 20-dimensional projective space.

Now consider the expression

$$Q(\alpha, \bar{\alpha}) = \int \alpha \wedge \bar{\alpha},$$

where again α is a holomorphic two-form. The integrand is locally

$$|f(z, w)|^2 \, \mathrm{d}z \wedge \mathrm{d}w \wedge \mathrm{d}\bar{z} \wedge \mathrm{d}\bar{w} = -|f(z, w)|^2 \, \mathrm{d}z \wedge \mathrm{d}\bar{z} \wedge \mathrm{d}w \wedge \mathrm{d}\bar{w}.$$

However, $(\mathrm{i}\, \mathrm{d}z \wedge \mathrm{d}\bar{z}) \wedge (\mathrm{i}\, \mathrm{d}w \wedge \mathrm{d}\bar{w})$ defines the natural orientation for the complex manifold, and so the integrand is positive. Thus, we have

$$Q(\alpha, \bar{\alpha}) > 0$$

for any nonzero element of $H^{2,0}$. This relation, which says that $H^{2,0}$ is a positive subspace for the Hermitian form $Q(\cdot, \bar{\cdot})$, is the analogue of the second bilinear relation for Riemann surfaces.

Via the correspondence of $H^2(S; \mathbf{C})$ defined by a choice of basis, we can transfer the bilinear and Hermitian forms discussed above to $\mathbf{C}^n$. The set of all isotropic, positive k-dimensional subspaces is the period domain for the given Hodge structures. It is an open subset D (defined by the second relation) of a closed submanifold X (defined by the first relation) of the Grassmannian:

$$D \subset X \subset \text{Grassmannian}.$$

The assertion made above is that the map from the t-disk to D defined by the period matrix is holomorphic.

We come now to the part of Hodge theory that is genuinely different in dimension two. This is the *Griffiths infinitesimal period relation* or *transversality relation* (see Griffiths, 1968), which in the present context asserts that $H^{2,0}(S_t)$ cannot vary arbitrarily as a subspace of $H^2(S_t; \mathbf{C})$. For this statement to have meaning, we must choose an identification of $H^2(S_t; \mathbf{C})$ with the fixed subspace $H^2(S_0; \mathbf{C})$. This we do in a topologically natural way, as we did in the case of Riemann surfaces. Fix a basis $c^1(0), \ldots, c^n(0)$ for $H^2(S_0)$ and let

$$c^i(t) = (\phi_t^*)^{-1} c^i(0),$$

where $\phi_t : S_0 \longrightarrow S_t$ is the trivializing diffeomorphism. The Griffiths relation asserts that

$$\left.\frac{\mathrm{d}}{\mathrm{d}t}\right|_{t=0} H^{2,0}(S_t) \subset H^2(S_0) + H^{1,1}(S_0).$$

To explain the meaning of this assertion, we need a definition of the derivative of a variable family of subspaces of a fixed vector space. The derivative of $H^{2,0}(S_t)$, identified as a subspace of $\mathbf{C}^n$, is defined as follows. Choose a basis $\alpha^1(t), \ldots, \alpha^k(t)$ for $H^{2,0}(S_t)$ so identified. Then set

$$\left.\frac{\mathrm{d}}{\mathrm{d}t}\right|_{t=0} H^{2,0}(S_t) := \text{Span of } \dot{\alpha}^1(0), \ldots, \dot{\alpha}^k(0) \text{ modulo } H^{2,0}(S_0).$$

One checks that the definition makes sense and that, if it is identically 0, then $H^{2,0}(S_t) = H^{2,0}(S_0)$ for all t close to 0. We can also interpret this procedure geometrically by viewing the variable subspaces $H^{2,0}(S_t)$ inside $H^2(S_0; \mathbf{C})$ as giving a curve in the Grassmannian, and what we defined above simply is the tangent space at the origin to this curve. See also Problem C.1.9 in Appendix C.

A priori, the derivative is simply a subspace of $H^2(S_0; \mathbf{C})$ modulo $H^{2,0}(S_0)$. Griffiths' relation asserts that it lies in the subspace $H^{2,0} + H^{1,1}$. This is a proper subspace of the cohomology if and only if there are nonzero holomorphic two-forms. For double planes, the relation is nontrivial if the degree of the branch locus is at least six.

At this stage we cannot give a complete proof of Griffiths' relation. However, we can explain the main idea. First, note that the relation makes a statement about a filtration of the cohomology. This is the Hodge filtration, defined by

$$F^2 = H^{2,0}, F^1 = H^{2,0} + H^{1,1}, F^0 = H^{2,0} + H^{1,1} + H^{0,2}.$$

Note that it satisfies the property

$$F^0 \supset F^1 \supset F^2,$$

i.e., is a decreasing filtration. The general form of the Griffiths relation is

$$\dot{F}^p \subset F^{p-1}.$$

Thus, differentiation moves the filtration infinitesimally by at most one step, as in the case of Frenet frames for space curves.

Now let us reconsider the cohomology of a variable family of double planes. Fixing a polynomial $a(x, y)$ but letting the branch locus vary in the one-parameter family defined by $f_t = f(x, y) + tg(x, y)$, we construct a one-parameter family of holomorphic two-forms,

$$\alpha_t = \frac{a\, dx \wedge dy}{f_t^{\frac{1}{2}}}.$$

The derivative of a family of cohomology classes is again a family of cohomology classes, and that derivative is given explicitly as

$$\dot{\alpha}_t = -\frac{1}{2}\frac{ga\, dx \wedge dy}{f_t^{\frac{3}{2}}}.$$

The second derivative is also a family of cohomology classes, given explicitly as

$$\ddot{\alpha}_t = \frac{1}{3}\frac{g^2 a\, dx \wedge dy}{f_t^{\frac{5}{2}}}.$$

These computations suggest that there is another filtration of the cohomology, given by order of the fractional pole. We know that F^2 is represented by classes with denominator $f^{1/2}$. The space of cohomology classes with denominator $f^{3/2}$ includes the space of classes with denominator $f^{1/2}$. Thus, it is natural to ask whether this space corresponds to F^1 and, likewise, whether the space with denominator $f^{5/2}$ corresponds to F^0, that is, to the full cohomology. In Section 3.2 we come back to a similar but simpler situation, the case of a smooth hypersurface of projective space. We then see that the analogues of these statements, with one technical adjustment (replace "cohomology" by "primitive cohomology") are all true: the pole and Hodge filtrations coincide for smooth varieties. In this context the Griffiths relation, which is the foundation for much of higher-dimensional Hodge theory, holds.

One application of the technique used to motivate the Griffiths relation is a calculation of the derivative of the period map. Consider, for example, the case of sextic branch curves. Then there is just one nonzero holomorphic two-form, up to constant multiples, namely

$$\alpha_t = \frac{dx \wedge dy}{f_t^{\frac{1}{2}}}.$$

Its derivative is

$$\dot{\alpha}_t = -\frac{1}{2}\frac{g \, dx \wedge dy}{f_t^{\frac{3}{2}}}.$$

However, the expression on the right is to be taken modulo $H^{2,0}$, that is, modulo α_t. Moreover, we consider these expressions not as differential forms, but as cohomology classes. Thus the derivative is 0 if and only if it is cohomologous to a form with denominator $f^{1/2}$. In short, the actual derivative is 0 if we can reduce the order of the denominator of the formal derivative.

How might the order of the pole be reduced? A natural way to proceed is to add the exterior derivative of

$$\frac{a \, dx + b \, dy}{f^{\frac{1}{2}}},$$

namely

$$-\frac{1}{2}\frac{(bf_x - af_y) \, dx \wedge dy}{f^{\frac{3}{2}}}.$$

Thus if g is in the ideal generated by the partial derivatives of f – the so-called Jacobian ideal – then the order of the pole can be reduced. In Problem 1.3.4 we give an application of this idea.

Problems

1.3.1 Show that the double plane $z^2 = f(x, y)$ is connected.

1.3.2 Consider the two-torus $T = E_1 \times E_2$, where $E_i = \mathbf{C}/(\mathbf{Z} + \tau_i\mathbf{Z})$, $\tau_i \in \mathfrak{h}$, $i = 1, 2$.

(a) Let $\{\delta_i, \gamma_i\}$, $i = 1, 2$ be a basis of $H_1(E_i)$. Using the Künneth formula, write down a basis for $H_2(T)$.

(b) The one-form dz defines a one-form ω_{E_1} on E_1 and ω_{E_2} on E_2 and hence gives a basis for the cotangent bundle Ω^1_T. There results a basis $\{\omega_1, \omega_2\}$ for $H^{1,0}(T)$. Express the Hodge decomposition of $H^2(T; \mathbf{C})$ in terms of these forms.

(c) Verify that the two cycles $E_1 \times e_2$ and $e_1 \times E_2$ have type $(1, 1)$ and give conditions on τ_1 and τ_2 guaranteeing there are more algebraic cycles on T. Give an example where the maximum number, 4, is attained.

1.3.3 Let $S \subset \mathbf{P}^3$ be a smooth algebraic surface. Let $\{x, y, z\}$ be affine coordinates and assume that $f = 0$ is an affine equation for S.

(a) Show that on the set where f_z is nonzero, a rational two-form on S without poles is given by

$$\frac{a(x, y, z)\, \mathrm{d}x \wedge \mathrm{d}y}{f_z(x, y, z)}, \quad \deg a \leq \deg f - 3.$$

Use this to give an explicit basis for $H^{2,0}(S)$.

(b) Calculate the Euler characteristic of the Fermat surface S_d given by $x^d + y^d + z^d + 1 = 0$ by projecting it onto the (x, y)-plane. Using the fact that S_d is simply connected, use this to calculate the Betti numbers of S_d. Using the fact that all smooth degree d surfaces are diffeomorphic (this will be proven in Section 4.1), determine the Hodge numbers of S.

(c) Now let $d = 3$ and let C be a smooth cubic surface. Show that $b_2(C) = h^{1,1}(S) = 7$ and hence S has no periods. However, C does have moduli: it is a classical fact (see e.g. Griffiths and Harris, 1978, p. 489) that C is isomorphic to the projective plane blown up in 6 general points and since 4 of them can be chosen fixed by the transitive action of PGL(3), there remain 2 variable points, i.e., 4 moduli.

1.3.4 Consider the family $X_0^6 + X_1^6 + X_2^6 + tX_0^2X_1^2X_2^2$ of sextic curves in $\mathbf{P}^2$ and let S_t be the corresponding family of double planes. Show that the corresponding period map is nontrivial.

1.4 Mixed Hodge Theory Revisited

We have already seen that Hodge theory can be generalized to apply not only to smooth projective varieties of any dimension, but also to singular varieties, open varieties, and to degenerations (the notion of a limit mixed Hodge structure). Although the general theory is elaborate and technically complex, it is possible to give a construction of Hodge theory for special singular varieties that introduces many of the main ideas and applies to sufficiently many examples to be of interest. To describe this construction, let us first consider the case of a singular projective variety X that is the union of two smooth projective components X_1 and X_2:

$$X = X_1 \cup X_2.$$

Let $X_{12} = X_1 \cap X_2$ be the intersection of these components, and let us suppose that this variety is also smooth. This situation is shown schematically in Fig. 1.19.

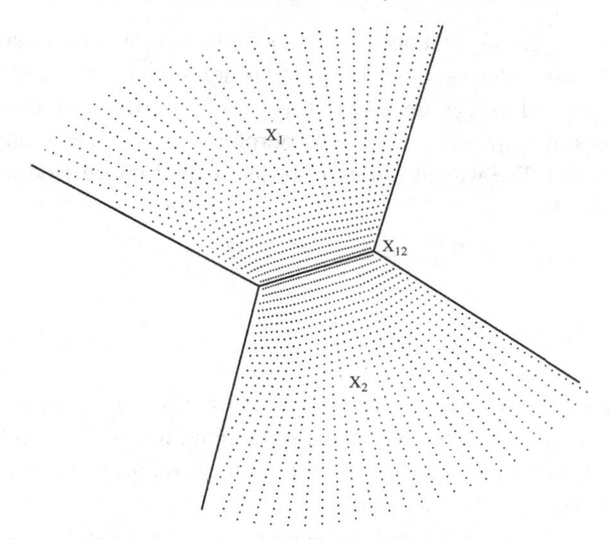

Figure 1.19 Smooth varieties intersecting.

One basic example is a double plane with affine equation $z^2 = f(x, y)^2$. In this case, that of a double branch curve, the variety X consists of two copies of projective two-space glued along the Riemann surface X_{12} defined by the affine equation $f(x, y) = 0$. Another fundamental example is given by the projective hypersurface $G_1(x)G_2(x) = 0$. Its two components are the hypersurfaces X_i defined by $G_i(x) = 0$, where $x = (x_0, \ldots, x_n)$. They meet along the set X_{12} defined by the two equations $G_1(x) = 0$ and $G_2(x) = 0$.

The natural tool for analyzing the cohomology of X is the Mayer–Vietoris sequence (see Appendix B):

$$\cdots H^{m-1}(X_1) \oplus H^{m-1}(X_2) \to H^{m-1}(X_{12}) \to H^m(X)$$
$$\to H^m(X_1) \oplus H^m(X_2) \to H^m(X_{12}) \cdots$$

From it we deduce the short exact sequence

$$0 \longrightarrow K \longrightarrow H^m(X) \longrightarrow Q \longrightarrow 0,$$

where

$$K = \text{cokernel } [H^{m-1}(X_1) \oplus H^{m-1}(X_2) \longrightarrow H^{m-1}(X_{12})] \tag{1.32}$$

and

$$Q = \text{kernel } [H^m(X_1) \oplus H^m(X_2) \longrightarrow H^m(X_{12})]. \tag{1.33}$$

The cohomology groups that occur in the definitions of K and Q carry Hodge structures, because the varieties in question are smooth and compact. The maps in (1.32) and (1.33) are morphisms of Hodge structure, because they are differences of maps induced by holomorphic maps. Hence K and Q carry Hodge structures. The weight filtration for the natural mixed Hodge structure on X is defined by

$$W_{m-1} = K, \quad W_m = H^m(X),$$

so that

$$W_m / W_{m-1} \cong Q$$

is an isomorphism of Hodge structures. To define the Hodge filtration, one needs a way of computing cohomology using differential forms. Then cohomology classes in the p-th level of the Hodge filtration are represented by differential forms which have at least p dz's.

Although we do not work out the details here, the basic idea is simple: a differential k-form on X is given by differential forms α_1 on X_1 and α_2 on X_2 that are compatible on X_{12}. By compatible we mean that $i_1^*\alpha_1 = i_2^*\alpha_2$, where $i_j : X_{12} \longrightarrow X_j$ is the inclusion. If we denote by $A^\bullet(X)$ the complex of such forms, then it sits in a short exact sequence of complexes

$$0 \longrightarrow A^\bullet(X) \longrightarrow A^\bullet(X_1) \oplus A^\bullet(X_2) \longrightarrow A^\bullet(X_{12}) \longrightarrow 0.$$

The associated long exact sequence of cohomology is the Mayer–Vietoris sequence for de Rham cohomology. This is the technical tool used to show that the given Hodge and weight filtrations define a mixed Hodge structure.

We give an example. Take the double plane X with affine equation $z^2 = g(x, y)^2$, where $g(x, y) = 0$ defines a smooth cubic curve C. The mixed Hodge structure of $H^2(X)$ has weights 1 and 2, and the associated Hodge structures are given as follows. Let A and B be two copies of $\mathbf{P}^2$, so that $X = A \cup B$ with $A \cap B = C$. Since $\mathbf{P}^2$ is simply connected, Eq. (1.32) yields

$$K \cong H^1(C).$$

Thus K is the Hodge structure of an elliptic curve. Using Eq. (1.33), we see that

$$Q \cong \mathbf{Z}(-1),$$

where a generator is given by a pair of compatible generators of $H^2(\mathbf{P}^2)$, one for A and one for B. Thus the mixed Hodge structure of X sits in an exact sequence

$$0 \longrightarrow H^1(C) \longrightarrow H^2(X) \longrightarrow \mathbf{Z}(-1) \longrightarrow 0,$$

where $\mathbf{Z}(-1)$ is the trivial Hodge structure of rank 1 and type $(1, 1)$. Unlike our previous examples, this mixed Hodge structure is split:

$$H^2(X) \cong H^1(C) \oplus \mathbf{Z}(-1).$$

Note that $H^2(X)$, viewed as a mixed Hodge structure, contains all the information needed to build the double plane X. Indeed, $H^2(X)$ determines $W_1 H^2(X)$, which is the Hodge structure of an elliptic curve isomorphic to C. Because the mixed Hodge structure is split, it contains no more information, which is as it should be.

For an example where the mixed Hodge structure does *not* split, consider the reducible quartic hypersurface $X = A \cup B \subset \mathbf{P}^3$, where A is a plane and B is a cubic surface. We assume A and B meet transversally, that is, with distinct tangent planes at each point of C. Thus C is a smooth cubic curve. The mixed Hodge structure in this case is also an extension of pure Hodge structures. A proof of this as a problem with hints is given below (Problem 1.4.1). However, unlike the mixed Hodge structure of the first example, it is not split. Although a full treatment of this phenomenon would take us too far afield, we can say what the basic idea is; details are available in Carlson (1980). Consider two skew lines L and L' which lie on the cubic surface. These lines meet C in points p and p'. The divisor class of $p - p'$ is given by a line integral on C that expresses part of the obstruction to splitting the mixed Hodge structure of X. A careful analysis of these obstructions shows that there is just enough information to build the quartic surface X from the mixed Hodge structure of $H^2(X)$.

Mapping Cones

Although very few singular varieties are of the form $A \cup B$, the ideas behind the Mayer–Vietoris sequence can be extended in a substantial yet simple way so as to treat far more general cases. One generalization is to consider arbitrary unions $A_1 \cup \cdots \cup A_n$ which are in "normal crossing form." These have local analytic equations of the form $z_1 \cdots z_k = 0$. Replacing the Mayer–Vietoris exact sequence by the Mayer–Vietoris spectral sequence (see e.g. Bott and Tu, 1982), one arrives at a general theory of varieties of this kind. Another generalization, which we discuss now, uses the idea of the mapping cone associated with a generalized cover. With this construction we can treat more complicated singularities.

Let us begin by re-examining the Mayer–Vietoris sequence from a more abstract point of view. To this end, consider a space X which can be written as the union of two subspaces A and B. Thus $X = A \cup B$, and $Y = A \cap B$ is the intersection. All of the relevant information is contained in the diagram

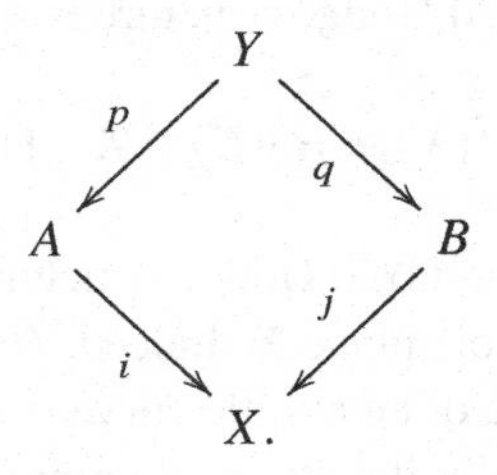

All the maps in the diagram are inclusions, and the data of these spaces and maps, specifically the diagram

$$K(p, q) = [A \overset{p}{\longleftarrow} Y \overset{q}{\longrightarrow} B],$$

are sufficient to define the Mayer–Vietoris sequence and so to compute the cohomology of X. We call $K(p, q)$ the *mapping cone* associated with the pair (p, q).

Associated with the geometric mapping cone is a mapping cone of cochain complexes. To describe it, suppose we are given a map of complexes

$$L^\bullet \overset{\phi}{\longrightarrow} M^\bullet.$$

Such a map defines a new complex $K(\phi)$, where

$$K^p = L^{p+1} \oplus M^p$$

and where the differential $\delta : K^p \longrightarrow K^{p+1}$ is given by

$$\delta(\ell, m) = (-\delta(\ell), -\phi(\ell) + \delta(m)).$$

The complex $(K(\phi), \delta)$ is usually called the (algebraic) *mapping cone* of ϕ, denoted Cone(ϕ). It fits in an exact sequence as we now explain. Clearly $M^\bullet$ is a subcomplex of $K^\bullet$. To ensure that the projection induces a morphism of complexes $K^p \longrightarrow L^{p+1}$, we introduce for any complex M the shifted complex $M[k]$ where $M[k]^p = M^{p+k}$ and where the differential is given by $\delta_{M[k]} = (-1)^m \delta_M$. Thus $H^p(M[k]) = H^{p+k}(M)$. Then $L^\bullet[1]$ is indeed a quotient complex of $K^\bullet$ and so we obtain an exact sequence of complexes

$$0 \longrightarrow M^\bullet \longrightarrow K^\bullet \longrightarrow L^\bullet[1] \longrightarrow 0.$$

Let us apply this construction to the map

$$C^\bullet(A) \oplus C^\bullet(B) \xrightarrow{p^* - q^*} C^\bullet(Y),$$

where $C^\bullet$ stands for the cochain complexes in the simplicial or singular setting (see Appendix B). The associated long exact sequence of cohomology for $\mathrm{Cone}(i^* - j^*)$ reads as follows:

$$\cdots \longrightarrow H^{p-1}(Y) \longrightarrow H^{p-1}(K^\bullet) \longrightarrow H^p(A) \oplus H^p(B) \xrightarrow{\delta} H^p(Y) \longrightarrow \cdots .$$

This sequence looks like the Mayer–Vietoris sequence, with $H^{p-1}(K^\bullet)$ playing the role of $H^p(X)$. Indeed, one can prove that this is so by comparing the above sequence with the Mayer–Vietoris sequence and using the five lemma.

To summarize, a cover of X by two sets A and B defines both a geometric mapping cone $K(i, j)$ and a cohomological one, $\mathrm{Cone}(i^* - j^*)$. The cohomology of the latter computes the cohomology of X, and it sits inside a Mayer–Vietoris sequence.

Observe now that the mapping cone and the associated long exact sequence of cohomology are defined for *any* diagram

$$[A \xleftarrow{p} Y \xrightarrow{q} B],$$

whether they arise from a cover of a space X or not. To make use of this level of generality, we need to know how to interpret the cohomology of the mapping cone $\mathrm{Cone}(p^* - q^*)$. To this end we define the *geometric realization* of $K(p, q)$ to be the following topological space. Form the disjoint union of A, $Y \times [0, 1]$, and B. Define an equivalence relation by identifying $(y, 0)$ with $p(y)$ and $(y, 1)$ with $q(y)$. The quotient space, which we write as $|K(p, q)|$, is by definition the geometric realization. The space for the mapping cone associated with the union $X = A \cup B$ is displayed in Fig. 1.20.

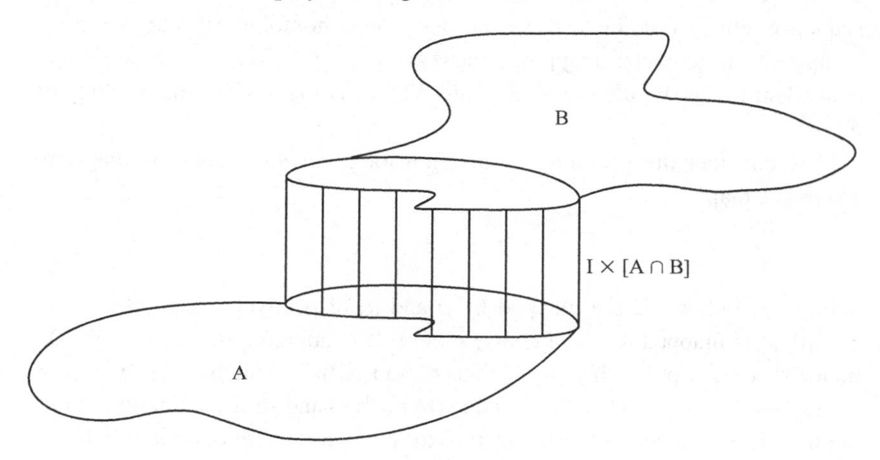

Figure 1.20 Mapping cone of a union.

From Fig. 1.21 we see that $|K(p, q)|$ is homotopy equivalent to X. Thus it is true in this case (and true in general) that[1]

$$H^{\bullet-1}(\mathrm{Cone}(p^* - q^*)) \cong H^\bullet(|K(p,q)|).$$

As our principal application, consider a singular variety X, and let Σ be its singular locus. For purposes of illustration, consider an algebraic curve with a node, as in Fig. 1.16. Let $p : \widetilde{X} \longrightarrow X$ be a resolution of singularities. This is a smooth algebraic variety and a surjective map p whose restriction to $X - p^{-1}(\Sigma)$ is an isomorphism. For our example this is illustrated in Fig. 1.17. A fundamental theorem of Hironaka (1964) states that a resolution of singularities always exists for projective algebraic varieties. Let $\widetilde{\Sigma} = p^{-1}(\Sigma)$, and consider the following, which we refer to as the *resolution diagram* of $p : \widetilde{X} \longrightarrow X$:

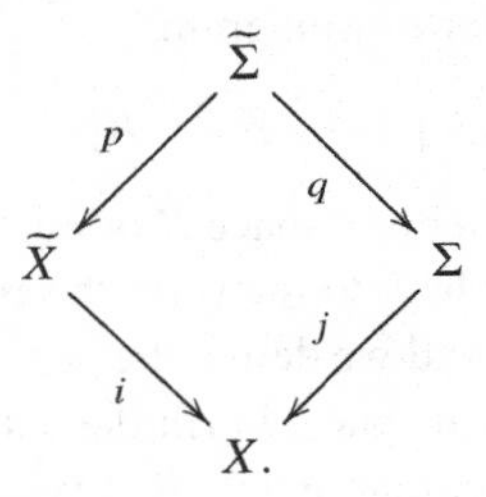

From a formal standpoint the resolution diagram is like the nerve of a cover of a space X by two open sets A and B, as in the Mayer–Vietoris diagram considered above. The mapping cone $K(p,q)$, the associated Mayer–Vietoris sequence, etc., are defined. But what does the cohomological mapping cone compute? The geometric mapping cone is given by Fig. 1.21, which is homotopy equivalent to the singular curve X. Thus $K(p,q)$ computes the cohomology of X.

Now consider the geometric realization for a general resolution diagram. There is a map

$$\epsilon : |K(p,q)| \longrightarrow X$$

defined as follows. Σ are mapped by p and q, respectively. A point (x,t) in $\widetilde{\Sigma} \times [0,1]$ is mapped to $i \circ p(x) = j \circ q(x)$. It is apparent that if $x \in X - \Sigma$, then $\epsilon^{-1}(x)$ is a point. If $x \in \Sigma$, then $\epsilon^{-1}(x)$ is the geometric realization of $[i^{-1}(x) \longleftarrow i^{-1}(x) \longrightarrow x]$. This is a cone over $i^{-1}(x)$ and so is contractible. Thus the map from the geometric realization to X is a map with contractible fibers and therefore is a homotopy equivalence.

[1] Note the shift by -1 on the left. To avoid this shift some authors prefer to use the complex $\mathrm{Cone}(f)[-1]$ instead of $\mathrm{Cone}(f)$.

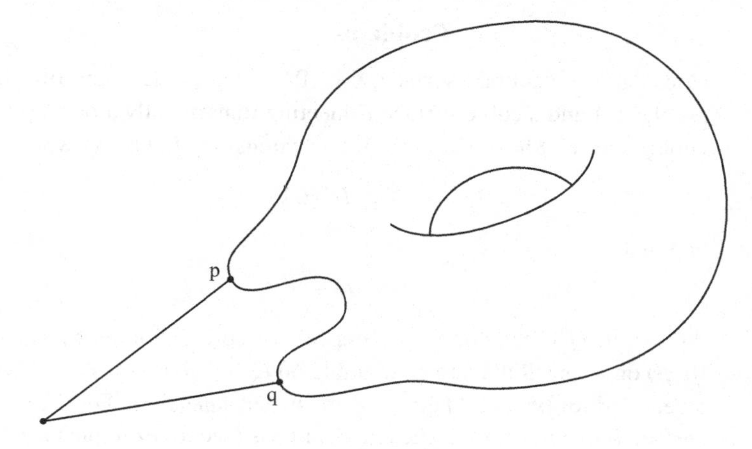

Figure 1.21 Geometric realization of the resolution of a double point.

From the preceding discussion, we see that the long exact sequence of cohomology associated with a resolution diagram reads as follows:

$$\cdots \longrightarrow H^{k-1}(\widetilde{X}) \oplus H^k(\Sigma) \longrightarrow H^{k-1}(\widetilde{\Sigma}) \longrightarrow H^k(X)$$
$$\longrightarrow H^k(\widetilde{X}) \oplus H^k(\Sigma) \longrightarrow H^k(\widetilde{\Sigma}) \longrightarrow \cdots .$$

Now consider the special case in which both Σ and $\widetilde{\Sigma}$ are smooth. Then each term in the above sequence, except for the cohomology of X, is the cohomology of a smooth projective variety and therefore carries a Hodge structure. As for X itself, note that its cohomology is computed from the mapping cone, the terms of which are complexes for smooth varieties. Thus, we can just as well formulate a de Rham theory, considering the mapping cone

$$[A^\bullet(\widetilde{X}) \oplus A^\bullet(\Sigma) \longrightarrow A^\bullet(\widetilde{\Sigma})].$$

With cohomology defined via differential forms, one defines the Hodge filtration in the usual way, as the part of cohomology represented by forms with at least p dz's. The weight filtration is defined as it was for the usual Mayer–Vietoris sequence. The result is a mixed Hodge structure on $H^k(X)$ with weights k and $k-1$. The long exact sequence for the resolution diagram of a singular algebraic curve gives the natural mixed Hodge structure and agrees with that described above. For more examples see Problems 1.4.1 and 1.4.2.

Problems

1.4.1 Consider the algebraic surface $X \subset \mathbf{P}^3$ which is the union of a hyperplane A and a cubic surface B meeting transversally along a plane cubic curve C. Show that $H^2(X)$ is an extension of Q by K, where

$$K \cong H^1(C),$$

and that

$$Q \cong \mathbf{Z}(-1)^7.$$

As a hint, Q is the space of pairs of degree-2 cohomology classes (α, β) on A and B that are compatible on C and $H^2(A) \cong \mathbf{Z}(-1)$, with a generator of type $(1, 1)$ given by the hyperplane class. For the cubic surface B one must use some classical facts (see for example Griffiths and Harris, 1978). First, on any smooth cubic surface there lie exactly 27 lines. These are projective lines, that is, Riemann spheres, defined as the intersection of two hyperplanes in projective three-space. The homology classes of these lines generate the second homology of the cubic surface. Furthermore, the rank of the second homology is 7.

1.4.2 Consider an algebraic surface X with one isolated singularity at a point p. Let $\widetilde{X}$ be a resolution of singularities, and denote the inverse image of p in the resolution by E. The inverse image, the so-called exceptional locus, is an algebraic curve. Let us assume that it is smooth. Discuss the mixed Hodge structure on X as an extension of a weight 1 Hodge structure by one of weight 2. Show that we get an ordinary Hodge structure if the singularity is an ordinary double point, but a genuine mixed Hodge structure if the singularity is an ordinary m-fold point with $m > 2$.

2

Cohomology of Compact Kähler Manifolds

The cohomology of a complex projective algebraic manifold has more structure than that of an ordinary compact manifold. For example (as we shall see), the first Betti number is even and the second Betti number is nonzero. There are several things which must be understood before this finer structure can be seen. First is the representation of cohomology classes by *harmonic forms*, that is, by solutions of Laplace's equation, $\Delta\phi = 0$. Second are the special properties of the Laplacian on an algebraic manifold. In fact, what is important is the *Kähler* condition, a kind of strong compatibility between the Riemannian and complex structures. The rest of the chapter is devoted to an exposition of the basic facts of harmonic theory on compact Kähler manifolds, up to and including the Lefschetz decomposition.

2.1 Cohomology of Compact Differentiable Manifolds

Recall that at every point $p \in M$ of a differentiable manifold M of dimension m, the tangent space $T_{M,p}$ is a real vector space of dimension m. The union when p moves over M can be seen as a manifold T_M of dimension $2m$ equipped with a natural, differentiable projection $\pi : T_M \to M$ giving T_M the structure of a vector bundle, the *tangent bundle*. We discuss the notion of a vector bundle in Appendix C. In our case, locally over a coordinate patch $(U, \{x_1, \ldots, x_m\})$, the tangent bundle receives a product-structure as follows. The vector fields $\partial/\partial x_i$, $i = 1, \ldots, m$ on U give a basis for the tangent space at every point in U and so the union of these becomes naturally identified with the product

$$T_M|U = \bigcup_{p\in U} T_{M,p} = U \times \left(\mathbf{R}\frac{\partial}{\partial x_1} + \cdots + \mathbf{R}\frac{\partial}{\partial x_m}\right),$$

and π is just projection onto the first factor. If you pass from U to another coordinate patch V, the Jacobian matrix provides the means to glue $T_M|U$ to $T_M|V$. This procedure defines a manifold structure on the resulting topological space T_M.

In a similar way, one defines the *cotangent bundle* $T^\vee_M$ and the bundle $\bigwedge^k T^\vee_M$ of k-forms starting from the cotangent spaces, respectively their k-th exterior powers. A smooth k-form is nothing but a global smooth section of the bundle of k-forms. We set

$$A^k(M) \stackrel{\text{def}}{=} \{\text{space of smooth } k\text{-forms on } M\}.$$

Any smooth map $f : N \to M$ induces pullback maps $f^* : A^k(M) \to A^k(N)$. The differentiable k-forms behave functorially with respect to differentiable mappings between manifolds. In addition, the exterior differentiation on the space A^k of (global) k-forms makes $A^\bullet$ into a complex, the *de Rham complex*. Its cohomology, the *de Rham cohomology*

$$H^\bullet_{\mathrm{dR}}(M) \stackrel{\text{def}}{=} \{\text{space of closed forms on } M\}/\{\text{space of exact forms on } M\}$$

is a differentiable invariant, functorially associated with differentiable manifolds. The wedge product of forms makes de Rham cohomology into a graded commutative algebra and it is natural to ask whether there is a relation between this invariant and the topological invariant given by the algebra of singular cohomology. This is the content of *de Rham's theorem*. To explain it, we make use of the fact that singular cohomology in our situation can also be calculated by means of differentiable simplices so that we can integrate without problems. This is explained in Warner (1983, Chapter 5) which includes details of what now follows. There is a natural map

$$\iota : H^k_{\mathrm{dR}}(M; k) \to H^k_{\mathrm{sing}}(M; k), \quad k = \mathbf{R}, \mathbf{C},$$

which is defined by letting a closed form act on a differentiable chain c by integration:

$$\iota(\phi)(c) = \int_c \phi.$$

That this is well defined on the level of cohomology classes is Stokes' theorem,

$$\iota(\mathrm{d}\psi)(c) = \int_c \mathrm{d}\psi = \int_{\partial c} \psi = 0,$$

and it turns out that the integration map is an isomorphism. For this result compactness is not required (paracompactness suffices). A further feature is the compatibility of the cup product of cochains and the wedge product of differential forms, leading up to the final result as follows.

Theorem 2.1.1 (de Rham's Theorem) *For a paracompact differentiable manifold M, integration gives an algebra isomorphism*

$$\boxed{\iota : H^\bullet_{\mathrm{dR}}(M;k) \xrightarrow{\sim} H^\bullet_{\mathrm{sing}}(M;k), \quad k = \mathbf{R}, \mathbf{C}.}$$

Harmonic theory provides a distinguished representative for each cohomology class. To describe how, fix a *Riemannian metric* for the underlying manifold M. This is by definition a smooth field g of inner products, one for each tangent space. It therefore defines a rule for measuring the lengths of and the angles between tangent vectors. As we have seen before, if $x_1, \ldots, x_m$ is a system of local coordinates on M, then the coordinate vector fields $\partial/\partial x_i$ give a basis for the tangent space at each point of the given neighborhood. Thus we can define a matrix-valued function with entries

$$g_{ij} = g\left(\frac{\partial}{\partial x_i}, \frac{\partial}{\partial x_j}\right).$$

The matrix (g_{ij}) is symmetric, positive definite, and gives a local formula for g:

$$g = \sum g_{ij}\, \mathrm{d}x_i\, \mathrm{d}x_j.$$

Riemannian metrics always exist. To see this, note that on $\mathbf{R}^m$ the usual dot product gives one, written

$$g = \sum \mathrm{d}x_i\, \mathrm{d}x_i$$

in the present notation. A global metric can be obtained by patching together local ones using a partition of unity and then using the fact that the set of positive definite symmetric bilinear forms on a vector space is convex. Alternatively, one may embed M in a suitable Euclidean space and use the induced metric: the inner product of two vectors tangent to M at p is the dot product of their images in Euclidean space.

Once a metric is defined on the tangent bundle, natural metrics are defined on all associated bundles, e.g., the cotangent bundle and its exterior powers. One way to define these is via orthonormal frames. Fix a local orthonormal frame $\{V_i\}$ for the tangent bundle, and let $\{\phi_i\}$ be the dual frame for the cotangent bundle, defined by the relations

$$\phi_i(V_j) = \delta_{ij}.$$

Define a metric on the cotangent bundle by declaring the ϕ_i's to be orthonormal, i.e., by imposing the condition

$$g(\phi_i, \phi_j) = \delta_{ij}.$$

Next let $\phi_I = \phi_{i_1} \wedge \cdots \wedge \phi_{i_k}$, where $I = (i_1, \ldots, i_k)$ is a strictly increasing multiindex, and define a metric on the exterior powers of the cotangent bundle by declaring these objects to be orthonormal, i.e., by imposing

$$g(\phi_I, \phi_J) = \delta_{IJ}.$$

Now assume that M is not only Riemannian, but oriented, and that the coordinates $x_1, \ldots, x_m$ in the given order are compatible with the orientation. Then the expression

$$\sqrt{\det(g_{ij})}\, \mathrm{d}x_1 \wedge \cdots \wedge \mathrm{d}x_m,$$

which is *a priori* defined only locally with reference to the given coordinate system, in fact defines a global m-form. This is the *volume form* of the manifold, which we write as $\mathrm{d}V$. If moreover M is *compact*, then one may define an inner product on the space $A^k(M)$ of globally defined k-forms:

$$(\phi, \psi) = \int_M g(\phi, \psi)\, \mathrm{d}V.$$

Given a positive inner product on $A^k(M)$ one has a norm $\|\phi\|^2 = (\phi, \phi)$. One may seek to minimize the norm of elements in a cohomology class $\{\phi + \mathrm{d}\psi\}$, where $\mathrm{d}\phi = 0$. Elements of minimal norm are unique if they exist. To show existence one completes $A^k(M)$ to a Hilbert space, the so-called space of L^2 k-forms. Then existence is trivial: the closure of is an affine subspace of a Hilbert space and so has a unique point closest to the origin. Nontrivial is the assertion that the minimum is a point of $A^k(M)$, not merely of its closure in L^2. This is called "regularity" and is a foundational result in the theory of elliptic differential equations, see Warner (1983, Chapter 7). To explain the connection with differential equations, we first note that the exterior derivative has an adjoint, namely a first-order partial differential operator

$$\mathrm{d}^* : A^k(M) \to A^{k-1}(M)$$

which is characterized by the identity

$$\boxed{(\mathrm{d}\phi, \psi) = (\phi, \mathrm{d}^* \psi).}$$

To see this, first define the *Hodge $*$-operator*. This is an algebraic operator defined pointwise on k covectors at a given point of the m-dimensional manifold M by the formula

$$* : \bigwedge^k T^\vee_{M,x} \to \bigwedge^{m-k} T^\vee_{M,x}, \quad \alpha \wedge *\beta = (\alpha, \beta)(\mathrm{d}V)_x, \quad \forall \alpha, \beta \in \bigwedge^k T^\vee_{M,x}.$$

A small calculation, which uses Stokes' theorem and compactness, shows that the operator $(-1)^{mk+1} * \mathrm{d} *$ indeed gives an adjoint for d

Suppose now that ϕ is of minimal norm in its cohomology class, so that

$$\|\phi\|^2 \leq \|\phi + t\, \mathrm{d}\psi\|^2$$

for all t. Expanding the right-hand side as $\|\phi\|^2 + 2t(\phi, \mathrm{d}\psi) + O(t^2)$, we see that $(\phi, \mathrm{d}\psi) = 0$, which implies that

$$(\mathrm{d}^* \phi, \psi) = 0$$

for all ψ. Consequently,

$$\mathrm{d}^* \phi = 0.$$

Thus, a minimal representative of a cohomology class is both closed and "co-closed." Moreover, it is a solution of

$$\Delta\phi = 0,$$

where

$$\Delta = \mathrm{d}\mathrm{d}^* + \mathrm{d}^* \mathrm{d}$$

is the self-adjoint operator associated with exterior differentiation. This is the Laplace equation for ϕ. As mentioned in the beginning of this chapter, the solutions of the Laplace equation are the *harmonic forms*. Thus, minimal representatives are harmonic. It is an easy exercise to show that harmonic forms are also minimal cohomology classes.

As noted above, a form which is both closed and co-closed is harmonic. The converse also holds. Indeed, if ϕ is harmonic, then

$$(\Delta\phi, \phi) = 0.$$

However,

$$(\Delta\phi, \phi) = (\mathrm{d}\mathrm{d}^* \phi, \phi) + (\mathrm{d}^* \mathrm{d}\phi, \phi) = \|\, \mathrm{d}^* \phi\|^2 + \|\, \mathrm{d}\phi\|^2,$$

which establishes the claim.

An immediate consequence of what has just been proved is that the spaces of harmonic, exact, and co-exact forms are mutually orthogonal, where by co-exact we mean an object of the form $\mathrm{d}^* \phi$. One of the main results of harmonic theory Warner (1983, Chapter 6) is that these spaces give an (orthogonal) direct sum decomposition:

Theorem (The harmonic-Hodge decomposition theorem) *Let M be a compact oriented Riemannian manifold. There is an orthogonal direct sum decomposition*

$$A^k(M) = \mathrm{H}^k_\Delta(M) \oplus \mathrm{d}A^{k-1}(M) \oplus \mathrm{d}^* A^{k+1}(M), \tag{2.1}$$

where $\mathrm{H}^k_\Delta(M)$ *denotes the space of harmonic forms. The latter is a finite dimensional vector space.*

Consequently, one can define an operator G which is 0 on the harmonic space and Δ^{-1} on its orthogonal complement. This is the so-called *Green's operator*. If H is orthogonal projection onto the harmonic space, then one has the identities

$$\Delta G = 1 - H,$$
$$G\Delta = 1 - H,$$

which assert that G is a pseudo-inverse for Δ. The Green's operator commutes with any self-adjoint operator commuting with the Laplacian. To show this, suppose that we have such an operator $T : A^p(M) \longrightarrow A^q(M)$. Let p_1, p_2 be the orthogonal projections onto $\mathrm{H}^p_\Delta(M)^\perp$ and $\mathrm{H}^q_\Delta(M)^\perp$, respectively. Because T commutes with the Laplacian, it sends $\mathrm{H}^p_\Delta(M)$ to $\mathrm{H}^q_\Delta(M)$, and because the orthogonal complement of the harmonic forms is the image of the Laplacian, T commutes with orthogonal projection onto these spaces, i.e., $T\circ p_1 = p_2\circ T$. Moreover, denoting the restriction of the inverse of the Laplacian to $\mathrm{H}^p_\Delta(M)^\perp$ and $\mathrm{H}^q_\Delta(M)^\perp$ by r_1 and r_2, respectively, we also have $T\circ r_1 = r_2\circ T$. So we obtain the desired commutation relation:

$$\begin{aligned} T\circ G = T\circ\Delta^{-1}\circ p_1 &= T\circ r_1\circ p_1 \\ &= r_2\circ T\circ p_1 = r_2\circ p_2\circ T = G\circ T. \end{aligned}$$

From this we find

$$\Delta G = \mathrm{d}\mathrm{d}^* G + \mathrm{d}^* \mathrm{d}G = \mathrm{d}G\,\mathrm{d}^* + \mathrm{d}^* G\,\mathrm{d}.$$

Combined with the pseudo-inverse property, this yields

$$\phi = H\phi + \mathrm{d}(G\,\mathrm{d}^*\phi) + \mathrm{d}^*(G d\phi), \tag{2.2}$$

in accord with the Hodge decomposition (2.1). Thus, if ϕ is a closed form, $\phi = H\phi + \mathrm{d}(G\,\mathrm{d}^*\phi)$ and ϕ is in the same cohomology class as $H\phi$. It is easy to see that $H\phi$ is the only harmonic form in its cohomology class (see Problems 2.2.1–2.2.4). So we have the following corollary.

Corollary 2.1.2 *Each de Rham class on a compact oriented Riemannian manifold contains a unique harmonic representative. In particular, the de Rham cohomology is finite dimensional.*

Problems

2.1.1 Let $U \xrightarrow{\mathrm{d}} V \xrightarrow{\mathrm{d}} W$ be a complex of finite-dimensional Hermitian vector spaces and denote by d^* the adjoint maps. Prove the Hodge decomposition in this case:

$$V = \mathrm{Ker}(\mathrm{d}\mathrm{d}^* + \mathrm{d}^*\mathrm{d}) \oplus \mathrm{d}U \oplus \mathrm{d}^* W.$$

2.1.2 Prove the uniqueness assertion in Corollary 2.1.2

2.1.3 Let $L = \mathrm{d}/\mathrm{d}x$ be the differentiation operator for functions on the real line, endowed with the inner product

$$(f, g) = \int_{-\infty}^{\infty} f(x)g(x)\,\mathrm{d}x.$$

Of course, one must consider functions for which this integral makes sense. Find the operator L^* such that $(Lf, g) = (f, L^*g)$. Then find $\Delta \stackrel{\text{def}}{=} L^*L$.

2.1.4 Let $M = S^1$ with the standard Euclidean metric. Functions on S^1 are identified with periodic functions $f(\theta)$ on the real line with period $\theta = 2\pi$. Compute the Laplacian on functions and on one-forms.

2.2 What Happens on Kähler Manifolds

Type Decomposition of Complex Forms

Recall that on a complex manifold M we have systems of local holomorphic coordinates $\{z_1, \ldots, z_m\}$. The associated vector fields $\partial/\partial z_i$ and $\partial/\partial \bar{z}_i$ give a local frame for the complexified tangent bundle and, at least in this local coordinate system, define a decomposition

$$T_M^{\mathbf{C}} = T'_M \oplus T''_M, \quad T''_M = \overline{T'_M}$$

of the complexified tangent bundle $T_M^{\mathbf{C}} = T_M \otimes \mathbf{C}$. In this decomposition T'_M is the *holomorphic tangent bundle*. Its sections are the fields spanned by the $\partial/\partial z_i$. Its complex conjugate is the *anti-holomorphic tangent bundle* T''_M. Sections of T''_M are spanned by the fields $\partial/\partial \bar{z}_i$. Because the functions which glue overlapping holomorphic coordinate systems together are themselves holomorphic, one sees that the local decompositions just defined are consistent and so define a global decomposition of the complexified tangent bundle. Smooth sections of T'_M and T''_M are called fields of type $(1, 0)$ and $(0, 1)$, respectively.

The decomposition of the complexified tangent bundle induces a decomposition of the bundles of complex-valued k-forms. To describe it, define a *basic form of type* (p, q) to be a form

$$\mathrm{d}z_I \wedge \mathrm{d}\bar{z}_J = \mathrm{d}z_{i_1} \wedge \cdots \wedge \mathrm{d}z_{i_p} \wedge \mathrm{d}\bar{z}_{j_1} \wedge \cdots \wedge \mathrm{d}\bar{z}_{j_q}.$$

A form of type (p, q) is a linear combination of these, where the coefficients are (say) C^∞ functions. Because the gluing functions are holomorphic, a form which is of type (p, q) in one coordinate system is also of the same type (p, q) in any other. Consequently, there is a direct sum decomposition

$$A^k(M; \mathbf{C}) = \bigoplus\nolimits_{p+q=k} A^{p,q}(M), \tag{2.3}$$

where $A^{p,q}(M)$ is the vector space of forms of type (p, q) on M and where $A^k(M; \mathbf{C}) = A^k(M) \otimes \mathbf{C}$ are the complex-valued k-forms. Note that

$$\overline{A^{p,q}(M)} = A^{q,p}(M).$$

We now ask whether this decomposition passes to cohomology. To make this precise, we let

$$H^{p,q}(M) \overset{\text{def}}{=} \{\text{classes representable by a closed } (p,q)\text{-form}\} \subset H^{p+q}_{\mathrm{dR}}(M)$$

so that obviously

$$\overline{H^{p,q}(M)} = H^{q,p}(M). \tag{$*$}$$

The decomposition we are after is

$$H^k_{\mathrm{dR}}(M; \mathbf{C}) = \bigoplus H^{p,q}(M)$$

and is called a *Hodge decomposition* for H^k. These might or might not exist. It is customary to employ the notation

$$h^{p,q}(M) = \dim H^{p,q}(M)$$

for the dimension of these spaces. These are the *Hodge numbers*.

Examples 2.2.1 (i) Consider a compact complex torus $M = \mathbf{C}^n/\Lambda$. A basis for the cohomology is given by the translation-invariant forms, i.e., the forms with constant coefficients relative to the basic (p, q)-forms. Consequently, the cohomology carries a Hodge decomposition.

(ii) For the case of compact Riemann surfaces, see Chapter 1. We saw there that we have

$$H^1(M; \mathbf{C}) = H^{1,0}(M) \oplus H^{0,1}(M),$$

where

$$H^{1,0}(M) = \{\text{classes of holomorphic one-forms}\}.$$

(iii) There are complex manifolds that admit no Hodge decomposition. Indeed, $(*)$ and the Hodge decomposition imply that the odd Betti numbers are even, e.g., $B_1 = h^{1,0} + h^{0,1} = 2h^{1,0}$. According to Example A.6.(vii) the Hopf manifold is homeomorphic to $S^1 \times S^{n-1}$ and so, by the Künneth formula, its first Betti number is 1. Thus neither the Hopf manifold nor any complex manifold homotopy equivalent to it supports a Hodge structure.

Kähler Manifolds

There is a large class of objects, the compact Kähler manifolds, for which a Hodge decomposition exists. It contains the projective algebraic manifolds. For the moment, we do not give a precise definition but rather point out an important consequence: the Laplace operator, which is an operator of degree 0 on forms, is in fact an operator of bidegree $(0, 0)$, i.e., maps a (p, q)-form to a (p, q)-form. Consequently, the (p, q) components of a harmonic form are harmonic. Thus, we do have a decomposition

$$\mathrm{H}^k_\Delta(M) = \bigoplus\nolimits_{p+q=k} \mathrm{H}^{p,q}_\Delta(M)$$

of the harmonic space. Moreover, the map $\mathrm{H}^k_\Delta \to H^k_{\mathrm{dR}}$ defined by sending a harmonic form to its de Rham class is an isomorphism and maps $\mathrm{H}^{p,q}_\Delta$ to $H^{p,q}$. It follows that the subspaces $H^{p,q}$ (which do not depend on the metric) decompose the de Rham cohomology.

Let us now see what is involved in proving that the Laplacian has bidegree (type) $(0, 0)$. In general an operator has bidegree (p, q) if it maps each $A^{r,s}$ to $A^{r+p,s+q}$. Thus the exterior derivative admits a decomposition

$$\mathrm{d} = \partial + \bar{\partial}$$

into an operator of bidegree $(1, 0)$ and a conjugate operator of bidegree $(0, 1)$. These are defined as the unique derivations that annihilate the basic (p, q)-forms and act on functions by

$$\partial f = \sum \frac{\partial f}{\partial z_i} \, \mathrm{d}z_i,$$

$$\bar{\partial} f = \sum \frac{\partial f}{\partial \bar{z}_i} \, \mathrm{d}\bar{z}_i.$$

The adjoint operators ∂^* and $\bar{\partial}^*$ have bidegrees $(-1, 0)$ and $(0, -1)$, respectively. Since $\Delta = \mathrm{d}\mathrm{d}^* + \mathrm{d}^*\mathrm{d}$, the Laplacian is in general a sum of operators of different types.

On a complex manifold there are two additional Laplace operators,

$$\Delta_\partial = \partial\partial^* + \partial^*\partial,$$
$$\Delta_{\bar\partial} = \bar\partial\bar\partial^* + \bar\partial^*\bar\partial,$$

each of which have bidegree (0, 0). To see how these relate to the ordinary Laplacian, we expand the definition:

$$\begin{aligned}\Delta &= \mathrm{d}\mathrm{d}^* + \mathrm{d}^*\mathrm{d}\\ &= (\partial + \bar\partial)(\partial^* + \bar\partial^*) + (\partial^* + \bar\partial^*)(\partial + \bar\partial)\\ &= (\partial\partial^* + \partial^*\partial) + (\bar\partial\bar\partial^* + \bar\partial^*\bar\partial) + (\partial\bar\partial^* + \bar\partial^*\partial) + (\bar\partial\partial^* + \partial^*\bar\partial).\end{aligned}$$

Thus,

$$\Delta = \Delta_\partial + \Delta_{\bar\partial} + \text{cross terms}.$$

The cross terms are the sum of one operator of bidegree $(1, -1)$ and another of bidegree $(-1, 1)$, in general nonzero. However, the (as yet unexplained) Kähler condition guarantees that the cross terms vanish and that

$$\Delta_\partial = \Delta_{\bar\partial}.$$

It follows that Δ has bidegree (0, 0) and the equality of the two Laplacians implies that the three Laplacians are proportional.

We come now to the definition of Kähler manifolds, for which we first review some notions from linear algebra. Fix a real vector space V and let J be a complex structure on V, i.e., a linear map $J : V \to V$ such that $J^2 = -1$. This of course implies that V is even-dimensional.

Definition 2.2.2 A *Hermitian metric* on (V, J) is given by a bilinear form $h : V \times V \to \mathbf{C}$ such that, writing h in real and imaginary parts as $h = g + i\omega$, one has

(i) h satisfies $h(Jx, Jy) = h(x, y)$ for all $x, y \in V$;
(ii) g is a euclidean metric and ω, the *metric form*, is a symplectic form, i.e., a non-degenerate skew-symmetric real form;
(iii) $\omega(x, y) = g(x, Jy)$ for all $x, y \in V$ – or, equivalently $g(x, y) = \omega(Jx, y)$.

Note that condition (iii) states that the metric form is uniquely determined by the underlying real metric g and the complex structure. Conversely, suppose that we have a symplectic vector space (V, ω) and a complex structure J such that

(iv) $\omega(Jx, Jy) = \omega(x, y)$ for all $x, y \in V$;
(v) $\omega(Jx, x) > 0$ if $x \neq 0$.

Then by condition (iii) in the above definition we get a metric g such that $g + \mathrm{i}\omega$ is a Hermitian metric on (V, J).

The eigenvalues of J are $\pm\mathrm{i}$. Thus $V_{\mathbf{C}} = V \otimes_{\mathbf{R}} \mathbf{C}$ splits into a direct sum

$$V_{\mathbf{C}} = V^+ \oplus V^-, \quad V^\pm = \{x \in V_{\mathbf{C}} \mid J(x) = \pm\mathrm{i}x\}.$$

Lemma 2.2.3 *Let there be given a Hermitian metric $h = g + \mathrm{i}\omega$ on (V, J). Then V^+ and V^- are isotropic with respect to $g_{\mathbf{C}}$, the $\mathbf{C}$-linear extension of g, as well as with respect to $\omega_{\mathbf{C}}$, the $\mathbf{C}$-linear extension of ω.*

Proof If x and y are in V^+ then

$$g(x, y) = g(Jx, Jy) = g(\mathrm{i}x, \mathrm{i}y) = -g(x, y).$$

So $g|V^+ = 0$. By condition (iii) in the definition we also have $\omega_{\mathbf{C}}(x, y) = 0$. The proof for V^- is analogous. □

We can reformulate this as follows. Consider any $\mathbf{R}$-linear complex valued form f on V. We define then

$$f^\pm(v) \stackrel{\text{def}}{=} f(v) \mp \mathrm{i} f(Jv).$$

Since $f^+(Jv) = \mathrm{i}J(v)$ the form f^+ is $\mathbf{C}$-linear while $f^- = \overline{f^+}$ is $\mathbf{C}$-anti-linear. It follows that the dual of $V_{\mathbf{C}}$ has a direct sum decomposition

$$\begin{aligned} V_{\mathbf{C}}^\vee &= \mathrm{Hom}_{\mathbf{R}}(V, \mathbf{C}) = (V^*)^{1,0} \oplus (V^*)^{0,1}, \\ f &= f^+ + f^-, \end{aligned}$$

where $(V^*)^{1,0} \stackrel{\text{def}}{=} \mathrm{Hom}_{\mathbf{C}}(V, \mathbf{C})$ and $(V^*)^{0,1} = \overline{(V^*)^{1,0}}$.

Now set

$$\bigwedge^{p,q} V_{\mathbf{C}}^\vee = \bigwedge^p (V^*)^{1,0} \otimes \bigwedge^q (V^*)^{0,1}.$$

Then we have a direct sum decomposition

$$\bigwedge^k V_{\mathbf{C}}^\vee = \bigoplus_{p+q=k} \bigwedge^{p,q} V_{\mathbf{C}}^\vee \tag{2.4}$$

and we say that elements in $\bigwedge^{p,q} V_{\mathbf{C}}^\vee$ are of type (p, q). Now ω is real valued, and since by Lemma 2.2.3 we have $\omega_{\mathbf{C}}(V^+, V^+) = 0$ and similarly for V^-, the decomposition (2.4) implies that the form $\omega_{\mathbf{C}}$ has type $(1, 1)$. Conversely, if ω is a real bilinear form of type $(1, 1)$ one has $\omega_{\mathbf{C}}(Jx, Jy) = \omega_{\mathbf{C}}(x, y)$ for all $x, y \in V_{\mathbf{C}}$. So being of type $(1, 1)$ is equivalent to condition (iv). In other words, we have the following.

Corollary 2.2.4 *Giving a Hermitian metric on (V, J) is equivalent to giving a triple (V, J, ω) with J a complex structure, ω a real and positive symplectic form such that $\omega_{\mathbf{C}}$ is of type $(1, 1)$ with respect to J.*

Now consider a complex manifold M. The underlying almost complex structure

$$J : T_M \to T_M$$

can be explicitly described as follows. If $z_k = x_k + \mathrm{i}y_k$ relates holomorphic and (distinguished) real coordinates, then we have

$$\frac{\partial}{\partial z_k} = \frac{1}{2}\left(\frac{\partial}{\partial x} - \mathrm{i}\frac{\partial}{\partial y}\right), \quad \frac{\partial}{\partial \bar{z}_k} = \frac{1}{2}\left(\frac{\partial}{\partial x} + \mathrm{i}\frac{\partial}{\partial y}\right).$$

The "usual" almost complex structure on the real tangent spaces is given by

$$J\left(\frac{\partial}{\partial x_k}\right) = \frac{\partial}{\partial y_k},$$
$$J\left(\frac{\partial}{\partial y_k}\right) = -\frac{\partial}{\partial x_k}.$$

Hence J acts on T', the holomorphic tangent bundle, as multiplication by $+\mathrm{i}$ and on T'', the anti-holomorphic tangent bundle by multiplication by $-\mathrm{i}$.

Suppose further that M carries a Riemannian metric g such that $g(Jx, Jy) = g(x, y)$. Such a manifold is called *Hermitian*, and carries, fiber by fiber, all the linear algebra structure discussed in the previous paragraph. In particular we can decompose the bundle of complex k-forms fiber by fiber into types,

$$\bigwedge^k T^*_M = \bigoplus_{p+q=k} \bigwedge^{p,q}_M,$$

which on the level of *sections* corresponds to the decomposition (2.3) of complex k-forms into types.

Also, we have a two-form ω, canonically associated with the metric, which is real, positive, and of type $(1, 1)$. We call it a Kähler form if it is closed.

Definition 2.2.5 A complex manifold M with Hermitian metric g is *Kähler* if the $(1, 1)$-form associated with g is closed. The metric g is called a *Kähler metric* and its associated form a *Kähler form.*

Remark 2.2.6 The condition that the Kähler form is closed is equivalent to the existence of a local holomorphic coordinate system $\{z_1, \ldots, z_m\}$ so that

$$g(z_1, \ldots, z_m) = \mathbf{1} + O(|z|^2),$$

i.e., all the linear terms in the Taylor expansion for g vanish. In other words, the metric osculates the Euclidean one up to order 2 (see Wells, 1980, Section 5.4).

Two fundamental examples of Kähler manifolds are the complex Euclidean and projective spaces, $\mathbf{C}^n$ and $\mathbf{P}^n$. The metric on the first is the "dot" product

$$h = \sum \mathrm{d}z_i \ \mathrm{d}\bar{z}_i,$$

whose associated $(1,1)$-form is

$$\omega = \mathrm{i} \sum \mathrm{d}z_i \wedge \mathrm{d}\bar{z}_i.$$

Remark Because ω is translation-invariant it defines a Kähler structure on any quotient torus $\mathbf{C}^n/\Lambda$. As an aside, note that not all complex tori are algebraic. See Problem 2.2.3(c).

For the Kähler structure on $\mathbf{P}^n$ it suffices to define the form ω, from which the Riemannian and Hermitian metrics g and h can be recovered. To this end we first define a real, closed, semipositive form on $\mathbf{C}^{n+1} - \{0\}$, namely,

$$\Omega = \mathrm{i}\partial\bar{\partial} \log ||z||^2, z = (z_0, \ldots, z_n).$$

It is real because in general the conjugate of $\mathrm{i}\partial\bar{\partial}\rho$, where ρ is real-valued, is

$$-\mathrm{i}\bar{\partial}\partial\rho = \mathrm{i}\partial\bar{\partial}\rho,$$

where we have used $\partial\bar{\partial} + \bar{\partial}\partial = 0$. It is closed because in general

$$\begin{aligned} \mathrm{d}\partial\bar{\partial} &= (\partial + \bar{\partial})\partial\bar{\partial} \\ &= \partial^2\bar{\partial} - \bar{\partial}^2\partial \\ &= 0. \end{aligned}$$

It is semipositive because for any collection $\{f_i\}$ of holomorphic functions, in the expression

$$\mathrm{i}\partial\bar{\partial} \log \sum |f_i|^2 = \mathrm{i} \sum h_{ij} \ \mathrm{d}z_i \wedge \mathrm{d}\bar{z}_j$$

the matrix (h_{ij}) is Hermitian semipositive. This requires a small calculation. To obtain the required form on $\mathbf{P}^n$, we observe that on the open set $U_i = \{z_i \neq 0\}$,

$$||z||^2 = |z_i|^2(1 + ||w||^2),$$

where w is the vector composed of the nonconstant ratios z_k/z_i. Moreover, on U_i, we have

$$\Omega = \mathrm{i}\partial\bar{\partial} \log(1 + ||w||^2). \qquad (*)$$

This is because

$$\partial\bar{\partial} \log |f|^2 = \partial\bar{\partial} \log(f) - \bar{\partial}\partial \log(\bar{f}) = 0$$

for any nowhere-zero holomorphic function f. Now view w as an affine coordinate vector on U_i, one of the basic open sets of $\mathbf{P}^n$. In each of these, Ω, according to what has just been stated, defines a real, closed $(1, 1)$-form, which one checks is positive, not merely semipositive. One also checks that these forms are consistent on overlaps and so define a global two-form ω with the required properties. This is almost evident because the relation $(*)$ has the same appearance in one coordinate system as it does in another. The metric just constructed is called the *Fubini–Study metric*.

The de Rham class of ω can be scaled to give the canonical generator of $H^2(\mathbf{P}^n)$ and so a multiple of the Fubini–Study metric gives an integral class. See Problem 2.2.1 for the scaling factor.

Now consider a Kähler manifold M and a complex submanifold N. The submanifold inherits a Riemannian metric from M and so has a natural Hermitian structure. Moreover, the canonical $(1, 1)$-form for M pulls back to give the canonical $(1, 1)$-form on N. Therefore N is, in a natural way, a Kähler manifold. From this we conclude that the Fubini–Study metric induces a Kähler metric on projective algebraic manifolds.

As we have mentioned earlier, the Laplace–Beltrami operator of a Kähler manifold is of type $(0, 0)$. To see why, let L denote the operator defined by multiplication against the Kähler form

$$L(\alpha) = \omega \wedge \alpha,$$

and let

$$\Lambda = L^*$$

be its formal adjoint. These operators are of type $(1, 1)$ and $(-1, -1)$, respectively. If A, B are endomorphisms of the module of complex-valued differential forms of pure degree a, b, respectively, we define the graded commutator

$$[A, B] = AB - (-1)^{ab} BA.$$

With this convention, the fundamental *Kähler identities* Wells (1980, Corollary 5.4.10) can be expressed as

$$\begin{aligned} \partial^* &= \mathrm{i}[\Lambda, \bar{\partial}], \\ \bar{\partial}^* &= -\mathrm{i}[\Lambda, \partial]. \end{aligned}$$

If you introduce the real operator

$$\mathrm{d}^c \stackrel{\text{def}}{=} -\mathrm{i}(\partial - \bar{\partial})$$

with formal adjoint $(\mathrm{d}^c)^*$, these can be rewritten as

$$[\Lambda, \mathrm{d}] = -(\mathrm{d}^c)^*.$$

The adjoint relation is also useful and reads as follows:

$$[L, d^*] = d^c.$$

Theorem (Consequences of the Kähler identities) *One has* $[\partial, \overline{\partial}^*] = 0$ *and hence* $\Delta_d = 2\Delta_{\overline{\partial}}$; *in particular*
(i) *the Laplacian is real;*
(ii) *the Laplacian preserves types; and*
(iii) *the L-operator preserves harmonic forms.*

Proof The commutation relation $\partial\overline{\partial}^* + \overline{\partial}^*\partial = 0$ follows from

$$\begin{aligned}-\mathrm{i}(\partial\overline{\partial}^* + \overline{\partial}^*\partial) &= \partial[\Lambda, \partial] + [\Lambda, \partial]\partial \\ &= \partial\Lambda\partial - \partial\Lambda\partial = 0.\end{aligned}$$

Then

$$\begin{aligned}\Delta_d &= (\partial + \overline{\partial})(\partial^* + \overline{\partial}^*) + (\partial^* + \overline{\partial}^*)(\partial + \overline{\partial}) \\ &= \Delta_\partial + \Delta_{\overline{\partial}}.\end{aligned}$$

Next observe that

$$\begin{aligned}\mathrm{i}\Delta_\partial &= -\partial[\Lambda, \overline{\partial}] - [\Lambda, \overline{\partial}]\partial \\ &= \partial\Lambda\overline{\partial} + \partial\overline{\partial}\Lambda - \Lambda\overline{\partial}\partial + \overline{\partial}\Lambda\partial \\ &= \overline{\partial}\Lambda\partial + \partial\overline{\partial}\Lambda - \Lambda\overline{\partial}\partial - \partial\Lambda\overline{\partial} \\ &= \overline{\partial}[\Lambda, \partial] + [\Lambda, \partial]\overline{\partial} \\ &= \mathrm{i}\Delta_{\overline{\partial}}.\end{aligned}$$

It follows that $\Delta_d = 2\Delta_{\overline{\partial}}$ as claimed and hence (i) and (ii) follow. Finally, to prove (iii) assume that α is harmonic so that $d\alpha = d^*\alpha = 0$. Consequently, $dL\alpha = d(\omega \wedge \alpha) = \omega \wedge d\alpha = 0$, and if one assumes that α has pure type (this is allowed because of (ii)), one also has $d^* L\alpha = -[L, d^*]\alpha = -d^c\alpha$ $= 0$. □

The discussion in the previous section now immediately gives the following.

Theorem 2.2.7 (The Hodge decomposition theorem) *Let M be a compact Kähler manifold. Let* $H^{p,q}(M)$ *be the space of cohomology classes represented by a closed form of type* (p, q). *There is a direct sum decomposition*

$$H^m_{\mathrm{dR}}(M; \mathbf{C}) = \bigoplus_{p+q=m} H^{p,q}(M).$$

This decomposition is induced by a decomposition on the level of harmonic forms:

$$H^m_{\mathrm{dR}}(M; \mathbf{C}) \xleftarrow{\simeq} \mathrm{H}^k_\Delta(M; \mathbf{C}) = \bigoplus_{p+q=m} \mathrm{H}^{p,q}_\Delta(M), \quad \mathrm{H}^{p,q}_\Delta(M) \xrightarrow{\simeq} H^{p,q}(M).$$

Moreover, $H^{p,q}(M) = \overline{H^{q,p}(M)}$. *In other words,* $H^m_{\mathrm{dR}}(M)$ *carries a real Hodge structure of weight m.*

Using the de Rham theorem 2.1.1, we conclude the following.

Corollary 2.2.8 *For a compact Kähler manifold M, the singular cohomology group* $H^m(M)$ *carries an integral weight m Hodge structure.*

As to homology, we can see (Facts B.2.1) that the homology $H_m(M; \mathbf{Q})$ is dual to the cohomology $H^m(M; \mathbf{Q})$. Thus we have the following.

Corollary 2.2.9 *The homology group* $H_m(M)$ *carries a natural Hodge structure of weight* $-m$.

As to functoriality, we already noted in Chapter 1 that for any holomorphic map $f : X \longrightarrow Y$ between smooth projective varieties, the map

$$f^* : H^m(Y) \longrightarrow H^m(X)$$

is a morphism of Hodge structures, and both the kernel and image of f^* are Hodge structures. This also holds of course when f is a holomorphic map between Kähler manifolds. A similar assertion then holds for the maps induced in homology and also for the Gysin maps. See Problem 2.2.5.

Problems

2.2.1 Determine the multiple of the Fubini–Study metric ω on $\mathbf{P}^n$ which gives the (positive) generator of $H^2(\mathbf{P}^n)$ by calculating its integral over a line.

2.2.2 Recall the notion of a connection in Riemannian geometry. If M is a manifold, an *affine connection* is an operator $\nabla : T_M \times T_M \to T_M$ which is $\mathbf{R}$-bilinear and for fixed vector fields Y the covariant derivative $\nabla_Y = \nabla(*, Y) : T_M \to T_M$ obeys the Leibniz rule $\nabla_Y(fX) = df(Y)X + f\nabla_Y(X)$. One says that ∇ has no *torsion* if

$$\nabla_X Y - \nabla_Y X = [X, Y].$$

If (M, g) is a Riemannian manifold, there is a unique affine connection without torsion which preserves the metric in the sense that $g(\nabla_Z X, Y) + g(X, \nabla_Z Y) = 0$ for all triples of vector fields (X, Y, Z) on M. This connection is the *Levi–Civita connection.* Prove that a Hermitian manifold (M, g) is Kähler if and only if the Levi–Civita connection commutes with the almost complex structure operator.

2.2.3 Let $z_1, \ldots, z_n$ be coordinates in $\mathbf{C}^n$ and consider a lattice $\Lambda = \mathbf{Z}\lambda_1 \oplus \cdots \oplus \mathbf{Z}\lambda_{2n}$. Place the generators λ_k as column vectors in a $n \times 2n$ matrix Ω. Suppose that a Hermitian metric on $\mathbf{C}^n$ is given by the positive definite Hermitian matrix H. Suppose that the corresponding translation-invariant two-form defines an integral cohomology class on the torus $\mathbf{C}^n/\Lambda$. Kodaira's embedding theorem (see Griffiths and Harris, 1978, p. 181) then ensures that the torus can be embedded in a projective space. Conversely, integral $(1,1)$-classes on a torus coming from translation-invariant $(1,1)$-classes on $\mathbf{C}^n$ must necessarily be given by $\sum h_{ik}\, dz_i \wedge d\bar{z}_k$ with a constant Hermitian positive definite matrix $H = (h_{ik})$. This leads to a necessary and sufficient criterion for Ω to be the period matrix of a projective torus, i.e., an abelian variety.

(a) Suppose that the real coordinates with respect to the lattice are given by $\{x_1, \ldots, x_{2g}\}$ and suppose that $\gamma = \sum g_{ij}\, dx_i \wedge dx_j$, with $G = (g_{ij})$ skew-symmetric, defines a real positive definite $(1,1)$-form. Express γ in terms of the coordinates z_j using the inverse of the $2n \times 2n$ matrix $\begin{pmatrix} \Omega \\ \bar{\Omega} \end{pmatrix}$ and show that we have

$$\begin{aligned} \mathrm{i}\bar{\Omega} G^{-1\mathsf{T}} \Omega &> 0, \\ \Omega G^{-1\mathsf{T}} \Omega &= 0. \end{aligned}$$

These are the two *Riemann conditions* or *Riemann bilinear relations*.

(b) Verify that the matrix

$$\Omega = \begin{pmatrix} 1 & 0 & \sqrt{-2} & \sqrt{-3} \\ 0 & 1 & \sqrt{-5} & \sqrt{-7} \end{pmatrix}$$

does not satisfy the Riemann relations.

(c) Show that one can make a change of coordinates for Λ so that the matrix G of (a) becomes

$$G = \begin{pmatrix} 0 & D \\ -D & 0 \end{pmatrix}, \quad D \overset{\text{def}}{=} \begin{pmatrix} d_1 & & 0 \\ & \ddots & \\ 0 & & d_r \end{pmatrix}, (d_i \in \mathbf{Z}).$$

Next, if the first g vectors of Λ are identified with the standard basis for $\mathbf{C}^n$, the period matrix thus normalized has as its first n columns the unit vectors and its last n columns form a matrix Z'; set $Z \overset{\text{def}}{=} Z'D$, and with our new G, express the Riemann conditions as

$$^{\mathsf{T}}Z = Z, \mathrm{Im}(Z) > 0.$$

Conclude that algebraic tori depend on $\frac{1}{2}n(n+1)$ continuous parameters and that there are thus plenty of nonalgebraic tori in dimensions > 1.

2.2.4 If X and Y are smooth projective varieties, then $H^k(X \times Y; \mathbf{Q})$ is, by the Künneth theorem (see Problem B.6 in Appendix B), the direct sum of the tensor products $H^m(X; \mathbf{Q}) \otimes H^n(Y; \mathbf{Q})$, where $m + n = k$. Show that the Hodge structure on $H^k(X \times Y)$ imposed by the Künneth theorem agrees with the natural Hodge structure.

2.2.5 Show that a holomorphic map $f : X \longrightarrow Y$ between Kähler manifolds induces a morphism of Hodge structures $f_* : H_k(X) \longrightarrow H_k(Y)$. Show that this is also true for the Gysin map $f_* : H^k(X)(-c) \longrightarrow H^{k+2c}(Y)$, $c = \dim Y - \dim X$.

2.3 How Lefschetz Further Decomposes Cohomology

In addition to the Hodge decomposition, there are two other structures inherent in the cohomology of a compact Kähler manifold. First is the *integral lattice*, which is the image of $H^k(M; \mathbf{Z})$ in $H^k(M; \mathbf{R})$. Second is the notion of *primitive cohomology*. To define it we must state the so-called "hard Lefschetz theorem".

Theorem (Hard Lefschetz theorem) *Let m be the complex dimension of M and define the Lefschetz operator $L : H^*(M) \longrightarrow H^*(M)$ by $L(x) = \omega \wedge x$. For M a compact Kähler manifold the k-th iterate is an isomorphism:*

$$L^k : H^{m-k}_{\mathrm{dR}}(M) \xrightarrow{\sim} H^{m+k}_{\mathrm{dR}}(M).$$

Sketch of the proof One of the modern proofs as given for instance in Griffiths and Harris (1978, pp. 118–122), uses representation theory on the level of harmonic forms. Let us explain this proof, leaving computational details aside.

The three operators L, its formal adjoint Λ, and the shifted counting operator H which equals $m - k$ on H^k form a basis of a Lie algebra isomorphic to $\mathfrak{sl}_2$

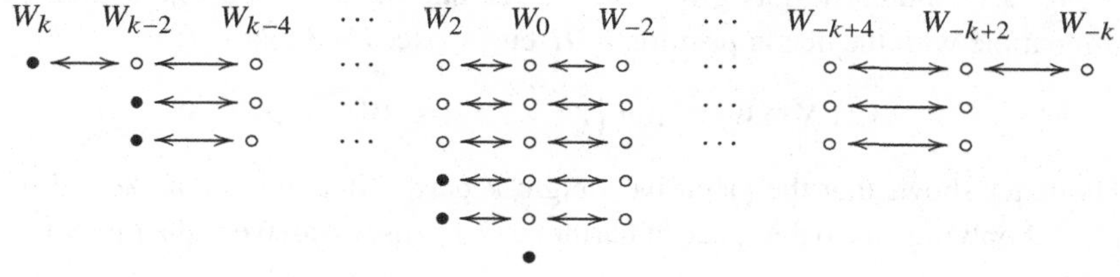

$W_k \quad W_{k-2} \quad W_{k-4} \quad \cdots \quad W_2 \quad W_0 \quad W_{-2} \quad \cdots \quad W_{-k+4} \quad W_{-k+2} \quad W_{-k}$

Figure 2.1 Lefschetz decomposition when k is even.

with its standard basis. More precisely, L, Λ, H correspond to

$$Y \stackrel{\text{def}}{=} \begin{pmatrix} 0 & 0 \\ 1 & 0 \end{pmatrix},$$

$$X \stackrel{\text{def}}{=} \begin{pmatrix} 0 & 1 \\ 0 & 0 \end{pmatrix},$$

$$H \stackrel{\text{def}}{=} \begin{pmatrix} 1 & 0 \\ 0 & -1 \end{pmatrix}.$$

Representations for $\mathfrak{sl}_2$ are direct sums of irreducible ones. It is easy to show that for such a V we have

$$V = V_m \oplus V_{m-2} \oplus \cdots \oplus V_{-m+2} \oplus V_{-m}$$

with V_k a one-dimensional eigenspace for H for the eigenvalue k. We have

$$Y^k : V_k \xrightarrow{\sim} V_{-k}.$$

So we can visualize V as a string of eigenspaces for H going from m to $-m$ connected from left to right by the Y-operators and in the opposite direction by the X-operators. It follows that an arbitrary representation W is the direct sum of several "strings" of the above type of varying lengths. But still Y^k establishes an isomorphism between the H-eigenspaces W_{-k} and W_k.

Applying this to the case where V is the space harmonic forms, we have that m is indeed the complex dimension, $W_{\pm k} = H_\Delta^{m \mp k}$, and Y^k becomes L^k, thereby proving the hard Lefschetz theorem. Note that the numbering gets reversed: instead of going from W_k to W_{-k}, the k-th power of the Y-operator becomes the operator L^k which goes from H^{m-k} to H^{m+k}.

Continuing with the case of an arbitrary $\mathfrak{sl}_2$-representation W, we now set $PW = \operatorname{Ker} X$ so that

$$W = PW \oplus YPW \oplus Y^2 PW \oplus \cdots . \tag{2.5}$$

In Fig. 2.1 , the black dots give PW. We see that the above decomposition is compatible with the decomposition in H-eigenspaces and that

$$\operatorname{Ker} X \cap W_k = \operatorname{Ker}\left(Y^{k+1} : W_k \to W_{-k-2}\right).$$

The latter shows that the primitive weight-k part, $PW \cap W_k$, is the kernel of Y^{k+1}. Applying this to the space of harmonic k-forms, we arrive at the following definition.

Definition 2.3.1 Let $m = \dim_{\mathbf{C}} M$. For $k \leq m$, the *primitive* $(m-k)$*-cohomology*, $H^{m-k}_{\text{prim}}(M; \mathbf{C})$, is the kernel of L^{k+1} acting on $H^{m-k}_{\text{dR}}(M; \mathbf{C})$. Equivalently, it is the kernel of the Λ-operator:

$$\begin{aligned} H^{m-k}_{\text{prim}}(M; \mathbf{C}) &= \operatorname{Ker}\left\{L^{k+1} : H^{m-k}_{\text{dR}}(M; \mathbf{C}) \to H^{m+k+2}_{\text{dR}}(M; \mathbf{C})\right\} \\ &= \operatorname{Ker}\left\{\Lambda : H^{m-k}_{\text{dR}}(M; \mathbf{C}) \to H^{m-k-2}_{\text{dR}}(M; \mathbf{C})\right\}. \end{aligned}$$

By convention, $H^k_{\text{prim}}(M; \mathbf{C}) = 0$ for $k > m$.

Note that formally one can define primitive cohomology with coefficients in a ring R whenever the Kähler class ω lives in $H^2(M; R)$. We denote this by $H^k_{\text{prim}}(M; R)$. One can for instance always take $R = \mathbf{R}$.

For example, for a Kähler surface, H^2_{prim} is the space of classes x satisfying $[\omega] \cup x = 0$. More geometrically, it is the space of classes dual to cycles that have intersection number 0 with a hyperplane in general position. This follows from the duality between the cup product of cohomology classes and the intersection product of cycles (Poincaré's original definition of cup product). Note also that in general $H^1_{\text{prim}} = H^1$, because it is the kernel of $L^m : H^1 \to H^{2m+1} = 0$.

The other part of the hard Lefschetz theorem, which is a direct consequence of the decomposition (2.5), asserts that there is a direct sum decomposition.

Theorem 2.3.2 (Lefschetz decomposition) *Let M be a compact Kähler manifold. Then there is a direct sum decomposition*

$$H^k_{\text{dR}}(M; \mathbf{C}) = H^k_{\text{prim}}(M; \mathbf{C}) \oplus LH^{k-2}_{\text{prim}}(M; \mathbf{C}) \oplus L^2 H^{k-4}_{\text{prim}}(M; \mathbf{C}) \cdots,$$

the Lefschetz decomposition. If, moreover, M is projective algebraic and the Kähler metric is induced by a suitable multiple of the Fubini–Study metric (see Problem 2.2.1), this decomposition is also valid when $R = \mathbf{Q}$.

To illustrate the theorem, consider, for example, the cohomology of an algebraic surface. For it we have

$$H^2(M; \mathbf{Q}) = H^2_{\text{prim}}(M; \mathbf{Q}) \oplus \mathbf{Q}[\omega],$$

where the second factor is generated by the Kähler class.

In Chapter 1 we saw that the cup product form for the middle cohomology has special properties with respect to holomorphic forms. For Riemann surfaces and double planes this gave rise to a special case of the *Riemann–Hodge bilinear relations*. We restrict ourselves to primitive cohomology. Indeed, on it we have a bilinear form b which satisfies the generalized Riemann–Hodge bilinear relations.

Theorem 2.3.3 *Let M be a Kähler manifold. Set*

$$b(\phi, \psi) \stackrel{\text{def}}{=} (-1)^{\frac{1}{2}k(k-1)} \int_M \phi \wedge \psi \wedge \omega^{m-k}, \quad \phi, \psi \in H^k(M; \mathbf{C}).$$

This form induces a polarization on primitive cohomology $H^k_{\text{prim}}(M; \mathbf{R})$, i.e.,[1] *$b$ is $(-1)^k$-symmetric (symmetric for k even and skew-symmetric for k odd); moreover, with $p + q = r + s = k$, it satisfies the following relations:*

$$\begin{aligned} &(1) \quad b(H^{p,q}, H^{r,s}) = 0 \ \text{ if } (r,s) \neq (q,p); \\ &(2) \quad \mathrm{i}^{p-q} b(u, \bar{u}) > 0, \quad u \in H^{p,q}_{\text{prim}}, \ \text{if } u \neq 0. \end{aligned}$$

In other words, introducing the Weil operator

$$\begin{aligned} C &: H^k(X; \mathbf{C}) \to H^k(X; \mathbf{C}) \\ C|H^{p,q} &= \text{multiplication by } \mathrm{i}^{p-q}, \end{aligned} \tag{2.6}$$

the form

$$h_C(\phi, \psi) \stackrel{\text{def}}{=} b(C\phi, \bar{\psi})$$

is positive definite Hermitian on primitive cohomology.

Remarks 2.3.4 (1) Note that $b(Cx, \bar{y}) = (-1)^k b(c, C\bar{y})$. Several people, e.g. Griffiths (1968, p. 574), Weil (1958, p. 77–78), use $Q = (-1)^k b$ for the polarizing form and then the second bilinear relation becomes $Q(x, C\bar{x}) > 0$ for $x \neq 0$.
(2) Since the Kähler form ω is a real closed form defining a class in $H^2(M; \mathbf{R})$, the definition of b (with $k = 2$) then gives

$$b(\omega, \omega) = -\int_M \omega^n < 0, \quad n = \dim_{\mathbf{C}} M.$$

However, $C\omega = \omega$, $\omega = \bar{\omega}$, and so $b(\omega, \omega) = h_C(\omega, \omega)$. Thus h_C is *not* a definite Hermitian form on the full cohomology $H^2(M; \mathbf{C})$.

Despite this, following Weil (1958, p. 77–78), one still can obtain a polarization on all of cohomology.

[1] See Definition 1.2.4.

Corollary 2.3.5

$$(-1)^r b(\phi,\psi) = (-1)^{r+\frac{1}{2}k(k-1)} \int_M \phi \wedge \psi \wedge \omega^{m-k}, \quad \phi,\psi \in L^r H^{k-2r}_{\text{prim}}(M)$$

defines a polarization on all of $H^k(M;\mathbf{R})$. If M is projective, the polarization can be defined as a $\mathbf{Q}$-valued form on $H^k(M;\mathbf{Q})$.

Observe that this checks with $b(\omega,\omega) < 0$ since here $r = 1$.

Sketch of the proof of Theorem 2.3.3 The first bilinear relation is elementary. If $(r,s) \neq (q,p)$, then the form $\phi \wedge \psi \wedge \omega^{m-k}$ either has too many dz's or too many $d\bar{z}$'s and therefore vanishes identically. The second relation lies deeper. We do not enter into the details of the proof. They can be found in Wells (1980, p.207), for instance. Instead, we illustrate it in a typical case: we shall show that the relation holds for the second cohomology of an algebraic surface. Consider first a holomorphic two-form given locally by $\phi = f\, dz \wedge dw$. Then

$$\begin{aligned} h_C(\phi,\phi) &= \int_M \phi \wedge \bar{\phi} \\ &= \int_M |f|^2 \mathrm{i}\, dz \wedge d\bar{z} \wedge \mathrm{i}\; dw \wedge d\bar{w} > 0. \end{aligned}$$

The sign is the correct one.

Consider next a primitive $(1,1)$-form ϕ, which by definition has the property $[\omega] \cup [\phi] = 0$. Since $H^{1,1}_{\text{prim}}$ is the complexification of a real vector space, we may assume that ϕ is *real*. However, one can say more. If we take ϕ to be harmonic then $L\phi = \omega \wedge \phi$ is harmonic. Moreover, by making a linear change of coordinates, we may also assume that

$$\omega = \mathrm{i}\; dz \wedge d\bar{z} + \mathrm{i}\; dw \wedge d\bar{w}$$

at the origin. Since any Hermitian bilinear form in $\mathbf{C}^2$ may be diagonalized with respect to another by a unitary change of coordinates, we may assume that

$$\phi = \mathrm{i}(a\; dz \wedge d\bar{z} + b\; dw \wedge d\bar{w})$$

at the origin. But then

$$\phi \wedge \omega = (a + b)\; dV$$

at the origin, where dV is the standard volume form on $\mathbf{C}^2$, i.e., $\mathrm{i}\, dz \wedge d\bar{z} \wedge \mathrm{i}\; dw \wedge d\bar{w}$. Consequently,

$$\phi = \mathrm{i}a(dz \wedge d\bar{z} - dw \wedge d\bar{w}),$$

from which we conclude that

$$\phi \wedge \bar{\phi} = -2|a|^2\, dV < 0$$

at the origin. Since the assertion regarding the sign is invariant under a linear

– even holomorphic – change of coordinates, the integrand is negative, which yields the claimed result. □

Bibliographical and Historical Remarks

Section 2.1 In Wells (1980), the Hodge decomposition theorem is fully proved. For another proof see Griffiths and Harris (1978).

Sections 2.2–2.3 The classic reference is Wells (1980). See also Bertin and Peters (2002); Weil (1958).

3

Holomorphic Invariants and Cohomology

We start this chapter by giving a step-by-step description of the de Rham groups and the Hodge filtration in algebraic terms, culminating in the filtered Čech–de Rham double complex. We postpone technical ingredients, like hypercohomology and spectral sequences, to Chapter 6.

By way of an example, the cohomology of hypersurfaces in projective space is treated in Section 3.2. We then introduce algebraic cycles and we state the Hodge conjecture for them. Related to cycles we also have the Abel–Jacobi maps whose targets are the intermediate Jacobians associated with cohomology of Kähler manifolds. This is treated in the final section, Section 3.6.

3.1 Is the Hodge Decomposition Holomorphic?

Holomorphic Forms

Consider a form of type $(p, 0)$. Its exterior derivative decomposes as

$$\mathrm{d}\phi = \partial\phi + \bar{\partial}\phi,$$

where the two parts have types $(p + 1, 0)$ and $(p, 1)$. Thus, if ϕ is closed, then $\partial\phi$ and $\bar{\partial}\phi$ vanish separately. If in local coordinates

$$\phi = \sum_I \phi_I \,\mathrm{d}z_I, \qquad (*)$$

then the condition $\bar{\partial}\phi = 0$ reads

$$\sum_{k,I} \frac{\partial \phi_I}{\partial \bar{z}_k} \,\mathrm{d}\bar{z}_k \wedge \mathrm{d}z_I = 0,$$

from which it follows that the coefficient functions are holomorphic. Consequently, a closed $(p, 0)$-form is a so-called *holomorphic p-form*, i.e., an expression of the form (*) with holomorphic coefficients. In general, the converse is not true because a holomorphic p-form may not be closed. Consider, for example, the form $z\, dw$ on $\mathbf{C}^2$. On a compact Kähler manifold, however, the converse does hold. One argument goes as follows. By hypothesis $\bar{\partial}\phi = 0$. Since ϕ has type $(p, 0)$, $\bar{\partial}^*\phi$ has type $(p, -1)$ and therefore vanishes. Consequently our form is a solution of the equation $\Delta_{\bar{\partial}}\phi = 0$. But all three Laplacians on a Kähler manifold are proportional, so that $\Delta_d\phi = 0$, hence $d\phi = 0$.

As a corollary of what we have done, we conclude that exact holomorphic forms are necessarily 0. To see this, consider a holomorphic p-form ϕ. Recall that the space of d-harmonic forms is orthogonal to the space of exact forms (this is the harmonic-Hodge decomposition theorem), so that $\phi = 0$. We summarize this as follows.

Proposition 3.1.1 *Any holomorphic p-form on a compact Kähler manifold is closed and it is 0 if and only if it is exact. Thus there is a natural injective homomorphism*

$$\{\text{holomorphic } p\text{-forms}\} \longrightarrow H^p_{\mathrm{dR}}$$

whose image is $H^{p,0}$.

The dimension of these spaces give invariants, generalizing the genus of a projective curve. Two of these play an important role in higher dimensions.

Definition 3.1.2 Let M be a complex manifold of dimension m. Then

(i) the dimension of the space of holomorphic m-forms is called the *geometric genus* of M and denoted by $p_g(M)$; and
(ii) the dimension of the space of holomorphic one-forms is called the *irregularity* $q(M)$ of M.

Proposition 3.1.1 suggests that perhaps other components of the Hodge decomposition have a "holomorphic description." This is indeed the case, and can be argued as follows. Begin with the identification of $H^{p,q}_{\mathrm{dR}}$ with the harmonic space $\mathrm{H}^{p,q}_{\Delta}$. Using the proportionality of Laplacians, substitute for the latter the space of $\Delta_{\bar{\partial}}$-harmonic (p, q) forms. By the analogue of the theorem relating the de Rham cohomology to the harmonic forms, identify it with the space of $\bar{\partial}$-closed forms modulo the space of $\bar{\partial}$-exact forms. This is the so-called Dolbeault cohomology:

$$H^{p,q}_{\bar{\partial}} = \frac{\{\bar{\partial}\text{-closed } (p, q) \text{ forms}\}}{\{\bar{\partial}\text{-exact } (p, q) \text{ forms}\}}.$$

We discuss below the sheaf theoretic proof of the de Rham theorem (see Section

6.3), and this proof can be imitated to identify the Dolbeault cohomology group with the group $H^q(\Omega^p)$, where Ω^p is the sheaf of holomorphic p-forms (see Problem 6.2.1). Since the latter can be described in terms of Čech cohomology (more is said about this below), it is a perfectly good holomorphic object.

Proposition 3.1.3 *We have*

$$\begin{aligned} H^{p,q} &\cong \mathrm{H}^{p,q}_{\triangle} \\ &\cong \mathrm{H}^{p,q}_{\triangle_{\bar\partial}} \\ &\cong H^{p,q}_{\bar\partial} \\ &\cong H^q(\Omega^p). \end{aligned}$$

Proof Most assertions have been shown already. We now show the third isomorphism. We have already seen the Green's operator for the operator $\triangle$ (see formula (2.2)). Now we let G stand for the Green's operator for $\triangle_{\bar\partial}$. For a $\bar\partial$-closed (p,q)-form α we then have

$$\alpha = H\alpha + \bar\partial\bar\partial^* G\alpha.$$

This proves the desired isomorphism. □

A more ambitious goal is to describe holomorphically all of the complex-valued cohomology. This too can be done using the notions of hypercohomology and double complex. All of the necessary foundational material can be found in Bott and Tu (1982). We start with a short digression on Čech cohomology.

Čech Cohomology

We first review the concept of a sheaf. A *sheaf* of abelian groups $\mathcal{F}$ on a topological space M consists of the following data. For each open set $U \subset M$ we have an abelian group $\mathcal{F}(U)$ and these are related by means of "restriction homomorphisms" $\mathcal{F}(V) \to \mathcal{F}(U)$, one for each inclusion $U \hookrightarrow V$ between open sets. When $U \subset V \subset W$ are inclusions between three open sets, the corresponding restriction homomorphisms should be compatible in the obvious sense. Moreover, a gluing property should hold: if an open set U is covered by open subsets $U_i, i \in I$, and "sections" $s_i \in \mathcal{F}(U_i)$ are given that are compatible under restrictions, there is a unique section $s \in \mathcal{F}(U)$ restricting the various sections s_i. A sheaf can have extra structure and given a ring R or a field k, it should be clear what is meant by a sheaf of rings, R-modules, or k-algebras.

Examples of sheaves include the constant sheaves G_M, where G is any abelian group. Other fundamental examples are the sheaves of germs of differentiable functions on a differentiable manifold or the sheaf of germs of holomorphic functions on a complex manifold. These sheaves of rings are also called the *structural sheaves*. For a complex manifold M this sheaf is denoted $\mathcal{O}_M$. We use the same symbol as for the trivial bundle. Indeed, for a holomorphic vector bundle E, we often use the notation $\mathcal{O}(E)$ to denote its sheaf of germs of holomorphic sections. As a variant of the latter, we also have the sheaf of germs of differentiable or holomorphic sections of a differentiable or holomorphic vector bundle, respectively. Below, we also consider coherent sheaves. See Section 7.2.

Fix some ring R. To define *Čech cohomology* with values in a sheaf $\mathcal{F}$ of R-modules, we start with an open cover $\mathfrak{U}$ of M. A collection $(U_0, \ldots, U_q)$ of members of $\mathfrak{U}$ with nonempty intersection is called a q-simplex $\sigma = \{0, \ldots, q\}$ and its support $|\sigma|$ is by definition $U_0 \cap \cdots \cap U_q$. The i-th face of σ is the $(q-1)$-simplex $\sigma^i = \{0, \ldots, i-1, i+1, \ldots, q\}$. The restriction induced by the inclusion of the corresponding supports is denoted

$$\rho_i : \mathcal{F}(U_0 \cap \cdots \cap U_q) \to \mathcal{F}(U_i \cap \cdots U_{i-1} \cap U_{i+1} \cap \cdots \cap U_q).$$

A *q-cochain* is a function f which assigns to any q-simplex σ an element $f(\sigma) \in \mathcal{F}(|\sigma|)$. This is the same as saying that f is an element of the *free product* of the R-modules $\mathcal{F}(|\sigma|)$ where σ runs over the q-simplices of $\mathfrak{U}$. This free product is again an R-module (with the obvious module-operations):

$$C^q(\mathfrak{U}, \mathcal{F}) \stackrel{\text{def}}{=} \prod_{\sigma \text{ a } q\text{-simplex of } \mathfrak{U}} \mathcal{F}(|\sigma|).$$

The coboundary homomorphism

$$\delta : C^q(\mathfrak{U}, \mathcal{F}) \to C^{q+1}(\mathfrak{U}, \mathcal{F})$$

is defined by

$$\delta f(\sigma) = \sum_{i=0}^{q+1} (-1)^i \rho_i f(\sigma^i)$$

defining the *Čech cochain complex* $C^\bullet(\mathfrak{U}, \mathcal{F})$. We set

$$H^q(\mathfrak{U}, \mathcal{F}) \stackrel{\text{def}}{=} H^q(C^\bullet(\mathfrak{U}, \mathcal{F})), \quad \check{H}^q(M, \mathcal{F}) \stackrel{\text{def}}{=} \varinjlim_{\mathfrak{U}} H^q(\mathfrak{U}, \mathcal{F}),$$

where the direct limit is taken over the set of coverings, partially ordered under the refinement relation. As an example, sections of $\mathcal{F}$ define zero-cocycles and conversely:

$$\left.\begin{aligned} \Gamma(\mathcal{F}) &= \{\alpha \in C^0(\mathfrak{U}, \mathcal{F}) \mid \delta\alpha = 0\} = Z^0(\mathfrak{U}, \mathcal{F}) \\ &= H^0(\mathfrak{U}, \mathcal{F}) = \check{H}^0(M, \mathcal{F}) \end{aligned}\right\}. \tag{3.1}$$

So for cohomology in degree 0 there is no need to pass to the limit. This is also true for higher-degree cohomology provided one uses special covers.

Theorem (Theorem of Leray, (Godement, 1964, p.209)) *Let* $\mathfrak{U} = \{U_i\}_{i \in I}$ *be an open cover of* M *which is acyclic in the sense that for any nonempty intersection* $U = U_{i_1} \cap U_{i_2} \cdots \cap U_{i_k}$ *of the cover we have*

$$\check{H}^p(U, \mathcal{F}) = 0 \textit{ for } p \geq 1.$$

Then the natural homomorphisms

$$\check{H}^p(\mathfrak{U}, \mathcal{F}) \to \check{H}^p(M, \mathcal{F})$$

are isomorphisms for all $p \geq 0$.

In general, if $\mathfrak{U}$ is such a cover, we call it a *Leray cover* for the sheaf $\mathcal{F}$.

Example 3.1.4 (i) All covers are Leray with respect to any of the sheafs $\mathcal{A}^k_M$ of k-forms on a compact differentiable manifold M. This follows, using partitions of unity. See Problem 3.1.2.

(ii) A Leray cover for constant sheaves is gotten by taking a cover $\{U_i\}$ by geodesically convex sets, thus ensuring that all intersections

$$U_I = U_{i_1} \cap \cdots \cap U_{i_p}$$

are convex and so contractible. For a discussion of the existence of such covers see Helgason (1962, Chapter 1, Theorem 9.9).

(iii) For a complex manifold M, one can find a Leray cover that works simultaneously for all holomorphic vector bundles $\mathcal{F}$, using open sets that are biholomorphic to balls. Indeed, these and their intersections are so-called *Stein manifolds* and for these, Cartan's famous "Théorème B" (see e.g. Grauert and Remmert, 1977) says that the cohomology groups $\check{H}^k(U, \mathcal{F})$ vanish for $k \geq 1$. In fact $\mathcal{F}$ can even be a coherent $\mathcal{O}_M$-module, but for the moment this is not needed. Coherent sheaves are briefly treated in Section 7.2.

Čech cohomology is a concrete incarnation of an axiomatic cohomology theory of sheaves $\mathcal{F}$ on topological spaces X. There is only one such theory and so we usually write $H^q(X, \mathcal{F})$ for the q-th cohomology group of $\mathcal{F}$. Čech cohomology only obeys the axioms for the class of paracompact spaces and continuous maps between them:

Proposition *For any paracompact topological space* X *and a sheaf* $\mathcal{F}$ *on it we have*

$$H^q(X, \mathcal{F}) = \check{H}^q(X, \mathcal{F}).$$

For a constant sheaf R *on a paracompact (topological) manifold there are canonical identifications between sheaf cohomology with values in R and the singular cohomology with values in R:*

$$H^q(X, R_X) = H^q(X; R).$$

For proofs, see the Bibliographic Remarks at the end of this chapter.

Hypercohomology and the Čech–de Rham Complex

After this digression we continue with our holomorphic picture. We choose a cover $\mathfrak{U}$ by geodesically convex open sets. These are Stein and so, as we saw above, such a cover is Leray not only with respect to constant sheaves and differentiable forms, but also with respect to holomorphic forms. Consider now the *double complex K* whose components are Čech cochains with values in the sheaves of holomorphic p-forms. Write this as

$$K^{p,q} = \mathscr{C}^q(\Omega^p),$$

and write the two differentials as

$$\mathrm{d}\colon K^{p,q} \longrightarrow K^{p+1,q},$$
$$\delta : K^{p,q} \longrightarrow K^{p,q+1},$$

where d is the exterior derivative of the complex $\Omega^\bullet$ and where δ is the Čech coboundary.

Let sK denote the total complex associated with K,

$$sK^m = \sum_{p+q=m} K^{p,q},$$

and let

$$D = \mathrm{d} + (-1)^p \delta.$$

One checks that $D^2 = 0$ so that sK is a complex relative to D. The hypercohomology group of the complex $\Omega^\bullet$ is by definition the cohomology of sK and is written

$$\mathbf{H}^m(\Omega^\bullet) \stackrel{\text{def}}{=} H^m(sK). \tag{3.2}$$

We claim that there is a natural map from $\mathbf{H}^\bullet(\Omega^\bullet)$ to $H^\bullet_{\mathrm{dR}}$ and that this map is an isomorphism. To construct the map, note that hypercohomology is a functor from complexes of sheaves to graded groups. Thus, if $\mathscr{A}^\bullet$ denotes the complex of C^∞ forms, the natural inclusion

$$\Omega^\bullet \longrightarrow \mathscr{A}^\bullet$$

induces a map

$$\mathbf{H}^\bullet(\Omega^\bullet) \to \mathbf{H}^\bullet(\mathscr{A}^\bullet).$$

Another functor from complexes of sheaves into graded groups is the cohomology of the associated complex of global sections, which we can write[1] as $R\Gamma$. Because the global sections are a subspace of the 0-cochains (see (3.1)), we have maps of complexes $\Gamma(\mathscr{A}^\bullet) \to s\mathscr{C}^\bullet(\mathscr{A}^\bullet)$ and hence there is a natural map

$$R\Gamma(\mathscr{A}^\bullet) \to \mathbf{H}(\mathscr{A}^\bullet).$$

The left-hand side, however, is a de Rham cohomology, so that we have a diagram

$$\mathbf{H}^\bullet(\Omega^\bullet) \longrightarrow \mathbf{H}^\bullet(\mathscr{A}^\bullet) \longleftarrow H^\bullet_{\mathrm{dR}}(M).$$

In Section 6.3, we use the technique of spectral sequences to show that both maps are isomorphisms and so, admitting this for the moment, we found a holomorphic description of the cohomology groups.

Theorem 3.1.5 *The de Rham cohomology of a compact Kähler manifold is canonically isomorphic to the hypercohomology of the holomorphic de Rham complex. Assuming we are given a cover by geodesically convex sets, this is the cohomology of a total complex associated with the double complex of Čech cochains with values in the holomorphic p-forms, the so-called Čech–de Rham complex.*

Because the complex cohomology can be described holomorphically, one might ask if the Hodge decomposition itself can be so described. This, it turns out, is too much. However, one can go part way. To do so, instead of the Hodge decomposition, look at the Hodge filtration which, we recall (see (1.23) in Chapter 1), is defined on cohomology by

$$F^p = \bigoplus\nolimits_{r \geq p} H^{r,s}. \tag{3.3}$$

Its graded quotients are isomorphic to the spaces of type (p, q), i.e.,

$$H^{p,q} \cong F^p / F^{p+1}.$$

Because

$$\begin{aligned} F^p \cap \overline{F^q} &= H^{p,q}, \\ H^k &= F^p \oplus \overline{F^{k-p+1}}, \end{aligned}$$

the Hodge decomposition can be recovered from the filtration.

Let us first define the *trivial filtration* $F^p K^\bullet$ of any complex $K^\bullet$ as the

[1] See Section 6.4 where we consider the derived functor mechanism.

complex one gets by forgetting everything before K^p (see Fig. 3.1). This is a subcomplex of $K^\bullet$. Turning to the Čech–de Rham complex, we can likewise introduce

$$F^p\mathscr{C}^\bullet(\Omega^\bullet) = \bigoplus\nolimits_{r\geq p}\mathscr{C}^\bullet(\Omega^r).$$

As one sees from Figure 3.1, this is a sub-double complex.

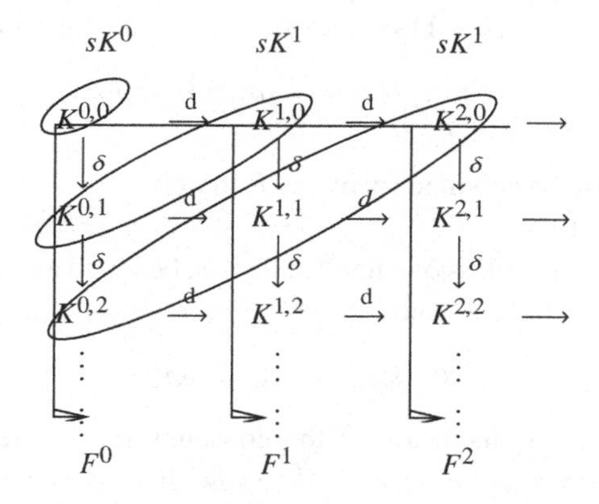

Figure 3.1 Double complex and the trivial filtration.

The image of its hypercohomology in the hypercohomology of the full Čech–de Rham complex defines a filtration. We shall see in Chapter 6, using the spectral sequence of Remark 6.3.4 which is associated with the F-filtration on our double complex, that it is the Hodge filtration (3.3). The crucial ingredient in proving this turns out to be the isomorphism $H^{p,q} \cong H^q(\Omega^p)$, established in Proposition 3.1.3. See Remark 6.3.4 for further discussion. With hindsight, we know that the meaning of all this is that our spectral sequence degenerates at the first stage, and it turns out that this property can be read off from the double complex directly. Degeneration simply means here that the differential of the associated double complex is strict with respect to the F-filtration (see Problem 6.1.2 in Chapter 6).

Summarizing, we have the following proposition.

Proposition 3.1.6 *On a compact Kähler manifold the Hodge filtration is the filtration induced by the trivial filtration $F^\bullet$ on the Čech–de Rham double complex. The total derivative D is strictly compatible with this filtration:*

$$F^p \cap \operatorname{Im}(D) = DF^p.$$

Since this is an entirely holomorphic description of the filtration, one might surmise that indeed it will vary holomorphically when we vary the complex structure. This turns out to be the case and will be pursued in Chapter 6.

Problems

3.1.1 Let M be a compact manifold with an open cover U_i, $i \in I$ and associated partition of unity τ_i. Let f be a k Čech cocycle with values in $\mathscr{A}^p_M$. Suppose that $k \geq 1$ and define a $(k-1)$-cochain g by giving its value on the $(k-1)$ simplex $\sigma = \{U_{j_0}, \ldots, U_{j_{k-1}}\}$ as follows:

$$g(\sigma) = \sum_{j \in I} \tau_j f(j\sigma), \quad j\sigma = \{U_j, U_{j_0}, \ldots, U_{j_{k-1}}\}.$$

Verify that the coboundary of g is f, thereby proving that $H^k(\mathfrak{U}, \mathscr{A}^p) = 0$ for $k \geq 1$.

3.1.2 The first Chern class of a line bundle can be found as follows. Consider the exponential sequence

$$0 \to \mathbf{Z} \xrightarrow{j} \mathscr{A}_M \xrightarrow{\mathbf{e}} \mathscr{A}^*_M \to 1,$$

where j is the inclusion of the constants in the sheaf of germs of C^∞-functions and $\mathbf{e}(f) = \mathrm{e}^{2\pi i f}$. Show that the coboundary map $\delta : H^1(M, \mathscr{A}^*_M) \to H^2(M; \mathbf{Z})$ is an isomorphism identifying the group of equivalence classes of C^∞ line bundles on M with the group $H^2(M; \mathbf{Z})$. Show that $\delta(L) = c_1(L)$. The fact that δ is an isomorphism shows that the first Chern class is a complete C^∞-invariant for line bundles.

3.2 A Case Study: Hypersurfaces

Suppose we start out with a projective manifold P such as projective space, whose cohomology is well known. We wish to compute the cohomology of a *smooth* hypersurface $X \subset P$ in terms of, say, its equation, and further global data on P. One way to do this is to express the cohomology of the complement as coming from a suitable complex of rational differential forms with poles along X and then link the cohomology of the complement $U = P - X$ to that of X. Of course, one can simply use the complex $\Omega^\bullet_U$ of holomorphic forms on U, but this complex is in general much too large. We want to replace it by smaller complexes defined by forms that are globally defined on P. We have two such complexes. The first is the complex

$$\Omega^\bullet_P(*X)$$

of meromorphic forms on P having at most poles along X (of any order). In general the order of the pole gets bigger under derivation so that the sheaves

$$\Omega^p(kX) = \text{ sheaf of } p\text{-forms with poles of order at most } k$$

do not yield subcomplexes. However, this is the case for the complex

$$\Omega_P^\bullet(\log X)$$

of meromorphic forms on P with at most logarithmic poles along X. These forms φ are defined locally: on an open set V where $f = 0$ is an equation for X we demand that $f\varphi$ and $f d\varphi$ extend to a holomorphic form on V.

Let us give a simple example: a smooth curve in the projective plane. Classically the holomorphic one-forms on such a curve are the residues of rational two-forms on the projective plane having a simple pole along the curve. Suppose for instance that the smooth curve X inside the projective plane P is given by a homogeneous equation $F = 0$. In general, "logarithmic poles" means more than just poles of order at most 1 (see Problem 3.2.1). But for the highest degree this is the case: $\Omega_P^2(\log X) = \Omega_P^2(X)$. This sheaf fits in the exact sequence

$$0 \to \Omega_P^2 \to \Omega_P^2(X) \xrightarrow{\text{Res}} \Omega_X^1 \to 0,$$

where the residue "Res" is defined as follows. In any affine chart centered at a point of X we choose coordinates (f, g) such that f is a local equation for the curve. For any local holomorphic two-form α having a pole of order at most 1 along X we then write

$$\alpha = \frac{\mathrm{d}f \wedge \beta}{f}.$$

Then $\text{Res}(\alpha)$ is the restriction of β to X. This is independent of the local equation because, if $f = uf'$ with u a nowhere vanishing function, then $\mathrm{d}f/f = \mathrm{d}f'/f' + \mathrm{d}u/u = (\mathrm{d}f' + f'u^{-1}\,\mathrm{d}u)/f'$, and so $\mathrm{d}f/f \wedge \beta = (\mathrm{d}f' + f'\gamma) \wedge \beta/f'$ for some holomorphic one-form γ, and both forms β and $\beta + f'\gamma$ restrict to the same form on the curve X. Globally, because $H^0(\Omega_P^2) = H^1(\Omega_P^2) = 0$, this gives an isomorphism

$$H^0(\Omega_P^2(X)) \xrightarrow{\text{Res}} H^0(\Omega_X^1)$$

that can be made totally explicit. Indeed, if z_0, z_1, z_2 are homogeneous coordinates, and if F has degree d, the isomorphism sends the rational form

$$\frac{A(z_0\,\mathrm{d}z_1 \wedge \mathrm{d}z_2 - z_1\,\mathrm{d}z_0 \wedge \mathrm{d}z_2 + z_2\,\mathrm{d}z_0 \wedge \mathrm{d}z_1)}{F}$$

to its residue along X. Here A is any homogeneous polynomial of degree $d - 3$

and so $H^0(\Omega^1_X)$ is isomorphic to the space of degree $(d-3)$-polynomials in z_0, z_1, z_2.

So we have recovered at least the holomorphic part of the de Rham cohomology. But this approach can be extended to rational forms on P having higher-order poles along X and then yields a description of the cohomology of X in terms of the cohomology of the complement of X in P. By means of a residue operation generalizing the one we just saw, one gets the remaining part of the de Rham cohomology. The goal of this section is to carry out this program.

We start with the first step: the calculation of the cohomology of the complement

$$j : U = P - X \hookrightarrow P.$$

To this end we look at the exact cohomology sequence for the pair (P, U). We replace the term $H^k(P, U)$ by $H^{k-2}(X)$ using the Thom isomorphism theorem. This requires some explanation. Let T be a tubular neighborhood of X in P. By the excision property (formula (B.2) in Appendix B), the cohomology of the pair (P, U) is the same as for the pair $(T, T{-}X)$. By the tubular neighborhood theorem Lang (1985, Chapter 4.5), T is diffeomorphic to a tubular neighborhood $T(N)$ of the zero-section inside the normal bundle N of X in P so that $H^k(T, T{-}X) = H^k(T(N), T(N) - \{\text{zero-section}\})$, which in turn is isomorphic to $H^{k-2}(X)$ by Thom's theorem (see Spanier, 1966, Section 5.7, Theorem 10).

We thus obtain a modified exact sequence, the *Gysin sequence*

$$\cdots \to H^{k-1}(U) \xrightarrow{r} H^{k-2}(X) \xrightarrow{i_*} H^k(P) \xrightarrow{j^*} H^k(U) \to \cdots .$$

The map r is a sort of topological residue map; in *rational cohomology* it turns out to be the transpose of the "tube over cycle map" $\tau : H_{k-2}(X; \mathbf{Q}) \to H_{k-1}(U; \mathbf{Q})$, which by definition associates with the class of a cycle c in X the class of the boundary in U of a suitably small tubular neighborhood of c. This follows when one analyzes the definition of the Thom isomorphism. The map i_* is the *Gysin map* associated with i, i.e., the map Poincaré dual to the map induced in homology by i (see also Problem B.1 in Appendix B).

Let us exploit this exact sequence in the case when $P = \mathbf{P}^{n+1}$ and for the only interesting rank, the one giving the middle cohomology $H^n(X)$. We easily see that for odd dimension the map r is an isomorphism, while for even n the map r is an injective homomorphism whose image is the primitive cohomology. Since for n odd, $H^n_{\text{prim}} = H^n$, in both cases we obtain an isomorphism

$$r : H^{n+1}(U) \xrightarrow{\sim} H^n_{\text{prim}}(X).$$

If we replace $\mathbf{P}^{n+1}$ by an arbitrary manifold we need to assume that X is *ample* in P. Then $U = P - X$ is affine and so by Theorem B.3.2 we have $H^{n+2}(U) = 0$ and thus

$$H^{n-1}(X) \xrightarrow{i_*} H^{n+1}(P) \xrightarrow{j^*} H^{n+1}(U) \xrightarrow{r} H^n(X) \xrightarrow{i_*} H^{n+2}(P) \to 0.$$

The image of r equals the kernel of the map i_* and it is called the *variable cohomology*

$$\boxed{H^n_{\text{var}}(X) = \operatorname{Ker}\left(i_* : H^n(X) \to H^{n+2}(P)\right)}$$

as opposed to the "fixed" cohomology

$$\boxed{H^n_{\text{fix}}(X) = i^* H^n(P)}$$

which is coming from P. In the case of projective space, variable cohomology coincides with primitive cohomology and indeed formula (3.2) remains valid provided we replace primitive cohomology by variable cohomology and if moreover we assume that $H^{n+1}_{\text{prim}}(P; \mathbf{Q}) = 0$ (this is automatic for projective space).

Proposition 3.2.1 *Let X be an ample smooth hypersurface of a projective manifold P of dimension $n+1$ and let $U = P - X$. Suppose that $H^{n+1}_{\text{prim}}(P; \mathbf{Q}) = 0$. Then the transpose of the "tube over cycle map" induces an isomorphism*

$$H^{n+1}(U) \xrightarrow[\cong]{r} H^n_{\text{var}}(X).$$

Proof We split

$$H^{n+1}(P; \mathbf{Q}) = H^{n+1}_{\text{prim}}(P; \mathbf{Q}) \oplus L_P H^{n-1}(P; \mathbf{Q}),$$

where L_P denotes the Lefschetz operator. The latter is multiplication with the fundamental class of a hyperplane section which we may assume to be given by the class of X. Then $i_* \circ i^* : H^{n-1}(P; \mathbf{Q}) \to H^{n+1}(P; \mathbf{Q})$ equals the Lefschetz operator L_P, and in particular the image of L_P is contained in the image of i_* and hence is killed by j^*. So the kernel of the residue map contains at most the image of primitive $(n+1)$-cohomology of the manifold P which is assumed to be 0. So r is injective and it remains to see that its image coincides with the variable cohomology. But this follows from the definition of variable cohomology if we use the Gysin sequence, since $\operatorname{Im}(r) = \operatorname{Ker}(i_*)$. □

The middle cohomology with *rational* coefficients actually can be shown to split (Problem 3.2.3).

Lemma 3.2.2 *The middle cohomology $H^n(X; \mathbf{Q})$ splits as follows:*

$$H^n(X; \mathbf{Q}) = H^n_{\text{var}}(X, \mathbf{Q}) \oplus H^n_{\text{fix}}(X; \mathbf{Q}).$$

This shows that $H^{n+1}(U;\mathbf{Q})$, which under our assumptions is equal to the variable cohomology, computes the interesting part of the middle cohomology of X. To continue with the former, recall that its cohomology is the hypercohomology of the complex $\Omega_U^\bullet$. However, on one hand it is the hypercohomology of the subcomplex $\Omega_P^\bullet(*X)$ of those meromorphic forms on P having at most poles along X. On the other hand, one can as well use the the subcomplex of those forms having at most a logarithmic pole along X, as follows.

Theorem 3.2.3 *The restriction map* $\Omega_P^\bullet(*X) \to \Omega_U^\bullet$ *and the inclusion map* $\Omega_P^\bullet(\log X) \hookrightarrow \Omega_P^\bullet(*X)$ *induce isomorphisms*

$$\mathbf{H}^p(\Omega_P^\bullet(\log X)) \xrightarrow{\sim} \mathbf{H}^p(\Omega_P^\bullet(*X)) \xrightarrow{\sim} \mathbf{H}^p(\Omega_U^\bullet) = H^p(U,\mathbf{C}).$$

The proof really becomes an exercise once we are familiar with the techniques of spectral sequences. See Problem 6.3.4 in Chapter 6.

The logarithmic complex naturally enters in the residue sequence

$$0 \to \Omega_P^\bullet \to \Omega_P^\bullet(\log X) \xrightarrow{\text{Res}} \Omega_X^{\bullet-1} \to 0,$$

where the residue map is defined locally on an open set $V \subset P$ on which X has an equation $f = 0$ by a formula which is slightly more complicated than that for curves. In fact, with local holomorphic coordinates $(x_0 = f, x_1, \dots, x_n)$ we write

$$\text{Res}(\alpha) = \eta|_X, if\alpha = \frac{df}{f} \wedge \eta + \eta',$$

where both η and η' only involve $(dx_1, \dots, dx_n)$. We leave it as an exercise (see Problem 3.2.2) to show that this map is well defined.

The long exact sequence in hypercohomology has the same terms as the Gysin sequence and morally should be the same. Indeed, this is almost the case.

Theorem 3.2.4 *We have a commutative diagram*

$$\begin{array}{ccccc}
H^{k+1}(U) & \xrightarrow{r} & H^k(X) & \xrightarrow{i_*} & H^{k+2}(P) \\
\downarrow & & \downarrow \tfrac{1}{2\pi i} & & \downarrow \\
H^{k+1}(U;\mathbf{C}) & \xrightarrow{\text{Res}} & H^k(X;\mathbf{C}) & \xrightarrow{\delta} & H^{k+2}(P;\mathbf{C}) \\
\downarrow \wr & & \downarrow \wr & & \downarrow \wr \\
\mathbf{H}^{k+1}(\Omega_P^\bullet(\log X)) & \xrightarrow{\text{Res}} & \mathbf{H}^k(\Omega_X^\bullet) & \xrightarrow{\delta} & \mathbf{H}^{k+2}(\Omega_P^\bullet).
\end{array}$$

Proof The maps in the diagram are self-explanatory. What has to be proved is the commutativity. The residue formula

$$\int_\gamma \operatorname{Res}(\alpha) = \frac{1}{2\pi \mathrm{i}} \int_{\tau(\gamma)} \alpha$$

for $\gamma \in H_k(X)$ and $\alpha \in H^{k+1}(U)$ shows that the map Res indeed is the transpose r of the "tube over cycle map" up to a factor of $2\pi\mathrm{i}$.

Showing that the connecting morphism in the long exact sequence is $2\pi\mathrm{i}$ times the Gysin map is left an an exercise (Problem 3.2.4). □

Let us now consider the preceding exact sequence of complexes degree by degree. The map i^* preserves the Hodge decomposition, and hence the adjoint i_* as well as δ is a homomorphism of degree (1, 1), and the Hodge components figure in the resulting exact sequence

$$\cdots \to H^q\left(\Omega_P^p\right) \to H^q\left(\Omega_P^p(\log(X)\right) \xrightarrow{\operatorname{Res}} H^q\left(\Omega_X^{p-1}\right) \to \cdots .$$

For $p+q = n$ the map Res, being the Hodge component of an isomorphism map, must itself be an isomorphism. It follows that the isomorphisms in cohomology respect the decomposition into Hodge components:

$$\begin{array}{ccc} H^{n+1}(U;\mathbf{C}) & \xrightarrow{\sim} & H^n_{\mathrm{var}}(X;\mathbf{C}) \\ \| & & \| \\ \bigoplus_{p+q=n+1} H^q(\Omega_P^p(\log X)) & \xrightarrow{\sim} & \left(\bigoplus_{p+q=n+1} H^q(\Omega_X^{p-1})\right) \cap H^n_{\mathrm{var}}(X). \end{array}$$

Therefore we can put a Hodge structure on $H^{n+1}(U)$ of weight $n+2$ by declaring the Hodge components of $H^{n+1}(U)$ to be the subspaces $H^q(\Omega_P^p(\log X))$ figuring on the left. Then the residue isomorphism becomes an isomorphism of Hodge structures of type $(-1,-1)$. In the next section we take a closer look at this.

For the moment we note that this description is not suitable for explicit calculations; the following "order of the pole filtration" on the larger complex $\Omega_P^\bullet(*X)$ of meromorphic forms with arbitrary poles along X turns out to be more manageable. We shall use the (standard) notation $\Omega_P^k(\ell X)$ for the sheaf of meromorphic k-forms on P having at most poles of order $\le \ell$ along X so that in particular $\Omega^{n+1}(X)$ and $\Omega^{n+1}(\log X)$ are the same. Then, the order of pole filtration is defined by:

$$F^p_{\mathrm{pole}}\left(\Omega_P^\bullet(*X)\right) = \Big(0 \to 0 \to \cdots \to 0 \to \Omega_P^p(X) \xrightarrow{d}$$
$$\to \Omega_P^{p+1}(2X) \xrightarrow{d} \cdots \xrightarrow{d} \Omega_P^{n+1}(n+2-p) \to 0\Big).$$

This subcomplex should be viewed as starting in degree p. Its hypercohomology groups $\mathbf{H}^{n+1}(F^p_{\mathrm{pole}}(\Omega_P^\bullet(*X))) \subset \mathbf{H}^{n+1}(\Omega_P^\bullet(*X))$ define a filtration on $H^{n+1}(U;\mathbf{C})$, the *order of pole filtration*. On the other hand, we have the trivial filtration both on the complex $\Omega^\bullet_P(*X)$ and on the logarithmic complex. By definition this gives the Hodge filtration on $H^{n+1}(U;\mathbf{C})$. It is functorial and so is compatible with the residue map. If we knew that the pole filtration gives the Hodge filtration on $H^{n+1}(U;\mathbf{C})$, the residue isomorphism then could be used to calculate the Hodge filtration on primitive cohomology by means of the pole filtration. Unfortunately, the pole filtration and the Hodge filtration do not automatically coincide. For this, one needs X to be sufficiently ample so that a certain vanishing property holds. The precise statement is as follows.

Theorem 3.2.5 *Let X be an ample smooth divisor in a projective manifold P of dimension $n+1$. Suppose that $H^{n+1}_{\rm prim}(P) = 0$ and that $H^i(P,\Omega^j_P(kX)) = 0$ for all $i, k \geq 1, j \geq 0$. The twisted residue map (which now is an isomorphism)*

$$\mathrm{Res} = \frac{1}{2\pi \mathrm{i}} r : H^{n+1}(U,\mathbf{C}) \xrightarrow{\sim} H^n_{\rm var}(X)$$

lowers the degree of the Hodge filtration by one. The pole order filtration on the left-hand side coincides with the Hodge filtration so that Res *induces isomorphisms from the graded pieces of the pole filtration on the left-hand side to the graded pieces of the Hodge filtration on the right-hand side:*

$$\frac{H^0\left(\Omega^{n+1}_P((n-p+1)X)\right)}{H^0\left(\Omega^{n+1}_P((n-p)X)\right) + dH^0\left(\Omega^n_P((n-p)X)\right)} \xrightarrow{\sim} H^{p,n-p}_{\rm var}(X).$$

Consequently, the order of pole filtration induces a pure Hodge structure of weight $(n+2)$ on $H^{n+1}(U)$ and r is an isomorphism of type $(-1,-1)$.

The proof of this theorem uses some further techniques from the theory of spectral sequences. The proof is outlined in Problem 6.3.4 in Chapter 6.

Example 3.2.6 Taking $P = \mathbf{P}^{n+1}$ and X a degree d hypersurface, Bott's vanishing theorem which we show below (Section 7.2) guarantees the required vanishing. We have already noted that there is no primitive cohomology for $\mathbf{P}^{n+1}$ and that variable and primitive cohomology coincide for hypersurfaces.

Let us pursue this example a little further. We are going to give an explicit description of the $(n+1)$-forms and the n-forms. Let $\{z_0,\ldots,z_{n+1}\}$ be a system of homogeneous coordinates on $\mathbf{P}^{n+1}$. We let S^k be the vector space of homogeneous polynomials of degree k in the variables $z_0,\ldots,z_{n+1}$, i.e., the degree k part of the graded ring

$$S = \mathbf{C}[z_0,\ldots,z_{n+1}].$$

The Euler vector field on $\mathbf{P}^{n+1}$ is denoted

$$E = \sum_{i=0}^{n+1} z_i \partial / \partial z_i,$$

and we introduce an $(n+1)$-form ω_{n+1} on $\mathbf{C}^{n+2}$ by

$$\omega_{n+1} = i_E(\mathrm{d}z_0 \wedge \cdots \wedge \mathrm{d}z_{n+1}) = \sum_\alpha (-1)^\alpha z_\alpha \cdot \mathrm{d}z_0 \wedge \ldots \widehat{\mathrm{d}z_\alpha} \cdots \wedge \mathrm{d}z_{n+1}.$$

Here i_E denotes contraction with the Euler vector field. Finally, we let $q : \mathbf{C}^{n+2} - \{0\} \to \mathbf{P}^{n+1}$ be the natural projection. We first state a general result, the proof of which is relegated to Problem 3.2.6 at the end of this section.

Lemma 3.2.7 *The k-form*

$$\tilde{\varphi} = \frac{A}{B} \mathrm{d}z_{i_1} \wedge \cdots \wedge \mathrm{d}z_{i_k}$$

on $\mathbf{C}^{n+2} - \{0\}$ *with A and B homogeneous polynomials satisfies*

$$\tilde{\varphi} = q^*(\text{rational form } \varphi \text{ on } \mathbf{P}^{n+1})$$

if and only if
(i) $i_E(\tilde{\varphi}) = 0$, *and*
(ii) φ *is homogeneous of degree* 0, *i.e.*, $\deg A + k - \deg B = 0$.

Proposition 3.2.8 (i) *Rational* $(n+1)$*-forms on* $\mathbf{P}^{n+1}$ *can be written as total degree-zero forms* $\frac{A}{B}\omega_{n+1}$.
(ii) *Any rational n-form on* $\mathbf{P}^{n+1}$ *is of the form*

$$\psi = \frac{1}{B} \sum_{i<j} (-1)^{i+j} A_{ij} \, \mathrm{d}z_0 \wedge \ldots \widehat{\mathrm{d}z_i} \cdots \wedge \widehat{\mathrm{d}z_j} \cdots \wedge \mathrm{d}z_{n+1},$$

where $A_{ij} = z_i A_j - z_j A_i$ *for some polynomial* A_i *of degree* $\deg(B) - n - 1$.

Proof Observe that we have an exact sequence

$$\Gamma(\mathbf{C}^{n+2}, \Omega^{k+1}) \otimes S^{q-1} \xrightarrow{i_E} \Gamma(\mathbf{C}^{n+2}, \Omega^k) \otimes S^q \xrightarrow{i_E} \Gamma(\mathbf{C}^{n+2}, \Omega^{k-1}) \otimes S^{q+1}.$$

It follows that $i_E(\omega_{n+1}) = 0$. Using the previous description of the rational forms on projective space, the description of the rational $(n+1)$-forms follows. The description of the rational n-forms follows in a similar fashion, using exactness of the sequence for $k = n$. □

Writing out $\mathrm{d}\psi$ in the preceding expression for an n-form with $B = F^l$ we find

$$\mathrm{d}\psi = \left(\frac{1}{F^{l+1}} l \cdot \sum A_j \partial F / \partial z_j - \frac{1}{F^l} \sum \partial A_j / \partial z_j \right) \omega_{n+1}$$

and hence we arrive at the following conclusion.

Proposition 3.2.9 *Any rational $(n+1)$-form φ on $\mathbf{P}^{n+1}$ having a pole of order l along a hypersurface $F = 0$ is a total degree-zero rational form $\frac{A\omega_{n+1}}{F^l}$ and*

$$\varphi = \mathrm{d}\psi + \text{ a rational form having pole order} \leq l - 1 \text{ along } F = 0$$

if and only if

$$A = \sum_j B_j \partial F/\partial z_j,$$

i.e., if and only if A belongs to the Jacobian ideal

$$\mathfrak{j}_F = (\partial F/\partial z_0, \ldots, \partial F/\partial z_{n+1}).$$

Recall that $S = \mathbf{C}[z_0, \ldots, z_{n+1}]$. We then put

$$R_F = \mathbf{C}[z_0, \ldots, z_{n+1}]/\mathfrak{j}_F.$$

Theorem 3.2.5 in this case then reads as follows.

Theorem 3.2.10 *Let X be a smooth hypersurface in $\mathbf{P}^{n+1}$ of degree d given by $F = 0$. Set $t(p) = d(n-p+1) - n - 2$. The residue isomorphism*

$$\mathrm{Res} : H^{n+1}(\mathbf{P}^{n+1} - X) \xrightarrow{\cong} H^n_{\mathrm{prim}}(X)$$

maps the subspace of $H^{n+1}(\mathbf{P}^{n+1} - X)$ represented by

$$\varphi = \frac{A}{F^{n+1-p}}\omega_{n+1}, \quad \deg A = t(p)$$

to $F^p H^n_{\mathrm{prim}}(X)$ and $\mathrm{Res}(\varphi) \in F^{p+1}H^n_{\mathrm{prim}}(X)$ if and only if $A \in \mathfrak{j}_F$. Hence we obtain an isomorphism

$$R_F^{t(p)} \xrightarrow{\sim} H^{p,n-p}_{\mathrm{prim}}(X).$$

Problems

3.2.1 Show that for a smooth curve C on a surface X the sheaves $\Omega^1_X(\log C)$ and $\Omega^1(C)$ are different.

3.2.2 Show that the residue as we have defined it (see (3.2)) is independent of the choice of coordinates and of the choice of the local equation of the hypersurface.

3.2.3 Give the proof of Lemma 3.2.2. As a hint, show first that the two summands are orthogonal by using the formula $\langle a, i^*b\rangle = \langle i_*a, b\rangle$. Then compare their dimensions using the Lefschetz hyperplane theorem for the fixed part and the Gysin sequence for the variable part.

3.2.4 Complete the proof of Theorem 3.2.4 as follows.

(a) Show that the definition of the logarithmic complex and the residue map also makes sense in the setting of smooth forms.

(b) Use this to establish the isomorphism

$$H^k_{\mathrm{dR}}(X;\mathbf{C}) = \frac{\{\alpha \in \Gamma\left(P, \mathcal{A}^{k+1}_P(\log X)\right) \mid \mathrm{d}\alpha \in \Gamma\left(P, \mathcal{A}^{k+2}_P\right)\}}{\mathrm{d}\Gamma\left(P, \mathcal{A}^k_P(\log X)\right) + \Gamma\left(P, \mathcal{A}^{k+1}_P\right)}.$$

(c) Use Leray's residue formula (see Leray, 1959):

$$\int_X \mathrm{Res}(\theta) = \frac{1}{2\pi \mathrm{i}} \int_P \mathrm{d}\theta$$

to show that for all closed $(2n-k)$-forms γ on P we have

$$\int_P \mathrm{d}\alpha \wedge \gamma = \langle \mathrm{d}\alpha, \gamma \rangle = 2\pi \mathrm{i} \langle i_* \beta, \gamma \rangle$$

$$= 2\pi \mathrm{i} \langle \beta, i^* \gamma \rangle = 2\pi \mathrm{i} \int_X \beta \wedge i^* \gamma.$$

(d) Deduce that

$$[d\alpha] = 2\pi \mathrm{i} \cdot i_*[\beta]$$

and show that this implies that the connecting morphism in the long exact sequence of the residue sequence is $2\pi \mathrm{i}$ times the Gysin map.

3.2.5 Let X be a hypersurface of degree d in $\mathbf{P}^{n+1}$. Consider the covering $\mathfrak{U}$ of $\mathbf{P}^{n+1}$ given by the open sets U_j which are the complement of the hypersurface $\partial F/\partial z_j = 0$, $(j = 0, \ldots, n+1)$. Let K_j be contraction with the vector field $\partial/\partial z_j$ and set $K_{j_0 \cdots j_p} = K_{j_0}, \ldots, K_{j_p}$.

(a) Show that for any polynomial A of degree $d(p+1) - n - 2$ the element $\lambda_p(A)$ is represented by the p-cocycle in $C^p(\mathfrak{U}, \Omega^{n-p}_X)$ given by

$$\alpha_{j_0 \cdots j_p} = \left. \frac{A K_{j_0 \cdots j_p} \omega_{n+1}}{\partial F/\partial z_{j_0} \cdots \partial F/\partial z_{j_p}} \right|_X .$$

(see Carlson and Griffiths, 1980, Section 3b).

(b) Show that for A of degree $d(p+1) - n - 2$ and B of degree $d(n+1-p) - n - 2$ the cup product of $\lambda_p(A)$ and $\lambda_{n-p}(B)$ in $H^{n,n}_{\mathrm{prim}}(X) \cong H^{n+1}(\omega_{n+1})$ is given by the n-cocycle

$$d \frac{AB\Omega_{\mathbf{P}^{n+1}}}{\partial F/\partial z_0 \cdots \partial F/\partial z_j}.$$

(see Carlson and Griffiths, 1980, Section 3c).

(c) Let $R = S/j_F$. Use (ii) to define an isomorphism

$$\lambda_\rho : R^\rho \xrightarrow{\sim} H^{n,n}(X)$$

such that the diagram

$$\begin{array}{ccc} R^{t(p)} \times R^{t(n-p)} & \xrightarrow{\text{multiplication}} & R^\rho \\ \downarrow \lambda_{t(p)} \times \lambda_{t(n-p)} & & \downarrow \lambda_\rho \\ H^{n-p,p}_{\text{prim}} \times H^{p,n-p}_{\text{prim}} & \xrightarrow{\text{cup-product}} & H^{n,n} \end{array}$$

is commutative up to sign.

3.2.6 Prove Lemma 3.2.7. Hint: one possibility is to make direct computations in standard affine coordinates. Alternatively and more conceptually, one could use the Lie derivative in conjunction with Cartan's formula which we treat below (see Definition 4.5.2 and Proposition 4.5.3). Indeed, one should show that the Lie derivative of $\tilde{\varphi}$ in the direction of the Euler vector field is equal to $(\deg \tilde{\varphi})\tilde{\varphi}$. By Cartan's formula it is also equal to $i_E \, d\tilde{\varphi} + d(i_E \tilde{\varphi})$ and so (i) and (ii) imply that the Lie derivative of $\tilde{\varphi}$ in the direction of the Euler field is 0. Verify that this is exactly the condition for a rational form on $\mathbf{C}^{n+2} - \{0\}$ to be the pullback of a form on projective space.

3.3 How Log-Poles Lead to Mixed Hodge Structures

In the situation given in Proposition 3.2.1, we were given a smooth hypersurface $X \subset P$ inside a projective manifold P which has no primitive cohomology in degree $n + 1$. We denoted the complement $P - X$ by U. The "tube over cycle map" gives an isomorphism

$$r : H^{n+1}(U; \mathbf{Q}) \to H^n_{\text{var}}(X; \mathbf{Q})$$

in that case. We also saw that the order of pole filtration induces a pure weight $(n + 2)$ Hodge structure on the left-hand side. Since the right-hand side has a pure Hodge structure of weight n, the map r shifts the weight by 2: we have a morphism of Hodge structures of type $(-1, -1)$. Behind this is the factor of $2\pi i$ which occurs in the residue map so that we have to make a Tate twist (see Example 1.2.6) on the left for compatibility on integral level; then the map $r : H^{n+1}(U) \otimes \mathbf{Z}(1) \to H^n(X)$ is actually a morphism of Hodge structures.

In the general situation, where P has nonzero primitive cohomology, the image of primitive cohomology of P inside $H^{n+1}(U; \mathbf{Q})$ clearly carries a (pure)

Hodge structure of weight $n+1$, because the restriction map is not supposed to shift the weight by naturality. So we have a situation in which we should have two different weights on a fixed vector space. Indeed, we have a mixed Hodge structure.

Example 3.3.1 In the introduction in the subsection starting on page 36 we saw that a punctured Riemann surface gives an example of a mixed Hodge structure: let C be a Riemann surface which is obtained by deleting m points $x_1, \ldots, x_m$ in a compact Riemann surface $\hat{C}$. Then we have the long exact sequence

$$0 \to H^1(\hat{C}, \mathbf{Z}) \to H^1(C, \mathbf{Z}) \to \mathbf{Z}^m \to H^2(\hat{C}, \mathbf{Z}) \to H^2(C, \mathbf{Z}) = 0.$$

From this, one can see that $H^1(C, \mathbf{Z})$ is an extension of $\mathrm{Ker}(\oplus_{i-1}^m \mathbf{Z} \to H^2(\hat{C}, \mathbf{Z}))$ by $H^1(\hat{C}, \mathbf{Z})$. The weight filtration on this free $\mathbf{Z}$-module $H^1(C, \mathbf{Z})$ is defined by $W_1 H^1(C, \mathbf{Z}) = H^1(\hat{C}, \mathbf{Z})$, and $W_2 H^1(C, \mathbf{Z}) = H^1(C, \mathbf{Z})$ is everything. The Hodge filtration comes from the residue sequence

$$0 \to \Omega^1_{\hat{C}} \to \Omega^1_{\hat{C}}(\log\{x_1, \ldots, x_m\}) \to \mathcal{O}_{\{x_1, \ldots, x_m\}} \to 0,$$

i.e., we have $F^0 H^1(C, \mathbf{C}) = H^1(C, \mathbf{C})$ and

$$F^1 H^1(C, \mathbf{C}) = H^0\left(\hat{C}, \Omega^1_{\hat{C}}(\log\{x_1, \ldots, x_m\}\right).$$

Finally, $F^2 H^1(C, \mathbf{C}) = 0$. Altogether we obtain an extension of (pure) Hodge structures

$$0 \to H^1(\hat{C}) \to H^1(C) \to \mathrm{Ker}\left(\bigoplus_{i=1}^m \mathbf{Z}(-1) \to H^2(\hat{C})\right) \to 0,$$

where $\mathbf{Z}(-1)$ is again the (-1)-st Tate twist.

In contrast to the case of pure Hodge structures, there is no canonical splitting of $H_{\mathbf{C}}$ into Hodge subgroups $H^{p,q}$ (see the previous example). However, it can be easily shown that we have a decomposition of the form

$$H_{\mathbf{C}} = \bigoplus\nolimits_{p,q} H^{p,q}$$

such that all elements in $H^{p,q}$ are in $F^p \cap W^{\mathbf{C}}_{p+q}$ and the summand $H^{p,q}$ is conjugate to $H^{q,p}$ modulo summands of the form $H^{r,s}$ where $r < p$ and $s < q$. Moreover, as shown by Deligne (1971b), these splittings are unique provided we refine the latter requirement about the conjugates. We do not really make use of this fact, but it forms the essential ingredient in the proof that morphisms of mixed Hodge structures are strict. Let us first give the definitions.

Definition 3.3.2 A *morphism of mixed Hodge structures* $f : H \to H'$ is a morphism of $\mathbf{Z}$-modules which is compatible with both filtrations. A morphism

$f : H \to H'$ is a morphism of mixed Hodge structures of type $(-m,-m)$, if under f the Hodge and weight filtrations behave like $f(F^k) \subset F^{k-m}$ and $f(W_k) \subset W_{k-2m}$. In other words, if one looks at $(2\pi i)^{-m} f : H \otimes \mathbf{Z}(m) \to H'$, then this is a morphism of mixed Hodge structures in the usual sense.

As stated above, morphisms of mixed Hodge structures behave particularly well with regard to the weight and Hodge filtration in that they strictly preserve these. Here we recall that a morphism between filtered vector spaces $f : (V,F) \to (V',F')$ is *strict* if not only f preserves the filtration, but also if $w \in \mathrm{Im}(f) \cap (F')^p$ then $w = f(v)$ with $v \in F^p$. We are not going to show this, but the original source is Deligne (1971b).

The basic idea behind introducing mixed Hodge structures is to impose a finer structure on the cohomology groups of noncompact or singular algebraic varieties that are preserved under simple functorial operations such as pullback under morphisms in cohomology or homology. Furthermore, they should be generalizations of the familiar Hodge structures on smooth projective varieties.

Let us return to the situation of Section 3.2 and define carefully the Hodge and weight filtration on $H^{n+1}(U)$. This is done best on the level of the log-complexes $\Omega_P^\bullet(\log X)$. We saw that we should use the order of the pole filtration. On the log-complex this amounts to the trivial filtration:

$$F^p\Omega_P^\bullet(\log X) = \left(0 \to \cdots \to 0 \to \Omega_P^p(\log X) \to \Omega_P^{p+1}(\log X) \to \cdots\right),$$

which we have defined in Section 3.1, inducing in cohomology

$$F^p H^{n+1}(U;\mathbf{C}) = \mathrm{Im}\left(\mathbf{H}^{n+1}(P, F^p\Omega_P^\bullet(\log X)) \to \mathbf{H}^{n+1}(P, \Omega_P^\bullet(\log X))\right).$$

To define the weight filtration, we proceed as follows again on the level of complexes: define

$$W_m\Omega_P^\bullet(\log X) = \begin{cases} 0 & \text{for } m < 0, \\ \Omega_P^\bullet & \text{for } m = 0, \\ \Omega_P^\bullet(\log X) & \text{for } m \geq 1. \end{cases}$$

This carries over to a filtration on cohomology via

$$W_{m+n+1}H^{n+1}(U;\mathbf{C}) = \mathrm{Im}\left(\mathbf{H}^{n+1}(P, W_m\Omega_P^\bullet(\log X)) \to \mathbf{H}^{n+1}(P, \Omega_P^\bullet(\log X))\right).$$

This filtration is defined over $\mathbf{Q}$, since the restriction map $H^\bullet(P) \to H^\bullet(U)$ is defined over $\mathbf{Q}$ so that the group $\mathbf{H}^{n+1}(P, \Omega_P^\bullet)$ carries a natural $\mathbf{Q}$ structure compatible with our definition. Note that our convention is such that the weight of an element in $\mathrm{Gr}^W_{m+n+1} H^{n+1}(U;\mathbf{Q})$ is exactly $m+n+1$, as it should be. The resulting sequence

$$\cdots \to H^{n+1}(P) \to H^{n+1}(U) \otimes \mathbf{Z}(1) \to H^n(X) \to \cdots$$

then becomes a sequence of mixed Hodge structures.

In Chapter 10 we shall encounter a more general situation, namely one in which $X \subset P$ is a simple normal crossing divisor. This is by definition a union of smooth hypersurfaces in P such that for each point of P there is a local coordinate patch $(U, z_1, \ldots, z_{n+1})$ such that X is locally given by the equation $z_1 z_2 \ldots z_k = 0$ for some $0 \leq k \leq n+1$ (here $n = \dim(P)$). So locally, k smooth branches of X meet transversally. In this case, the sheaf $\Omega^1_P(\log X)$ is defined as the subsheaf of $\Omega^1_P(*X)$ that is generated by Ω^1_P and all forms $\mathrm{d}z_i/z_i$ for all local equations $\{z_i = 0\}$ of components of X. The higher wedge powers $\Omega^p_P(\log X)$ are defined as the p-th exterior powers of $\Omega^1_P(\log X)$. The cohomology of U with complex coefficients is defined to be the hypercohomology group $\mathbf{H}^*(P, \Omega^\bullet_P(\log X))$. There is again a natural subcomplex $F^p\Omega^\bullet_P(\log X)$ of $\Omega^\bullet_P(\log X)$ given by the trivial filtration, and it induces the Hodge filtration on hypercohomology in the same way as before.

The weight filtration on $H^k(U, \mathbf{Q})$ is a bit more difficult to define and involves some residue calculus. In this setting the weight filtration on the complex of log-poles that keeps track of the number of $\mathrm{d}z_i/z_i$ possibly has more than two nontrivial steps. Indeed, the subsheaves $W_m\Omega^p_P(\log X)$ of those differential forms that are of the form

$$\alpha \wedge \frac{\mathrm{d}z_{i(1)}}{z_{i(1)}} \wedge \ldots \wedge \frac{\mathrm{d}z_{i(m)}}{z_{i(m)}}, m \leq n+1, \ \alpha \text{ holomorphic},$$

give a filtration with at most as many steps as we have branches of X that locally meet. This is a generalization of what we previously had, which becomes apparent on writing the weight filtration as

$$W_m\Omega^p_P(\log X) = \begin{cases} 0 & \text{for } m < 0, \\ \Omega^{p-m}_P \wedge \Omega^m_P(\log X) & \text{for } 0 \leq m \leq p, \\ \Omega^p_P(\log X), & \text{for } m \geq p. \end{cases}$$

This yields an increasing filtration $W_\bullet$ on the whole complex $\Omega^\bullet_P(\log X)$ and therefore on the cohomology by using the same shift as before,

$$W_{m+k}H^k(U, \mathbf{C}) = \mathrm{Im}\left(\mathbf{H}^k(W_m\Omega^\bullet_P(\log X)) \to \mathbf{H}^k(\Omega^\bullet_P(\log X))\right)$$

so that the elements have weights $\leq m+k$.

With some more work, the weight filtration is seen to be already defined over $\mathbf{Q}$. First, there are multiple residue maps

$$\mathrm{res} : \mathrm{Gr}^W_m \Omega^p_P(\log X) \cong i_*\Omega^{p-m}_{X(m)},$$

which are all isomorphisms under our assumptions. Here $i : X(m) \hookrightarrow X$ denotes the inclusion of the disjoint union of the m-fold intersections $X_{i(1)} \cap$

$\ldots \cap X_{i(m)}$ of X. Then the weight pieces W_m on cohomology sit in successive extensions of cohomology groups defined by the $X(m)$ and are therefore defined over $\mathbf{Q}$.

A more general situation is that of a smooth, quasi-projective manifold U by definition the complement of a Zariski closed subset inside an *arbitrary* projective variety. One can replace this compactification by a smooth projective manifold P with complement a simple normal crossing divisor X so that we are in the situation above. The associated mixed Hodge structure on $H^k(U, \mathbf{C})$ using this particular pair (P, X) does not depend on the choices (see Deligne, 1971b). Because the number of intersecting components of X is bounded by $\dim(U)$, the description of the weight filtration as given above immediately leads us to the following conclusion.

Lemma 3.3.3 *The weights i of the mixed Hodge structure on the cohomology group $H^k(U, \mathbf{C})$ of a smooth quasi-projective variety U are bounded in the range $k \leq i \leq k + \dim(U)$.*

We also can define a mixed Hodge structure on the cohomology of a pair (X, A) with X and A smooth-projective or quasi-projective. We do not explain this in detail here, but we present the necessary ingredients. First, we interpret $H^\bullet(X, A)$ as the cohomology of the mapping cone of the natural restriction homomorphism $S^\bullet(X) \to S^\bullet(A)$ between the complexes of singular cohomology. This concept was introduced in Chapter 1. We recall its definition here.

Definition 3.3.4 Let $h : A^\bullet \to B^\bullet$ be a morphism of complexes of sheaves on X. Then the *mapping cone* of h is defined as the complex of sheaves

$$\mathrm{Cone}^n(h) = A^{n+1} \oplus B^n, \quad d_C = (-d_A, -f + d_B).$$

With this definition, it is easy to check that d_C really gives a differential with $d_C^2 = 0$ and that we have a short exact sequence of complexes

$$0 \to B^\bullet \to \mathrm{Cone}^\bullet(h) \to A^\bullet[1] \to 0.$$

Because we have smooth algebraic manifolds, we can calculate their complex cohomology also using the holomorphic de Rham complex. So we have the de Rham isomorphisms $H^i(X; \mathbf{C}) = \mathbf{H}^i(X, \Omega_X^\bullet)$ and $H^i(A; \mathbf{C}) = \mathbf{H}^i(X, i_*\Omega_X^\bullet)$, where $i : A \hookrightarrow X$ is the inclusion. The five lemma together with the long exact sequences for singular cohomology, as well its de Rham version, show that we indeed may compute $H^\bullet(X, A; \mathbf{C})$ as the hypercohomology of the cone complex of $h : \Omega_X^\bullet \to i_*\Omega_A^\bullet$. The trivial filtration on this cone complex does not give the correct Hodge filtration, unless X and A are projective. In the general case, one has to replace $\Omega_X^\bullet$ and $\Omega_A^\bullet$ by their logarithmic cousins, obtained by compactifying X and A suitably. The map h gets replaced by its logarithmic version and the correct Hodge filtration is the trivial filtration on the cone of this new map. The point here is that this latter cone still computes $H^\bullet(X, A)$.

3.4 Algebraic Cycles and Their Cohomology Classes

Among the elementary consequences of the Kähler condition is a further topological restriction, namely, that all of the even Betti numbers in the obvious range are nonzero. To see this, observe that in local holomorphic coordinates

$$\omega^n = \text{const.} \cdot \det(h_{rs}) \mathrm{i}\, \mathrm{d}z_1 \wedge \mathrm{d}\bar{z}_1 \wedge \cdots \wedge \mathrm{i}\, \mathrm{d}z_n \wedge \mathrm{d}\bar{z}_n,$$

where the constant is positive and the wedge product is the positively oriented volume form on $\mathbf{C}^n$ with its euclidean structure. Since (h_{rs}) is positive, so is the determinant, and so

$$\int_M \omega^n > 0.$$

We conclude that the cohomology class of ω^n and, *a fortiori*, its lower exterior powers are nonzero. This establishes what was claimed. Indeed, we have the following proposition.

Proposition 3.4.1 *For a compact Kähler manifold M we have*

$$h^{p,p}(M) > 0$$

for $0 \le p \le \dim M$.

In any case, this gives a more elementary proof that a manifold homotopy equivalent to the Hopf manifold, which has $B_2 = 0$, carries no Kähler structure (see the examples in Section 2.2.1).

This reasoning shows also that the fundamental homology class of any n-dimensional complex submanifold $i : N \hookrightarrow M$ is of pure Hodge type. We recall that this class $i_*[N]$ is defined in Appendix B for any oriented submanifold of M. To begin with, we claim that it is nonzero. In fact, we know that the Kähler class of M restricts to the Kähler class of N. From calculus we have

$$\int_{i_*[N]} \omega^n = \int_N i^* \omega^n.$$

However, the last integral is positive by the argument in the previous paragraph, so that $i_*[N] \neq 0$, as required. Next, we adapt the argument to prove the assertion about the Hodge type. To explain this, we use the dual Hodge decomposition on the complexified homology, as defined at the end of Section

2.2. Recall that a homology class c has type (r, s) if $\phi(c) = 0$ whenever ϕ has type (p, q) with $(p + r, q + s) \neq (0, 0)$. We thus have a Hodge decomposition

$$H_\ell(M; \mathbf{C}) = \bigoplus_{p+q=-\ell} H_{p,q}(M),$$

and we have the following proposition.

Proposition 3.4.2 *Let $i : N \hookrightarrow M$ be a compact complex submanifold of a compact Kähler manifold M. The fundamental homology class $i_*[N]$ is of type $(-n, -n)$, $n = \dim N$.*

Proof Let ϕ be a closed (p, q)-form and consider the integral

$$\int_N i^* \phi.$$

Locally, the integral is a sum of integrals

$$\int f \, \mathrm{d}z_I \wedge \mathrm{d}\bar{z}_J,$$

where the z_i's are now holomorphic coordinates on N. If $p > n$ then $\mathrm{d}z_I = 0$, and if $q > n$ then $\mathrm{d}\bar{z}_J = 0$, from which the assertion follows. □

Let us pause for a moment to show that the Poincaré duality isomorphism respects the Hodge decomposition:

Proposition 3.4.3 *Poincaré duality induces an isomorphism of Hodge structures*

$$H^\ell(M)(2m) \xrightarrow{\sim} H_{2m-\ell}(M).$$

Proof We have just shown that the fundamental class of M has pure type $(-m, -m)$. Poincaré duality is induced by the cap product

$$H_{2m}(M) \times H^{2m-\ell}(M) \to H_\ell(M),$$

which respects Hodge structures (it is dual to the cup product) and so the Poincaré duality isomorphism maps classes of type (p, q) in $(2m-\ell)$-cohomology to classes of type $(p - m, q - m)$ in ℓ-homology, in accord with the weight $p + q - 2m = -\ell$ of the Hodge structure on H_ℓ. The Tate twist corrects the shift produced by cap product with the fundamental class. □

Let us further generalize the notion of fundamental class in case N is a subvariety of a complex manifold M, possibly with singularities. There still exists a fundamental homology class $[N] \in H_{2n}(N)$. For instance, one can use resolution of singularities $\sigma : \tilde{N} \to N$ (see Hironaka, 1964) and push the fundamental class of $\tilde{N}$ to N. To test whether the resulting class is nonzero in the homology of M it suffices to show that the integral of some global form over an open dense subset of N is nonzero. Taking for this the regular locus N_{reg} of N we find

$$\int_{i_*[N]} \omega^n = \int_N i^*\omega^n = \int_{N_{\text{reg}}} i^*\omega^n > 0.$$

In a similar vein, the preceding proposition can be shown for any irreducible algebraic subvariety:

Addendum Let M be a complex projective manifold and $i : N \hookrightarrow M$ an irreducible algebraic subvariety. The fundamental homology class $i_*[N]$ is of type $(-n, -n)$, $n = \dim N$.

We can generalize this a bit more by introducing the following fundamental concepts.

Definition 3.4.4 An *n-cycle* on an algebraic variety M is a finite formal linear combination of irreducible algebraic subvarieties of dimension n with integral coefficients. These form the group $\mathfrak{Z}_n(M)$ of algebraic n-cycles. The cycle class map

$$\text{cl} : \mathfrak{Z}_n(M) \to H_{2n}(M)$$

is obtained by extending linearly the map that with a subvariety $i : N \hookrightarrow M$ associates its class $i_*[N]$. Its image is the group of n-dimensional *algebraic classes*.

So we see that the algebraic cycle classes are of pure Hodge type $(-n, -n)$. We showed this for curves on surfaces in Section 1.3, where we discussed the famous *Hodge conjecture*, one of the million-dollar millennium prize questions.

Conjecture (Hodge conjecture) *All classes of type $(-n, -n)$ in $H_{2n}(M; \mathbf{Q})$ are algebraic.*

Some remarks are in order. First of all, the stronger conjecture with integral coefficients in general is false. Atiyah and Hirzebruch (1962) have given a counterexample. Secondly, to this date the conjecture is known in general only for $n = 1$ and $n = m - 1$. We treat this now in more detail.

As we observed in Chapter 1, the case of divisors is due to Lefschetz and Hodge. This the content of the Lefschetz theorem on $(1, 1)$-classes. The proof by Kodaira and Spencer can be found in Griffiths and Harris (1978, p.163). It is based on the long exact sequence in cohomology for the so-called exponential sequence on M

$$0 \to 2\pi \mathrm{i}\mathbf{Z} \to \mathcal{O}_M \xrightarrow{\exp} \mathcal{O}_M^* \to 0,$$

where by definition $\exp(f) = e^f$. The group $H^1(\mathcal{O}_M^*)$ classifies isomorphism classes of line bundles on M:

$$H^1(\mathcal{O}_M^*) = \operatorname{Pic} M.$$

The right-hand side also is the group of classes of divisors ($(m-1)$-dimensional algebraic cycles) under linear equivalence. What is important is that linearly equivalent divisors have the same fundamental class in $H_{2m-2}(M)$. Applying Poincaré duality, this yields the same classes in $H^2(M)$ as well. By abuse of language we also call this element in $H^2(M)$ the "cycle class" of the linear equivalence class of the divisor under consideration. A crucial computation reveals that the coboundary map $H^1(\mathcal{O}_M^*) \to H^2(M)$ can be identified with this new cycle class map. The exact sequence associated with the preceding exponential sequence then shows that the image of this map, the algebraic $(2m-2)$-classes, is the kernel of

$$\pi : H^2(M) \to H^2(\mathcal{O}_M).$$

By Dolbeault's theorem (Proposition 3.1.3) $H^q(\Omega_M^p)$ can be identified with the classes of (p, q)-forms on M and so the kernel of π consists of integral classes of type $(2, 0) + (1, 1)$. Because an integral class in particular is real, the $(2, 0)$-part must vanish and hence that class must be of pure type $(1, 1)$. The Hodge conjecture for integral algebraic $(m-1)$-classes thus follows.

The Hodge conjecture for algebraic one-cycles then follows from this and the hard Lefschetz theorem (Theorem 2.3): because the isomorphism $H^2 \xrightarrow{\sim} H^{2m-2}$ from this theorem preserves rational pure-type classes as well as the algebraic cycles, the Hodge conjecture thus follows.

Motived by this, we give a special name to the integral $(1, 1)$-classes; they form the *Néron–Severi group*

$$\mathsf{NS}(M) = \left\{c \in H^2(M) \mid c \otimes 1 \in H^{1,1}(M) \subset H^2(M) \otimes_{\mathbf{Z}} \mathbf{C}\right\}, \tag{3.4}$$

whose rank is called the *Picard number*.

We end this section by giving a few special cases where the Hodge conjecture is true. For a more extensive discussion of the Hodge conjecture, see Lewis (1991).

Examples 3.4.5 (i) The conjecture holds for projective spaces, because all cohomology groups have rank 1 and are generated by classes of linear subspaces. Likewise, it is true for Grassmannians, since the cohomology is generated by the so-called Schubert cycles (see Griffiths and Harris, 1978, p. 196),
(ii) Based on the previous examples we can construct more interesting ones. These are based on the computation of the (co)homology of the blowup. See

Example C.2.3. So let $M \subset N$ be a compact submanifold of a compact complex manifold N, say of codimension d, and let $\sigma : \mathrm{Bl}_M N \to N$ be the blowup with center N. Then, putting $k^- = \max\{0, k-d+1\}$, the even-dimensional homology is given by

$$H_{2k}(\mathrm{Bl}_N M) \xrightarrow{\sim} H_{2k}(M) \oplus H_{2k-2}(N) \oplus \cdots \oplus H_{2k^-}(N).$$

In particular, if the Hodge conjecture is true for M and it is true for N, it is also true for $\mathrm{Bl}_N M$. Particular cases include projective spaces blown up in points, curves, and surfaces.

(iii) Continuing with the previous example, we can find an interesting particular case, the so-called *unirational* fourfold M. By definition this means that there exists a dominant rational map $\mathbf{P}^4 \dashrightarrow M$. To show that the Hodge conjecture is true, one first resolves the indeterminacies of the rational map by blowing up M (see Hironaka, 1964),

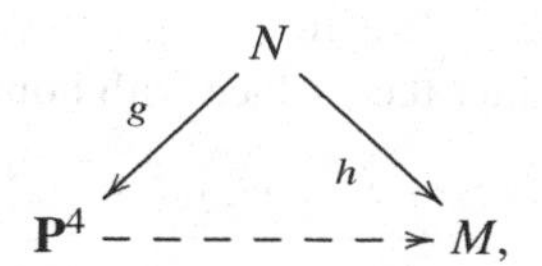

where g is a succession of blowups in points, curves, and surfaces, and h is a generically finite morphism of degree say d. The Hodge conjecture only needs to be checked for codimension two-cycles. The previous example shows that it is true for N. Next, one has to look at $h : N \longrightarrow M$. Let us consider a topological 4-cycle x on M. Since h is generically finite, the inverse image of a suitable cycle homologous to x gives a 4-cycle y on N. Suppose that the class $[x] \in H_4(M)$ has type $(-2,-2)$. Then the same is true for the class $[y] \in H_4(N)$, which therefore is an algebraic class, say $[y] = \mathrm{cl}\, Z$. On the level of cycles the induced homomorphism $\mathfrak{Z}_2(N) \xrightarrow{h_*} \mathfrak{Z}_2(M)$ is defined as follows. For an irreducible cycle Z set $h_* Z = 0$ if Z is mapped to a lower- dimensional cycle and otherwise set $h_* Z = (\deg h \mid Z) \cdot h(Z)$. It is not too hard to show that this is compatible with the cycle class maps. One therefore gets $\mathrm{cl}(h_* Z) = d \cdot [x]$ and so the $\mathbf{Q}$-Hodge conjecture holds.

Problem

3.4.1 The goal of this exercise is to give an explicit example of a surface with Picard number 1. The Lefschetz–Noether theorem, which we show below (Section 7.5), states that the *generic* surface of degree ≥ 4 has Picard number 1, but the point is to construct one *explicitly*! Suppose a cyclic group G of order n is acting on a surface $S \subset \mathbf{P}^3$

through projective transformations. We want to investigate the induced action on the Néron–Severi group of S. We first recall that for every divisor m of n there is one irreducible rational representation

$$W_m = \mathbf{Q}[t]/(\Phi_m(t))$$

of G, where $\Phi_m(t)$ is the m-th cyclotomic polynomial and the generator of G acts as multiplication with t. For every n-th root of unity α there is likewise one irreducible representation

$$U_\alpha = \mathbf{C}[t]/(t - \alpha).$$

The crucial observation is that $\mathsf{NS}(S)\otimes\mathbf{Q}$ is a rational G-representation and so is its orthogonal complement $T(S)_\mathbf{Q}$. The complexification of the latter contains $H^{2,0} + H^{0,2}$, and so knowledge of the action on two-forms at least shows what kind of irreducible G-representations $T(S)_\mathbf{Q}$ should contain. This in turn gives upper bounds for the Picard number. Now consider the surface with homogeneous equation

$$z_0^5 + z_1 z_2^4 + z_2 z_3^4 + z_3 z_1^4 = 0,$$

which admits an action of a group of order $n = 65$ as follows:

$$(z_0, z_1, z_2, z_3) \mapsto (\eta^{13} z_0, \eta^{-16} z_1, \eta^4 z_2, \eta^{-1} z_3), \eta = \exp(2\pi i/65).$$

(a) Show that this gives an action.
(b) Using the explicit representation for $H^{2,0}$ by residues, show that $T(S)_\mathbf{Q}$ contains $W_{65} + W_5$, a space of dimension 52, while $B_2(S) = 53$. Conclude that $\mathsf{NS}(S)$ has rank 1. For more examples of this kind see Shioda (1981).

3.5 Tori Associated with Cohomology

With Hodge structures H of odd weight $2p - 1$ one associates a natural compact complex torus, its "Jacobian,"

$$JH = H_\mathbf{C}/(F^p + H_\mathbf{Z}).$$

The compactness follows from the observation that the map of vector spaces $H \longrightarrow H_\mathbf{C}$ induces an isomorphism $H \longrightarrow H_\mathbf{C}/F^p$ where both sides are viewed as real vector spaces. This is a consequence of the fact that

$$H_\mathbf{C} = F^p \oplus \overline{F^p}.$$

Consequently one has an induced isomorphism

$$H/H_{\mathbb{Z}} \xrightarrow{\simeq} JH. \qquad (*)$$

The prototype is the classical Jacobian which is $JH^1(S)$, S a Riemann surface. More generally, by definition (2.6), the Weil operator C of a Hodge structure H of weight $2p-1$ satisfies $C^2 = -\operatorname{id}$ and hence $H_{\mathbb{C}}$ is the direct sum of the $\pm$i-eigenspaces, say

$$H_{\mathbb{C}} = H^+ \oplus H^-, \; C|H^\pm = \pm\mathrm{i}, \quad \bar{H}^+ = H^-, \qquad (3.5)$$

that is, C defines a weight-one Hodge structure, say (H, C) and Weil's Jacobian is its Jacobian, $J(H, C)$. This notion has been modified by Griffiths as follows.

Definition 3.5.1 Let X be a compact Kähler manifold. *Griffiths' intermediate Jacobian* $J^p(X)$ is the Jacobian for the cohomology group $H^{2p-1}(X)$ with its usual Hodge structure.

Griffiths' Jacobian varies holomorphically with its defining Hodge structure (see Section 4.5), while Weil's Jacobian in general does not.

Suppose now that the Hodge structure is polarized by the bilinear form b. It defines a skew-form on the real vector space H, hence yielding a translation-invariant (and therefore closed) two-form ω on the quotient torus (recall $(*)$). Explicitly, we have

$$\omega(u, v) = -\mathrm{i}b(u, \bar{v}). \qquad (**)$$

Its integral on the one-cycle defined by a lattice vector is an integer, so its cohomology class is integral. Because $T_0 JH = H_{\mathbb{C}}/F^p \cong \overline{F^p}$, one can identify the space $\overline{F^p}$ with the space of translation-invariant holomorphic vector-fields. Thus the first bilinear relation, $b(F^p, F^p) = 0$, implies that the $(0, 2)$-component of ω vanishes. Because it is real, the $(2, 0)$-component vanishes as well, and so ω is of type $(1, 1)$. The second bilinear relation and $(**)$ imply that ω has sign $-\mathrm{i}^{s-r+1} = (-1)^{r+p+1}$ on $H^{r,s}$ and thus we have (see (3.5))

$$\omega > 0 \text{ on } H^- = H^{p-1,p} \oplus H^{p-3,p+2} \oplus \cdots \subset \overline{F^p H},$$
$$\omega < 0 \text{ on } H^+ = H^{p-2,p+1} \oplus H^{p-4,p+3} \oplus \cdots \subset \overline{F^p H}.$$

It follows that there is no natural polarization on JH unless $p = 1$ or if the Hodge structure is very special. In the geometric situation it follows that $J^1(X)$ is a polarized variety, it is the ordinary Jacobian. Moreover, $J^n(X)$ is a polarized torus, the *Albanese torus*, which happens to be the dual of the Jacobian torus. Note however that the Weil Jacobian *does* get a natural polarization and so it is an abelian variety.

What is significant is that in these cases ω is the first Chern class of a positive line bundle. By the Kodaira embedding theorem, sections of some high power of this bundle embed the underlying torus as a closed complex submanifold of some projective space (see Griffiths and Harris, 1978, p. 181). Thus in these special cases we get abelian varieties, meaning that they carry the additional structure of an algebraic manifold. For example, the Jacobian of H^1 is always an abelian variety, while Griffiths' intermediate Jacobian of H^3 is an abelian variety only in the special case $H^{3,0}(M) = 0$.

Let us see how our definitions relate to the classical theory, where the Jacobian is defined as

$$\mathrm{Jac}(M) = H^{1,0}(M)^\vee / H_1(M; \mathbf{Z})$$

for any compact Riemann surface M. The resulting torus is designed so as to admit a natural map

$$u : M \longrightarrow \mathrm{Jac}(M),$$

the *Abel–Jacobi map*. To define it, fix a base point p, then map an arbitrary point q to the functional on abelian differentials (holomorphic one-forms) given by the line integral, i.e., to the functional

$$\phi \mapsto \int_p^q \phi.$$

This functional is well defined up to addition of a functional defined by integrating over a loop based at p, so that the class of the functional $\int_p^q$ in $\mathrm{Jac}(M)$ is well defined. To relate $\mathrm{Jac}(M)$ to $JH^1(M)$, observe that the natural pairing

$$H^1(M; \mathbf{Z}) \times H^1(M; \mathbf{Z}) \to \mathbf{Z},$$

given by the cup product followed by integration over the fundamental class, allows one to identify duals of subspaces of H^1 with quotients of H^1. In the case at hand $H^{1,0}$ is isotropic, and so its dual is naturally isomorphic with $H^1(M; \mathbf{C})/H^{1,0}(M)$. It follows that $\mathrm{Jac}(M) = JH^1(M)$.

In the case of an elliptic curve, say

$$\mathscr{E}_\lambda = \left\{ y^2 = x(x-1)(x-\lambda) \right\} + \text{ point at infinity}, \lambda \neq 0, 1, \infty,$$

the Abel–Jacobi map yields an isomorphism

$$u : \mathscr{E}_\lambda \xrightarrow{\cong} JH^1(\mathscr{E}_\lambda).$$

This was explained in Chapter 1.

Since the Hodge structure determines the Jacobian up to isomorphism, it determines the original elliptic curve up to isomorphism. This observation is the prototype for the next result.

Theorem (Torelli's theorem) *A compact Riemann surface is determined up to isomorphism by the polarized Hodge structure on its first cohomology.*

This is not difficult to see in the case of genus 2. The image of the Abel–Jacobi map is another Riemann surface M', possibly singular. In fact it is smooth and isomorphic to M. On the other hand, the polarization ω determines a line bundle L. Line bundles with fixed Chern class on a torus are parametrized by the torus itself, so the torus acts on the set of such line bundles by translation. If the representative line bundle is chosen with care, it has a nontrivial holomorphic section. A classical fact is that the zero set of this section is M'. Thus there is a prescription for constructing a Riemann surface isomorphic to M from the Jacobian, viewed with its structure of projective algebraic variety. This structure is completely determined by H^1 viewed as a *polarized* Hodge structure. The proof of the general result is more difficult (see for instance Griffiths and Harris, 1978, p. 35).

Problems

3.5.1 Show that for a cubic threefold X the intermediate Jacobian is polarized.

3.5.2 Let X be a complex projective manifold. Suppose that there exists a rationally defined sub-Hodge structure $H \subset H^{p-1,p}(X) \oplus H^{p,p-1}(X)$ contained in $H^{2p-1}(X;\mathbf{C})$. So some sublattice $H_{\mathbf{Z}} \subset H^{2p-1}(X)$ generates H as a real vector space. Prove that the torus $H/H_{\mathbf{Z}}$ is an abelian variety.

3.5.3 Shioda has constructed examples of hypersurfaces of odd dimension for which there are no nontrivial rationally defined Hodge structures as in the previous exercise (see Shioda, 1985). This illustrates a theorem of Noether–Lefschetz type proven by Griffiths (1968) stating that this is true for any "generic" smooth hypersurface of projective space of sufficiently high degree. We are presenting here the simplest example, that of a sextic threefold $X \subset \mathbf{P}^4$ with equation

$$z_0z_1^5 + z_1z_2^5 + z_2z_3^5 + z_3z_4^5 + z_4z_0^5 = 0.$$

The cyclic group of prime order 521 acts on it as follows. Choose a primitive m-th root of unity σ, where $m = 5^5 + 1 = 6 \cdot 521$ and define

$$g : (\ldots, z_i, \ldots) \mapsto \left(\ldots, \sigma^{(-5)^{4-i}} z_i, \ldots\right).$$

(a) Show that X is smooth.

(b) Show that g acts on X and that it induces a projective action of the cyclic group G of order 521.
(c) Show that $B_3(X) = 520$ and determine the Hodge numbers.
(d) Calculate the action of g on $H^{3,0}(X)$ and conclude that the representation of G on $H^3(X;\mathbf{Q})$ is the unique nontrivial irreducible representation of G (it was called W_{521} in Problem 3.4.1).
(e) Conclude that there cannot be a nontrivial rational sub-Hodge structure inside $H^{2,1}(X) \oplus H^{1,2}(X)$.

3.6 Abel–Jacobi Maps

Let us see how our definition of the intermediate Jacobian relates to geometry. There is a dual view of the intermediate Jacobian. The cup product pairing establishes a duality between $H^{2p-1}(M)$ and $H^{2m-2p+1}(M)$ with the property that F^k is annihilated by F^{m-k+1} (a form having at least k dz's gives 0 when wedged against one having at least $m-k+1$ dz's). So this duality induces an isomorphism between H^{2p-1}/F^p and the dual of F^{m-p+1}. Moreover, if you compose the duality with the Poincaré duality isomorphism you get an embedding

$$\alpha : H_{2m-2p+1}(M;\mathbf{Z})/\text{torsion} \longrightarrow H^{2m-2p+1}(M)^\vee,$$

which is just integration. So we find

$$\boxed{J^p(M) \cong \left(F^{m-p+1}H^{2m-2p+1}(M)\right)^\vee / \alpha\left(H_{2m-2p+1}(M;\mathbf{Z})\right).}$$

This description generalizes the description of Jac(M) as given in the previous section. Using this, we can define the *Abel–Jacobi map* on algebraic cycles of (complex) codimension p which are homologous to 0. If $Z = \partial\Gamma$ is such a cycle, integrating closed forms over Γ defines an element of $F^{m-p+1}H^{2m-2p+1}(M)^\vee$, and a different choice of Γ' with $Z = \partial\Gamma'$ determines the same element in the intermediate Jacobian $J^m(M)$ because $\Gamma - \Gamma'$ gives a period integral. One needs to verify furthermore that different closed forms representing the same class yield the same integral over Z. So we have to show $\int_\Gamma \alpha = \int_\Gamma \alpha'$ if α and α' are two closed forms in the same cohomology class. But in Chapter 6 we prove that the spectral sequence associated with the double complex $(A^{\bullet,\bullet}, \partial, \bar\partial)$ degenerates at E_1, which implies that $\mathrm{d} = \partial + \bar\partial$ strictly respects the type decomposition. This is similar to what we did above (see Proposition 3.1.6). It implies that the form β for which $\mathrm{d}\beta = \alpha - \alpha'$ may be assumed to have at least $m-p+1$ dz's and so $\int_\Gamma \mathrm{d}\beta = \int_Z \beta = 0$.

One may also use upper indices to grade cycles by codimension and then the Abel–Jacobi map can be written as a linear map:

$$\mathcal{Z}^p_{\text{hom}}(M) \stackrel{\text{def}}{=} \{\text{ codim } p\text{-cycles on } M \text{ homologous to } 0\} \xrightarrow{u} J^p(M).$$

The left-hand side usually is an enormous group. For instance, for zero-cycles, it is just the group of formal sums $\sum n_P P$, $P \in M$ and $\sum n_P = 0$. To reduce the size, one can impose equivalence relations. There is another reason to do this: the process of intersecting a given cycle with other cycles is not always well behaved. For instance, intersecting a codimension d variety with a codimension e variety does not always produce a cycle with the expected codimension $d + e$. A sufficient geometric condition for this to happen is that the cycles meet transversally: each component of the cycle should meet each component of the other cycle in smooth points, and at such intersection points the tangent spaces intersect in a subspace of the expected codimension. Any good equivalence relation should obey the "moving lemma" which says that, given a finite set of subvarieties, any cycle can be replaced by an equivalent one which meets all of the given varieties transversally. An obvious example of such an equivalence relation is that of homological equivalence induced by the cycle class map $\text{cl} : \mathcal{Z}_p(M) \to H_{2p}(M)$.

The classical moving lemma (see Roberts, 1972) states that this property also holds for *rational equivalence* and *algebraic equivalence*. These are defined as follows. We say that two cycles belong to the same algebraic family parametrized by a variety S if they occur as fibers over s of a cycle Z on $M \times S$ that is equidimensional over S and meets the fiber $M \times \{s\}$ transversally in the cycle Z_s. Belonging to the same algebraic family generates by definition algebraic equivalence. Restricting the base manifolds to rational curves we get the definition of rational equivalence. For divisors this coincides with the notion of linear equivalence.

The quotient of cycles modulo rationally equivalence by definition is the n-th *Chow group*

$$\text{Ch}_n(M) \stackrel{\text{def}}{=} \mathcal{Z}_n(M)/\{\text{cycles rationally equivalent to } 0\}.$$

Using upper indices to grade by codimension, $\text{Ch}^n(M)$ denotes the equivalence classes of codimension n-cycles.

Next note that two cycles which are algebraically equivalent have the same homology class: choose any path γ connecting s_1 and s_2 and let Z_γ be the union of the cycles $(M \times t) \cap Z$, for $t \in \gamma$. Then the boundary of Z_γ is precisely $\pm(Z_1 - Z_2)$. It follows that the cycle class map passes to Chow groups:

$$\text{cl} : \text{Ch}^d(M) \to H^{2d}(M).$$

As an example, the kernel of the class map for divisors is the Picard torus. This follows from the exponential sequence. The image is the group of algebraic $(m-1)$-cycle classes, the *Néron–Severi group* $\mathsf{NS}(M) \subset H^2(M)$ we encountered previously (3.4). So we have an exact sequence

$$0 \to \mathrm{Pic}^0(M) \to \mathrm{Ch}^1(M) = \mathrm{Pic}(M) \to \mathsf{NS}(M) \to 0$$

describing what happens with divisors.

As a second example, let us look at zero-cycles. Because $H^{2m}(M) \cong \mathbf{Z}$, a zero-cycle has class 0 if and only if it has degree 0. Because a zero-cycle is supported by a curve, and because on a curve cycles are algebraically equivalent to 0 if and only they have degree 0, this is also true on M.

On the other end of the spectrum we have the Abel–Jacobi map which is defined for cycles homologous to 0. We see below (Corollary 4.7.2) that the Abel–Jacobi map varies holomorphically in families. Because a rational curve cannot map nonconstantly to a torus, the Abel–Jacobi map factors over rational equivalence as well. This motivates the introduction of

$$\boxed{\mathrm{Ch}^d_{\mathrm{hom}}(M) \stackrel{\mathrm{def}}{=} \frac{\{\text{codimension } d \text{ cycles homologous to } 0\}}{\{\text{cycles rationally equivalent to } 0\}},}$$

so that the Abel–Jacobi map now becomes

$$u : \mathrm{Ch}^d_{\mathrm{hom}}(M) \to J^d(M).$$

The kernel of the Abel–Jacobi map can be nontrivial, contrary to what we know for divisors. For example, for zero-cycles the right-hand side is the Albanese torus, and Mumford (1969b) showed that even for surfaces the kernel can be large. Indeed, if the surface admits a nontrivial holomorphic two-form, no curve can be found on the surface that supports the kernel up to rational equivalence. In other words, the kernel is larger than any of the Jacobians associated with curves on the surface.

As seen above, zero-cycles are algebraically equivalent to 0 if and only if their degree is 0, but this ceases to be true for higher-dimensional cycles. The first examples of this phenomenon were given by Griffiths (1969). We treat these in Chapter 10.

Problems

3.6.1 Verify that the moving lemma implies that the direct sum of the Chow groups becomes a graded ring under the intersection product, the Chow ring.

3.6.2 Prove that the Abel–Jacobi map is an isomorphism for divisors.

3.6.3 Consider the Fermat surfaces $X_0^d + \cdots + X_3^d = 0$ in $\mathbf{P}^3$. Determine the lines on those for $d = 3, 4$ and find out when two lines have the same homology class.

Bibliographical and Historical Remarks

In Section 3.1 we continued explaining some of the basic results of the subject. For the techniques of sheaves, Čech cohomology, and double complexes, see Griffiths and Harris (1978); Wells (1980).

The residue calculus leading to the explicit representation of the primitive cohomology of smooth hypersurfaces as explained in Section 3.2 is due to Griffiths (1969).

For the basic results in Section 3.3, see Deligne (1971b, 1974).

The results in Section 3.4 have been known for a long time. The foundational theory of algebraic cycles is due to Chow, Lefschetz, Poincaré, Samuel, van der Waerden, Weil, and others. One of the highlights of this early development is the link with topology known as the Lefschetz Theorem on (1, 1)-classes. It was preceded by the first proof that the Picard number is finite (see Atiyah and Hodge, 1955); this proof uses residue calculus, somewhat similar in spirit to what we did in Section 3.2. Sir William Hodge (1952) posed a bold generalization of this result, known as the Hodge conjecture, which remains one of the famous unsolved conjectures. For developments on the Hodge conjecture see Lewis (1991).

A new breakthrough in the theory of algebraic cycles came from P. Griffiths. The generalization of the classical construction of the Jacobian for Riemann surfaces (due to Abel and Jacobi) as explained in Section 3.5 is due to him (see Griffiths, 1968). Using his version of the Abel–Jacobi maps, he was able to show a spectacular nonclassical phenomenon: there exist cycles that are homologically but not algebraically equivalent to zero (see Griffiths, 1969). We return to this discovery in Chapter 10.

For generalizations of Weil's intermediate Jacobian and applications to mathematical physics see Müller-Stach et. al. (2012a).

4

Cohomology of Manifolds Varying in a Family

In this chapter we study how the Hodge structure of X_t varies with a parameter t. First one has to define precisely how the X_t glue together. This is captured in the notion of a "smooth family." In Chapter 1 we studied by way of an example the Legendre family and its associated period map which takes values in the upper half-plane. To generalize this, we introduce Griffiths' period domains, which serve as targets for period maps in the general case. Period maps are holomorphic and "horizontal", as we shall see, and this leads to the notion of abstract variations of Hodge structures, the central theme of this book.

4.1 Smooth Families and Monodromy

A *smooth family* of compact complex manifolds consists of a proper surjective holomorphic map between complex manifolds which is everywhere of maximal rank. So, if $f : X \to S$ is a smooth family, all fibers $X_t = \pi^{-1}(t)$ are smooth compact complex manifolds. We say that f is a *smooth projective family* if, moreover, X is a submanifold of a projective fiber bundle P such that f is the restriction of the projection onto the second factor:

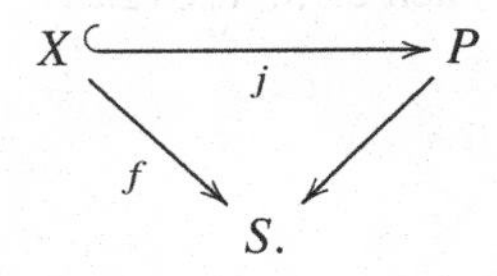

Example 4.1.1 (i) Consider the family of algebraic surfaces in $\mathbf{P}^3$ given by the homogeneous equation

$$Z_0^4 + Z_1^4 + Z_2^4 + Z_3^4 + tZ_0Z_1Z_2Z_3 = 0.$$

For all but finitely many $t \in \mathbf{C}$, to be determined as an exercise, the resulting X_t are smooth and projective.

(ii) Take $A_\Omega = \mathbf{C}^g/\mathbf{Z}^{2g}\Omega$, where Ω is a matrix varying in $\mathbf{C}^{2g}$, i.e., we consider the columns as giving a basis for a lattice in $\mathbf{C}^g$.

We next need two basic results from differential topology. Let $p : X \longrightarrow S$ be a map of differentiable manifolds which is *submersive*, i.e., such that $\mathrm{d}p$ is of maximal rank. Let ξ be a vector field on S that is nowhere zero. We claim that there is a vector field ξ' on X such that $\mathrm{d}p(\xi'_x) = \xi_{p(x)}$ for all x. To see this, we show first that the assertion holds locally. Begin by choosing a neighborhood of $p(x)$ small enough so that $\xi = \partial/\partial s_1$ for some system of coordinates $s_1, \ldots, s_k$. Next choose, using the implicit function theorem, a neighborhood of x and coordinates $x_1, \ldots, x_n$ such that $x_i = s_i \circ p$ for $i \leq k$. Then $\partial/\partial x_1$ is the required lift. For the global version, cover X by open sets U_α which admit local lifts ξ'_α, let $\{\rho_\alpha\}$ be a partition of unity subordinate to this cover, and let ξ' be the sum of the $\rho_\alpha \xi'_\alpha$.

The second result states that we can integrate any smooth vector field ξ' on a manifold X. Intuitively, think of a particle, starting at time 0 following a path which at every moment t is tangent to the vector field. This defines the *smooth flow* $\phi_t : U_t \to X$ of the vector field. Here U_t is the (open) subset of X of points for which the flow extends to all times $< t$. This subset may be strictly smaller than X. The semigroup property, $\phi_t \circ \phi_{t'} = \phi_{t+t'}$, on the subset where it makes sense, should also be intuitively clear. For a rigorous proof of these assertions see Warner (1983, pp. 37–38).

An important special case of the preceding construction is the following. We assume that $p : X \to S$ is *evenly submersive*. This means that each $s \in S$ admits an open neigborhood W such that $p^{-1}W$ is covered by open sets $U \subset X$ diffeomorphic to $V \times W$, V open in X_s, and such that $p|U$ coincides with the projection onto the second factor.

Let $\gamma : [0, T] \longrightarrow S$ be a curve whose velocity field $\xi = \dot{\gamma}$ has unit length. Let ξ' be a lift of the velocity field to the part of X lying above the image of γ, and let ϕ_t be the corresponding flow. This flow is defined for all points of $X_{\gamma(0)}$ and for all times $t \in [0, T]$. Moreover,

$$\phi_t : X_{\gamma(0)} \longrightarrow X_{\gamma(t)}$$

is a diffeomorphism. All of this follows because ϕ_t is compatible with the flow associated with $\dot{\gamma}$, which after time t maps $\gamma(0)$ to $\gamma(t)$. More precisely, if $x \in X_{\gamma(t)}$ and $\tilde{\gamma} : [t, t+\epsilon] \to X$ is an integral curve of the vector field ξ' starting in x at time t, its projection is an integral curve of ξ starting at $p(x)$ and so, by unicity of solutions to systems of ordinary differential equations with given initial data, must coincide with $\gamma|[t, t+\epsilon]$ for small enough ϵ. By compactness of $[0, T]$ we can thus find an integral curve $\tilde{\gamma} : [0, T] \to X$ projecting to γ, and because we can do this for any point $x \in X_{\gamma(0)}$, the assertion follows. Note that this argument is valid without assuming that $X_{\gamma(0)}$ is compact.

If $S = \Delta^k$ is now a polydisk centered at the origin 0, the preceding construction can be used to inductively trivialize the family differentiably. Indeed, for $k = 1$ this follows from what was just said. At the same time, if $p : X \to \Delta^k$ is trivial over the polydisk Δ^{k-1} given by the first $k-1$ coordinates, the flow associated with a lift of the last coordinate proves that X is diffeomorphic to a product $p^{-1}\Delta^{k-1} \times \Delta$. Thus we have the following.

Theorem 4.1.2 (Ehresmann's theorem) *An evenly submersive smooth, not necessarily proper map is locally differentiably trivial over the base.*

Continuing with the preceding discussion, note first that the lift of the velocity vector field of our curve γ is not unique, and two different lifts give different flows, but they are isotopic, inducing homotopic diffeomorphisms between X_0 and $X_{\gamma(T)}$. More generally, we may deform γ leaving the endpoints fixed. So, let $\Gamma : [0, T] \times [0, 1] \longrightarrow S$ be a homotopy between $\gamma(t) = \Gamma(t, 0)$ and $\gamma'(t) = \Gamma(t, 1)$ such that $\Gamma(0, s)$ and $\Gamma(T, s)$ are constant. Then one can extend the lift ξ' to a lift of $\partial/\partial t$ over the image of Γ. Fixing s, this gives a flow $\phi_t^s : X_0 \to X_T$ depending continuously on s that has the properties $\phi_t^0 = \phi_t$ and $\phi_t^1 = \phi_t'$. So the diffeomorphisms ϕ_T and ϕ_T' associated with γ and γ', respectively, are homotopic.

It follows from these two remarks that the effect of ϕ_T on (co)homology does not depend on the choice of the lifting vector field and depends only on the homotopy class of γ. In particular, for a simply connected base S, this provides an explicit trivialization of the bundle whose fiber at $s \in S$ is the cohomology group $H^w(X_s)$. We say that this bundle is a *local system* of **Z**-modules.

For a projective family we can locally view the X_t, $t \in [0, T]$ as embedded in the same projective space. Fixing a hyperplane H, we can always assume that the vector field ξ' lifting a given vector field ξ in the base space is tangent along $X_t \cap H$ for all $t \in [0, T]$. This implies that the flow sends $X_0 \cap H$ to $X_t \cap H$ for $t \in [0, T]$, and in particular, primitive cohomology is preserved under the flow and thus primitive cohomology also yields locally constant systems.

If S is no longer simply connected, we have to look at the effect of the fundamental group. So, look at a closed loop based at $o \in S$, so that $T = 1$ and $\gamma(0) = \gamma(1) = o$. Recall that the product $\gamma * \sigma$ of two such loops γ and σ is the loop based at o obtained by *first* following γ and *then* σ, both with double speed. If ϕ_t is the flow associated with γ and ψ_t that associated with σ, by the semigroup property of the flow associated with $\gamma * \sigma$ we find that this flow is given by

$$\begin{cases} \phi_{2t}, & 0 \leq t \leq \frac{1}{2}, \\ \psi_{2t-1} \circ \phi_1, & \frac{1}{2} \leq t \leq 1, \end{cases}$$

so that one obtains a representation of the fundamental group in *cohomology*, as follows.

Definition 4.1.3 The *monodromy representation* in cohomology associates with each element $[\gamma] \in \pi_1(S, o)$ the unique automorphism in cohomology induced by any of the diffeomorphisms $\phi_1 : X_o \to X_o$ induced by a lift to X of the tangent vector field to any of the loops γ representing $[\gamma]$.[1] We write this as

$$\rho : \pi_1(S, o) \longrightarrow \mathrm{GL}(H^\bullet(X_o)).$$

Note that this representation preserves the cup product because this is a purely topological invariant. Also, in cohomology it preserves the hyperplane class and hence the Lefschetz decomposition and polarization.

For instance, the monodromy representation on the first cohomology group takes values in the group of integer matrices preserving a nondegenerate integral skew-form, i.e., takes values in a lattice in $G \cong \mathrm{Sp}(n, \mathbf{R})$. For the second cohomology of an algebraic surface, on the other hand, the receiving group is a lattice in an indefinite orthogonal group $G \cong \mathrm{SO}(2p, q)$.

4.2 An Example: Lefschetz Fibrations and Their Topology

In Section 3.2 we considered a single ample hypersurface X of a fixed manifold P. In this section, we think of P as being embedded in some projective space and we consider the projective space of all hyperplane sections of P. Those hyperplane sections that are smooth give an interesting projective family whose monodromy group has been studied by Lefschetz.

Let P be a fixed $(n+1)$-dimensional projective manifold embedded in a given projective space $\mathbf{P}^N$. We let $(\mathbf{P}^N)^\vee$ be the dual projective space, i.e., the space of hyperplanes of $\mathbf{P}^N$. The *dual variety* $P^\vee$ of P consists of hyperplanes that are tangent to P at some point and forms an irreducible subvariety of $(\mathbf{P}^N)^\vee$. The complement $U = (\mathbf{P}^N)^\vee - P^\vee$ forms the base manifold of the tautological family $X_u, u \in U$ of the smooth hyperplane sections of P. Let us fix a base point $o \in U$ and let us set $X = X_o$.[2] Recall in Section 3.2 we have shown that

[1] Here $\gamma(0) = o$ so that X_o is the fiber of $X \to S$ over this point.

[2] Keep in mind that this is different from the notation used in the previous section where X stands for the total space of the family.

there is an orthogonal splitting

$$H^n(X;\mathbf{Q}) = H^n_{\mathrm{var}}(X) \oplus H^n_{\mathrm{fixed}}(X),$$

with

$$H^n_{\mathrm{var}}(X;\mathbf{Q}) = \mathrm{Ker}(i_{n*} : H^n(X;\mathbf{Q}) \to H^{n+2}(P;\mathbf{Q})),$$
$$H^n_{\mathrm{fixed}}(X;\mathbf{Q}) = \mathrm{Im}(i_n^* : H^n(P;\mathbf{Q}) \to H^n(X;\mathbf{Q})).$$

The theory of Lefschetz pencils describes the first summand in terms of the monodromy of the tautological family. A *Lefschetz pencil* is a family of hyperplane sections parametrized by a projective line ℓ which meets the dual variety $P^\vee$ transversally in a finite set of points $u_1, \ldots, u_M$. The hyperplane sections corresponding to u_j have precisely one ordinary double point p_j (see Problem 4.2.1). The hyperplanes in $\mathbf{P}^N$ corresponding to this pencil ℓ all have a codimension-two subspace in $\mathbf{P}^N$ in common, meeting P in the base locus of the pencil. Let us denote by $\tilde{P}$ the blowup of P in this base locus. We get a natural fibration

$$\tilde{P} = \{(m,t) \in P \times \ell \mid m \in X_t\} \longrightarrow \ell,$$

the Lefschetz fibration associated with the Lefschetz pencil. Over $U \cap \ell$, this Lefschetz fibration is smooth, and if we choose our base point o in $U \cap \ell$, the monodromy associated with this family factors over the monodromy representation of the tautological family

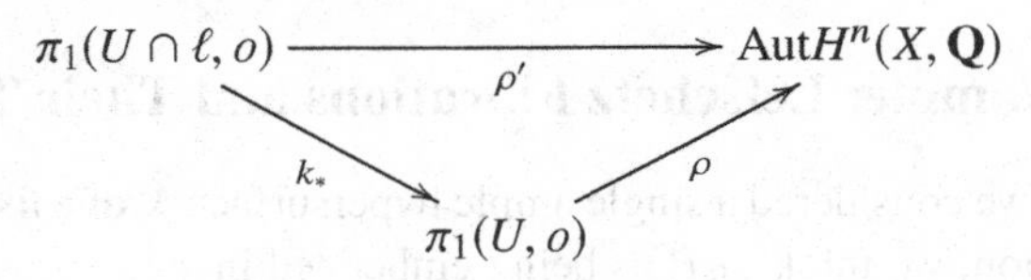

where $k : U \cap \ell \to U$ is the inclusion. By a theorem of van Kampen (1933), the map k_* is surjective and so the monodromy group of our tautological family is the same as the monodromy group for the smooth part of any Lefschetz fibration. This monodromy group has been studied classically by Lefschetz (1924); for a modern treatment see Lamotke (1981). We start out by taking a special set of generators for the fundamental group $\pi_1(U \cap \ell, o)$. We take small mutually nonintersecting circles σ_j of radius ϵ winding once positively about their center u_j, and paths ℓ_j from o to the point u_j^ϵ on the circle σ_j which is closest to o. See Fig. 4.1. The circles are chosen to be small enough so that $|\sigma_j| \cup |\ell_j|$ meets $|\sigma_k| \cup |\ell_k|$ for $k \neq j$ only in the base point o. We then let λ_j be the class of the path obtained by first following ℓ_j, then σ_j, and then going back to o along ℓ_j. Finally we let $T_j = \rho'(\lambda_j)$ be their images in the monodromy group. These elements are called the *Picard–Lefschetz transformations*.

Now we return to what happens near the double point p_j over the critical value t_j. In a suitable small coordinate neighborhood W of p_j in M with coordinates $(z_1 = x_1 + \mathrm{i}y_1, \ldots, z_{n+1} = x_{n+1} + \mathrm{i}y_{n+1})$, the Lefschetz fibration is given by $(z_1, \ldots, z_{n+1}) \mapsto \sum_{i=1}^{n+1} z_i^2$. So, if we write $\vec{x} = (x_1, \ldots, x_{n+1})$, $\vec{y} = (y_1, \ldots, y_{n+1})$, the fiber over a real point ϵ^2 is given by the equations

$$(\vec{x}, \vec{y}) = 0,$$
$$\|\vec{x}\|^2 - \|\vec{y}\|^2 = \epsilon^2.$$

So the set of points where $\vec{y} = 0$, the real points of this set, is a real n-sphere, the *vanishing cycle* associated with p_j. When ϵ tends to 0, the vanishing cycle indeed vanishes, i.e., shrinks to p_j. Observe that a differentiable trivialization along ℓ_j gives a corresponding cycle in X. Its class is the vanishing cycle in $H_n(X)$, and any cocycle δ_j whose class in $H^n(X)$ is Poincaré dual to the class of this vanishing cycle is called the *vanishing cocycle* associated with p_j.

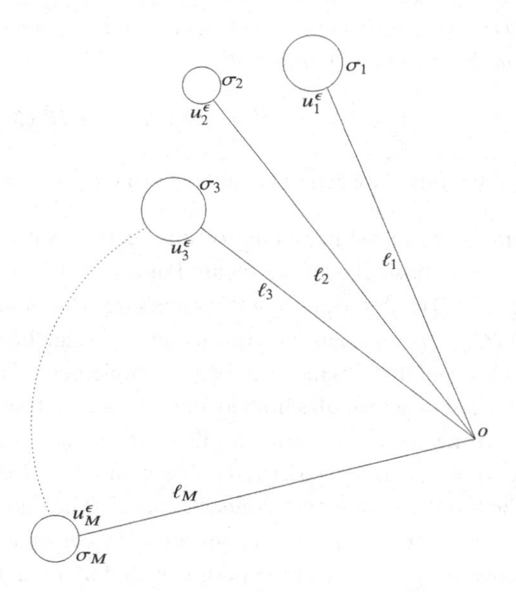

Figure 4.1 Special Generators.

We state without proof the following important result about these.

Theorem 4.2.1 *The vanishing cocycle δ_j has the following intersection behavior:*

$$\begin{cases} (\delta_j, \delta_j) = 0 & \text{for } n \text{ odd}, \\ (\delta_j, \delta_j) = (-1)^{n/2} 2 & \text{for } n \text{ even}. \end{cases}$$

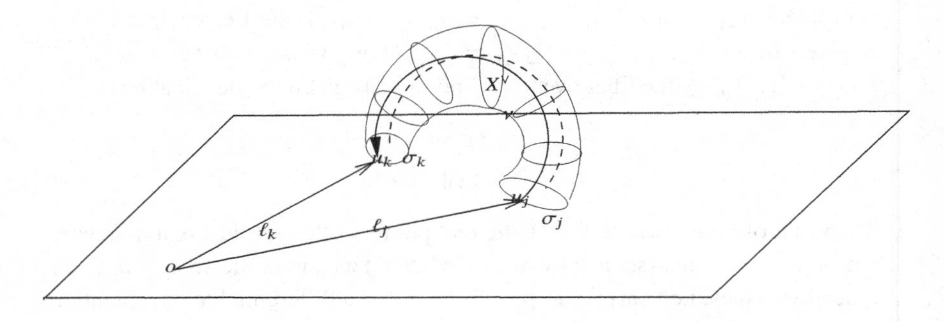

Figure 4.2 Connecting u_j and u_k.

The associated transformation in cohomology is trivial except in degree n where it is given by the Picard–Lefschetz formula:

$$T_j(v) = v + (-1)^{\frac{1}{2}(n+1)(n+2)}(v, \delta_j)[\delta_j], \quad v \in H^n(X; \mathbf{Q}).$$

For details of the proof we refer to Lamotke (1981).

By the definition from the beginning of this section, variable cohomology classes are precisely those classes which are Poincaré dual to classes of cycles in X bounding in P. The description of the vanishing cocycles shows that these give classes in $H^n_{\mathrm{var}}(X)$. Lefschetz has shown that they span the variable part. As for the fixed part, recall that it is the orthogonal complement of the variable part, and the Picard–Lefschetz formulas immediately show that the monodromy fixes this part and that the converse is true. So this part is the π_1-trivial summand. The variable part turns out to be an irreducible π_1-module. Let us give a proof of this fact. The proof uses the path-connectedness of the dual variety $P^\vee$.

Indeed, as shown in Fig. 4.2 we can connect u_j and u_k by a path in this dual variety and hence we can connect the points u_j^ϵ and u_k^ϵ by a path ν in a small tubular neighborhood of $P^\vee$ in U. So σ_j and $\nu * \sigma_k * \nu^{-1}$ are homotopic, and if w is the path obtained by following first ℓ_k, then ν^{-1}, and then ℓ_j^{-1}, the loop λ_j is homotopic to $w^{-1} * \lambda_k * w$. It follows that $T_j = \rho(w) \cdot T_k \cdot \rho(w)^{-1}$. The Picard–Lefschetz formula implies that

$$(v, \delta_j)\rho(w)\delta_j = (\rho(w)v, \delta_k)\delta_k.$$

Since the cup product pairing is nondegenerate for a nonzero vanishing class of a cocycle δ_j, for some v we must have $(v, \delta_j) \neq 0$ and so $\rho(w)\delta_j = c\delta_k$. So

$$\begin{aligned}(v, \delta_j)\rho(w)\delta_j &= (\rho(w)v, \delta_k)\delta_k \\ &= c^{-2}(\rho(w)v, \rho(w)\delta_j)\rho(w)\delta_j \\ &= c^{-2}(v, \delta_j)\rho(w)\delta_j,\end{aligned}$$

which implies that $c = \pm 1$ and the vanishing cocycles are mutually conjugate up to sign.

Now the orthogonal complement of the space spanned by the vanishing cycles is the π_1-trivial part. So, if $0 \neq v \in H^n_{\text{var}}(X)$ we can always find some vanishing cocycle δ_j with $(v, \delta_j) \neq 0$. Hence, if $W \subset H^n_{\text{var}}(X)$ is a nonzero π_1-subspace, we infer from $T_j v \in W$, that $\delta_j \in W$, and hence all vanishing cocycles belong to W and so $W = H^n_{\text{var}}(X)$. Let us summarize these last results.

Theorem 4.2.2 *The variable part*

$$H^n_{\text{var}}(X; \mathbf{Q}) = \text{Ker}\,(i_{n*} : H^n(X; \mathbf{Q}) \to H^{n+2}(P; \mathbf{Q}))$$

coincides with the subspace of $H^n(X; \mathbf{Q})$ spanned by the classes of the vanishing cocycles. These are mutually conjugate up to sign. The space they span is an irreducible module for the monodromy of the tautological family. Its orthogonal complement

$$H^n_{\text{fixed}}(X; \mathbf{Q}) = \text{Im}\,(i^*_n : H^n(P; \mathbf{Q}) \to H^n(X; \mathbf{Q}))$$

coincides with the classes left invariant by the monodromy of the tautological family.

Problem

4.2.1 Let X be a smooth subvariety of $\mathbf{P}^N$ spanning $\mathbf{P}^N$. Let $X^\vee$ be its dual variety. Show that this is an irreducible proper subvariety of the dual projective space. Show by a local calculation that if a line of this space meets $X^\vee$ transversally at a point, the corresponding hypersurface section has exactly one ordinary double point. Deduce the existence of Lefschetz pencils.

4.3 Variations of Hodge Structures Make Their First Appearance

We can now explain more precisely what we mean by "variation of Hodge structure." For each homotopy class of paths from the base point o to an arbitrary point t there is a diffeomorphism $\phi_t : X_o \longrightarrow X_t$ that identifies the cohomology of an arbitrary fiber with that of the central one:

$$\phi_t^* : H^w(X_t) \longrightarrow H^w(X_o).$$

Now consider the Hodge decompositions on $H^w(X_o; \mathbf{C})$ obtained by pulling back the decompositions on $H^w(X_t; \mathbf{C})$, i.e., the decompositions defined by

$$\phi_t^* H^{p,q}(X_t).$$

These give a family or "variation" of Hodge structures on a fixed vector space, namely $H^w(X_o; \mathbf{C})$; it is this object we now study.

We state the basic first result, the **Hodge numbers are constant in families.**

Proposition 4.3.1 *The spaces $H^{p,q}(X_t)$ (although varying with t) all have the same dimension. So we have a locally constant C^∞ vector bundle, given fiberwise by $H^{p,q}(X_t; \mathbf{C})$.*

Proof To see this, note first that these Hodge numbers for $p + q = w$ sum up to a constant number, the w-th Betti number of the fiber, which is constant because the family is differentiably locally trivial as noted above. To show that each of the numbers $h^{p,q}(X_t)$ is constant, it is therefore sufficient to show that they behave in an *upper-semicontinuous* fashion:

$$h^{p,q}(X_t) \leq h^{p,q}(X_o), \text{ for } |t - o| < \epsilon.$$

Recall (Theorem 2.2.7) that the Hodge component $H^{p,q}(X)$ can also be described as the space of harmonic (p, q)-forms, i.e. as the kernel of the Laplacian operator acting on complex forms of type (p, q). So, we may apply the following general result about elliptic operators depending on parameters (see Wells, 1980, Chapter 4, Theorem 4.13).

Theorem *Let E and F be smooth complex vector bundles over a differentiable manifold and let $D_t : C^\infty(E) \to C^\infty(F)$ be an elliptic operator depending smoothly on t. Then the dimension of the kernel depends upper-semicontinuously on t.*

Some explanation is in order. A differential operator D between two vector bundles E and F over a manifold M is by definition a global operator that in a trivializing coordinate chart $U \subset M$ is given by a differential operator D_U in the usual sense. Suppose that the highest-order part of D_U is an expression of the form $\sum_{|I|=k} a_I(x)\,\partial^k/\partial x_I$, where

$$a_I(x) = \begin{pmatrix} a_I^{1,1}(x) & \cdots & a_I^{1,s}(x) \\ \vdots & & \vdots \\ a_I^{r,1}(x) & \cdots & a_I^{r,s}(x) \end{pmatrix}, \quad a_I^{\alpha,\beta}(x)) \in C^\infty(U),\ r = \operatorname{rank} E,\ s = \operatorname{rank} F.$$

Using auxiliary variables ξ_j instead of the tangent vectors $\partial/\partial x_j$, $j = 1, \ldots, m$. the symbol of $D_U = D|U$ is the $r \times s$ polynomial matrix

$$\sigma(D_U) = \left(\sigma(D_U)^{\alpha,\beta}\right) \stackrel{\text{def}}{=} \left(\sum_{|I|=k} a_I^{\alpha,\beta}(x)\xi^I\right), \text{ where } \xi^I = \xi_1^{i_1}\cdots\xi_m^{i_m}.$$

If one interprets the ξ^I, $|I| = k$, as local sections of the bundle $\operatorname{Sym}^k T_M$, the symbol transforms as a section of $\operatorname{Sym}^k T_M \otimes \operatorname{Hom}(E, F)$ and so makes sense globally. One says that a differential operator is *elliptic* if its symbol is invertible. An example is the Laplacian acting on functions; then $E = F$ is the trivial bundle and its symbol is multiplication by the polynomial $\sum_{j=1}^m \xi_j^2$.

In our situation, using the diffeomorphism ϕ_t, the Laplacian, acting on the $A^{p,q}(X_t)$, the space of forms of type (p, q) on X_t, can be viewed as an operator $D_t : C^\infty(E) \to C^\infty(E)$, where E is the bundle of (p, q)-forms on X_o. Indeed, with $k = p + q$, the composition of the following complex bundle maps

$$\begin{array}{ccccccc} (p,q)\text{-forms} & \hookrightarrow & k\text{-forms} & \xrightarrow{\phi_t^*} & k\text{-forms} & \twoheadrightarrow & (p,q)\text{-forms} \\ \text{on } X_t & & \text{on } X_t & & \text{on } X_o & & \text{on } X_o, \end{array}$$

is an isomorphism for small t.

The operator D_t depends smoothly on the parameter t and we can apply the preceding result: the dimension of the kernel, $h^{p,q}(X_t)$, thus depends upper-semicontinuously on the parameter. □

Note that if the base manifold S is simply connected, then the construction just given is single-valued. If not, then it is multi-valued but with the various isomorphisms ϕ_t^* related by the action of the monodromy representation.

Because we need polarized Hodge structures in order to have a nice quotient under the monodromy group, we now repeat the preceding constructions with primitive cohomology instead of full cohomology. As a consequence of Theorem 4.1.2, we have seen that the primitive cohomology groups are also locally constant vector bundles, and the Hodge filtration induces on these a filtration satisfying the two bilinear relations from Theorem 2.3.3, which we recall here:

(1) $b(H^{p,q}, H^{r,s}) = 0$ if $(r, s) \neq (q, p)$;

(2) $h_C > 0$, (where $C|H^{p,q} = \mathrm{i}^{p-q}$ and $h_C(u, v) = b(Cu, \bar{v})$).

The first relation, in terms of the Hodge filtration $F^w \subset F^{w-1} \subset \cdots \subset F^0$, is equivalent to the condition

$$(F^p)^\perp = F^{w-p+1},$$

where the orthogonal complement with respect to the bilinear form b is taken. The second relation says that the real structure on the flag is such that the Hermitian form $h_{\mathbf{C}}$ in fact becomes a unitary metric on $H^w(X_t, \mathbf{C})$.

4.4 Period Domains Are Homogeneous

In Chapter 1 we saw that the variable Hodge filtration reflects itself in the period map. The target generalizing the upper half-plane will be a *period domain* parametrizing the set of polarized Hodge structures of weight w on a fixed real vector space H with fixed Hodge numbers. Such a Hodge structure is determined by specifying a flag $F^w \subset F^{w-1} \cdots \subset F^0$ of fixed type satisfying the above two bilinear relations.[3]

Let us first consider the **first bilinear relation**. It is a *closed* condition which says that we need roughly half of this flag. The variety $\check{D}$ parametrizing flags of the above type satisfying the first bilinear relation can therefore be identified with the flag manifold of flags $F^w \subset \cdots \subset F^v$, $v = [w + 1/2]$ of given type in the complexified space $H_{\mathbf{C}} = H \otimes \mathbf{C}$.[4] For example, if $w = 2$, then $(F^2)^\perp = F^1$, so that $F^2 = H^{2,0}$ determines the Hodge structure, and $\check{D}$ is the Grassmannian of b-isotropic $h^{2,0}$-planes in $H_{\mathbf{C}}$.

In addition to being an algebraic manifold, $\check{D}$ turns out to be homogeneous for the action of a complex Lie group $G_{\mathbf{C}}$, namely, the group of linear automorphisms of $H_{\mathbf{C}}$ which preserve the polarization b. We prove this below. But first we explain what goes on in an example of weight 1 and $h^{1,0} = 1$. Then $G_{\mathbf{C}} = \mathrm{Sp}(1, \mathbf{C}) \cong \mathrm{SL}(2, \mathbf{C})$. The subgroup B fixing a given filtration is the group of upper triangular matrices. The manifold $D = G_{\mathbf{C}}/B$ is the group theorist's definition of $\mathbf{P}^1$. In any case, we see that $\check{D} \cong G_{\mathbf{C}}/B$ is indeed a smooth compact manifold. We can make this more explicit. For odd weight we have an even dimensional vector space H, say of dimension $2g$, and then

$$G_{\mathbf{C}} \simeq \mathrm{Sp}(g, \mathbf{C}) \overset{\text{def}}{=} \left\{ g \in \mathrm{GL}(2g, \mathbf{C}) \,\middle|\, {}^{\mathsf{T}}g J_g g = J \right\}, \quad J_g = \begin{pmatrix} 0_g & \mathbf{1}_g \\ -\mathbf{1}_g & 0_g \end{pmatrix}.$$

[3] Although we generally restrict ourselves to effective Hodge structures on H, i.e., those for which $H_{\mathbf{C}} = F^0$, more general Hodge structures come up later in Chapter 15 ff.

[4] We refer to Examples A.6.(v) and (vii) for an introduction to Grassmannians and flag manifolds.

If the weight is even, in a suitable basis the complexified form b becomes $\mathbf{1}_n$, $n = \dim H$ and the connected group preserving this form becomes

$$G_{\mathbf{C}} \simeq \mathrm{SO}(n, \mathbf{C}).$$

Proposition 4.4.1 *The variety $\check{D}$ classifying Hodge decompositions of weight w with fixed Hodge numbers $h^{p,q}$ and which obeys the first bilinear relation is a flag manifold of type $(f_w, \ldots, f_v)$, $f_p = \dim F^p$, $v = \left[\frac{w+1}{2}\right]$. It can also be identified with the homogeneous manifold*

$$\check{D} = G_{\mathbf{C}}/B,$$

where $G_{\mathbf{C}} = \mathrm{Aut}^0\, H_{\mathbf{C}}, b)$ and B is the subgroup fixing a reference flag $F^w \subset \cdots \subset F^v$ with F^v a maximal b-isotropic subspace of $H_{\mathbf{C}}$.

Proof We first consider the case of weight 1 so that we have a totally isotropic subspace $F^1 = (F^1)^{\perp} \subset H_{\mathbf{C}}$ with $\dim F^1 = g$. Start now with a basis $\{f_1, \ldots, f_g\}$ for F^1. Since b is nondegenerate, we can find vectors $\{f'_1, \ldots, f'_g\}$ with

$$b(f_i, f'_j) = \delta_{ij}.$$

We may replace f'_j by a suitable linear combination $h_i = f'_i + \sum B_{ij} f_j$ such that $\{f_1, \ldots, f_g, h_1, \ldots, h_g\}$ becomes a symplectic basis for $H_{\mathbf{C}}$. Since any two symplectic bases are related by a complex symplectic isometry of $H_{\mathbf{C}}$ this shows that any two isotropic subspaces of $H_{\mathbf{C}}$ of dimension g are related by such an isometry.

This construction can be adapted to arbitrary odd weight $w = 2m - 1$, by constructing a symplectic basis adapted to the Hodge filtration

$$F^{2m-1} \subset \cdots \subset F^m = (F^m)^{\perp} \cdots \subset (F^{2m-1})^{\perp} \subset H_{\mathbf{C}}.$$

Next, consider the case of weight 2 so that we have a Hodge flag $F^2 \subset (F^2)^{\perp} = F^1 \subset H_{\mathbf{C}} = F^0$. First choose an arbitrary basis $\{f_1, \ldots, f_\ell\}$ for F^2 and complement it to a basis $\{f_1, \ldots, f_\ell, h_1, \ldots, h_h\}$ for $F^1 = (F^2)^{\perp}$ such that $b(h_i, h_j) = \delta_{ij}$. This is possible since b, being nondegenerate, induces a nondegenerate quadratic form on $(F^2)^{\perp}/F^2$. For the same reason we can construct vectors $k'_1, \ldots, k'_\ell$ orthogonal to the vectors h_j such that $b(k'_i, f_j) = \delta_{ij}$. We then can replace each k'_i by a suitable linear combination $k_i = k'_i + \sum_j C_{ij} f_j$ in order that the k_i span an isotropic subspace. The basis $\{f_1, \ldots, f_\ell, h_1, \ldots, h_h, k_1, \ldots, k_\ell\}$ is called a *b-adapted basis* for the Hodge flag. The form b takes the shape

$$\begin{pmatrix} 0_\ell & 0_{h\times\ell} & \mathbf{1}_\ell \\ 0_{\ell\times h} & \mathbf{1}_h & 0_{\ell\times h} \\ \mathbf{1}_\ell & 0_{h\times\ell} & 0_\ell \end{pmatrix}.$$

Two b-adapted bases for two given Hodge flags are obviously related by an isometry, say φ of $(H_{\mathbb{C}}, b)$. By interchanging f_1 and k_1 or replacing h_1 by $-h_1$ in the first basis (these operation preserves the Hodge flag) one can force $\det(\varphi) = 1$. Again, this construction can be modified for arbitrary even weight $2m$ by constructing a suitable basis adapted to the Hodge flag:

$$F^{2m} \subset \cdots \subset F^{m+1} \subset (F^{m+1})^{\perp} \subset \cdots \subset (F^{2m})^{\perp} \subset H_{\mathbb{C}}. \qquad \square$$

Next we consider the **second bilinear relation**. As a positivity condition it defines an open subset D of $\check{D}$. This is the set of polarized structures we seek. As an open subset of a complex manifold it has a natural complex structure. We have more.

Proposition 4.4.2 *The period domain D classifying the Hodge filtration $F^{\bullet}$ of fixed dimension $f^p = \dim F^p$ satisfying both bilinear relations is an open subset of $\check{D}$, the* compact *manifold classifying Hodge filtrations satisfying only the first relation,*[5] *and it is a homogeneous manifold*

$$D \cong G/V,$$

where $G = \mathrm{Aut}(H, b) \cap \mathrm{SL}(H)$ and $V = G \cap B$, the subgroup fixing a reference flag satisfying both bilinear relations.

The proof makes use of a concept that will come back several times, that of a Hodge frame.

Definition 4.4.3 Let $(H, b, F^{\bullet})$ be a polarized weight w Hodge structure and let h_C be the associated Hermitian metric. A *Hodge frame* consists of an h_C-unitary basis for $H_{\mathbb{C}}$ such that

- it is adapted to the Hodge filtration $F^w \subset F^{w-1} \cdots \subset F^1 \subset F^0 = H_{\mathbb{C}}$; in particular, it provides bases for all Hodge summands $H^{p,q}$;
- the conjugate of the resulting basis for $H^{p,q}$ forms part of the Hodge frame; it is a basis for $H^{q,p}$. If $p = q = \frac{1}{2}w$ (so w is even) this means that the Hodge frame restricted to $H^{p,p}$ is real.

Sketch of proof It suffices to show that $\mathrm{Aut}(H, b)$ acts transitively on Hodge structures satisfying both bilinear relations.

We consider only the case of weight 1 and 2 and, apart from a few side remarks, we leave the proof for arbitrary weight to the reader. The proof essentially consists of showing the existence of a Hodge frame. First consider

[5] The manifold $\check{D}$ is called the *compact dual* of D.

weight 1. We use the Hermitian metric $(x, y) \mapsto ib(x, \bar{y})$ to choose a unitary basis $\{f_1, \ldots, f_g\}$ for F^1. This means

$$ib(f_\alpha, \bar{f}_\beta) = \delta_{\alpha\beta}, \quad \alpha, \beta = 1, \ldots, g.$$

Since $H_\mathbf{C} = F^1 \oplus \bar{F}^1$ the set $\{f_1, \ldots, f_g, \bar{f}_1, \ldots, \bar{f}_g\}$ is a basis for $H_\mathbf{C}$. This provides us in this case with a Hodge frame. We are going to use this frame as follows.

We have $b(f_\alpha, f_\beta) = 0$, since F^1 is b-isotropic and hence, since b is real, we also have $b(\bar{f}_\alpha, \bar{f}_\beta) = 0$. It follows that a suitable complex linear combination of these gives a real symplectic basis for (H, b). Take for instance $\left\{\frac{1}{\sqrt{2}}(f_1 + \bar{f}_1), \ldots, \frac{1}{\sqrt{2}}(f_g + \bar{f}_g), \frac{\mathrm{i}}{\sqrt{2}}(f_1 - \bar{f}_1), \ldots, \frac{\mathrm{i}}{\sqrt{2}}(f_g - \bar{f}_g)\right\}$. It follows that the linear map which sends a unitary basis for F^1 to a unitary basis for another totally b-isotropic subspace of $H_\mathbf{C}$ satisfying the two bilinear relations is indeed a real symplectic transformation of H. This proof adapts without problem to the case of arbitrary odd weight.

Next, consider the case of weight 2: one has a polarized Hodge flag $F^2 \subset (F^2)^\perp \subset H_\mathbf{C}$. As in the weight 1 case, choose a basis $\{f_1, \ldots, f_k\}$ for F which is h_C-unitary. This time $h_C(u, v) = -b(u, \bar{v})$ and the same complex linear combination of the unitary basis $\{f_1, \ldots, f_k, \bar{f}_1, \ldots, \bar{f}_k\}$ as the one used in the weight 1 case gives a real basis for the real subspace $F^2 \oplus \bar{F}^2 \subset H$ for which the form b becomes $-\mathrm{id}_{2k}$. The second bilinear relation tells us that form b restricted to its orthogonal complement is an inner product. Choose any b-orthonormal real basis for it, say $\{g_1, \ldots, g_\ell\}$. The resulting frame thus is adapted to the Hodge flag and the polarizing form b takes the form $\mathrm{diag}(-\mathrm{id}_{2k}, \mathrm{id}_\ell)$. For any other polarized Hodge flag do the same. So, by construction there is a $g \in G$ sending the first frame to the new frame, i.e., G acts transitively on polarized Hodge flags.

In general, for weight $w = 2m$ this method provides a real basis for $H^{p,q} \oplus H^{q,p}$ $(p \neq q)$ such that b has signature $(-1)^{p-m}$ while on $H^{m,m}$ it is positive definite. So this also shows that now G acts transitively on polarized Hodge frames for any even weight. □

We now investigate the structure of G in more detail. First the case of **odd weight**. Then b is anti-symmetric and so G is a symplectic group isomorphic to $\mathrm{Sp}(n, \mathbf{R})$ with $n = \dim_\mathbf{R} H$. The subgroup V then is a product of unitary groups. We see this as follows. A Hodge frame for a reference Hodge structure in this case can be constructed from a unitary basis for each of the $H^{p,q}$ with $p < q$. Any $g \in V$ acts on such a basis unitarily and since it is real it does the same on the conjugate bases for the $H^{q,p}$. This shows that $V \simeq \prod_{p<q} \mathrm{U}(h^{p,q})$.

Next, look at **even weight** $w = 2m$. Then $G = \mathrm{SO}(r, s)$ for suitable r and s, to be determined in what follows. The form b has sign $(-1)^{p-m}$ on the real space

W_p underlying $H^{p,q} + \overline{H^{q,p}}$. So, adjusting the signs by $(-1)^m$, we get positive signs for even p and negative signs for odd p. Now the component $H^{m,m}$ of the Hodge decomposition of the reference Hodge structure is real and therefore $g \in V$ preserves this component orthogonally. Thus in addition to the factors that are unitary groups $\mathrm{U}(h^{p,q})$ for each $p < w/2$ we have an extra orthogonal group for the "middle" component if present. Note that for any integer k the unitary group $\mathrm{U}(k)$ embeds in $\mathrm{SO}(2k)$ via $A + \mathrm{i}B \mapsto \begin{pmatrix} A & -B \\ B & A \end{pmatrix}$ so that the extra component from V thus also must have determinant 1. Recall from Appendix D that the groups $\mathrm{Sp}(h; \mathbf{R})$, $\mathrm{SO}(n)$ and $\mathrm{U}(n)$ are connected and hence the corresponding domains are so. However, $G = \mathrm{SO}(s,t)$ has two connected components whenever $s, t > 0$ and these components are preserved by the action of V so that $D = G/V$ also has two components. We summarize our results as follows.

Proposition 4.4.4 *Let $D = G/V$ be a period domain for weight w Hodge structures with Hodge numbers $h^{p,q}$, $p = 0, \ldots, w$. The group G is an algebraic real matrix group which is simple as a real Lie group.*

(i) *For odd weight $w = 2m + 1$, the form b is anti-symmetric and we have*

$$\boxed{\begin{aligned} &G \simeq \mathrm{Sp}(n, \mathbf{R}), \quad \dim H = 2n, \\ &V = \prod_{p \le m} \mathrm{U}(h^{p,q}). \end{aligned}}$$

In this case D is connected and non-compact.

(ii) *In the even-weight case $w = 2m$, the form b is symmetric and then*

$$\boxed{\begin{aligned} &G \simeq \mathrm{SO}(s,t),\ s = \sum_{p \text{ even}} h^{p,q},\ t = \sum_{p \text{ odd}} h^{p,q}, \\ &V = \prod_{p<m} \mathrm{U}(h^{p,q}) \times \mathrm{SO}(h^{m,m}). \end{aligned}}$$

In this case, if $s, t > 0$, the domain D consists of two isomorphic connected components and if $s = 0$ or $t = 0$, the domain D is connected and compact.

Examples 4.4.5 (i) For weight 1 structures with Hodge decomposition

$$H = H^{1,0} \oplus H^{0,1}$$

the period domain is

$$D = \mathrm{Sp}(g, \mathbf{R})/\mathrm{U}(g),$$

where $g = \dim H^{1,0}$. This is one of the classical Hermitian symmetric domains (see Helgason, 1962, Chapter 10, Section 5.3), namely Siegel's upper half-space which we already encountered in Chapter 1 (see (1.19)):

$$\mathfrak{h}_g = \{Z \text{ a } g \times g \text{ matrix} \mid Z \text{ symmetric with a positive imaginary part}\}.$$

(See Problem 4.4.3). In the next section we explain this using period matrices.

(ii) For weight 2 structures with Hodge decomposition

$$H^2 = H^{2,0} \oplus H^{1,1} \oplus H^{0,2}$$

we have

$$D = \mathrm{SO}(2a, b)/\mathrm{U}(a) \times \mathrm{SO}(b),$$

where

$$a = \dim H^{2,0} \text{ and } b = \dim H^{1,1}.$$

In this case, however, a connected component $\mathrm{SO}^0(2a, b)/\mathrm{U}(a) \times \mathrm{SO}(b)$ of D is not usually a Hermitian symmetric domain.

To explain why, we recall that symmetric spaces are of the form G/K where G is a semisimple Lie group and K is a maximal compact subgroup. The maximal compact subgroup of $\mathrm{SO}^0(2a, b)$ is $\mathrm{SO}(2a)\times\mathrm{SO}(b) = \mathrm{U}(a)\times\mathrm{SO}(b)$. Except in the case $a = 1$, this does not give a symmetric space (see Helgason, 1962, Chapter 10, Section 5.3). For $a = 1$ we have another classical Hermitian domain (see Problem 4.4.3). In Problem 4.4.3 we show that the period domain in this case consists of two isomorphic components, each isomorphic to the Hermitian symmetric domain IV_{20}.

The same domain comes up in geometry as the period domain of $K3$ surfaces. Recall that a $K3$ surface by definition has trivial canonical bundle and trivial first Betti number. In order to apply our considerations, we use a non-trivial fact: $K3$ surfaces are all Kähler as shown by Siu (1980). We have seen (see Eq. (1.31)) that the Hodge vector of a $K3$ surface is $(1, 20, 1)$. This was established only for the example of a sextic double plane but follows from a little surface geometry, as we now explain. The Riemann–Roch theorem in the guise of Noether's formula (see e.g. Barth et al., 1993, Chap. I.5) tells us that $h^{2,0} - h^{1,0} + 1 = \frac{1}{12}(K^2 + e)$ and so, since K is trivial and $h^{1,0} = 0, h^{2,0} = 1$, the Euler number $e = 24$ which forces $b_2 = 22$ and hence the Hodge numbers.

The usual polarization on the total cohomology comes in two parts: the one coming from the intersection form on the primitive cohomology and a second part which is minus the intersection form on the line spanned by the class of a Kähler form. It is customary *not* to use this polarization, but the intersection form. The domain classifying such Hodge structures is then also called a period domain. The advantage of this approach is that the intersection form is integral on *integral* cohomology and, since it is purely topological, it is preserved by monodromy.

The "classical" period domains we have introduced so far would have been gotten by fixing the Kähler class and considering the primitive cohomology; this is usually only done in the projective situation when one can take as a Kähler class the cohomology class of an ample divisor, which is integral. The signature of the intersection form on primitive cohomology being (2, 19), this gives a domain of type IV_{19}. If we consider projective $K3$ surfaces with a fixed ample class, the monodromy then fixes this class and acts discretely on the latter domain.

Coming back to the full cohomology, we have $a = 1$ and $b = 20$ in this case and the signature of the intersection form in this case is (3, 19). So we get a domain of type IV_{20} instead.

$$S \text{ is a } K3 \text{ surface} \implies \text{conn. comp. of its period domain} \simeq IV_{20}.$$

The preceding examples give Hermitian symmetric spaces. This is not always the case but we do have a fibration to the "associated symmetric space," namely

$$\omega : G/V \longrightarrow G/K$$

with fiber K/V. The fibers are complex submanifolds but the projection is not holomorphic, even when G/K is a Hermitian symmetric domain (see Problem 4.4.2). Whether or not G/K is Hermitian symmetric depends on the parity of the weight. We see below in the examples that G/K always has a complex structure if the weight is odd and, apart from one exceptional case, never has a complex structure if the weight is even.

Example 4.4.6 (i) For even-weight Hodge structures, say of weight $w = 2m$, we have seen that $D = \mathrm{SO}(s,t)/\prod_{p<m} \mathrm{U}(h^{p,q}) \times \mathrm{SO}(h^{m,m})$ with s the sum of the Hodge numbers $h^{p,q}$ with p even and t the sum of the remaining ones. The maximal compact subgroup of G is $(\mathrm{O}(s)\times \mathrm{O}(t)) \cap \mathrm{SO}(s+t)$ and G/K is the set of real s-planes in H on which b is positive. For $s = 1$ this is a Hermitian symmetric space and then ω is the map identifying the two isomorphic connected components (see Problem 4.4.3), but otherwise this is never true (see Helgason, 1962, Chapter 10, Section 5.3). The fibration ω sends the Hodge flag to the real subspace of H for which the complexification is the direct sum of the Hodge components $H^{p,q}$ with p even.

(ii) For odd-weight Hodge structures, say of weight $w = 2m + 1$, we have $D = \mathrm{Sp}(n, \mathbf{R})/\prod_{p\le m} \mathrm{U}(h^{p,q})$, $2n = \dim H$. Here $K = \mathrm{U}(n)$ and G/K is the Grassmannian of complex n-planes W in $H_{\mathbf{C}}$, isotropic with respect to b and for which $ib(w, \bar{w}) > 0, w \in W - \{0\}$. This is a complex manifold and it is indeed one of the Hermitian symmetric domains (see Problem 4.4.3). The fibration ω sends the Hodge flag to the direct sum of the Hodge components $H^{p,q}$ with p even.

We come back to this fibration in Section 12.5 where we relate it to the splitting of the tangent bundle.

Problems

4.4.1 Fill in the details of the proofs of Propositions 4.4.1 and 4.4.2.

4.4.2 Show that $\omega : G/V \longrightarrow G/K$ in general is not holomorphic and describe its fibers both group-theoretically and geometrically in terms of flag manifolds. Compare your answer with Griffiths (1983, Proposition 8.16.).

4.4.3 (a) Show that the group $\mathrm{SO}(s,t)$ has two connected components but that $D(s,t) \stackrel{\text{def}}{=} \mathrm{SO}(s,t)/\mathrm{S}(\mathrm{O}(s) \times \mathrm{O}(t))$ is connected and in fact isomorphic to $\mathrm{SO}^0(s,t)/\mathrm{SO}(s) \times \mathrm{SO}(t)$.

(b) Equip the set $M_{2,t}(\mathbf{R})$ of 2 by t matrices $X = \begin{pmatrix} X_1 \\ X_2 \end{pmatrix}$ with a complex structure given by $J\begin{pmatrix} X_1 \\ X_2 \end{pmatrix} = \begin{pmatrix} -X_2 \\ X_1 \end{pmatrix}$ (this complex structure is the one obtained from identifying the space $\mathbf{C}^t$ of row vectors of length t with the real space $M(2,t)$ in the obvious way). Show that $\mathrm{SO}^0(2,t)$ acts transitively on

$$(IV)_t \stackrel{\text{def}}{=} \left\{ X \in M_{2,t}(\mathbf{R}) \;\middle|\; X^{\mathsf{T}}X - \mathbf{1}_2 < 0 \right\}, \qquad (4.1)$$

(meaning that the left-hand side is negative definite). By identifying the isotropy group at the origin, show that this set can be identified with $D(2,t)$. As such it inherits a complex structure from the complex structure defined above. This is by definition a domain of type IV.

(c) Consider the period domain of even-weight Hodge structures where all Hodge numbers $h^{p,q}$ for p even are 0 except either the middle one which must be equal to 2: $h^{w/2,w/2} = 2$, or two excentric symmetric ones both equal to 1: $h^{p,q} = h^{q,p} = 1$, $p \neq w/2$. Show that this period domain is the subset of projective space $\mathbf{P}^{n-1}$, $n = \dim H$ of lines $X = [(X_1, \ldots, X_n)]$ in $H_{\mathbf{C}}$ satisfying the two bilinear relations $X^{\mathsf{T}}X = 0$ and $X^{\mathsf{T}}\bar{X} > 0$ and that ω is the identity. Compare the above description with the description given in (b).

(d) Show that $\mathrm{Sp}(n, \mathbf{R})/\mathrm{U}(n)$ is $\mathfrak{h}_n$ by showing that $\mathrm{Sp}(n, \mathbf{R})$ acts transitively on $\mathfrak{h}_n$ and that the isotropy group at $\mathrm{i}\mathbf{1}_n$ is $\mathrm{U}(n)$.

4.5 Period Maps

Once the period domains are defined, the construction made previously in Chapter 1 generalizes. Fix a marking

$$m : H^w_{\text{prim}}(X_o)/\text{torsion} \xrightarrow{\simeq} \mathbf{Z}^n$$

for the primitive cohomology of a reference fiber at $o \in S$ and choose a path reaching from o to a general point t and pick a diffeomorphism $\phi_t : X_o \to X_t$ determined by the path. Then

$$F^\bullet_t \overset{\text{def}}{=} m_{\mathbf{C}} \circ \phi_t^* F^\bullet H^w_{\text{prim}}(X_t; \mathbf{C})$$

defines a Hodge filtration on $\mathbf{C}^n$. In the case of a simply connected, e.g., local, base manifold it is single-valued but in general it is multi-valued. The multi-valuedness, however, is controlled by the monodromy representation. Thus, we obtain a single-valued map

$$\mathcal{P} : S \longrightarrow \Gamma\backslash D,$$

where Γ is the image of the monodromy representation.

Let us discuss the structure of the quotient $\Gamma\backslash D$. First, this quotient is a Hausdorff topological space. This is because Γ is discrete in G, and V is compact, so the action on D is *properly discontinuous*, meaning that if K_1 and K_2 are two compact subsets of D, then at most finitely many Γ-translates of K_1 will meet K_2. Indeed, the inverse image $\tilde{K}_i \subset G$, $i = 1, 2$ of the set K_i under the projection $G \to G/V = D$ is compact, as well as the product $L = \tilde{K}_2 \cdot \tilde{K}_1^{-1}$. But if $\gamma K_1 \cap K_2 \neq \varnothing$, we have $\gamma \in L$ and thus γ belongs to a discrete compact and hence finite set.

From this it follows that locally $\Gamma\backslash D$ is either a complex manifold or, at a point x fixed by Γ, a quotient of a ball by a finite group. The group acts on the ring of germs of functions at x. The subring of invariant functions is finitely generated (this is a classical result; see for instance Eisenbud (1994, pp. 29–31)) by functions $f_1, \ldots, f_N$ with finitely many relations among them, thus defining what is called an analytic subspace of some open neighborhood of the origin in $\mathbf{C}^N$. We say that $\Gamma\backslash D$ is a complex space. Complex spaces will be treated in more detail in Section 5.2.

As with the period map for curves of genus g (see Chapter 1), we can describe the period map explicitly in terms of matrices, as follows.

Proposition 4.5.1 *Consider the Hodge structure on the integral primitive cohomology $H^w_{\text{prim}}(X)$ of a projective manifold X and let $\{\gamma_j\}$ be a rational basis. Let $w = 2m$ or $w = 2m + 1$ and let $\{\omega_i\}$ be a basis for F^m adapted to the filtration $F^w \subset F^{w-1} \cdots \subset F^m$ in the sense that subsets of it give bases for each of the subspaces F^p, $p = w, \ldots, m$. Then the Hodge structure is determined by the row space of the matrix of periods*

$$Z_{ij} = \int_{\gamma_j} \omega_i .$$

We now aim to show that in general the period map is holomorphic. Note that this makes sense, since both S and $\Gamma \backslash D$ are complex analytic spaces. To begin the proof, we note first that the question is a local one, so we may assume that the domain of the period map is a contractible open set and thus that there is no monodromy so that the target of the period map is just D. The complex structure on D is given via its description as an open subset of an algebraic submanifold of a product of Grassmannians. In accord with the discussion above, the "coordinates" in each of these Grassmannians are given as the row space of a matrix M^p of integrals as given in Proposition 4.5.1, where now the ω_i give a basis of F^p. The tangent space at F^p of the corresponding Grassmannian is canonically isomorphic to $\mathrm{Hom}(F^p, H/F^p)$, where $H = H^w_{\text{prim}}(X)$ (see Problem C.1.9). To prove that a map $\gamma : D \to \mathrm{Gr}(f^p, H)$ is holomorphic at $o \in D$, $\gamma(o) = F^p$, we must show that the antiholomorphic tangent vector $\partial/_{\bar{\partial} z}\gamma(z) \mid_o$ vanishes. So, if $\tau(z)$ is the row space of $F^p(z)$, we must show that $\frac{\partial}{\bar{\partial} z}\tau(z) \mid_o \subset F^p$. In our case we must therefore show that

$$\text{row space of } \frac{\partial M^p}{\partial \bar{z}} \subset \text{row space of } M^p .$$

Equivalently, we must show that for $\omega \in F^p$ we have

$$\frac{\partial \omega}{\partial \bar{z}} \in F^p,$$

where the differentiation is that of a cohomology class.

To explain this last notion, consider the homology classes $\gamma_i(t) = \phi_{t*}\gamma_i$ obtained by pushing a fixed basis from the homology of the central fiber to nearby fibers via the flow considered above. We regard these classes as well as the classes $\gamma^i(t)$, forming the dual basis $\{\gamma^i(t)\}$, as constant. A (smooth) family of cohomology classes $\omega_t \in H^w(X_t)$ is given by

$$\omega_t = \sum_i A_i(t)\gamma^i(t),$$

where

$$A_i(t) = \int_{\gamma_i(t)} \omega_t$$

are smooth functions. Thus derivatives of ω_t are defined by derivatives of the periods $A_i(t)$.

To compute these, suppose first we are given a family X_t where t is a real parameter. Such families arise, for example, by pulling back an algebraic family to the interval $(-\epsilon, \epsilon)$ via a curve η. Derivatives with respect to t are therefore of the form

$$\frac{\mathrm{d}}{\mathrm{d}t}\int_{(\phi_t)_*\gamma} \omega_t = \frac{\mathrm{d}}{\mathrm{d}t}\int_{\gamma} \phi_t^*\omega_t = \int_{\gamma} \frac{\mathrm{d}}{\mathrm{d}t}\phi_t^*\omega_t.$$

The last equality makes sense because $\phi_t^*\omega_t$ is a curve in the space of w-forms on X_o and therefore has a velocity vector. Formally, we first consider the family ω_t as a smooth w-form on the total space and then we take the so-called *Lie derivative*, defined as follows.

Definition 4.5.2 The *Lie derivative* of a w-form ω on the total space X in the direction of the tangent vector field ξ of a curve $\eta : (-\epsilon, \epsilon) \to S$ is defined by

$$\mathscr{L}_{\xi'}\omega \overset{\text{def}}{=} \lim_{t\to 0} \frac{\phi_t^*(\omega|X_t) - \omega|X_0}{t},$$

where ϕ_t is the flow defined by a lift ξ' of ξ.

In our case ω restricts to a closed form on the fibers. Explicitly, if

$$f : X \to S$$

is our family, the *bundle of relative one-forms* by definition is the bundle

$$T_X^\vee / f^* T_S^\vee,$$

and its w-th exterior power $\bigwedge^w \left(T_X^\vee / f^* T_S^\vee\right)$ is the *bundle of relative w-forms.* Its sections are the relative w-forms. By construction these always can be lifted (nonuniquely) to forms on the total space. Choosing local coordinates $(t, x = (x_2, \dots, x_n))$ on X such that $f(t, x) = t$, such a relative w-form ω_t can be locally written as

$$\omega_t = \sum_{|I|=w} a_I(t, x)\, \mathrm{d}x_I$$

with lifting to a global w-form on X which locally is of the form

$$\omega = \omega_t + \beta \wedge \mathrm{d}t.$$

If you introduce the derivation along the x-direction,

$$d\omega = d_x\,\omega \quad (\text{mod } dt),$$

the condition that the relative form ω_t restricts to closed forms on the fibers just means $d_x\,\omega = 0$. From the definition of the Lie derivative it follows that $\mathcal{L}_{\xi'}$ commutes with this operator and so $\mathcal{L}_{\xi'}$ acts on closed relative forms. Similarly, it acts on exact relative forms. So we get an induced action on relative cohomology classes of fibers.

We need to see that this last action does not depend on the choice of the lifting ξ' of ξ. This can be done by means of the homotopy formula which relates the Lie derivative to the contraction $i_{\xi'}$ with the vector field ξ'. This contraction is the unique derivation linear with respect to the C^∞ functions such that on one-forms we have

$$i_{\xi'}\phi = \phi(\xi').$$

Thus

$$i_{\xi'} : A^w(X) \longrightarrow A^{w-1}(X),$$

where for a product of one-forms,

$$i_{\xi'}(\phi_1 \wedge \cdots \wedge \phi_w) = \sum_i (-1)^{i-1}(i_{\xi'}\phi_i)\,\phi_1 \wedge \cdots \wedge \widehat{\phi_i} \wedge \cdots \wedge \phi_w.$$

The formula we are after is the following.

Lemma 4.5.3 (Cartan's formula) *We have* $\mathcal{L}_{\xi'}\omega = d(i_{\xi'}\omega) + i_{\xi'}\, d\omega$.

This homotopy formula, which we prove below, implies our claim: for any vector field η tangent along the fibers, $i_\eta\, d_t\,\omega + d_t\, i_\eta\omega = 0$ and so $\mathcal{L}_\eta\omega = d_x\, i_\eta\omega$, an exact form along the fibers. So the action of $\mathcal{L}_{\xi'}$ on relative cohomology classes does not depend on the lift ξ' of ξ and we get the following.

Corollary–Definition 4.5.4 *Lie differentiation preserves forms which restrict to closed, respectively exact forms along the fibers. The induced action on relative cohomology classes does not depend on the choice of the lift of ξ. This is the definition of the* Gauss–Manin connection*:*

$$\boxed{\nabla_\xi[\omega|X_o] \stackrel{\text{def}}{=} [\mathcal{L}_{\xi'}\omega].}$$

Remarks 4.5.5 1. The Gauss–Manin connection is a connection on the bundle E whose fiber at t is the vector space $H^w_{\text{prim}}(X_t)$. Note that a section in this bundle is a family ω_t of w-forms, and the Gauss–Manin connection evaluates to 0 on such a family precisely if $\phi_t^*\omega_t$ is constant. These are the *flat sections*. So if we

trivialize our bundle E by means of the flow, constant sections are exactly the flat sections and the Gauss–Manin connection can also be defined by these (the extension to arbitrary sections of E being uniquely given by the Leibniz rule).
2. The Picard–Fuchs equation (1.11) for the Legendre family from Chapter 1 is in fact equivalent to the Gauss–Manin connection for this family. See Problem 4.5.1.

Proof of Lemma 4.5.3 We establish the formula for vector fields that are 0 nowhere even though it holds in general. In this case, which is general enough for our purposes, one may apply the implicit function theorem to choose local coordinates $x_1, \ldots, x_n$ so that $\xi = \partial/\partial x_1$. Thus ϕ_t is a translation by t units along the x_1-axis. Now any differential form is a sum of terms that can be written as $f\,\mathrm{d}x_1 \wedge \mathrm{d}x_J$ or $f\,\mathrm{d}x_J$, where $\mathrm{d}x_J = \mathrm{d}x_{j_1} \wedge \cdots \wedge \mathrm{d}x_{j_w}$ is free of $\mathrm{d}x_1$, i.e., satisfies $j_k \neq 1$ for all k. It suffices to establish the formula for each kind of term separately. For the first kind we have

$$
\begin{aligned}
\mathscr{L}_\xi(f\,\mathrm{d}x_1 \wedge \mathrm{d}x_J) &= \frac{\partial f}{\partial x_1}\,\mathrm{d}x_1 \wedge \mathrm{d}x_J,\\
\mathrm{d}(i_\xi f\,\mathrm{d}x_1 \wedge \mathrm{d}x_J) = \mathrm{d}(f\,\mathrm{d}x_J) &= \sum_{k\notin J} \frac{\partial f}{\partial x_k}\,\mathrm{d}x_k \wedge \mathrm{d}x_J,\\
i_\xi\,\mathrm{d}(f\,\mathrm{d}x_1 \wedge \mathrm{d}x_J) &= -\sum_{k\notin J\cup\{1\}} \frac{\partial f}{\partial x_k}\,\mathrm{d}x_k \wedge \mathrm{d}x_J.
\end{aligned}
$$

For the second kind we have

$$
\begin{aligned}
&\mathscr{L}_\xi(f \wedge \mathrm{d}x_J) = \frac{\partial f}{\partial x_1}\,\mathrm{d}x_1 \wedge \mathrm{d}x_J,\\
&\mathrm{d}(i_\xi f \wedge \mathrm{d}x_J) = 0,\\
&i_\xi\,\mathrm{d}(f\,\mathrm{d}x_J) = i_\xi \sum_{k\notin J} \frac{\partial f}{\partial x_k}\,\mathrm{d}x_k \wedge \mathrm{d}x_J = \frac{\partial f}{\partial x_1}\,\mathrm{d}x_1 \wedge \mathrm{d}x_J.
\end{aligned}
$$

In both cases these computations establish the required formula. □

We use the foregoing to compute the derivative

$$
\frac{\partial}{\partial \bar z}\int_{(\phi_z)_*\gamma} \omega = \frac{\partial}{\partial \bar z}\int_\gamma \phi_z^*\omega = \int_\gamma \frac{\partial}{\partial \bar z}\phi_z^*\omega,
$$

where z is now a complex parameter. This requires some explanation. Recall that in the complex plane we have

$$
\begin{aligned}
\frac{\partial}{\partial z} &= \frac{1}{2}\left(\frac{\partial}{\partial x} - \mathrm{i}\frac{\partial}{\partial y}\right),\\
\frac{\partial}{\partial \bar z} &= \frac{1}{2}\left(\frac{\partial}{\partial x} + \mathrm{i}\frac{\partial}{\partial y}\right),
\end{aligned}
$$

where $z = x+iy$ is the usual complex coordinate. Thus $\partial/\partial z$ and $\partial/\partial\bar{z}$ are *complex* linear combinations of *real* tangent vector fields. The Gauss–Manin connection is defined via differentiation with respect to flows and so makes sense for real vector fields. However, we can extend it to complex-valued fields by the rule

$$\nabla_{U+iV}\omega = \nabla_U\omega + \mathrm{i}\nabla_V\omega.$$

Now consider a path $\gamma : [0, 1+\epsilon] \longrightarrow S$ from o to z which passes horizontally through z at time $t = 1$ with unit speed, so that its velocity vector at that instant is $\partial/\partial x$. Then we have

$$\frac{\partial}{\partial x}\int_{(\phi_z)_*\gamma}\omega = \int_\gamma \mathcal{L}_U\phi_z^*\omega,$$

where U is a lifting of $\partial/\partial x$. Similarly, we find that

$$\frac{\partial}{\partial y}\int_{(\phi_z)_*\gamma}\omega = \int_\gamma \mathcal{L}_V\phi_z^*\omega$$

for a lifting V of $\partial/\partial y$. Thus

$$\frac{\partial}{\partial\bar{z}}\int_{(\phi_z)_*\gamma}\omega = \int_\gamma \mathcal{L}_{\bar{Z}}\phi_z^*\omega,$$

where $\bar{Z}$ is a lifting of $\partial/\partial\bar{z}$ to a vector field of type (0, 1). This kind of lifting is possible because one can choose compatible holomorphic coordinates on the base and total space, with real analytic coordinates associated in the usual way with the complex analytic ones. Finally comes the key observation. The operator d applied to a form does not decrease the number of $\mathrm{d}z$'s or $\mathrm{d}\bar{z}$'s. The operator $i(\bar{Z})$, on the other hand, decreases the number of $\mathrm{d}\bar{z}$'s by one although it leaves the number of $\mathrm{d}z$'s unchanged. Consequently, if ω has at least p $\mathrm{d}z$'s to begin with, its Lie derivative with respect to $\bar{Z}$ also has that many $\mathrm{d}z$'s. Because a class in F^p is represented by a closed form that is a linear combination of forms having at least p $\mathrm{d}z$'s, the Lie derivative $\mathcal{L}_{\bar{Z}}$ thus preserves F^p. The relation we obtain, written more telegraphically as

$$\frac{\partial F^p}{\partial\bar{z}} \subset F^p,$$

is precisely the assertion to be proved.

Theorem 4.5.6 *The period map is holomorphic.*

We note the following obvious, but very useful, consequence.

Corollary 4.5.7 *Intermediate Jacobians vary holomorphically in families.*

This is clear, since the intermediate Jacobian of the fiber X_t is the dual of $F^{m-p+1}H^{2m-2p+1}(X_t)$ divided out by a (locally constant) lattice, an isomorphic image of $H_{2m-2p+1}(X_t, \mathbf{Z})$ modulo torsion.

The argument to prove the theorem in fact shows more, namely, that

$$\frac{\partial F^p}{\partial z} \subset F^{p-1}.$$

The point is that contraction with a lift Z of $\partial/\partial z$ to a field of type $(1,0)$ can decrease the number of dz's, but only by one unit. In terms of the Gauss–Manin connection we thus have the following.

Corollary 4.5.8 (Griffiths' transversality condition) *The Gauss–Manin connection evaluated in directions ξ of type* $(1,0)$ *(for instance along holomorphic vector fields) decreases the Hodge degree by at most one:*

$$\nabla_\xi F^p \subset F^{p-1}.$$

As we saw in Chapter 1, this is the Griffiths infinitesimal period relation, or third bilinear relation; we noted in Chapter 1 that it was not observed classically because it is vacuous for variations of weight 1 structures, e.g., for the period map of a family of curves. Its importance will become clear in the course of the book. We find, for example, that even though general period domains are not bounded domains, they behave in many ways as if they were for holomorphic maps satisfying Griffiths' condition. For example, there is a version of the Schwarz lemma for such maps.

Remark 4.5.9 Maps satisfying Griffiths' condition are sometimes called horizontal. *Horizontal maps* are indeed nonvertical with respect to the fibration $G/V \longrightarrow G/K$ discussed above. For us horizontality has a more general meaning as we see later (Section 12.5) and we shall then see that Griffiths' transversality implies horizontality in that sense.

Problem

4.5.1 This problem serves to show the equivalence of the geometric Gauss–Manin connection and the classical Picard–Fuchs equation.

(a) Recall the Picard–Fuchs equation for the Legendre family, $\mathscr{E}_\lambda$:

$$\lambda(\lambda - 1)\pi'' + (2\lambda - 1)\pi' + \frac{1}{4}\pi = 0.$$

The pair (π, π') gives a basis for $H^1(\mathscr{E}_\lambda)$. Use the Picard–Fuchs equation to calculate the Gauss–Manin connection in the λ-direction.

(b) Show that conversely the action of the Gauss–Manin-connection in the λ-direction gives a second order ordinary differential equation for π and that this must be the Picard–Fuchs equation.
(c) Generalize this to arbitrary families.

4.6 Abstract Variations of Hodge Structure

Period mappings, which abound in nature, motivate the definition of a variation of Hodge structure. Such an object consists of a number of parts, beginning with a base space S on which a local system of real or complex vector spaces is given. This means that the associated bundle E in the sense of Appendix C.4 has locally constant transition functions. Equivalently, this bundle comes from a representation $\rho : \pi(S, o) \to \mathrm{GL}(E_o)$ of the fundamental group. To be explicit, let $\tilde{S}$ be a universal cover of S. Then the local system on S associated with ρ is the set of equivalence classes

$$\{[\tilde{s}, e] \mid (\tilde{s}, e) \in \tilde{S} \times E_o, \quad (\tilde{s}, e) \sim (\tilde{s} \cdot g, \rho_{g^{-1}}(e))\}$$

with $g \in \pi(S, o)$ acting as a covering transformation on $\tilde{S}$. Its associated vector bundle E has a canonical flat connection, say ∇: locally E is trivialized by a constant frame and in this frame $\nabla = \mathrm{d}$. This is well defined on overlaps, since the transition functions are locally constant.

Any local system of complex vector spaces gives rise to a natural holomorphic vector bundle $E_{\mathbb{C}}$, because its defining transition functions are locally constant and therefore *a fortiori* holomorphic.

The next ingredient in the definition of a variation of Hodge structure is a system of holomorphic subbundles F^p which, fiber-by-fiber, defines a real Hodge structure of fixed weight w. To say this differently, we first digress on the complexified flat connection

$$\nabla : E_{\mathbb{C}} \to \mathcal{A}^1_S(E_{\mathbb{C}}),$$

where $\mathcal{A}^1_S(E_{\mathbb{C}})$ denotes the bundle of complex-valued one-forms with values in $E_{\mathbb{C}}$. This bundle then can be split into its $(1, 0)$ and $(0, 1)$-components giving a type decomposition for the connection as well:

$$\nabla = \nabla' + \nabla''.$$

The holomorphicity condition is equivalent to the condition that $\nabla_{\bar{\zeta}} s$ is a section of F^p for any section s of F^p, where $\bar{\zeta}$ is a vector field of type $(0, 1)$. This can be abbreviated as

$$\nabla'' : F^p \longrightarrow F^p \otimes \mathcal{A}^{0,1}_S.$$

In addition, we require that *Griffiths' transversality condition* holds, i.e., that

$$\nabla' : F^p \longrightarrow F^{p-1} \otimes \mathscr{A}_S^{1,0}. \tag{$*$}$$

Thus, if s is a section of F^p and ζ is a vector field of type $(1,0)$, then $\nabla_\zeta s$ is a section of F^{p-1}.

We now summarize what we have so far.

Definition 4.6.1 A (real) *variation of Hodge structure* consists of a triple (E, ∇, F^p), where E is a locally constant real vector bundle on a complex base manifold S, ∇ is the canonical flat connection on E, and F^p is a filtration of the holomorphic bundle $E_{\mathbf{C}} \stackrel{\text{def}}{=} E \otimes \mathbf{C}$ by holomorphic subbundles which fiber-by-fiber give Hodge structures and which satisfy Griffiths' transversality condition (i.e., $(*)$ above).

By Remark 4.5.5, in the geometric situation the canonical flat connection is exactly the Gauss–Manin connection. Observe also that in the geometric setting there is more structure: the bundle E actually came from cohomology with real coefficients and so (at least the full cohomology bundle) has an integral structure. This leads to the concept of a variation of Hodge structure with an integral structure: such data is given by a locally constant system $E_{\mathbf{Z}}$ of (free) $\mathbf{Z}$-modules. Replacing $\mathbf{Z}$ by $\mathbf{Q}$ in the foregoing we arrive at a *rational structure* $E_{\mathbf{Q}}$.

One may in addition require that the variation of Hodge structure be polarized.

Definition 4.6.2 A *polarization* of the variation of Hodge structure E consists of an underlying rational structure $E_{\mathbf{Q}}$ and a nondegenerate bilinear form

$$b : E_{\mathbf{Q}} \times E_{\mathbf{Q}} \longrightarrow \mathbf{Q}_S,$$

where $\mathbf{Q}_S$ is the trivial $\mathbf{Q}$-valued local system on S. This form must in addition have the following properties:

(i) b induces a polarization fiber-by-fiber, i.e., satisfies the first and second bilinear relations;

(ii) the generalization of the rule for differentiating the dot product of vector-valued functions holds, i.e.,

$$db(s, s') = b(\nabla s, s') + b(s, \nabla s') \quad \text{(i.e., } b \text{ is flat)}.$$

Now any polarized variation of Hodge structure defines a monodromy representation. For this notion we refer to Theorem C.4.3:

$$\rho : \pi_1(S, o) \to G_{\mathbf{Q}} = \operatorname{Aut}(E_{\mathbf{Q},o}, b),$$

where $o \in S$ is a base point. Its image

$$\Gamma = \rho(\pi_1(S, o)) \subset G_{\mathbf{Q}},$$

the monodromy group, has to be factored out in order to have a well-defined map

$$\mathcal{P} : S \to \Gamma\backslash D.$$

Such a map is *locally liftable and holomorphic* in the sense that for every point $s \in S$ the map $\mathcal{P}$ restricted to a small polydisk U about s lifts to a holomorphic map $\tilde{\mathcal{P}} : U \to D$, i.e., $\mathcal{P}|U$ is just $\tilde{\mathcal{P}}$ followed by the natural quotient map $D \to \Gamma\backslash D$. Any such lift obeys Griffiths' infinitesimal period relation. Conversely, holomorphic maps $S \to \Gamma\backslash D$ that are locally liftable and whose local lifts satisfy Griffiths' transversality condition define a variation of Hodge structure with monodromy contained in Γ. This is just a way of rephrasing the definition of a variation of Hodge structure. We summarize.

Lemma–Definition 4.6.3 *Let S be a complex manifold, $D = G/V$ a period domain classifying Hodge structures of given weight and Hodge numbers, and $\Gamma \subset G$ a subgroup; then giving a variation of Hodge structure with monodromy contained in Γ is equivalent to giving a locally liftable and holomorphic map*

$$S \to \Gamma\backslash D$$

whose local lifts satisfy Griffiths' transversality condition. Such maps will be called period maps.

Note that Γ need not be discrete and so $\Gamma\backslash D$ is not necessarily an analytic space. However, if we have an *integral* polarized variation of Hodge structure (such as the ones coming from algebraic families), then Γ is a discrete subgroup of G, $\Gamma\backslash D$ is an analytic space, and the period map itself is an analytic map between analytic spaces.

Contrary to what happens for curves, in general the fiber of the period map may consist of more than one point. Later on we discuss this more fully when we discuss the Torelli problems in Section 5.7 and Chapter 8. Furthermore, Griffiths' transversality condition implies that the period map rarely is surjective.

4.7 The Abel–Jacobi Map Revisited

We refer to §3.5 for the definition of the Abel–Jacobi map. In this section we want to study the behavior of the Abel–Jacobi map when the cycle varies in a family. We need the concept of a *relative cycle*. This is a cycle

$$Z \subset S \times M$$

which intersects each fiber $(\{s\} \times M)$ in a cycle Z_s of the correct codimension $p = \dim S + \dim M - \dim Z$ in M. We then say that the relative cycle defines *a family of codimension p cycles* on M parametrized by S. If all members Z_s of this family are homologous to 0, there is an induced Abel–Jacobi map $u^p_Z : S \to J^p(M)$. It is holomorphic.

Lemma 4.7.1 *Let Z be a family of codimension p-cycles homologous to 0 on M.*

(i) *The Abel–Jacobi map*

$$u^p_Z : S \to J^p(M)$$

is holomorphic.

(ii) *If we identify the tangent space at any point of the intermediate Jacobian with*[6] $H_{\mathbb{C}}/F^p \cong \overline{F^p H^{2p-1}(M)}$ *(an isomorphism of complex vector spaces) we have at any point* $o \in S$ *an inclusion*

$$(u^p_Z)_* T_{S,o} \subset H^{p-1,p}(M) \subset \overline{F^p H^{2p-1}(M)}.$$

Proof Suppose $Z_s = \partial \Gamma_s$. We consider a small neighborhood U in S of the point s in which a vector field ξ is given. We lift it in the obvious way to a vector field $\tilde{\xi}$ on $U \times M$. Let p_1, p_2 be the projections onto U and M, respectively, and let ϕ be a closed $(2m - 2p + 1)$-form on the m-dimensional manifold M. The Lie derivative $\mathcal{L}_{\tilde{\xi}} = i_{\tilde{\xi}} \circ \mathrm{d} + \mathrm{d} \circ i_{\tilde{\xi}}$ in the direction of $\tilde{\xi}$ is obviously 0 when calculated for $p_2^* \phi$. So, if d_ξ denotes derivation in the direction of ξ, we have

$$0 = \int_{\Gamma_s} \mathcal{L}_{\tilde{\xi}} p_2^* \phi = \mathrm{d}_\xi \int_{\Gamma_s} \phi + \int_{Z_s} i_\xi \phi,$$

where the last equality follows from Stokes' theorem. Now suppose that ϕ is of type $(m - p + 1, m - p) + (m - p + 2, m - p - 1) + \cdots$ and ξ has type $(0, 1)$. Then the form $i_\xi \phi$ has no components of type $(m - p, m - p)$ and hence the last integral vanishes and so $\mathrm{d}_\xi \int_{\Gamma_s} \phi = 0$ for all closed forms ϕ. This implies that the map u^p_Z is holomorphic.

Similarly, if ξ is of type $(1, 0)$, we have

$$[\phi] \in F^{m-p+2} H^{2m-2p+1}(M) \implies \int_{Z_s} i_\xi \phi = 0.$$

To see that this implies the second assertion, write f for the functional of integration over Γ_s and note that for any closed form ϕ on M we have

$$\langle \mathrm{d}_\xi f, \phi \rangle = \mathrm{d}_\xi \langle f, \phi \rangle = - \int_{Z_s} i_\xi \phi.$$

[6] Recall that $H_{\mathbb{C}} = H^{2p-1}(M; \mathbb{C})$.

It follows that $d_\xi f$ vanishes on $F^{m-p+2}H^{2m-2p+1}(M)$ and so the differential of u_Z maps the holomorphic tangent map to the annihilator of this space in $F^pH^{2p-1}(M)$, i.e., $H^{p-1,p}(M)$. □

Corollary 4.7.2 *The Abel–Jacobi map is zero on cycles rationally equivalent to zero and hence factors over the Chow groups.*

Proof Any holomorphic map $\mathbf{P}^1 \to J^p(X)$ must be constant. □

Problem

4.7.1 The subtorus $J_a(M) \subset J(M)$ is by definition the largest complex subtorus whose complex tangent space at 0 is contained in $H^{p-1,p}$.

(a) Show that $T^{\mathbf{C}}_{J_a(M),0}$ is the complexification of the largest rationally defined sub-Hodge structure of $H^{2p-1}(M)$ contained in $H^{p-1,p} \oplus H^{p,p-1}$. Deduce that $J_a(M)$ is a principally polarized variety.

(b) Show that the image of the Abel–Jacobi map is always contained in $J_a(M)$.

(c) Deduce that for Shioda's examples from Problem 3.5.3 the image of the Abel–Jacobi map is trivial.

Bibliographical and Historical Remarks

Theorem 4.1.2, which states the local triviality of families in the differentiable setting, is sometimes attributed to Ehresmann (1947).

Section 4.2 is from the classic book by Lefschetz (1924); for a modern approach that stays as topological as possible, see Lamotke (1981).

Sections 4.3–4.7 explain parts of Griffiths' original work (see Griffiths (1968, 1970)).

5

Period Maps Looked at Infinitesimally

We start out with a compact complex manifold. The set of complex structures that are close to the given complex structure is completely described by Kuranishi's theorem, which we state after introducing a few auxiliary concepts. Next, we look more closely at the derivative of the period map and come back to the previously studied example of hypersurfaces. The derivative of the period map comes with multilinear algebra invariants, formalized in the notion of infinitesimal variation of Hodge structure introduced in Section 5.5; we give a few examples to show how this structure can be used to reconstruct geometric objects. The last sections are devoted to the infinitesimal Torelli problem; in Section 5.7 we investigate some interesting counterexamples due to Kynev and Todorov.

5.1 Deformations of Compact Complex Manifolds

In Chapter 4 we considered smooth families of compact complex manifolds. In this chapter we fix a compact complex manifold X_o and view it as isomorphic to a special fiber over a point o in a smooth base manifold S of such a family $f : X \to S$ which then is said to be a *deformation* of X_o. Two deformations $f : X \to S$ and $g : Y \to T$ are considered to be isomorphic if there exists a fiber-preserving biholomorphic map between X and Y that induces the identity on X_o (remember that the identification of the fiber over the base point is part of the data). The biholomorphic map induced on the base spaces is not required to be the identity when $S = T$.

We see below that it is natural to allow more general base spaces than complex manifolds, but for the moment let us suppose that both S and X are manifolds.

We saw in Chapter 4 that any fiber X_t of such a deformation is differentiably isomorphic to X_o. We proved this by choosing a smooth path γ from o to t and lifting the velocity vector field over $f^{-1}\gamma \subset X$. This we did in two steps; the first is local and can also be used to lift holomorphic vector fields on the base to holomorphic vector fields on an open set in X. The second step involved a partition of unity and cannot be used in the holomorphic setting. And indeed, we may have nontrivial changes in the complex structure when we go from X_o to X_t. The idea is that by shrinking S if necessary, we can cover X by open coordinate neighborhoods $\{U_j\}$, $j \in I$ over which we can lift a given germ of a local holomorphic vector field v near o, say to $\tilde{v}_j$. Then the differences $\tilde{v}_i - \tilde{v}_j$ in the intersections $U_i \cap U_j$ define a Čech cocycle with values in the sheaf Θ_{X_o} of germs of holomorphic vector fields on the fiber X_o. This defines the *Kodaira–Spencer class* $\theta(v) \in H^1(X_o, \Theta_{X_o})$. One can verify (see Problems 5.1.1 and 5.1.2) that this class does not depend on the actual liftings. It is clear that $\theta(v)$ depends linearly on the tangent direction v so that we have obtained a linear map

$$\rho : T_{S,o} \to H^1(X_o, \Theta_{X_o}), \quad \textit{(Kodaira–Spencer map)}$$

which measures the infinitesimal changes of the complex structure near o in the various tangent directions.

The next question is whether this indeed reflects actual changes in the complex structure so that a vanishing Kodaira–Spencer map means that we have locally a trivial family, i.e., a family isomorphic to the product family $X_o \times S$. This is in general not true, due to jumping phenomena for the dimensions of the vector spaces $H^1(X_t, \Theta_{X_t})$ when t varies. But if these dimensions are locally constant near o, then ρ adequately detects deviations of the product structure near o. For further discussion as well as details related to this entire section, see Kodaira et al. (1958).

Returning to deformations, one can ask if there is a maximal one from which all others can be derived in some standard way. The relevant notions here are those of an induced deformation and of completeness. To start with the former, let $\phi : (T, o) \to (S, o)$ be a holomorphic map preserving base points. Then the fiber product over S of our deformation $f : X \to S$ and $\phi : T \to S$ yields a new deformation of X_o, the *pullback family* or *family induced by* ϕ. As to the second notion, a deformation of X_o is called *complete* if any other deformation of X_o can be induced from it. If the inducing map is unique, we call the deformation *universal*. If only its derivative at the base point is unique it is called *versal*. Versal families are unique up to isomorphism, at least locally around the base point.

There is a very useful criterion for completeness whose proof can be found in Kodaira and Spencer (1962):

Theorem *A smooth family of compact complex manifolds with a surjective Kodaira–Spencer map is complete at the base point. In particular, if the Kodaira–Spencer map is a bijection, the family is versal.*

Finally, a natural question is whether versal or universal deformations always exist. The answer turns out to be no if we are restricted to families whose base is a manifold. But there is one special case that can be treated within the framework of manifolds. This is another result due to Kodaira, Nirenberg, and Spencer (see Kodaira et al., 1958).

Theorem 5.1.1 *Suppose that $H^2(X, \Theta_X) = 0$. Then a versal deformation for X exists whose Kodaira–Spencer map is an isomorphism.*

It is not too hard to show that if, moreover, $H^0(X, \Theta_X) = 0$, this versal family is universal for X.

Example 5.1.2 If C is a curve of genus $g > 1$ the preceding theorem implies that the base of the versal deformation is smooth and has dimension $\dim H^1(C, \Theta_C) = \dim H^0(C, \Omega_C^{\otimes 2}) = 3g - 3$ and that it is universal.

Example 5.1.3 Let X be a smooth hypersurface in $\mathbf{P}^{n+1}$ of degree d. We show below that except for the case of quartic surfaces in $\mathbf{P}^3$, a versal family with smooth base can be obtained from the family of all smooth hypersurfaces of degree d in $\mathbf{P}^{n+1}$ if it is restricted to a slice S_F transversal to the PGL-orbit of the base point F in the parameter space. One can show that $H^0(X, \Theta_X) = 0$ if $n \geq 2, d \geq 3$ (Problem 5.1.5). So in these cases the versal family is universal.

Problems

5.1.1 The goal of this exercise is to find a different expression for $\theta_{ij}(v)$, the Kodaira–Spencer class. Assume that S is a ball in $\mathbf{C}^s$ with coordinates $t = (t_1, \ldots, t_s)$ centered at 0, the special point. Assume further that X is covered by coordinate charts U_j with coordinates (z_j, t), $z_j = (z_j^1, \ldots, z_j^n)$. In overlaps $U_j \cap U_k$ we have

$$z_j^\alpha = f_{jk}^\alpha(z_k, t), \alpha = 1, \ldots, n.$$

So the functions $f_{jk}^\alpha(z_k, t)$ for $t = 0$ describe how the restrictions of the charts to $t = 0$ glue together to the special fiber and for any small but fixed t they give a nearby fiber. We fix a vector field

$$v = \sum_{\beta=1}^{n} v_\beta \frac{\partial}{\partial t_\beta}.$$

The infinitesimal change of these gluing data in this direction is measured by

$$\frac{\partial f_{ij}^{\alpha}(z_j,t)}{\partial v} \stackrel{\text{def}}{=} \sum_{\beta=1}^{n} v_\beta \left.\frac{\partial f_{ij}^{\alpha}(z_j,t)}{\partial t_\beta}\right|_{t=0}.$$

Show that in fact

$$\theta_{ij}(v,t) = \sum_{\alpha=1}^{n} \frac{\partial f_{ij}^{\alpha}(z_j,t)}{\partial v}\frac{\partial}{\partial z_i^{\alpha}} = \tilde{v}_i - \tilde{v}_j.$$

This shows that the definition of the Kodaira–Spencer class does not depend on the actual liftings, provided we fix the coordinates.

5.1.2 Verify that the definition of the Kodaira–Spencer class does not depend on the choice of coordinates by showing that another choice of coordinates yields another cocycle which differs from the first one by a coboundary.

5.1.3 Let D be a smooth hypersurface of M and consider infinitesimal deformations of D within M.

(a) Show that these are classified by $H^0(\nu_{D/M})$, where $\nu_{D/M}$ is the normal bundle of D in M.

(b) Let $D \subset S \times M$ be a hypersurface such that projection onto the first factor gives a deformation $D \to S$ of a hypersurface D_o, $o \in S$. In analogy with the Kodaira–Spencer map, use (a) to define a *characteristic map* $\sigma : T_{S,o} \to H^0(\nu_{D/M})$.

(c) Let $\delta : H^0(\nu_{D/M}) \to H^1(\Theta_D)$ be the coboundary map in cohomology of the exact sequence of vector bundles

$$0 \to \Theta_D \to \Theta_M|D \to \nu_{D/M} \to 0.$$

Show that the map $\delta \circ \sigma$ is the Kodaira–Spencer map of the deformation $D \to S$.

5.1.4 Show that for any deformation $f : X \to S$ there is an exact sequence of vector bundles

$$0 \to \Theta_f \to \Theta_X \to f^*\Theta_S \to 0,$$

where the first bundle – defined by this sequence – has the property that its restriction to a fiber X_s is the tangent bundle Θ_{X_s} of that fiber. Restrict this sequence to the fiber over the base point and consider the coboundary map $T_{S,o} = (\Theta_f)_o \to H^1(X_o, \Theta_{X_o})$. Show that this is the Kodaira–Spencer map.

5.1.5 Show that $H^0(X, \Theta_X) = 0$ for any smooth hypersurface in $\mathbf{P}^{n+1}$ of degree $d > n + 2$. (Apply Serre duality and the Nakano vanishing theorem (consult Griffiths and Harris, 1978). In fact this is also true whenever $n \geq 2, d \geq 3$ (see e.g. Kodaira et al., 1958, Lemma 14.2). It follows that the group of automorphisms of X is finite in this range.

5.2 Enter: the Thick Point

As we have said before, the problem of existence of versal deformations cannot be resolved without more general parameter spaces. Indeed, one needs complex spaces. Let us recall their definition. As with manifolds, one starts with suitable local models, the analytic subspaces of some open subset $U \subset \mathbf{C}^n$. Such an *analytic subspace* is defined by specifying finitely many analytic functions $f_1, \ldots, f_n$ in U. These define first of all a set, namely the locus $V \subset U$ where these functions vanish. Second, they define a subsheaf $\mathcal{J}$ of the sheaf $\mathcal{O}_U$ of the holomorphic on U by taking in each point the ideal generated by their germs at that point. Thus, V is the support of this sheaf and the analytic functions on V are then the sections of the quotient sheaf $\mathcal{O}_U/\mathcal{J}$. In fact, the data of $\mathcal{J}$ alone suffices, since it determines the set V, but the converse is not true!

Having described our local models, one may glue these to a *complex space* X. An economical way of doing this is by giving X as a *ringed space* upon specifying a subsheaf of the sheaf of complex-valued functions on X. This subsheaf should form a sheaf of $\mathbf{C}$-algebras and in every local model as above, it should be isomorphic to $\mathcal{O}_U/\mathcal{J}$. Summarizing, a complex space is a Hausdorff ringed space locally isomorphic (as a ringed space) to an analytic subset $(V, \mathcal{O}_V)$ of an open subset in $\mathbf{C}^n$, i.e., $\mathcal{O}_V = \mathcal{O}_U/\mathcal{J}$ for some subsheaf of ideals $\mathcal{J} \subset \mathcal{O}_U$ with support V.

Example 5.2.1 For every $n \in \mathbf{N}$ we have the *n-th order thick point* $\mathbf{O}_n$ given by the subset of $\mathbf{C}$ defined by the ideal (z^n) in the ring of one-variable convergent power series. The point $\mathbf{O}_2$ plays a natural role in deformation theory as we see below.

To generalize the notion of deformation to the case where the basis is any complex space, we recall that a holomorphic map $f : X \to Y$ is defined to be a submersion if X is locally diffeomorphic to a product of an open set in $\mathbf{C}^n$ and an open subset in the base such that via this isomorphism f is just a projection onto the second factor. Note that, as a consequence, all fibers of f are manifolds. In the case in which X and Y are both manifolds, f is a submersion precisely when it is a submersion in the usual sense, i.e., the derivative of f is everywhere surjective so that we find back the notion of a smooth family. This motivates the extended definition that follows.

Definition 5.2.2 (i) A *family of manifolds over* S or *with base* S is a submersive surjective holomorphic map $X \longrightarrow S$ between complex spaces with connected fibers. We say that we have a family of *compact* manifolds if f happens to be proper. A family is called *a deformation* of a given manifold X_o if $o \in S$ and the fiber over o is isomorphic with the manifold X_o with which we started. (This isomorphism is part of the data of a deformation.) If X_o is compact, the family is required to be proper.

(ii) If $f : X \longrightarrow S$ is a family of manifolds over S and $\phi : T \longrightarrow S$ is a holomorphic map, the induced family is the fiber product of f and ϕ considered as a family over T.

As before, we can now introduce induced deformations and complete, versal and universal families. The Kodaira–Spencer map now gets a particularly elegant description using deformations over the thick point $\mathbf{O}_2$. Indeed, the set of isomorphism classes of deformations over $\mathbf{O}_2$ can be identified with the cohomology group $H^1(X_o, \Theta_{X_o})$ (see Problem 5.2.1). Because every tangent vector ζ to the base point o of a given deformation $f : X \longrightarrow S$ can be seen as a morphism of the thick point $\mathbf{O}_2$ to (S, o), we can consider the induced deformation over the thick point, and this yields an element $\kappa(f)(\zeta)$ in $H^1(X_o, \Theta_{X_o})$, the *Kodaira–Spencer class* of the deformation. Taken together, as before, they define the *Kodaira–Spencer map*

$$\boxed{\rho = \kappa(f) : T_{S,o} \longrightarrow H^1(X_o, \Theta_{X_o}).}$$

One can verify that this map coincides with our earlier definition if S is a manifold (see Problems 5.2.1 and 5.1.3). Now we can formulate the desired existence theorem due to Kuranishi. See also Douady (1984/85).

Theorem 5.2.3 (Kuranishi's theorem (Kuranishi, 1965)) *For any compact complex manifold* X *there exists a versal deformation with a bijective Kodaira–Spencer map. If* $H^0(X, \Theta_X) = 0$*, such a deformation can be chosen to be universal. The base* S *of the deformation is the fiber over the origin of a holomorphic map between neighborhoods of the origin in* $H^1(X, \Theta_X)$ *and* $H^2(X, \Theta_X)$ *with a* 0 *derivative at the origin. In particular* S *is smooth precisely when this map is* 0.

Note that the existence theorem of the previous section is the special case when $H^2(X, \Theta_X) = 0$, in which case the deformation space S is smooth.

The versal deformation of X whose existence is claimed in the preceding theorem is appropriately called the *Kuranishi family* or the *Kuranishi deformation* of X, and the parameter space is called its *Kuranishi space*. In fact, only the germ of the Kuranishi family around the marked fiber is well defined. The same holds for the Kuranishi space: only its germ at the marked point is well defined.

There are instances in which the Kuranishi space is smooth even when $H^2(X, \Theta_X)$ does not vanish, as in the following example.

Example 5.2.4 Suppose that dim $X = 3$ and that K_X is trivial. Such manifolds are called *Calabi–Yau manifolds*. By Serre duality, the obstruction space $H^2(X, \Theta_X)$ is dual to $H^1(\Omega^1_X)$ which is certainly not 0 when X is Kähler. But by Tian (1986) the Kuranishi space is smooth in this case.

One can go further and look at deformations of complex spaces themselves. In the case of compact spaces, Grauert (1974) and and Palamodov (1976) developed such a theory. For isolated singularities of a not necessarily compact space, such deformations were examined by Donin (1978), Grauert (1972), and Pourcin (1974). Later Bingener (1987) gave a nice uniform treatment.

Problems

5.2.1 A deformation $\mathbf{X}_2 \to \mathbf{O}_2$ of X can be described as follows. Choose an open cover $\mathfrak{U} = \{U_j\}$ of X by polydisks and let the transition functions in $U_j \cap U_k$ be given by

$$z_j^\alpha = f_{jk}^\alpha(z_k), \alpha = 1, \ldots, n.$$

Then the transition functions for $\mathbf{X}_2$ in $U_j \cap U_k$ are of the form

$$z_j^\alpha = f_{jk}^\alpha(z_k) + t g_{jk}^\alpha(z_k). \qquad (*)$$

Let $\widetilde{(\partial/\partial t)_j}$ be the vector field $\partial/\partial t$ on $U_j \times \mathbf{O}_2$. Show that

(a) $\widetilde{(\partial/\partial t)_j} - \widetilde{(\partial/\partial t)_k} = \sum_\alpha g_{jk}^\alpha(z_k)\partial/\partial z_j^\alpha$ is a one-cocycle on $\mathfrak{U}$ with values in Θ_X;

(b) every cocycle as in (a) defines a deformation of X using transition functions as in (*);

(c) cohomologous cocycles define isomorphic deformations;

(d) $H^1(\Theta_X)$ classifies isomorphism classes of deformations of X over the thick point;

(e) if $X \to S$ is a one-parameter deformation of X_o with $\partial/\partial t$ a tangent at the base point, the Kodaira–Spencer class for $\partial/\partial t$ coincides with the previously defined Kodaira–Spencer class.

5.2.2 Assume that $H^0(X, \Theta_X) = 0$. Show that any isomorphism of a universal deformation of X inducing the identity on the base and the fiber over the base point is the identity. Deduce that the group of holomorphic automorphisms of X acts faithfully as a group of isomorphisms of a universal deformation of X.

5.3 The Derivative of the Period Map

In this section we interpret the derivative of the period map of a geometric variation as a linear algebra construction involving the Kodaira–Spencer map of the underlying deformation.

We start, however, with an abstract variation of Hodge structures over a suitably small pointed manifold (S, o) with period map

$$\mathcal{P} : S \to D,$$

where D is a period domain for, say, polarized weight-w Hodge structures on a free $\mathbf{Z}$-module $H_{\mathbf{Z}}$ of finite rank endowed with a $(-1)^w$-symmetric bilinear form b. We let

$$\bigoplus_{p+q=w} H_F^{p,q}$$

be the Hodge decomposition corresponding to the base point $o \in S$. We claim that this induces a decomposition on the Lie algebra $\mathfrak{g}$ of endomorphisms of $H = H_{\mathbf{Z}} \otimes \mathbf{R}$:

$$\mathfrak{g} \stackrel{\text{def}}{=} \operatorname{End}(H, b).$$

The proof uses the following way to rephrase the skew-symmetry of b:

$$\mathfrak{g} = \operatorname{Ker}(s : \operatorname{End}(H) \to \operatorname{End}(H)), \quad s(\xi) = \xi + {}^{\mathsf{T}}\xi. \tag{5.1}$$

Here the transpose is defined by the usual equation

$$b(\xi u, v) = b(u, {}^{\mathsf{T}}\xi v) \text{ for all } u, v \in H,$$

which determines the transpose completely since b is nondegenerate. We can now state and prove the above claim in a more precise fashion.

Lemma 5.3.1 (i) *The choice of a reference Hodge structure in D defines a real weight 0 Hodge structure on* $\operatorname{End}(H)$.
(ii) *The homomorphism s from* (5.1) *is a morphism of Hodge structures and hence $\mathfrak{g}$ inherits the Hodge structure on* $\operatorname{End}(H)$*; in particular, its $(p, -p)$-component is given by*

$$\mathfrak{g}^{p,-p} = \left\{\xi \in \mathfrak{g}_{\mathbf{C}} \mid \xi(H^{r,s}) \subset H^{r+p,s-p} \text{ for all } (r, s)\right\}.$$

Proof (i) The space $\mathrm{End}(H)$ gets a weight 0 Hodge structure as a consequence of the "functorial package" (see the discussion in the first chapter just after the definition of a Hodge structure (Definition 1.2.4)).
(ii) Suppose that ξ has weight $(-r, r)$ and $u \in H^{p+r,q-r}$. Then $\xi u \in H^{p,q}$ and by the first bilinear relation $b(\xi u, v) = 0$ unless $v \in H^{q,p}$. On the other hand we have that $b(u, {}^{\mathsf{T}}\xi v) = 0$ unless ${}^{\mathsf{T}}\xi v \in H^{q-r,p+r}$. It follows that ${}^{\mathsf{T}}\xi$ has also weight $(-r, r)$. So $\xi \mapsto {}^{\mathsf{T}}\xi$ is a morphism of Hodge structures. Then s is a morphism of Hodge structures and hence $\mathfrak{g}$ carries a real Hodge structure. □

The subspace $\mathfrak{g}^{-1,1}$ corresponds to the tangent directions for which Griffiths' transversality holds, i.e., we have

$$\mathrm{d}\mathcal{P}_o : T_{S,o} \longrightarrow T^{-1,1}_{D,F} = \mathfrak{g}^{-1,1}.$$

Before we describe this derivative, let us recall that we have seen that on S the Hodge filtration bundles $\mathcal{F}^p$ are holomorphic subbundles of the bundle $\mathcal{H}$ but that the subbundles $\mathcal{H}^{p,q}$ are in general only C^∞ subbundles. The bundles themselves still have a holomorphic structure via the isomorphism

$$\mathcal{H}^{p,q} \simeq \mathrm{Gr}^p_{\mathcal{F}} \mathcal{H}.$$

Indeed, the right-hand side as a quotient of the holomorphic bundle $\mathcal{F}^p$ by the holomorphic subbundle $\mathcal{F}^{p+1}$ has a natural holomorphic structure. The left-hand side with this holomorphic structure, is called the *Hodge (p, q)-bundle*. The *total Hodge bundle* by definition is then

$$\mathcal{H}_{\mathrm{Hdg}} \stackrel{\mathrm{def}}{=} \mathrm{Gr}_{\mathcal{F}} \mathcal{H} = \bigoplus_p \mathrm{Gr}^p_{\mathcal{F}} \mathcal{H} \simeq \bigoplus_{p+q=w} \mathcal{H}^{p,q}.$$

This bundle, with its holomorphic structure is the underlying bundle of a so-called Higgs bundle, to be studied in greater detail later: see Sections 13.1–13.5. The restriction of $\mathrm{d}\mathcal{P}(o)$ on the Hodge bundle components can be described as follows.

Lemma 5.3.2 *The* Gauss–Manin connection *of the local system underlying the variation of Hodge structure induces $\mathcal{O}_S$-linear maps on the Hodge bundles:*

$$\sigma^p : \mathrm{Gr}^p_{\mathcal{F}} \mathcal{H} \to \mathrm{Gr}^{p-1}_{\mathcal{F}} \mathcal{H} \otimes \Omega^1_S.$$

Put $F = \mathcal{P}(o)$. For each tangent vector t of S at o the induced linear map between the fibers at o of the total Hodge bundle $\mathcal{H}_{\mathrm{Hdg}}$ is an endomorphism of type $(-1, 1)$ and we have

$$\mathrm{d}\mathcal{P}_o(t) = \bigoplus_p \sigma^p(t) \in \mathrm{End}^{-1,1} \mathcal{H}_{\mathrm{Hdg}}.$$

Proof By Griffiths' transversality (Cor. 4.5.8), the canonical flat connection ∇ maps any local holomorphic section s of $\mathcal{F}^p$ near o to a section of $\mathcal{F}^{p-1} \otimes \Omega^1_S$. Let f be the germ at o of a holomorphic function. The Leibniz rule shows that $\nabla(fs) = f\nabla(s)$ modulo $\mathcal{F}^p$ and so ∇ becomes $\mathcal{O}_S$-linear as a map from $\mathrm{Gr}^p_{\mathcal{F}}\,\mathcal{H}$ to $\mathrm{Gr}^{p-1}_{\mathcal{F}}\,\mathcal{H} \otimes \Omega^1_S$. So we get a holomorphic bundle map which visibly is of type $(-1, 1)$. Since the derivative of the period map is given by ∇ acting on $\mathrm{Gr}_{\mathcal{F}}\,\mathcal{H}$ the final assertion follows. □

Let us interpret this in case we have a variation that comes from the primitive rank w cohomology of a projective family over a base manifold (S, o). We let the fiber over the base point be X. Recall that any tangent vector t at the base point determines a *Kodaira–Spencer class* $\kappa(t) \in H^1(X, \Theta_X)$. Contraction with $\kappa(t)$ defines a linear map $\Omega^p_X \to \Omega^{p-1}_X$ and hence a linear map

$$\delta^p : H^q_{\mathrm{prim}}(\Omega^p_X) \to H^{q+1}_{\mathrm{prim}}(\Omega^{p-1}_X).$$

We can identify $H^q_{\mathrm{prim}}(X, \Omega^p_X)$ and the fiber at o of $\mathcal{F}^p/\mathcal{F}^{p+1}$.

Lemma 5.3.3 *Consider the variation of Hodge structure over a contractible manifold (S, o) associated with the primitive cohomology of the fibers of a smooth projective family over S. Let X be the fiber over o. The derivative at o of the period map in the direction t is the cup product with the Kodaira–Spencer class $\kappa(t) \in H^1(X, \Theta_X)$, i.e.,*

$$\delta^p(t) = \sigma^p_F(t),$$

where F corresponds to the fiber at o of $\mathcal{F}^\bullet$.

Proof The Kodaira–Spencer class $\kappa(t)$ can be computed in a Čech covering $\{U_j\}$ of X by first taking lifts $\tilde{t}_j$ of v and then taking the class of the one-cocycle given by $\tilde{t}_j - \tilde{t}_k$ on $U_j \cap U_k$ (see Problem 5.2.1). Now the map $\sigma^p(t) : \mathcal{H}^{p,q}_F \to \mathcal{H}^{p-1,q+1}_F$ induced by the Gauss–Manin connection can be computed (see Problem 5.3.1) and we find that on the level of cocycles we get the cup product with the one-cocycle $\{\tilde{t}_j - \tilde{t}_k\}$. □

Combining the two lemmas we find the following.

Theorem 5.3.4 *Let X be the fiber over o of a projective family over a contractible base manifold S. The cup product with the Kodaira–Spencer class $\kappa(v)$ defines linear maps*

$$\delta_p(t) = \lambda_p(\kappa(t)) : H^q_{\mathrm{prim}}(X, \Omega^p_X) \to H^{q+1}_{\mathrm{prim}}(X, \Omega^{p-1}_X), \quad p + q = w.$$

These are the nonzero components of the endomorphism $\delta(t)$ of $H^w_{\mathrm{prim}}(X, \mathbf{C})$

corresponding to the derivative of the period map $\mathcal{P} : S \to D$ in the direction of v at o. So there is a commutative diagram

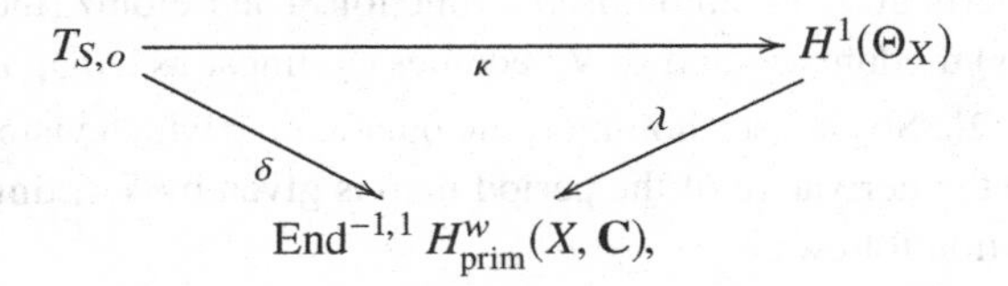

$$\begin{array}{ccc} T_{S,o} & \xrightarrow{\kappa} & H^1(\Theta_X) \\ & \delta \searrow \quad \swarrow \lambda & \\ & \mathrm{End}^{-1,1} H^w_{\mathrm{prim}}(X, \mathbf{C}), & \end{array}$$

in which λ comes from the cup product.

Problem

5.3.1 Complete the proof of Lemma 5.3.3 by showing that the map $\sigma^p(t) : \mathcal{H}_F^{p,q} \to \mathcal{H}_F^{p-1,q+1}$ is the cup product with $\kappa(t)$. Hint: do a Čech computation using an acyclic cover $\{U_j\}$ of X.

5.4 An Example: Deformations of Hypersurfaces

In Section 3.2, we saw that the middle cohomology of a hypersurface splits into a fixed part and a variable part. We now consider the hypersurface $X = X_o \subset P$ as a fiber of the family of all smooth hypersurfaces parametrized by, for example, S. The fixed cohomology for the varying fibers forms a constant subsystem inside the local system consisting of the middle cohomology of the fibers. These classes therefore are killed by the Gauss–Manin connection. Look now at the germ of the family around $o \in S$ and consider the associated infinitesimal variation of Hodge structures

$$\delta : T = T_{S,o} \longrightarrow \mathrm{End}\, H^n(X).$$

We consider the subspace of $H^n(X)$ consisting of classes that are infinitesimally fixed in the directions corresponding to T:

$$H^n_T(X) \stackrel{\mathrm{def}}{=} \{\alpha \in H^n(X) \mid \delta(t)\alpha = 0, \forall t \in T\}.$$

Clearly this subspace contains the fixed classes. In Chapter 7, we give sufficient conditions so that the converse holds (see Theorem 7.5.1). These conditions can be shown to hold for surfaces of degree $d \geq 4$ in projective three-space, and the theorem then implies that for "generic" (in a sense to be specified) such surfaces the only Hodge classes, i.e., the rational classes of type $(1, 1)$, are those coming from the hyperplane sections. In other words, the only divisors on a generic surface of degree $d \geq 4$ are the hyperplane sections and their multiples.

To start with, we need to specify the family of hypersurfaces. This we do by fixing a very ample line bundle L on P, and we put

$$S = \{(s) \in |L| \mid \text{the corresponding divisor } X_s \subset P \text{ is smooth}\}.$$

The incidence variety $I_S = \{(s, y) \in S \times P \mid y \in X_s\}$ is the total space of the *tautological family* $p : I_S \to S$, where p is the projection onto the first factor.

We recall that the tangent space of the projective space $|L| = \mathbf{P}H^0(P, L)$ at a point $* = (s_o) \in |L|$ can be identified with

$$T = T_{S,*} = H^0(P, L)/\mathbf{C} \cdot (s_o).$$

Below we assume that $H^1(P, \mathcal{O}_P) = 0$ so that the exact sequence

$$0 \to \mathcal{O}_P \xrightarrow{s_o} L \to L \mid X \to 0$$

identifies the right-hand side with

$$H^0(L|X) = H^0(\nu_{X/P}).$$

To compute the action on cohomology, we use the fact that it is given by the cup product with the Kodaira–Spencer class, inducing maps $H^{p,n-p}(X) \to H^{p-1,n-p+1}(X)$. On the other hand, viewing $H^0(P, L)$ as the space of rational functions on P having a pole of order ≤ 1 along X, the cup product with $t \in T = H^0(P, L)/\mathbf{C} \cdot (s_o)$ defines maps $H^0(\Omega_P^{n+1}(kX)) \to H^0(\Omega_P^{n+1}((k+1)X))$.

Suppose now that L is sufficiently ample so that the following vanishing result holds:

$$H^i(P, \Omega_P^j(L^{\otimes k})) = 0, \forall i, k \geq 1, j \geq 0. \qquad (*)$$

In this case the two actions match. To express this, we recall that the residue isomorphisms

$$\frac{H^0(\Omega_P^{n+1}((n-p+1)X))}{H^0(\Omega_P^{n+1}((n-p)X)) + dH^0(\Omega_P^n((n-p)X))} \xrightarrow[\sim]{\text{Res}} H_{\text{var}}^{p,n-p}(X)$$

induce surjective maps

$$r_p : H^0(P, \Omega_P^{n+1}(q+1)(X)) \to H_{\text{var}}^{p,q}(X), \quad p + q = n.$$

Proposition 5.4.1 *Assume that the vanishing condition* (*) *holds. Let*

$$T = H^0(P, \mathcal{O}_P(X))/\mathbf{C} \cdot (s_o),$$

the tangent space at $X = \{s_o = 0\}$ *to the base of the tautological family. There is a commutative diagram*

$$\begin{array}{ccc} T \otimes H^0(P, \Omega_P^{n+1}((n-p+1)X)) & \xrightarrow{\text{cup}} & H^0(P, \Omega_P^{n+1}((n-p+2)X)) \\ \downarrow{\scriptstyle \kappa \otimes r_p} & & \downarrow{\scriptstyle r_{p-1}} \\ H^1(X, \Theta_X) \otimes H^{p,n-p}_{\text{var}}(X) & \xrightarrow{\text{cup}} & H^{p-1,n-p+1}_{\text{var}}(X). \end{array}$$

Proof We deform X in the direction of $g \in H^0(P, \mathcal{O}_P(X))$. So we consider the first-order deformation $\mathbf{X} \subset P \times \mathbf{O}_2$ which in local affine charts U_i covering P is given by $f_i + tg_i = 0, t^2 = 0$. Let $\alpha_0 \in H^0(\Omega_P^{n+1}((n-p+1)(X))$ be given in U_i by ω_i / f_i^{n-p+1}, ω_i a holomorphic $(n+1)$-form on U_i. Then there is a global form $\alpha_t \in H^0(\Omega_P^{n+1}((n-p+1)X_t))$ which in U_i is given by $\omega_i/(f_i + tg_i)^{n-p+1}$ and hence $\mathrm{d}/\mathrm{d}t\ \left(\omega_i/(f_i + tg_i)^{n-p+1}\right)\big|_{t=0} = (-1)\cdot (n-p+1)\cdot g_i \cdot \omega_i / f_i^{n-p+2}$ modulo lower-order poles. Tracing through the numerical factors in the definition of the maps r_d we find from this expression $\nabla_{\partial/\partial g}(r_p(\alpha_0)) = r_{p-1}(g \cdot \alpha_0)$, proving commutativity of the diagram. □

Example 5.4.2 (Hypersurfaces in projective spaces) Let $P = \mathbf{P}^{n+1}$, and fix a system of homogeneous coordinates on $\mathbf{P}^{n+1}$. Let X be the hypersurface given by a homogeneous polynomial F of degree d. Now $S = U_{n,d}$, the parameter space for the tautological family of degree d hypersurfaces in $\mathbf{P}^{n+1}$. Finally, we let j_F be the Jacobian ideal of F (ideal generated by the partial derivatives of F). It follows that $T = S^d/\mathbf{C}\cdot F$, where S^d is the vector space of degree-d homogeneous polynomials on $\mathbf{P}^{n+1}$. Indeed, the tangent space to $\mathbf{P}(S^d)$ at X is the set of equivalence classes of curves $F + tG$ where G and G' give the same tangent if and only if G and G' differ by a multiple of F:

$$T = T_{(U_{d,n},[X])} \xrightarrow{\sim} S^d/\mathbf{C}\cdot F.$$

The tangents at X to the $\mathrm{GL}(n+2)$-orbit through X correspond to $\mathrm{j}_F/\mathbf{C}\cdot F$. To see this, consider the one-parameter group g_t of linear transformations sending e_j to $e_j + te_i$ and fixing all other e_k. Since $\mathrm{d}/\mathrm{d}t(\,g_t{\circ}F)\big|_{t=0} = z_j\partial F/\partial z_i$, it follows that the tangent vectors to $\mathrm{GL}(n+2)F$ at F span j_F^d.

To identify the action of T on middle cohomology, we need the Kodaira–Spencer map for our family. Recall (see Problem 5.1.3) that one has a characteristic map $T_{(U,[X])} \to H^0(\nu_{X/P})$. In our case, this map is the identity. In Problem 5.1.3 we showed that the coboundary δ from the exact sequence defining the normal bundle $\nu_{X/P}$ composed with the characteristic map is the Kodaira–Spencer map for the tautological family over U. So the Kodaira–Spencer map is the coboundary map

$$T = H^0(\nu_{X/P}) \xrightarrow{\kappa} H^1(X, \Theta_X).$$

Because $H^1(\mathcal{O}_{\mathbf{P}^{n+1}}) = 0$, we have an identification $T = H^0(X, \nu_{X/\mathbf{P}^{n+1}})$, and the Kodaira–Spencer map is the map $\kappa : S^d \longrightarrow H^1(\Theta_X)$ which associates with $G \in S^d$ the infinitesimal deformation of X with equation $F + tG = 0$, $t^2 = 0$. The image T of this map consists of those infinitesimal deformations that preserve the polarization.

Suppose next that X is not a quartic surface in $\mathbf{P}^3$. Then the exact sequence defining the normal bundle

$$0 \longrightarrow \Theta_X \longrightarrow \Theta_{\mathbf{P}^{n+1}}|X \longrightarrow \nu_{X/\mathbf{P}^{n+1}} \longrightarrow 0$$

induces a surjective coboundary map $\delta : H^0(\nu_{X/\mathbf{P}^{n+1}}) \longrightarrow H^1(\Theta_X)$. To see this, we first note that the vector space $H^1(\Theta_{\mathbf{P}^{n+1}})$ is Serre dual to $H^n(\Omega^1_{\mathbf{P}^{n+1}}(-n-2))$ which vanishes by Bott's vanishing theorem, which we discuss below (see Theorem 7.2.3). Likewise $H^2(\Theta_{\mathbf{P}^{n+1}}(-d)) = 0$ unless $n = 2, d = 4$. Using the exact sequence

$$0 \longrightarrow \Theta_{\mathbf{P}^{n+1}}(-d) \longrightarrow \Theta_{\mathbf{P}^{n+1}} \longrightarrow \Theta_{\mathbf{P}^{n+1}}|X \longrightarrow 0,$$

it follows that $H^1(\Theta_{\mathbf{P}^{n+1}}|X) = 0$, and using the defining sequence for the normal bundle as given above, we conclude that the coboundary map δ is onto as required. Consequently, we can describe the Kodaira–Spencer map in this case by means of the exact sequence

$$0 \longrightarrow \mathrm{j}^d_F \longrightarrow S^d \xrightarrow{\kappa} H^1(\Theta_X) \longrightarrow 0. \tag{5.2}$$

It follows that the Kodaira–Spencer map restricted to the slice $V_F \subset \mathbf{P}(S^d)$ through the base point F transversal to the $\mathrm{GL}(n+2)$-orbit (see also Example 5.1.3) is an isomorphism. By Kodaira's completeness theorem (see Section 5.1) we have a versal family. Since $H^0(\Theta_X) = 0$, this family is even universal (see Example 5.1.3).

Proposition 5.4.1 can now be applied to projective space $P = \mathbf{P}^{n+1}$ using the results from Section 3.2. We find the following.

Lemma 5.4.3 *Let $t(p) = d(n-p+1) - n - 2$ and let*

$$\lambda_p : R_F^{t(p)} \xrightarrow{\sim} H^{p,n-p}_{\mathrm{prim}}(X)$$

be the residue map followed by multiplication with $(n-p)!(-1)^{n-p}$. The following diagram commutes:

$$\begin{array}{ccc}
R^d_F \otimes R^{t(p)}_F & \xrightarrow{\text{multiplication}} & R^{t(p-1)}_F \\
\Big\downarrow{\scriptstyle \lambda_d \otimes \lambda_{t(p)}} & & \Big\downarrow{\scriptstyle \lambda_{t(p-1)}} \\
T \otimes H^{p,n-p}_{\mathrm{prim}}(X) & \xrightarrow{\text{cup product}} & H^{p-1,n-p+1}_{\mathrm{prim}}(X).
\end{array}$$

5.5 Infinitesimal Variations of Hodge Structure

In this section we axiomatize the notion of the derivative of the period map in the abstract setting, obtaining what is called an infinitesimal variation of Hodge structure. We do this by taking as ingredients of this structure two consequences of transversality. One we have already seen; it is the horizontality of the period map. The second consequence is a certain commutativity property.

Lemma 5.5.1 *$d\mathcal{P}(T)$ is an abelian subspace of $\mathfrak{g}$.*

Proof Let v and v' be two complex vector fields in a neighborhood of s in S and let ∇ be the flat connection given by the variation of Hodge structures. The tangent vector $dp(v(s))$ can be viewed as the endomorphism of $(H_{\mathbf{C}}, b)$ corresponding to the endomorphism $\nabla_{v(s)}$ of the fiber at s of the complex local system underlying the variation of Hodge structure. The curvature tensor for ∇ at s evaluated on the vectors $v(s), v'(s)$ is $[\nabla_v, \nabla_{v'}](s) - \nabla_{[v,v']}(s)$. But ∇ is a flat connection and so (see Appendix C, Problem C.3.1) we get $[\nabla_v, \nabla_{v'}](s) = \nabla_{[v,v']}(s)$. So, if $d\mathcal{P}(v) = w, d\mathcal{P}(v') = w'$, we have $[w(F), w'(F)] = d\mathcal{P}([v, v'](F)) \in \mathfrak{g}^{-1,1}$. On the other hand, the bracket on $\mathfrak{g}$ has the property that $[\mathfrak{g}^{-1,1}, \mathfrak{g}^{-1,1}] \subset \mathfrak{g}^{-2,2}$, as follows directly from the definition. It follows that $[w(F), w'(F)] = 0$, i.e., the image of T is an abelian subspace of $\mathfrak{g}$. □

Thus we arrive at the following definition.

Definition 5.5.2 An infinitesimal variation of Hodge structure (IVHS) consists of:

(i) a polarized weight-w Hodge structure H on a free $\mathbf{Z}$-module of finite rank $H_{\mathbf{Z}}$ endowed with a $(-1)^w$-symmetric bilinear form b;
(ii) a complex vector space T;
(iii) a complex linear map $\delta : T \longrightarrow \mathrm{End}(H_{\mathbf{Z}} \otimes_{\mathbf{Z}} \mathbf{C}, b) = \mathfrak{g}$ such that
 (a) $\delta(T) \subset \mathfrak{g}^{-1,1}$, where we give $\mathfrak{g}$ the natural Hodge structure induced by H (see Proposition 5.3.1),
 (b) $\delta(T)$ is an abelian subspace of $\mathfrak{g}$, i.e., for all $t, t' \in T$, the endomorphisms $\delta(t) \circ \delta(t')$ and $\delta(t') \circ \delta(t)$ of $H_{\mathbf{C}}$ are equal and define a map $\delta^2 : S^2T \to \mathrm{End}^{-2,2}(H_{\mathbf{C}})$.

If we have a real polarized Hodge structure on a finite-dimensional real vector space H the above definition still makes sense and we speak of an infinitesimal variation of *real* Hodge structures.

We show now how standard multilinear algebra operations can be applied to an infinitesimal variation of Hodge structure that eventually give back geometric data. To be specific, in an obvious way we get linear maps

$$\delta^r : S^r T \to \mathrm{End}^{-r,r}(H_{\mathbf{C}})$$

for all r. Observe that the image of δ^r does not consist of b-skew endomorphisms, but rather of maps that are $(-1)^r$-symmetric under b. Now b induces a perfect pairing between the Hodge components $H^{a,w-a}$ and $H^{w-a,a}$. So, if these components are at the extreme ends of the Hodge decomposition, and say $a \geq w - a$ then with $r = 2a - w$, we have $\delta^r : S^r T \to \mathrm{End}^{-r,r}(H_{\mathbf{C}}) = \mathrm{Hom}(H^{a,w-a}, H^{w-a,a}) = H^{w-a,a} \otimes H^{w-a,a}$.

Lemma 5.5.3 *The image of δ^r $(r = 2a - w)$ actually is contained in the subspace of the symmetric homomorphisms, i.e., we have*

$$\delta^r : S^r T \longrightarrow \mathrm{Hom}^{\mathrm{sym}}(H^{a,w-a}, H^{w-a,a}) = S^2 H^{w-a,a}.$$

Proof We have to show that for an r-tuple of vectors $(t_1, \ldots, t_r) \in S^r T$, the linear map $A = \delta^r(t_1, \ldots, t_r) \in \mathrm{Hom}((H^{w-a,a})^\vee, H^{w-a,a})$ has the property that $l(A(m)) = m(A(l))$ for all $l, m \in H^{w-a,a}$. However, we have $l(A(m)) = b(l, A(m)) = (-1)^w b(A(m), l)$ just by $(-1)^w$-symmetry of b. Because $r = 2a - w$ we have $(-1)^w = (-1)^r$ and so the fact that A is $(-1)^r$-symmetric implies that $(-1)^w b(A(m), l) = b(m, A(l)) = m(A(l))$. □

Let us now apply this construction in a geometric situation. Suppose now that $r = w = n$ and that the infinitesimal variation comes from the variation of Hodge structure on the primitive n-cohomology of a projective family of n-dimensional manifolds over a contractible base manifold S. As before, we fix $s \in S$, let X be the fiber of the family over s, and we set $T = T_{S,s}$. We have seen that the derivative of the period map fits into a commutative diagram

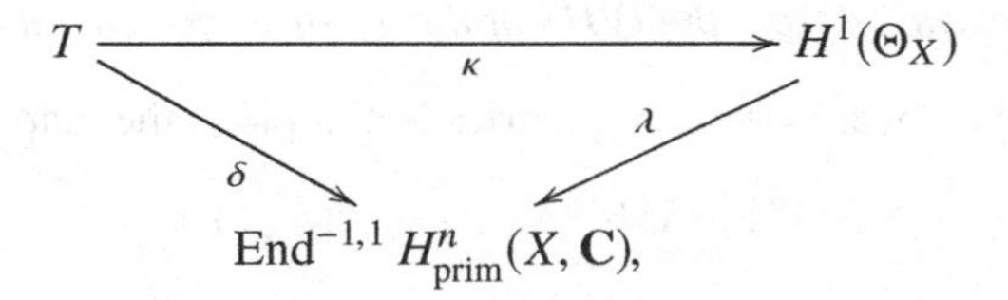

$$\begin{array}{ccc} T & \xrightarrow{\kappa} & H^1(\Theta_X) \\ & \delta \searrow \quad \swarrow \lambda & \\ & \mathrm{End}^{-1,1} H^n_{\mathrm{prim}}(X, \mathbf{C}), & \end{array}$$

where the components λ come from the cup products

$$H^1(\Theta_X) \to \mathrm{Hom}\left(H^{n-j}(\Omega_X^j), H^{n-j+1}(\Omega_X^{j-1})\right).$$

Because the natural maps

$$\underbrace{H^1(\Theta_X) \otimes \cdots \otimes H^1(\Theta_X)}_{n} \to H^n(\underbrace{\Theta_X \otimes \cdots \otimes \Theta_X}_{n}) \to H^n(\Lambda^n \Theta_X)$$

are both antisymmetric, iterating the Kodaira–Spencer map, we get a linear map

$$\kappa^n : S^nT \to H^n(\Lambda^n\Theta_X) = H^n\left(K_X^{-1}\right),$$

which fits into a commutative diagram

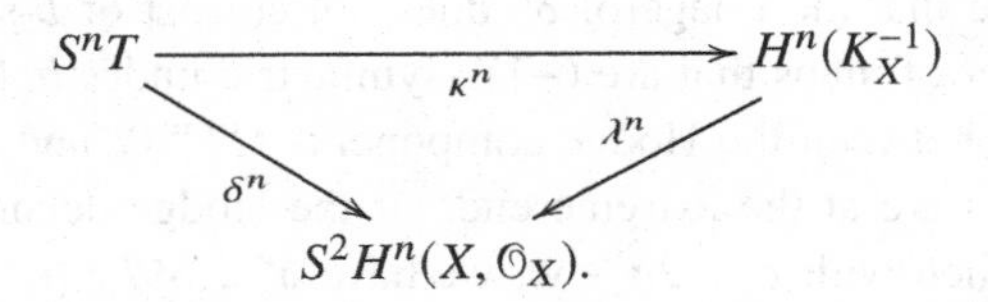

To interpret this geometrically, we dualize this diagram and find

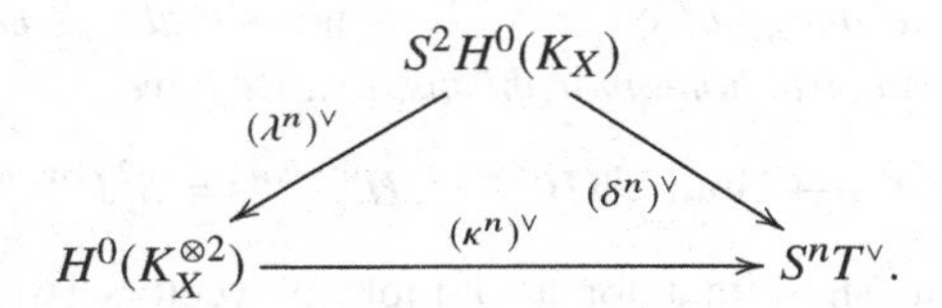

Let us call $(\kappa^n)^\vee$ the *n-th iterate of the dual Kodaira–Spencer map* and $(\delta^n)^\vee$ *the n-th iterate of the codifferential*. The geometric interpretation we have in mind is as follows.

Lemma 5.5.4 *Let X be an n-dimensional projective manifold for which the rational map defined by the canonical system*

$$\varphi_{|K_X|} : X \longrightarrow \mathbf{P}H^0(K_X)^\vee$$

is an embedding. Suppose that the quadrics cut out a complete linear system on the image of X and that this image is the intersection of all quadrics that pass through the image. Assume moreover that X is the fiber of a projective family for which the n-th iterate of the dual Kodaira–Spencer map is injective. Then X can be reconstructed from the IVHS of the given projective family.

Proof Since the linear system of quadrics is complete, the map

$$(\lambda^n)^\vee : S^2H^0(K_X) \to H^0(K_X^{\otimes 2})$$

is surjective. The kernel is formed by the quadrics through the canonical image of X. Since the n-th iterate of the dual Kodaira–Spencer map is injective, this kernel is the same as the kernel for the n-th iterate of the codifferential, as we can see from the previous commutative diagram. So the IVHS gives the system of quadrics through the canonical image of X, whose intersection is X. □

Example 5.5.5 Let X be a curve which is nonhyperelliptic, is not trigonal, and is not a plane quintic. Then a result of Enriques–Petri (see Saint-Donat,

1975) tells us that the canonical image of X is cut out by quadrics. The system of quadrics cuts out a complete linear system on the canonical curve by Noether's theorem, c.f., loc. cit. Because $n = 1$ the condition on the dual of the Kodaira–Spencer mapping just says that the Kodaira–Spencer mapping should be surjective, so any complete deformation of X yields an IVHS from which one can recover the curve.

Problem

5.5.1 The map of Lemma 5.5.3 induces a map

$$\phi : T \to \mathrm{Hom}^{\mathrm{sym}}(H^{a,w-a}, H^{w-a,a})$$

by $\phi(t) = \delta^r(t, \ldots, t)$. Let $\Sigma_{a,k}$ be the locus in $\mathbf{P}(T)$ where ϕ has rank $\leq k$. Consider the variation of Hodge structure given by the Kuranishi deformation of a curve C of genus $g \geq 5$. Show that $\Sigma_{0,1}$ is the bicanonical image of C in $T = \mathbf{P}H^1(\Theta_C) = \mathbf{P}H^0(K_C^{\otimes 2})^\vee$ (see also Griffiths and Harris, 1983).

5.6 Application: A Criterion for the Period Map to be an Immersion

Let us formulate the infinitesimal Torelli problem as follows.

Problem (Infinitesimal Torelli problem) Let X be the fiber over 0 of a projective deformation over a contractible base manifold S with the property that the Kodaira–Spencer map is injective. Decide whether the period map $\mathcal{P} : S \to D$ for the primitive cohomology group in degree w is an immersion at 0.

If the parameter space of the Kuranishi family (see Section 5.2) for X is smooth, as is the case for all examples, we have to consider this question for the Kuranishi family only.

Because $\mathcal{P}(s)$, $s \in S$, reflects the Hodge decomposition of $H^w_{\mathrm{prim}}(X_s, \mathbf{C})$ and because the Kodaira–Spencer map, which measures infinitesimal changes in the complex structure, is injective, we can loosely state this problem as follows: determine whether the infinitesimal change of complex structure at o is faithfully reflected in the infinitesimal change of the Hodge decomposition at o.

Because $\kappa(f) : T_{S,o} \to H^1(X, \Theta_X)$ is injective, we can identify $T_{S,o}$ with a subspace of $H^1(X, \Theta_X)$. Theorem 5.3.4 then yields the following criterion.

Theorem 5.6.1 *Let X be a smooth projective manifold whose Kuranishi space is smooth and of dimension* $\dim H^1(X, \Theta_X)$*. The period map for the Kuranishi family associated with primitive cohomology in degree w is injective precisely when the cup product map*

$$T_{S,o} \xrightarrow{\kappa(f)} H^1(\Theta_X) \longrightarrow \bigoplus_{p+q=w} \operatorname{Hom}\big(H^q(\Omega_X^p), H^{q+1}(\Omega_X^{p-1})\big)$$

is injective.

Examples 5.6.2 (i) Consider a family of curves. The dual of the cup product map is the product map

$$H^0(\Omega^1) \otimes H^0(\Omega^1) \to H^0((\Omega^1)^{\otimes 2}),$$

and this map is surjective unless $g > 2$ and the curve is hyperelliptic. It follows that the infinitesimal Torelli theorem holds for curves of genus 2 and for non-hyperelliptic curves of higher genus.
(ii) Let X be an n-dimensional variety with a trivial canonical bundle. Then contraction of local holomorphic vector fields with local holomorphic n-forms gives an isomorphism $\Theta_X \xrightarrow{\sim} \Omega_X^{n-1}$. It follows that the cup product mapping $H^0(\Omega_X^n) \otimes H^1(\Theta_X) \to H^1(\Omega_X^{n-1})$ is an isomorphism and so is the induced mapping $H^1(\Theta_X) \to \operatorname{Hom}(H^0(\Omega_X^n), H^1(\Omega_X^{n-1}))$. It follows that the position of $H^{n,0}(X)$ inside $H^n_{\text{prim}}(X)$ gives local moduli for X. Examples include abelian varieties, Calabi–Yau manifolds such as $K3$ surfaces, and hypersurfaces of degree $n+2$ in $\mathbf{P}^{n+1}$. See Example 5.2.4.

Other examples are given in later chapters when we have developed more techniques.

5.7 Counterexamples to Infinitesimal Torelli

Trivial Counterexamples

The simplest counterexamples are provided by surfaces having nontrivial moduli but without holomorphic one- and two-forms. We have seen such an example in Chapter 1, Exercise 1.3.3(c): the cubic surface in projective 3-space has 4 moduli but no periods. Later we shall nevertheless find a Hodge theoretic period domain. See Section 17.3.

A nonrational example of this kind was given by Enriques (1914) and still bears his name. As we shall show, several nonrational surfaces can be constructed as quotients of complete intersection surfaces in some projective space by a cyclic group of projective transformations which preserves the intersection. Among these we shall find Enriques' surfaces.

Let us first say a few words about properties of a quotient of a complex manifold X by a cyclic group G generated by a holomorphic automorphism $g : X \to X$ of prime order p. For simplicity, assume that the fixed locus is a (possibly empty) smooth divisor $R \subset X$. Then $Y = X/G$ has the structure of a complex manifold with a holomorphic projection $\pi : X \to Y$. This can be seen by choosing local holomorphic coordinates $\{s, t_1, \ldots, t_k\}$ centered at a point of R such that g is given by

$$(s, t_1, \ldots, t_k) \mapsto (\rho_p s, t_1, \ldots, t_k), \quad \rho_p = \exp(2\pi \mathrm{i}/p).$$

Then R is given by $s = 0$ and $(s^p, t_1, \ldots, t_k)$ can be taken as local coordinates on Y and hence $\pi : X \to Y$ is locally given by

$$\pi : (s, t_1, \ldots, t_k) \to (s^p, t_1, \ldots, t_k).$$

To investigate the relation between the tangent sheaves T_X and T_Y, we best use direct images. Although these will be treated more extensively later in Section 6.4, in our setting, the direct image $\pi_* \mathcal{F}$ of a sheaf on X is easy to describe since π is a finite morphism. If $y \in Y$ is not a branch point, say $\pi^{-1} y = \{x_1, \ldots, x_p\}$ the stalk of $\pi_* \mathcal{F}$ at y is the direct sum of the stalks of $\mathcal{F}$ at the x_k. If y is a branch point and $\pi^{-1} y = x$, we have $(\pi_* \mathcal{F})_x \simeq \mathcal{F}_y$. Moreover, if for instance $\mathcal{F} = \mathcal{O}_X, T_X$ or Ω^k_X, then G acts on each stalk of $\pi_* \mathcal{F}$. Also, in this case the direct image has a natural structure of an $\mathcal{O}_Y$-module.

With this in mind, the above calculation is seen to imply that there is a short exact sequence of $\mathcal{O}_Y$-modules

$$0 \to (\pi_* T_X)^G \xrightarrow{\pi_*} T_Y \to \mathcal{O}_D(D) \to 0, \quad D = \pi_* R, \tag{5·3}$$

where the first map is induced by the tangent map. To see this, since $g_*(\partial/\partial s) = \rho_p \cdot \partial/\partial s$, the G-invariant holomorphic vector fields are generated by $s\partial/\partial s$, $\partial/\partial t_1, \ldots, \partial/\partial t_k$. On the other hand, setting $t_0 = s^p$, one of the local coordinates on Y, one has $\pi_*(s\partial/\partial s) = p t_0 \partial/\partial t_0$, hence π_* is injective with cokernel isomorphic to $\mathcal{O}_D(D)$. Indeed, t_0 is a coordinate transversal to D and $t_0 \partial/\partial t_0$ is a local section of $\mathcal{J}_D \cdot \mathcal{O}_Y(T_Y)$ and so the cokernel of π_* is a locally free sheaf on D with $\left.\dfrac{\partial}{\partial y_0}\right|_D$ a local generator. This is a local section of $\mathcal{O}_D(D)$.

The same local calculation can also be used to investigate the relation between the canonical bundles of X and Y. Since

$$\pi^*(t_0 \wedge \mathrm{d}t_1 \wedge \cdots \wedge \mathrm{d}t_k) = p s^{p-1} \, \mathrm{d}s \wedge \mathrm{d}t_1 \wedge \cdots \wedge \mathrm{d}t_k \tag{5·4}$$

we deduce

$$\pi^* K_Y = K_X \otimes R^{p-1}. \tag{5·5}$$

Let us apply this to the following examples of surfaces. Recall that

$$p_g(X) = h^{d,0}(X) = h^{0,d}(X) = \dim H^0(K_X), \quad d = \dim X,$$
$$q(X) = h^{1,0}(X) = h^{0,1}(X) = \dim H^0(\Omega^1_X).$$

Examples 5.7.1 1. **Godeaux surfaces** Consider the Fermat quintic surface $F_5 = \{x_0^5 + x_1^5 + x_2^5 + x_3^5 = 0\}$ in $\mathbf{P}^3$. The group $\mathbf{Z}/5\mathbf{Z}$ generated by the projective transformation $(x_0, x_1, x_2, x_3) \xrightarrow{g} (x_0, \rho_5 x_1, \rho_5^2 x_2, \rho_5^3 x_3)$ acts on $\mathbf{P}^3$ with fixed locus the four coordinate points $(1, 0, 0, 0)$, $(0, 1, 0, 0)$, $(0, 0, 1, 0)$ and $(0, 0, 0, 1)$. Since these do not lie on the Fermat, the action restricts to a free action on F_5. Put $S = F_5/(\mathbf{Z}/5\mathbf{Z})$. It is a surface with K_S ample: by (5.5) the canonical bundle equals $\pi^* K_S = K_{F_5}$, which, by Examples C.1.6 in Appendix C, is just the hyperplane bundle. It follows that S is not a rational surface. We claim:

$$p_g(S) = q(S) = 0.$$

By Lefschetz' hyperplane theorem B.3.3 we have $2q(F_5) = b_1(F_5) = 0$ and so $q(S) = 0$. To calculate $p_g(S)$, we recall that the two-forms on F_5 are residues of rational three-forms on $\mathbf{P}^3$ with poles along F_5; the latter are of the form

$$L \frac{\omega_3}{x_0^5 + x_1^5 + x_2^5 + x_3^5}, \qquad L \text{ linear homogeneous.}$$

The two-forms on S come from $(\mathbf{Z}/5\mathbf{Z})$-invariant differential forms. Since ω_3 is the ρ_5-character space, we seek for linear forms in the ρ_5^4-character space, but there are none, whence the claim.

Next, consider moduli. Instead of the Fermat quintic, we may take any g-invariant quintic. A local deformation of the above Godeaux surface can be obtained by adding another monomial from the g-invariant collection, e.g. $x_0^3 x_2 x_3$. This corresponds to a direction in the tautological family of quintic surfaces on $\mathbf{P}^3$ which has a nonzero Kodaira–Spencer class, since it does not belong to the Jacobian ideal of the Fermat quintic. Here we use the description (5.2) of the Kodaira–Spencer map for hypersurfaces in projective space.

This shows that the Godeaux surface has nontrivial moduli, but trivial period map.

2. **Enriques surfaces** We now start with a complete intersection X of three quadrics in $\mathbf{P}^5$. Repeatedly applying the canonical bundle formula (C.1) we see that $K_X = \mathcal{O}_X$ and since $b_1(X) = 0$, again by repeated application of the Lefschetz hyperplane theorem, the surface X is a $K3$ surface. Consider now the involution on $\mathbf{P}^5$ given by

$$\iota : (x_0, x_1, x_2, x_3, x_4, x_5) \mapsto (x_0, x_1, x_2, -x_3, -x_4, -x_5).$$

The fixed locus of ι consists of two planes: the plane $x_0 = x_1 = x_2 = 0$ and the plane $x_3 = x_4 = x_5 = 0$. The quadrics invariant under ι are given by equations of the form

$$Q(x_0, \ldots, x_6) = Q'(x_0, x_1, x_2) + Q''(x_3, x_4, x_5) = 0.$$

For generic choices of three such quadrics Q, the resulting intersection X is smooth and does not contain the fixed locus of ι. The surface $Y = X/\langle\iota\rangle$ is therefore smooth.

Claim *The surfaces Y are Enriques surfaces, by definition surfaces with invariants*

$$p_g(Y) = q(Y) = 0$$

that have a $K3$ surface as their universal cover.

Let us prove the assertion about the invariants. As before $b_1(X) = 0$ implies that $2q(Y) = b_1(Y) = 0$. To show that $p_g(Y) = 0$ it suffices to show that a nonzero two-form on X must be anti-invariant under ι. Any holomorphic two-form on X is a repeated residue of a rational five-form on $\mathbf{P}^5$, this time with poles along 3 quadrics whose intersection is X. But such a form is always anti-invariant under ι.

The above construction can be seen to depend on 10 moduli; indeed Enriques surfaces have 10 moduli (see e.g. Barth et al., 1993, VIII.20) and so this yields indeed a surface with moduli but no period map. Nevertheless, as is shown there, it possible to study moduli through the period map of its universal cover, a $K3$ surface.

The Kynev Examples

Kynev surfaces are examples with $p_g = 1, q = 0$ and hence have a nontrivial period domain: these are the domains we called IV_b with $b = h^{1,1} = b_2 - 2$. See Eq. (4.1) in Problem 4.4.3.

Let us now proceed to the construction of these surfaces. Recall that $\rho_3 = \exp(2\pi i/3)$; consider the projective transformation

$$g : (x_0, x_1, x_2, x_3) \mapsto (x_0, \rho_3 x_1, \rho_3^2 x_2, -x_3)$$

and let $G = \langle g\rangle \simeq \mathbf{Z}_2 \times \mathbf{Z}_3$. The G-invariant polynomials are generated by $x_0, x_1x_2, x_1^3, x_2^3, x_3^2$; we are especially interested in a subfamily of G-invariant sextics

$$T \stackrel{\text{def}}{=} \mathbf{C}\text{-span of } \{H_1, H_2, H_3, H_4\},$$
$$H_1 = x_0^4 x_3^2, \quad H_2 = x_0^4 x_1 x_2, \quad H_3 = x_0^2 x_3^4, \quad H_4 = x_1 x_2 x_3^4,$$

and certain deformations of the Fermat sextic:

$$X_{t_1,\ldots,t_4} = \{F_{\underline{t}} \stackrel{\text{def}}{=} x_0^6 + x_1^6 + x_2^6 + x_3^6 + \sum_{j=1}^{4} t_j H_j = 0\}.$$

The quotient surfaces

$$Y_{\underline{t}} = X_{\underline{t}}/G$$

belong to a class of surfaces introduced by Kynev (1977) as counterexamples to infinitesimal Torelli. They are indeed called *Kynev surfaces*. Let us first determine their invariants. It suffices to do this for X_0.

Claim (i) *The surface* $X = X_0$ *has the following invariants:* $h^{2,0}(X) = 10, h^{1,0}(X) = 0, K_X^2 = 24, e(X) = 108.$
(ii) *The transformation g has no fixed points,* g^2 *has* 6 *fixed points:*

$$p_\pm = (1, 0, 0, \pm\rho_{12}),$$
$$q_\pm = (1, 0, 0, \pm\rho_{12}^3),$$
$$r_\pm = (1, 0, 0, \pm\rho_{12}^5),$$

and these are pairwise permuted by the involution g^3.
(iii) *The fixed locus of the involution* g^3 *is the smooth curve* $\{x_3 = 0\}$ *on which G acts freely by a group of order* 3.
(iv) *The surface* $Y = X/G$ *has precisely* 3 *singular points coming from* $p_\pm, q_\pm, r_\pm$*; these points are of type*[1] A_2. *Let* $\tilde{Y}$ *be the minimal desingularization of* Y.
(v) *The surface* $\tilde{Y}$ *has invariants* $h^{2,0}(\tilde{Y}) = 1,\ h^{1,0}(\tilde{Y}) = 0,\ K_{\tilde{Y}}^2 = 1,\ e(\tilde{Y}) = 23.$

Proof As before, by Lefschetz' hyperplane theorem $b_2(X) = 2q(X) = 0$ and the adjunction formula gives $K_X = 2H$, where H is the hyperplane bundle. So $K_X^2 = 4 \deg X = 24$ and $h^{2,0}(X)$ is the number of independent quadrics in $\mathbf{P}^3$, which is 10. The Noether formula

$$h^{2,0} - h^{1,0} + 1 = \frac{1}{12}(K^2 + e)$$

gives $e(X) = 108$.

The assertions about the fixed points can be easily verified and so the calculation of the Euler characteristic of $\tilde{Y}$ is a consequence; indeed, the curve $C = \{x_3 = 0\}$ is a sextic plane curve with $e(C) = -18$ and its image $\tilde{C}$ in $\tilde{Y}$ has

[1] This means that the minimal resolution of the singularity forms two smooth rational curves intersecting transversally in a point.

Euler number $\frac{-18}{3} = -6$ so that

$$e(\tilde{Y}) = \frac{1}{6}(e(X) - e(C) - 6) + e(\bar{C}) + 3e(A_2)$$
$$= \frac{1}{6}(108 - (-18) - 6) + (-6) + 3 \cdot 3 = 23.$$

To determine the spaces of holomorphic one- and two-forms of $\tilde{Y}$ just note that these come from the g-invariant forms on X which we know: there are no holomorphic one-forms and the holomorphic two-forms on Y are residues of G-invariant rational forms

$$\frac{Q}{F_{\underline{t}}}\omega_3, \deg Q = 2.$$

Since $g^*F_{\underline{t}} = F_{\underline{t}}$, $g^*\omega_3 = -\omega_3$ the only G-invariant form ω comes from $Q = x_0x_3$. Since g^3 is an involution fixing the curve $\{x_3 = 0\}$ on the quotient, the form ω no longer has a zero along this curve: see the formula (5.4). Here, $p = 2$ and $\pi^*(dt_0 \wedge dt_1) = s \cdot ds \wedge dt_1$ which has a simple zero along the ramification divisor $(x_3 = 0)$ but none on $Y_{\underline{t}}$. □

Let us now show, using Theorem 5.6.1 that this family indeed gives a counterexample.

Claim (i) *There is a one-dimensional subfamily of* $\tilde{Y}_{\underline{t}}$ *of effective deformations;* (ii) *the period map for this family is trivial.*

Proof (i) We first calculate the Kodaira–Spencer map of X at $H = 0$ in the direction H_j employing the tools from Section 3.2. This direction is represented by the monomial H_j which does not belong to $\mathfrak{j}_F$. So by (5.2) the deformation in this direction is not trivial. Let us next show that there are also directions for which the Kodaira–Spencer map for $\tilde{Y}$ remains nonzero at $t = 0$. Let $R = \{x_3 = 0\}$ be the ramification locus of the involution g^3. The transformation g^2 has only isolated fixed points and these don't interfere with what follows. Now we use the exact sequence (5.3) from which we get

$$0 \to H^0(\mathcal{O}_D(D)) \xrightarrow{\delta} H^1(X, T_X)^G \to H^1(T_{\tilde{Y}}).$$

The vector space on the left is generated by the linear forms x_0, x_1, and x_2. So dim $H^0(\mathcal{O}_D(D) = 3$ and hence the four-dimensional subspace $T \subset H^1(X, T_X)^G$ has an image in $H^1(T_{\tilde{Y}})$ of dimension ≥ 1, i.e., we get an induced deformation of $\tilde{Y}$ depending on 1 parameter.
(ii) Next, we show that the derivative of the period map in this direction is zero. Recall that $\omega \in H^{2,0}(X)$ is the G-invariant holomorphic two-form on X, unique up to a multiplicative constant. Now invoke the commutative diagram

from Lemma 5.4.3. The Jacobi ideal $\mathfrak{j}_F$ is generated by $(x_0^5, \ldots, x_3^5)$ and hence $x_0x_3 \cdot T \in \mathfrak{j}_F$ so that $T \cup \omega = 0$. But then the derivative of the period map for $\tilde{Y}$ in the directions given by T is identically zero. □

Remarks 1. The description above is Kynev's original description of the surfaces (see Kynev, 1977) . As one easily checks, there are 13 linearly G-invariant sextics that are not projectively equivalent and these give a ten-dimensional family of Kynev surfaces. The subfamily invariant under ι has dimension 4; they are also invariant under the involution that interchanges x_1 and x_2. The quotient of these Kynev surfaces by the group H generated by the 2 involutions is $\mathbf{P}^2$ and in this way we get a description as a bi-double cover of $\mathbf{P}^2$ with covering group H. Todorov (1980) has shown that the general such bi-double cover Y has 12 moduli. One of the involutions, say ι, in the covering group of $Y \longrightarrow \mathbf{P}^2$ gives a $K3$ surface Z which depends on 10 moduli. The branch curve of ι depends on 2 moduli and the corresponding double covers Y all have the same periods. The full moduli space of surfaces with these invariants is of dimension 18 (see Catanese, 1979, 1980; Todorov, 1981).
2. Todorov (1981) gave further counterexamples to Torelli. These are surfaces of general type having invariants $p_g = 1$, $q = 0$ but with $2 \leq K^2 \leq 8$ (see also Morrison, 1988). All of them are constructed as double covers of nodal $K3$ surfaces branched in some curve and several of the nodes. As in the Kynev case, the double covers have varying moduli when the branch curve moves but the period vector of the resulting surfaces remains unchanged. This shows that the period map has positive dimensional fibers.

Relation with Grothendieck's Period Conjecture

There is more geometry to the Kynev surfaces. They admit an extra involution ι induced by $x_0 \mapsto -x_0$ and the quotient by it, or rather its minimal resolution of singularities, is a $K3$ surface. To see this, first note that the involution ι extends to $\tilde{Y}$ and fixes the unique canonical curve C, the image of $\{x_0 = 0\} \subset Y$. Each A_2-configuration on $\tilde{Y}$ is preserved by ι; the two rational curves of such a configuration are separately preserved but not pointwise fixed and hence the intersection point must be fixed and on each component there is one more fixed point. So in total there are 9 isolated fixed points and the surface $\tilde{Y}/\langle\iota\rangle$ hence has 9 singularities of type A_1. Its minimal desingularization $\tilde{Z}$ (which is also the minimal desingularization of Z) is indeed a $K3$ surface. This is so, because the form ω is invariant under g as well as ι and the involutions g^3 and ι make that ω no longer has zeroes on the image of the curves $\{x_0 = 0\}$ and $\{x_2 = 0\}$ and so the canonical bundle has become trivial. This can also be done for the members of the one-dimensional family $\tilde{Y}_t$ of Kynev surfaces with trivial period map which we constructed above. Since the induced period map for the $K3$ surface is also trivial, the resulting deformation of $K3$ surfaces is trivial.

Summarizing, we have a nontrivial family $\tilde{Y}_t$ of Kynev surfaces and a trivial family $\tilde{Z}_t$ of $K3$ surfaces which are linked as follows:

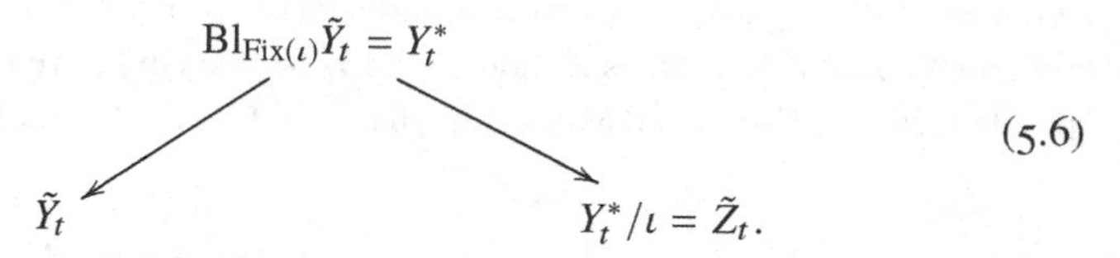

(5.6)

We have first blown up the 9 fixed points of ι, extended the involution to the resulting surface Y_t^* and then formed the quotient.

Let us now look at the period matrix for the families $\tilde{Y}_t$ and $\tilde{Z}_t$. Since in both cases there is essentially one holomorphic two-form, the period matrix is actually a *period vector* of length $b = b_2$ which gives a well defined point in the corresponding projective space, indeed in the subset given by the Riemann bilinear relations. We have seen that $e(\tilde{Y}) = 23$ and hence $b_2(\tilde{Y}) = 21$. Consequently, the period domain for $H^2(\tilde{Y})$ is of type IV_{19}. For surfaces, periods of holomorphic two-forms over algebraic cycles vanish, only periods over the transcendental cycles matter. This implies that the period vector belongs to subvarieties of these domains cut out by certain linear sections.

In our case we consider only the algebraic cycles coming from blowing up the 9 isolated fixed points of ι acting on $\tilde{Y}_t$. These span a lattice $\mathsf{E} \subset H^2(Y_t^*; \mathbf{Z})$ with orthogonal complement isomorphic to $H^2(\tilde{Y}_t; \mathbf{Z})$. Their images in $\tilde{Z}_t$ span the lattice $\overline{\mathsf{E}} \subset H^2(\tilde{Z}_t; \mathbf{Z})$. Hence a natural identification $\mathsf{E} = \overline{\mathsf{E}}$ which leads to the commutative diagram

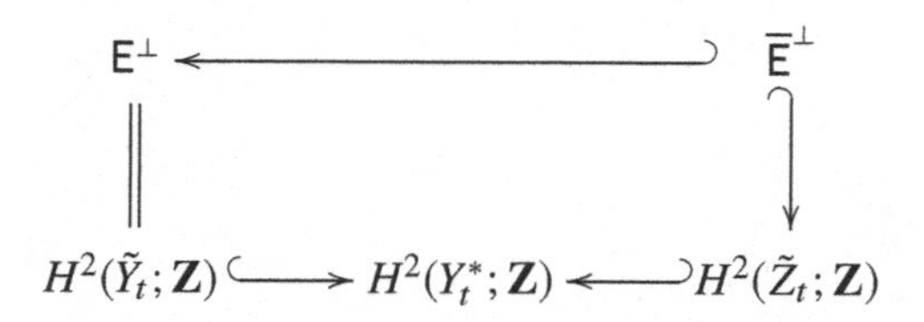

of abelian groups. Next, recall that the classes of algebraic cycles on a surface S generate the Néron–Severi lattice inside $H^2(S; \mathbf{Z})$ (see (3.4)) and the transcendental lattice is its orthogonal complement. For a surface with $h^{2,0}(S) = 1$ this lattice has a Hodge structure with $h^{2,0} = 1$. More generally, the orthogonal complement of any negative definite lattice such as $\overline{\mathsf{E}}$ has such Hodge structure and we can look at the period vector in such an orthogonal complement. Note that $\overline{\mathsf{E}}^\perp$ is a lattice of rank $22 - 9 = 13$. The surfaces $\tilde{Y}_t$ for varying t have the same period vector which belongs to the subdomain $D(\overline{\mathsf{E}}^{\perp}) \subset D(\mathsf{E}^{\perp})$ of type IV_{11}, where it is the period vector for the isomorphic $K3$ surfaces $\tilde{Z}_t$.

The preceding phenomenon is related to the so called *period conjecture of Grothendieck*. This conjecture in a crude and imprecise form states that varieties with the same periods should be related by correspondences. Indeed, for the Kynev examples, (5.6) gives a correspondence of $\tilde{Y}_t$ for varying t with the mutually isomorphic $K3$ surfaces $\tilde{Z}_t$ and so the $\tilde{Y}_t$ for varying t are related by correspondences. See Huber and Müller-Stach (2017, Ch. 12) for background on this.

Bibliographical and Historical Remarks

In Sections 5.1 and 5.2 we collected what we need from the general theory of deformations of complex spaces. Further historical comments can be found in these sections. In Section 5.3–5.6, the infinitesimal study of the period map is initiated following Griffiths (1968) and Carlson et al. (1983).

The property that the image of the period map lands in an abelian subspace has been employed to estimate ranks of period maps; see Carlson et al. (1989); Carlson and Toledo (1989a), and references therein. See also Section 14.3.

Part TWO

ALGEBRAIC METHODS

6

Spectral Sequences

Spectral sequences form an efficient algebraic tool for keeping track of the effect in cohomology of extra structures on the complexes used to calculate it. More specifically, this tool is extremely useful when dealing with a filtration on the level of complexes such as the trivial filtration on the Čech–de Rham complexes introduced in Section 3.1. One of our results will be of crucial importance in Chapter 10. It describes how to associate a spectral sequence with an exact sequence of vector bundles. We apply it here to prove an abstract de Rham theorem, thereby giving the promised sheaf-theoretic proof of the classical de Rham theorem. We also return to the Hodge filtration which we treat from a more abstract point of view. In the last but final section we present another way to define abstract sheaf cohomology, namely using the language of derived functors. As an application we give the algebraic interpretation of the Gauss–Manin connection as proposed by Katz and Oda.

6.1 Fundamental Notions

A *spectral sequence* consists of a sequence (E_r, d_r), $r = 0, 1, \dots$ of bigraded groups $E_r = \bigoplus_{p,q \in \mathbf{Z}} E_r^{p,q}$ with homomorphisms

$$d_r : E_r^{p,q} \longrightarrow E_r^{p+r,q-r+1}$$

such that

(i) $d_r \circ d_r = 0$ and
(ii) $E_{r+1}^{p,q} = \operatorname{Ker} d_r^{p,q} / \operatorname{Im} d_r^{p-r,q+r-1}$.

Observe that d_r increases the total degree $n = p + q$ by one so that for fixed r we have a complex (E_r^n, d_r), and E_{r+1} computes the cohomology of this complex.

A spectral sequence arises when one has a filtered complex $(K^\bullet, \mathrm{d})$ of groups, say with a decreasing filtration $F^\bullet$. Setting

$$Z_r^{p,q} = \left\{a \in F^p K^{p+q} \mid da \in F^{p+r} K^{p+q+1}\right\},$$

one defines the terms $E_r^{p,q}$ by

$$E_r^{p,q} \stackrel{\text{def}}{=} \frac{Z_r^{p,q}}{\mathrm{d}Z_{r-1}^{p-r+1,q+r-2} + Z_{r-1}^{p+1,q-1}}.$$

The differential d then induces d_r. Successive quotients of the filtration F frequently occur in this setting, so we introduce the p-th graded part

$$\mathrm{Gr}_F^p(K^n) = F^p K^n / F^{p+1} K^n.$$

The first term in the spectral sequence is

$$E_0^{p,q} = \mathrm{Gr}_F^p(K^{p+q})$$

and so

$$E_1^{p,q} = H^{p+q}\left(\mathrm{Gr}_F^p(K^\bullet)\right).$$

If $d_r = 0$ for $r \geq k$ we say that *the spectral sequence degenerates* at $E_k = E_\infty$. If on each K^n the filtration $F^\bullet$ has finite length, for fixed (p, q), the groups $E_r^{p,q}$ remain the same from a certain index on. An easy calculation then identifies the limit as

$$E_\infty^{p,q} = \mathrm{Gr}_{F_\infty}^p H^{p+q}(K^\bullet),$$

where, with $i : F^p(K^\bullet) \hookrightarrow K^\bullet$ the inclusion, we have

$$F_\infty^p(H^n(K^\bullet, d)) = \mathrm{Im}\left(H^n(F^p(K^\bullet)) \xrightarrow{H^n(i)} H^n(K^\bullet)\right).$$

Such filtrations are called *biregular filtrations.* One says that the *spectral sequence abuts* to $H^\bullet(K^\bullet, d)$. This is commonly denoted

$$E_r^{p,q} \Longrightarrow H^{p+q}(K^\bullet, d).$$

The next easily verified result is used to compare spectral sequences for related filtered complexes.

Lemma 6.1.1 *If $f : K^\bullet \to L^\bullet$ is a filtered homomorphism between complexes, there is an induced homomorphism $E(f_r)$ between the spectral sequences. If $E(f_r)$ is an isomorphism for $r = r_0$, it is an isomorphism for $r \geq r_0$ as well. In particular, if the filtrations are biregular, the spectral sequences abut to isomorphic groups.*

Often, the use of diagrams is very helpful; one exhibits for some j the nonzero E_j-terms in a (p, q)-lattice and all derivatives to and from these to see the process of abutment. A typical example is the existence of the so-called *edge-homomorphisms*. To explain these, suppose that we have a finite first quadrant spectral sequence in the sense that the only nonzero terms $E_1^{p,q}$ occur when $0 \leq p \leq N$ and $0 \leq q \leq M$. The reader is encouraged to visualize the diagram and see how the derivatives behave. At the left edge there are no incoming arrows and so $E_2^{0,q} = \operatorname{Ker}\left\{d_1 : E_1^{0,q} \to E_1^{1,q}\right\} \subset E_0^{0,q}$, and the situation is similar for the higher groups $E_k^{0,q}$. So, because in this case $E_\infty^{0,q}$ is a graded quotient of all of $H^q(K^\bullet)$, we get our first edge-homomorphism:

$$e^q : H^q(K^\bullet) \twoheadrightarrow E_\infty^{0,q} \subset E_1^{0,q} = H^q(\operatorname{Gr}^0 K^\bullet). \tag{6.1}$$

Similarly, looking at the right-hand edge $(N, 0)$, we obtain a surjection $E_1^{N,0} \twoheadrightarrow E_\infty^{N,0}$, and because the latter is the smallest graded quotient, it is in fact a subspace of $H^N(K^\bullet)$. This yields the second edge-homomorphism:

$$f^N : H^N(F^N K^\bullet) = E_1^{N,0} \twoheadrightarrow E_\infty^{N,0} \subset H^N(K^\bullet).$$

Below we give another example of this (see Fig. 6.1).

Spectral sequences are used here mostly in the context of double complexes. As we saw in Section 3.1, a double complex (living in the first quadrant) consists of a bigraded group $K^{\bullet,\bullet} = \bigoplus_{p,q\in\mathbf{Z}} K^{p,q}$ and differentials $\mathrm{d}: K^{p,q} \to K^{p+1,q}$ and $\delta : K^{p,q} \to K^{p,q+1}$, satisfying $\mathrm{d}^2 = \delta^2 = 0, \mathrm{d}\delta = \delta\,\mathrm{d}$ The associated total complex is $sK^n = \bigoplus_{p+q=n} K^{p,q}$ with differential $D = \mathrm{d}+(-1)^p\delta$. It has two obvious filtrations,

$$'F^p = \bigoplus_{r\geq p} K^{r,s},$$

$$''F^q = \bigoplus_{s\geq q} K^{r,s},$$

and these then induce filtrations $'F$ and $''F$ on the associated total complex. The associated spectral sequences are denoted by $'E_r^{p,q}$ and $''E_r^{p,q}$. Because $H^{p+q}(K^{p,\bullet}, D) = H^q(K^{p,\bullet}, \delta)$, and a similar relation holds for $H^{p+q}(K^{\bullet,q}, D)$, we find:

$$'E_1^{p,q} = H^q(K^{p,\bullet}, \delta), \qquad ''E_1^{p,q} = H^p(K^{\bullet,q}, \mathrm{d}),$$

$$'E_2^{p,q} = H^p(H^q(K^{\bullet,\bullet}, \delta), \mathrm{d}), \qquad ''E_2^{p,q} = H^p(H^q(K^{\bullet,\bullet}, \mathrm{d}), \delta).$$

Problems

6.1.1 Prove that the spectral sequence for a biregularly filtered complex $(K^\bullet, d\, F^\bullet)$ yields an exact sequence

$$0 \to E_2^{1,0} \to H^1(K^\bullet) \to E_2^{0,1} \xrightarrow{d_2} E_2^{2,0} \to H^2(K^\bullet).$$

6.1.2 Show that the spectral sequence degenerates at E_1 if and only if the derivative is strictly compatible with the filtration, i.e., $F^p \cap \operatorname{Im} d = d(F^p)$.

6.2 Hypercohomology Revisited

Recall from Section 3.1 that we defined hypercohomology of the holomorphic de Rham complex $\Omega_M^\bullet$ on a compact complex manifold M as the cohomology of the total complex associated with the Čech–de Rham complex.

Suppose that we now start out with an arbitrary complex of sheaves on a topological space X,

$$(\mathcal{K}^\bullet, d) = (\mathcal{K}^0 \xrightarrow{d} \mathcal{K}^1 \xrightarrow{d} \mathcal{K}^2 \xrightarrow{d} \cdots),$$

and suppose that we are given a Leray cover $\mathfrak{U} = \{\mathfrak{U}_i\}_{i \in I}$ of X. Recall that this means that we can compute Čech cohomology directly from the cover. We can now form the double complex consisting of Čech cochains $\mathscr{C}^\bullet(\mathfrak{U}, \mathcal{K}^\bullet)$ with differential d coming from the complex of sheaves and Čech differential δ. In this case the cohomology of the associated total complex by definition is the hypercohomology group of the complex of sheaves we started with. Without assuming the cover is Leray, we have to modify the definition slightly.

Definition 6.2.1 The *hypercohomology group* of the complex of sheaves $\mathcal{K}^\bullet$ is the direct limit of the total complexes associated with $\mathscr{C}^\bullet(\mathfrak{U}, \mathcal{K}^\bullet)$:

$$\mathbf{H}^p(\mathcal{K}^\bullet, d) \stackrel{\text{def}}{=} \varinjlim_{\mathfrak{U}} H^p(s\mathscr{C}^\bullet(\mathfrak{U}, \mathcal{K}^\bullet)).$$

To compute this hypercohomology, one uses the two spectral sequences of the double complex $\mathscr{C}^\bullet(\mathfrak{U}, \mathcal{K}^\bullet)$ that have E_1-terms,

$$'E_1^{p,q} = \varinjlim_{\mathfrak{U}} H^q(\mathscr{C}^\bullet(\mathcal{K}^p), \delta) = H^q(X, \mathcal{K}^p)$$

and

$$''E_1^{p,q} = \varinjlim_{\mathfrak{U}} H^p(\mathscr{C}^q(\mathfrak{U}, \mathcal{K}^\bullet), d),$$

respectively.

In the first spectral sequence the derivative comes from d and in the second sequence it comes from δ. Hence, the E_2-terms are given by

$$'E_2^{p,q} = H^p(H^q(X, \mathcal{K}^\bullet), \mathrm{d})$$

and

$$''E_2^{p,q} = H^p(X, H^q(\mathcal{K}^\bullet, \mathrm{d})),$$

respectively, where the right-hand side is the Čech cohomology of the cohomology *sheaf* associated with the complex $\mathcal{K}^\bullet$.

For reasons that become apparent below, we introduce the term *de Rham cohomology groups* for

$$H^p_{dR}(X, \mathcal{K}^\bullet) \stackrel{\text{def}}{=} {'E_2^{p,0}} = H^p(H^0(X, \mathcal{K}^\bullet), \mathrm{d}).$$

Any global section of $\mathcal{K}^p$ is a 0-Čech cocycle, and thus if it is d-closed, it is a cocycle in the total complex $s\mathcal{C}^\bullet(\mathfrak{U}, \mathcal{K}^\bullet)$ and so one gets an associated element in the hypercohomology, whereby defining the linear map

$$h_{dR} : H^p_{dR}(\mathcal{K}^\bullet) \longrightarrow \mathbf{H}^p(X, \mathcal{K}^\bullet).$$

Remark 6.2.2 If the complex of sheaves is a trivial complex with $\mathcal{K}^q = 0$ for $q \neq s$, the n-th hypercohomology of this complex simply coincides with $H^{n-s}(X, \mathcal{K}^s)$.

Remark 6.2.3 A map $j : \mathcal{A}^\bullet \to \mathcal{B}^\bullet$ between complexes induces maps between the hypercohomology groups of the complexes. If j induces an isomorphism on the level of cohomology sheaves (in which case we say that j is a *quasi-isomorphism*), it induces an isomorphism in hypercohomology, because the second spectral sequences are isomorphic (use Lemma 6.1.1).

A special case of the preceding situation arises when the complex of sheaves $(\mathcal{K}^\bullet, d)$ we started with in fact is an exact sequence, so that the cohomology sheaves vanish. So the E_2-terms of the second spectral sequence are all 0 and so *a fortiori* it abuts to 0. This then also holds for the first spectral sequence $'E_1^{p,q} = H^q(X, \mathcal{K}^p)$. If these vanish in a certain range, we can compute the higher degree terms of the spectral sequence at the borders of this range, which thus must be zero also, yielding specific information. For instance, by inspecting Fig. 6.1, we obtain the following lemma which will be of use later (see the proof of Theorem 7.4.1).

Lemma 6.2.4 *Suppose that $\mathcal{K}^\bullet$ is an exact complex in degrees ≥ 0. Assume that $H^i(X, \mathcal{K}^p) = 0$ for all $p > 0$ and $i = 1, \ldots, q-1$. Then the derivative d_{q+1}*

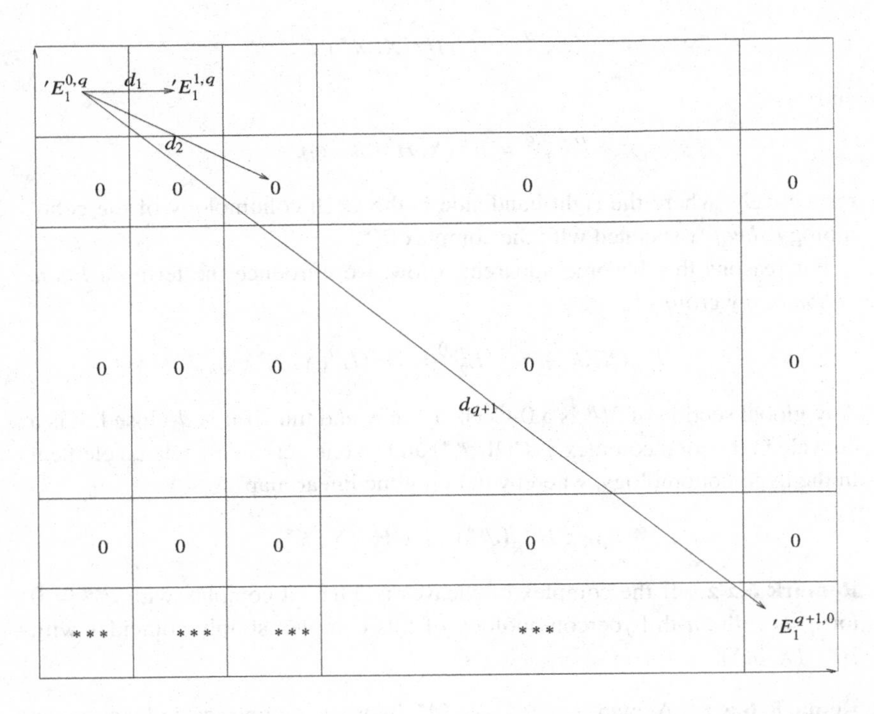

Figure 6.1 Spectral sequence of an exact complex.

induces an isomorphism

$$\operatorname{Ker}\left(H^q(X,\mathcal{K}^0)\xrightarrow{H^q(\mathrm{d})} H^q(X,\mathcal{K}^1)\right),$$

$$\xrightarrow{\cong} H^{q+1}_{\mathrm{dR}}(\mathcal{K}^\bullet,\mathrm{d}) = \frac{\operatorname{Ker}\left(\mathrm{d}\colon H^0(X,\mathcal{K}^{q+1})\to H^0(X,\mathcal{K}^{q+2})\right)}{\operatorname{Im}\left(\mathrm{d}\colon H^0(X,\mathcal{K}^{q})\to H^0(X,\mathcal{K}^{q+1})\right)}.$$

We next formulate an abstract version of de Rham's theorem with the de Rham complex $\mathcal{A}^\bullet$ replaced by any complex.

Theorem (Abstract theorem of de Rham) *Let X be a topological space and let $\mathcal{K}^\bullet$ be an exact complex of sheaves on X. We have a canonical identification*

$$H^p(X,\mathcal{F}) = \mathbf{H}^p(X,\mathcal{K}^\bullet),\ \ \textit{where}\ \ \mathcal{F} \overset{\text{def}}{=} \operatorname{Ker}\left\{\mathrm{d}\colon \mathcal{K}^0\to\mathcal{K}^1\right\}.$$

If, in addition, for all q and all $p>0$ we have $H^p(X,\mathcal{K}^q)=0$, then we also *have an isomorphism*

$$\mathbf{H}^p(X, \mathcal{K}^\bullet) \xrightarrow{\sim} H^p_{\mathrm{dR}}(X, \mathcal{K}^\bullet) \stackrel{\mathrm{def}}{=} H^p(\Gamma(X, \mathcal{K}^\bullet), \mathrm{d}).$$

Proof The higher cohomology sheaves are all 0 and so the second spectral sequence degenerates at E_2 and we have

$$''E_2^{p,0} = H^p(X, \mathcal{F}) \xrightarrow{\cong} \mathbf{H}^p(X, \mathcal{K}^\bullet).$$

For the second assertion, since $'E_2^{p,q} = H^p(H^q(X, \mathcal{K}^\bullet), \mathrm{d}) = 0$ for $q > 0$ and $'E_2^{p,0} = H^p(X, \mathcal{F})$, the first spectral sequence degenerates at E_2 and this yields the stated identification. □

Now, if $X = M$ is a differentiable manifold and $\mathcal{K}^\bullet = \mathcal{A}^\bullet_M$ is the de Rham complex, the last assertion is precisely the isomorphism $H^n_{\mathrm{dR}}(M) \cong \mathbf{H}^n(\mathcal{A}^\bullet)$ which is one half of the proof of Theorem 3.1.5. Combined with the first assertion, we get the *de Rham isomorphism*

$$H^n_{\mathrm{dR}}(M) \cong H^n(M; \mathbf{R}).$$

If we take the holomorphic de Rham complex, this does not work, because in general $H^p(M, \Omega^q_M)$ does not vanish for $p > 0$. There is one special case worth mentioning, where hypercohomology can be replaced by de Rham cohomology.

Example 6.2.5 Recall (Example 3.1.4), that a *Stein variety* is defined as the of zero set of finitely many holomorphic functions defined in an open subset of $\mathbf{C}^n$. We further recall (as mentioned there without proof) that the higher cohomology groups for a vector bundle over a Stein variety vanish. In particular, if X is a Stein *manifold*, $H^n(X; \mathbf{C})$ can thus be computed as the cohomology of the complex of global sections for the holomorphic de Rham complex.

In general, because of the *holomorphic* Poincaré lemma, we only get

$$\mathbf{H}^n(M, \Omega^\bullet_M) \cong H^n(M; \mathbf{C}).$$

This suffices, however, to complete the proof of Theorem 3.1.5.

One can also obtain directly an identification of the left-hand side with the de Rham group as follows. Consider the injection $\Omega^\bullet_M \hookrightarrow \mathcal{A}^\bullet_M \otimes \mathbf{C}$ of the holomorphic de Rham complex into the usual de Rham complex. On the level of cohomology sheaves this induces an isomorphism: for the sections one gets the constants, and the higher cohomology sheaves vanish (by the holomorphic and the usual Poincaré lemmas, respectively). Since quasi-isomorphisms induce isomorphisms between the corresponding hypercohomology groups this indeed establishes the existence of a natural isomorphism

$$\mathbf{H}^n(M, \Omega^\bullet_M) \cong H^n_{\mathrm{dR}}(M; \mathbf{C}).$$

Problem

6.2.1 If M is a complex manifold, consider the Dolbeault complex

$$(\mathcal{A}^{p,\bullet}, \bar{\partial}) = (\mathcal{A}^{p,0} \xrightarrow{\bar{\partial}} \mathcal{A}^{p,1} \xrightarrow{\bar{\partial}} \cdots).$$

Show that it is quasi-isomorphic to the complex Ω^p placed in degree p. Establish the Dolbeault isomorphism

$$H^q(M, \Omega^p) \cong H^{p,q}_{\bar{\partial}}(X) = H^q_{\mathrm{dR}}(\mathcal{A}^{p,\bullet}, \bar{\partial}).$$

6.3 The Hodge Filtration Revisited

For additional applications we need spectral sequences in hypercohomology associated with a filtered complex of sheaves on a topological space. This is completely analogous to what we did for filtered complexes of groups.

We start with a filtered complex of sheaves on a topological space X, say $(\mathcal{K}^\bullet, \mathrm{d})$ with decreasing filtration $F^\bullet$. This means that $F^p\mathcal{K}^\bullet$ is a subcomplex of $\mathcal{K}^\bullet$. A morphism between complexes of sheaves $f : (\mathcal{K}^\bullet, \mathrm{d}) \to (\mathcal{L}^\bullet, \mathrm{d}')$ is a *filtered quasi-isomorphism* if f preserves the filtration and $\mathrm{Gr}(f)$ is a quasi-isomorphism.

Examples 6.3.1 (i) The *canonical filtration*

$$F^{\mathrm{can}}_p(\mathcal{K}^\bullet) = (\mathcal{K}^0 \to \cdots \to \mathcal{K}^{p-1} \to \mathrm{Ker}\,\mathrm{d} \to 0 \to 0 \cdots).$$

This is an increasing filtration and its graded complex Gr^F_p is filtered quasi-isomorphic to the complex having zeroes everywhere except in degree p, where the p-th cohomology sheaf $H^p(\mathcal{K}^\bullet)$ is located.
(ii) The *trivial filtration*

$$F^p_{\mathrm{triv}}(\mathcal{K}^\bullet) = (0 \to 0 \to \cdots \to 0 \to \mathcal{K}^p \to \mathcal{K}^{p+1} \to \cdots),$$

i.e., everything in degree less than p is replaced by 0. This is a decreasing filtration.

Suppose next that we have a filtered complex $\mathcal{L}^\bullet$ that has the property that the sheaves $\mathrm{Gr}^p_F \mathcal{L}^\bullet$ have no higher cohomology groups and is filtered quasi-isomorphic to $\mathcal{K}^\bullet$. Then by the abstract de Rham theorem, its de Rham groups compute the hypercohomology of $\mathcal{L}^\bullet$. The filtration F induces a filtration on these de Rham groups and hence on the hypercohomology groups. The spectral sequence for this filtration has as its $E_1^{p,q}$-term $H^{p+q}_{dR}(X, \mathrm{Gr}^p_F \mathcal{L}^\bullet)$. Since $\mathrm{Gr}^p_F \mathcal{L}^\bullet$ has no higher cohomology, applying the abstract de Rham theorem once more shows that this E_1-term is equal to the hypercohomology group $\mathbf{H}^{p+q}(X, \mathrm{Gr}_F^p \mathcal{K}^\bullet)$. Combining everything, we find the spectral sequence

$$E_1^{p,q} = \mathbf{H}^{p+q}\left(X, \mathrm{Gr}_F^p \mathcal{K}^\bullet\right) \Longrightarrow \mathbf{H}^{p+q}(X, \mathcal{K}^\bullet).$$

We do this now for two complexes on a Kähler manifold M, namely $\mathcal{K}^\bullet = \Omega_M^\bullet$ and $\mathcal{L}^\bullet = \mathcal{A}_M^\bullet \otimes \mathbf{C}$. The canonical embedding

$$\Omega_M^\bullet \xrightarrow{j} \mathcal{A}_M^\bullet \otimes \mathbf{C} \tag{6.2}$$

is, as we have seen, a quasi-isomorphism. The decomposition into types of the sheaf complex $\mathcal{A}^\bullet \otimes \mathbf{C}$ gives the filtered complex

$$F^p(\mathcal{A}^\bullet \otimes \mathbf{C}) = \bigoplus\nolimits_{r \geq p} \mathcal{A}_M^{r, \bullet - r}.$$

This indeed induces the Hodge filtration on de Rham cohomology in the following way. The inclusion of F^p in the full complex induces a homomorphism on the level of hypercohomology,

$$\mathbf{H}^k(F^p(\mathcal{A}^\bullet \otimes \mathbf{C})) \to \mathbf{H}^k(\mathcal{A}^\bullet \otimes \mathbf{C}) = H_{\mathrm{dR}}^k(M; \mathbf{C}),$$

and its images are the classes representable by forms of type $(p, k-p) + (p+1, k-p-1) + \cdots + (k, 0)$. This is exactly the p-th level of the Hodge filtration, because we have

$$H^k(M, \mathbf{C}) = \bigoplus\nolimits_{p+q=k} H^{r,s}(M), \quad F^p H^k(M, \mathbf{C}) = \bigoplus\nolimits_{r \geq p} H^{r,k-r}(M).$$

The homomorphism j (see (6.2) above) becomes a filtered homomorphism provided we put the trivial filtration on the de Rham complex. Then $\mathrm{Gr}^p(j)$ gives the Dolbeault complex

$$0 \to \Omega_M^p \to \mathcal{A}_M^{p,0} \xrightarrow{\bar\partial} \to \mathcal{A}_M^{p,1} \xrightarrow{\bar\partial} \cdots.$$

By Dolbeault's lemma, this is exact, and so j induces a quasi-isomorphism on the level of graded complexes. So the E_1-term of the first spectral sequence, which computes the hypercohomology of the graded complex, is just the de Rham cohomology of the preceding complex, i.e., $'E_1^{p,q} = H^q(M, \Omega_M^p)$. The first spectral sequence of hypercohomology (viewed as coming from the trivial filtration) therefore reads

$$'E_1^{p,q} = H^q\left(M, \Omega_M^p\right) \Longrightarrow \mathbf{H}^{p+q}(\Omega_M^\bullet) = H_{\mathrm{dR}}^{p+q}(M; \mathbf{C}).$$

This is the *Hodge to de Rham spectral sequence*. Consider now the filtration on the abutment.

Definition 6.3.2 The *trivial filtration* on $H^k_{\mathrm{dR}}(M;\mathbf{C})$ is given by

$$\boxed{F^p_{\mathrm{triv}} H^k_{\mathrm{dR}}(M;\mathbf{C}) = \mathrm{Im}\left(\mathbf{H}^k\left(F^p_{\mathrm{triv}}\Omega^\bullet_M\right) \xrightarrow{\alpha_p} \mathbf{H}^k(\Omega^\bullet_M)\right),} \tag{6.3}$$

where the map α_p is induced by the inclusion map.

Of course, in view of the fact that the map (6.2) is a filtered quasi-isomorphism, this filtration is the filtration induced by the F-filtration (6.3), i.e., the Hodge filtration. Let us show that the spectral sequence in fact degenerates. To summarize, we have the following.

Proposition 6.3.3 *Let M be a compact Kähler manifold. The trivial filtration coincides with the Hodge filtration and the Hodge to de Rham spectral sequence degenerates at E_1.*

Proof By Proposition 3.1.3 we have a canonical isomorphism $H^{r,s}(M) \cong H^s(M,\Omega^r_M)$, and so

$$\sum_{p+q=k} \dim {}'E_1^{p,q} = \sum_{p+q=k} \dim H^{p,q} = \dim H^k(M;\mathbf{C}),$$
$$= \sum_{p+q=k} \dim E_\infty^{p,q},$$

which implies that the spectral sequence degenerates at E_1 (since E_{r+1} is a subquotient of E_r). □

Remark 6.3.4 The Čech–de Rham complex with its trivial filtration (see Theorem 3.1.5 and the discussion following it) also gives rise to a spectral sequence

$$E_1^{p,q} = H^q(M,\Omega^p) \Longrightarrow H^{p+q}_{\mathrm{dR}}(M;\mathbf{C}).$$

The trivial filtration on the abutment is the same as the trivial filtration (6.3) and hence, by the preceding proposition, is equal to the Hodge filtration. The same proof now shows that the spectral sequence above degenerates, as we claimed in a previous chapter (Proposition 3.1.6).

Problems

6.3.1 Show that the canonical filtration gives the spectral sequence

$$E_1^{p,q} = H^{2p+q}(X, H^{-p}(\mathcal{K}^\bullet, d)) \Longrightarrow \mathbf{H}^{p+q}(X,\mathcal{K}^\bullet),$$

which is nothing but the second spectral sequence in hypercohomology up to renumbering $E_r^{p,q} \to E_{r+1}^{2p+q,-p}$.

6.3.2 Show that the trivial filtration leads to

$$E_1^{p,q} = H^q(X, \mathcal{K}^p) \Longrightarrow \mathbf{H}^{p+q}(X, \mathcal{K}^\bullet),$$

which from E_2 on is just the first spectral sequence in hypercohomology.

6.3.3 Suppose we have a spectral sequence with $E_1^{p,q} = 0$ unless $p \leq 0, q \geq 0$ (a so-called fourth quadrant spectral sequence). Fix some natural number i. Suppose that $E_1^{p,q} = 0$ unless $q \leq i$ or $q = n$ and suppose that $E_\infty^{0,i} = 0$. Prove that d_n induces an isomorphism $E_2^{0,i} \cong E_2^{-n,i-n-1}$.

6.3.4 Let $X \subset P$ be a smooth hypersurface.

(a) Show that the inclusion $\kappa : (\Omega_P^\bullet(\log X), F_{\text{triv}}) \hookrightarrow (\Omega_P^\bullet(*X), F_{\text{pole}})$ is a filtered quasi-isomorphism. One should prove this by a calculation in suitable local coordinates. Use Lemma 6.2.4 to deduce that κ induces an isomorphism on the level of hypercohomology if $H^i(P, \Omega_P^j(kX)) = 0$ for all $i, k \geq 1, j \geq 0$.

(b) Show that κ is a quasi-isomorphism and hence induces an isomorphism in cohomology.

(c) Show that the restriction $\rho : \Omega_P^\bullet(*X) \longrightarrow \Omega_U^\bullet$ induces isomorphisms in cohomology. Hint: look at the proof of Theorem 3.2.4. It uses this isomorphism, but if we only know that the diagram commutes (this is proven there without assuming that ρ induces isomorphisms), the five-lemma can be used instead to show the isomorphism.

6.4 Derived Functors

There is a way to define the hypercohomology groups of *any* complex of sheaves $\mathcal{F}^\bullet$ on M without passing to Čech-cohomology. This is motivated by the definition (3.2) of the hypercohomology for the holomorphic de Rham complex. The constant sheaf $\mathbf{C}_M$ on a complex manifold M is the kernel of the derivation $\Omega_M^0 \to \Omega_M^1$ and so prolongs to give the holomorphic de Rham complex:

$$0 \to \mathbf{C}_M \to (\Omega_M^0 \xrightarrow{\text{d}} \Omega_M^2 \to \cdots).$$

This complex has no higher cohomology sheaves; this is equivalent to the holomorphic Poincaré lemma; by definition, it is an acyclic resolution of $\mathbf{C}_M$. Next, by the Dolbeault lemma, each of the sheaves Ω_M^p admit an acyclic resolution

$\mathscr{A}_M^{p,\bullet}$ and then one forms the double complex

$$\begin{array}{ccccc} A^{0,0}(M) & \xrightarrow{\bar\partial} & A^{0,1}(M) & \xrightarrow{\bar\partial} & \cdots \\ \downarrow \partial & & \downarrow \partial & & \\ A^{1,0}(M) & \xrightarrow{\bar\partial} & A^{1,1}(M) & \xrightarrow{\bar\partial} & \cdots \\ \downarrow \partial & & \downarrow \partial & & \\ \vdots & & \vdots & & \end{array}$$

with the global (p,q)-forms $A^{p,q}(M)$ as constituents.

Recall that for a double complex K, the associated single complex has been denoted sK. See page 99. Then, from what we have seen there, the following canonical identifications can be made:

$$H^p(M;\mathbf{C}) = \mathbf{H}^p(M,\Omega_M^\bullet) = H^p(sA^{\bullet,\bullet}(M)).$$

We could use the second equality as a definition for hypercohomology of the complex of sheaves $\Omega_M^\bullet$. This suggest how to do this for any complex of sheaves provided we have a way to associate a canonical acyclic resolution with any sheaf $\mathscr{F}$ on M. Godement (1964, II.4.3) has found such a resolution whose construction is quite simple; we refer to Problem 6.4.1. In what follows we shall denote this resolution simply by

$$C^\bullet(\mathscr{F}): \text{ the } \textit{Godement resolution}.$$

Fix a ring R. To define the hypercohomology groups for a complex of sheaves $\mathscr{F}^\bullet$ of R-modules on a topological space M, form for each p its Godement resolution $C^\bullet(\mathscr{F}^p)$. These, for varying p, by functoriality fit naturally in a double complex $\mathscr{K}^{\bullet,\bullet} = C^\bullet(\mathscr{F}^\bullet)$. Then put

$$\mathbf{H}^q(M,\mathscr{F}^\bullet) \stackrel{\text{def}}{=} H^q(s\Gamma(M,\mathscr{K}^{\bullet,\bullet})).$$

This definition is indeed equivalent to our previous definition and has the merit that it is clearly completely canonical and functorial. The special case where $\mathscr{F}^\bullet$ is one sheaf placed in degree 0 then provides an alternative definition for sheaf cohomology:

$$H^q(M,\mathscr{F}) = H^q(\Gamma(M,C^\bullet\mathscr{F})).$$

The above construction is a special case of the formation of a derived functor, namely the one for the functor "taking sections over M". This can be phrased differently. Define the total derived section functor by

$$R\Gamma(M,\mathscr{F}^\bullet) \stackrel{\text{def}}{=} s\Gamma(M,C^\bullet(\mathscr{F}^\bullet)).$$

It associates with a complex of sheaves of R-modules on M a complex of vector spaces in a canonical way. The cohomology of this complex then is its hypercohomology:

$$\mathbf{H}^q(M, \mathcal{F}^\bullet) = H^q R\Gamma(M, \mathcal{F}^\bullet).$$

Let us apply this to the theory of spectral sequences. So, starting with a complex $(\mathcal{F}^\bullet, \delta)$ of sheaves on M, let $(C^\bullet \mathcal{F}^p, \mathrm{d})$ be the Godement resolution complex. The double complex of sections can be denoted by

$$(\Gamma(M, C^\bullet \mathcal{F}^\bullet), \mathrm{d}\, \delta)\,.$$

For the first spectral sequence we use the differential δ first, which yields the term $C^p(M, H^q(\mathcal{F}^\bullet))$; here the cohomology is a *sheaf* on M. Hence, fixing q and calculating d-cohomology we get the usual cohomology groups of this sheaf and the first spectral sequence reads

$$'E_2^{p,q} = H^p(M, H^q(\mathcal{F}^\bullet)) \Longrightarrow \mathbf{H}^{p+q}(M, \mathcal{F}^\bullet). \tag{6.4}$$

We can use a similar procedure to define for instance the derived direct image functor associated with a continuous map $f : M \to N$. If $\mathcal{F}$ is a sheaf of R-modules on M, recall that the direct image $f_*\mathcal{F}$ is defined as follows. For any open subset $U \subset N$, consider $\Gamma(f^{-1}U, \mathcal{F})$. This defines a presheaf on N and the associated sheaf is the direct image sheaf $f_*\mathcal{F}$. The resulting functor f_* is left exact. Now set:

$$\begin{aligned} Rf_*(\mathcal{F}^\bullet) &\stackrel{\text{def}}{=} f_* s[C^\bullet \mathcal{F}^\bullet], \\ \mathbf{R}^q f_*(\mathcal{F}^\bullet) &\stackrel{\text{def}}{=} H^q\left(f_* s[C^\bullet \mathcal{F}^\bullet]\right). \end{aligned}$$

The first line defines the total direct image functor which associates in a canonical way with a complex of sheaves of R-modules on M a complex of sheaves of R-modules on N; the second line defines the q-th direct image of $\mathcal{F}^\bullet$ as a *sheaf* of R-modules on N. From this it follows also that $\mathbf{R}^q f_*\mathcal{F}^\bullet$ is the sheaf associated with the presheaf

$$U \mapsto \mathbf{H}^q(f^{-1}U, \mathcal{F}^\bullet). \tag{6.5}$$

Similarly, as for the global section functor, the direct image functor can be applied to a single sheaf instead of a complex of sheaves and this yields the usual higher direct images (see Eq. (6.5)):

$$R^q f_*\mathcal{F} = H^q\left(f_* C^\bullet \mathcal{F}\right).$$

Next, consider the spectral sequence (6.4) for the complex of sheaves $G^\bullet = f_*C^\bullet(\mathcal{F})$ on M. Then by the above, $H^q(G^\bullet) = R^q f_*\mathcal{F}$. On the other hand,

$\mathbf{H}^n(N, G^\bullet) = H^n(\Gamma(N, f_* C^\bullet(\mathcal{F})) = H^n(\Gamma(M, C^\bullet(\mathcal{F}))) = H^n(M, \mathcal{F})$ so (6.4) becomes:

$$E_2^{p,q} = H^p(N, R^q f_* \mathcal{F}) \implies H^{p+q}(M, \mathcal{F}), \tag{6.6}$$

which is the *Leray spectral sequence*. As to functoriality, we have

$$R(f \circ g)_* \mathcal{F}^\bullet \xrightarrow[\sim]{\text{qis}} Rf_*(Rg_* \mathcal{F}^\bullet).$$

Problems

6.4.1 Let $\mathcal{F}$ be a sheaf of abelian groups on a topological space M. A "discontinuous section" over an open subset U of M consists of a collection of germs $\{a_x \in \mathcal{F}_x \mid x \in U\}$. The set of all such sections is denoted $C^0\mathcal{F}(U)$. Varying U, we obtain a presheaf $C^0\mathcal{F}$ which is in fact a sheaf. By definition it comes equipped with an injective homomorphism $\mathcal{F} \hookrightarrow C^0\mathcal{F}$. One inductively defines

$$\begin{aligned} \mathfrak{Z}^0\mathcal{F} &= \mathcal{F}, \\ \mathfrak{Z}^p\mathcal{F} &\stackrel{\text{def}}{=} C^{p-1}\mathcal{F}/\mathfrak{Z}^{p-1}\mathcal{F}, \\ C^p\mathcal{F} &\stackrel{\text{def}}{=} C^0\left(C^{p-1}\mathcal{F}/\mathfrak{Z}^{p-1}\mathcal{F}\right). \end{aligned}$$

(a) Show that the sheaves $C^p\mathcal{F}$ are *flabby* or *flasque*, i.e., any section over an open set extends to the entire space X.

(b) Show that there are natural maps $\mathrm{d}\colon C^p\mathcal{F} \longrightarrow \mathfrak{Z}^{p+1}\mathcal{F} \hookrightarrow C^{p+1}\mathcal{F}$ which fit into a complex $C^\bullet(\mathcal{F})$ resolving $\mathcal{F}$, the *Godement resolution.*

(c) Show functoriality by showing that any morphism of sheaves $f : \mathcal{F} \longrightarrow \mathcal{G}$ induces a morphism of complexes $C^\bullet(f)$ between the respective Godement resolutions.

(d) Prove that $H^p(M, \mathcal{F}) = H^p(\Gamma(M, C^\bullet\mathcal{F}))$.

6.4.2 Show, given a continuous map $f : M \longrightarrow N$, that there are natural identifications $\mathbf{H}^p(M, \mathcal{F}^\bullet) = \mathbf{H}^p(N, Rf_*\mathcal{F}^\bullet)$.

6.4.3 Show, again using the Godement resolution, that for any short exact sequence $0 \longrightarrow \mathcal{F} \xrightarrow{f} \mathcal{G} \xrightarrow{g} \mathcal{H} \longrightarrow 0$ of sheaves on M of R-modules, there is a long exact sequence of sheaves of R-modules on N:

$$\cdots \longrightarrow \mathbf{R}^q f_*\mathcal{F} \xrightarrow{\mathbf{R}^q f_*} \mathbf{R}^q f_*\mathcal{G} \xrightarrow{\mathbf{R}^q g_*} \mathbf{R}^q f_*\mathcal{H} \xrightarrow{\delta} \mathbf{R}^{q+1} f_*\mathcal{F} \longrightarrow \cdots.$$

6.5 Algebraic Interpretation of the Gauss–Manin Connection

We want to give an alternative treatment of the Gauss–Manin connection in the situation where $f : X \to S$ is a smooth projective family. Recall (Remark 4.5.5) that we have characterized this connection as the unique flat connection of the local system E of cohomology groups $H^w(X_t; \mathbf{C})$.

The first step is to write E as a direct image sheaf, i.e., $E = R^w f_* \mathbf{C}_X$. This is justified by the fact that

$$(R^w f_* \mathbf{C}_X)_t = \lim_U H^w(f^{-1}U; \mathbf{C}) = H^w(X_t; \mathbf{C}),$$

where we take the limit over smaller and smaller open neighborhoods U of t; the rightmost equality holds, since for a small enough contractible such neighborhood this cohomology group is just $H^w(X_t; \mathbf{C})$.

Consider next the complex of sheaves of relative complex differential forms $\mathscr{A}^\bullet_{X/S,\mathbf{C}}$. If we restrict this complex to X_t we just get the de Rham complex and hence its hypercohomology gives $H^*(X_t; \mathbf{C})$. Consider the direct image functor $\mathbf{R}^w f_* \mathscr{A}^\bullet_{X/S,\mathbf{C}}$, a sheaf on S whose stalk at t is the k-th de Rham cohomology of X_t, i.e., $E = \mathbf{R}^w f_* \mathscr{A}^\bullet_{X/S,\mathbf{C}}$. Instead, we can work with holomorphic forms: just replace $\mathscr{A}^\bullet_{X/S}$ by $\Omega^\bullet_{X/S}$. Concluding, we have

$$R^w f_* \mathbf{C}_X = \mathbf{R}^w f_* \Omega^\bullet_{X/S}.$$

As a second step, we want to recover the variation of Hodge structure algebraically. Note that the trivial filtration $F_{\rm triv}$ on the complex $\Omega^\bullet_{X/S}$ induces in a canonical way a filtration on the direct image:

$$\mathscr{F}^p = F^p \mathbf{R}^w \Omega^\bullet_{X/S} \stackrel{\rm def}{=} \operatorname{Im}\left(\mathbf{R}^w f_* F^p_{\rm triv} \Omega^\bullet_{X/S} \to \mathbf{R}^w f_* \Omega^\bullet_{X/S}\right). \tag{6.7}$$

We claim that this gives the Hodge filtration bundles defining the usual variation of Hodge structure on E. To see this, we just look at the fiber at s and note by Proposition 6.3.3 that this indeed gives the Hodge filtration on $H^w(X_s; \mathbf{C})$.

Now we can give the description of the Gauss–Manin connection due to Katz and Oda (1968). The exact sequence

$$0 \to f^*\Omega^1_S \to \Omega^1_X \to \Omega^1_{X/S} \to 0 \tag{6.8}$$

defines the relative holomorphic cotangent bundle. Assume now for simplicity[1] that the base S is a curve. Then this sequence fits in an exact sequence of complexes

$$0 \to f^*\Omega^1_S \otimes \Omega^\bullet_X[-1] \to \Omega^\bullet_X \to \Omega^\bullet_{X/S} \to 0.$$

Next, apply the direct image functor. As we have seen (Problem 6.4.3), given a

[1] See also Problem 6.5.2.

short exact sequence of sheaves of R-modules, this gives a long exact sequence for the derived functors in which the connecting homomorphism

$$\mathbf{R}^w f_* \Omega^\bullet_{X/S} \to \mathbf{R}^w f_* (f^* \Omega^1_S \otimes \Omega^\bullet_X) = \Omega^1_S \otimes \mathbf{R}^w f_* \Omega^\bullet_X$$

induces

$$\nabla : E \to E \otimes \Omega^1_S,$$

where the last identification, $E = R^w f_* \mathbf{C}_X = \mathbf{R}^w f_* \Omega^\bullet_X$, uses the fact that $\Omega^\bullet_X$ is a resolution of $\mathbf{C}_X$. It can be shown that this is indeed the same as the Gauss–Manin connection. First, one shows that the connection is flat and second, one shows that the flat sections are exactly the locally constant cohomology classes. For details we refer to Bertin and Peters (2002, §2.C). It follows quite directly from this and (6.7) that Griffiths' transversality holds, i.e.

$$\nabla : \mathcal{F}^p \to \mathcal{F}^{p-1} \otimes \Omega^1_S. \tag{6.9}$$

As a consequence of this approach we also have a nice geometric description of the specific Hodge bundle $\mathcal{F}^w$ when w is the fiber dimension. Note that the fiber of this bundle at $s \in S$ is the Hodge component $H^{w,0}(X_s) = H^0(X_s, \Omega^w_{X_s})$. It is customary to call $\Omega^w_{X/S}$ the *relative canonical bundle* and denote its associated locally free sheaf, the *relative canonical sheaf*, as

$$\omega_{X/S} = \Omega^w_{X/S} = \mathcal{O}_X(K_X \otimes f^* K_S^{-1}), \quad w = \dim X_s, \tag{6.10}$$

where we leave the verification of the last equality to the reader (Problem 6.5.1). Then the Hodge bundle $\mathcal{F}^w$ can be given as

$$\mathcal{F}^w = f_* \omega_{X/S}.$$

Later we also need the logarithmic version. So, let $\Sigma \subset S$ be a normal crossing divisor and set $D = f^{-1}\Sigma$. We assume that D is a *reduced* normal crossing divisor. We say that $f : (X, D) \to (S, \Sigma)$ is a *normal crossing degeneration*. First we define the relative log-one forms by means of the exact sequence

$$f^* \Omega^1_S(\log \Sigma) \to \Omega^1_X(\log D) \to \Omega^1_{X/S}(\log D) \to 0$$

and then put $\Omega^p_{X/S}(\log D) = \bigwedge^p \Omega^1_{X/S}(\log D)$. The differential of this complex comes from the complex $\Omega^1_X(\log D)$. We leave it as an exercise (see Problem 6.5.4) to show that $\Omega^1_{X/S}(\log D)$ is locally free of rank equal to the fiber dimension and that $(\Omega^p_{X/S}(\log D), d)$ is a complex. Next, we consider

$$E = \mathbf{R}^w f_* \Omega^\bullet_{X/S}(\log D).$$

This comes with a logarithmic connection

$$\nabla : E \to E \otimes \Omega^1_S(\log \Sigma)$$

as before. The Hodge filtration on the restriction of $E \otimes \mathcal{O}_S$ to $S - \Sigma$ is induced by the trivial filtration on $\Omega^\bullet_{X/S}(\log D)$ and it then follows from the above description of the Gauss–Manin connection that the connection behaves with respect to the filtering bundles: just as in the case of a smooth family (see (6.9)):

$$\nabla : \mathcal{F}^p \to \mathcal{F}^{p-1} \otimes \Omega^1_S(\log \Sigma). \tag{6.11}$$

This is the logarithmic version of Griffiths' transversality.

The above description for the bundle $\mathcal{F}^w$, where $w = \dim X_s$, has a logarithmic version as well:

$$\mathcal{F}^w = f_*(\omega_{X/S}), \quad \omega_{X/S} \stackrel{\text{def}}{=} \Omega^w_{X/D}(\log D). \tag{6.12}$$

Problems

6.5.1 Prove the second equality in equation (6.10).

6.5.2 Let $0 \to \mathcal{G} \to \mathcal{H} \to \mathcal{F} \to 0$ be a short exact sequence of complex vector bundles on a complex manifold X. Then $\wedge^k \mathcal{H}$ has a natural filtration, the *Koszul filtration*, given by

$$F^p(\wedge^k \mathcal{H}) = \operatorname{Im}(\wedge^p \mathcal{H} \otimes \wedge^{k-p} \mathcal{G} \to \wedge^k \mathcal{H}).$$

Apply this to the sequence (6.8), make

$$0 \to \operatorname{Gr}^1_F \to F^1/F^2 \to \operatorname{Gr}^0_F \to 0$$

explicit, and write the connecting homomorphism in the resulting sequence for hypercohomology. Compare this with the case where $\dim S = 1$.

6.5.3 Recalling that $E = R^w f_* \mathbf{C}_X = \mathbf{R}^w f_* \Omega^\bullet_{X/S}$, and that $\mathcal{F}^\bullet$ is the filtration induced by the trivial filtration on $\Omega^\bullet_{X/S}$, show that

$$\operatorname{Gr}^p_{\mathcal{F}} E = R^{w-p} f_* \Omega^p_{X/S}.$$

Indication: the fiber of $\Omega^p_{X/S}$ at $s \in S$ equals $\Omega^p_{X_s}$; from this find the fiber of $R^{w-p} f_* \Omega^p_{X/S}$ at $s \in S$ and deduce the result.

6.5.4 Let $f : (X, D) \to (S, \Sigma)$ be a normal crossing degeneration. Show that local coordinates $(x_1, \ldots, x_n)$ on X centered at a point of D can be found so that

$$f(x_1, \ldots, x_n) = (y_1, \ldots, y_m), \quad y_j = \prod_{i \in \pi(j)} x_i,$$

where the $\pi(k)$ are distinct subsets of $\{1, \ldots, n\}$.

Bibliographical and Historical Remarks

Spectral sequences originate from algebraic topology and were introduced by Koszul and Serre in the 1950s. The book by Godement (1964) provides further details.

The treatment of Hodge theory through degeneration of spectral sequences is relatively new. It opened the way for an algebraic treatment of the Hodge decomposition (see Deligne and Illusie, 1987).

The use of derived functors was absent in a previous edition of the book. We use it here because it gives an efficient way to treat the the Leray spectral sequence as well as the Gauss–Manin connection in an algebraic manner. The latter has been suggested by Katz and Oda (1968).

7

Koszul Complexes and Some Applications

The concept of regularity forms a bridge between geometric properties of linear systems and the vanishing of cohomology groups of related sheaves. Koszul complexes intervene in the calculation of regularity. These are treated in the first two sections of this chapter, and regularity is treated in the third section. One of the first applications is a cheap proof of Bott's vanishing theorem.

In projective geometry, quotients of a polynomial ring by an ideal generated by homogeneous polynomials play a dominant role. Clearly, these quotients are graded. An important special case arises when we take the ideal of the partial derivatives of a homogeneous polynomial defining a smooth hypersurface in projective space. Because this quotient is related to the middle cohomology group of the corresponding smooth hypersurface it merits an algebraic study. One of the central results concerning such finite dimensional quotients is Macaulay's theorem which is used here repeatedly.

We then develop a few more sophisticated tools related to the multiplicative structure of the quotient rings. One of these is Donagi's symmetrizer lemma, which plays an important role in the variational approach to Torelli properties of hypersurfaces in projective space (see Chapter 8). Second, we have a Koszul-type short exact sequence (Theorem 7.3.5) which is applied to Noether–Lefschetz-type problems (Section 7.5).

7.1 The Basic Koszul Complexes

We first explain a basic multilinear algebra construction. Start with a complex vector space

$$V = \mathbf{C}e_0 \oplus \mathbf{C}e_2 \oplus \cdots \oplus \mathbf{C}e_r$$

of dimension r and the polynomial ring

$$\boxed{S = \mathbf{C}[X_0, \ldots, X_r]}$$

in $r+1$ variables. We consider S with its natural grading, i.e., the X_j all have degree 1 and S^q denotes the subspace of homogeneous polynomials in degree q. If we write $S(-k)$ this means that the homogeneous polynomials of degree m have been placed in degree $m-k$. This is useful if we want to preserve gradings. For instance, we shift S appropriately to define a degree-preserving contraction operator

$$\begin{array}{lll} \bigwedge^k V \otimes S(a) & \xrightarrow{i_k} & \bigwedge^{k-1} V \otimes S(a+1), \\ e_{i_1} \wedge \cdots \wedge e_{i_k} \otimes P & \longmapsto & \sum_{j=1}^k (-1)^{j-1} e_{i_1} \wedge \cdots \wedge \widehat{e_{i_j}} \cdots \wedge e_{i_k} \otimes X_{i_j} P. \end{array} \tag{7.1}$$

These fit in an exact complex, as follows.

Proposition 7.1.1 *The* Koszul complex

$$K^\bullet(X_0, \ldots, X_n) \stackrel{\text{def}}{=} \Big(0 \to \bigwedge^r V \otimes S(-r) \xrightarrow{i_r} \bigwedge^{r-1} V \otimes S(-r+1) \xrightarrow{i_{r-1}} \cdots$$
$$\cdots \to V \otimes S(-1) \to S\Big)$$

is exact and degree-preserving; in particular the short sequences

$$\bigwedge^{p+1} V \otimes S^{q-1} \xrightarrow{i_{p+1}} \bigwedge^p V \otimes S^q \xrightarrow{i_p} \bigwedge^{p-1} V \otimes S^{q+1}$$

are exact for $p \geq 1$.

Proof Consider the map $D_p : \bigwedge^p V \otimes S \to \bigwedge^{p+1} \otimes S$ defined by $D_p(e_{i_1} \wedge \cdots \wedge e_{i_p} \otimes P) = \sum_{k=0}^r e_k \wedge e_{i_1} \cdots \wedge e_{i_p} \otimes \partial P/\partial X_k$. This gives a homotopy operator for the Koszul complex in the sense that

$$i_{p+1} \circ D_p + D_{p-1} \circ i_p = p + q \qquad \text{on } S^q.$$

This implies in particular that the complex is exact. □

Of course we can use similar contraction maps when instead of S we have a graded S-module M. This time we do get a complex, but it need no longer be exact. It is called the *Koszul complex* for M,

$$0 \to \bigwedge^r V \otimes M(-r) \xrightarrow{i_r} \bigwedge^{r-1} V \otimes M(-r+1) \xrightarrow{i_{r-1}}$$
$$\cdots \to V \otimes M(-1) \to M \to 0,$$

and its cohomology groups are the *Koszul cohomology groups*:

$$\boxed{\mathcal{K}_{p,q}(M,V) \stackrel{\text{def}}{=} \frac{\operatorname{Ker}\left(i_p^{q-p} : \bigwedge^p V \otimes M^{q-p} \to \bigwedge^{p-1} V \otimes M^{q-p+1}\right)}{\operatorname{Im}\left(i_{p+1}^{q-p-1} : \bigwedge^{p+1} V \otimes M^{q-p-1} \to \bigwedge^p V \otimes M^{q-p}\right)}.}$$

The Koszul complex of Proposition 7.1.1 is a minimal free resolution of the S-module $\mathbf{C}$. The definition of the latter is as follows.

Definition 7.1.2 A *minimal free resolution* of an S-module M is a graded exact sequence

$$\cdots \to F^p \xrightarrow{A_p} \cdots \to F^1 \xrightarrow{A_1} F^0 \to M \to 0,$$

where $F^p = \bigoplus_q S(-q)^{n(p,q)}$ and where the homomorphism A_p is given by a matrix whose nonzero entries are homogeneous polynomials of positive degree.

Minimal free resolutions exist. A proof can be found in Matsumura (1989, Section 19). Although such resolutions are not unique, we have the following theorem.

Theorem 7.1.3 *The dimension of $\mathcal{K}_{p,q}(M,V)$ equals the number $n(p,q)$ of copies of $S(-q)$ at place p of a minimal free resolution of M.*

Proof The second Koszul sequence is obtained by tensoring a minimal free resolution of $\mathbf{C} = S/S^+$, $S^+ = \bigoplus_{q>0} S^q$ with the S-module M. By definition, its cohomology groups coincide with the torsion modules $(\mathrm{Tor}_p^S(\mathbf{C}, M))^q$. Now Tor is symmetric in its arguments (see Problem 7.1.2 below). So we could as well start with a free resolution of M, tensor it with $\mathbf{C}$, and then take cohomology groups. Because all nonzero entries of the matrices A_p have positive degree, in the tensored complex all maps (except for the last one) are 0; this proves the theorem. □

The numbers $n(p,q)$ found from a minimal free resolution determine an important invariant of our S-module.

Definition 7.1.4 The *regularity* of an S-module M is the number

$$m(M) \stackrel{\text{def}}{=} \max\{q - p \mid n(p,q) \neq 0\}.$$

To get a feeling for this, consider first the case of a 0-regular module. Because F^p can only consist of a number of copies of $S(-p)$ and because the entries of the matrices A_p are homogeneous polynomials of positive degree, only linear polynomials occur. Similarly, for an m-regular module, only homogeneous polynomials up to degrees $m+1$ can occur.

For future reference we want to record the following consequence of this definition.

Lemma 7.1.5 *If M or N is free, the regularity of $M \otimes N$ is equal to the sum of the regularities of M and N.*

The proof is left as an exercise (Problem 7.1.3 below).

Problems

7.1.1 Koszul sequences exist in a slightly more general framework. Let S be any commutative ring and M a finitely generated S-module. Given any ordered sequence of elements $x_0, \ldots, x_r$ in S, taking for V a free S-module with generators $e_0, \ldots, e_r$, one can now write a Koszul sequence as before:

$$K^{\bullet}(x_0, \ldots, x_r, M) \stackrel{\text{def}}{=} [0 \to \textstyle\bigwedge^n V \otimes M \stackrel{i_r}{\longrightarrow} \textstyle\bigwedge^{n-1} V \otimes M \to \cdots \to M \to 0].$$

In general this will not be exact. For special choices of the x_j, however, the sequence turns out to be almost exact. These are the so-called M-regular sequences: for all $j = 0, \ldots, r$, the element x_j is not a zero-divisor in $M/(x_1 M + \cdots + x_{j-1} M)$. For instance, if $M = S = \mathbf{C}[Z_0, \ldots, Z_r]$, the coordinates $Z_0, \ldots, Z_r$ form a regular sequence. Show inductively that $H^i(K^{\bullet}(x_0, \ldots, x_r, M)) = 0$ for $i < n$ and also that $H^n(K^{\bullet}(x_0, \ldots, x_r, M)) = M/(x_0 M + \cdots + x_r M)$.

7.1.2 Show that $\mathrm{Tor}(N, M) = \mathrm{Tor}(M, N)$ by considering the spectral sequence for the double complex $F_p \otimes G_q$, where $F_{\bullet}$ and $G_{\bullet}$ are minimal free resolutions for N and M, respectively.

7.1.3 Prove Lemma 7.1.5.

7.2 Koszul Complexes of Sheaves on Projective Space

As before, we let $S = \mathbf{C}[X_0, \ldots, X_r]$ be the ring of polynomials in $r+1$ variables and we let $\mathbf{P} = \mathbf{P}^r$ be a projective space with homogeneous coordinates $X_0, \ldots, X_r$. Any graded S-module M determines a sheaf on projective space (in the Zariski topology) which we denote by $\mathcal{S}(M)$ and which is defined as follows. Over a Zariski-open U, we let

$$\mathcal{S}(M)(U) = \left\{ \frac{m}{P} \;\middle|\; m \in M, P \in S, P(x) \neq 0, \forall x \in U, \deg m = \deg P \right\}.$$

Taking $M = S$ yields the structure sheaf, and $M = S(k)$ yields the locally free sheaf $\mathcal{O}_{\mathbf{P}}(k)$. Concerning these, we have Serre's fundamental result, as follows.

Theorem *The cohomology groups $H^i(\mathbf{P}^r, \mathcal{O}(k))$ vanish in the range $i = 1, \ldots, r-1$. Also $H^0(\mathbf{P}^r, \mathcal{O}(k)) = 0$ for $k < 0$ and $H^n(\mathbf{P}^r, \mathcal{O}(k)) = 0$ for $k > -r-1$. Moreover, there is a natural isomorphism $S^k \cong H^0(\mathbf{P}^r, \mathcal{O}(k))$ inducing a degree-preserving isomorphism*

$$S(k) \cong \textstyle\bigoplus_{d=0}^{\infty} H^0(\mathbf{P}^r, \mathcal{O}(d+k)).$$

Direct sums of copies of $\mathcal{O}_{\mathbf{P}}$ are called free $\mathcal{O}_{\mathbf{P}}$-modules. In general, $\mathcal{S}(M)$ is always a sheaf of $\mathcal{O}_{\mathbf{P}}$-modules, in short an $\mathcal{O}_{\mathbf{P}}$-module. If M is of finite rank, the resulting $\mathcal{O}_{\mathbf{P}}$-module is a coherent sheaf. Let us recall the definition.

Definition 7.2.1 A sheaf of $\mathcal{O}_{\mathbf{P}}$-modules $\mathcal{F}$ is *coherent* if it is locally (in the Zariski topology) a cokernel of a morphism between free $\mathcal{O}_{\mathbf{P}}$-modules.

The resulting functor $\mathcal{S}$ is an exact functor, i.e., it transforms exact sequences of S-modules into exact sequences of $\mathcal{O}_{\mathbf{P}}$-modules. As an application we have the exact Koszul sheaf sequence (recall that $V = \mathbf{C}e_0 \oplus \cdots \oplus \mathbf{C}e_r$):

Lemma 7.2.2 (i) *The Koszul sheaf sequence*

$$\mathcal{K}^\bullet(\mathbf{P}^r) = \Big(0 \to \bigwedge^{r+1} V \otimes \mathcal{O}_{\mathbf{P}^r}(-r-1) \xrightarrow{i_{r+1}} \cdots \\ \cdots \to V \otimes \mathcal{O}_{\mathbf{P}^r}(-1) \to \mathcal{O}_{\mathbf{P}^r} \to 0\Big) \quad (7.2)$$

is exact. The maps i_k are defined as before (formula (7.1)).

(ii) *The kernel of the map i_p is naturally isomorphic to $\Omega^p_{\mathbf{P}^r}$.*

For (ii), see Problem 7.2.1 below. It provides the familiar description of the (co-)tangent bundle of projective space by means of the *Euler sequence*.

An important corollary of these last remarks is Bott's vanishing theorem (see Bott, 1957).

Theorem 7.2.3 (Bott's vanishing theorem) *The cohomology groups $H^q(\mathbf{P}^r, \Omega^p_{\mathbf{P}^r}(k))$ vanish unless*

(i) $p = q$ *and* $k = 0$,
(ii) $q = 0$ *and* $k > p$, *or*
(iii) $q = r$ *and* $k < -r + p$.

Proof As we have seen, the sheaves of holomorphic p-forms on $\mathbf{P}^r$ are described in Lemma 7.2.2 as the kernel of a partial Koszul sequence. Tensor this with $\mathcal{O}(k)$. Because $H^i(\mathcal{O}(k)) = 0$ unless $i = 0$ and $k \geq 0$, or $i = r$ and $k \leq -r - 1$, the conditions of Lemma 6.2.4 are fulfilled and it follows that $H^q(\Omega^p_{\mathbf{P}^r}(k))$ is equal to

$$H^q_{\mathrm{dR}}\left(\bigwedge^{p-\bullet} V \otimes \mathcal{O}_{\mathbf{P}^r}(-p + k + \bullet)\right).$$

This cohomology group is equal to the cohomology group of the short exact sequence $\bigwedge^{p-q+1} V \otimes S^{-p+k+q-1} \to \bigwedge^{p-q} V \otimes S^{-p+k+q} \to \bigwedge^{p-q-1} V \otimes S^{-p+k+q+1}$ and this group is 0 either for trivial reasons or because of Proposition 7.1.1 unless $p = q$ and $k = 0$. □

Now, conversely, given a coherent $\mathcal{O}_{\mathbf{P}}$-module $\mathcal{F}$ on $\mathbf{P}$, one may form the associated graded S-module

$$M_t(\mathcal{F}) \overset{\text{def}}{=} \bigoplus\nolimits_{d \geq t} H^0(\mathbf{P}, \mathcal{F}(d)),$$

and one can show that for t large enough this is a finitely generated S-module whose associated sheaf is the sheaf with which we started. For instance, Serre's theorem, quoted above, implies that $M_0(\mathcal{O}(k)) = S(k)$ with associated sheaf $\mathcal{O}(k)$. Minimal free resolutions of $M_0(\mathcal{F})$ sheafify to what we call a *minimal free resolution* of the sheaf $\mathcal{F}$:

$$0 \to \mathcal{F}_s \to \cdots \to \mathcal{F}_1 \to \mathcal{F}_0 \to \mathcal{F} \to 0,$$

where $\mathcal{F}_p = \bigoplus_q \mathcal{O}(-q)^{n(p,q)}$ and where the maps are given by matrices whose nonzero entries are homogeneous polynomials of positive degree. The numbers $n(p, q)$ are the dimensions of the Koszul cohomology groups $\mathcal{K}_{p,q-p}(\mathcal{F}, \mathcal{O}_{\mathbf{P}^r}(1))$ where

$$\begin{aligned} &\mathcal{K}_{p,q}(\mathcal{F}, \mathcal{O}_{\mathbf{P}^r}(1)) \overset{\text{def}}{=} \text{the cohomology of} \\ &\quad \textstyle\bigwedge^{p+1} V \otimes H^0(\mathcal{F}(q-1)) \to \bigwedge^p V \otimes H^0(\mathcal{F}(q)) \to \bigwedge^{p-1} V \otimes H^0(\mathcal{F}(q+1)). \end{aligned}$$

Problems

7.2.1 We let $W \overset{\text{def}}{=} V^\vee$ and we identify $\bigwedge^p V$ with the trivial vector bundle Ω^p_W of p-forms on W. Let $z_0, \ldots, z_r$ be coordinates in W, i.e., a basis for V. We let $E = \sum_{i=0}^r z_i \partial/\partial z_i$ be the Euler vector field and $i_E : \Omega^p_W \otimes S^q V \to \Omega^{p-1}_W \otimes S^{q+1} V$ be the contraction with the Euler vector field. Show that there is an exact sequence

$$\begin{aligned} 0 \to \Omega^{r+1}_W \otimes S^\bullet V(-r-1) \xrightarrow{i_E} \Omega^r_W \otimes S^\bullet V(-r) \xrightarrow{i_E} \\ \cdots \to \Omega^1_W \otimes S^\bullet V(-1) \to \mathcal{O}_W \otimes S^\bullet V \to 0. \end{aligned}$$

7.2.2 (a) Let $\mathcal{L}$ be a holomorphic line bundle over a complex manifold M and let $s_1, \ldots, s_n$ be n independent sections of $\mathcal{L}$ spanning a subspace V of $H^0(M, \mathcal{L})$ that have as a common zero locus an analytic subvariety Z of M of codimension k. Such a subvariety is locally defined as the zero locus of a regular sequence of germs of holomorphic functions. See, for instance, Matsumura (1989, Theorem 17.4). Define a Koszul complex (starting in degree $-n$)

$$0 \to \textstyle\bigwedge^n (V \otimes \mathcal{L}^\vee) \to \bigwedge^{n-1} (V \otimes \mathcal{L}^\vee) \to \cdots \to V \otimes \mathcal{L}^\vee \to \mathcal{O}_M \to 0$$

in the obvious way. Show that it is exact in degrees $< -(n-k)$.

If $n = k$ the entire sequence is exact provided we replace $\mathcal{O}_M$ by $\mathcal{O}_M \to \mathcal{O}_Z$. Hint: use Problem 7.1.1.

(b) Let $\mathcal{E}$ be a holomorphic vector bundle of rank n over a complex manifold M and let s be a holomorphic section of $\mathcal{E}$ with zero locus an analytic subspace Z of M of codimension k. Define a Koszul complex (starting in degree $-n$)

$$0 \to \bigwedge^n \mathcal{E}^\vee \to \bigwedge^{n-1} \mathcal{E}^\vee \cdots \to \mathcal{E}^\vee \to \mathcal{O}_M \to 0,$$

where each map is obtained by contraction with s. Show that it is exact in degrees $< -(n-k)$. If $n = k$ and, moreover, s meets the zero section of $\mathcal{E}$ transversally (so that Z is a reduced set of points) the entire sequence is exact provided we replace $\mathcal{O}_M$ by $\mathcal{O}_M \to \mathcal{O}_Z$.

7.3 Castelnuovo's Regularity Theorem

In the first section we introduced the concept of regularity of a graded S-module M. It can be calculated by means of Koszul cohomology. This leads to a first concept of regularity for coherent sheaves on $\mathbf{P} = \mathbf{P}^r$. In the applications we compute regularity from Koszul sequences related to the geometry of the situation. We then apply a central theorem which asserts that the regularity is the minimal number m for which the groups $H^i(\mathbf{P}^r, \mathcal{F}(m-i))$ all vanish for $i > 0$. It is this vanishing result which turns out to be extremely useful. This result presupposes the existence of a minimal number m with the said properties. That this is true is the assertion of Castelnuovo's regularity theorem.

Let us now introduce the two basic concepts making use of the previously introduced concept of regularity for S-modules (Definition 7.1.4):

Definition 7.3.1 (i) The *regularity* of $\mathcal{F}$ is the regularity of the S-module

$$\bigoplus_q H^0(\mathbf{P}^r, \mathcal{F}(q)).$$

(ii) $\mathcal{F}$ is *m- regular* if $H^i(\mathbf{P}^r, \mathcal{F}(m-i)) = 0$ for all $i > 0$.

For use in subsequent sections we need an immediate consequence of this definition and Lemma 7.1.5.

Lemma 7.3.2 *If $\mathcal{F}$ and $\mathcal{G}$ are coherent sheaves on $\mathbf{P}^r$ and one of them is locally free, the regularity of $\mathcal{F} \otimes \mathcal{G}$ is the sum of the regularities of $\mathcal{F}$ and $\mathcal{G}$.*

We next investigate the relationship between these two notions. According to the outline above, we first state and prove the following.

Theorem 7.3.3 (Castelnuovo's regularity theorem) *If $\mathcal{F}$ is m-regular, it is also $(m+i)$-regular.*

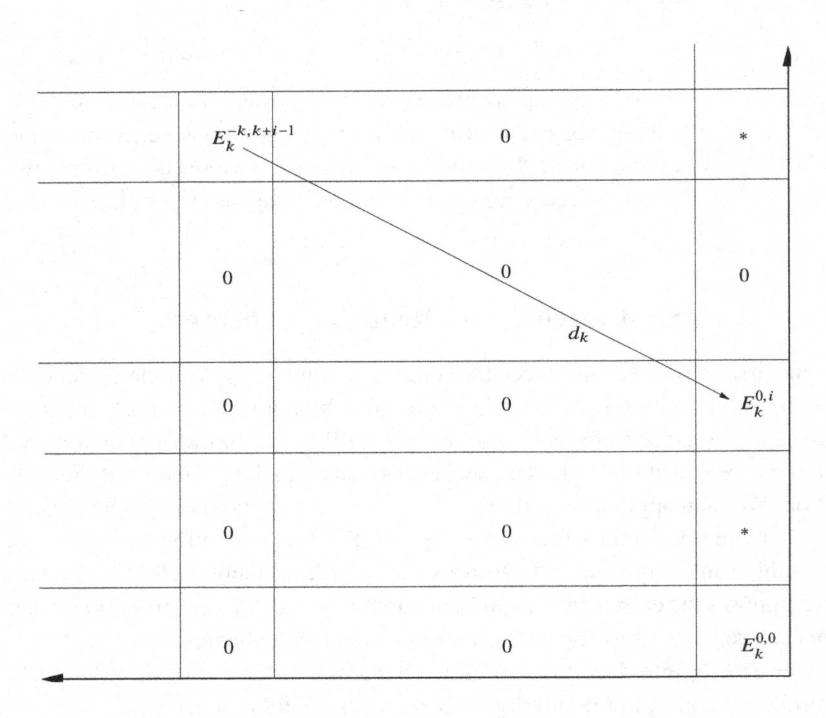

Figure 7.1 Spectral sequence $E_k^{p,q}$.

Proof Fix some integer $i > 0$. Consider the exact sequence obtained from the Koszul sequence, Eq. (7.2), upon tensoring it by $\mathcal{F}(m+1-i)$. It can be viewed as a complex which has nonzero terms only in degrees $-r-1, -r, \ldots, 0$. In degree 0 we have $\mathcal{F}(m+1-i)$. We want to apply Lemma 6.2.4 and so we consider the E_1-term of the spectral sequence associated with the Koszul sequence:

$$E_1^{p,q} = \bigwedge^{-p} V \otimes H^q(\mathbf{P}^r, \mathcal{F}_{\mathbf{P}^r}(m+1-i+p)).$$

By construction $E_1^{p,q} = 0$ unless $q \geq 0, p \leq 0$. This gives a so-called fourth quadrant spectral sequence. We must prove that $E_1^{0,i} = H^i(\mathbf{P}^r, \mathcal{F}(m+1-i)) = 0$

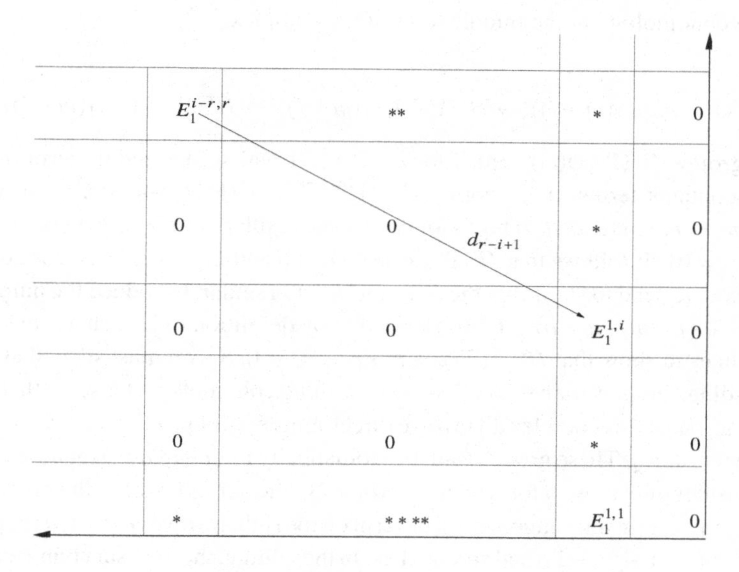

Figure 7.2 Spectral sequence $E_1^{p,q}$.

for $i > 0$. Since $\mathcal{F}$ is m-regular it follows that

$$E_1^{-k,k+i-1} = \bigwedge^k V \otimes H^{k+i-1}(\mathcal{F}(m-(k+i-1))) = 0.$$

Hence $d_k : E_k^{-k,k+i-1} \to E_k^{0,i}$ is the zero-map for $k \geq 1$. The maps going out of $E_k^{0,i}$ are trivially 0 and so $E_1^{0,i} = E_\infty^{0,i} = 0$. The last equality follows since by Lemma 6.2.4 the spectral sequence converges to 0; see Fig. 7.1. □

From this theorem it follows that a minimal m exists, for which $\mathcal{F}$ is m-regular.

Theorem 7.3.4 *The regularity of $\mathcal{F}$ is equal to* $\min\{m \mid \mathcal{F}$ *is m-regular*$\}$.

Proof We consider a minimal free resolution $\mathcal{F}_\bullet$ of $\mathcal{F}$ as a complex with terms in degrees ≤ 1, where $\mathcal{F}$ is placed in degree 1 and $\mathcal{F}_p$ in degree $-p$. Tensor this minimal free resolution by $\mathcal{O}_{\mathbf{P}^r}(m-i)$ and consider the spectral sequence of the resulting exact sequence (Lemma 6.2.4):

$$E_1^{p,q} = \begin{cases} H^q(\mathbf{P}^r, \mathcal{F}_{-p}(m-i)) & \text{for } p \leq 0, \\ H^q(\mathbf{P}^r, \mathcal{F}(m-i)) & \text{for } p = 1, \\ 0 & \text{otherwise.} \end{cases}$$

Because $H^k(\mathbf{P}^r, \mathcal{O}(l)) = 0$ for $0 < k < r$, the only nonzero differential coming into $E_1^{1,i}, i > 0$ is d_{r-i+1} (see Fig. 7.2) and so $H^i(\mathbf{P}^r, \mathcal{F}(m-i))$ is isomorphic to the cohomology at the middle term of the complex,

$$H^r(\mathbf{P}^r, \mathcal{F}_{r-i+1}(m-i)) \to H^r(\mathbf{P}^r, \mathcal{F}_{r-i}(m-i)) \to H^r(\mathbf{P}^r, \mathcal{F}_{r-i-1}(m-i)).$$

The groups $H^r(\mathbf{P}^r, \mathcal{O}(l))$ vanish for $l > -r-1$ and so the middle term itself only contains terms of the form $\bigoplus_q \bigoplus^{n(r-i,q)} H^r(\mathbf{P}^r, \mathcal{O}(-q+m-i))$ with $q > m-i+r$. This *a fortiori* holds for $m = \mu$, the regularity of $\mathcal{F}$: $\mu = \max\{q-p \mid n(p,q) \neq 0\}$. It follows that $H^i(\mathbf{P}^r, \mathcal{F}(\mu - i)) = 0$ and the sheaf $\mathcal{F}$ is μ-regular.

Next, we need to show that the $\mathcal{F}$ is not $(\mu-1)$-regular. Introduce the number $\pi = \max\{p \mid n(p, \mu+p) \neq 0\}$ (observe that by definition of μ such a π exists). It suffices to show that $H^{r-\pi}(\mathcal{F}(\mu+\pi-r-1)) \neq 0$. So we must show that the preceding complex for $r-i = \pi$ has nonvanishing cohomology. Observe that the three terms of this complex all involve direct sums of groups $H^r(\mathcal{O}(\mu+\pi-q-r-1))$ for certain q. These groups vanish automatically for $\mu+\pi-q > 0$ and we have to consider only those q for which $q \geq \mu+\pi$. By the definition of μ, the numbers $n(\pi+1, q)$ within this range are 0. The term on the right involves $n(\pi-1, q)$ copies of $H^r(\mathcal{O}(\mu+\pi-q-r-1))$ and so vanishes. In the middle, the only surviving terms in the range $q \geq \mu+\pi$ are those with $q = \mu+\pi$, and on the left we have the two possibilities $q = \mu+\pi$ and $q = \mu+\pi+1$. The definition of π forces $n(\pi+1, \mu+\pi+1)$ to be 0 and only the term with $q = \mu+\pi$ survives. So the sequence reduces to

$$\bigoplus^{n(\pi+1,\mu+\pi)} H^r(\mathcal{O}(-r-1)) \to \bigoplus^{n(\pi,\mu+\pi)} H^r(\mathcal{O}(-r-1)) \to 0.$$

The map on the left is induced by multiplication by a degree zero polynomial and so by minimality it must be 0. Consequently, the cohomology of the complex has dimension $n(\pi, \mu+\pi) \neq 0$. □

We can now prove the main technical result.

Theorem 7.3.5 *Let $W \subset H^0(\mathbf{P}^r, \mathcal{O}_{\mathbf{P}^r}(d))$ be a base point free linear subspace of codimension c. The cohomology at the middle term of the complex*

$$\textstyle\bigwedge^{p+1} W \otimes H^0(\mathbf{P}^r, \mathcal{O}(k-d)) \to \bigwedge^{p} W \otimes H^0(\mathbf{P}^r, \mathcal{O}(k)) \to \bigwedge^{p-1} W \otimes H^0(\mathbf{P}^r, \mathcal{O}(k+d))$$

vanishes if

$$k \geq d + p + c.$$

Proof Consider the graded ring

$$S^\bullet W = \bigoplus_{k=0}^{\infty} S^k W.$$

Because $W \subset H^0(\mathbf{P}^r, \mathcal{O}(d))$, the vector space

$$N \stackrel{\text{def}}{=} \bigoplus_q H^0(\mathbf{P}^n, \mathcal{O}_{\mathbf{P}^r}(k + qd))$$

can be considered as a graded $S^\bullet W$-module. Its Koszul sequence, after twisting it suitably, reads

$$\cdots \to \bigwedge^{p+1} W \otimes \mathcal{O}(k-d) \xrightarrow{j_{p+1}} \bigwedge^{p} W \otimes \mathcal{O}(k) \xrightarrow{j_p} \bigwedge^{p-1} W \otimes \mathcal{O}(k+d) \to \cdots,$$

and its p-th de Rham group (a Koszul group of the module N) is exactly the group we want to investigate. We study it by means of the spectral sequence of the exact complex, considered in Lemma 6.2.4. The exact complex we have in mind is obtained from the above complex by replacing it on the left by the kernel of the map j_{p+1}. Introduce the vector bundle $\mathcal{E}$ by means of the exact sequence

$$0 \to \mathcal{E} \to W \otimes \mathcal{O}_{\mathbf{P}^r} \to \mathcal{O}_{\mathbf{P}^r}(d) \to 0,$$

where the surjectivity on the right just means that W has no base points. Then the desired exact complex starts with the vector bundle

$$\operatorname{Ker} j_{p+1} = \bigwedge^{p+1} \mathcal{E}(k-d),$$

and from the spectral sequence for this exact complex (Lemma 6.2.4) we derive that the desired cohomology group is isomorphic to

$$H^1\left(\mathbf{P}^r, \bigwedge^{p+1} \mathcal{E}(k-d)\right).$$

Let us first deal with the case of $W = H^0(\mathbf{P}^r, \mathcal{O}(d))$. The defining sequence for $\mathcal{E}$ immediately gives that

$$H^i(\mathcal{E}(1-i)) = 0 \text{ for all } i > 0$$

and thus $\mathcal{E}$ is 1-regular. Using Lemma 7.3.2 it follows that $\mathcal{E}^{\otimes(p+1)}$ and hence the subbundle $\bigwedge^{p+1} \mathcal{E}$ is $(p+1)$-regular. We conclude that

$$H^1\left(\bigwedge^{p+1} \mathcal{E}(k-d)\right) = 0 \text{ provided } k - d \geq p,$$

which proves the theorem for $c = 0$.

To prove the theorem for arbitrary c we consider a complete flag of subspaces

$$W = W_c \subset W_{c-1} \subset \cdots \subset W_1 \subset W_0 = H^0(\mathcal{O}(d)).$$

The subspace W_i defines vector bundles for $\mathscr{E}_i$ like W for $\mathscr{E} = \mathscr{E}_c$.

We just proved that $\bigwedge^p \mathscr{E}_0$ is p-regular for any $p \geq 1$. We want to show inductively that $\bigwedge^p \mathscr{E}_i$ is $(p+i)$-regular for all $p, i \geq 0$. This suffices for the proof in the general case: for $i = c$ and $q = 1$ we get

$$H^1\left(\bigwedge^{p+1} \mathscr{E}(p+c)\right) = 0 \text{ for all } p > 0$$

and so

$$H^1\left(\mathbf{P}^r, \bigwedge^{p+1} \mathscr{E}(k-d)\right) = 0 \text{ for } k-d \geq p+c \text{ and for all } p > 0.$$

Let us now prove the desired regularity. For the induction step, we observe that by definition there are exact sequences of vector bundles on $\mathbf{P}^r$:

$$0 \to \mathscr{E}_i \to \mathscr{E}_{i-1} \to \mathcal{O} \to 0$$

leading to

$$0 \to \bigwedge^{p+1} \mathscr{E}_i \to \bigwedge^{p+1} \mathscr{E}_{i-1} \to \bigwedge^p \mathscr{E}_i \to 0.$$

Now, tensor this by $\mathcal{O}(p+i-q)$ for any $q > 0$ and consider the long exact sequence in cohomology:

$$\cdots \to H^q\left(\bigwedge^{p+1} \mathscr{E}_{i-1}(p+i-q)\right) \to H^q\left(\bigwedge^p \mathscr{E}_i(p+i-q)\right)$$
$$\to H^{q+1}\left(\bigwedge^{p+1} \mathscr{E}_i(p+i-q)\right) \to \cdots.$$

Using ascending induction on i, the term on the left vanishes, the case $i = 1$ reducing to the p-regularity for $\mathscr{E}_0$. Next, descending induction on p gives that the term on the right vanishes, the case $p > \operatorname{rank} \mathscr{E}_i$ being automatic. So we find that the middle term vanishes for all i, i.e., $\bigwedge^p \mathscr{E}_i$ is $(p+i)$-regular. □

Corollary 7.3.6 *Let $S = \mathbf{C}[z_0, \ldots, z_r]$ and consider W as a subspace of S^d.*

(i) *The multiplication map*

$$W \otimes S^k \longrightarrow S^{d+k}$$

is surjective provided $k \geq c$.

(ii) *Consider the complex*

$$\bigwedge^2 W \otimes S^{k-d} \longrightarrow W \otimes S^k \longrightarrow S^{d+k},$$

where the first map is defined by sending $F_1 \wedge F_2 \otimes G$ to $F_1 \otimes F_2 G - F_2 \otimes F_1 G$ and the second map is multiplication. This complex is exact provided $k > d + c$.

Problems

7.3.1 Show that the structure sheaf of $\mathbf{P}^r$ has regularity 0. Determine the regularity of the line bundles $\mathcal{O}(k)$ on $\mathbf{P}^r$.

7.3.2 Calculate the regularity of $\Omega^p(k)$ on $\mathbf{P}^r$.

7.4 Macaulay's Theorem and Donagi's Symmetrizer Lemma

In geometric applications finite dimensional quotients of the polynomial ring $S = \mathbf{C}[z_0, \ldots, z_r]$ often occur. For instance, if F is a homogeneous polynomial, its *Jacobian ideal* $\mathfrak{j}_F$ is the ideal generated by the partial derivatives, and if F defines a smooth hypersurface the quotient $S/\mathfrak{j}_F$ is finite dimensional. Below, we see its geometric meaning as the middle cohomology of the corresponding hypersurface. This explains the interest of this sort of ring.

If R is any finite-dimensional quotient of S, the subspace of highest degree is called the *socle* of R. These turn out to be one-dimensional for a special type of quotient – *Gorenstein rings* – which one obtains by taking the quotient of S by an ideal generated by a regular sequence $\{F_0, \ldots, F_r\}$ of $r+1$ polynomials. We recall that this means that for $i = 1, \ldots, r$ the polynomial F_i is not a zero-divisor in the quotient $S/(F_0, \ldots, F_{i-1})$. The standard example is the sequence of the partial derivatives of a smooth hypersurface.

Macaulay's theorem describes multiplication as a self-dual operation, a common feature of all Gorenstein rings. We give a geometric proof of this theorem using a technique similar to the one employed in the proof of Bott's vanishing theorem.

Theorem 7.4.1 (Macaulay's theorem) *Let*

$$S = \mathbf{C}[z_0, \ldots, z_r]$$

and suppose there is given an ordered $(r+1)$*-tuple of homogeneous polynomials* $F_j \in S$ *of degree* d_j, $j = 0, \ldots, r$ *that form an S-regular sequence. Let* $\mathfrak{i}$ *be the ideal in S they generate and put*

$$R \stackrel{\text{def}}{=} S/\mathfrak{i}.$$

Finally, set $\rho \stackrel{\text{def}}{=} \sum_{i=0}^{r} d_j - (r+1)$.

(i) *The grading on S given by degree induces a grading on R and* $R^k = 0$ *for* $k \geq \rho + 1$, *whereas* $\dim_{\mathbf{C}} R^\rho = 1$.

(ii) *Multiplication in S induces a perfect pairing*

$$R^k \otimes R^{\rho-k} \longrightarrow R^\rho \cong \mathbf{C}.$$

Proof Introduce the rank $(r+1)$-bundle $\mathscr{E} \stackrel{\text{def}}{=} \mathcal{O}(d_0) \oplus \cdots \oplus \mathcal{O}(d_r)$ and the section s of $\mathscr{E}$ defined by the $(r+1)$-tuple $(F_0, \ldots, F_r)$. Because this $(r+1)$-tuple is a regular sequence, it follows that the section s vanishes nowhere, i.e., has codimension $r+1$ in $\mathbf{P}^r$. The Koszul sequence we get by repeated contraction with s then gives an exact complex which we let start in degree 0 (compare Problem 7.2.2 where it would start in degree $-r-1$):

$$K^\bullet \stackrel{\text{def}}{=} \left(0 \to \bigwedge^{r+1} \mathscr{E}^\vee \to \bigwedge^r \mathscr{E}^\vee \to \cdots \to \mathscr{E}^\vee \to \mathcal{O} \to 0\right).$$

Because $H^j(\mathbf{P}^r, \mathcal{O}(l)) = 0$ for $1 \le j \le r-1$, one can apply Lemma 6.2.4 to the complex $K^\bullet \otimes \mathcal{O}(k)$. So for each $k \in \mathbf{Z}_{\ge 0}$ we have an isomorphism:

$$\begin{aligned}\operatorname{Ker}\left(H^r(\mathcal{O}(-r-1+k-\rho)) \xrightarrow{\alpha_k} \textstyle\bigoplus_j H^r(\mathcal{O}(-r-1+k+d_j-\rho))\right) \\ \xrightarrow{\cong} \operatorname{Coker}\left(\textstyle\bigoplus_j H^0(\mathcal{O}(k-d_j)) \xrightarrow{\beta_k} H^0(\mathcal{O}(k))\right).\end{aligned}$$

Both maps α_k and β_k are induced by multiplication with $(F_0, \ldots, F_r)$. If we apply Serre duality to the kernel of α_k we see that α_k is the transpose of $\beta_{\rho-k}$ and we obtain an isomorphism between the dual of the cokernel of the map

$$\textstyle\bigoplus_j S^{\rho-k-d_j} \xrightarrow{(F_0,\ldots,F_r)} S^{\rho-k}$$

and the cokernel of the map

$$\textstyle\bigoplus_j S^{k-d_j} \xrightarrow{(F_0,\ldots,F_r)} S^k.$$

The first cokernel is $R^{\rho-k}$ and the second cokernel is R^k. First, this (or the previous isomorphism) proves part (i) and it also shows that R^k and $R^{\rho-k}$ are naturally dual to each other. What is not clear, however, is that this duality is given by multiplication of polynomials. But below we show (see Lemma 7.4.2) that the multiplication map $R^1 \times R^k \to R^{k+1}$ induces an injective map $R^k \to \operatorname{Hom}(R^k, R^{k+1})$ whenever $k < \rho$, from which (ii) then follows. □

Lemma 7.4.2 *If $k < \rho$, $G \in S^k$ and $Gz_j \in \mathfrak{i}$, $j = 0, \ldots, r$, then $G \in \mathfrak{i}$.*

Proof We use induction on the number r of variables, the result being obvious for $r = 0$. Suppose $r > 0$. After a linear change of variables $(z_0, F_1, \ldots, F_r)$ will also be a regular sequence. Write $Gz_j = \sum_k H_{jk} F_k$ with H_{jk} of degree $k+1-d_j$. Then $0 = G(z_0 z_j - z_j z_0) \equiv (z_0 H_{j0} - z_j H_{00}) F_0 \bmod (F_1, \ldots, F_r)$. As F_0 is not a zero-divisor modulo $(F_1, \ldots, F_r)$, we conclude $z_0 H_{j0} - z_j H_{00} \in (F_1, \ldots, F_r)$ and so $z_j H_{00} \in (z_0, F_1, \ldots, F_r)$. Working modulo z_0, identifying $S/(z_0)$ with $\mathbf{C}[z_1, \ldots, z_r]$, and observing that $\deg H_{00} = k+1-d_0 < \sum_{j=1}^r d_j - r$, we can apply induction and conclude that $H_{00} \in (z_0, F_1, \ldots, F_r)$. Write now $H_{00} = H_0 z_0 +$

polynomial in $z_1, \dots z_r$. Because $Gz_0 \equiv H_{00}F_0 \bmod (F_1, \dots, F_r)$ it follows that $z_0(G - H_0F_0)$ belongs to the ideal $(F_1, \dots, F_r)$. Again, because $z_0, F_1, \dots, F_r$ is a regular sequence, we get $G - H_0F_0 \in (F_1, \dots, F_r)$, so $G \in \mathfrak{i}$. □

Corollary 7.4.3 *Multiplication in S induces symmetric bilinear maps*

$$R^i \times R^j \longrightarrow R^{i+j}$$

which are nondegenerate in each factor as long as $i, j \geq 0$ *and* $i + j \leq \rho$.

We shall employ Macaulay's theorem on various occasions. The first application is the proof of Donagi's symmetrizer lemma (see Donagi, 1983).

Theorem 7.4.4 (Donagi's symmetrizer lemma) *Let*

$$\mathrm{Sym}(R^a, R^b) = \{h \in \mathrm{Hom}(R^a, R^b) \mid$$
$$h(P_1)P_2 = h(P_2)P_1 \text{ in } R^{a+b} \text{ for all } P_1, P_2 \in S^a\}$$

and set $\rho \stackrel{\text{def}}{=} \sum_{i=0}^{r} d_j - (r + 1)$.
The natural map

$$R^{b-a} \longrightarrow \mathrm{Sym}(R^a, R^b)$$

is an isomorphism provided $a > 0$, $\rho > a + b$, *and* $\rho \geq b + \max_j d_j$.

Proof The result can be rephrased by saying that the Koszul complex

$$0 \to R^{b-a} \to (S^a)^\vee \otimes R^a \to \textstyle\bigwedge^2 (S^a)^\vee \otimes R^{a+b}$$

is exact. By Macaulay's theorem 7.4.1, this complex is dual to the Koszul complex

$$\textstyle\bigwedge^2 S^a \otimes R^{\rho-a-b} \to S^a \otimes R^{\rho-b} \to R^{\rho-b+a} \to 0.$$

The surjectivity at the right is a consequence of the surjectivity of multiplication of polynomials (here we use that $\rho - b > a > 0$). Exactness at the middle term follows from the following auxiliary result with $d = a, W = S^a$, and $k = \rho - b$. □

Proposition 7.4.5 *Suppose there is given an ordered* $(r + 1)$*-tuple of homogeneous polynomials* $F_j \in S = \mathbf{C}[z_0, \dots, z_r]$ *of degree* d_j, $j = 0, \dots, r$ *which forms an S-regular sequence. Let* $\mathfrak{i}$ *be the ideal in S they generate and as before, let* $R = S/\mathfrak{i}$. *Let* $W \subset H^0(\mathcal{O}_{\mathbf{P}^r}(d)) = S^d$ *be a linear system of codimension c. Consider the complex*

$$\textstyle\bigwedge^2 W \otimes R^{k-d} \to W \otimes R^k \to R^{d+k}.$$

If W is base-point free this complex is exact for those k for which $k > d + c$ *and* $k \geq c + \max d_j$.

Proof Consider the commutative diagram

$$\begin{array}{ccccccc} \bigwedge^2 W \otimes S^{k-d} & \longrightarrow & W \otimes S^k & \longrightarrow & S^{d+k} & & \\ \downarrow & & \downarrow & & \downarrow & & \\ \bigwedge^2 W \otimes R^{k-d} & \longrightarrow & W \otimes R^k & \longrightarrow & R^{d+k} & \longrightarrow & 0. \end{array}$$

Corollary 7.3.6 implies that the upper sequence is exact and the multiplication map $W \otimes S^c \longrightarrow S^{c+d}$ is surjective. Since $\mathfrak{i}^k = S^c\, \mathfrak{i}^{k-c}$ (by assumption $\mathfrak{i}$ has generators in degree $\leq k - c$), this last fact implies that the map $W \otimes \mathfrak{i}^k = W \otimes S^c \mathfrak{i}^{k-c} \longrightarrow \mathfrak{i}^{d+k}$ is surjective. A simple diagram chase then proves exactness at the middle of the upper sequence. □

Problem

7.4.1 Give an elementary proof of Donagi's symmetrizer lemma (Corollary 7.4.4) by giving an alternative proof of the first half of Proposition 7.4.5 in the case $W = S^k$ as follows (see Donagi and Green, 1984). We have to show exactness of

$$\bigwedge^2 S^d \otimes R^{k-d} \xrightarrow{\alpha} S^d \otimes R^k \xrightarrow{\beta} R^{d+k},$$

with $\alpha(F_1 \wedge F_2 \otimes G) = F_1 \otimes F_2 G - F_2 \otimes F_1 G$, and β is the multiplication map. Elements of $\operatorname{Im} \alpha$ are spanned by

$$z^I \otimes z^{I'+L} - z^{I'} \otimes z^{I+L}, |I| = |I'| = a, |L| = b - a.$$

Use this to show that, first,

$$z^I \otimes z^J \equiv z^{I'} \otimes z^{J'} \mod \operatorname{Im} \alpha$$

if $I + J = I' + J'$ and $J' + I \geq 0$, and second, that

$$z^I \otimes z^{K-I} \equiv z^{I'} \otimes z^{K-I'} \mod \operatorname{Im} \alpha$$

if there exists a J with $|J| = b, J \leq K$, and $J - I' \geq 0, J - I \geq 0$. Use the hypothesis $b > a$ to show that

$$z^I \otimes z^{K-I} \equiv z^{I'} \otimes z^{K-I'} \mod \operatorname{Im} \alpha$$

if I and I' differ by a permutation of two elements – and more generally whenever $K \geq I$ and $K \geq I'$. Using standard multi-index notation, let

$$\sum_{|I|=a, |J|=b} c_{IJ} z^I \otimes z^J \in \operatorname{Ker} \beta, c_{IJ} \in \mathbf{C}.$$

Then one has

$$\sum_{I+J=K} c_{IJ} = 0 \text{ for all } K \text{ with } |K| = a + b.$$

Use this together with the previous equation to show that

$$\sum_{|I|=a,|J|=b} c_{IJ} z^I \otimes z^J \text{ with } I + J = K \text{ fixed}$$

belongs to $\operatorname{Im} \alpha$.

7.5 Applications: The Noether–Lefschetz Theorems

Recall the description of the infinitesimal variation of Hodge structure for hypersurfaces as given in Section 5.4. The setting is as follows. We fix a projective manifold P of dimension $n + 1$ and a very ample line bundle L on it which satisfies the vanishing condition

$$H^i(P, \Omega^j_P(L^{\otimes k})) = 0, \quad \forall i, j \geq 1, k \geq 0. \tag{$*$}$$

The sections of L defining smooth hypersurfaces give a tautological family over $\mathbf{P}H^0(P, L)$ whose tangent space at $X = \{s = 0\}$ can be identified with

$$T = H^0(P, \mathcal{O}(X))/\mathbf{C} \cdot (s).$$

Now Proposition 5.4.1 implies that the product map

$$T \otimes H^{p,n-p}_{\text{var}}(X) \longrightarrow H^{p-1,n-p+1}_{\text{var}}(X)$$

is surjective provided the multiplication map

$$H^0(\mathcal{O}_P(X)) \otimes H^0(\Omega^{n+1}_P((n-p+1)X)) \to H^0(\Omega^{n+1}_P((n-p+2)X)) \qquad (**)_p$$

is surjective as well. This fact will play a crucial role in the proof of the infinitesimal Noether theorem, as we now show. But we first recall that a cohomology class α is called *infinitesimally fixed* in our situation, if the Kodaira–Spencer image of α vanishes. Equivalently, this means that $\alpha \cup t = 0$ for all $t \in T$.

Theorem 7.5.1 (Infinitesimal Noether–Lefschetz theorem) *Assume that L is a very ample line bundle on a smooth $(n + 1)$-dimensional projective manifold P. Suppose that L satisfies the vanishing conditions $(*)$ as well as the condition that the products $(**)_p$ are surjective for $p = 1, \ldots, n$. Then infinitesimally fixed classes are fixed. In other words, $H^n_T(X; \mathbf{Q}) = H^n_{\text{fixed}}(X; \mathbf{Q})$. Here T is the tangent space to the tautological family defined by sections of the line* *bundle L at the point corresponding to X, and* $H^n_T(X)$ *is the subspace of classes annihilated by the action of the Gauss–Manin connection in the directions of T.*

Proof We shall show that for any p the intersection of $H^{p,n-p}_T(X)$ and $H^n_{\mathrm{var}}(X)$ consists of $\{0\}$. Assume that $\alpha \in H^{p,n-p}_{\mathrm{var}}(X)$ is infinitesimally fixed. As explained just before the assertion of the proposition, the assumption that the product maps are surjective implies that any variable class β of type $(n-p,p)$ is a linear combination of elements of the form $t \cup \gamma$ with $t \in T$ and γ a variable class of type $(n-p+1, p-1)$. With $\langle -, - \rangle$ the intersection form on $H^n(X)$ we have $\langle \alpha, \beta \rangle = \langle \alpha, t \cup \gamma \rangle = -\langle t \cup \alpha, \gamma \rangle = 0$ since α is infinitesimally fixed. Hence we conclude that α is orthogonal to $H^{n-p,p}_{\mathrm{var}}(X)$. By Lemma 3.2.2, variable and fixed cohomology are orthogonal with respect to one another and so α is orthogonal to all of $H^{n-p,p}(X)$. However, because α was supposed to be of type $(p, n-p)$, this implies that $\alpha = 0$. □

Corollary 7.5.2 *Let* $n = 2m$. *As before, assume that L is sufficiently ample so that the conditions of the previous theorem hold. For s outside of a countable union of proper subvarieties of S, the parameter space of the tautological family, we have for the corresponding hypersurface* $X = X_s$:

$$H^{m,m}(X; \mathbf{Q}) = \mathrm{Im}\{H^{m,m}(P; \mathbf{Q}) \to H^{2m}(X; \mathbf{Q})\}.$$

Proof We work on the universal cover $\tilde{S}$ of S. On it the local system formed by the n-dimensional primitive cohomology groups of the corresponding hypersurfaces is constant and for every constant cohomology class α we can look at the analytic variety $\tilde{S}_\alpha = \{s \in \tilde{S} \mid \alpha \in H^{m,m}(X_s)\}$ (see Problem 7.5.2). If $\tilde{S}_\alpha = \tilde{S}$ the class α is constant and hence infinitesimally fixed. The theorem implies that α is the restriction of a class in P. If this is true for all classes we are done. Otherwise, we leave out the proper subvarieties which are the images in S of the varieties $\tilde{S}_\alpha$ for which $\tilde{S}_\alpha \neq \tilde{S}$. □

Example 7.5.3 Take $P = \mathbf{P}^3$. The required vanishing in cohomology is a special case of Bott's vanishing theorem (7.2.3) and the demand on surjectivity of the product mapping

$$H^0(\mathcal{O}(d)) \otimes H^0(\mathcal{O}(p+1)d-4) \to H^0(\mathcal{O}(p+2)d-4)$$

for $p = 0, 1$ follows as soon as $d \geq 4$. Recall that the Lefschetz's theorem on $(1,1)$-classes says that a class in $H^2(X)$ is of type $(1,1)$ precisely when it is the first Chern class of a divisor (see Griffiths and Harris, 1978). Such classes form the Picard group

$$\mathrm{Pic}(X) = H^1(\mathcal{O}^*_X) \cong H^{1,1}(X) \cap H^2(X),$$

because $\mathrm{Pic}^0(X) = 0$ and $H^2(X)$ has no torsion. The smooth degree-d surfaces form a Zariski-open subset U_d of the projective space of all such surfaces. We obtain the classical Noether–Lefschetz theorem, as follows.

Theorem 7.5.4 (Noether–Lefschetz theorem) *Assume that $d \geq 4$. There exists a locus NL_d, which is a countable union of proper closed subvarieties of U_d, such that the Picard group of any surface X in the complement of this set is generated by $\mathcal{O}_X(1)$,* i.e., *each curve on X is a complete intersection of X with another surface.*

The set

$$\boxed{NL_d = \{X \in U_d \mid X \text{ smooth and } \mathrm{Pic}(X) \text{ not generated by } \mathcal{O}_X(1)\}}$$

is called the *Noether–Lefschetz locus*.

The Bundle of Principal Parts and the Second Noether–Lefschetz Theorem

Before we can state the second main theorem we have to explain a useful but not widely known tool from deformation theory.

Definition 7.5.5 Let $\Delta \subset X \times X$ be the diagonal with ideal sheaf $\mathcal{I}_\Delta$. Let p_1 and p_2 be the projections of $X \times X$ onto the first and second factor. Let L be any line bundle on X.

(i) The *bundle of principal parts* of L is the bundle

$$P^1(L) \stackrel{\text{def}}{=} (p_1)_* \left(p_2^* L \otimes \mathcal{O}_{X\times X}/\mathcal{I}_\Delta^2\right).$$

(ii) Dually we have the bundle

$$\Sigma_L \stackrel{\text{def}}{=} \mathrm{Hom}(P^1(L), L) = P^1(L)^\vee \otimes L.$$

The exact sequence

$$0 \to \mathcal{I}_\Delta/\mathcal{I}_\Delta^2 \to \mathcal{O}_{X\times X}/\mathcal{I}_\Delta^2 \to \mathcal{O}_\Delta \to 0$$

yields an exact sequence of $\mathcal{O}_X$-modules

$$0 \to \Omega_X^1 \otimes L \to P^1(L) \to L \to 0. \tag{7.3}$$

From this one can see that $P^1(L)$ is locally free. In fact, we first write $P^1(L)$ differentiably as a direct sum

$$P^1(L) = L \oplus \Omega_X^1 \otimes L.$$

Then the $\mathcal{O}_X$-module structure is given by

$$f(s, \eta) = (fs, f\beta + df \otimes s).$$

Now choose a coordinate cover $\{U_i\}, i \in I$ of X which is also a trivializing cover for L. Let $\phi_i : L|U_i \to \mathcal{O}_{U_i}$ be the corresponding trivialization. This also gives a trivialization $\phi_i : \Omega^1_X \otimes L|U_i \to \Omega^1_X|U_i$. Using this, we define a map

$$j^1 : \mathcal{O}_X(L) \to P^1(L),$$

which over U_i is given by $s \mapsto (s, \phi_i^{-1}(d\phi_i s))$. In other words, we send s to its 1-jet. What remains to be verified is that j^1 is $\mathcal{O}_X$-linear. This we leave to the reader. We note that this does *not* give a holomorphic splitting of the sequence, because the splitting we started out with is only a smooth splitting. In fact, the extension class of the sequence (7.3) is known to be $2\pi\mathrm{i}$ times the first Chern class of L. See Problem 7.5.1 below.

The bundle Σ_L fits into the exact sequence

$$0 \to \mathcal{O}_X \to \Sigma_L \to \Theta_X \to 0. \tag{7.4}$$

This gives rise to

$$\cdots \to H^1(\mathcal{O}_X) \to H^1(\Sigma_L) \to H^1(\Theta_X) \to \cdots .$$

This relates to infinitesimal deformations of smooth pairs (X, D) where D is a divisor on X with normal bundle L. As explained for instance in Horikawa (1976, Appendix) these deformations are in one-to-one correspondence with

$$\mathbf{H}^1(\Theta_X \to i_*\nu_{D/X}) = H^1(\Theta_X(-\log D)),$$

where $i : D \hookrightarrow X$ is the embedding. The relevant diagram is now

$$\begin{array}{ccccccccc} & & & & \Theta_X(-\log D) & = & \Theta_X(-\log D) & & \\ & & & & \downarrow & & \downarrow & & \\ 0 & \to & \mathcal{O}_X & \to & \Sigma_L & \to & \Theta_X & \to & 0 \\ & & \| & & \downarrow & & \downarrow & & \\ 0 & \to & \mathcal{O}_X & \to & \mathcal{O}_X(D) & \to & L & \to & 0. \end{array}$$

The cohomology diagram then shows the following. Interpreting $H^0(L)$ as the deformations coming from the sections of L, leaving X fixed, and also assuming $H^1(L) = 0$, the space $H^1(\Sigma_L)$ is the quotient of the full deformation space by this subspace.

We now have made enough preparations to state and prove the second main result of this section.

Theorem 7.5.6 (Explicit Noether–Lefschetz theorem) *For $d \geq 3$, every component of the Noether–Lefschetz locus NL_d has codimension $\geq d - 3$.*

Proof A portion of the long exact sequence associated with (7.4) reads as follows:

$$H^1(\Sigma_L) \xrightarrow{\alpha} H^1(T_X) \xrightarrow{\beta} H^2(\mathcal{O}_X).$$

By the preceding description of the extension class the coboundary map β is multiplication with $c_1(L)$. Now specialize to X a surface in three-space of degree d. To identify the first space we form

$$\widetilde{NL_d} = \{(X, L) \mid X \in NL_d, L \in \mathrm{Pic}(X)\}.$$

Here we consider $\widetilde{NL_d}$ as a subset of the local system over U_d whose stalk over $[X]$ is $H^2(X)$. The group $G = \mathrm{SL}(3, \mathbf{C})$ acts with at most finite stabilizers, and $H^1(\Sigma_L)$ is the Zariski tangent space at $([X], L)$ of the quotient by G of $\widetilde{NL_d}$. The middle space is of course equal to the Zariski tangent space at $[X]$ to the moduli space, and α is the map induced by projection (forgetful map)

$$\begin{aligned} p_2 : \widetilde{NL_d}/G &\to NL_d/G, \\ ([X], L) &\mapsto [X]. \end{aligned}$$

Let

$$Z_{[X],L} \stackrel{\text{def}}{=} \text{the irreducible component of } \widetilde{NL_d}/G \text{ containing } ([X], L).$$

Then

$$T = \operatorname{Ker}\beta = \operatorname{Im}\alpha = \text{Zariski tangent space at } [X] \text{ to } p_2(Z_{[X],L}).$$

Assume now that $[X]$ belongs to the Noether–Lefschetz locus and choose $L \in \mathrm{Pic}(X)$ primitive. Now $T \otimes c_1(L)$ maps to 0 in $H^2(X, \mathcal{O}_X)$. Dually, $T \otimes H^0(\Omega^2_X)$ maps to the annihilator of $c_1(L)$ in $H^{1,1}_{\mathrm{prim}}(X)$. In particular, this map is not surjective. We have identified the cup product $H^1(T_X) \otimes H^{2,0}(X) \to H^{1,1}_{\mathrm{prim}}(X)$ as the multiplication map $R^d \otimes R^{d-4} \to R^{2d-4}$ (see Lemma 5.4.3). Let $\tilde{T}$ be the preimage of T in S^d. Then the multiplication map

$$\tilde{T} \otimes S^{d-4} \to S^{2d-4}$$

is not surjective. Because X is nonsingular, the degree-d piece of the Jacobian ideal j_F is base-point free ($X = \{F = 0\}$) and so is $\tilde{T}$. If $d - 4 \geq \operatorname{codim}\tilde{T}$ this map would have to be surjective by Corollary 7.3.6 (i). This contradiction proves the theorem. □

Problems

7.5.1 Using that $c_1(L)$ is the class given by the 1-cocycle $\{\frac{1}{2\pi i} \operatorname{dlog} \phi_{ij}\}$ (here the ϕ_{ij} are the transition functions for L) prove that the extension class of the sequence (7.3) is equal to $2\pi i c_1(L)$ (see Atiyah, 1957, Proposition 12).

7.5.2 Show that the subset $\tilde{S}_\alpha$ of points in $\tilde{S}$ where a rational class $\alpha \in H^{2m}(X)$ is of type (m, m) is indeed an analytic subvariety of $\tilde{S}$.

7.5.3 Show that the bundle of principal parts for $\mathcal{O}_{\mathbf{P}^n}(1)$ is the direct sum of $(n+1)$ copies of this bundle. Hint: compare the exact sequence (7.4) with the Euler sequence.

Bibliographical and Historical Remarks

We first go through the sections.

Section 7.1: Koszul complexes in connection with Torelli problems were used for the first time in Lieberman et al. (1977). Later Green made abundant use of these in his study of the Torelli problems (see Green, 1984a,b).

A basic reference for Section 7.2 is the fundamental article by Serre (1955).

Sections 7.3, 7.4: Mumford (1970) introduced the concept of Castelnuovo-regularity. See also Mumford (1963, Lecture 14) where other proofs of the results in Section 7.3 can be found. The notion of regularity still plays an important role today. Macaulay's theorem is stated in Macaulay (1916). Griffiths (1969) used it for the first time in algebraic geometry to compute the cohomology of hypersurfaces in projective space. Since then it has turned out to be one of the basic tools from commutative algebra used to understand the geometry of period maps, especially for hypersurfaces.

Section 7.5: The infinitesimal Noether–Lefschetz theorem was proved in Carlson et al. (1983). A modern proof of the classical Noether–Lefschetz theorem can be found in Hartshorne (1975). The proof given here of the explicit version is due to Green (1988). The discussion of principal parts is based on Atiyah (1957).

We finish mentioning some relevant developments in the 1990s:

- In Ciliberto et al. (1988) it is shown that the Noether–Lefschetz locus NL_d contains infinitely many components for $d \geq 4$ and that the union of these components is Zariski-dense in U_d. For a simpler proof see Kim (1991).

- Concerning special components, Green (1989a) and Voisin (1988a) have shown that for $d \geq 5$, a component V of NL_d has codimension $d - 3$ if and only if V is the base locus of the family of surfaces of degree d containing a line. Under the same degree restrictions, each component V of NL_d not of codimension $d - 3$ has codimension $\geq 2d - 7$, and equality holds precisely for the base of a family of degree-d hypersurfaces containing a conic (see Voisin, 1989b, 1990).
- Harris conjectured that there are at most finitely many special components for fixed d, but Voisin (1991) found counterexamples in degrees $d = 4s$, s large.
- For three-dimensional hypersurfaces in $\mathbf{P}^4$ one can pose similar problems for the one-cycles. It is not true, however, that a curve on a generic hypersurface of large enough degree is cut out by a surface as conjectured in Griffiths and Harris (1985). Voisin (1989a) found counterexamples for large-degree curves. For very small degrees there is an affirmative result by Wu (1990a). The Noether–Lefschetz problem for threefolds with an ample line bundle on it has been treated in Joshi (1995).
- Generalizations to even-dimensional hypersurfaces in $\mathbf{P}^n$ were found by Voisin and Green (see Green, 1988).
- One can also pose the Noether–Lefschetz problem for dependency loci of sections in vector bundles and of vector bundle homomorphisms. This problem has been studied in Ein (1985) and Spandaw (1994, 2002).
- For other publications concerning the Noether–Lefschetz problem see Amerik (1998); Ciliberto and Lopez (1991); Cox (1990); Debarre and Laszlo (1990); Lopez (1991); Spandaw (1992, 1996); Szabo (1996); and Xu (1994).
- The locus where a rational cohomology class of degree $2p$ stays of type (p, p) not only is *complex analytic*, but even *algebraic*. This theorem, proven in Cattani et al. (1995) is strong evidence for the Hodge conjecture.

8

Torelli Theorems

The classical Torelli theorem states that a compact Riemann surface of genus g, $g \geq 1$, up to isomorphism is determined by its Jacobian (as a polarized abelian variety), or, equivalently, by its polarized Hodge structure on the first cohomology. We discussed this in Section 3.5.

For higher-dimensional varieties this leads to the *global Torelli problem* for a given class of compact Kähler manifolds of dimension n: suppose that for two manifolds X and X' from this set you have an isomorphism $\varphi : H^n(X, \mathbf{Z}) \to H^n(X', \mathbf{Z})$ between the integral cohomology groups that preserves the cup product pairing and respects the Hodge decomposition. Is it true that X and X' are isomorphic? Is it even true that φ is induced by an isomorphism of manifolds (strong Torelli problem)?

An easier problem is the *infinitesimal Torelli problem* for a given manifold where we ask whether the period map for the Kuranishi family for the given manifold is locally an immersion. It turns out that this problem can be reduced to a cohomological question. The latter can be settled in several important cases, either directly or by means of Koszul complexes. This we do in the first Section. In Section 8.3 we prove Donagi's theorem that the Hodge structure of a degree-d hypersurface in projective $(n + 1)$-space generically determines the equation of the hypersurface for most pairs (n, d). This is the prototype of a *generic Torelli theorem*.

In Section 8.4 we discuss a reformulation of this theorem in the framework of moduli spaces.

8.1 Infinitesimal Torelli Theorems

Recall the criterion for infinitesimal Torelli stated in Section 5.6: infinitesimal Torelli holds for the Kuranishi family for X with respect to primitive

w-cohomology whenever the composite map

$$T_{S,0} \to H^1(\Theta_X) \xrightarrow{\text{cup product}} \bigoplus_{p+q=w} \mathrm{Hom}\big(H^q(\Omega_X^p), H^{q+1}(\Omega_X^{p-1})\big)$$

is injective, where S is the Kuranishi space.

We start by discussing the case of hypersurfaces in projective space.

Example 8.1.1 Let X be a smooth degree-d hypersurface in $\mathbf{P}^{n+1}$. We computed the derivative of the period map in Lemma 5.4.3. We see that infinitesimal Torelli holds if the product

$$R_F^d \times R_F^{t(p)} \to R_F^{t(p-1)}$$

is nondegenerate in the first factor for some p between 1 and n. Macaulay's theorem (or rather its Corollary 7.4.3) tells us that this product map is nondegenerate in each factor as long as $0 \le t(p-1) \le \rho = (n+2)(d-2)$. Working this out we find that the only exceptions are quadric and cubic curves and cubic surfaces.

We now describe a more sophisticated approach to the infinitesimal Torelli problem which has given the most far-reaching results to date. Start with a general set-up. Let X be a compact complex manifold, L a line bundle on X, V a subspace of $H^0(X, L)$ of dimension k, and $\mathcal{F}$ any locally free sheaf on X. We want to see how Koszul complexes can be used to measure the failure of the injectivity of the cup product mapping

$$H^q(X, \mathcal{F}) \to \mathrm{Hom}(V, H^q(X, L \otimes \mathcal{F})).$$

This is clearly related to the cohomological formulation of the infinitesimal Torelli problem.

We start by forming the Koszul complex

$$K^\bullet = \left(0 \to \textstyle\bigwedge^k (V \otimes L^\vee) \xrightarrow{i_k} \bigwedge^{k-1}(V \otimes L^\vee) \to \cdots \to V \otimes L^\vee \xrightarrow{i} \mathcal{O}_X \to 0\right),$$

where $i(v \otimes f) = f(v)$ and $i_p(\alpha_1 \wedge \cdots \wedge \alpha_p) = \sum_j (-1)^{j-1} i(\alpha_j)\alpha_1 \wedge \cdots \wedge \widehat{\alpha_j} \wedge \cdots \wedge \alpha_p$. This Koszul complex can be written as $K^\bullet = L^\bullet \otimes \bigwedge^k (V^\vee \otimes L)$, with

$$L^\bullet = \left(0 \to \mathcal{O}_X \xrightarrow{\mathfrak{d}} V^\vee \otimes L \to \cdots \to \textstyle\bigwedge^k (V^\vee \otimes L) \to 0\right).$$

We need a description of $\mathfrak{d}$. The complexes $K^\bullet$ and $L^\bullet$ are dual to each other via the natural pairing

$$\textstyle\bigwedge^i V \otimes L^\vee \bigotimes \bigwedge^{k-i} V^\vee \otimes L \longrightarrow \mathcal{O}.$$

In particular, the derivative $\mathfrak{d}$ in $L^\bullet$, like the last derivative in $K^\bullet$, comes from the evaluation map $H^0(X, L) \otimes \mathcal{O}_X \to L$. We now tensor $L^\bullet$ with any locally free sheaf $\mathcal{F}$ and consider the complex $H^q(X, L^\bullet \otimes \mathcal{F})$. Its cohomology group in degree 0 is nothing but the kernel of the cup product map $H^q(X, \mathcal{F}) \to \mathrm{Hom}(V, H^q(X, L \otimes \mathcal{F}))$, as follows from the description of the derivative d_0. This cohomology group is also the term $'E_2^{0,q}$ of the first spectral sequence for the hypercohomology of the complex $L^\bullet \otimes \mathcal{F}$, as introduced in Section 6.2. We first show how a consideration of the second spectral sequence leads to vanishing of the hypercohomology.

Lemma 8.1.2 *If no component of the base locus of $V \subset H^0(X, L)$ has codimension $\leq \ell$, the hypercohomology groups $\mathbf{H}^q(L^\bullet \otimes \mathcal{F})$ vanish for $q < \ell$ and $\mathcal{F}$ any locally free sheaf on X.*

Proof The base locus having codimension $\leq \ell$, the Koszul complex $K^\bullet$ is exact up to degree ℓ (see Problem 7.2.2). Hence, the complex $L^\bullet \otimes \mathcal{F}$ is also exact up to this degree. The second spectral sequence

$$''E_2^{p,q} = H^q(X, H^p(L^\bullet \mathcal{F})) \Longrightarrow \mathbf{H}^{p+q}(X, L^\bullet \otimes \mathcal{F})$$

then shows that the hypercohomology vanishes for $p + q < \ell$. □

Now we look at the first spectral sequence, which is more relevant for our problem, as we indicated above.

$$'E_2^{p,q} = H^p(H^q(X, L^\bullet \otimes \mathcal{F}), \mathrm{d}) \Longrightarrow \mathbf{H}^{p+q}(X, L^\bullet \otimes \mathcal{F}).$$

We have an exact sequence (Problem 7.1.1 in Chapter 7)

$$0 \to E_2^{1,0} \to \mathbf{H}^1 \to E_2^{0,1} \to E_2^{2,0} \to \mathbf{H}^2,$$

which translates into

$$0 \to H^1_{\mathrm{dR}}(L^\bullet \otimes \mathcal{F}) \xrightarrow{h_{\mathrm{dR}}} \mathbf{H}^1(L^\bullet \otimes \mathcal{F}) \to \mathrm{Ker}[H^1(\mathcal{F}) \to \mathrm{Hom}(V, H^1(\mathcal{F} \otimes L))]$$
$$\to H^2_{\mathrm{dR}}(L^\bullet \otimes \mathcal{F}) \xrightarrow{h_{\mathrm{dR}}} \mathbf{H}^2(L^\bullet \otimes \mathcal{F}).$$

From Lemma 8.1.2 we conclude the following.

Proposition 8.1.3 *If the base locus of $V \subset H^0(X, L)$ does not contain a divisor, there is an injection*

$$\mathrm{Ker}\left[H^1(\mathcal{F}) \to \mathrm{Hom}(V, H^1(\mathcal{F} \otimes L))\right] \hookrightarrow H^2_{\mathrm{dR}}(L^\bullet \otimes \mathcal{F}).$$

This is an isomorphism if the base locus of V has no components of codimension ≥ 2.

We derive the following corollary.

Corollary 8.1.4 *Suppose that the base locus of $V \subset H^0(X, L)$ does not contain a divisor. Then, the cup product map*

$$H^1(\mathcal{F}) \to \mathrm{Hom}(V, H^1(\mathcal{F} \otimes L))$$

is injective whenever the second de Rham group of the complex $L^\bullet \otimes \mathcal{F}$ vanishes.

We apply this to the situation where the canonical bundle is of the form $\Omega_X^n = L^{\otimes \ell}$, $\ell > 0$ and we arrive at the following result from Lieberman et al. (1977).

Theorem 8.1.5 *Let X be a compact complex manifold with $K_X = L^\ell$, $\ell > 0$. Assume that no component of the base locus of $H^0(X, L)$ has codimension 1. Then the cup product map*

$$H^1(\Theta_X) \to \mathrm{Hom}\big(H^0(X, \Omega_X^n), H^1(\Omega_X^{n-1})\big)$$

is injective whenever the de Rham groups $H^2_{\mathrm{dR}}(L^\bullet \otimes (\Theta_X \otimes L^i))$ vanish for $i = 0, \ldots, \ell - 1$.

Proof Suppose that $H^2_{\mathrm{dR}}(L^\bullet \otimes (\Theta_X \otimes L^i)) = 0$, $i = 0, \ldots, \ell - 1$. Assume that there exists a nonzero $\alpha \in H^1(\Theta_X)$ such that $\sigma \cdot \alpha = 0$ in $H^1(\Omega_X^{n-1})$ for all $\sigma \in H^0(X, L^\ell)$. Pick now j minimal $(1 \le j \le \ell)$ such that multiplication with $H^1(L^j)$ kills α in $H^1(\Theta_X \otimes L^j)$ and fix $\tau \in H^0(X, L^{j-1})$ with $\alpha \cdot \tau \neq 0$. Such an element τ defines a nonzero element in $\mathrm{Ker}\big(H^1(\Theta_X \otimes L^{j-1}) \to \mathrm{Hom}(H^0(L), H^1(\Theta_X \otimes L^j))\big)$ and so $H^2_{\mathrm{dR}}(L^\bullet \otimes (\Theta_X \otimes L^{j-1})) \neq 0$, contradicting our assumption. □

Corollary 8.1.6 *Let X be a compact projective manifold of dimension n with $K_X = L^\ell, \ell > 0$ and such that no component of the base locus of $H^0(K_X)$ is a divisor. If $H^0(\Omega_X^{n-1} \otimes L) = 0$, the infinitesimal Torelli theorem holds with respect to the n-forms.*

Proof We have $H^0(\Theta_X \otimes L^{\ell+1}) = H^0(\Omega_X^{n-1} \otimes L) = 0$. Fix a nonzero section of L. It defines an injection $\Theta_X \otimes L^j \to \Theta_X \otimes L^{\ell+1}$ for $j < \ell$; hence $H^0(\Theta_X \otimes L^{\ell+1}) = 0$ implies the vanishing of all the groups $H^0(\Theta_X \otimes L^j) = 0$ with $j < \ell + 1$; hence $H^2_{\mathrm{dR}}(L^\bullet \otimes [\Theta_X \otimes L^i]) = 0$ for $i < \ell$. □

As a first application, we look at complete intersections in projective space.

Theorem 8.1.7 *Let X be a smooth complete intersection of dimension n in $\mathbf{P}^{n+r}$. Assume that $k = 1$ or $k < 0$. If $p + q \le n - 1$ and $(q, p) \neq (0, 0)$, we have $H^q(\Omega_X^p(k)) = 0$. Consequently, by Corollary 8.1.6, when, moreover, K_X is ample, infinitesimal Torelli holds.*

Proof Induction on r. For $r = 0$ this follows from Bott's vanishing theorem 7.2.3. Assume that the assertion is true for X and we wish to show it for

$Y := X \cap H$, where H is a hypersurface of degree $d \geq 2$ meeting X transversally. From the exact sequence defining the conormal bundle $\mathcal{O}_Y(-d)$ of Y in X,

$$0 \to \mathcal{O}_Y(-d) \to \Omega^1_X|Y \to \Omega^1_Y \to 0,$$

we derive the exact sequences

$$0 \to \Omega^{q-1}_X(-d+k) \to \Omega^q_X(k)|Y \to \Omega^q_Y(k) \to 0$$

and we see that $H^q(\Omega^p_Y(k)) = 0$ provided $H^{q+i}(\Omega^{p-i}_X(k-di)|Y) = 0$ for $i = 0, 1$. The exact sequences

$$0 \to \Omega^{p-i}_X(k-d(i+1)) \to \Omega^{p-i}_X(k-di) \to \Omega^{p-i}_X(k-di)|Y \to 0$$

show that for any integer i with $i = 0, \ldots, p$ this group vanishes if

$$H^{q+i}\big(\Omega^{p-i}_X(k-di)\big) = 0 = H^{q+i+1}\big(\Omega^{p-i}_X(k-d(i+1))\big). \qquad (*)$$

One immediately checks that we are in the correct range of indices in order to apply the induction hypothesis and we conclude that $(*)$ holds. □

When K_X is no longer ample, infinitesimal Torelli is almost always true: by Flenner (1986, Theorem 2.1) infinitesimal Torelli holds for complete intersections X of dimension n and degree d in $\mathbf{P}^{n+r}$ *except in the following cases* where it fails:

(i) $r = 1, d = 3, n = 2$ (cubic surfaces);
(ii) X an even-dimensional complete intersection of two quadrics.

Similar techniques can be applied to complete intersections in weighted projective space $\mathbf{P}(e_0, \ldots, e_{n+r})$ (the quotient of $\mathbf{C}^{n+r+1} - 0$ by the action of $\mathbf{C}^*$ given by $t \cdot (x_0, \ldots, x_{n+r}) = (t^{e_0}x_0, \ldots, t^{e_{n+r}}x_{n+r})$). Here, as we have seen in Section 5.7, infinitesimal Torelli is false in general. On the positive side we have the following.

Theorem 8.1.8 *Let $Y = \mathbf{P}(1, 1, d_2, \ldots, d_{n+r})$ and let $z_0, \ldots, z_{n+r}$ be homogeneous coordinates. Let X be an n-dimensional smooth complete intersection in Y $(n \geq 2)$ with ample canonical bundle and assume that $z_0 = z_1 = 0$ defines a codimension 2 subvariety of X. Then infinitesimal Torelli holds for X.*

Proof This follows from Corollary 8.1.6. We take $L = \mathcal{O}_X(1)$ and observe that the linear system $|L|$ contains the pencil $\lambda z_0 + \mu z_1$, which by assumption has no fixed locus of codimension 1 on X. The required vanishing theorem can be shown as in the previous example. See Usui (1976) for details (see also Problem 8.1.2). □

The first statement covers the case of m-fold coverings of $\mathbf{P}^n$ with ample canonical bundle. One takes $e_0 = \cdots = e_n = 1, e_{n+1} = d$ and X a hypersurface of degree $(dm, \ldots, dm, m)$ and chooses the homogeneous coordinates carefully.

Problems

8.1.1 Give an elementary proof that $H^p(L^\bullet) = 0$ if the base locus of V has no component of codimension $\leq p + 1$ in case $p = 0, p = 1$. (This is all we need for the application in Theorem 8.1.5.)

8.1.2 Let $Y = \mathbf{P}(1, 1, d_2, \ldots, d_{n+r})$ and let X be an n-dimensional smooth complete intersection in Y ($n \geq 2$). Show that $H^0(\Omega_X^{n-1}(1)) = 0$.

8.2 Global Torelli Problems

Let S be a manifold over which we have a real variation of Hodge structure with monodromy group Γ and period map

$$\mathcal{P} : S \to \Gamma/D.$$

Let us recall the original version of the Torelli problem.

Problem (The global Torelli problem) Is it true that within a given class of varieties the periods classify its members (up to isomorphism)?

This should be stated more precisely since the periods depend on the choice of a basis for the cohomology and Hodge filtration. What we mean here is that up to these choices the period matrix uniquely determines the variety up to isomorphism. In other words, given an isomorphism of polarized Hodge structures $H^w(X) \simeq H^w(Y)$, is is true that X and Y are isomorphic?

The classical Torelli theorem (see Torelli, 1913) states that this is indeed the case for smooth projective curves. One can rephrase this theorem by saying that for such a curve C, the isomorphism class of the polarized Hodge structure on $H^1(C)$ determines C up to isomorphism, or, equivalently, that the jacobian $J(C)$ together with its polarization determines C, again up to isomorphism. Let us next discuss some other important examples.

Examples 8.2.1 (i) First of all, note that giving a weight 1 Hodge structure on a real vector space H is the same as giving its Weil operator, which in this case is an almost complex structure. Conversely, such an almost complex structure J defines a decomposition of $H \otimes \mathbf{C}$ in two complex conjugate eigenspaces

and hence a weight 1 Hodge structure. If $H = H_{\mathbf{Z}} \otimes \mathbf{R}$ and one has an integral polarized weight 1 Hodge structure on H, the torus $H/H_{\mathbf{Z}}$ with the complex structure given by the Hodge structure is an abelian variety: the polarization of the Hodge structure is a polarization for the complex torus. The Torelli problem then becomes trivial: polarized abelian varieties up to isomorphism correspond exactly to isomorphism classes of polarized weight 1 Hodge structures.
(ii) Let us consider $K3$ surfaces. Abelian varieties of dimension 2 are related to $K3$ surfaces and indeed for these the Torelli problems are related. Recall that a $K3$ surface is a simply connected Kähler surface S with K_S trivial. In Section 1.3 we considered the example of a double plane branched in a smooth sextic curve. There are many more examples, like the smooth quartic surfaces in $\mathbf{P}^3$ or complete intersections of three quadrics in $\mathbf{P}^5$. For this discussion, another class of $K3$ surfaces is more relevant; the so-called *Kummer surfaces*. These are defined to be the minimal resolutions of quotients of two-dimensional complex tori T by the natural involution ι sending x to $-x$. Such quotients acquire precisely 16 ordinary double points in the two-torsion points of T. Since T need not be algebraic, neither does $T/\langle\iota\rangle$ or its minimal resolution $\mathrm{Km}(T)$. So there are indeed non-algebraic $K3$ surfaces. Coming back to algebraic Kummer surfaces, how can these be linked to the Torelli theorem for the corresponding abelian variety when the former, as any $K3$ surface, have no H^1 at all? Here one should keep in mind that for complex tori T the total cohomology is generated by $H^1(T;\mathbf{Z})$ and for Kummer surfaces $H^2(T;\mathbf{Z})$ is naturally embedded in $H^2(\mathrm{Km}(T);\mathbf{Z})$. Indeed, this idea forms a principal step towards proving a Torelli theorem for $K3$ surfaces which we now formulate. For a proof as well as for further details and historical background of what we are about to discuss, we refer to Barth et al. (1993, Ch. VIII).

Theorem *Let S and S' be two $K3$ surfaces. Suppose that there exists a* Hodge isometry, *i.e., an isomorphism $H^2(S';\mathbf{Z}) \to H^2(S;\mathbf{Z})$ of Hodge structures preserving the cup product pairing. Then S and S' are isomorphic.*

Here we recall that all $K3$ surfaces are Kähler so that the cohomology does have a Hodge structure. Also note that a Hodge isometry is not the same as a morphism of polarized Hodge structures: for the latter we have to fix a Kähler class and split off the primitive cohomology, but then we lose the $\mathbf{Z}$-structure in general. Compare Example 4.4.5(ii).

The above theorem can be used to describe various moduli spaces for $K3$ surfaces. Since this plays some role in other parts of the book, let us briefly take this up. First remark that for any $K3$ surface S Serre duality and the fact that K_S is trivial imply that $h^j(\Theta_S) = h^{2-j}(\Omega^1_S)$. In particular, $H^2(S,\Theta_S) = 0$

which, as we have seen (Theorem 5.1.1), implies that the Kuranishi space of S is smooth of the expected dimension $h^1(\Theta_S) = h^1(\Omega^1_S) = 20$. So one expects a moduli space M of dimension 20. As we have seen, the period domain D for $K3$ surfaces is the domain IV_{20}, of the same dimension. Since local Torelli holds, the moduli space M locally around a point $[S]$ maps isomorphically to a neighborhood of $\mathcal{P}(S)$ in D, where $\mathcal{P}$ is the period map for the Kuranishi family for S. To make this precise we first need to establish the isomorphism class of the lattice $H^2(S; \mathbf{Z})$ equipped with the intersection pairing q_S. Here surface theory comes to our aid: q_S is isometric to the unique even lattice[1] of signature (3, 19). Let us call it the *K*3-*lattice* Λ. A *marking* for S is the choice of an isometry $H^2(S; \mathbf{Z}) \xrightarrow{\sim} \Lambda$. Via this marking the lattice Λ receives a Hodge structure and so

$$D = \left\{[x] \in \mathbf{P}(\Lambda \otimes \mathbf{C}) \mid x^2 = 0, \quad x^2 > 0\right\}$$

parametrizes marked $K3$ surfaces. The orthogonal group of Λ permutes the markings. It does not act discretely on D. This changes for algebraic $K3$ surfaces. You find them back inside D as follows: fix a primitive vector[2] $\ell \in \Lambda$ with $\ell^2 = 2d > 0$. Then surfaces for which ℓ corresponds to an ample class (under the marking) have periods on the hyperplane orthogonal to ℓ, say

$$D_\ell = D \cap \ell^\perp \subset \mathbf{P}(\Lambda \otimes \mathbf{C}).$$

The group Γ_ℓ of isometries of Λ fixing ℓ this time does operate properly discontinuously on D and the quotient space

$$M_\ell = \Gamma_\ell \backslash D_\ell$$

turns out to be the quasi-projective 19-dimensional moduli spaces M_ℓ for algebraic $K3$ surfaces with a polarization of class ℓ. Roughly speaking, the period map is an isomorphism for these. Of course, some care should be exercised; the precise statement is a bit more involved. We refer to Barth et al. (1993, Chapter VIII, § 22) for details.

(ii) A *hyperkähler manifold* is a simply connected Kähler manifold with a *unique* holomorphic two-form which is everywhere nondegenerate. Here we recall that a two-form σ on a complex manifold X is nondegenerate in a point $p \in X$ if the bilinear skew form that σ defines on T_pX is nondegenerate. This forces dim X to be even, say dim $X = 2n$. Also, the n-fold wedge product of σ then gives a nowhere zero section of the canonical bundle: K_X is trivial. If $n = 2$ one finds back the definition of a $K3$ surface. The uniqueness of σ implies that $h^{2,0}(X) = h^{0,2}(X) = 1$ and so the period domain resembles the one

[1] A lattice with quadratic form q is even if $q(x, x) \in 2\mathbf{Z}$ for all x in the lattice.

[2] This means that ℓ is not a nontrivial multiple of an element from the lattice Λ.

for $K3$ surfaces. The polarization which we usually employ is only defined on primitive cohomology: using a Kähler form ω it is defined by

$$b(u,v) = \int_X u \wedge v \wedge \omega^{2n-2}, \quad u, v \in H^2_{\text{prim}}(X).$$

This form of course can be extended to all of $H^2(X)$ by declaring the class of ω orthogonal to $H^2_{\text{prim}}(X)$. However, it is not clear that we obtain a form which is rational on rational cohomology. To remedy this, one replaces b by the Beauville–Bogomolov form q_X; it is integral on all of $H^2(X;\mathbf{Z})$ and has signature $(3, b_2(X) - 3)$. It is defined as[3] (see Beauville, 1983).

$$\begin{aligned}
q_X(u,v) &= q_0(u,v) + \frac{1}{2}(q_1(u,v) + q_1(v,u)),\\
q_0(u,v) &= \frac{n}{2}\int_X \sigma^{n-1} \wedge \bar{\sigma}^{n-1} \wedge u \wedge v,\\
q_1(u,v) &= (1-n)\cdot \int_X \sigma^{n-1} \wedge \bar{\sigma}^{n} \wedge u \cdot \int_X \sigma^{n} \wedge \bar{\sigma}^{n-1} \wedge v.
\end{aligned}$$

Hence, we obtain a lattice

$$\Lambda(X) = (H^2(X,\mathbf{Z}), q_X).$$

Contrary to $K3$ surfaces, the lattices $\Lambda(X)$ need not be isometric to a fixed one when X varies over the class of hyperkähler manifolds. Still, one can fix an isometry class of an integral lattice, say Λ of signature $(3, b-3)$ with which we associate a period domain

$$D_\Lambda = \{x \in \mathbf{P}(\Lambda_{\mathbf{C}}) \mid q(x) = 0,\ q(x,\bar{x}) > 0\},$$

a connected manifold of dimension $b-2$. A marking is an isometry $\Lambda(X) \xrightarrow{\simeq} \Lambda$ and one lets M_Λ be the moduli space of Λ-marked hyperkähler manifolds. In general it may have many components. Fix one and continue to denote it M_Λ. Clearly, there is a holomorphic period map

$$p_\Lambda : M_\Lambda \to D_\Lambda.$$

This map is a local biholomorphic map (Beauville, 1983) but there is also a global statement.

Theorem (Huybrechts, 1999; Verbitsky, 2013) *The period map p_Λ is a generically injective and surjective holomorphic map.*

Remark It is known that p_Λ is in general *not* injective although it is in the bimeromorphic sense.

[3] In Beauville (1983) it is explained that this form restricts on primitive cohomology to a multiple of b.

Apart from these examples the global Torelli problem is wide open and so one might ask for a less stringent demand. This leads to the following version of the problem.

Problem (The generic Torelli problem) Is it true that the generic fiber[4] of the period map for a given projective family consists of points corresponding to isomorphic varieties?

If U is a connected and simply connected neighborhood of $s \in S$ we can locally lift $\mathcal{P}$ to a period map $\tilde{\mathcal{P}} : U \to D$. The associated infinitesimal variation of Hodge structure at a point $s \in S$ is nothing but the derivative of the period map at s. Now we can formulate our third global problem.

Problem (The variational Torelli problem) Can one reconstruct the generic fiber in a projective family up to isomorphism from its infinitesimal variation of Hodge structure?

Clearly, if variational Torelli holds, generic Torelli holds.

Example 8.2.2 Consider a projective family of curves of genus ≥ 5 which is everywhere locally complete (i.e., the Kodaira–Spencer map is surjective). We saw in Chapter 5 that we can reconstruct the curve from its infinitesimal variation, at least if C is not hyperelliptic, trigonal, or a plane quintic. This is generically the case in any complete family if the genus is ≥ 5. So generic Torelli holds in this case.

We want to reformulate the variational Torelli problem in a more geometric and suggestive fashion by introducing the prolongation of the period map. If the rank of the lifted period map is constant and equal to r and $\mathbf{G}(r, TD)$ is the Grassmannian bundle over D of dimension r subspaces of the tangent bundle of D we define the prolongation $\widetilde{P\mathcal{P}} : U \to \mathbf{G}(r, TD)$ by letting $\widetilde{P\mathcal{P}}(s)$ be the subspace $d\mathcal{P}(T_{S,s})$ of $T_{D,\mathcal{P}(s)}$. If the period map only has generically rank r, we can use these local liftings to construct the prolongation over the dense set of S where the period map has rank r. Then we divide out by the natural left-action of the monodromy group on $\mathbf{G}(r, TD)$ to obtain a meromorphic *prolongation of the period map*:

$$P\mathcal{P} : S \dashrightarrow \Gamma\backslash\mathbf{G}(r, TD).$$

Suppose now that the infinitesimal (real) variation of Hodge structure is entirely determined by giving the image of the tangent map. This is the case, for instance, if the infinitesimal Torelli theorem is true. Then the variational Torelli problem can be reformulated as follows.

[4] By "generic" we always mean outside a proper analytic subset.

Problem (The variational Torelli problem (bis)) Is it true that the generic fiber of the prolongation map of the period map for a given projective family consists of points corresponding to isomorphic varieties?

It is clear that this form of variational Torelli implies generic Torelli ("principle of prolongation").

In applications to projective families which lead to moduli spaces we need a more sophisticated version of the principle of prolongation, which we now formulate.

Theorem 8.2.3 (Principle of prolongation) *Let X and Y be complex manifolds, not necessarily connected, and let $f : X \to Y$ be an analytic map of constant rank r. Let $G(r, T_Y)$ be the bundle of Grassmannians of r-dimensional subspaces of the tangent bundle T_Y and let $F : X \to G(r, Y)$ be the prolongation of f, i.e., $F(x) = (f(x), df(T_{X,x}))$. There is a thin subset T of X consisting of fibers of f (i.e., $T = f^{-1}f(T)$ is contained in a countable union of proper analytic subsets of X) such that outside of T the fibers of f and F coincide.*

Proof Introduce the subvarieties

$$\Delta_f = \{(x, x') \in X \times X \mid f(x) = f(x')\},$$
$$\Delta_F = \{(x, x') \in X \times X \mid F(x) = F(x')\}$$

of $X \times X$ and let p_1, p_2 be the two projections of $X \times X$ to X. We then set

$$E := \Delta_f - \Delta_F, \quad T = p_1(E).$$

We claim that E has codimension $> r$ in $X \times X$ and, since f has rank r, any component of T is a proper subvariety of X. This would prove the proposition. To prove the claim, let $(x, x') \in E$ and $y = f(x) = f(x')$. We can choose small complex neighborhoods U of x and U' of x' such that

$$f : U \to V \subset Y, \quad f : U' \to V' \subset Y$$

are submersions onto r-dimensional submanifolds V, V' of Y in a neighborhood of y. Because $df_x \neq df'_{x'}$, it follows that $\dim_y(V \cap V') < r$. Because $g = f \circ p_1|\Delta_f$ has fibers of codimension $2r$ in $X \times X$ and because $g(E \cap U \times U') \subset V \cap V'$, it follows that E has codimension $> r$ in $X \times X$ at (x, x'). This settles the claim and completes the proof of the proposition. □

8.3 Generic Torelli for Hypersurfaces

Using the same notation as before, we let $U_{d,n}$ be the base space of the universal family of smooth hypersurfaces of degree d in $\mathbf{P}^{n+1} = \mathbf{P}(V^\vee)$, where V has a basis $\{e_0, \ldots, e_{n+1}\}$ and the dual basis $\{z_0, \ldots, z_{n+1}\}$ for $V^\vee$ gives homogeneous coordinates on $\mathbf{P}^{n+1}$. Set

$$S := \mathbf{C}[z_0, \ldots, z_{n+1}].$$

Any smooth hypersurface X of degree d in $\mathbf{P}^{n+1}$ can be given by a polynomial $F \in S^d$ with the property that the generators $\partial F/\partial z_0, \ldots, \partial F/\partial z_{n+1}$ of the Jacobian ideal $\mathfrak{j}_F$ form a regular sequence. Let

$$R_F := S/\mathfrak{j}_F.$$

Let us translate Macaulay's theorem (Theorem 7.4.1) in the present situation. The number ρ in this case equals $(d-2)(n+2)$. We find the following.

Lemma 8.3.1 *The multiplication map*

$$R_F^a \times R_F^b \longrightarrow R_F^{a+b}$$

is nondegenerate if $a + b \leq (d-2)(n+2)$.

Let us first explain how we can apply Donagi's symmetrizer method to data coming from the infinitesimal variation of Hodge structure. We have computed the cup product map $T \times H^{n-p,p}_{\text{prim}} \longrightarrow H^{n-p-1,p+1}_{\text{prim}}$, where T is the subspace of $H^1(\Theta_X)$ corresponding to the infinitesimal deformations of X that stay inside the projective space (Proposition 5.4.1). From Lemma 8.3.1 it follows that the infinitesimal variation of Hodge structure at F gives us vector spaces W^i plus isomorphisms $\mu_i : R_F^i \xrightarrow{\sim} W^i$ for $i = d, t(p) = d(n-1+p)-n-2, t(p+1) = t(p)+d$ and a bilinear map $B_{t(p),d} : W^d \times W^{t(p)} \longrightarrow W^{d+t(p)}$, which make the following diagram commutative:

$$\begin{array}{ccc} R_F^d \times R_F^{t(p)} & \xrightarrow{\text{multiplication}} & R_F^{t(p)+d} \\ \downarrow{\scriptstyle \mu_d \times \mu_{t(p)}} & & \downarrow{\scriptstyle \mu_{t(p)+d}} \\ W^d \times W^{t(p)} & \xrightarrow{B_{t(p),d}} & W^{t(p)+d}. \end{array}$$

The symmetrizer construction now comes into play. It starts from a bilinear map $B : U \times V \longrightarrow W$ between vector spaces. Introduce

$$\operatorname{Sym}(U, V) = \{P \in \operatorname{Hom}(U, V) \mid B(u, P(u')) = B(u', P(u)) \text{ for all } u, u' \in U\}$$

and the natural bilinear map

$$B_- : \operatorname{Sym}(U, V) \times U \longrightarrow V$$

given by $B_-(P, u) = P(u)$. We refer to $(\operatorname{Sym}(U, V), B_-)$ as the *symmetrizer* of B.

Using this construction, let us define a sequence of bilinear maps

$$B_{a_i,b_i} : W^{a_i} \times W^{b_i} \longrightarrow W^{a_i+b_i}$$

inductively, starting with $a_1 = t(p)$, $b_1 = d$. The pairs (a_i, b_i) follow the Euclidean algorithm, beginning with the pair (a_1, b_1). For each $i \geq 1$ we define $(W^{a_{i+1}}, B_{a_{i+1},b_{i+1}})$ as the symmetrizer of B_{a_i,b_i}. The sequence stops when $a_i = b_i = e$, where

$$e = \gcd(d, n+2) = \gcd(t(p), n+2).$$

We can compare this construction with the multiplication maps between the corresponding graded pieces of the ring R_F. To do so, we observe that Donagi's symmetrizer lemma (Lemma 7.4.4) can be rephrased in this special case in the language of symmetrizers, as follows.

Lemma 8.3.2 *Consider the multiplication map $R_F^a \times R_F^b$. The natural map $R_F^{b-a} \longrightarrow \mathrm{Sym}(R_F^a, R_F^b)$ is an isomorphism provided $a > 0, (d-2)(n+2) > \max(a+b, b+d)$.*

Corollary 8.3.3 *Assume that $t(p) < d$ and $(d-2)(n-1) \geq 3$. Moreover, if $d = 3$ we should have $n \geq 5$. There are commutative diagrams*

$$\begin{array}{ccc} R_F^{a_i} \times R_F^{b_i} & \xrightarrow{\text{multiplication}} & R_F^{a_i+b_i} \\ \Big\downarrow \mu_{a_i} \times \mu_{b_i} & & \Big\downarrow \mu_{a_i+b_i} \\ W^{a_i} \times W^{b_i} & \xrightarrow{B_{a_i,b_i}} & W^{a_i+b_i}, \end{array}$$

where the vertical arrows are isomorphisms.

Proof By induction, the maps μ_{b_i} and $\mu_{a_i+b_i}$ are defined, W^{a_i} is a subset of $\mathrm{Hom}(W^{b_i}, W^{a_i+b_i})$, and B_{a_i,b_i} is just evaluation, so that μ_{a_i} is uniquely defined by $\mu_{a_i}(P)u = \mu_{a_i+b_i}(P \cdot \mu_{b_i}^{-1}(u))$ for any $P \in R_F^{a_i}$ and $u \in W^{b_i}$. Observe that the numerical condition on $t(p)$, n, and d implies that $(d-2)(n+2) > 2d$, and because W^{a_i} is the symmetrizer of $W^{a_i} \times W^{a_i+b_i} \longrightarrow W^{2a_i+b_i}$, commutativity of the diagram implies that μ_{a_i} is an isomorphism. At the last stage we have to verify that the two vector spaces W^e occurring are actually the same. This follows by repeating the construction once more and observing that R_F^0 and W^0 are one-dimensional and so can be identified by choosing a generator. □

Now we are ready to state and prove our main theorem.

Theorem (Torelli Theorem for projective hypersurfaces) *A smooth hypersurface X of degree d in $\mathbf{P}^{n+1}$ can be reconstructed up to projective equivalence* *from the image of the derivative of the period map at X, except possibly in the following two cases:*

$$d \text{ is a divisor of } 2(n+2);$$
$$d = 3,\ n = 2 \text{ (cubic surfaces).}$$

Proof **Step 1**: reduction to $(d-2)(n-1) \geq 3$. Observe that hyperplanes and quadrics have no moduli, so we can assume that $d \geq 3$; for plane cubic curves the result is trivial, whereas it is false (and excluded) for cubic surfaces, and known for threefolds (Clemens and Griffiths, 1972). For plane curves of degree 4 or more the result follows from the Torelli theorem plus the uniqueness of a nonsingular g^2_d (see Accola, 1979). Finally the case of quartic surfaces is a special case of the global Torelli theorem for $K3$ surfaces (see Barth et al., 1993, Chapter VIII).

For the remainder of the proof we take p maximal so that $H^{p,n-p} \neq 0$, and then $t := t(p)$ is as small as possible:

$$n + 2 = (n - p + 1)d + t, 0 \leq t \leq d - 1.$$

It follows that we can apply Corollary 8.3.3. Recall that $e = \gcd(d, n+2)$ and so $e < d-1$ because d is not a divisor of $n+2$ and $d > 2$. So $R^e = S^e$ and we have an isomorphism $\mu_e : S^e \longrightarrow W^e$ which we would like to be able to reconstruct.
Step 2: reconstruction of μ_e if $2e < d - 1$. In this case $R^{2e} = S^{2e}$ and we can go back to the diagram of Corollary 8.3.3 with $a_i = b_i = e$. Formally, we have the following lemma.

Lemma *Let V be a vector space and let*

$$B : S^2(S^t V^\vee) \longrightarrow S^{2t} V^\vee$$

be the multiplication map. Let $s : \mathbf{P}(V) \longrightarrow \mathbf{P}(S^t V)$ *be the Veronese embedding. The ideal of the image of s is generated by quadrics, i.e., by* Ker B.

Proof Ker B is the system of quadrics through the Veronese image. We need to show that any point in the intersection of these quadrics also belongs to the Veronese image. If $z_0, \ldots, z_n$ is a basis for $V^\vee$, the monomials $z_I, |I| = t$ give a basis for $S^t V^\vee$ and Ker B is spanned by $z_I z_J - z_K z_L, |I| + |J| = |K| + |L| = 2t$. Let p be any point in the intersection of these quadrics; then there is a unique point $q \in \mathbf{P}(V)$ with $z_i(q) = (z_{0\ldots 0i}(p)), i = 0, \ldots, n$. □

Corollary *Up to an automorphism of* S^1*, the isomorphism* $\mu_e : S^e \longrightarrow W^e$ *can be recovered from the infinitesimal variation of Hodge structure.*

Proof The map $B_{e,e}$ determines the Veronese Y in $\mathbf{P}(W^e)^\vee$ and so, if we make a choice for the identification $\mathbf{P}V^\vee = Y$, we again find μ_e as the inverse of the restriction map

$$W^e = H^0(\mathbf{P}((W^e)^\vee), \mathcal{O}(1)) \xrightarrow{\sim} H^0(Y, \mathcal{O}(1)) \xrightarrow{\sim} H^0(\mathbf{P}(V^\vee), \mathcal{O}(e)) = S^e. \quad \square$$

Step 3: reconstruction of μ_e if $2e \geq d - 1$. Observe that the numerical restrictions imply that now either $e = 1$ or $d = 2e$. In the first case, we can fix μ_e arbitrarily and we are done. So we assume that $e > 2$ and $d = 2e$. We now first use the existence of the commutative diagram of Corollary 8.3.3 in the case of $a_i = e, b_i = 2e$ together with the previous diagram. These give a diagram

$$\begin{array}{ccccc} \bigwedge^2 R_F^e \otimes R_F^e & \to & R_F^e \otimes R_F^{2e} & \to & R_F^{3e} \\ \downarrow \wedge^2\mu_e \otimes \mu_e & & \downarrow \mu_e \otimes \mu_{2e} & & \downarrow \mu_{3e} \\ \bigwedge^2 W^e \otimes W^e & \to & W^e \otimes W^{2e} & \to & W^{3e}. \end{array}$$

Observe that we can apply Corollary 7.3.6 with $k = 2e, d = e, c = 1$. It follows that we can identify those hyperplanes $\mu_e(V)$ of W^e which correspond to subsystems V of $|\mathcal{O}_{\mathbf{P}}^{n+1}(e)|$ which have a (unique) base point. The points in $\mathbf{P}(W^e)^\vee$ corresponding to these hyperplanes form the Veronese and as before we can recover the isomorphism μ_e.

Step 4: reconstruction of the Jacobian ideal.

Using μ_e we want to go back to the diagrams of Corollary 8.3.3 and this time we reverse the direction of the induction so that it goes up in order to reconstruct the maps μ_{a_i}. So we apply decreasing induction on i to get the commutative diagrams

$$\begin{array}{ccc} S^{a_i} \times S^{b_i} & \xrightarrow{\text{multiplication}} & S^{a_i+b_i} \\ \downarrow \mu_{a_i} \times \mu_{b_i} & & \downarrow \mu_{a_i+b_i} \\ W^{a_i} \times W^{b_i} & \xrightarrow{B_{a_i,b_i}} & W^{a_i+b_i}, \end{array}$$

where each $\mu_{a_i+b_i}$ is induced from μ_{a_i}, μ_{b_i} and is surjective. Indeed, the existence of the maps μ and the natural quotient maps $S^{a_i} \to R_F^{a_i}$ guarantees that $\mu_{a_i+b_i}$ is well defined.

Going all the way to $i = 1$ we construct the map μ_d, whose kernel is precisely the Jacobian ideal in degree d. We then find the Jacobian ideal in degree $d - 1$ by Macaulay's theorem.

Step 5: Completion of the proof: reconstruction of F from the Jacobian ideal. We prove the following.

Claim *Let $F, G \in S^d, d \geq 2$ and assume that their partial derivatives span the same subspace of S^{d-1}. Then F and G are related by a linear transformation of the coordinates.*

Let us prove this claim. We may assume that the rank of the subspace spanned by the partial derivatives of F is maximal, equal to $n+1$. For, if $\sum_j \alpha_j \partial F/\partial z_j = 0$ is any relation, F is constant along lines parallel to the vector $(\alpha_0, \ldots, \alpha_{n+1})$ and we can make a linear change of variables to change these relations into $\partial F/\partial z_j = 0, j = k+1, \ldots, n+1$, so that F (and hence G) depends only on the first $k+2$ variables, and span a subspace of rank $k+2$.

The assumption on the rank now implies that for some invertible $n+2$ by $n+2$ matrix B we have

$$\begin{pmatrix} \partial G/\partial z_0 \\ \vdots \\ \partial G/\partial z_{n+1} \end{pmatrix} = B \begin{pmatrix} \partial F/\partial z_0 \\ \vdots \\ \partial F/\partial z_{n+1}, \end{pmatrix}$$

which we abbreviate as $\nabla G = B \circ \nabla F$.

Whenever $B_1(t) := (\mathbf{1} - t(\mathbf{1} - B))$ is invertible, we put $B(t) = d^{-1}(\mathbf{1} - B) \cdot B_1(t)^{-1}$ and let ${}^{\mathsf{T}}B(t)$ be its transpose. Choose a path $\gamma(t)$ from 0 to t along which $B_1(t)$ is invertible and set $C(t) = \exp \int_{\gamma(t)} {}^{\mathsf{T}}B(\tau)d\tau$. Finally, define $\Phi(z,t) = (1-t)F(z) + tG(z)$. We claim

$$\Phi(C(t)z, t) = F(z) \text{ for all } t. \tag{8.1}$$

If we can show this, it will follow that $F(z) = G(C(1)z)$, i.e., F and G are related by a linear transformation and we are done.

It suffices to show that $\partial/\partial t\{\Phi(C(t)z,t)\} = 0$, because $C(0) = \mathbf{1}$ and $\Phi(z,0) = F(z)$. Setting $x = C(t)z$ we calculate

$$\begin{aligned} \partial/\partial t\{\Phi(x,t)\} &= -F(x) + G(x) \\ &\quad + \sum_j [(1-t)\partial F/\partial z_j(x) + t\partial G/\partial z_j(x)] \circ (C'(t)z)_j. \end{aligned}$$

Euler's equation gives $F(x) = d^{-1}{}^{\mathsf{T}}x \circ \mathbf{1} \circ \nabla F(x)$ and $G(x) = d^{-1}{}^{\mathsf{T}}x \circ B \circ \nabla F(x)$. Because $C'(t)z = {}^{\mathsf{T}}B(t) \circ C(t)z = {}^{\mathsf{T}}B(t)x$, we find

$$\partial/\partial t\{\Phi(x,t)\} = {}^{\mathsf{T}}x[-d^{-1}(\mathbf{1} - B) + B(t) \cdot (\mathbf{1} - t(\mathbf{1} - B))] \circ \nabla F(x).$$

This vanishes by the definition of $B(t)$. This proves (8.1) and hence the Claim. □

8.4 Moduli

We have seen that any deformation of a given compact complex manifold can be pulled back from the Kuranishi family which is the best substitute one has for a universal family. Even if this family is everywhere universal (such as for curves or for hypersurfaces) its base does not parametrize the isomorphism classes of the fibers. This is because of the presence of nontrivial automorphisms which act nontrivially on the Kuranishi family. Let us illustrate this with an example.

Example 8.4.1 (Elliptic curves) Here we continue the discussion started in Chapter 1, Remark 1.1.8. The group $\mathbf{Z}^2$ acts on the product of the upper half-plane $\mathfrak{h}$ and the plane $\mathbf{C}$ as follows:

$$(n, m)\circ(\tau, z) = (\tau, z + n + m\tau).$$

The quotient U is a complex manifold and the projection onto $\mathfrak{h}$ realizes it as a family of elliptic curves. The group $\mathrm{SL}(2, \mathbf{Z})$ acts on the upper half-plane $\mathfrak{h}$ in the usual manner:

$$\begin{pmatrix} a & b \\ c & d \end{pmatrix}(\tau) = \frac{a\tau + b}{c\tau + d}.$$

We identify each fiber over τ with $\mathbf{R}^2$ by taking the ordered basis $\{\tau, 1\}$. In this way, any $\gamma \in \mathrm{SL}(2, \mathbf{R})$ defines a linear isomorphism from the fiber over τ to the fiber over $\gamma(\tau)$. It is obvious that this defines an action of $\mathrm{SL}(2, \mathbf{Z})$ on $\mathfrak{h} \times \mathbf{C}$ which descends to U. Since two elliptic curves U_τ and $U_{\tau'}$ are isomorphic if and only $\tau = \gamma(\tau')$ for some $\gamma \in \mathrm{SL}(2, \mathbf{Z})$, the quotient variety $\mathrm{SL}(2, \mathbf{Z})\backslash\mathfrak{h}$ (the j-line) represents the isomorphism classes of elliptic curves. However, elliptic curves all have the nontrivial automorphism $x \mapsto -x$, and so $\mathrm{SL}(2, \mathbf{Z})\backslash U$ is not a family of elliptic curves and we have no universal family.

In the preceding example, the j-line $\mathbf{P}^1$ parametrizes in a natural fashion the isomorphism classes of elliptic curves, and for any family of elliptic curves over a base B, the map which sends $s \in S$ to the j-invariant of the fiber over s is a holomorphic map $j : B \longrightarrow \mathbf{P}^1$. This can be formalized as follows. Suppose one has a given set $\mathcal{M}$ of complex manifolds together with an equivalence relation $\sim$, e.g., curves of genus g, with isomorphism between them, hypersurfaces of given degree in a fixed projective space under projective equivalence, the set of complex structures on a given differentiable manifold with equivalence complex-analytic isomorphism, etc.

The corresponding *moduli problem* is to find a complex space M whose points correspond uniquely to the equivalence classes of objects in $\mathcal{M}$. Moreover, if one has a family F of objects in $\mathcal{M}$ over a complex space S, the map $\phi(F) : S \longrightarrow M$ sending a point $s \in S$ to the point in M corresponding to the equivalence class of the fiber F_s of F over s should be analytic. Ideally, M should have a universal family of objects in $\mathcal{M}$ over it and in this case we call M a *fine moduli space*. Usually this is too much to ask for, as we saw in the previous example and one has to be content with a weaker notion, that of a *coarse moduli space*. To define the latter formally, we first have to recall what a natural transformation $\mathbf{T} : \mathbf{F}_1 \longrightarrow \mathbf{F}_2$ is between functors $\mathbf{F}_1, \mathbf{F}_2 : \mathbf{C} \longrightarrow \mathbf{D}$ relating the same categories in a contravariant way: it is a collection of morphisms $\mathbf{T}(S) : \mathbf{F}_1(S) \longrightarrow \mathbf{F}_2(S)$, one for each object S in $\mathbf{C}$, such that for every morphism $f : S \longrightarrow S'$ we have $\mathbf{T}(S') \circ \mathbf{F}_1(f) = \mathbf{F}_2(f) \circ \mathbf{T}(S)$.

Next, we must assume that we can extend the notion of the given equivalence relation to families of varieties in $\mathcal{M}$ in such a way that equivalent families induce equivalent families when we pull them back. So we can now consider the set $\mathbf{F}(S)$ of equivalence classes of families over S with objects on $\mathcal{M}$. These define a functor

$$\mathbf{F} : \{\text{analytic spaces}\} \longrightarrow \{\text{sets}\}\,.$$

The set of morphisms from S to an analytic space M will be denoted by $\mathbf{H}_M(S)$. These define a second functor $\mathbf{H}_M$ between the same categories. The maps $\phi(F) : S \longrightarrow M$ we introduced above define a natural transformation

$$\Phi : \mathbf{F} \longrightarrow \mathbf{H}_M\,.$$

If this is a bijection, we say that the *functor* $\mathbf{F}$ *is represented* by M and in this case we have a fine moduli space. The family over M corresponding to the identity morphism of M can be seen to have the obvious universality property we required for a fine moduli space. The definition of a coarse moduli space runs as follows.

Definition 8.4.2 A *coarse moduli space* for $(\mathcal{M}, \sim)$ consists of an analytic space M whose points correspond bijectively to $(\mathcal{M}/\sim)$ such that

(i) for any family F of manifolds in $\mathcal{M}$ parametrized by an analytic space S the map $\phi(F) : S \longrightarrow M$ sending a point $s \in S$ to the point in M corresponding to the equivalence class of the fiber F_s of F over s is a morphism of analytic spaces, and
(ii) for any analytic space N and any natural transformation $\Psi : \mathbf{F} \longrightarrow \mathbf{H}_N$ the map on points $\mu : M \longrightarrow N$ induced by $\Psi(\text{point})$ is a morphism.

The usual definition is stated entirely in categorical terms. For completeness we give it here.

Definition 8.4.3 (Alternative definition) A *coarse moduli space* for $(\mathcal{M}, \sim)$ consists of an analytic space M together with a natural transformation $\Phi : \mathbf{F} \longrightarrow \mathbf{H}_M$ such that

(i) $\Phi(\text{point}) : \mathbf{F}(\text{point}) = (\mathcal{M}/\sim) \longrightarrow \mathbf{H}_M(\text{point}) = M$ is bijective, and
(ii) for any analytic space N and any natural transformation $\Psi : \mathbf{F} \longrightarrow \mathbf{H}_N$ there exists a natural transformation $\chi : \mathbf{H}_M \longrightarrow \mathbf{H}_N$ such that $\Psi = \chi \circ \Phi$.

To see the equivalence, note that $\Phi(S)(F)$ is the morphism which on points coincides with the map we denoted by $\phi(F)$. Moreover, μ defines the required natural transformation upon setting $\chi(S)(\phi) = \mu \circ \phi$. Conversely, one finds again μ since $\mu = \chi(M)(\mathrm{id}_M)$. The advantage of this definition is that it can be easily modified to work within the categories of projective varieties, schemes, etc.

There is a framework in which moduli spaces can be seen to exist, namely when we consider equivalence classes defined by the action of an affine algebraic group G, e.g., $G = \mathrm{GL}(n, \mathbf{C})$, $G = \mathrm{SL}(n, \mathbf{C})$ on a variety H, over which we have a family in which every manifold in $\mathcal{M}$ appears and which is locally complete at every point. The construction of a coarse moduli space for the original moduli problem then can be reduced to the problem of whether a good quotient for the G-action on H exists, as we shall see now. Let us introduce the following.

Definition 8.4.4 A *categorical quotient* of H by G is a pair (M, ϕ) consisting of a variety M and a morphism $\phi : H \longrightarrow M$ which is constant on G-orbits and which has the obvious universal property with respect to morphisms $\psi : H \longrightarrow N$ constant on G-orbits: there should exist a unique morphism $\chi : M \longrightarrow N$ such that $\chi \circ \phi = \psi$. If in addition fibers of ϕ consist of exactly one orbit we call (M, ϕ) an *orbit space*.

Clearly, a categorical quotient is determined up to isomorphism. The connection between quotients and coarse moduli spaces can be stated as follows.

Definition 8.4.5 (Construction principle) Suppose there is a family $U \longrightarrow H$ of manifolds in $\mathcal{M}$ in which every manifold of $\mathcal{M}$ appears and which is locally complete at each point of H. Assume that the equivalence relation is induced by the action of the group G, i.e., $h \sim h'$ if and only if h and h' are in the same G-orbit. An orbit space for the group action on H then gives a coarse moduli space for $(\mathcal{M}, \sim)$.

Proof This is rather straightforward; we outline the argument as given in Newstead (1978, pp. 40–41). We observe that there is a one-to-one correspondence between morphisms $\phi : H \longrightarrow M$ which are constant on G-orbits and natural transformations $\Phi : \mathbf{F} \longrightarrow \mathbf{H}_M$.

Given Φ, for each family f over H we set $\phi = \Phi(H)(f)$. It is constant on orbits by our assumptions.

Conversely, if we are given ϕ and a family $f' : U' \longrightarrow H'$, local completeness of $U \longrightarrow H$ implies that one can cover H' by open sets over which g is induced by a morphism to H. These morphisms can differ by the action of G on overlaps, but after composing with ϕ they glue to a morphism $\Phi(H')(f')$. This defines the natural transformation Φ.

It is easy to check that the two correspondences $\phi \longrightarrow \Phi$ and $\Phi \longrightarrow \phi$ are inverse bijections.

Next, because our quotient is an orbit space, condition (i) in the alternative definition 8.4.3 is satisfied. Finally, a categorical quotient also satisfies the second property. □

It is, however, hard to see directly whether categorical quotients exist, except in some special cases, which we discuss now. For simplicity, we let $G = \mathrm{SL}(n, \mathbf{C})$ act linearly on a complex affine space $\mathbf{C}^n$ in such a way that it maps a given affine subvariety H into itself. The coordinate ring $R(H)$ of H is acted upon by G. By a theorem of Hilbert, the invariant ring is finitely generated (for a modern proof see Dieudonné and Carrell, 1971) and hence is the coordinate ring of some affine variety M. The quotient map $\phi : H \longrightarrow M$ is a surjective G-equivariant morphism which has the property that for any open U in M the pullback via ϕ induces an isomorphism

$$\phi^* : \mathcal{O}(U) \xrightarrow{\sim} \mathcal{O}(\phi^{-1}U)^G$$

from the regular functions on U to the G-invariant functions on $\phi^{-1}(U)$. Moreover, ϕ sends closed G-invariant Zariski-closed sets to Zariski-closed sets; if these are disjoint, their images remain disjoint.

For proofs consult Newstead (1978, Chapter 3.3). In fact, the proof holds for reductive groups, a class of groups we need later in Chapter 15 and whose definition is recalled in Appendix D.

Any pair (M, ϕ) satisfying the preceding properties is called a *good quotient* and if it is an orbit space (see Definition 8.4.4) it is called a *geometric quotient* or a *GIT-quotient*, after the monograph by Mumford (1965). The important fact concerning these is the following.

Lemma 8.4.6 (Newstead 1978, Chapter 1.3) *A good quotient (M, ϕ) is a categorical quotient of H by G. Moreover,*

(i) *the ϕ-images of two points coincide if and only if the Zariski closures of their G-orbits intersect,*
(ii) *if the action of G is closed, then (M, ϕ) is a geometric quotient, and*
(iii) *if $H' := \{h \in H \mid Gh$ is Zariski-closed and* $\dim Gh$ *is maximal$\}$, then $M' = \phi(H')$ is open in M, and M' together with the restriction of ϕ to H' is a geometric quotient for H'.*

So for linear actions of $\mathrm{SL}(n, \mathbf{C})$ on affine varieties, geometric quotients exist. We want to see whether this is also true for projective varieties. For simplicity, let us look at a linear action of $G = \mathrm{SL}(n + 1, \mathbf{C})$ on a complex vector space V

of dimension $n+1$. Let $z_0, \ldots, z_n$ be a basis for $V^\vee$ and consider

$$R(G) := \mathbf{C}[z_0, \ldots, z_n]^G,$$

the ring of invariant polynomials. Again, it is finitely generated, say by the homogeneous polynomials $P_0, \ldots, P_m$ of degrees $d_0, \ldots, d_m$, respectively. Consider the map $z := (z_0, \ldots, z_n) \mapsto (P_0(z), \ldots, P_m(z))$ sending a point in V to a point in $\mathbf{C}^{m+1}$. To make it well defined on $P(V)$ we have to replace the target by *weighted projective space* $\mathbf{P}(d_0, \ldots, d_m)$, i.e., the quotient of $\mathbf{C}^{m+1} - \{0\}$ by the action of $\mathbf{C}^*$ given by $t(y_0, \ldots, y_m) = (t^{d_0} y_0, \ldots, t^{d_m} y_m)$. Still, a problem remains; the polynomials $P_0, \ldots, P_m$ could be 0 in all points of V. An easy example of this situation is given by the standard action of $\mathrm{SL}(n, \mathbf{C})$ on $\mathbf{C}^n$, where the only invariant polynomials are constants. In view of this, let us give the following definition.

Definition 8.4.7 A point $z \in \mathbf{P}(V)$ is called

(i) *semistable* if there is a nonconstant invariant homogeneous polynomial which does not vanish at z, and
(ii) *stable* if it is semistable and if in addition Gz is Zariski-closed and has dimension equal to $\dim G$.

So, if we set

$$\mathbf{P}_{ss}(V) := \{z \in \mathbf{P}(V) \mid z \text{ semistable}\},$$

the map

$$\pi : \mathbf{P}(V) \dashrightarrow \mathbf{P}(d_0, \ldots, d_m)$$

given by $z := (z_0, \ldots, z_n) \mapsto (P_0(z), \ldots, P_m(z))$ is a rational map, defined on $\mathbf{P}_{ss}(V)$ with image a certain closed subvariety M of $\mathbf{P}(d_0, \ldots, d_m)$. It is not difficult to show the following Lemma (see e.g. Newstead, 1978, Ch. 3.4).

Lemma 8.4.8 *The morphism $\pi : \mathbf{P}_{ss}(V) \to M$ is a good quotient of $\mathbf{P}_{ss}(V)$. The image of the stable points $\mathbf{P}_s(V)$ in $\mathbf{P}(V)$ under π forms a Zariski-open subset M_s of M which is a geometric quotient for $\mathbf{P}_s(V)$.*

This applies, for instance, to families of hypersurfaces, as in the following example.

Example 8.4.9 Let V be a complex vector space of dimension $n+2$. Hypersurfaces in $\mathbf{P}(V^\vee)$ of degree d correspond in a one-to-one fashion to points of $\mathbf{P}(S^d V)$. The singular hypersurfaces F correspond to a $SL(V)$-invariant hypersurface given by the vanishing of the discriminant of F, a polynomial in the coefficients of F. It follows that the complement $U_{d,n}$ is an $SL(V)$-invariant affine subvariety of $\mathbf{P}(S^d V)$, which decomposes into *closed* orbits. Moreover, if $n \geq 1$ and $d \geq 3$ the stabilizer of a point is finite. Indeed, the automorphism group of the corresponding hypersurface is finite, as we saw in Problem 5.1.5. It follows that smooth hypersurfaces are always stable points for the action of $SL(V)$ and a geometric quotient exists. Over $\mathbf{P}(S^d V)$ we have the obvious universal family of hypersurfaces $\{(z, F) \in \mathbf{P}(V^\vee) \times U_{d,n} \mid F(z) = 0\}$, so this geometric quotient $U_{d,n}/SL(V)$ is a coarse moduli space for hypersurfaces of degree d under projective equivalence. Because there are hypersurfaces with nontrivial groups of automorphisms, this cannot be a fine moduli space.

We end this section with a reformulation of the generic Torelli problem in the language of moduli spaces. We are going to rephrase the Torelli problem for the situation in which we have a coarse moduli space M for a set $\mathcal{M}$ of smooth projective varieties with an equivalence relation $\sim$. Assume now:

(i) $\mathcal{M}$ is the set of fibers of an everywhere locally complete projective family $f : U \to H$ over a connected manifold H with a relative ample line bundle.
(ii) An affine group G acts fiberwise on the family f, and the given relative ample bundle is (up to isomorphism) preserved by the G-action. Furthermore, the equivalence relation is induced by the action of the group G, i.e., $h \sim h'$ if and only if h and h' are in the same G-orbit.
(iii) A categorical quotient H/G exists.

From our previous discussion it follows that $M = H/G$ is a coarse moduli space for $(\mathcal{M}, \sim)$. Because it is a categorical quotient, the period map for the family $f : U \to H$ descends to a holomorphic map $\mathcal{P}_M : M \to \Gamma\backslash D$. By abuse of notation we shall call this map a period map as well. So we arrive at the following.

Problem (Reformulation of the (generic) Torelli problem) In the preceding setup, is it true that the period map $\mathcal{P}_M : M \to \Gamma\backslash D$ is (generically) injective (i.e., outside a proper analytic subset of M)?

We can now state the main result of this section.

Theorem 8.4.10 *If variational Torelli (bis) holds for a family $f : U \to H$ as above, then generic Torelli holds for any discrete subgroup $\Gamma' \subset G_{\mathbf{R}}$ containing the monodromy group Γ, i.e., the map*

$$\mathcal{P}'_M : M \to \Gamma'\backslash D$$

induced by $\mathcal{P}_M$ is generically injective.

Proof First, we can shrink H so that the resulting orbit space $M = H/G$ is smooth, the quotient map $q : H \longrightarrow M$ has everywhere maximal rank, and, finally, the lifted period map $\tilde{\tau} : \tilde{H} \longrightarrow D$ from the universal cover $\tilde{H}$ of U to D has constant rank r. So the map

$$t : \tilde{H} \times \Gamma' \to D$$

defined by $t(h, \gamma) = \gamma(\tilde{\tau} h), h \in \tilde{H}$ has a prolongation

$$T : \tilde{H} \times \Gamma' \to \mathbf{G}(r, TD).$$

The assumption that variational Torelli holds for $U \longrightarrow H$ just says that the fibers of T are contained in $V \times \Gamma'$, where V is the lift of any G-orbit in H to the universal cover of H.

Because q has maximal rank, there are local sections to q over a contractible open set, and because the period map $\mathcal{P}$ is locally liftable, it follows that $\mathcal{P}'_M$ is locally liftable as well and leads to a holomorphic map $\tilde{\mathcal{P}}' : \tilde{M} \longrightarrow D$, where $\tilde{M}$ is the universal cover of M. Because the fibers of q consist of isomorphic polarized varieties, we have

$$\mathrm{d}\tilde{\tau}(T_{\tilde{H},h}) = \mathrm{d}\tilde{\mathcal{P}}(T_{\tilde{M},m}),$$

where $h \in \tilde{H}$ and m is any point in $\tilde{M}$ above the image of h under the composition of the natural projection $\tilde{H} \longrightarrow H$ and q. It follows that $\tilde{\mathcal{P}}'$ also has constant rank and that the infinitesimal variation of Hodge structure at h is determined by m.

Now consider the map

$$t' : \tilde{M} \times \Gamma' \to D$$

defined by $t'(m, \gamma) = \gamma(\tilde{\mathcal{P}}'(m))$ with prolongation

$$T' : \tilde{M} \times \Gamma' \to \mathbf{G}(r, TD).$$

It follows that the fibers of T' are contained in (fiber of $\tilde{M} \longrightarrow M) \times \Gamma'$. By the principle of prolongation this is true for t' as well, at least outside a thin set, i.e., the map $\mathcal{P}'_M : M \longrightarrow \Gamma' \backslash D$ is injective outside a thin set. We need to see that this set is actually analytic. A proof of this result must be postponed to the last part of the book after we have discussed more differential geometry. In fact, by Corollary 13.7.6 it follows that the period map $\mathcal{P}'$ can be extended to a proper map over a partial compactification of M. □

Example 8.4.11 Let us take for our family the family of all smooth hypersurfaces of degree d in projective space $\mathbf{P}^{n+1}$. It follows that the generic Torelli theorem for this family follows from the already proven variational version.

Remark In Cox et al. (1987) the reader can find a stronger version of the main theorem of this section. No assumption about the existence of a coarse moduli space for the family $U \to H$ is needed. Considering the equivalence relation R given by identifying polarized isomorphic varieties and using results due to Grothendieck it is shown that R is constructible. Then the existence of a good quotient follows as in the case of a group action. We have chosen to work in the setup of groups because it is more elementary and it is all we need for the applications.

Bibliographical and Historical Remarks

We comment briefly on the subjects of this chapter.

Torelli problems: Various proofs of the classical Torelli theorem are known. See, for instance Griffiths and Harris (1978). A precise version of the Torelli theorem for $K3$ surface has been formulated by Andreotti and Weil (independently). A proof for algebraic $K3$s has been given by Piateckii-Shapiro and Shafarevich (1971), while Burns and Rapoport (1975) extended this to Kähler $K3$ surfaces. Siu (1983) proved that all $K3$ surfaces are Kähler. For references and more details, see Barth et al. (1993, Chap VIII). See also the more recent monograph Huybrechts (2016). For an application of the Torelli theorem for $K3$ surfaces to Torelli problems of surfaces of general type see Usui (1982).

For background on hyperkähler manifolds and the proof the Torelli theorem for these, see Huybrechts (2012). Gross et al. (2003), which contains chapters on Calabi–Yau manifolds as well as hyperkähler manifolds, is highly recommended as a source for applications to mathematical physics.

Infinitesimal Torelli theorems: The first proofs of the infinitesimal Torelli theorem for complete intersections in $\mathbf{P}^n$ use vanishing results for the twists of the bundle of k-forms on $\mathbf{P}^n$. See Peters (1975, 1976a,b) and Usui (1976).

For m-fold covers of $\mathbf{P}^n$ with an ample canonical bundle the first results can be found in Kiĭ (1973), preceding the more general results in Peters (1976b). For nonample canonical bundles there are results by Saito (1986) and Wehler (1986). Konno (1985) shows that the infinitesimal Torelli theorem holds for certain types of m-cyclic covers of the Hirzebruch surfaces $\Sigma_r = \mathbf{P}(\mathcal{O}_{\mathbf{P}^1} \oplus \mathcal{O}_{\mathbf{P}^1}(r))$. His proof corrects some of the statements in Peters (1976b).

Flenner (1986) also uses Koszul-type arguments in the case of zero loci of sections of vector bundles. In case of a line bundle this becomes an infinitesimal Torelli theorem for smooth hypersurfaces in the linear system $|\mathcal{O}(k)|$ on Y for k large enough. This is a result previously obtained by Green (1984a). Kiĭ (1978) has obtained an infinitesimal Torelli statement using different methods. It implies the infinitesimal Torelli property for a surface having an elliptic pencil with nonconstant j-invariant and without multiple fibers. For one or two multiple fibers, infinitesimal Torelli is false; the period map even has positive-dimensional fibers (see Chakiris, 1980). Saito (1983) gives a fairly complete account of the infinitesimal Torelli theorem for elliptic surfaces. He also treats the case of an elliptic fibration over a curve of higher genus and he gives a list of counterexamples to the infinitesimal Torelli theorem in this situation.

Using Kiĭ's main theorem, Konno (1986) showed that the infinitesimal Torelli theorem holds for certain smooth complete intersections of simply connected homogeneous Kähler manifolds Y with $b_2(Y) = 1$. Wehler (1988) shows that the infinitesimal Torelli theorem holds for cyclic covers X of a homogeneous rational manifold of rank 1 with K_X ample.

For manifolds with many one-forms one also has infinitesimal Torelli theorems. See Reider (1988), and Peters (1988) for higher dimensions (see also Tu, 1983).

Variational Torelli: The introduction of variational methods in the study of global Torelli problems started with the seminal papers Carlson et al. (1983); Griffiths (1983), and Griffiths and Harris (1983). See also the exposition by Peters and Steenbrink (1983), which we largely have followed here.

Variational methods have been applied successfully by Donagi (1983); Green (1984b); Donagi and Tu (1987); Konno (1991) and Ivinskis (1993).

The basic symmetrizer lemma is due to Donagi (1983); see also the more elementary proof by Donagi and Green (1984). For elliptic surfaces see also Cox and Donagi (1986).

With much more elaborate methods, Voisin (1999a) treated a missing case of Torelli, namely the case of the quintic threefold.

Moduli: The categorical approach to moduli problems is due to Grothendieck. For historical references the reader may consult the classic book Mumford (1965). In this book geometric invariant theory as introduced by Hilbert takes its modern form as presented here. For a more recent treatment of moduli problems, see Viehweg (1995).

9

Normal Functions and Their Applications

In this chapter we introduce infinitesimal methods in order to study algebraic cycles and explain several results of Griffiths (see Griffiths, 1968, 1969, 1970; Green, 1989b; Voisin, 1988b). The study started in Section 3.4 is continued in Section 9.1 of this chapter, where we introduce the concept of a normal function and its infinitesimal invariants. In Section 9.2 we investigate the Griffiths group of hypersurface sections. We finish this chapter with a proof of the theorem of Green (1989b) and Voisin (1988b) on the image of the Abel–Jacobi mapping for hypersurfaces.

9.1 Normal Functions and Infinitesimal Invariants

This section introduces normal functions for families of algebraic cycles and Griffiths' idea of differentiating them. In Section 3.4 we saw that all algebraic subvarieties and therefore all algebraic cycles have cohomology classes, i.e., there are natural morphisms $\mathrm{cl}^p : \mathrm{Ch}^p(X) \to H^{2p}(X, \mathbf{Z})$ for every $p \geq 0$. If $\mathrm{Ch}^p(X)_{\mathrm{hom}}$ denotes the kernel of cl^p, the group of homologically trivial cycles, then we have already defined the Abel–Jacobi map in Section 3.6,

$$u^p : \mathrm{Ch}^p(X)_{\mathrm{hom}} \to J^p(X),$$

where

$$J^p(X) = \frac{H^{2p-1}(X, \mathbf{C})}{F^p + H^{2p-1}(X, \mathbf{Z})}$$

is the complex torus associated with the Hodge structure $H^{2p-1}(X)$ in Section 3.6. We recall the definition of u^p: if $Z \in Z^p(X)_{\mathrm{hom}}$, write $Z = \partial\Gamma$ for some piecewise differentiable $(2n - 2p + 1)$-chain Γ on X. Integration over Γ defines a linear functional on $\mathcal{A}_X^{2n-2p+1}$ via

$$\omega \mapsto \int_\Gamma \omega.$$

We have shown that

$$\left\{ \int_\Gamma - \right\} \in \frac{F^{n-p+1} H^{2n-2p+1}(X, \mathbf{C})^*}{H_{2n-2p+1}(X, \mathbf{Z})^*}$$

is well defined. By Poincaré duality the latter group equals $J^p(X)$ for any compact Kähler manifold X.

Now, if $f : X \to S$ is a smooth projective family, where S is any complex manifold, then we can construct a bundle of complex tori over S by taking the p-th *intermediate Jacobian* in each fiber and gluing them together. More precisely, we define the torus bundle $\mathcal{J}^p_{X/S}$ over S by the condition that its sheaf of holomorphic sections is given by

$$\frac{R^{2p-1} f_* \mathbf{C} \otimes \mathcal{O}_S}{F^p R^{2p-1} f_* \mathbf{C} \otimes \mathcal{O}_S + R^{2p-1} f_* \mathbf{Z}}.$$

Note that $\mathcal{J}^p_{X/S}$ is more than a torus bundle. Its total space is indeed a complex manifold, smooth and proper over S with a holomorphic projection map. It is called the family of intermediate Jacobians associated with $f : X \to S$.

Definition 9.1.1 Let $f : X \to S$ be a smooth projective family and $W \in Z^p(X)$ an algebraic cycle which has the property that its intersection with all fibers X_t for $t \in S$ is again of codimension p and homologous to 0. Then the section $t \mapsto \nu(t) = u^p(W \cap X_t)$ is called the *normal function* associated with the analytic family of cycles W_t.

Normal functions associated with families of algebraic cycles are clearly continuous sections, but we will see now that they are in fact holomorphic maps. Furthermore, they satisfy an additional differential equation, called horizontality.

Definition 9.1.2 A holomorphic section $\nu : S \to \mathcal{J}^p_{X/S}$ is called an *abstract normal function* if every lifting $\tilde{\nu} \in \Gamma(U, R^{2p-1} f_* \mathbf{C} \otimes \mathcal{O}_S)$ over any open subset $U \subset S$ is *horizontal*, i.e., satisfies the differential equation

$$\nabla \tilde{\nu} \in \Omega^1_U \otimes F^{p-1} R^{2p-1} f_* \mathbf{C} \otimes \mathcal{O}_S.$$

Note that this means that $\nabla \tilde{\nu}$ satisfies a system of first-order differential equations. Usually this property is stated by saying that $\nabla \tilde{\nu}$ vanishes in the quotient vector bundle $\Omega^1_U \otimes R^{2p-1} f_* \mathbf{C} \otimes \mathcal{O}_S / F^{p-1}$ or, equivalently, that ν vanishes under

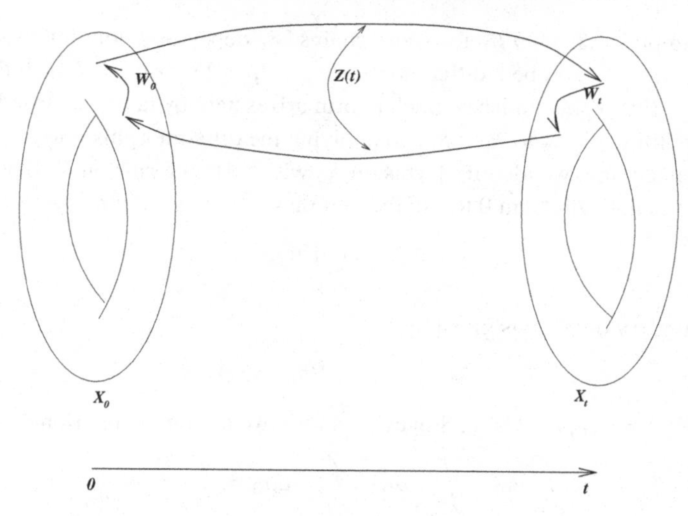

Figure 9.1 Deformation of a cycle.

the natural map

$$\nabla : \mathcal{J}^p_{X/S} \to \Omega^1_S \otimes R^{2p-1} f_* \mathbf{C} \otimes \mathcal{O}_S / F^{p-1},$$

which is well defined because $\nabla(F^p) \subset F^{p-1} \otimes \Omega^1_S$.

Theorem 9.1.3 *Under the above assumptions, every normal function associated with a family of algebraic cycles is an abstract normal function, i.e., it is holomorphic and satisfies the horizontality condition.*

Proof We have already proven such a statement in Lemma 4.7.1 for trivial families. This theorem was first proved by Ph. Griffiths. We give two proofs here: first a classical proof (essentially the one of Griffiths) and then, in Section 10.1, another one using Deligne cohomology and the Leray spectral sequence. One advantage of the second proof is that it leads to higher-dimensional generalizations (see the historical remarks at the end of this chapter).

The assertions are local and depend on one chosen tangent vector in S, so we may assume that S is the one-dimensional unit disk throughout the proof. The map $f : X \to S$ is diffeomorphic to the product $X_0 \times S$: Consider a lift $\tilde{\xi}$ of a tangent vector $\xi = \frac{\partial}{\partial t}$ of S in the origin; as shown in Section 4.1, this can be used to trivialize the family differentiably. We also recall the notion of the Lie derivative from Definition 4.5.2.

Write Z_t for the restriction of a cycle $\mathfrak{Z} \in Z^p(X)$ to a cycle Z_t of codimension p on X to X_t and we assume that the intersection is defined for all $t \in S$.

By assumption, $Z_t = \partial \mathscr{W}_t$ for some chains $\mathscr{W}_t$ depending continuously on t (see Fig. 9.1). Let α_0 be a differentiable $(2n - 2p + 1)$-form on X such that its Lie derivative $\mathscr{L}_\xi \alpha_0$ vanishes. Such a form arises here by taking a closed form on X_0, pulling it back to $X_0 \times S$, and applying the diffeomorphism to X. In this way of extending, we identify forms on X_0 with certain forms on X. If γ is the standard radial path from 0 to t in the unit disk S, then we define

$$\mathscr{W}(t) = \bigcup_{s \in \gamma} \mathscr{W}_s.$$

The boundary of $\mathscr{W}(t)$ is given by

$$\partial \mathscr{W}(t) = \mathscr{W}_t - \mathscr{W}_0 + Z(t),$$

where $Z(t) = \cup_{s \in \gamma} Z_s$. Using Stokes' theorem, we obtain the relation

$$\int_{\mathscr{W}_t} \alpha_0 = \int_{\mathscr{W}_0} \alpha_0 - \int_{Z(t)} \alpha_0 + \int_{\mathscr{W}(t)} d\alpha_0 = \int_{\mathscr{W}_0} \alpha_0 - \int_{Z(t)} \alpha_0.$$

Let α be a holomorphic section of the Hodge bundle $F^{n-p+1} R^{2n-2p+1} f_* \mathbf{C} \otimes \mathscr{O}_S$ which coincides with α_0 when restricted to the fiber X_0. The Taylor series of α can be written as

$$\alpha = \alpha_0 + t\alpha_1 + \cdots .$$

Therefore,

$$\frac{\partial}{\partial t}\left(\int_{\mathscr{W}_t} \alpha\right)(0) = \frac{\partial}{\partial t}\left(\int_{\mathscr{W}_t} \alpha_0\right)(0) + \int_{\mathscr{W}_0} \alpha_1$$

and

$$\frac{\partial}{\partial \bar{t}}\left(\int_{\mathscr{W}_t} \alpha\right)(0) = \frac{\partial}{\partial \bar{t}}\left(\int_{\mathscr{W}_t} \alpha_0\right)(0).$$

Now, as in the proof of Lemma 4.7.1, for any β with vanishing Lie derivative $\mathscr{L}_\xi \beta$, we have that

$$\frac{\partial}{\partial t}\left(\int_{\mathscr{W}_t} \beta\right)(0) = -\int_{Z_0} i_\xi \beta$$

as well as

$$\frac{\partial}{\partial \bar{t}}\left(\int_{\mathscr{W}_t} \beta\right)(0) = -\int_{Z_0} i_{\bar{\xi}} \beta.$$

In particular, since Z_0 is an algebraic subvariety, both integrals vanish if $i_\xi \beta$ or $i_{\bar{\xi}} \beta$ has no components of type $(n-p, n-p)$. In particular, if $\beta = \alpha_0 \in F^{n-p+1}$, this implies that

$$\frac{\partial}{\partial \bar{t}}\left(\int_{\mathscr{W}_t} \alpha\right)(0) = \frac{\partial}{\partial \bar{t}}\left(\int_{\mathscr{W}_t} \alpha_0\right)(0) = -\int_{Z_0} i_{\bar{\xi}} \alpha_0 = 0,$$

i.e., ν is holomorphic. The computation of $\partial/\partial t\,(\int_{\mathcal{W}_t}\alpha)(0)$ shows in a similar way that ν is horizontal. □

Now we turn to the so-called *infinitesimal invariants of normal functions*, first invented by Griffiths (1983) and later extended by Green (1989b) and Voisin (1988b).

In order to define those, let us assume that we have a smooth, proper morphism $f : X \longrightarrow S$, where S is Stein, for example a ball or a polydisk. We denote by $\mathcal{H}$ the flat vector bundle

$$\mathcal{H} \stackrel{\text{def}}{=} R^{2p-1}f_*\mathbf{C}\otimes\mathcal{O}_S,$$

and by $\mathcal{F}^i \subseteq \mathcal{H}$ the i-th Hodge subbundle.

Recall the definition of ∇ from Section 4.6. Flatness of ∇ means just that $\nabla\circ\nabla = 0$ which leads to the following definition.

Definition 9.1.4 The complex

$$\boxed{\mathcal{K}^\bullet_{X/S}(p) : \mathcal{F}^p \xrightarrow{\nabla} \mathcal{F}^{p-1}\otimes\Omega^1_S \xrightarrow{\nabla} \mathcal{F}^{p-2}\otimes\Omega^2_S \xrightarrow{\nabla} \cdots}$$

of sheaves on S is called the *p-th Koszul complex* associated with $f : X \longrightarrow S$.

The letter $\mathcal{K}$ stands for *Koszul* and we see below why the Koszul complexes are going to appear. Now let $\nu : S \longrightarrow \mathcal{J}^p_{X/S}$ be an arbitrary abstract normal function for the family $f : X \longrightarrow S$.

Lemma 9.1.5 *If $\tilde{\nu} \in \Gamma(U, \mathcal{H}|_U)$ is any local lifting of ν over $U \subset S$, then $\nabla\tilde{\nu} \in \Omega^1_U\otimes\mathcal{F}^{p-1}/\mathrm{Im}(\nabla(\mathcal{F}^p))$ is well defined independently of the lifting, and provides a section $\delta(\nu) \in H^0(U, \mathcal{H}^1(\mathcal{K}^\bullet_{X/S}(p)))$ of the first cohomology sheaf of $\mathcal{K}^\bullet_{X/S}(p)$.*

Proof Two liftings ν_1 and ν_2 differ by

$$\nu_1 - \nu_2 = \nu_f + \nu_{\mathbf{Z}},$$

where ν_f is a section of $\mathcal{F}^p$ and $\nu_{\mathbf{Z}}$ is a section of $R^{2p-1}f_*\mathbf{Z}$. However, $\nu_{\mathbf{Z}}$ is locally constant, hence $\nabla(\nu_{\mathbf{Z}}) = 0$ and $\nabla(\nu_f) \in \mathrm{Im}(\nabla(\mathcal{F}^p))$, i.e., $\nabla(\nu_1) = \nabla(\nu_2)$ in $\Omega^1_U \otimes \mathcal{F}^{p-1}/\mathrm{Im}(\nabla(\mathcal{F}^p))$. □

What we have just done is to make explicit the coboundary map in the long exact sequence for hypercohomology associated with the short exact sequence of complexes

$$0 \to \mathcal{K}^\bullet_{X/S}(p) \to \mathcal{H}\otimes\Omega^\bullet_S \to \mathcal{Q}^\bullet_{X/S}(p) \to 0$$

in the case where S is a Stein manifold (e.g., a polydisk) so that all liftings exist.

Here $\mathcal{Q}^\bullet_{X/S}(p)$ is the complex of sheaves defined as

$$\mathcal{Q}^a_{X/S}(p) = \Omega^a_S \otimes \mathcal{H}/\mathcal{F}^{p-a}.$$

Therefore, if we have a class ν together with an intermediate lifting over S (forgetting only the integral part)

$$\hat{\nu} \in H^0\big(S, \mathcal{Q}^0_{X/S}(p)\big),$$

then $\delta(\nu) \in H^0(S, \mathcal{H}^1(\mathcal{K}^\bullet_{X/S}(p)))$ is the image of $\hat{\nu}$ under the coboundary map. By definition of the coboundary map, this is $\nabla(\tilde{\nu})$ for a further lifting $\tilde{\nu} \in \Gamma(S, \mathcal{H}|_S)$ of $\hat{\nu}$, and hence of ν. We can refine $\delta(\nu)$ into a series of infinitesimal invariants as follows. Observe first that the complexes $\mathcal{K}^\bullet_{X/S}(p) = \mathcal{F}^{p-\bullet} \otimes \Omega^\bullet_S$ have a natural filtration by subcomplexes

$$F^k\mathcal{K}^\bullet_{X/S}(p) \overset{\text{def}}{=} \mathcal{F}^{p+k-\bullet} \otimes \Omega^\bullet_S, \quad k \geq 0.$$

Recall from Section 6.1 that a filtration on a complex defines a spectral sequence. In this case, the filtration of $\mathcal{K}^\bullet_{X/S}(p)$ by the subcomplexes $\Omega^\bullet_S \otimes \mathcal{F}^{p+k-\bullet}$ defines the spectral sequence

$$E_1^{a,b}(p) = \mathcal{H}^{a+b}(\mathrm{Gr}^a_F\, \mathcal{K}^\bullet_{X/S}(p)) \Longrightarrow \mathcal{H}^{a+b}(\mathcal{K}^\bullet_{X/S}(p)).$$

Since $\delta(\nu)$ is a section of $\mathcal{H}^1(\mathcal{K}^\bullet_{X/S}(p))$ we are particularly interested in the resulting filtration on this sheaf. It has successive quotients given by $E_\infty^{a,1-a}(p)$, $a = 0, \ldots, p$. For each $s \in S$ the images of the values at s of $\delta(\nu)$ in the stalks at s of the first quotient sheaf $F^0/F^1 = E_\infty^{0,1}(p)$ define a section, and hence a well-defined invariant

$$\boxed{\delta_1(\nu) \in H^0(S, \mathcal{H}^1(\mathrm{Gr}^0_F\, \mathcal{K}^\bullet_{X/S}(p))) = H^0(S, E_\infty^{0,1}(p)),}$$

called the *first infinitesimal invariant of Griffiths*. Because $E_\infty^{0,1}(p) \subset E_1^{0,1}(p)$, we may view $\delta_1(\nu)$ as a section of either one.

The second, third, and higher invariants are defined as follows. Suppose that $\delta_1(\nu) = 0$. This means that $\delta(\nu) \in F^1$ and one consider its image in F^1/F^2 yielding the second invariant

$$\delta_2(\nu) \in H^0(S, E_\infty^{1,0}(p)),$$

and inductively, if $\delta_1(\nu), \ldots, \delta_k(\nu)$ vanish, then there is a well-defined $(k+1)$-st infinitesimal invariant

$$\boxed{\delta_{k+1}(\nu) \in H^0(S, E_\infty^{k,1-k}(p)).}$$

The following definition was first proposed by Griffiths for $k = 0$ (see Griffiths, 1983) and later extended by Green for all k (see Green, 1989b).

Definition 9.1.6 The invariants $\delta_{k+1}(\nu) \in H^0(S, E_\infty^{k,1-k}(p))$ are called the *(higher) infinitesimal invariants of the normal function* ν.

The following property motivates the definition of infinitesimal invariants.

Proposition 9.1.7 *A locally constant normal function* $\nu : S \to \mathcal{J}^p_{X/S}$ *has locally liftings* $\tilde{\nu}$ *to flat sections of* $R^{2p-1}f_*\mathbf{C}$*. Equivalently, its infinitesimal invariant* $\delta(\nu)$ *vanishes, or, what is the same,* all *infinitesimal invariants* $\delta_i(\nu)$ *vanish for* $i \geq 1$.

Proof If ν is locally constant, then clearly $\delta(\nu) = 0$, because for local liftings $\delta(\nu) = \nabla\tilde{\nu} = 0$. Conversely, if $\delta(\nu) = 0$, one may write $\tilde{\nu} = \nabla(\nu_f)$ locally over U for some $\nu_f \in H^0(U, \mathcal{F}^p)$. Changing the lifting $\tilde{\nu}$ over U by ν_f creates a flat lifting. The second equivalence is obvious: by definition of the spectral sequence above, $\delta(\nu) = 0$ if and only if $\delta_i(\nu) = 0$ for all $i \geq 1$. □

Example 9.1.8 We have discussed the Hodge structures of weight 1 underlying abelian varieties in Section 4.3 in the examples. Normal functions and infinitesimal invariants for families of abelian varieties are very instructive models for us: let $f : X \to S$ be a universal family of abelian varieties equipped with some level structure, i.e., S is a finite cover of $\mathcal{A}_g$. Let V be the vector space $H^{1,0}(A)$ for some reference point $0 \in S$ corresponding to A. It is a representation of the Lie algebra $\mathfrak{sl}_g(\mathbf{C})$ in a natural way. Because f is universal, we can use the matrix description of Siegel's upper half-plane to see that $\Omega^1_{S,0} = \mathrm{Sym}^2(V)$. Also, all Hodge groups $H^{a,b}(A)$, using V and its dual $V^\vee$, are given as

$$H^{a,b}(A) = \textstyle\bigwedge^a V \otimes \bigwedge^b V^\vee.$$

All vector spaces occurring in the fibers of the Koszul complexes at the point 0 are therefore representations of the Lie algebra $\mathfrak{sl}_g(\mathbf{C})$. It is also true (see Problems 9.1.1–9.1.2 below) that the map induced by the Gauss–Manin connection is a morphism of representations. As a consequence, the cohomology of the Koszul complexes can be computed as a sum of irreducible representations of $\mathfrak{sl}_g(\mathbf{C})$. In Problem 9.1.2 the reader is invited to do this for $g \leq 3$. For higher dimensions this computation is related to Lie algebra cohomology and explained in detail in the work of Fakhruddin (1996).

Problems

9.1.1 Prove that the Koszul complexes for universal families of abelian varieties with any level structure are complexes of representations for the

Lie algebra $\mathfrak{sl}(g;\mathbf{C})$. Hint: use the example above and prove that cup product maps are maps of representations.

9.1.2 Show that the stalk of $\mathscr{H}^1(\mathcal{K}^\bullet_{X/S}(2))$ at any point $A \in \mathcal{A}_3$ is isomorphic to $\mathrm{Sym}^4(V)$ where $V = H^{1,0}(A)$ as above (what happens for $g \leq 2$?). Hint: decompose everything into irreducible representations.

9.2 The Griffiths Group of Hypersurface Sections

The quotient of cycles of codimension p which are homologically equivalent to 0 modulo those algebraically equivalent to 0 (see Section 3.6 for definitions) is called the *Griffiths group*:

$$\mathrm{Gr}^p(X) \stackrel{\text{def}}{=} \frac{\mathrm{Ch}^p_{\text{hom}}(X)}{\mathrm{Ch}^p_{\text{alg}}(X)}. \tag{9.1}$$

Griffiths first proved in 1969 that there exist threefolds for which this group (for $p = 2$) is nontrivial. In particular, homological and algebraic equivalence become different for cycles of codimension at least 2. This phenomenon can still be detected by using Abel–Jacobi invariants. For higher codimensions we see below as a corollary of Nori's theorem that the Abel–Jacobi invariants are in general not detecting the cycles faithfully. Here we present an infinitesimal method that works for sufficiently ample hypersurface sections of a four-dimensional smooth projective variety Y. In Theorem 9.3.1 we show that the Abel–Jacobi map is torsion for threefolds of degree ≥ 6 in $\mathbf{P}^4$. However, for quintic threefolds the statement is false.

Theorem 9.2.1 (Griffiths, 1969) *Let Y be a smooth quintic fourfold containing two-planes P_1 and P_2 such that the cohomology class of the difference $[P_1] - [P_2] \in H^{2,2}_{\text{prim}}(Y,\mathbf{Q})$ is a nonzero primitive $(2,2)$-form. Then the intersection of $P_1 - P_2$ with a very general hyperplane section X is homologically equivalent to 0, but no multiple of it is algebraically equivalent to 0.*

Proof We give a sketch of a slight modification of the original proof of Griffiths. We encourage the reader to work out the details (see Problem 9.3.1 below). To start, choose a Lefschetz pencil $(X_t)_{t\in\ell}$ of hyperplane sections of Y over a line ℓ. The total space $\tilde{Y}$ is a blowup of Y along the base locus of the pencil. We denote the structure map by $\bar{f} : \tilde{Y} \to \ell$. Denote by $U \subset \ell$ the set of points where the sections X_t are smooth and let $Y_U \stackrel{\text{def}}{=} f^{-1}(U)$ be the inverse image of U under $\bar{f}$. Then $f : Y_U \to U$ becomes the restriction of $\bar{f}$ to Y_U. Fix a base point $o \in U$. The cycles $(P_1 - P_2) \cap X_t$ induce a normal function ν. The image of ν under the map

$$H^0(U, \mathcal{J}^2_{Y_U/U}) \to H^1(U, R^3 f_* \mathbf{Z})$$

induced by the sequence defining intermediate Jacobians

$$0 \to R^3 f_* \mathbf{Z} \to R^3 f_* \mathbf{C} \otimes \mathcal{O}_U / F^2 \to \mathcal{J}^2_{Y_U/U} \to 0$$

is given by a cohomology class in $H^1(U, R^3 f_* \mathbf{Z})$ which in the remainder of the proof will be identified with $[P_1] - [P_2]$. We identify $H^1(U, R^3 f_* \mathbf{Q})$ with the subspace

$$L^1 H^4(Y_U, \mathbf{Q}) = \mathrm{Ker}(H^4(Y_U, \mathbf{Q}) \to H^4(X_o, \mathbf{Q}))$$

via the Leray spectral sequence. It is easy to check that the class of $[P_1] - [P_2] \in H^{2,2}_{\mathrm{prim}}(Y, \mathbf{Q})$ restricts to a nonzero primitive cohomology class in $L^1 H^4(Y_U, \mathbf{Q})$ because the kernel

$$\mathrm{Ker}(H^4(\tilde{Y}, \mathbf{Q}) \to H^4(Y_U, \mathbf{Q}))$$

consists of cohomology classes with support in the singular fibers $\tilde{Y} - Y_U$, which map to 0 in primitive cohomology (they are multiples of $\omega = c_1(L)$). Furthermore, this class in $L^1 H^4(Y_U, \mathbf{Q})$ can be identified with the class in $H^1(U, R^3 f_* \mathbf{Q})$ obtained above after tensoring with $\mathbf{Q}$: this can be checked by choosing a covering of U by simply connected sets U_i and choosing topological cochains ζ_i such that $\partial \zeta_i = P_1 - P_2$ over U_i. The Poincaré dual classes ζ^i define a Čech cocycle $\zeta^{ij} \in H^1(U, R^3 f_* \mathbf{Q})$ that is equal to both classes. Therefore, we conclude that the normal function cannot be lifted to a flat section of $R^3 f_* \mathbf{C} \otimes \mathcal{O}_U / F^2$ and therefore also not to a section of $R^3 f_* \mathbf{C} \otimes \mathcal{O}_U$ over U. This proves that the Abel–Jacobi map is already nontorsion for a general $t \in U$. It remains to be shown that the cycles $(P_1 - P_2) \cap X_t$ are not in general algebraically equivalent to 0. However, $H^3(X_t)$ has no fixed part, because $H^3(Y) = 0$, and the action of the monodromy group $\pi_1(U, o)$ on $H^3(X_o, \mathbf{Q})$ is irreducible. If now all cycles $(P_1 - P_2) \cap X_t$ were algebraically equivalent to 0, then their Abel–Jacobi classes would be contained in the subspace $H^{2,1}(X_o) \oplus H^{1,2}(X_o)$, which contradicts the irreducibility of the monodromy action and the fact that $H^{3,0}(X_o) = \mathbf{C}$, as the canonical bundle of X_o is trivial. □

In this chapter we also prove the following very similar theorem. Notice that for the proof of this theorem one could also use the same topological methods as in the previous proof, but we want to present our new infinitesimal methods which also apply to many other cases.

Theorem 9.2.2 *Let Y be a smooth, projective fourfold with $H^{1,2}(Y) = 0$ and L a sufficiently ample line bundle on Y. Assume that Y carries a nonzero primitive*

(2, 2)-form which arises from an algebraic cycle Z. Then the intersection of Z with a very general hypersurface section X in L is homologically equivalent to 0*, but no multiple of it is algebraically equivalent to* 0*. In other words, $Z \cap X$ defines a nontorsion element in the Griffiths group of X.*

The literature contains several variations of Theorem 9.2.2 (Griffiths, 1969; Müller-Stach, 1992; Nori, 1993; Voisin, 1988b; Wu, 1990b). An explanation of the notion of *sufficient ampleness*, which is used frequently, is necessary here. In general, given an ample line bundle L, we mean by such a statement that the conclusion holds whenever L is replaced by a sufficient power L^m of itself. There is also the notion of *very general*: a point in a moduli space S is called very general with respect to some property if the assertion of this property holds for all moduli points in S outside a countable union of proper analytic subsets. This is supposed to be a variant of the notion of *general*, for which one takes out a finite union of proper analytic subsets.

But before we begin the proof, we need another simpler definition of the first infinitesimal invariant $\delta_1(\nu)$. Let Y be a four-dimensional projective manifold and L an ample line bundle whose general section is smooth. Let X be a smooth hypersurface section of Y in the linear system defined by L. Recall that $\delta_1(\nu)$ is a section of a vector bundle over some parametrizing set of hypersurface sections and that the value at the point o (corresponding to X) belongs to the middle cohomology of

$$H^1(X,\Omega_X^2) \xrightarrow{\sigma} H^2(X,\Omega_X^1)\otimes\Omega_{S,o}^1 \xrightarrow{\sigma} H^3(X,\mathcal{O}_X)\otimes\Omega_{S,o}^2, \qquad (9.2)$$

where σ is induced by the Gauss–Manin connection. The alternative description of this group starts as follows. There is an exact sequence

$$0 \to \Omega_X^1 \otimes L^{-1} \to \Omega_Y^2|_X \to \Omega_X^2 \to 0$$

obtained by taking the second exterior power in the conormal bundle sequence. It induces a long exact sequence

$$\cdots \to H^1(X,\Omega_X^2) \to H^2(X,\Omega_X^1 \otimes L^{-1}) \to H^2(X,\Omega_Y^2|_X) \to H^2(X,\Omega_X^2) \to \cdots.$$

The natural map $H^2(Y,\Omega_Y^2) \to H^2(X,\Omega_Y^2|_X)$ sends primitive classes to 0 in $H^2(X,\Omega_X^2)$, thereby inducing a natural map

$$\alpha : H^{2,2}_{\mathrm{prim}}(Y) \to \frac{H^2(X,\Omega_X^1\otimes L^{-1})}{\mathrm{Im}(H^1(X,\Omega_X^2))}.$$

We have in addition a canonical map

$$^{\mathsf{T}}m : H^2(X,\Omega_X^1\otimes L^{-1}) \to H^2(X,\Omega_X^1)\otimes H^0(X,L|_X)^\vee$$

induced by the multiplication map

$$m : H^2(X, \Omega_X^1 \otimes L^{-1}) \otimes H^0(X, L|_X) \to H^2(X, \Omega_X^1).$$

We obtain therefore, by composition, a map

$$\beta : H^{2,2}_{\mathrm{prim}}(Y) \to \frac{H^2(X, \Omega_X^1) \otimes H^0(X, L|_X)^\vee}{\mathrm{Im}(H^1(X, \Omega_X^2))}.$$

Now we look at the associated family $f : \mathcal{X} \to S$ of smooth hypersurface sections defined by L. Here, by definition, S is an open subset of $\mathbf{P}H^0(Y, L)$ which we can identify with an open subset of $H^0(X, L|_X)$. This implies that the tangent space to S at any point is isomorphic to $H^0(X, L|_X)$. Let us write T for it. We can now make the connection with the previous definition of $\delta_1(\nu)$ by going back to (9.2). Indeed, one has $\sigma = {}^{\mathsf{T}}m \circ \delta$; in other words, the diagram

$$\begin{array}{ccc} H^1(X, \Omega_X^2) & \xrightarrow{\delta} & H^2(X, \Omega_X^1 \otimes L^{-1}) \\ & \searrow^{\sigma} & \downarrow {}^{\mathsf{T}}m \\ & & H^2(X, \Omega_X^1) \otimes T^* \end{array}$$

is commutative. We leave this verification to the reader. It then follows that the target of β contains the cohomology group of the sequence (9.2) which is the value at o of the first infinitesimal invariant $\delta_1(\nu)$. Indeed, the following result, which we state without proof, links the map β to the first infinitesimal invariant.

Lemma 9.2.3 (Voisin, 1988b) *If Z is an algebraic cycle on Y that has a primitive cohomology class of type $(2, 2)$, then the first infinitesimal invariant $\delta_1(\nu)$ associated with the normal function ν of the family of cycles $Z \cap X_s$ is equal to $\beta(\mathrm{cl}^2(Z))$.*

We draw a first consequence.

Corollary *In the situation above, if $H^{1,2}(Y) = 0$ and L is sufficiently ample, then the associated infinitesimal invariant $\delta_1(\nu)$, and therefore $\delta(\nu)$, is nonzero.*

Proof In view of Lemma 9.2.3, we only need to prove that β is injective. The map $H^2(Y, \Omega_Y^2) \to H^2(X, \Omega_Y^2|_X)$ is injective because $H^2(Y, \Omega_Y^2 \otimes L^{-1}) = 0$ for L sufficiently ample. Furthermore, the map

$$H^2(X, \Omega_X^1 \otimes L^{-1}) \to H^2(X, \Omega_X^1) \otimes H^0(X, L|_X)^\vee$$

is injective because, as we shall presently show, its Serre dual, the multiplication map

$$H^1(X, \Omega_X^2) \otimes H^0(X, L|_X) \to H^1(X, \Omega_X^2 \otimes L), \tag{9.3}$$

is surjective if L is sufficiently ample. It follows that β is injective.

So it remains to prove our claim that the map (9.3) is surjective under the condition that $H^{1,2}(Y) = 0$ and L is sufficiently ample. To show this, first remark that since X is a threefold, one has $\Omega_X^2 \cong T_X \otimes K_X$. Consequently,

$$H^1(X, \Omega_X^2 \otimes L) = H^1(X, T_X \otimes K_X \otimes L).$$

Now look at the exact sequences

$$0 \to T_X \to T_Y|_X \to \mathcal{O}_X(L) \to 0$$

and

$$0 \to T_Y \otimes L^{-1} \to T_Y \to T_Y|_X \to 0.$$

Twisting the first one with $K_X \otimes L = K_Y|_X \otimes L^2$ yields

$$0 \to T_X \otimes K_X \otimes L \to T_Y|_X \otimes K_X \otimes L \to K_X \otimes L^2 \to 0,$$

so that we obtain as part of the long exact sequence,

$$\cdots \to H^0(X, K_X \otimes L^2) \to H^1(X, T_X \otimes K_X \otimes L) \to H^1(X, T_Y \otimes K_X \otimes L) \to \cdots.$$

The second sequence twisted with $K_Y|_X \otimes L^2$ shows that the last term vanishes for L sufficiently ample, because for the same reason $H^1(Y, T_Y \otimes K_Y \otimes L^2)$ and $H^2(Y, T_Y \otimes K_Y \otimes L)$ also both vanish. It follows that $H^1(X, \Omega_X^2 \otimes L)$ is a quotient of $H^0(Y, K_Y \otimes L^3)$. In a similar way, one shows that $H^1(X, \Omega_X^2)$ is a quotient of $H^0(Y, K_Y \otimes L^2)$ under the assumption that $H^{1,2}(Y) = 0$. Hence it remains to be proven that the multiplication map

$$H^0(Y, K_Y \otimes L^2) \otimes H^0(Y, L) \to H^0(Y, K_Y \otimes L^3)$$

is surjective for L sufficiently ample. Consider the product $Y \times Y$ with the two projections p_1, p_2. The line bundle $\mathcal{L} \stackrel{\text{def}}{=} p_1^*(K_Y \otimes L^2) \otimes p_2^*(L)$ is sufficiently ample on $Y \times Y$, because L is sufficiently ample. If $\Delta \subset Y \times Y$ denotes the diagonal, then we have the exact sequence

$$0 \to I_\Delta \to \mathcal{O}_{Y \times Y} \to \mathcal{O}_Y \to 0$$

because Δ is isomorphic to Y. The multiplication map above coincides with the restriction map $H^0(Y \times Y, \mathcal{L}) \to H^0(\Delta, \mathcal{L}_\Delta)$. Therefore, it is enough to show that $H^1(Y \times Y, I_\Delta \otimes \mathcal{L}) = 0$, which follows from the fact that $\mathcal{L}$ is sufficiently ample. □

Now we are in a position give the proof of Theorem 9.2.2.

Proof of Theorem 9.2.2 Let ν be the normal function associated with the family of intersections $Z \cap X$. We have already shown that $\delta(\nu)$ is nonzero (see the previous corollary). This implies that $Z \cap X$ has a nontrivial image in the intermediate Jacobian of X for very general X in $|L|$. It remains to be shown that $Z \cap X$ is not algebraically equivalent to 0 for very general X. By Lemma 4.7.1(2), it is enough to prove that $J^2(X)$ has no subtorus with tangent space contained in $H^{2,1}(X) \oplus H^{1,2}(X)$ for very general X. Note that $H^{3,0}_{\text{var}}(X) \neq 0$ because, as L is sufficiently ample, the dimension of $H^0(X, K_X)$ tends to infinity for sufficiently ample L by the adjunction formula, whereas the dimension of the fixed part of the cohomology stays bounded. Therefore, $H^{2,1}(X) \oplus H^{1,2}(X)$ is a proper subspace of $H^3_{\text{var}}(X)$, and it remains to be shown that $H^3_{\text{var}}(X)$ has no subspaces that are invariant under the monodromy action. This can be done as in the proof of Theorem 9.2.1 in this chapter with a slight modification. By the irreducibility of the monodromy action (see Theorem 4.2.2), there are no $(1,2)$-classes that remain type $(1,2)$ in our deformation, because by the assumption $H^{1,2}(Y) = 0$, all classes we consider are variable cohomology classes. This implies that there is no such nontrivial subtorus on a very general member of our deformation. □

The literature contains slightly more general statements, such as the following theorem (see Müller-Stach, 1992).

Theorem 9.2.4 *Let Y be a smooth fourfold. Then $H^{2,2}_{\text{prim}}(Y, \mathbf{Q})$ is a subquotient of the Griffiths group $\mathrm{Gr}^2(X) \otimes \mathbf{Q}$ for any very general sufficiently ample hypersurface section X of Y.*

The reader is also encouraged to read the work of Wu (1990b).

Problem

9.2.1 Show that the Fermat quintic fourfold always contains a pair of planes that satisfy the conditions of Theorem 9.2.1.

9.3 The Theorem of Green and Voisin

In this section we prove a theorem of Green and Voisin (see Green, 1989b) stating that the image of the Abel–Jacobi map for codimension m-cycles on a very general smooth hypersurface of degree $d \geq 2+4/(m-1)$ in $\mathbf{P}^{2m}_{\mathbf{C}}$ is torsion for $m \geq 2$. This theorem has been a strong motivation for many subsequent results. In the proof we make use of our earlier Koszul theoretic computations.

If $f : X \to S$ is the universal family of smooth hypersurface sections of degree d in $\mathbf{P}^{2m}$, then one can prove that the complex of sheaves $\mathcal{K}^\bullet_{X/S}(m)$ is exact in degrees 0 and 1 for all $d \geq 2 + 4/(m-1)$. Using this fact, one can prove the following theorem of Green and Voisin.

Theorem 9.3.1 *If $X \subset \mathbf{P}^{2m}$, $m \geq 2$ is a very general smooth hypersurface section of degree $d \geq 2 + 4/(m-1)$, then the image of the Abel–Jacobi map of X is contained in the torsion points of the intermediate Jacobian $J^m(X)$ of X.*

Remark 9.3.2 As explained in Section 9.2, by *very general* we mean that there exists a countable union (as opposed to *general* by which we mean a finite union) of proper analytic subsets of the linear system of hypersurfaces of degree d such that the statement holds for all X in the complement.

Proof Let $f : X \to S$ be the universal family of smooth hypersurface sections of degree d in $\mathbf{P}^{2m}$. If one has a cycle that exists on a very general fiber of f then, by the definition of very general, it exists as a cycle in $\mathrm{Ch}^m(X)$ after replacing S by a finite unramified covering $T \to S$. The reason is that over S there exists a family of relative Chow varieties parametrizing the effective cycles on X_s. Over the complement of a countable union of proper subvarieties of S, these form an algebraic fiber bundle by the countability of the number of components of Chow varieties (see Harris, 1992). Therefore, any point in this family can be extended to a multisection of the family of relative Chow varieties over this complement. This defines T, if we observe that any cycle may be written as the difference of effective ones.

The Koszul complexes $\mathcal{K}^\bullet_{X/S}(m)$ and in particular their homology do not change under finite unramified base changes as above, so that we may assume that S is in fact the universal family of smooth hypersurface sections of degree d in $\mathbf{P}^{2m}$ for the following argument. By Proposition 9.1.7, it is clear that the vanishing of the cohomology of every complex

$$H^{m+k,m-1-k}_{\mathrm{prim}}(X_t) \to \Omega^1_{S,t} \otimes H^{m+k-1,m-k}_{\mathrm{prim}}(X_t) \to \Omega^2_{S,t} \otimes H^{m+k-2,m-k+1}_{\mathrm{prim}}(X_t)$$

at the middle position (for all $k \geq 0$) implies that every infinitesimal invariant arising from any cycle of codimension m is 0. By duality this is equivalent to the statement that the complexes

$$\bigwedge^2 T_{S,t} \otimes H^{m-k+1,m+k-2}_{\mathrm{prim}}(X_t) \to T_{S,t} \otimes H^{m-k,m+k-1}_{\mathrm{prim}}(X_t) \to H^{m-k-1,m+k}_{\mathrm{prim}}(X_t)$$

are exact in the middle. Using the theory of Jacobian rings (see Macaulay's lemma 7.4.1), and the isomorphisms $R_F^{t(p)} = H^{p,2m-1-p}_{\mathrm{pr}}(X_t)$ with $t(p) = d(2m-p) - 2m - 1$ of Lemma 5.4.3), this comes down to proving the exactness of the complexes of vector spaces

$$\bigwedge^2 W \otimes R^{d(m+k-1)-2m-1} \longrightarrow W \otimes R^{d(m+k)-2m-1} \longrightarrow R^{d(m+k+1)-2m-1}$$

in the middle, where $W = \mathbf{P}H^0(\mathbf{P}^{2m}, \mathcal{O}_{\mathbf{P}^{2m}}(d))$. This, in turn, we have already shown in Proposition 7.4.5 as long as $d(m+k) - 2m - 1 > d$ for all $k \geq 0$, i.e., $d \geq 2 + 4/(m-1)$. Therefore, under the condition $d \geq 2 + 4/(m-1)$, every normal function that arises can be lifted locally to a constant section of the Hodge bundle $R^{2m-1} f_* \mathbf{C} \otimes \mathcal{O}_S$. Such local liftings are even unique up to sections of $R^{2m-1} f_* \mathbf{Z}$, because locally there are no flat sections in $\mathcal{F}^m$ by the injectivity of the sequences

$$H^{m+k,m-1-k}_{\text{prim}}(X_t) \longrightarrow \Omega^1_{S,t} \otimes H^{m+k-1,m-k}_{\text{prim}}(X_t)$$

for all $k \geq 0$ (using Proposition 7.4.5 again).

To finish the proof of the theorem, we observe that the primitive cohomology of hypersurfaces has strong additional properties connected with the Picard–Lefschetz formalism. We have the following claim.

Claim: If the normal function ν has local constant liftings, which are unique up to sections of $R^{2m-1} f_* \mathbf{Z}$, then ν is torsion as a holomorphic section of the bundle of intermediate Jacobians.

Let $\tilde{\nu}$ be a locally constant lifting of ν in a neighborhood of $0 \in S$. Take any loop γ with base point 0. By parallel transport along γ we obtain a new lifting $\hat{\nu}$ of ν locally around 0 with the property that

$$\rho(\gamma)(\tilde{\nu}(0)) - \tilde{\nu}(0) = \hat{\nu}(0) - \tilde{\nu}(0) \in H^{2m-1}(X_0, \mathbf{Z}).$$

Here $\rho : \pi(S, 0) \longrightarrow H^{2m-1}(X_0, \mathbf{C})$ is the monodromy representation associated with the local system $R^{2m-1} f_* \mathbf{C}$.

Thus it remains to be proven that if a class $a \in H^{2m-1}(X_0, \mathbf{C})$ has the property that $\rho(\gamma)(a) - a \in H^{2m-1}(X_0, \mathbf{Z})$ for all $\gamma \in \pi_1(S, 0)$, then some integer multiple of a is in $H^{2m-1}(X_0, \mathbf{Z})$.

We prove this using the Picard–Lefschetz formalism, which we have outlined in Section 4.2. In order to do this, one chooses a Lefschetz pencil of hyperplane sections through X_0, parametrized by a line ℓ in the projective space of all degree-d hypersurfaces in $\mathbf{P}^{2m}$. Over some nonempty $U = \ell \cap U$ the fibers are smooth and we have vanishing cocycles δ_j associated with each point $t_j \in \ell - U$. Recall that these are all conjugate under the monodromy group of the universal family of smooth degree d hypersurfaces and that these thus span an irreducible submodule of $H^{2m-1}(X_0; \mathbf{Q})$. On the other hand, they also span the so-called variable cohomology group, which is the whole group because we are in odd degree.

Then, according to the Picard–Lefschetz formula, the image of γ_i under the monodromy representation ρ satisfies

$$\rho(\gamma_i)(a) - a = \pm(a, \delta_i)\delta_i.$$

As the δ_i generate $H^{2m-1}(X_0; \mathbf{Q})$, by Poincaré duality over $\mathbf{Q}$, we obtain $a \in H^{2m-1}(X_0; \mathbf{Q})$. □

Problems

9.3.1 Show that the direct image of Ω^p_Z of a cyclic Galois covering $f : Z \to \mathbf{P}^3$ of degree m branched along a smooth divisor B of degree $m \cdot d$ in $\mathbf{P}^3$ is given by

$$f_*\Omega^p_Z = \Omega^p_{\mathbf{P}^3} \oplus \bigoplus_{i=1}^{m-1} \Omega^p_{\mathbf{P}^3}(\log B) \otimes L^{-i},$$

where $L = \mathcal{O}(d)$. Prove a similar formula for $f_*\Theta_Z$. Hint: study the eigenspace decomposition analogous to $f_*\mathcal{O}_Z = \bigoplus_{i=0}^{m-1} L^{-i}$ under the cyclic group by using the group action on the pushdown of the sheaves of differential forms. You should also consult Müller-Stach (1994).

9.3.2 Use the previous exercise to compute $H^{p,3-p}(Z)_{\text{prim}}$ and $H^1(Z, \Theta_Z)$ in terms of logarithmic forms on $\mathbf{P}^3$.

9.3.3 Show that, for $m \geq 2$ and $d \geq 5$, the Koszul complexes $\mathcal{K}^\bullet_{Z/S}(2)$ for the universal family of cyclic coverings of $\mathbf{P}^3$ are exact in degrees 0 and 1. Conclude and state a result analogous to the theorem of Green and Voisin. What happens for $d = 4$ and $m = 2$?

Bibliographical and Historical Remarks

As we have seen, Abel–Jacobi maps are not surjective in general. This follows from the Green–Voisin result, but also from the countability of the Griffiths group observed by Griffiths, because intermediate Jacobians are uncountable objects. Mumford (1969b) and Roĭtman (1974), on the other hand, proved that the Albanese maps for zero-cycles are in general far from being injective. The first instance where this occurs is on a complex projective surface with geometric genus $p_g > 0$. A famous conjecture of Bloch asserts that the Albanese map for complex projective surfaces with $p_g = 0$ should be injective. This was verified in many cases but remains open in general. See Voisin (1994b) for results about surfaces in $\mathbf{P}^3$ using higher generalizations of Abel–Jacobi maps.

Normal functions have also been studied by means of Picard–Fuchs equations in del Angel and Müller-Stach (2002). There the idea is that to show that a normal function is not contained in the integral lattice at a general point, it must be shown that the normal function does not satisfy the same differential equation as the periods of the given family. Of course this approach is very much related to Griffiths' idea of infinitesimal invariants.

Several authors have studied the *Griffiths group* of Calabi–Yau threefolds. Griffiths (1969) first proved that some quintic threefolds have a nontrivial (and nontorsion) Griffiths group. Later Clemens showed that a very general quintic threefold has an infinite-dimensional Griffiths group, using degeneration techniques. Voisin (1992) reproved this theorem using Noether–Lefschetz loci and infinitesimal invariants. Paranjape (1991) and Bardelli and Müller-Stach (1994) showed that a very general Calabi–Yau threefold obtained as a complete intersection of two cubic hypersurfaces in $\mathbf{P}^5$ has infinite-dimensional Griffiths group $\mathrm{Gr}^2(X)$ (and also studied other examples). The computation relies on the fact that a cycle arising from the intersection with a primitive nonzero cohomology class of type (2, 2) on a fourfold has a nontrivial infinitesimal invariant on the very general hyperplane section X. Then a monodromy argument due to H. Clemens (1977) (which was also applied to study quintic threefolds) is used to show that there is a countable number of linearly independent classes. Voisin (1994a, 2000) (using Noether–Lefschetz loci) has shown that every nonrigid very general Calabi–Yau threefold has infinite-dimensional Griffiths groups $\mathrm{Gr}^2(X)$.

Collino and Pirola (1995) have computed the *infinitesimal invariant* of the cycle $C - C^-$ on a Jacobian of a genus 3 curve C. The value turns out to be precisely the quartic equation of the canonical image of C inside $\mathbf{P}^2$ interpreted as a section of the appropriate Koszul bundle. Using this they were able to show that the Griffiths groups of Jac(C) for very general C is nontorsion (we know that it is even infinite-dimensional by a theorem of Bardelli (1989) and Nori (1989). M. Nori and N. Fakhruddin have generalized these results by computing the cohomology of the local systems of primitive cohomology groups of the universal family of Hodge structures over the Siegel space. For example, Fakhruddin (1996) obtains the following results.

Theorem *The Griffiths groups of codimension* 3 *and* 4 *of the very general abelian variety of dimension* 5 *are of infinite rank.*

Theorem *The Griffiths groups of codimension p cycles on the very general Prym variety of dimension $g \geq 5$ is nontorsion for $3 \leq p \leq g - 1$.*

10

Applications to Algebraic Cycles: Nori's Theorem

Deligne cohomology is a tool that makes it possible to unify the study of cycles through an object that classifies extensions of (p, p)-cycles by points in the p-th intermediate Jacobian (which is the target of the Abel–Jacobi map on cycles of codimension p). This is treated in Section 10.1 with applications to normal functions.

Before giving the proof of Nori's theorem in Section 10.6, we need some results from mixed Hodge theory. These are proven in Section 10.2 where we also state different variants of the theorem. Sections 10.3 and 10.4 treat a local-to-global principle and an extension of the method of Jacobian representations of cohomology which are both essential for the proof. We finish the chapter with some applications of Nori's theorem and discuss the conjectured filtrations on the Chow groups to which these lead.

10.1 A Detour into Deligne Cohomology with Applications

Here we introduce Deligne cohomology in the form first defined by P. Deligne. We illustrate its connections to intermediate Jacobians and explain its functorial properties. Then we give the second proof of Theorem 9.1.3. The results of this section are not needed for an understanding of the rest of the chapter, and some readers may want to skip this section and read it later. In the historical remarks at the end of the chapter we point out some further directions where Deligne–Beilinson cohomology becomes more important. Deligne cohomology was defined first by P. Deligne and later extended by A. Beilinson. We use here mainly the original version of Deligne without growth conditions for noncompact spaces. Beilinson later imposed such growth conditions in order to get a more functorial theory. The extension of Beilinson has been worked out in detail in Esnault and Viehweg (1988).

Definition 10.1.1 Let X be a Kähler manifold. Define the *analytic Deligne complex* on X by

$$\boxed{\mathbf{Z}_{\mathcal{D}}(p) : (2\pi \mathrm{i})^p \mathbf{Z} \to \mathcal{O}_X \to \cdots \to \Omega_X^{p-1}.}$$

This is a complex of sheaves in the analytic topology and we put the first sheaf $(2\pi\mathrm{i})^p\mathbf{Z}$, which is a constant subsheaf of $\mathbf{C}$, in degree 0. Hence the last sheaf, Ω_X^{p-1}, sits in degree p. We denote by

$$H^{2p}_{\mathcal{D}}(X, \mathbf{Z}(p)) \stackrel{\text{def}}{=} \mathbf{H}^{2p}(X, \mathbf{Z}_{\mathcal{D}}(p)) \tag{10.1}$$

the $2p$-th hypercohomology of this complex and call it the *Deligne cohomology group* of X.

The Deligne complex $\mathbf{Z}_{\mathcal{D}}(p)$ sits in an exact sequence

$$0 \to \Omega_X^{<p}[-1] \to \mathbf{Z}_{\mathcal{D}}(p) \to \mathbf{Z}(p) \to 0,$$

where $\Omega_X^{<p}$ is the complex

$$0 \to \mathcal{O}_X \to \Omega_X^1 \to \cdots \to \Omega_X^{p-1}$$

and the symbol $[-1]$ denotes a shift of degree 1 in a complex. As a consequence, there exist long exact sequences of cohomology groups

$$\begin{aligned} \cdots &\to H^{2p-1}(X, \mathbf{Z}(p)) \to \mathbf{H}^{2p-1}\left(X, \Omega_X^{<p}\right) \to H^{2p}_{\mathcal{D}}(X, \mathbf{Z}(p)) \\ &\to H^{2p}(X, \mathbf{Z}(p)) \to \mathbf{H}^{2p}\left(X, \Omega_X^{<p}\right) \to H^{2p+1}_{\mathcal{D}}(X, \mathbf{Z}(p)) \to \cdots. \end{aligned}$$

If X is a compact Kähler manifold, then we know by the degeneration of the Hodge-to-de Rham spectral sequence from Section 6.3 that $\mathbf{H}^i(X, \Omega_X^{<p}) \cong H^i(X, \mathbf{C})/F^i$, or equivalently that

$$F^p H^i(X, \mathbf{C}) = \operatorname{Ker}(H^i(X, \mathbf{C}) \to \mathbf{H}^i(X, \Omega_X^{<p})).$$

In particular, we have a short exact sequence

$$\boxed{0 \to J^p(X) \to H^{2p}_{\mathcal{D}}(X, \mathbf{Z}(p)) \to H^{p,p}(X, \mathbf{Z}) \to 0,}$$

where $H^{p,p}(X, \mathbf{Z}) \stackrel{\text{def}}{=} \{\alpha \in H^{2p}(X, \mathbf{Z}) \mid \alpha_{\mathbf{C}} \in F^p\}$. Here, $\alpha_{\mathbf{C}}$ denotes the image of α under the change-of-coefficients map $H^{2p}(X, \mathbf{Z}) \to H^{2p}(X, \mathbf{C})$. Note that $H^{p,p}(X, \mathbf{Z})$ contains by definition all torsion classes in $H^{2p}(X, \mathbf{Z})$, because for them $\alpha_{\mathbf{C}} = 0$ and is therefore contained in F^p.

Example 10.1.2 In the case of $p = 1$, all the sequences just described are very familiar objects: $\mathbf{Z}_{\mathcal{D}}(1) = ((2\pi\mathrm{i})\mathbf{Z} \to \mathcal{O}_X)$, a complex starting in degree 0. It is well known that we have an exponential sequence

$$0 \to (2\pi\mathrm{i})\mathbf{Z} \to \mathcal{O}_X \xrightarrow{\exp} \mathcal{O}_X^* \to 0$$

on every complex manifold. Therefore, we obtain a natural quasi-isomorphism of complexes

$$\exp : \mathbf{Z}_{\mathcal{D}}(1) \xrightarrow{\cong} \mathcal{O}_X^*[-1].$$

This implies isomorphisms

$$H^i_{\mathcal{D}}(X, \mathbf{Z}(1)) \cong H^{i-1}(X, \mathcal{O}_X^*)$$

for all $i \geq 0$. For example, we obtain in this way that

$$H^2_{\mathcal{D}}(X, \mathbf{Z}(1)) \cong H^1(X, \mathcal{O}_X^*) \cong \mathrm{Pic}(X)$$

is equal to the Picard group of X, and

$$H^1_{\mathcal{D}}(X, \mathbf{Z}(1)) \cong H^0(X, \mathcal{O}_X^*)$$

is the group of global holomorphic units on X. Other interesting examples are the groups $H^2_{\mathcal{D}}(X, \mathbf{Z}(2))$: they parametrize all isomorphism classes of pairs (L, ∇) on a compact Kähler manifold X, where L is a holomorphic line bundle and ∇ is a holomorphic connection on L (see Problem 10.1.1 below). The interested reader will find more details about Deligne cohomology in Esnault and Viehweg (1988), in particular a description of its product structure.

Now let us explain the main technique of this section: the use of Deligne cohomology when we deform a given variety X_o in a family. This is related to normal functions in the following form. If $f : X \longrightarrow S$ is any projective family (so in particular f has smooth projective fibers) and $\alpha \in H^{2p}_{\mathcal{D}}(X, \mathbf{Z}(p))$ is a global class in Deligne cohomology which is homologous to zero on all fibers, then α gives rise to a normal function because $\alpha|_{X_t} \in H^{2p}(X_t, \mathbf{Z}(p))$ has a zero image in $H^{p,p}(X_t, \mathbf{Z})$ and therefore lives in $J^p(X_t)$:

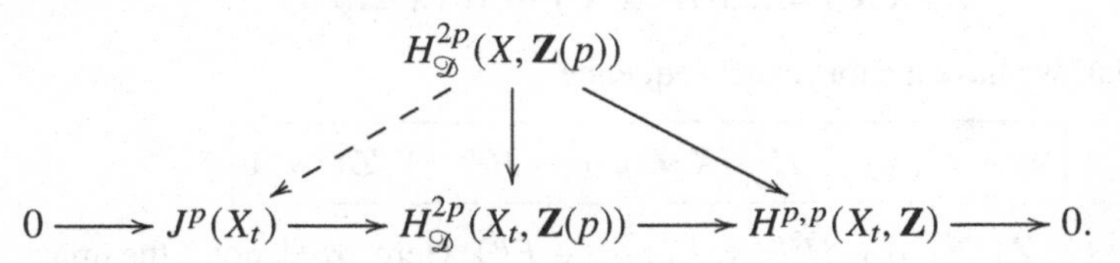

So, even if α is not the class of an algebraic cycle, it nevertheless defines a normal function. In fact, we prove a slightly more general version of Theorem 9.1.3, as follows.

Theorem 10.1.3 *Under the above assumptions, every normal function arising from a Deligne cohomology class which is homologous to zero on all fibers is an abstract normal function, i.e., it is holomorphic and satisfies the horizontality condition.*

This is slightly more general than the previous way of stating it, because every actual cycle on X gives rise to a Deligne class (see Esnault and Viehweg, 1988; Gillet, 1984) but not necessarily vice versa.

Proof of the Theorem. If $f : X \to S$ is a submersion between complex manifolds with fibers projective manifolds of dimension, say d, one has the exact sequence for the relative cotangent bundle:

$$0 \to f^*\Omega^1_S \to \Omega^1_X \to \Omega^1_{X/S} \to 0.$$

The sheaf $\Omega^1_{X/S}$ which appears on the right fits naturally in a complex

$$\Omega^\bullet_{X/S} = \left[0 \to \mathcal{O}_X \to \Omega^1_{X/S} \to \cdots \to \Omega^d_{X/S}\right].$$

This sequence comes up in the graded parts of the filtration

$$F^r = \operatorname{Im}[f^*\Omega^r_S \wedge \Omega^{<p}_X[-r] \to \Omega^{<p}_X] \tag{10.2}$$

on the complex $\Omega^{<p}_X$, where we recall that this complex plays a role in the definition of the Deligne cohomology. Indeed, we have

$$\operatorname{Gr}^r_F = f^*\Omega^r_S \wedge \Omega^{<p}_{X/S}[-r],$$

where the meaning of the complex $\Omega^{<p}_{X/S}$ should be clear. Letting $\mathcal{F}^\bullet$ be the Hodge filtration on the bundle $\mathcal{H}^n = R^n f_*\mathbf{C} \otimes \mathcal{O}_S$, we observe that $\mathbf{R}^n f_*\Omega^{<p}_{X/S}$ is quasi-isomorphic to $\mathcal{H}^n/\mathcal{F}^p$. If we apply the direct image functor to the right hand side, we get the E_1-term of the spectral sequence associated with the filtration induced by F on the complex $\mathbf{R}f_*\Omega^{<p}_X$. This implies that we can write the spectral sequence as

$$E_1^{a,b}(p) \stackrel{\text{def}}{=} \mathbf{R}^{a+b} f_* Gr^a_F = \Omega^a_S \otimes \mathcal{H}^b/\mathcal{F}^{p-a} \Longrightarrow \mathbf{R}^{a+b} f_*\Omega^{<p}_X. \tag{10.3}$$

We may shrink S to prove the assertion; therefore we assume that S is a polydisk for the rest of the proof. From what we wrote in Section 6.5, we know that under this identification, the differentials $d_1^{a,b} : E_1^{a,b} \to E_1^{a+1,b}$ are induced by the Gauss–Manin connection:

$$d_1^{a,b} = \nabla : \Omega^a_S \otimes_{\mathcal{O}_S} \mathcal{H}^b/\mathcal{F}^{p-a} \to \Omega^{a+1}_S \otimes_{\mathcal{O}_S} \mathcal{H}^b/\mathcal{F}^{p-a-1}. \tag{10.4}$$

Now let us look at Deligne cohomology, as defined above. We have already seen that there exists a long exact sequence of cohomology groups

$$\begin{aligned} \cdots H^{2p-1}(X, \mathbf{Z}(p)) &\to \mathbf{H}^{2p-1}\left(X, \Omega^{<p}_X\right) \\ &\to H^{2p}_{\mathcal{D}}(X, \mathbf{Z}(p)) \to H^{2p}(X, \mathbf{Z}(p)) \to \mathbf{H}^{2p}(X, \Omega^{<p}_X) \cdots . \end{aligned}$$

If $\alpha \in H^{2p}_{\mathcal{D}}(X, \mathbf{Z}(p))$ is a Deligne cohomology class that is homologous to 0 on all fibers of f, then the normal function associated with it is obtained by the following procedure (also explained in Section 9.2).

First, note that because α is homologous to 0 on all fibers, it maps to 0 in $H^{2p}(X, \mathbf{Z}(p)) = H^0(S, R^{2p} f_* \mathbf{Z}(p))$ after possibly shrinking S. Therefore it lifts noncanonically to a class $\tilde{\alpha} \in \mathbf{H}^{2p-1}(X, \Omega_X^{<p})$. Next, we combine the Leray spectral sequence for f and the spectral sequence (10.3) for $a + b = 2p - 1$. Note that

$$E_1^{0,2p-1}(p) = \mathbf{R}^{2p-1} f_* \Omega_{X/S}^{<p} = \mathcal{H}^{2p-1}/\mathcal{F}^p.$$

The normal function ν_α associated with α is a section of this bundle (in fact of a subbundle). One checks that it is the image of $\tilde{\alpha}$ under an "edge homomorphism" $\mathbf{R}^{2p-1} f_* \Omega_X^{<p} \to E_1^{0,2p-1}(p)$ as given in Chapter 6, Eq. (6.1), but after taking global sections. This morphism can be written as a composition

$$\mathbf{H}^{2p-1}\left(X, \Omega_X^{<p}\right) \xrightarrow{e^{2p-1}} H^0\left(S, \mathbf{R}^{2p-1} f_* \Omega_X^{<p}\right) \xrightarrow{r_*} H^0\left(S, E_1^{0,2p-1}(p)\right).$$

The first map is induced by a similar edge homomorphism $H^{2p-1}(X, \mathbf{C}) \to H^0(S, Rf_*^{2p-1}\mathbf{C})$ for the Leray spectral sequence (6.6) from Section 6.4, after truncating the complex $\Omega_X^\bullet$ which resolves the constant sheaf $\mathbf{C}$. The second map is induced by the natural projection $\Omega_X^{<p} \to \Omega_{X/S}^{<p}$ coming from the projection $\Omega_X^1 \twoheadrightarrow \Omega_{X/S}^1$. This description obviously shows that the normal function is horizontal.

Since, by description of the edge-homomorphism, ν_α is a section of the bundle given by $E_\infty^{0,2p-1}$, all derivatives of the spectral sequence vanish; in particular $d_1(\nu_\alpha) = 0$. So by (10.4) the normal function is holomorphic as well. □

Problems

10.1.1 Show that the group $H^2_{\mathcal{D}}(X, \mathbf{Z}(2))$ classifies holomorphic line bundles with a holomorphic connection by relating the group to the complex $\mathcal{O}_X^* \to \Omega_X^1$ given by the $d \log$ map.

10.1.2 Prove that for X smooth, we have in the Zariski topology $H^i(X, \mathcal{O}_X^*) = 0$ for $i \geq 2$. What happens in the analytic topology? Hint: give a flasque resolution of $\mathcal{O}_X^*$ using the sheaf of meromorphic functions on X.

10.2 The Statement of Nori's Theorem

In this section we state Nori's connectivity theorem in the form of an effective version due to Jan Nagel (2002). It gives the best bounds in the theorem of Nori that are known to date. They may still not be optimal, but in the case where $Y = \mathbf{P}^{n+1}$ they are known to be sharp. The proof grew out of Nagel's thesis and builds on unpublished work of Green and Müller-Stach (1996). Nagel was able to invoke the original approach in a very clever way in order to obtain such effective bounds. This theorem implies Griffiths' theorem about the Abel–Jacobi maps (see Theorem 9.2.1) as well as the Noether–Lefschetz theorem and the theorem of Green and Voisin (Theorem 9.3.1). See Section 10.7 for a treatment of those applications. Before we state the theorem, let us fix some notation.

Notation 10.2.1 Let $(Y, \mathcal{O}_Y(1))$ be a polarized variety of dimension $n + 1 + r$. For any multi-index $\underline{d} = (d_0, \ldots, d_r)$, let $X \subset Y$ be a complete intersection given by a global section of the vector bundle

$$E = \mathcal{O}_Y(d_0) \oplus \cdots \oplus \mathcal{O}_Y(d_r).$$

We introduce

$$\begin{aligned} S &\stackrel{\text{def}}{=} \mathbf{P}H^0(Y, E), \\ \Delta &\subseteq S \text{ the discriminant locus}, \\ X_S &\subset Y_S = Y \times S \text{ the universal complete intersection} \\ &\qquad \text{of type } \underline{d} \text{ in } Y, \\ U &\stackrel{\text{def}}{=} S - \Delta. \end{aligned}$$

In the affine setting we have likewise

$$\begin{aligned} V &\stackrel{\text{def}}{=} H^0(Y, E), \\ \Delta' &\subseteq V \text{ the affine discriminant locus}, \\ U' &\stackrel{\text{def}}{=} V - \Delta'. \end{aligned}$$

We consider *submersive base changes* $p : T \to S$, i.e., the induced map on tangent spaces is surjective at every point of T. The total spaces pull back to

$$\begin{aligned} Y_T &\stackrel{\text{def}}{=} Y \times T, \\ X_T &\stackrel{\text{def}}{=} X_S \times_S T. \end{aligned}$$

The original theorem of Nori (1993) does not include effective bounds on degrees and can be stated as follows using the relative cohomology groups of (Y_T, X_T).

Theorem 10.2.2 (Nori, 1993) *For every positive integer $c \leq n$ there exists a natural number $N = N(Y, \mathcal{O}_Y(1), c)$ such that*

$$H^{n+k}(Y_T, X_T; \mathbf{Q}) = 0, \textit{ for all } k \leq c$$

and every submersion $p : T \to S$ provided that

$$\min(d_0, \dots, d_r) \geq N.$$

From this formulation one sees immediately that the theorem is a generalization of the Lefschetz hyperplane theorem, stated as Theorem B.3.3. Indeed, first of all, using induction, it implies a similar theorem for complete intersections of Y. Now we apply the Leray spectral sequence to both morphisms, $Y_T \to T$ and $X_T \to T$. Then for $i < n$ the direct images $R^i f_* \mathbf{Z}$ for both morphisms are isomorphic by the generalized version of the Lefschetz hyperplane theorem. Note here that the fibers may be singular, but the complements $Y - X$ are always smooth so that we may apply the theorem. In the Leray spectral sequence the E_2-terms are then isomorphic and hence so are the E_∞-terms, so the natural restriction map $H^i(Y_T, \mathbf{Z}) \to H^i(X_T, \mathbf{Z})$ is an isomorphism for $i < n$ and injective for $i = n$. In other words, the relative cohomology groups $H^i(Y_T, X_T; \mathbf{Z})$ vanish for $i \leq n$. Nori's theorem extends the range of coefficients for the vanishing to $i \leq 2n$ but only for $\mathbf{Q}$-coefficients. One can indeed construct examples where the theorem fails with $\mathbf{Z}$-coefficients for $n < i \leq 2n$ (see Problem 10.2.1 below).

Paranjape (1994) showed later that for the necessary bounds one has the estimate $N(Y, \mathcal{O}_Y(1), c) \leq m_Y + n + c + 1$ with $m_Y = \max\{m_j - j - 1 : 0 \leq j \leq \dim Y\}$, where m_j is the Castelnuovo regularity of Ω^j_Y; see also Braun and Müller-Stach (1996) and Ravi (1993) for bounds and explicit computations of m_Y that were obtained earlier. Nagel's version of Nori's theorem can be stated in the following way (recall that $U = S - \Delta$).

Theorem 10.2.3 (Nagel, 2002) *Fix an integer $c \leq n$ and let $p : T \to U$ be a submersion. Then*

$$H^{n+k}(Y_T, X_T; \mathbf{Q}) = 0$$

vanishes for every $0 \leq k \leq c$ provided the following conditions are satisfied: assume that $d_0 \geq \cdots \geq d_r$ and

$$(C): \sum_{a=\min(c,r)}^{r} d_a \geq m_Y + \dim(Y) - 1,$$

$$(C_i): \sum_{a=i}^{r} d_a + (\mu - c + i) d_r \geq m_Y + \dim(Y) + c - i,$$

for all i with $0 \leq i \leq \min(c-1, r)$. Here, for n, c as above, μ is defined as the constant $\mu = [\frac{n+c}{2}]$.

We devote the remainder of this section to reducing the proof to a statement about a filtration which is coarser than the Hodge filtration. In order to explain its formulation, we advise the reader to review the mixed Hodge theory discussed in Section 3.3. In particular, it is necessary to understand Hodge filtrations on relative cohomology groups in more detail.

To do this, let Y be a complex algebraic manifold and $i : X \hookrightarrow Y$ a smooth irreducible subvariety. Assume that $Y \subset \overline{Y}$ and $X \subset \overline{X}$ are smooth algebraic compactifications with simple normal crossing boundary divisors

$$D \stackrel{\text{def}}{=} \overline{Y} - Y,$$
$$D' \stackrel{\text{def}}{=} D \cap \overline{X}.$$

The logarithmic de Rham complex $\Omega^\bullet_{\overline{Y}}(\log D)$ together with the Hodge and weight filtrations on its hypercohomology define a mixed Hodge structure on $V_{\mathbf{C}} = H^i(Y, \mathbf{C})$. We now use the description of relative cohomology as a mapping cone, as explained in Section 3.3:

$$H^i(Y, X) = \mathbf{H}^i(Y, \operatorname{Ker}(\Omega^\bullet_Y \to i_* \Omega^\bullet_X)).$$

In order to define the mixed Hodge structure of Deligne on $H^i(Y, X)$, we need both the Hodge and weight filtrations. Here is a description of the Hodge filtration $F^\bullet$ on $H^i(Y, X)$: look at the commutative diagram

$$\begin{array}{ccccccccc} 0 & \to & i_*\Omega^\bullet_{\overline{X}}(\log D')[-1] & \to & C(\beta) & \to & \Omega^\bullet_{\overline{Y}}(\log D) & \to & 0 \\ & & \downarrow & & \downarrow & & \downarrow & & \\ 0 & \to & i_*\Omega^\bullet_X[-1] & \to & C(\alpha) & \to & \Omega^\bullet_Y & \to & 0, \end{array}$$

where

$$C(\alpha) \stackrel{\text{def}}{=} \operatorname{Cone}(\alpha : \Omega^\bullet_Y \to i_*\Omega^\bullet_X) = \operatorname{Ker}(\Omega^\bullet_Y \to i_*\Omega^\bullet_X)$$

and

$$C(\beta) \stackrel{\text{def}}{=} \operatorname{Cone}(\beta : \Omega^\bullet_{\overline{Y}}(\log D) \to i_*\Omega^\bullet_{\overline{X}}(\log D')).$$

Note that because both cone complexes are quasi-isomorphic, they both compute $H^\bullet(Y, X)$. But only the second complex can be used to obtain the correct Hodge filtration coming from the trivial filtration F^p (see Section 6.3):

$$F^p H^i(Y, X) \stackrel{\text{def}}{=} \operatorname{Im}(\mathbf{H}^i(F^p C(\beta)) \to \mathbf{H}^i(C(\beta))).$$

See Deligne and Dimca (1990) for a proof.

The trivial filtration on the first cone complex, or rather on the complex

$$\Omega^\bullet_{Y,X} = \operatorname{Ker}\left(\Omega^\bullet_Y \to i_*\Omega^\bullet_X\right),$$

which is quasi-isomorphic to it, gives a second filtration which was used by

Nori in his proof. This is his *G-filtration*:

$$G^p H^i(Y,X) = \mathrm{Im}(\mathbf{H}^i(F^p\Omega^\bullet_{Y,X}) \to \mathbf{H}^i(\Omega^\bullet_{Y,X})).$$

It follows immediately from the commutative diagram above that

$$F^p H^i(Y,X) \subset G^p H^i(Y,X), \tag{10.5}$$

because the vertical maps preserve the corresponding filtrations.

We do not explain the weight filtration here, because it is sufficient for our purposes to note that the maps in the long exact sequence associated with the pair (Y, X) are strict morphisms of mixed Hodge structures. These properties all enter in the following auxiliary but important lemma.

Lemma 10.2.4 *If $X \subset Y$ are smooth, quasi-projective varieties as above and $k \geq 0$ is an arbitrary integer, then*
(i) $\mathrm{Gr}^i_W H^{n+k}(Y,X) = 0$ *if* $i \leq n+k-2$;
(ii) *if* $F^k H^{n+k}(Y,X) = 0$ *for all* $k \leq c$ *and some* $c \leq n$, *then* $H^{n+k}(Y,X;\mathbf{Q}) = 0$ *for all* $k \leq c$.

Proof The assertion about the graded pieces follows from the fact that for any smooth quasi-projective variety Z the weights of $H^n(Z)$ are $\geq n$ by Lemma 3.3.3 and by the long exact sequence

$$\cdots \to H^{n+k-1}(X) \to H^{n+k}(Y,X) \to H^{n+k}(Y) \to \cdots$$

using the strictness of morphisms of mixed Hodge structures. For the second assertion, assume that one of the Hodge numbers $h^{p,q}$ of $H^{n+k}(Y,X)$ is non-zero. By assumption we have then $p \leq k-1$ and (by symmetry) $q \leq k-1$. Hence, if $k \leq n$, we have $p+q \leq 2k-2 \leq n+k-2$. This contradicts our first assumption. □

Note that in this proof we have used only the inequality $p \leq [\frac{n+k}{2}] - 1$, because then, by symmetry, $p+q \leq 2[\frac{n+k}{2}] - 2 \leq n+k-2$. So instead of looking at F^k, we could have looked at the vanishing of $F^{b_k} H^{n+k}(Y,X) = 0$ for some integer $k \leq b_k \leq [\frac{n+k}{2}]$. Nori's choice is $b_k = k$, in fact, but one can also take other sequences b_k to prove the theorem. Our choice is as follows. Fix n and c as above. Then we set

$$\boxed{b_k \stackrel{\text{def}}{=} k + \left[\frac{n-c}{2}\right] \quad \forall k \leq c \implies k \leq b_k \leq \left[\frac{n+k}{2}\right].}$$

In the next sections we shall employ the following principle.

Lemma 10.2.5 (Reduction to the G-filtration) *Suppose that*

$$G^{b_k} H^{n+k}(Y_T, X_T; \mathbf{Q}) = 0.$$

Then

$$H^{n+k}(Y_T, X_T; \mathbf{Q}) = 0, \text{ for all } k \le c.$$

Proof This follows directly from Lemma 10.2.4 and the inclusion $F^p H^i(Y, X) \subseteq G^p H^i(Y, X)$ explained above (10.5). □

By what we said so far, this Lemma implies Nori's Theorem 10.2.2 provided we can show that $G^{b_k} = 0$. In the next section we use these ideas but follow a different route to obtain Jan Nagel's effective version.

Problems

10.2.1 Find a counterexample to Nori's theorem stated over $\mathbf{Z}$. Hint: construct quotients of smooth hypersurfaces in projective space that carry free and properly discontinuous actions of finite groups. Then form the quotient and take such varieties for Y.

10.2.2 Show that condition (C_i) is a consequence of the inequality

$$(\mu - c + r + 1)d_r \ge m_Y + \dim(Y) + c.$$

10.3 A Local-to-Global Principle

In this technical section we explain how, using the first reduction to a statement about the G-filtration as explained in the previous section, we can further reduce Nori's theorem to a vanishing assertion about certain E_∞-terms of a spectral sequence with E_1-term given by

$$E_1^{p,q}(b) = \Omega^p_{T,t} \otimes H^{p+q}(Y, \Omega^{b-p}_{Y,X_t}). \tag{10.6}$$

So Nori's statement about global cohomology is reduced to a "local" sheaf-theoretic vanishing result. This reduction in addition uses Nori's condition (N_c).

Definition 10.3.1 Let $p : T \to U$ be a submersion and c a positive integer. We say that *Nori's condition* (N_c) *holds for the pair* (Y_T, X_T), if

$$R^a(\mathrm{pr}_T)_* \Omega^b_{Y_T, X_T} = 0,$$

for all (a, b) with $a + b \le n + k$, $b \ge b_k = k + [\frac{n-c}{2}]$, $k \le c$.

To be precise, what we are after is this.

Proposition 10.3.2 *Suppose that* $E_\infty^{p,q}(b) = 0$ *for all* (p, q, b) *for which* $p + q + b \le n + k$, $b \ge b_k$ *and* $k \le c$. *Then condition* (N_c) *holds. Moreover* $G^{b_k} H^{n+k}(Y_T, X_T) = 0$ *– and hence* $H^{n+k}(Y_T, X_T, \mathbf{Q}) = 0$ *for all* $k \le c$.

Before we show this, we first prove that the property (N_c) has very good base-change properties; this will be used in subsequent sections.

Lemma 10.3.3 *Let $g : T \to U$ be a submersion. Then*

(1) *if (N_c) holds for (Y_U, X_U), then (N_c) also holds for (Y_T, X_T);*
(2) *if g is surjective and (N_c) holds for (Y_T, X_T), then (N_c) also holds for (Y_U, X_U).*

Proof We may assume that T is also affine by looking at an affine cover of T. All exact sequences of vector bundles then split over T. In particular, for relative cotangent sheaves, one gets a direct sum decomposition (see Problem 10.3.1)

$$\Omega^b_{Y_T,X_T} = \bigoplus_{p+q=b} \mathrm{pr}_Y^* \Omega^p_{Y_U,X_U} \otimes \mathrm{pr}_T^* \Omega^q_{T/U}.$$

Apply now $g_* R^a(\mathrm{pr}_T)_*$ to get

$$g_* R^a(\mathrm{pr}_T)_* \Omega^b_{Y_T,X_T} = \bigoplus_{p+q=b} g_* \Omega^q_{T/U} \otimes R^a(\mathrm{pr}_U)_* \Omega^p_{Y_U,X_U}.$$

From this formula, the assertions follow immediately. □

Our next job is to explain the spectral sequence (10.6). By our assumption, $\mathrm{pr}_T : X_T \to T$ is a submersion. This implies that we have an exact sequence

$$0 \to \mathrm{pr}_T^* \Omega^1_T \to \Omega^1_{X_T} \to \Omega^1_{X_T/T} \to 0.$$

On the complex $\Omega^\bullet_{X_T}$ this induces an increasing filtration $L^\bullet$, which is similar to the F-filtration defined by (10.2):

$$L^p \Omega^\bullet_{X_T} \stackrel{\text{def}}{=} \mathrm{Im}\left(\mathrm{pr}_T^* \Omega^p_T \otimes \Omega^\bullet_{X_T}[-p] \to \Omega^\bullet_{X_T}\right).$$

Because $Y_T = Y \times T$, there is a natural split exact sequence

$$0 \to \mathrm{pr}_T^* \Omega^1_T \to \Omega^1_{Y_T} \to \mathrm{pr}_Y^* \Omega^1_Y \to 0.$$

It induces a filtration $L^\bullet$ on $\Omega^\bullet_{Y_T}$ in the same way as L^p did. Define now

$$\Omega^\bullet_{(Y_T,X_T)/T} \stackrel{\text{def}}{=} \mathrm{Ker}\left(\Omega^\bullet_{Y_T/T} \to i_* \Omega^\bullet_{X_T/T}\right).$$

This induces a filtration

$$L^p \Omega^\bullet_{Y_T,X_T} = \mathrm{Ker}\left(L^p \Omega^\bullet_{Y_T} \to L^p i_* \Omega^\bullet_{X_T}\right)$$

with graded pieces

$$\mathrm{Gr}^p_L \, \Omega^\bullet_{Y_T,X_T} \cong \mathrm{pr}_T^* \Omega^p_T \otimes \Omega^\bullet_{(Y_T,X_T)/T}[-p].$$

The induced filtration on $\Omega^\bullet_{Y_T,X_T}|_{Y \times \{t\}}$ creates a spectral sequence

$$E_1^{p,q}(b) \Rightarrow H^{p+q}\left(Y, \Omega^b_{Y_T,X_T}|_{Y \times \{t\}}\right)$$

with $E_1^{p,q}(b) \cong \Omega^p_{T,t} \otimes H^{p+q}(Y, \Omega^{b-p}_{Y,X_t})$.

We are now ready to give the proof of Proposition 10.3.2.

Proof of Proposition 10.3.2. To start, if

$$E_\infty^{p,q}(b) = 0,$$

for all $(p, q, b, k) \in \mathbf{N}^4$ with $p + q + b \leq n + k, b \geq b_k, k \leq c$, then Nori's condition (N_c) holds for (Y_T, X_T), because the vanishing holds in all fibers of the direct image sheaves.

By the reduction Lemma 10.2.5, it is indeed enough to prove that

$$G^{b_k} H^{n+k}(Y_T, X_T) = 0.$$

Now we use the Leray spectral sequence for hypercohomology with $g = \mathrm{pr}_T$:

$$E_2^{p,q} = H^p\big(T, \mathbf{R}^q g_* \Omega^\bullet_{Y_T,X_T}\big) \Longrightarrow \mathbf{H}^{p+q}\big(\Omega^\bullet_{Y_T,X_T}\big) = H^{p+q}(Y_T, X_T).$$

This spectral sequence can be constructed with the methods of Chapter 6 (see Problem 10.3.2). Using this spectral sequence for the filtered complex $F^{b_k}\Omega^\bullet_{Y_T,X_T}$ instead, we get the spectral sequence

$$E_2^{p,q} = H^p\big(T, \mathbf{R}^q g_* F^{b_k}\Omega^\bullet_{Y_T,X_T}\big) \Longrightarrow \mathbf{H}^{p+q}\big(F^{b_k}\Omega^\bullet_{Y_T,X_T}\big).$$

We see that to show $G^{b_k} H^{p+q}(Y_T, X_T) = 0$, it suffices to prove that

$$\mathbf{R}^q g_* F^{b_k}\Omega^\bullet_{Y_T,X_T} = 0 \quad \forall q \leq n + k.$$

But $\mathbf{R}^q g_* F^{b_k}\Omega^\bullet_{Y_T,X_T}$ has a filtration such that the graded pieces are subquotients of $R^a g_* \Omega^b_{Y_T,X_T}$. However, these groups vanish as long as Nori's condition (N_c) holds. □

Problems

10.3.1 Verify the claim

$$\Omega^b_{Y_T,X_T} = \bigoplus\nolimits_{p+q=b} \mathrm{pr}_Y^* \Omega^p_{Y_U,X_U} \otimes \mathrm{pr}_T^* \Omega^q_{T/U}$$

in the proof of Lemma 10.3.3.

10.3.2 Construct formally the Leray spectral sequence for hypercohomology used in the proof of Proposition 10.3.2.

10.4 Jacobi Modules and Koszul Cohomology

At this point we have seen that we need to investigate the spectral sequence with E_1-term

$$E_1^{p,q}(b) = \Omega^p_{T,t} \otimes H^{p+q}(Y, \Omega^{b-p}_{Y,X_t}).$$

We do this for hypersurfaces in Y. Later, in Section 10.6, we shall explain how also complete intersections can be handled.

The idea is to compare the above spectral sequence with a spectral sequence which is directly related to Jacobian representations for the Hodge summands in the cohomology of smooth, very ample hypersurfaces in arbitrary ambient varieties. This spectral sequence comes from the double complex $B^{\bullet,\bullet}(b)$ (depending, as our spectral sequence above, on an auxiliary parameter b) we introduce below as (10.7). As a double complex, it has two associated spectral sequences, the first of which we shall compare with $E_1^{p,q}(b)$. Both spectral sequences will play a role. The actual comparison involves duality and will be done in the next section (§10.5).

The formalism we develop here was first defined in Green and Müller-Stach (1996) and is used below in the proof of Nori's theorem. First, we need to explain the general set-up. Let L be an ample line bundle on a smooth projective variety Y. Then we have the bundle of principle parts $P^1(L)$ sitting in the exact sequence

$$0 \to \Omega^1_Y \otimes L \to P^1(L) \to L \to 0$$

with extension class $\pm(2\pi \mathrm{i}) \cdot c_1(L)$, defined in Section 7.5. Dualizing and tensoring with L yields an exact sequence

$$0 \to \mathcal{O}_Y \to \Sigma_L \to \Theta_Y \to 0, \quad \Sigma_L \stackrel{\text{def}}{=} P^1(L)^\vee \otimes L.$$

Example 10.4.1 If $Y = \mathbf{P}^n$ is projective space and $L = \mathcal{O}(1)$, then the Euler sequence coincides with the sequence involving Σ_L, so we get that $\Sigma_L = \mathcal{O}(1)^{n+1}$. The reason is that the map $\Sigma_L \to \Theta_Y$ is given by $x_i \mapsto \partial/\partial x_i$ in this case and hence coincides with the definition given in Section 7.5.

We set

$$\begin{aligned} V &\stackrel{\text{def}}{=} H^0(Y, L), \\ M_L &= \text{the kernel of the evaluation map } V \otimes \mathcal{O}_Y \to L. \end{aligned}$$

For the remainder of the proof s is an element of V, and X denotes a hypersurface section $X = \{s = 0\}$ inside Y. Recall also that we have defined a natural map $H^0(Y, L) \to H^0(Y, P^1(L))$ that sends any section s to its jet $j^1(s)$, i.e., the image of s under the natural map $H^0(Y, L) \to H^0(Y, P^1(L))$. We are ready to introduce the associated Jacobi module.

Definition 10.4.2 Let $\mathcal{F}$ be a locally free sheaf on Y and $s \in V$. We define

$$R_{Y,s}(\mathcal{F}) \stackrel{\text{def}}{=} \operatorname{Coker}(H^0(Y, \mathcal{F} \otimes \Sigma_L \otimes L^{-1}) \xrightarrow{j^1(s)} H^0(Y, \mathcal{F})),$$

The $R(\mathcal{O}_Y)$-module

$$R_*(\mathcal{F}) = \bigoplus\nolimits_{p \geq 0} R_{Y,s}(\mathcal{F} \otimes L^p)$$

is called the *Jacobi module* of $\mathcal{F}$. It depends on the choice of the section $s \in V$.

Example 10.4.3 In the case $Y = \mathbf{P}^n$ and $L = \mathcal{O}(d)$, the Jacobian ring

$$R_*(K_Y \otimes L) = \bigoplus\nolimits_{q \geq 0} R_{Y,s}(K_Y \otimes L^{q+1})$$

coincides with Griffiths' Jacobian ring $\oplus_p R_{(q+1)d-n-1}$, i.e., the quotient of the polynomial ring modulo the Jacobian ideal of s. This follows from the local description of $j^1(s)$ in Section 7.5.

Now we can define the double complex (of vector spaces) we alluded to at the start of this section:

$$B^{-p,q}(b) \stackrel{\text{def}}{=} \bigwedge^p V \otimes H^0(Y, K_Y \otimes \bigwedge^{b-q} \Sigma_L \otimes L^{q+1-p}), \quad p \leq q \leq b. \quad (10.7)$$

This is a second quadrant complex starting at the diagonal:

$$\begin{array}{ccccc}
\bigwedge^b V \otimes H^0(K_Y \otimes L) \to & \bigwedge^{b-1} V \otimes H^0(K_Y \otimes L^2) & \cdots & H^0(K_Y \otimes L^{b+1}) \to 0 \\
& \uparrow & & \uparrow \\
& \bigwedge^{b-1} V \otimes H^0(K_Y \otimes \Sigma_L \otimes L) & \cdots & H^0(K_Y \otimes \Sigma_L \otimes L^b) \to 0 \\
& & & \uparrow \\
& & \ddots & \vdots \\
& & & \uparrow \\
& & & H^0(K_Y \otimes \Sigma_L \otimes L) \to 0.
\end{array}$$

The vertical differentials are induced by the contraction maps with $j^1(s)$ explained above, whereas the horizontal differentials are given by the Koszul maps. Let

$$B^i(b) = \bigoplus\nolimits_{p=0}^{b} B^{-p,i+p}(b)$$

be the associated total complex. As a consequence, we get two spectral sequences of vector spaces converging to the total cohomology:

$$'E_1^{r,s}(b) = H^s(B^{r,\bullet}(b)) \Longrightarrow H^{r+s}(B^\bullet(b)),$$
$$''E_1^{r,s}(b) = H^s(B^{\bullet,r}(b)) \Longrightarrow H^{r+s}(B^\bullet(b)).$$

To interpret these spectral sequences we go back to the definition we have given of Koszul cohomology in Section 7.2. Obviously this notion can be generalized when we replace $\mathbf{P}^r$ by Y and $\mathcal{O}(1)$ by the line bundle L. We get the *Koszul cohomology group*

$$\mathcal{K}_{p,q}(\mathcal{F}, L^{q-1}) \stackrel{\text{def}}{=} \text{the cohomology of the 3-term complex}$$
$$\bigwedge^{p+1} V \otimes H^0(Y, \mathcal{F} \otimes L^{q-1}) \to \bigwedge^p V \otimes H^0(Y, \mathcal{F} \otimes L^q) \to \bigwedge^{p-1} V \otimes H^0(Y, \mathcal{F} \otimes L^{q+1}).$$

Lemma 10.4.4 *We have*

$${}'E_1^{-p,b}(b) = \bigwedge^p V \otimes R_{Y,s}(K_Y \otimes L^{b-p+1})$$

(for $q = b$) and

$${}''E_1^{p,-q}(b) = \mathcal{K}_{q,p-q+1}(K_Y \otimes \bigwedge^{b-p} \Sigma_L, L).$$

Proof The first identity follows from the definition of $R_{Y,s}(K_Y \otimes L^{b-p+1})$ and the fact that ${}'E_1^{-p,q}(b) = 0$ for $q > b$. For the second identity we note that

$$B^{-q,p}(b) = \bigwedge^q V \otimes K_Y \otimes \bigwedge^{b-p} \Sigma_L \otimes L^{p+1-q}.$$

Hence, by definition, we immediately get

$${}''E_1^{p,-q}(b) = \mathcal{K}_{q,p-q+1}(K_Y \otimes \bigwedge^{p-b} \Sigma_L, L). \qquad \square$$

We later need the following criterion for the vanishing of Koszul cohomology.

Lemma 10.4.5 *If $H^1(Y, \bigwedge^{p+1} M_L \otimes \mathcal{F} \otimes L^{q-1}) = 0$, then $\mathcal{K}_{p,q}(\mathcal{F}, L) = 0$.*

Proof It is a standard fact that, if we have a short exact sequence of vector bundles

$$0 \to A \to B \to C \to 0,$$

where C is a line bundle, then there is a long exact sequence

$$0 \to \bigwedge^{p+1} A \otimes C^{q-1} \to \bigwedge^{p+1} B \otimes C^{q-1} \to \bigwedge^p B \otimes C^q \to \cdots$$
$$\cdots \to B \otimes C^{p+q-1} \to C^{p+q} \to 0.$$

We apply this to the sequence

$$0 \to M_L \to V \otimes \mathcal{O}_Y \to L \to 0$$

and tensor with the locally free sheaf $\mathcal{F}$. Then the wedge powers $\bigwedge^{p+1} M_L$ of M_L sit in the long exact sequences

$$0 \to \bigwedge^{p+1} M_L \otimes \mathcal{F} \otimes L^{q-1} \to \bigwedge^{p+1} V \otimes \mathcal{F} \otimes L^{q-1} \to$$
$$\to \bigwedge^p V \otimes \mathcal{F} \otimes L^q \to \cdots \to \mathcal{F} \otimes L^{p+q} \to 0.$$

Therefore the vanishing of $H^1(Y, \bigwedge^{p+1} M_L \otimes \mathcal{F} \otimes L^{q-1}) = 0$ implies that the sheaf map

$$\bigwedge^{p+1} V \otimes \mathcal{F} \otimes L^{q-1} \to \operatorname{Ker}\left(\bigwedge^p V \otimes \mathcal{F} \otimes L^q \to \bigwedge^{p-1} V \otimes \mathcal{F} \otimes L^{q+1}\right)$$

is surjective on the H^0-level, i.e., the Koszul complex is exact at the position $\bigwedge^p V \otimes H^0(Y, \mathcal{F} \otimes L^q)$. □

10.5 Linking the Two Spectral Sequences Through Duality

To compare the spectral sequence for $E_1^{p,q}(b)$ with the one for $'E_1^{-p,q}(b)$ we need to make a connection between the wedge products of the space $\Omega^1_{T,t}$ that figures in the first sequence and the wedge products of V which come up in the second one. This goes as follows. Note that $T = V - \Delta'$ is an open subset of the vector space $V = H^0(Y, L)$ of sections of L. Therefore, the tangent space to T at $t \in T$ is canonically isomorphic to V itself at each point:

$$T_{T_t} \cong V. \tag{10.8}$$

So, clearly, some duality is at play here.

To put this in proper perspective, it is necessary to also understand the local structure of the sheaves $\Omega^p_{Y,X}$ and $\Omega^q_Y(\log X)$ if $X \subset Y$ is a smooth divisor. Let us choose local coordinates $z_0, \ldots, z_n$ in Y such that X is given by the locus where $z_0 = 0$. Then a local section $\sum a_I \, \mathrm{d}z_I \in \Omega^p_Y$ is contained in $\Omega^p_{Y,X}$ if and only if all coefficients a_I with $0 \notin I$ are in the ideal $I_X = (z_0)$. In other words, a local basis of $\Omega^p_{Y,X}$ is given by

$$\begin{aligned} &\mathrm{d}z_I, && 0 \in I,\\ z_0\, &\mathrm{d}z_I, && 0 \notin I. \end{aligned}$$

As we already know, the sheaf $\Omega^q_Y(\log X)$ is generated by differential forms that are either holomorphic or contain a copy of $\mathrm{d}z_0/z_0$. Therefore, we may pair both sheaves and get a map

$$\Omega^p_{Y,X} \otimes \Omega^q_Y(\log X) \to \Omega^{p+q}_Y(\log X)(-X) \subseteq \Omega^{p+q}_Y.$$

We note that the sheaf $\Omega^p_{Y,X}$ coincides with $\Omega^p_Y(\log X)(-X)$ in this situation. By comparing the local bases for both sheaves for $q = \dim(Y) - p$, one obtains a nondegenerate pairing with values in K_Y and hence a duality isomorphism

$$K_Y \otimes (\Omega^p_{Y,X})^\vee \cong \Omega_Y^{\dim(Y)-p}(\log X).$$

Combining this with Serre duality and using our previous notation $X = X_t$, we have that the dual of

$$\Omega^p_{T,t} \otimes H^{p+q}(Y, \Omega^{b-p}_{Y,X_t})$$

is naturally isomorphic to

$$\bigwedge^p V \otimes H^{n+1-p-q}\big(Y, \Omega_Y^{n+1-b+p}(\log X_t)\big).$$

Now recall the spectral sequence

$$E_1^{p,q}(b) = \Omega^p_{T,t} \otimes H^{p+q}\big(Y, \Omega^{b-p}_{Y,X_t}\big) \Longrightarrow H^{p+q}\big(Y \times \{t\}, \Omega^b_{Y_T,X_T}|_{Y\times\{t\}}\big)$$

defined in Section 10.3. Combining it with the preceding discussion, we see that its dual spectral sequence is

$$E_1^{-p,n+1-q}(b) = \bigwedge^p T_{T,t} \otimes H^{n+1-q-p}\big(Y, \Omega_Y^{n+1+p-b}(\log X_t)\big)$$

or, equivalently after renumbering $n + 1 - q$ to q,

$$E_1^{-p,q}(b) = \bigwedge^p T_{T,t} \otimes H^{q-p}\big(Y, \Omega_Y^{n+1+p-b}(\log X_t)\big).$$

We use the same notation for E_1 and its dual because it should be clear from the context which is meant.

The main result of this section states what we are after, namely that this spectral sequence is related to the sequence starting with $'E_1^{p,q}(b)$.

Lemma 10.5.1 (Green and Müller-Stach, 1996) *Suppose that the condition*

$$H^\alpha\big(Y, \Omega_Y^{n+2-\beta-\gamma} \otimes L^{b-p+1-\beta}\big) = 0$$

is satisfied for all $\alpha > 0$, $0 \le \beta \le b - p$ and $0 \le \gamma \le 1$. Then, for $T = V - \Delta'$, where $\Delta' \subset V = H^0(Y, L)$ is the discriminant locus of singular sections, there is an isomorphism of groups

$$h : {'E}_1^{-p,n+1-q}(b) \xrightarrow{\sim} E_1^{-p,n+1-q}(b)$$

compatible with the differentials d_1.

Proof Let us start out by recalling the identification (10.8) so that we have

$$E_1^{-p,n+1-q}(b) = \bigwedge^p V \otimes H^{n+1-q-p}(Y, \Omega^{n+1+p-b}(\log X_t)).$$

We first describe a resolution of the sheaf $\Omega^{n+1+p-b}(\log X_t)$. Let us start by observing that $\Theta_Y(-\log X_t)$, the sheaf dual to $\Omega^1_Y(\log X_t)$, sits in an exact sequence

$$0 \to \Theta_Y(-\log X_t) \to \Sigma_L \to L \to 0,$$

which follows from the commutative diagram

$$\begin{array}{ccccccccc} & & & & \Theta_Y(-\log X_t) & = & \Theta_Y(-\log X_t) & & \\ & & & & \downarrow & & \downarrow & & \\ 0 & \to & \mathcal{O}_Y & \to & \Sigma_L & \to & \Theta_Y & \to & 0 \\ & & \| & & \downarrow & & \downarrow & & \\ 0 & \to & \mathcal{O}_Y & \to & L & \to & N_{X_t|Y} & \to & 0. \end{array}$$

The short exact sequence gives rise to a resolution (as in the proof of Lemma 10.4.5).

$$\begin{aligned} 0 \to \textstyle\bigwedge^{b-p} \Theta_Y(-\log X_t) \to \bigwedge^{b-p} \Sigma_L \to \bigwedge^{b-p-1} \Sigma_L \otimes L \\ \to \cdots \to L^{b-p} \to 0. \end{aligned}$$

Using the duality isomorphisms

$$\textstyle\bigwedge^{i} \Theta_Y(-\log X_t) \cong \Omega_Y^{n+1-i}(\log X_t) \otimes K_Y^{-1} \otimes L^{-1}$$

and tensoring with $K_Y \otimes L$, we get our desired resolution

$$\begin{aligned} 0 \to \Omega_Y^{n+1-b+p}(\log X_t) \to K_Y \otimes \textstyle\bigwedge^{b-p} \Sigma_L \otimes L \\ \to \cdots \to K_Y \otimes L^{b-p+1} \to 0. \end{aligned}$$

The main point of the proof is that our assumption implies that the sheaves in this resolution are acyclic, i.e.,

$$H^{\alpha}(Y, K_Y \otimes \textstyle\bigwedge^{\beta} \Sigma_L \otimes L^{b-p+1-\beta}) = 0$$

for all $\alpha > 0$ and $0 \leq \beta \leq b - p$. In fact, by using the isomorphisms

$$\textstyle\bigwedge^{i} \Sigma_L \cong \bigwedge^{n+2-i} \Sigma_L^{\vee} \otimes K_Y^{-1}$$

(following from $\det(\Sigma_L) = K_Y^{-1}$) and the short exact sequence

$$0 \to \Omega_Y^{n+2-\beta} \to \textstyle\bigwedge^{n+2-\beta} \Sigma_L^{\vee} \to \Omega_Y^{n+2-\beta-1} \to 0$$

together with the condition

$$H^{\alpha}\left(Y, \Omega_Y^{n+2-\beta-\gamma} \otimes L^{b-p+1-\beta}\right) = 0$$

for all $\alpha > 0$, $0 \leq \beta \leq b - p$ and $0 \leq \gamma \leq 1$, acyclicity follows:

$$H^{\alpha}\left(Y, \textstyle\bigwedge^{n+2-\beta} \Sigma_L^{\vee} \otimes L^{b-p+1-\beta}\right) = 0.$$

This implies that our E_1-term comes from the de Rham cohomology at place $(n + 1 - p - q)$ of the resolution:

$$E_1^{-p,n+1-q}(b) = \textstyle\bigwedge^{p} V \otimes H_{\mathrm{dR}}^{n+1-p-q}\left(K_Y \otimes \bigwedge^{b-(p+\bullet)} \Sigma_L \otimes L^{1+\bullet}\right).$$

On the other hand, $'E_1^{-p,n+1-q}(b)$ comes from the de Rham cohomology group at place $(n+1-q)$ of the complex $K_Y \otimes \bigwedge^{b-\bullet} \Sigma_L \otimes L^{1-p+\bullet}$ upon tensoring with $\bigwedge^p V$. Comparing this with the previous calculation shows the existence of the desired morphism h, which is an isomorphism of vector spaces.

In order to prove the compatibility of the d_1-maps, it is enough to note that d_1 is the cup product with the extension e of the short exact sequence

$$0 \to \Theta_Y(-\log X_t) \to \Sigma_L \to L \to 0.$$

Now, $'d_1^{-p,n+1-q}(b)$ is induced by the cup-product maps with $j^1(s)$:

$$H^0\left(Y, \bigwedge^{b+q-n-1} \Sigma_L \otimes L^{n+2-p-q}\right) \to H^0\left(Y, \bigwedge^{b+q-n-2} \Sigma_L \otimes L^{n+3-p-q}\right).$$

Under the natural isomorphism constructed above, the cup-product maps correspond to the multiplication maps using the same argument as in Lemma 5.4.3. □

Remark Now let $p = 0$ and use Lemma 10.4.4. Then, assuming the vanishing assumption of Lemma 10.5.1, we can translate the isomorphism of E_1-terms as follows. It gives a *generalized Jacobian representation*.

$$H^p\left(Y, \Omega_Y^{n+1-p}(\log X_t)\right) \cong R_{Y,s}\left(K_Y \otimes L^{p+1}\right).$$

10.6 A Proof of Nori's Theorem

In this section we prove Nori's connectivity theorem (10.2.2) in the form of Theorem 10.2.3. As a first step, we consider hypersurfaces in Y since we want to use the isomorphism of spectral sequences of Section 10.5. In that section we have identified the E_1-terms of our two spectral sequences, but we want to identify the E_∞-terms, because then the vanishing of $'E_\infty^{-p,n+1-q}(b)$ (which we are going to show) would imply the vanishing of $E_\infty^{-p,n+1-q}(b)$ and hence property (N_c). However, if two spectral sequences agree on the E_1-level with compatible differentials, they still need not converge to the same limit. We have to show, moreover, that both spectral sequences are obtained from filtered complexes and that the isomorphism between the E_∞-terms is coming from a filtered morphism of both complexes (compare with Chapter 6). This has been done in Nagel (2002) by defining a filtered quasi-isomorphism between two filtered complexes whose associated spectral sequences are closely related to the spectral sequences we consider. One of those complexes appears in Green and Müller-Stach (1996).

We shall assume this highly technical result. It allows us to deduce Nori's theorem in the special case of hypersurfaces, which, as we shall see, is enough for our purposes. The upshot of all of this is as follows.

Proposition 10.6.1 (Hypersurfaces) *Assume that for all $k \le c$ and for all $t \in U' = H^0(Y, L) - \Delta'$ the following properties hold:*

(1) $E_1^{-x,y}(b) \cong {}'E_1^{-x,y}(b)$ *for all x, y such that $0 \le x \le b$ and $b - k + 1 \le y \le b$.*
(2) ${}''E_1^{x,-y}(b) = 0$ *for all x, y such that $x, y \ge 0$ and $b + 1 - k \le x - y \le b$.*

Then for every submersion $g : T \to U$, we have

$$H^{n+k}(Y_T, X_T) = 0$$

for all $k \le c$.

Proof By Proposition 10.3.2 it is sufficient to show that $E_\infty^{p,q}(b) = 0$ for all p, q, b, k with $p+q+b \le n+k$, $b \ge b_k$, and $k \le c$. For the dual spectral sequence this is equivalent to the vanishing of $E_\infty^{-p,n+1-q}(b)$ and, by our assumption (1), to the vanishing of ${}'E_\infty^{-p,n+1-q}(b)$ in the appropriate range of indices (one checks this by considering all incoming and outgoing differentials). Now, as ${}'E_1^{p,q}(b)$ and ${}''E_1^{p,q}(b)$ converge to the same limit, condition (2) guarantees that ${}'E_\infty^{n+1-p-q}(b)$ and ${}''E_\infty^{n+1-p-q}(b)$ both vanish. However, ${}'E_\infty^{-p,n+1-q}(b)$ is a direct summand of ${}'E_\infty^{n+1-p-q}(b)$ and hence also vanishes for all p, q, b, k with $p + q + b \le n + k$, $b \ge b_k$, and $k \le c$. □

Tools For Complete Intersections

As promised, the next step is to reduce to the hypersurface case. We employ the notation from (10.2.1). Furthermore, we set

$\mathbf{P} \stackrel{\text{def}}{=} \mathbf{P}(E)$ = the projective space bundle of hyperplanes in the vector bundle
$$E = \mathcal{O}_Y(d_0) \oplus \cdots \oplus \mathcal{O}_Y(d_r)$$

with the projection map

$$\pi : \mathbf{P} \to Y$$

and the tautological line bundle $\mathcal{O}_{\mathbf{P}(E)}(1)$. We assume that $d_0 \ge \cdots \ge d_r$ for the rest of the section.

The key observation for us, when looking at sheaf cohomology, is that

$$H^0(\mathbf{P}(E), \mathcal{O}_{\mathbf{P}(E)}(1)) = H^0(Y, E),$$

so that every section $s \in H^0(Y, E)$ defining X_s corresponds to a section $\tilde{s} \in H^0(\mathbf{P}(E), \mathcal{O}_{\mathbf{P}(E)}(1))$ and its zero locus $\tilde{X}_s = V(\tilde{s}) \subset \mathbf{P}(E)$. This construction can be done uniformly for the whole universal family of complete intersections, so that one gets a subset $\tilde{X}_S \subset \mathbf{P}_S$ corresponding to the inclusion $X_S \subset Y_S = Y \times S$. Here $\mathbf{P}_S$ is defined as $\mathbf{P}(\mathscr{E})$, where $\mathscr{E}$ is the vector bundle

$$\mathrm{pr}_Y^* E \otimes \mathrm{pr}_S^* \mathcal{O}_S(1)$$

on Y_S. This bundle has a tautological section whose zero locus defines $X_S \subset \mathbf{P}_S$. If $g : T \longrightarrow S$ is a submersion, then this extends to a section $\tilde{X}_T \subset \mathbf{P}_T = g^*\mathbf{P}(\mathscr{E})$. We denote by $\pi_T : \mathbf{P}_T \longrightarrow Y_T$ the extension of π to T.

Remark 10.6.2 In passing, we note that $\mathbf{P}_T$ is only isomorphic to $\mathbf{P}\times T$ if $\mathcal{O}_T(1)$ is trivial on T. This is, for example, the case when $T = U'$, the complement of the affine discriminant in $V = H^0(Y, \mathscr{E})$, because then $\mathcal{O}_V(1)$ is already trivial on V.

Lemma 10.6.3 *The map π induces an isomorphism*

$$(\pi_T)_* : H^{k+2r}(\mathbf{P}_T, \tilde{X}_T) \cong H^k(Y_T, X_T)$$

for all $k \geq 0$.

Proof The map

$$\pi_T : \mathbf{P}_T - \tilde{X}_T \rightarrow Y_T - X_T$$

is a fiber bundle with fibers isomorphic to the contractible space $\mathbf{C}^r$, because on each fiber $\mathcal{O}(1)$ restricts to a linear section of projective space. Therefore, the cohomology groups with compact support $H_c^{k+2r}(\mathbf{P}_T - \tilde{X}_T)$ and $H_c^k(Y_T - X_T)$ coincide. By topological duality (Theorem B.3.1) the assertion follows. □

For every $t \in T$ this induces an isomorphism

$$H^{k+2r}(\mathbf{P}, \tilde{X}_t) \cong H^k(Y, X_t).$$

Now π_* is a morphism of mixed Hodge structures, because the weight and Hodge filtrations are compatible under π. We conclude that

$$H^p\left(Y, \Omega^q_{Y,X_t}\right) \cong H^{p+r}\left(\mathbf{P}, \Omega^{q+r}_{\mathbf{P},\tilde{X}_t}\right).$$

As a consequence of this whole construction, we get a new set of spectral sequences

$$\tilde{E}_1^{p,q}(b+r) \stackrel{\text{def}}{=} \Omega^p_{T,t} \otimes H^{p+q}\left(\mathbf{P}, \Omega^{b+r-p}_{\mathbf{P},\tilde{X}_t}\right) \Longrightarrow H^{p+q}\left(\mathbf{P}, \Omega^{b+r}_{\mathbf{P},\tilde{X}_t}|_{Y\times\{t\}}\right)$$

for every $t \in T$. Associated with it are its dual spectral sequence $\tilde{E}_1^{-p,q}(b+r)$ and the spectral sequences $'\tilde{E}_1^{-x,y}(b+r)$ and $''\tilde{E}_1^{-x,y}(b+r)$ obtained by using $\mathcal{O}_{\mathbf{P}(E)}(1)$ instead of L as before and constructing in analogy a double complex $\tilde{B}^{-x,y}$ obtained by replacing (Y, L) with $(\mathbf{P}, \mathcal{O}_{\mathbf{P}}(1))$. Let us now summarize what we have to show, finally, in order to obtain the main result, which is as follows.

Proposition 10.6.4 (Complete intersections) *Assume that for all $k \le c$ and for all $t \in U' = H^0(Y, E) - \Delta'$ the following properties hold:*

(i) $\tilde{E}_1^{-x,y+r}(b+r) \cong {}'\tilde{E}_1^{-x,y+r}(b+r)$ *for all r, s such that $0 \le x \le b$ and $b-k+1 \le y \le b$;*

(ii) ${}''\tilde{E}_1^{-x,y+r}(b+r) = 0$ *for all x, y such that $x \ge 0$, $y \ge r$ and $b+1-k \le x-y \le b$;*

Then for every submersion $g : T \to U$, we have

$$H^{n+k}(Y_T, X_T) = 0$$

for all $k \le c$.

Proof All these statements follow from Proposition 10.6.1 applied to the pair $(\mathbf{P}, \mathcal{O}_{\mathbf{P}}(1))$ together with Lemma 10.6.3 and Proposition 10.3.2. Here we should recall Remark 10.6.2: $\mathbf{P}_U$ need not be isomorphic to the product $\mathbf{P} \times U$ in general unless $\mathcal{O}_T(1)$ is trivial on T. However, Lemma 10.3.3 can be adapted to this more general situation (see Nagel, 2002, Remark 2.11). □

Final Steps of the Proof of Nori's Theorem

To finish the proof of Theorem 10.2.3, we need to show that the vanishing conditions

$$(C): \sum_{a=\min(c,r)}^{r} d_a \ge m_Y + \dim(Y) - 1$$

and

$$(C_i): \sum_{a=i}^{r} d_a + (\mu - c + i)d_r \ge m_Y + \dim(Y) + c - i$$

for all i with $0 \le i \le \min(c-1, r)$ imply the conditions of Proposition 10.6.4.

Before we state the next, crucial lemma, we introduce some more notation

$$\begin{cases} M_E & \stackrel{\text{def}}{=} & \operatorname{Ker}\left(V \xrightarrow{\text{ev}} E\right), \\ M_{\mathcal{O}(1)} & \stackrel{\text{def}}{=} & \operatorname{Ker}\left(V \otimes \mathcal{O}_{\mathbf{P}} \xrightarrow{\text{ev}} \mathcal{O}_{\mathbf{P}}(1)\right), \end{cases}$$

where "ev" denotes evaluation.

Lemma 10.6.5 *We have:*

(i) $H^i(\mathbf{P}, \Omega^j_{\mathbf{P}}(k)) = 0$ *if*

$$H^{i+t}\left(Y, \Omega^u_Y \otimes \textstyle\bigwedge^v E \otimes \operatorname{Sym}^w E\right) = 0$$

for all (t, u, v, w) *such that* $0 \leq u \leq j$, $u + v = j + t + 1$, $v + w = k$, *and* $0 \leq t \leq r - j + u$;

(ii) $H^1(\mathbf{P}, \bigwedge^i M_{\mathcal{O}(1)} \otimes \Omega^j_{\mathbf{P}}(k)) = 0$ *if*

$$H^{s+t+1}\left(Y, \bigwedge^e M_E \otimes \bigwedge^f E \otimes \Omega^u_Y \otimes \bigwedge^v E \otimes \mathrm{Sym}^w E\right) = 0$$

for all (e, f, s, t, u, v, w) *such that* $s, t \geq 0$, $e + f = i + s + 1$, $0 \leq e \leq i$, $u + v = j + t + 1$, *and* $v + w = k - s - 1$;

(iii) $H^i(Y, \Omega^j_Y \otimes \bigwedge^c M_E \otimes \mathcal{O}_Y(k)) = 0$ *if* $i \geq 1$ *and* $k + i \geq m_j + c$ *(recall that* m_j *is the regularity of* Ω^j_Y*).*

Proof We only describe the principal reduction method without keeping track of the inequalities that have to be satisfied. This is left to the reader as an easy exercise.

(i) Using the exact sequence

$$0 \to \pi^*\Omega^1_Y \to \Omega^1_{\mathbf{P}} \to \Omega^1_{\mathbf{P}/Y} \to 0,$$

and taking j-th powers, we reduce to the vanishing of groups of the type $H^i(\mathbf{P}, \pi^*\Omega^u_Y \otimes \Omega^{j-u}_{\mathbf{P}/Y} \otimes \mathcal{O}_{\mathbf{P}}(k))$. On the other hand, $\Omega^1_{\mathbf{P}/Y}$ has a resolution

$$0 \to \Omega^1_{\mathbf{P}/Y} \to \pi^*E(-1) \to \mathcal{O}_{\mathbf{P}} \to 0.$$

Taking wedge powers, we obtain a resolution of $\Omega^{j-u}_{\mathbf{P}/Y}$. Therefore, we are reduced to the vanishing of groups of the sort

$$H^{i+t}(\mathbf{P}, \pi^*\Omega^u_Y \otimes \pi^* \bigwedge^{j-u+t+1} E \otimes \mathcal{O}_{\mathbf{P}}(k - j + u - t - 1)) = 0.$$

Now we apply the Leray spectral sequence and the projection formula to get down to the vanishing of $H^{i+t}(Y, \Omega^u_Y \otimes \bigwedge^v E \otimes \mathrm{Sym}^w E) = 0$ for all (t, u, v, w) such that $0 \leq u \leq j$, $u + v = j + t + 1$, $v + w = k$, and $0 \leq t \leq r - j + u$.

(ii) The bundles $\pi^* M_E$ and $M_{\mathcal{O}(1)}$ are sitting in the exact sequence

$$0 \to \pi^* M_E \to M_{\mathcal{O}(1)} \to \Omega^1_{\mathbf{P}/Y}(1) \to 0.$$

This becomes evident by noting that

$$\mathrm{Ker}(\pi^* E \to \mathcal{O}_{\mathbf{P}}(1)) \cong \Omega^1_{\mathbf{P}/Y}(1)$$

and using the definitions of M_E and $M_{\mathcal{O}(1)}$ together with the snake lemma. Taking the i-th wedge power and the exact sequence above for $\Omega^j_{\mathbf{P}}$, we get that

$$H^1(\mathbf{P}, \bigwedge^i M_{\mathcal{O}(1)} \otimes \Omega^j_{\mathbf{P}}(k)) = 0$$

if

$$H^{s+t+1}(Y, \bigwedge^e M_E \otimes \bigwedge^f E \otimes \Omega^u_Y \otimes \bigwedge^v E \otimes \mathrm{Sym}^w E) = 0$$

for all (e, f, s, t, u, v, w) such that $s, t \geq 0$, $e + f = i + s + 1$, $0 \leq e \leq i$, $u + v = j + t + 1$, and $v + w = k - s - 1$.

(iii) We give the proof for $Y = \mathbf{P}^{n+1}$ only. The general case can be found in Nagel (2002). The argument uses a reduction to projective space given in Paranjape (1994, Lemma 2.4). For projective space we use the fact that $m(M_E) = 1$ (see Problem 10.6.2) and that one has the inequalities $m(F \otimes G) \leq m(F) + m(G)$ whenever either F or G is a vector bundle (see Section 7.3). It follows that the regularity of $\Omega_Y^j \otimes \bigwedge^c M_E \otimes \mathcal{O}_Y(k)$ is $m_j + c + k$, and the assertion follows. □

It remains to be shown how the conditions (C) and (C_i) imply (i) and (ii) in Proposition 10.6.4. Let (Y, L) be the pair $(\mathbf{P}, \mathcal{O}_\mathbf{P}(1))$ and look at Lemma 10.5.1. It implies that we need to show that

$$H^{y+r-x-i-j}\left(\mathbf{P}, \Omega_\mathbf{P}^{n+2-b+x+i-z} \otimes \mathcal{O}_\mathbf{P}(i+1)\right) = 0$$

for all pairs i, j such that $0 \leq i \leq y + r - x - j - 1$ and $0 \leq j, z \leq 1$. Using Lemma 10.6.5 (i), this amounts to proving that

$$H^{y+r-x-i-j+t}\left(Y, \Omega_Y^u \otimes \bigwedge^v E \otimes \mathrm{Sym}^w(E)\right) = 0$$

for an appropriate range of indices. This can be checked by using the decomposition $E = \oplus \mathcal{O}(d_i)$ into line bundles and computing regularities. In this way, one can check that condition (C) is sufficient to imply all these conditions (see Problem 10.6.1). This proves condition (i).

To prove (ii), we have to show that $''\tilde{E}_1^{x,-y+r}(b+r)$ vanishes for certain values of x, y. However, by Lemma 10.4.4, this group is isomorphic to

$$\mathcal{K}_{y-r,x-y+r+1}(\bigwedge^{n+2-b+x} \Sigma^\vee_{\mathbf{P},\mathcal{O}(1)}, \mathcal{O}(1)).$$

By Lemma 10.4.5, this group will vanish once we have

$$H^1\left(\mathbf{P}, \bigwedge^{y-r+1} M_{\mathcal{O}(1)} \otimes \bigwedge^{n+2-b+x} \Sigma^\vee_{\mathbf{P},\mathcal{O}(1)} \otimes \mathcal{O}_\mathbf{P}(x-y+r)\right) = 0.$$

This, in turn, is a consequence of

$$H^1\left(\mathbf{P}, \bigwedge^{y-r+1} M_{\mathcal{O}(1)} \otimes \Omega_\mathbf{P}^{n+2-b+x-z} \otimes \mathcal{O}_\mathbf{P}(x-y+r)\right) = 0$$

by using the exact sequences

$$0 \to \Omega_\mathbf{P}^p \to \bigwedge^p \Sigma^\vee_{\mathcal{O}(1)} \to \Omega_\mathbf{P}^{p-1} \to 0.$$

Again using Lemma 10.6.5(ii), this follows from the condition

$$H^{s+t+1}\left(Y, \bigwedge^e M_E \otimes \bigwedge^f E \otimes \Omega_Y^u \otimes \bigwedge^v E \otimes \mathrm{Sym}^w(E)\right) = 0$$

for the seven-tuple of integers (e, f, s, t, u, v, w) satisfying $e + f = y - r + s + 2$, $0 \leq e \leq y - r + 1$, $u + v = n + r + 3 - b + x - z + t$, and $v + w = x - y + r - s - 1$. This can be checked (see Problem 10.6.1) by splitting E into the line bundles $\mathcal{O}(d_i)$ and a regularity computation. It follows that this vanishing is a consequence of condition (C_i).

Problems

10.6.1 Complete the regularity computation at the end of the proof above.

10.6.2 Make the conditions (C) and (C_i) explicit for the case where Y itself is a complete intersection in some projective space or is projective space itself by computing m_Y with suitable exact sequences. Hint: use the methods explained in the proof of Theorem 8.1.7.

10.6.3 Assume that Y is a variety that satisfies the property that $H^i(Y, \Omega^j_Y(k)) = 0$ for all $i > 0$, $k > 0$, and $j \geq 0$. Show that then the assumption of Lemma 10.5.1 is satisfied and write the conditions (C) and (C_i) in this case. Find examples for such Y.

10.7 Applications of Nori's Theorem; Filtrations on Chow Groups

In this last section we explain two applications of Nori's theorem and then introduce the conjectures of Bloch and Beilinson concerning filtrations of Chow groups which give an indication of what Chow groups should look like. We relate this to Lefschetz-type conjectures by Hartshorne and Nori. Finally, we indicate how infinitesimal methods can be used in order to show that the graded quotients of such filtrations are nontrivial (see Historical Remarks below).

Noether–Lefschetz Revisited

Our first application is a common generalization of the theorems of Noether and Lefschetz and of Green and Voisin (see Chapter 9, Theorem 9.3.1). The proof once again uses Deligne cohomology with logarithmic growth conditions (see Esnault and Viehweg, 1988). For every $Z \in \mathrm{Ch}^p(X)$, one can define a Deligne cycle class $c_p(Z) \in H^{2p}_{\mathcal{D}}(X_T, \mathbf{Q}(p))$. If U is compact, this definition of Deligne cohomology agrees with the previous one from Section 10.1 and for our purposes we only need to know that this new form of Deligne cohomology has functorial behavior with respect to morphisms.

Theorem 10.7.1 (Green and Müller-Stach, 1996) *Let $(Y, \mathcal{O}_Y(1))$ be a smooth* *and projective polarized variety of dimension $n+r+1$, $i : X \hookrightarrow Y$ a very general complete intersection of dimension n and multidegree $(d_0, \ldots, d_r)$ with d_i satisfying the conditions of Theorem 10.2.3. Furthermore, assume that $0 \le p \le n-1$. Then:*

$$\operatorname{Image}\big(c_p : \operatorname{Ch}^p(X) \otimes \mathbf{Q} \to H^{2p}_{\mathscr{D}}(X, \mathbf{Q}(p))\big)$$
$$\subseteq \operatorname{Image}\big(i^* : H^{2p}_{\mathscr{D}}(Y, \mathbf{Q}(p)) \to H^{2p}_{\mathscr{D}}(X, \mathbf{Q}(p))\big).$$

Proof If we have a cycle on a very general member $X = X_o$ of the universal family $X_S \to S$, then we have a finite unramified morphism $p : T \to S$ and a cycle $Z \in \operatorname{Ch}(X_T)$ that restricts to the given cycle on a fiber in the preimage $p^{-1}(o)$. One looks at the commutative diagram

$$\begin{array}{ccc} H^{2p}_{\mathscr{D}}(Y_T, \mathbf{Q}(p)) & \to & H^{2p}_{\mathscr{D}}(X_T, \mathbf{Q}(p)) \\ \downarrow & & \downarrow \\ H^{2p}_{\mathscr{D}}(Y, \mathbf{Q}(p)) & \to & H^{2p}_{\mathscr{D}}(X, \mathbf{Q}(p)), \end{array}$$

where the horizontal maps are the canonical maps induced by the inclusion $i : X \hookrightarrow Y$ and the vertical maps are restriction maps to the fiber over $t = o$. Because $p < n$, we have $2p < 2n$ and hence Nori's theorem applies. It gives

$$H^j(Y_T; \mathbf{Q}) \xrightarrow{\simeq} H^j(X_T; \mathbf{Q}), \quad j = 2p-1, 2p.$$

Now use that by Esnault and Viehweg (1988, Cor. 2.10) the Deligne cohomology groups of quasi-projective manifolds U (with $\mathbf{Q}$-coefficients) are determined by the exact sequence

$$\cdots \to H^{2p-1}(U, \mathbf{C})/F^p \to H^{2p}_{\mathscr{D}}(U, \mathbf{Q}(p)) \to H^{2p}(U, \mathbf{Q}) \to \cdots,$$

and that morphisms of mixed Hodge structures are strict (see Section 3.3). Hence, from the diagram

$$\begin{array}{ccccccccc} H^{2p-1}(Y_T,\mathbf{Q}) & \to & H^{2p-1}((Y_T,\mathbf{C})/F^p & \to & H^{2p}_{\mathscr{D}}((Y_T,\mathbf{Q}(p)) & \to & H^{2p}((Y_T,\mathbf{Q}) & \to & H^{2p}((Y_T,\mathbf{C})/F^p \\ \| \simeq & & \| \simeq & & \downarrow & & \| \simeq & & \| \simeq \\ H^{2p-1}(X_T,\mathbf{Q}) & \to & H^{2p-1}(X_T,\mathbf{C})/F^p & \to & H^{2p}_{\mathscr{D}}(X_T,\mathbf{Q}(p)) & \to & H^{2p}(X_T,\mathbf{Q}) & \to & H^{2p}(X_T,\mathbf{C})/F^p \end{array}$$

we deduce that $H^{2p}_{\mathscr{D}}(Y, \mathbf{Q}(p)) \xrightarrow{\simeq} H^{2p}_{\mathscr{D}}(X, \mathbf{Q}(p))$. This immediately implies that the Deligne cycle class of Z in $H^{2p}_{\mathscr{D}}(X_T, \mathbf{Q}(p))$ comes from a Deligne class in $H^{2p}_{\mathscr{D}}(Y_T, \mathbf{Q}(p))$. By the commutativity of the diagram this implies that the restriction of Z to X_o comes from a cohomology class in $H^{2p}_{\mathscr{D}}(Y, \mathbf{Q}(p))$. □

Example 10.7.2 If $Y = \mathbf{P}^4$, $r = 0$, and $d = d_0 \ge 6$, then this theorem implies that the image of the Abel–Jacobi map $\operatorname{Ch}^2_{\text{hom}}(X) \to \mathcal{J}^2(X)$ for very general X is contained in the image of $\mathcal{J}^2(Y)$ modulo torsion. However, the latter torus is trivial since $\mathbf{P}^4$ has no odd cohomology. Therefore, we obtain the theorem of Green and Voisin. On the other hand, if $Y = \mathbf{P}^3$, $r = 0$, and $X \subset Y$ is a very general hypersurface of degree $d \geq 4$, then the theorem implies that the image of $\mathrm{Ch}^1(X) \to H^2(X, \mathbf{Q})$ is the same as the image of $H^2(Y, \mathbf{Q})$, which is of rank one. Hence we deduce the Noether–Lefschetz theorem.

Remark 10.7.3 With more care, one can show that the image of c_p : $\mathrm{Ch}^p(X) \to H^{2p}_{\mathscr{D}}(X, \mathbf{Q}(p))$ is contained in the image of

$$c_p i^* = i^* c_p : \mathrm{Ch}^p(Y) \to H^{2p}_{\mathscr{D}}(X, \mathbf{Q}(p))$$

obtained by restricting cycles to X and then taking their cohomology classes, at least if one factors out the largest abelian subvariety $J^p_0(X)$ inside $J^p(X)$ whose classes are not parametrized by algebraic cycles (conjectured to be zero by Grothendieck's generalized Hodge conjecture). In other words, one can construct a cycle on Y which has the same Deligne class as the given cycle modulo $J^p_0(X)$ when restricted to X. See Green and Müller-Stach (1996) for details.

Applications to the Griffiths Group

Our second application is due to M. Nori himself. It is a generalization of the theorem of Griffiths saying that there are quintic threefolds which have codimension-2 cycles homologically that are not algebraically equivalent to 0. Also this proof requires a little bit of Deligne cohomology.

Theorem 10.7.4 *Let Y be a smooth projective variety of dimension $n + r + 1$ and $D_0, \ldots, D_r$ be very general hypersurfaces of sufficiently large degrees. Let $X = Y \cap D_0 \cap \ldots \cap D_r$ be a very general complete intersection of dimension n. Assume that Z is a codimension $p < n$ algebraic cycle on Y such that its cohomology class in $H^{2p}_{\mathrm{prim}}(Y, \mathbf{Q})$ is nonzero. Then the restriction of Z and of every multiple of it to X is not algebraically equivalent to* 0.

In other words, the Griffiths group of such a variety is nontrivial, where we recall (9.1) that the Griffiths group is defined as

$$Gr^p(X) = \frac{\mathrm{Ch}^p_{\mathrm{hom}}(X)}{\mathrm{Ch}^p_{\mathrm{alg}}(X)}.$$

Proof We argue by contradiction. The cycle Z induces for every finite unramified morphism $p : T \to S$ a cycle $W_T \in \mathrm{Ch}^p(X_T)$ by pulling it back to Y_T and restricting to X_T, in short $W_T = i^* \mathrm{pr}_Y^*(Z)$. Assume now that it is algebraically equivalent to 0 on a general fiber of $X_T \to T$.

By the definition of algebraic equivalence, this implies that there is a family of algebraic curves $C_T \to T$ and a correspondence $\Gamma \subset X_T \times_T C_T$ such that W_T is in the image of the induced map $\Gamma_* : \mathrm{Pic}^0(C_T) \to \mathrm{Ch}^p_{\mathrm{hom}}(X_T)$, which is obtained by pulling back an element from $\mathrm{Pic}^0(C_T)$ to $\mathrm{Ch}^*(X_T \times C_T)$, intersecting it with Γ and pushing forward to X_T. The projection map $C_T \to T$ is a submersion (after possibly shrinking T) and therefore Nori's theorem implies that

$$H^{2p+1}(Y_U, X_U; \mathbf{Q}) = 0$$

for $U \stackrel{\mathrm{def}}{=} C_T$ and $p < n$. If we use Deligne cohomology instead of Chow groups, we obtain that the Deligne cohomology class of $W_U \in H^{2p}_{\mathcal{D}}(X_U, \mathbf{Q}(p))$ is in the image of $H^{2p}_{\mathcal{D}}(Y_U, \mathbf{Q}(p))$. Note that $X_U = X_T \times_T C_T$ so that Γ is a cycle on X_U.

By Nori's theorem we conclude that the class of $\mathrm{pr}_Y^*(Z) \in H^{2p}_{\mathcal{D}}(Y_T, \mathbf{Q}(p))$ is in the image of the correspondence $\Gamma_* : \mathrm{Pic}^0(C_T) \to H^{2p}_{\mathcal{D}}(Y_T, \mathbf{Q}(p))$, because both classes coincide after applying i^*, which is an isomorphism. Therefore, this argument shows that if we restrict to the fiber over a point $o \in T$, we get that the cohomology class of $Z|_{X_o}$ is in the image of the correspondence $\Gamma(0)_* : \mathrm{Pic}^0(C_o) \to H^{2p}_{\mathcal{D}}(Y \times \{o\}, \mathbf{Q}(p))$. Therefore, it has to be 0, because correspondences map degree zero cycles to cycles that are homologous to 0. This contradicts the fact that the cohomology class was nonzero. This argument also works for any multiple of the class because we take $\mathbf{Q}$-coefficients. □

Example 10.7.5 Let Y be a smooth six-dimensional quadric in $\mathbf{P}^7$. By Theorem 10.2.3 we see that if $r = 1$ and Z = difference of two Schubert three-planes, then neither $Z \cap D_0 \cap D_1$ with $\deg(D_i) \geq 5$ nor even any multiple of it is algebraically equivalent to zero on $D_0 \cap D_1$ for very generally chosen hypersurfaces D_i. Note that the corresponding intermediate Jacobian $J^3(X)$ is 0 here, so that the cycles cannot be detected by the Abel–Jacobi map.

This example of Nori was the first example of a cycle that is not of dimension 0 and contained in the kernel of the Abel–Jacobi map. Recall that for a smooth, projective complex variety of dimension n the Abel–Jacobi map

$$\psi^n : \mathrm{Ch}^n_{\mathrm{hom}} \to \mathrm{Alb}(X)$$

is always surjective (by the universal property of the Albanese map) and has an infinite dimensional uniquely torsion-free kernel $T(X)$ if the Hodge number $h^{n,0} \neq 0$ (Roitman's theorem; a complete proof can be found in Green et al., 1994).

Correspondences and Motives

Correspondences between two smooth projective varieties X and Y are just cycles on the product $X \times Y$ up to rational equivalence and up to torsion; more precisely, a degree s *correspondence* from X to Y is an element of

$$\boxed{\operatorname{Corr}^s(X,Y) \stackrel{\text{def}}{=} \operatorname{Ch}^{\dim(X)+s}(X\times Y)\otimes \mathbf{Q}.}$$

This depends on the order in which we take the two factors. When we switch them, the same cycle defines the transposed correspondence. For instance, any morphism $\varphi : X \to Y$ defines a degree-zero correspondence in $\operatorname{Corr}^0(Y,X)$ by taking the transpose of its graph. So, a correspondence should be viewed as a contravariant generalization of a morphism. Like these, any correspondence $f \in \operatorname{Corr}(X,Y)$ induces additive homomorphisms on the Chow groups and in cohomology (but now covariantly),

$$f_* : \operatorname{Ch}(X) \to \operatorname{Ch}(Y),$$
$$f_* : H(X) \to H(Y),$$

both defined by the formula $(\mathrm{pr}_Y)_*(\mathrm{pr}_X^* Z \cdot f)$, where "pr" denotes projection. Correspondences can be composed like morphisms:

$$\begin{array}{ccc} \operatorname{Corr}(X,Y)\times\operatorname{Corr}(Y,Z) & \longrightarrow & \operatorname{Corr}(X,Z), \\ (f,g) & \longmapsto & g\bullet f \stackrel{\text{def}}{=} \mathrm{pr}_{XZ}^*\left\{\mathrm{pr}_{XY}^* f\cdot \mathrm{pr}_{YZ}^*(g)\right\}. \end{array}$$

So we can speak of projectors $p \in \operatorname{Corr}^0(X,X)$ which by definition satisfy $p\bullet p = p$.

We can now enlarge the category of complex smooth projective varieties by adding projectors, thereby obtaining *effective motives* (X,p). By definition, a morphism $(X,p)\to(Y,q)$ is a correspondence from X to Y of the form $q\circ f\circ p$, with $f \in \operatorname{Corr}^0(X,Y)$. We again find the varieties by taking (X,id), but the morphisms behave contravariantly, because we have to identify them with the transpose of their graphs. *Pure motives* (X,p,n) are obtained by adding an integer n to the effective motive (X,p), which serves as a bookkeeping device for the morphisms $(X,p,n)\to(Y,q,m)$. This time they are of the form $q\circ f\circ p$, with $p \in \operatorname{Corr}^{m-n}(X,Y)$. The motive of a variety X is then denoted $h(X) = (X,\mathrm{id},0)$.

Example 10.7.6 (i) The disjoint union defines an addition, making motives into an abelian monoid. Direct product defines a tensor product for motives.

(ii) Let X be a smooth n-dimensional projective variety. The two trivial projectors

$$\pi_0 = x\times X,$$
$$\pi_{2n} = X\times x$$

define the motives $h^0(X) = (X,\pi_0)$ and $h^{2n}(X,\pi_{2n})$, and because $\pi^+ = \mathrm{id} - \pi_0 - \pi_{2n}$ is a projector as well, the motive of the variety X breaks up as a direct sum

$$h(X) = h^0(X) \oplus h^+(X) \oplus h^{2n}(X).$$

(iii) For curves, the motive h^+ carries all the information that its jacobian has: morphisms $h^+(C) \longrightarrow h^+(C')$, where C, C' are curves, are exactly the homomorphisms $\mathrm{Jac}(C) \longrightarrow \mathrm{Jac}(C')$ up to isogenies.

Filtrations on the Chow Groups

For this section we refer to Murre et al. (2013) where the reader finds background and references for what here is merely outlined.

Bloch (1980) first formulated a conjecture about the existence of a filtration

$$\mathrm{Ch}^p(X) = F^0 \supseteq F^1 \supseteq F^2 \supseteq \cdots \supseteq F^{p+1} = 0$$

on Chow groups of smooth projective varieties which was later extended and axiomatized by A. Beilinson. As a first step he conjectured that, in dimension $n = 2$, $F^2\,\mathrm{Ch}^2(X)$ is modulo torsion given by the kernel of the Albanese map. $F^1\,\mathrm{Ch}^p(X)$ should always be equal to $\mathrm{Ch}^p_{\mathrm{hom}}(X)$.

We first sketch Murre's approach to the existence of such a filtration (see Murre, 1993). It depends on several conjectures.

Conjectures (Murre) (i) *The Chow–Künneth conjecture:*
Every smooth projective variety X admits a Chow–Künneth decomposition *in the sense that for every integer $i = 0, \ldots, 2n = 2\dim X$ there exists a projector $\pi_i \in \mathrm{Corr}^0(X, X)$ with the following properties:*

(a) $\pi_i \bullet \pi_j = 0$ *for* $i \neq j$*;*

(b) *the sum of the projectors decomposes the diagonal*

$$\pi_0 + \cdots + \pi_{2n} = \Delta(X);$$

(c) *the class of π_i in cohomology is the Künneth component of the diagonal* $\Delta_{2n-i,i}(X) \in H^{2n-i}(X) \otimes H^i(X)$.

Assuming this conjecture, put

$$\mathrm{Ch}^i(X) = (X, \pi_i, 0).$$

(ii) *The projectors $\pi_0, \ldots, \pi_{j-1}$, as well as the projectors $\pi_{2n}, \ldots, \pi_{2j+1}$, operate trivially on* $\mathrm{Ch}^j(X)_{\mathbf{Q}}$.

Assuming this conjecture as well, define

$$F^\nu = \mathrm{Ker}\,\pi_{2j} \cap \mathrm{Ker}\,\pi_{2j-1} \cap \cdots \cap \mathrm{Ker}\,\pi_{2j-\nu+1}$$

so that

$$\begin{aligned} F^0 &= \mathrm{Ch}^j(X)_{\mathbf{Q}}, \\ F^1 &= \mathrm{Ker}(\pi_{2j}), \\ F^2 &= \mathrm{Ker}(\pi_{2j-1}|F^1) = \mathrm{Ker}\,\pi_{2j} \cap \mathrm{Ker}\,\pi_{2j-1}, \end{aligned}$$

etc.

(iii) *The filtration $F^\bullet$ does not depend on the choice of the projectors π_i.*
(iv) *F^1 consists of the cycles homologically equivalent to* 0.

Jannsen (2000) has shown that a filtration satisfying these properties is unique and that Murre's conjectures are equivalent to the following.

Conjectures 10.7.7 (Bloch–Beilinson) *There exists a filtration $F^\bullet$ on $\mathrm{Ch}^p(X)_{\mathbf{Q}}$ such that*

(i)

$$\begin{aligned} F^0\,\mathrm{Ch}^p(X)_{\mathbf{Q}} &= \mathrm{Ch}^p(X)_{\mathbf{Q}}, \\ F^1\,\mathrm{Ch}^p(X)_{\mathbf{Q}} &= \mathrm{Ch}^p(X)_{\mathrm{hom},\mathbf{Q}}. \end{aligned}$$

(ii)

$$F^r\,\mathrm{Ch}^p(X)_{\mathbf{Q}} \cdot F^s\,\mathrm{Ch}^q(X)_{\mathbf{Q}} \subset F^{r+s}\,\mathrm{Ch}^{p+q}(X)_{\mathbf{Q}}.$$

(iii) *$F^\bullet$ is respected by f^*, f_* for morphisms f between smooth projective varieties.*
(iv) *Assuming the algebraicity of the Künneth components $\Delta_{2n-i,i}(X)$, the ν-th graded piece of the filtration $F^\bullet$ only depends on the motive $h^{2j-\nu}(X)$.*
(v) $F^{p+1}\,\mathrm{Ch}^p(X)_{\mathbf{Q}} = 0$.

There is a more refined version of conjecture (iv), the so-called Beilinson formula,

$$\mathrm{Gr}^i_F\,\mathrm{Ch}^p(X)_{\mathbf{Q}} = \mathrm{Ext}^i_{MM_{\mathbf{C}}}(\mathbf{Q}, h^{2p-i}(X)(p)),$$

where $MM_{\mathbf{C}}$ is the hypothetical category of mixed motives. The formula means that $\mathrm{Gr}^i_F\,\mathrm{Ch}^p(X)_{\mathbf{Q}}$ should be the Ext^i-group in that category. Such a category does not exist yet, but several attempts have been made, the most successful one up to now being the Voevodsky triangulated category of motives (see Voevodsky et al., 2000; Levine, 1998, and Mazza et al., 2006). For a completely different approach, suggested by Nori, see Huber and Müller-Stach (2017). For a precise formulation of Beilinson's formula and more discussions on equivalence relations and motives, see Jannsen (2000).

Several candidates for the "Bloch–Beilinson filtration" have been proposed by various people. For instance, Shuji Saito (1996) has defined a filtration on

Chow groups such that it has many good properties (except the vanishing of $F^{p+1}\operatorname{Ch}^p(X)$) and is known to be contained in the searched for filtration by the result of Asakura and Saito (1998). It is known Jannsen (2000) that if such filtrations exist at all, they coincide with the filtration of Saito.

Abelian varieties serve as a wonderful example to illustrate these filtrations. Beauville (1986a) and Bloch (1979) began to study the Chow groups of abelian varieties in the 1980s and found out that one can use the Fourier transform to define a filtration on each $\operatorname{Ch}^p(X)$ by setting

$$F^\nu \operatorname{Ch}^p(A)_\mathbf{Q} \stackrel{\text{def}}{=} \bigoplus\nolimits_{s\geq\nu} \operatorname{Ch}^p_s(A),$$

where $\operatorname{Ch}^p_s(A) \subseteq \operatorname{Ch}^p(A)_\mathbf{Q}$ is the subgroup of cycles where n^* (the pullback of multiplication by n) acts with the eigenvalue n^{2p-s}. It is a conjecture of Beauville that the filtration defined in this way has the properties of a Bloch–Beilinson filtration, in particular that $\operatorname{Ch}^p_s(A) = 0$ for $s < 0$. Deninger and Murre (1991) and Künnemann (1994) extended the formalism of Bloch and Beauville to so-called Grothendieck motives and proved many interesting results. See also Miller et al. (2007).

Several authors have shown that the kernel of the Abel–Jacobi map (and hence conjecturally F^2) is nontrivial for very general abelian varieties over the complex numbers: Ceresa's work (1983) has shown that the cycle $C - C^- \in \operatorname{Ch}^2(A)$ on a very general abelian threefold $A = \operatorname{Jac}(C)$ is nonzero modulo algebraic equivalence. Later Bardelli (1989) proved that these cycles and their isogenic images give rise to an infinite-dimensional Griffiths group as in the case of quintic threefolds. Nori (1993) and Fakhruddin (1996) have generalized that approach, applying infinitesimal methods to abelian four- and fivefolds (see Historical Remarks in Chapter 9).

These filtrations (in particular through the last axiom) give rise to many expected properties of Chow groups. For example, they suggest that there should be Lefschetz theorems for Chow groups in the range where they hold for cohomology.

Conjecture (Hartshorne) *Let $X \subset Y$ be a smooth ample hyperplane section. Then the restriction maps*

$$i^* : \operatorname{Ch}^p(Y) \otimes \mathbf{Q} \to \operatorname{Ch}^p(X) \otimes \mathbf{Q}$$

are isomorphisms for $2p \leq \dim(X) - 1$.

The point here is that the deeper layers of the filtration in Chow groups are controlled (in the sense of Beilinson's formula) by parts of the cohomology. As an example where the conjecture holds, you may just look at $p = 1$, when $\dim(X) \geq 3$. On the other hand, we have the Noether–Lefschetz theorems; for example, if $Y = \mathbf{P}^3$ and $X \subset Y$ is a hypersurface, then $\mathrm{Ch}^1(X) = \mathrm{Pic}(X)$ is not always isomorphic to $\mathrm{Ch}^1(Y) = \mathbf{Z}$, for example when $d = 3$, $\mathrm{Ch}^1(X) = \mathbf{Z}^7$. But, as we know already, if d is sufficiently large (≥ 4), then the Lefschetz theorem holds, provided X is sufficiently general in the moduli space. This conjecture could therefore be extended as follows.

Conjecture (Nori) *Let $X \subset Y$ be a general complete intersection of multidegree $(d_1, \ldots, d_r)$ such that all d_i are sufficiently large and X corresponds to a very general point in the moduli space. Then the restriction maps*

$$i^* : \mathrm{Ch}^p(Y) \otimes \mathbf{Q} \to \mathrm{Ch}^p(X) \otimes \mathbf{Q}$$

are isomorphisms for $p \leq \dim(Y) - 1$.

It is easy to see that this conjecture is sharp in the sense that for $p = \dim(Y)$ there are many counterexamples, due to Mumford (1969b) and Roĭtman (1974). For example, by Mumford's theorem, since a hypersurface of degree $d \geq 4$ in $\mathbf{P}^3$ has positive geometric genus, its second Chow group $\mathrm{Ch}^2(X)$ is a group of infinite $\mathbf{Q}$-rank (consult e.g. Bloch, 1980), whereas Ch^2 of projective space $\mathbf{P}^3$ is of rank 1.

We do not want to finish this chapter without giving some indication of current developments and possible directions for future research in this area.

S. Bloch (1986) has defined *higher Chow groups* $\mathrm{Ch}^p(X, n)$ as a generalization of the classical Chow groups. To define it, we need the algebraic n-simplex

$$\boxed{\Delta^n \stackrel{\text{def}}{=} \left\{t_0, \ldots, t_n) \in \mathbf{A}^{n+1} \,\middle|\, \sum t_i = 1\right\}}$$

and its faces, obtained by setting some of the coordinates t_j equal to zero. One then defines

$$\begin{aligned} z^p(X, n) = \{ \oplus \mathbf{Z} \cdot Y \quad | \quad & Y \subset X \times \Delta^n, \quad \operatorname{codim} Y = p; \\ & \operatorname{codim}(Y \cap (X \times \text{face of } \Delta^n)) = p.\} \end{aligned}$$

In other words, this is the free abelian group $z^p(X, n)$ on integral algebraic subvarieties of codimension p in $X \times \Delta^n$ meeting all its "faces"properly. This definition also makes sense for arbitrary base fields instead of $\mathbf{C}$.

The alternating sum of restriction maps to faces of Δ^n defines a chain complex

$$\cdots \to z^p(X, n+1) \xrightarrow{\delta_{n+1}} z^p(X, n) \xrightarrow{\delta_n} z^p(X, n-1) \to \cdots$$

similar to the complex defining singular homology. Its homology groups are the higher Chow groups

$$\mathrm{Ch}^p(X, n) \stackrel{\text{def}}{=} \frac{\mathrm{Ker}(z^p(X, n) \to z^p(X, n-1))}{\mathrm{Im}(z^p(X, n+1) \to z^p(X, n))}.$$

These are related to the higher algebraic K-theory of Quillen (1973) in the same way as classical Chow groups are related to Grothendieck's $K_0(X)$ via a Riemann–Roch theorem and Chern characters (see Grothendieck, 1958).

At this point recall the definition of the Deligne cohomology (10.1). For every quasi-projective complex variety X there are Chern class maps

$$c_{p,n} : \mathrm{Ch}^p(X,n) \longrightarrow H^{2p-n}_{\mathscr{D}}(X,\mathbf{Z}(p))$$

which generalize Griffiths' Abel–Jacobi maps. Many results proved above can be easily generalized to higher Chow groups. For example, one can obtain the following theorem.

Theorem 10.7.8 (Müller-Stach, 1997) *Let $(Y,\mathscr{O}_Y(1))$ be a smooth and projective polarized variety of dimension $n+h$, with $X \subset Y$ a general complete intersection of dimension n, and multidegree $(d_1,\ldots,d_h)$ with $\min(d_i)$ sufficiently large. Furthermore, assume that $1 \le p \le n$. Then:*

$$\mathrm{Image}(c_{p,1} : \mathrm{Ch}^p(X,1)\otimes\mathbf{Q} \to H^{2p-1}_{\mathscr{D}}(X,\mathbf{Q}(p))) \\ \subseteq \mathrm{Image}(i^* : H^{2p-1}_{\mathscr{D}}(Y,\mathbf{Q}(p)) \to H^{2p-1}_{\mathscr{D}}(X,\mathbf{Q}(p))).$$

Corollary 10.7.9 *Let $X \subset \mathbf{P}^3$ be a general hypersurface of degree $d \ge 5$. Then the Chern class $\mathrm{Ch}^2(X,1)\otimes\mathbf{Q} \to H^3_{\mathscr{D}}(X,\mathbf{Q}(2))$ has image isomorphic to $H^3_{\mathscr{D}}(\mathbf{P}^3,\mathbf{Q}(2)) \cong \mathbf{C}/\mathbf{Q}(1)$.*

This corollary shows that the so-called *Hodge $\mathscr{D}$-conjecture* of Beilinson fails in general (see Voisin (1995)). It is quite easy to extend these results to other higher Chow groups $\mathrm{Ch}^p(X,m)$ (see Nagel, 2002).

Jan Nagel (1997) has worked out several other special cases of this theorem.

Theorem 10.7.10 *Let $X = V(d_0,\ldots,d_r) \subset \mathbf{P}^{2m+r}$ be a very general complete intersection of degrees $d_0 \ge \cdots \ge d_r$ such that*

$(C_0)\ \sum_{i=0}^r d_i + (m-1)d_r \ge 2m+r+3,$

$(C_1)\ \sum_{i=0}^r d_i + md_r \ge 2m+r+2.$

Then the image of the Abel–Jacobi map on $\mathrm{Ch}^{m+1}(X,1)\otimes\mathbf{Q}$ is contained in the image of $\mathrm{Ch}^{m+1}(\mathbf{P}^{2m+r},1)\otimes\mathbf{Q}$.

Here the exceptional cases where the theorem does not apply are, for example, quartic surfaces in $\mathbf{P}^3$, in which we know from Müller-Stach (1997) that the image of $\mathrm{Ch}^2(X,1)$ is much bigger than the image of $\mathrm{Ch}^2(\mathbf{P}^3,1)$. Other examples have been obtained by Collino (1997) and Gordon and Lewis (1998).

Bibliographical and Historical Remarks

First a remark about the historical development of the *connectivity theorem*: Nori's first (unpublished) proof of the connectivity theorem (as described in a letter to M. Green) used Koszul cohomology. At the same time, in Green and Müller-Stach (1996), Koszul methods were used in an attempt to prove results, which were later proved directly via Nori's theorem in Green and Müller-Stach (1996). Jan Nagel (1997) was able to invoke the original approach in a very clever way in order to obtain the connectivity theorem together with effective bounds. The original method of Nori does not immediately give effective bounds.

There have been several attempts to generalize normal functions and infinitesimal invariants as well as intermediate Jacobians to the setting of the *Beilinson/Bloch filtrations*. Such objects should be targets of cycle class maps from the graded pieces of the filtrations on (higher) Chow groups. In Green and Müller-Stach (1996), the complexes which involve Griffiths' infinitesimal invariants are related to a filtration on Deligne cohomology of the total space of a family. The Griffiths' infinitesimal invariants live in the first cohomology group. However, at that time it was not clear what to do with the higher cohomology groups.

Later, Voisin (1994b) used the second cohomology to study 0-cycles on surfaces in $\mathbf{P}^3$ and finally Asakura and Saito (1998) have worked out a theory of higher normal functions and higher infinitesimal invariants building on that. On the other hand, Green (1996) has defined a higher Albanese object $J_2^2(X)$ for any complex projective surface X together with a higher Abel–Jacobi map $\psi_2^2 : \mathrm{Ch}^2(X)_{\mathrm{alb}} \to J_2^2(X)$ which is defined on the subgroup of 0-cycles that have a trivial Albanese image. $J_2^2(X)$ is defined as a quotient of $H^2(X, \mathbf{Z})/NS(X) \otimes \mathbf{R}/\mathbf{Z} \otimes \mathbf{R}/\mathbf{Z}$ by the image of cycles that are trivial as cycles on X. The map ψ_2^2 is obtained by putting a 0-cycle Z on a curve C and contracting the Abel–Jacobi class of Z on C versus the extension class of the mixed Hodge structure of the pair (X, C) in a certain precise way. Voisin has recently shown in Voisin (1999b) that ψ_2^2 is in general neither surjective nor injective but sometimes has an infinite-dimensional image. Green and Griffiths (see Green, 1998) have also introduced so-called arithmetic Gauss–Manin complexes to study cycles by sort of deforming their field of definition. This has been taken up and generalized by Asakura (1999) and Saito (2001).

Part THREE

DIFFERENTIAL GEOMETRIC METHODS

11

Further Differential Geometric Tools

If a manifold and a vector bundle admit holomorphic structures, the choice of a Hermitian metric on the bundle leads to a unique metric connection, the Chern connection. Its curvature is of type $(1, 1)$ and so the Chern classes of such bundles have type (p, p). We end the first section with a Bochner-type principle, the principle of plurisubharmonicity, which is used in Chapter 13. In Section 11.2, we study how curvature behaves on subbundles as well as on quotient bundles. Next comes a section on principal bundles. This section becomes important in Chapter 12. We finish this chapter with a section on totally geodesic submanifolds. This will be used in Chapter 17 on special subvarieties of Mumford–Tate varieties.

11.1 Chern Connections and Applications

Let E be a $\mathbf{C}^r$-bundle on a manifold M. Recall that a connection on E is an operator

$$\nabla : \mathscr{A}^0(E) \to \mathscr{A}^1(E)$$

which first of all is linear in the sense that for two sections s_1, s_2 of E, one has $\nabla(s_1 + s_2) = \nabla s_1 + \nabla s_2$, and which, moreover, obeys the Leibniz rule $\nabla(fs) = \mathrm{d}f \otimes s + f\nabla(s)$ for a function f and a section s of E. Before we can explain the notion of a metric connection in the case where E is equipped with a Hermitian metric h we explain some useful notation. First we define a sort of metric contraction operator

$$\begin{aligned} h : \mathscr{A}^k(E) \times \mathscr{A}^\ell(E) &\to \mathscr{A}^{k+\ell}(M), \\ (\alpha \otimes s, \beta \otimes t) &\mapsto h(\alpha \otimes s, \beta \otimes t) \stackrel{\text{def}}{=} h(s,t)\, \alpha \wedge \bar{\beta}. \end{aligned} \tag{11.1}$$

If $\sigma \in \mathcal{A}^k(E)$ and $\tau \in \mathcal{A}^\ell(E)$, we then set

$$\overline{h(\sigma,\tau)} = h(\bar{\sigma},\bar{\tau}).$$

This definition makes sense, because for real forms (σ,τ) we then have $\overline{h(\sigma,\tau)} = h(\sigma,\tau)$.

For later reference, we also need the following formula:

$$h(\sigma,\tau) = (-1)^{k\cdot\ell}\,\overline{h(\tau,\sigma)}. \tag{11.2}$$

Using this notation, we say that a connection ∇ on E is metric if

$$\mathrm{d}h(s,t) = h(\nabla s,t) + h(s,\nabla t) \quad \forall s,t \in \mathcal{A}^0(E).$$

For metric connections, we then have

$$\mathrm{d}h(\sigma,\tau) = h(\nabla\sigma,\tau) + (-1)^k h(\sigma,\nabla\tau), \quad \sigma \in \mathcal{A}^k(E), \tau \in \mathcal{A}^\ell(E). \tag{11.3}$$

These metric connections always exist by a standard partition of unity argument. Moreover, we have the following.

Lemma 11.1.1 *If M is a complex manifold and (E,h) is a Hermitian holomorphic vector bundle, we can split a metric connection into types and there is a unique metric connection ∇_h with $\nabla_h^{0,1} = \bar\partial$. Its curvature F_{∇_h} is of type $(1,1)$ and hence $c_k(E)$ is of type (k,k).*

Proof If $s_j^U, j = 1,\ldots,r$ is a local *holomorphic* frame with connection matrix A_U, and with $h_U = (h(s_i^U, s_j^U))$ the corresponding Gram matrix, the condition that ∇_h is metric means

$$\mathrm{d}h_U = A_U{\circ}h_U + h_U{\circ}{}^{\mathsf{T}}\bar{A}_U. \tag{11.4}$$

Since $\nabla_h^{0,1} = \bar\partial$, the connection matrix A_U relative to the given holomorphic frame only contains forms of type $(1,0)$. Comparing sides in (11.4) we see that $A_U = \partial h_U{\circ}h_U^{-1}$. Conversely, one checks that this formula indeed defines a connection, proving existence. Using the formula $F_U = \mathrm{d}A_U + A_U \wedge A_U$ for the curvature, one checks that

$$F_U = -\partial\bar\partial h_U{\circ}h_U^{-1} + \partial h_U{\circ}h_U^{-1} \wedge \bar\partial h_U{\circ}h_U^{-1}.$$

This is visibly of type $(1,1)$. □

Definition 11.1.2 The unique metric connection $\nabla = \nabla_h$ on a Hermitian holomorphic vector bundle (E,h) with $D^{0,1} = \bar\partial$ is called the *Chern connection* on (E,h) with curvature F_{∇_h}.

We paraphrase this notion in terms of (0, 1)-connections, i.e., operators $\overline{\partial}_E : \mathscr{A}^0(E) \to \mathscr{A}^{0,1}(E)$ satisfying the Leibniz rule. In general, any connection ∇ on any (not necessarily holomorphic) complex vector bundle E can be split into types $\nabla = \nabla^{1,0} + \nabla^{0,1}$, and $\nabla^{0,1}$ is an example of such a (0, 1)-connection. As in the case of connections, these extend to operators $\overline{\partial}_E : \mathscr{A}^k(E) \to \mathscr{A}^{k+1}(E)$, and the composition $\overline{\partial}_E \circ \overline{\partial}_E : \mathscr{A}^0(E) \to \mathscr{A}^{0,2}(E)$ is linear over the C^∞-functions. If $\overline{\partial}_E$ is the (0, 1)-part of a connection, this composition is the (0, 2)-part of the curvature. We say that $\overline{\partial}_E$ is *integrable* if $\overline{\partial}_E \circ \overline{\partial}_E = 0$. Indeed, if E happens to be holomorphic there is a canonical integrable (0, 1)-connection $\overline{\partial}_E$. Note that the Leibniz rule implies that such partial connections are local, and now $\overline{\partial}_E$ can be characterized by saying that $\overline{\partial}_E s = 0$ for a local section s if and only if s is holomorphic. Indeed, choosing a local holomorphic frame, sections of E are vector-valued functions s and we demand that the operator $\overline{\partial}_E$ act component-wise as $\overline{\partial}$. This is independent of the choice of trivialization because another holomorphic trivialization leads to $g(s)$ with g a *holomorphic* matrix so that indeed $\overline{\partial}(gs) = \overline{\partial}g(s) + g(\overline{\partial}s) = 0$. The fact that, conversely, an integrable (0, 1)-connection on a bundle defines a holomorphic structure on it is a nontrivial result – the Koszul–Malgrange integrability theorem – (see Donaldson and Kronheimer, 1990, Chapter 2).

If we started out with a Hermitian (but not necessarily holomorphic) bundle (E, h) and a metric connection ∇, the (2, 0)-part of the curvature F_∇ is the Hermitian transpose of the (0, 2)-part and so in this case E is integrable if and only if F_∇ has type (1, 1). If we started out with a holomorphic vector bundle E, this is the case if and only if ∇ is the Chern connection. Indeed, the complex structure on E being uniquely defined by $\nabla^{0,1}$, we must have $\nabla^{0,1} = \overline{\partial}$, the usual anti-holomorphic derivative. Summarizing, we have the following.

Theorem 11.1.3 *Suppose that E is a complex vector bundle over a complex manifold with a connection ∇. Then E admits a unique holomorphic structure for which $\nabla^{0,1} = \overline{\partial}$ if and only if $\nabla^{0,1}$ is integrable. If, moreover, E is equipped with a Hermitian metric and ∇ is a metric connection, this is equivalent to saying that the curvature of ∇ has pure type* (1, 1). *In this case $\nabla = \nabla_h$, the Chern connection with respect to the Hermitian holomorphic structure on E.*

Example 11.1.4 If $r = 1$ we have a line bundle and the Gram matrix is just a function h_U and the preceding formula for the curvature becomes

$$F_U = -\partial\overline{\partial} \log h_U,$$

and hence the Chern form γ_1 is given by

$$\gamma_1 = -\frac{1}{2\pi \mathrm{i}} \partial\overline{\partial} \log h_U.$$

In particular, if h_U is a metric on the tangent bundle T_M, on the line bundle $\det T_M$ we have the metric $\det h_U$. If M is one-dimensional and ω_M is the metric $(1,1)$-form, the curvature form F_U in U of the metric connection on the tangent bundle and the Gaussian curvature K_h are related as follows:

$$\frac{1}{2\mathrm{i}} F_U = -\frac{1}{2\mathrm{i}} \partial\bar{\partial} \log h_U = -K_h \omega_M | U.$$

By definition, the left-hand side of this expression is the Ricci curvature of the metric and so the formulas can be summarized neatly as follows:

$$\operatorname{Ric} \omega_M = -\text{Gaussian curvature} \cdot \omega_M.$$

We finish this section by giving a basic calculation, which plays an important role below.

Proposition 11.1.5 *For any holomorphic section s of a Hermitian holomorphic vector bundle (E, h) over a complex manifold M we have:*

$$\partial\bar{\partial} h(s, s) = h(\nabla_h s, \nabla_h s) - h(F_{\nabla_h} s, s).$$

Proof Simplifying notation, we write ∇ instead of ∇_h. For any two sections s and t of E we have $\mathrm{d}h(s,t) = h(\nabla s, t) + h(s, \nabla t)$. Suppose that s is holomorphic. Then ∇s is of type $(1,0)$ and hence the $(0,1)$-part of $\mathrm{d}h(s,s)$, is equal to $\bar{\partial} h(s,s) = h(s, \nabla s)$. Differentiating once more, we find (using (11.3)) $\partial\bar{\partial} h(s,s) = \mathrm{d}\bar{\partial} h(s,s) = h(\nabla s, \nabla s) + h(s, F_{\nabla_h} s)$. Because by (11.2) we have $h(s, F_{\nabla_h} s) = \overline{h(F_{\nabla_h} s, s)}$ and because the latter form is purely imaginary we get $h(s, F_{\nabla_h} s) = -h(F_{\nabla_h} s, s)$ as desired. □

To interpret this, we first recall that a $(1,1)$-form ω is *(semi-)positive*, written $\omega > 0$, (respectively $\omega \geq 0$) if in local holomorphic coordinates $z_1, \ldots, z_m$ we have

$$\omega = \mathrm{i} \sum h_{ij} \, \mathrm{d}z_i \wedge d\bar{z}_j,$$

where (h_{ij}) is a positive definite (resp. semi-positive definite) Hermitian matrix.[1] Equivalently: for any vector field ξ of type $(1,0)$, we have $\mathrm{i}\omega(\xi, \bar{\xi}) > 0$, resp. ≥ 0. Observe that this is coherent with the usual notion of positivity in dimension 1: in a local coordinate $z = x + \mathrm{i}y$, the $(1,1)$-form $\frac{1}{2}\mathrm{i}\, \mathrm{d}z \wedge \mathrm{d}\bar{z} = \mathrm{d}x \wedge \mathrm{d}y$ equals the standard volume form and is positive.

In our case, the form $\mathrm{i}h(\nabla_h s, \nabla_h s)$ is positive semidefinite and so Proposition 11.1.5 states that the real $(1,1)$-form $\mathrm{i}\partial\bar{\partial} h(s,s)$ is the sum of a positive semidefinite form plus the form $-\mathrm{i}h(F_{\nabla_h} s, s)$. If the latter is positive semidefinite as well, the result is a *plurisubharmonic* $(1,1)$-form in the sense that in

[1] The matrix (h_{ij}) being Hermitian is equivalent to ω being a real form.

local coordinates $\{z_1, \ldots, z_m\}$ the Hessian matrix is positive semidefinite:

$$\left(\frac{\partial^2 h(s,s)}{\partial z_j \partial \bar{z}_j}\right) \geq 0.$$

It is well known that such functions obey a maximum principle (see for instance Lelong, 1957).

The previous calculations can be summarized as follows.

Proposition 11.1.6 (Principle of plurisubharmonicity) *Let (E, h) be a Hermitian holomorphic vector bundle over a compact complex manifold and let s be a holomorphic section of E. Suppose that the $(1, 1)$-form $(F_{\nabla_h} s, s)$ is negative semidefinite. Then s is a horizontal section, i.e., $\nabla_h s = 0$.*

Remark 11.1.7 (1) Plurisubharmonic functions played a central role in the development of analysis of several complex variables, such as the Levi problem, and in the study of Stein manifolds (see Gunning and Rossi, 1965, Chapter 9). (2) One may extend the above principle to noncompact manifolds. Indeed, there is such a principle for *bounded* harmonic functions on any complex manifold.

Problem

11.1.1 Let (E, h) be a Hermitian holomorphic vector bundle and ∇_h the Chern connection. Suppose that $S \subset E$ is a C^∞-complex subbundle preserved by ∇_h. Prove that S and its orthogonal complement in E are holomorphic subbundles of E.

11.2 Subbundles and Quotient Bundles

Let M be a differentiable manifold, (E, h) a Hermitian complex vector bundle over M, and ∇ a metric connection.

For a subbundle $S \subset E$, the quotient bundle $Q = E/S$ admits a natural identification with $S^\perp$. We can then give both the induced metric. In this way, we get a C^∞ metric decomposition

$$(E, h) = (S, h|_S) \oplus (Q, h|_Q). \tag{11.5}$$

We define a connection ∇_S on S by projection, that is, for any section s of S we decompose $\nabla(s)$ into an S-valued and a Q-valued form. The S-valued form by definition is $\nabla_S(s)$. It is straightforward to see that this is a connection, even a metric connection. A similar story holds for Q.

Pursuing this a bit further, we thus write

$$\nabla s = \nabla_S s + \sigma(s),$$

where $\sigma : \mathscr{A}^0(S) \longrightarrow \mathscr{A}^1(Q)$ is called the *second fundamental form* of S in E. We remark that σ is $\mathscr{A}^0$-linear, because it is a difference of connections. This means that the second fundamental form can be viewed as a vector bundle homomorphism between differentiable bundles.

Suppose now M is a complex manifold, E is a holomorphic vector bundle, and ∇ is the Chern connection on E. Then ∇_S is the Chern connection on S, and similarly for ∇_Q. Note that the splitting (11.5) is differentiable but in general not holomorphic.

The relation between the connections can be examined by a choice of orthonormal frames for S and Q and by looking at the connection matrix A_∇ written out in block form with respect to these frames:

$$\begin{array}{cc} S \longrightarrow \\ Q \longrightarrow \end{array} \left(\begin{array}{c|c} A_S & -{}^{\mathsf{T}}\bar{\sigma} \\ \hline \sigma & A_Q \end{array}\right), \qquad \begin{array}{cc} \uparrow & \uparrow \\ S & Q \end{array} \tag{11.6}$$

where σ is the second fundamental form of S in E.

Lemma 11.2.1 *For the curvature of the subbundle and quotient bundle we have*

$$F(\nabla_S) = F_{\nabla_h}|_S + {}^{\mathsf{T}}\bar{\sigma} \wedge \sigma,$$
$$F(\nabla_Q) = F_{\nabla_h}|_Q + \sigma \wedge {}^{\mathsf{T}}\bar{\sigma}.$$

This follows by writing out the block-decomposition for $A_\nabla \wedge A_\nabla + dA_\nabla$ as follows:

$$\begin{pmatrix} (A_S \wedge A_S + dA_S) - {}^{\mathsf{T}}\bar{\sigma} \wedge \sigma & *** \\ *** & (A_Q \wedge A_Q + dA_Q) - \sigma \wedge {}^{\mathsf{T}}\bar{\sigma} \end{pmatrix}.$$

We now introduce an extension of the notion of positivity for forms, recalled at the end of the previous section, to E-valued forms.

Definition 11.2.2 $F \in A^{1,1}(\operatorname{End} E)$ is said to be *positive*, written $F > 0$, if for all nonzero C^∞ sections s of E the real $(1,1)$-form $\mathrm{i}h(Fs, s)$ is positive (see page 312), i.e.,

$$\mathrm{i}h(Fs, s)(\xi, \bar{\xi}) > 0, \quad \xi \text{ a nonzero } (1,0)\text{-vector}.$$

Likewise, we write $F \geq 0$ if positive semidefinite is meant.

Now, since $\sigma \wedge {}^{\mathsf{T}}\bar{\sigma} \geq 0$ (see Problem 11.2.1), we have the following useful corollary.

Corollary 11.2.3 *In the notation of Lemma 11.2.1, we have*

(i) $F(\nabla_S) \leq F_{\nabla_h}|_S$ *(curvature decreases on subbundles),*
(ii) $F(\nabla_Q) \geq F_{\nabla_h}|_Q$ *(curvature increases on quotient bundles).*

Example 11.2.4 *The tautological bundle on the Grassmann manifold $G(n, d)$ of $(d + 1)$-planes in $\mathbf{C}^{n+1}$*

Recall that this tautological (sub)bundle is defined as follows. Its total space is

$$S = \left\{(v, W) \in \mathbf{C}^{n+1} \times \mathbf{G}(n, d) \mid v \in W\right\} \subset \mathbf{C}^{n+1} \times \mathbf{G}(n, d),$$

and its projection $S \to \mathbf{G}(n, d)$ comes from the projection onto the second factor. The unitary group $\mathrm{U}(n+1)$ acts from the left on the Grassmann manifold and it preserves the universal subbundle. Let $\langle\, , \rangle$ denote the standard Hermitian metric on $\mathbf{C}^{n+1}$. It induces a Hermitian metric on the trivial $\mathbf{C}^{n+1}$-bundle and hence it gives S a $\mathrm{U}(n + 1)$-equivariant Hermitian metric. We want to give the Chern connection on S explicitly. To do this, we first introduce the *Stiefel manifold*:

$$\mathrm{St}(n + 1, d + 1) = \left\{\text{unitary } (d + 1)\text{-frames in } \mathbf{C}^{n+1}\right\}.$$

The unitary group $\mathrm{U}(n + 1)$ acts transitively from the right on $(d + 1)$-frames and the map

$$\lambda : \mathrm{St}(n + 1, d + 1) \to \mathbf{G}(n, d),$$

sending a $(d + 1)$ frame to the space it spans is a $\mathrm{U}(n + 1)$-equivariant map. Using the standard basis F_0 of $\mathbf{C}^{n+1}$ as a reference $(n + 1)$-frame, we identify $U(n + 1)$ with the set of unitary $(n + 1)$-matrices. Now map $g \in \mathrm{U}(n + 1)$ to the $(d + 1)$-frame formed by the first $(d + 1)$ vectors of the gF_0, thus defining

$$\mu : \mathrm{U}(n + 1) \to \mathrm{St}(n + 1, d + 1).$$

The induced maps $\mu^* : A^\bullet(\mathrm{St}(n + 1, d + 1)) \to A^\bullet(\mathrm{U}(n + 1))$ and $\lambda^* : A^\bullet(\mathbf{G}(n, d)) \to A^\bullet(\mathrm{St}(n+1, d+1))$ are injective and will be used to identify forms on the Grassmann and Stiefel manifold with forms on the unitary group $\mathrm{U}(n+1)$.

On the latter we have the matrix-valued *Maurer–Cartan form*

$$\omega = g^{-1}\,\mathrm{d}g \in A^1\left(\mathrm{U}(n + 1), \mathfrak{u}(n + 1)\right),$$

which defines in fact a $\mathrm{U}(n + 1)$-equivariant one-form on the Stiefel manifold $\mathrm{St}(n + 1, d + 1)$ of $(d + 1)$-frames. This serves an an example for a connection on a principal bundle, to be introduced in the next section. For now we show how to use it to get the Chern connection of S.

We first rewrite ω in terms of frames $F = \{f_0, \ldots, f_n\}$, as follows. Consider f_β as the β-th column of the unitary matrix $g = (g_{\alpha\beta})$. The matrix-valued one-form $\langle \mathrm{d}f_\alpha, f_\beta\rangle$ coincides with the form ${}^{\mathsf{T}}\bar{g}\,\mathrm{d}g = \omega$.

Next, we consider a Zariski-open subset $U \subset \mathbf{G}(n, d)$ over which we have a unitary $(n+1)$-frame field $F : U \longrightarrow \mathrm{U}(n+1)$. The first $(d+1)$ vectors of this frame field define a frame field for the bundle $S|U$, and the matrix-valued one-form on U given by

$$\omega' = F^*\left(\langle \mathrm{d}f_\alpha, f_\beta\rangle\right)_{\alpha,\beta=0,\ldots,n}$$

then defines a connection, as we show now. Two frame fields are related by a map $g : U \longrightarrow \mathrm{U}(n+1)$, and a small computation shows that in the frame gF the form becomes $\omega' + g^{-1}\,\mathrm{d}g$, proving our claim. The connection is clearly a metric connection, and because ω' is holomorphic it must be the Chern connection. Its curvature is computed in Problem 11.2.5.

Problems

11.2.1 Show that for any complex column vector $\vec{z}$, the matrix $\vec{z} \cdot {}^{\mathsf{T}}\bar{\vec{z}}$ is a positive semidefinite matrix, and use this to show that $\sigma \wedge {}^{\mathsf{T}}\bar{\sigma} \geq 0$ with σ the second fundamental form.

11.2.2 Suppose that the second fundamental form vanishes identically. If S is a holomorphic subbundle, so is Q, yielding a holomorphic splitting

$$E = S \oplus S^{\perp}.$$

11.2.3 Suppose that (E, h) is a Hermitian holomorphic vector bundle with the property that the curvature F of the Chern connection is seminegative $F \leq 0$ on a subbundle S in the sense that for all sections s of S the $(1,1)$-form $h(Fs, s)$ is negative semidefinite. Prove that holomorphic sections of E are necessarily flat.

11.2.4 Prove the *Maurer–Cartan equation*

$$\mathrm{d}\omega + \omega \wedge \omega = 0,$$

which should be read as $\mathrm{d}\omega_{\alpha\beta} + \sum_\gamma \omega_{\alpha\gamma} \wedge \omega_{\gamma\beta} = 0$.

11.2.5 Prove that the curvature of the Chern connection on the tautological subbundle $S \longrightarrow \mathbf{G}(n, d)$ is given by

$$F_{\alpha,\beta} = -\sum_{r=d+1}^{n} \omega'_{\alpha r} \wedge \overline{\omega'}_{\beta r}$$

by calculating $\mathrm{d}\omega' + \omega' \wedge \omega'$ using the Maurer–Cartan equations. This shows not only that the curvature is a (1, 1)-form, but also that it is seminegative in accordance with Corollary 11.2.3.

11.3 Principal Bundles and Connections

In this important section we explain how one can directly introduce the concept of connection on a principal bundle and how this is related to our previous notion of a connection on any of the associated vector bundles. Any local section in the principal bundle, together with a basis for the vector space on which the associated bundle is built, gives a frame for the latter. We are giving formulas for the connection (Proposition 11.4.3) and curvature form (Proposition 11.4.5) in terms of this frame, the representation giving the vector bundle and the fundamental form (11.8) defined by the connection on the principal bundle. These formulas are crucial in later chapters.

For basic facts about Lie groups and Lie algebras we refer to Appendix D. Recall in particular that a *representation* of a group G (in a vector space V) is a homomorphism $\rho : G \to \mathrm{GL}(V)$. In what follows one particular representation plays a special role, the *adjoint representation* of a Lie group G on its Lie algebra $\mathfrak{g}$. We repeat here the definition from Example D.2.2. For a fixed $g \in G$, conjugation by g defines the isomorphism $G \xrightarrow{\sim} G$ given by $x \mapsto gxg^{-1}$ whose derivative is the adjoint homomorphism $\mathrm{Ad}(g) : \mathfrak{g} \to \mathfrak{g}$. For varying g this defines the adjoint representation

$$\left.\begin{array}{ccc} G & \xrightarrow{\mathrm{Ad}} & \mathrm{End}\,\mathfrak{g} \\ g & \mapsto & \mathrm{Ad}(g). \end{array}\right\} \tag{11.7}$$

Let us next introduce the notion of a principal bundle over a differentiable manifold M.

Definition 11.3.1 Let V be a Lie group. A V*-principal bundle* over M consists of a differentiable manifold P equipped with a free *right* V-action and a differentiable map $\pi : P \longrightarrow M$ such that the fiber over $m \in M$ consists of one V-orbit. Moreover, π is locally trivial in the sense that each point $m \in M$ admits an open neighborhood $U \subset M$ with a V-equivariant diffeomorphism $p^{-1}U \xrightarrow{\sim} U \times V$.

Examples 11.3.2 (i) The set LM of linear frames of the tangent bundle of M. By linear frame we mean "basis for a tangent space." The group $V = \mathrm{GL}(n, \mathbf{R})$ acts on the right by change of basis, and clearly $LM/V \cong M$. The choice of an orientation M selects from LM the bundle L^+M of oriented frames, which has group $\mathrm{GL}^+(n, \mathbf{R})$, the set of matrices with positive determinant. The further choice of a Riemannian metric selects from L^+M the set of oriented orthonormal frames. This is a principal bundle P with group $\mathrm{SO}(n)$.

(ii) The Stiefel manifold of unitary $(d+1)$-frames in $\mathbf{C}^{n+1}$ (see the end of the previous section) is a principal $\mathrm{U}(n+1)$-bundle.

(iii) If G is a Lie group with subgroup V acting from the right, this defines the principal bundle $G \longrightarrow G/V$. The manifold G/V is called *homogeneous space*. A particular type of homogeneous space is going to play a dominant role.

Definition 11.3.3 Let G be a matrix Lie group and $V \subset G$ a closed subgroup. We say that (G, V) is a *reductive pair* or that $D = G/V$ is a *reductive domain* if there is a decomposition

$$\boxed{\mathfrak{g} = \mathfrak{v} \oplus \mathfrak{m}, \quad [\mathfrak{v}, \mathfrak{m}] \subset \mathfrak{m}.}$$

Equivalently, there is an $\mathrm{Ad}(V)$-invariant splitting $\mathfrak{g} = \mathfrak{v} \oplus \mathfrak{m}$.

The equivalence follows from

$$\left.\frac{\mathrm{d}}{\mathrm{d}t}\right|_{t=0} \mathrm{Ad}(e^{t\xi})\eta = \mathrm{ad}(\xi)\eta = [\xi, \eta].$$

We see in Chapter 12 that period domains are examples of such spaces.

Suppose we are given a principal bundle $\pi : P \longrightarrow M$. There is a natural bundle of *vertical vectors*, written T^{vert}, defined to be the kernel of $d\pi$. Note that it is V-invariant. A connection, from our point of view, is simply a V-invariant complementary bundle T^{hor}, called the *horizontal bundle*.

Definition 11.3.4 A *connection* in a principal V-bundle P is a V-invariant splitting

$$T_P = T^{\mathrm{vert}} \oplus T^{\mathrm{hor}}.$$

Example 11.3.5 The principal bundle $G \to G/V$ over a reductive homogeneous space G/V always has a natural connection in the sense just defined. At the identity element of G, the splitting is $\mathfrak{g} = \mathfrak{v} \oplus \mathfrak{m}$. At other points, it is gotten by pushing this decomposition forward by left-translation, i.e., by the differential of the map $L_g(x) = gx$. Because the splitting of $\mathfrak{g}$ is $\mathfrak{v}$-invariant, the resulting splitting is V-invariant,

With a connection on a principal bundle P we associate a $\mathfrak{v}$-*valued fundamental one-form* for P, say ω, which is uniquely given by demanding

$$\begin{aligned} \omega(\xi) &= 0, \quad \xi \in T_p^{\mathrm{hor}}, \\ \omega(X_p^*) &= X, \quad X \in \mathfrak{v}, \end{aligned} \tag{11.8}$$

where X^* is the vertical vector field on P given by

$$X^*_p = \left.\frac{\mathrm{d}}{\mathrm{d}t}\right|_{t=0} (p \cdot \exp(tX)),$$

with $\exp : \mathfrak{v} \longrightarrow V$ the exponential map. Since by varying X, we can cover all of the vertical tangent space T_p^{vert} at p, this really defines ω. This last property can be expressed using Maurer–Cartan forms.

Definition 11.3.6 Suppose that V is a Lie group with Lie algebra $\mathfrak{v}$. Its *Maurer–Cartan form* is the unique left-invariant $\mathfrak{v}$-valued form ω_V on V whose value at $e \in V$ on $X \in T_e V = \mathfrak{v}$ equals X. In particular, the value of ω_V on the left-invariant vector field X^* defined by $X \in \mathfrak{v}$ is constant and equals X.

Example If $V \subset \mathrm{GL}(n)$ is a matrix group the coordinate functions restrict to holomorphic functions on V and the Maurer–Cartan form can be identified with the matrix-valued one-form

$$\omega_V = v^{-1}\,\mathrm{d}v.$$

The following lemma collects a few useful properties of the form ω_V.

Lemma 11.3.7 1. *ω_V is V-equivariant under the adjoint representation (see (11.7)) of V on $\mathfrak{v}$ given by*

$$\mathrm{Ad}(v)\xi = v\xi v^{-1},\ v \in V,\ \xi \in \mathfrak{v}.$$

2. *A V-equivariant $\mathfrak{v}$-valued one-form ω on P satisfies*

$$\omega(X^*) = X,$$

if and only if

$$V_p^*\omega = \omega \quad \text{where } V_p : V \longrightarrow P \text{ is the orbit map } v \mapsto p\cdot v$$

at all points p.
3. *If this is the case, it is the fundamental one-form of a unique connection on P defined by setting*

$$T^{\mathrm{hor}}_{P,p} \stackrel{\mathrm{def}}{=} \mathrm{Ker}\left\{\omega : T_{P,p} \longrightarrow \mathfrak{v}\right\}.$$

Proof 1. We see this as follows: with R_v the right action, we have $R_v^*\omega_V = \mathrm{Ad}(v^{-1})\omega_V$, which means that the value of ω_V at $p \cdot v$, evaluated on $(R_v)_*(\xi) \in T_{P,p\cdot v}$ equals $v^{-1}\omega_{V,p}(\xi)v$.
2. Since $[V_p]_*(L_v)_*X = X^*_{p\cdot v}$, equation (11.8) then states that ω at the point $p \cdot v$ takes the value X on this vector. The converse is clear.
3. We leave this to the reader. □

Before we introduce the curvature associated with a connection form ω we need some notation. For any tangent vector $X \in T_{P,p}$ we write

$$X = X^{\mathrm{vert}} + X^{\mathrm{hor}}, \quad X^{\mathrm{vert}} \in T^{\mathrm{vert}}_{P,p}, \quad X^{\mathrm{hor}} \in T^{\mathrm{vert}}_{P,p}$$

and for any $\mathfrak{v}$-valued k-form $\alpha \in A^k(P, \mathfrak{v})$ we introduce

$$\alpha^{\mathrm{hor}}(X_1, \ldots, X_h) \stackrel{\mathrm{def}}{=} \alpha(X_1^{\mathrm{hor}}, \ldots, X_k^{\mathrm{hor}}).$$

The curvature form can now be defined.

Definition 11.3.8 The *curvature form* Ω associated with $\omega = \omega_V$ is the $\mathfrak{v}$-valued two-form $\Omega \stackrel{\mathrm{def}}{=} (\mathrm{d}\omega)^{\mathrm{hor}}$.

Recall the convention

$$[\omega_1, \omega_2](X, Y) \stackrel{\mathrm{def}}{=} [\omega_1(X), \omega_2(Y)] - [\omega_1(Y), \omega_2(X)] \in \mathfrak{v} \tag{11.9}$$

whenever $\omega_1, \omega_2 \in \mathcal{A}^1(\mathfrak{v})$ and $X, Y \in T_{P,p}$. Using this shorthand, a more useful expression for the curvature can be derived.

Lemma 11.3.9 *The curvature can be given by*

$$\Omega = \mathrm{d}\omega + \frac{1}{2}[\omega, \omega].$$

Proof Let X, Y be two vectors tangent to P. Then

$$\Omega(X, Y) = \mathrm{d}\omega(X^{\mathrm{hor}}, Y^{\mathrm{hor}}).$$

On the other hand, by equation (11.9),

$$\mathrm{d}\omega(X, Y) + \frac{1}{2}[\omega, \omega](X, Y) = \mathrm{d}\omega(X, Y) + [\omega(X), \omega(Y)].$$

So we need to establish the equality

$$\mathrm{d}\omega(X^{\mathrm{hor}}, Y^{\mathrm{hor}}) = \mathrm{d}\omega(X, Y) + [\omega(X), \omega(Y)].$$

By linearity, we may reduce X and Y to both horizontal, both vertical or mixed. In the first case, the desired equality is trivial. In the second case, we may assume that vertical vectors X, Y extend to the fundamental vector fields X^*, Y^*. On the other hand, for vector fields X^*, Y^* we can employ the usual formula for $\mathrm{d}\omega$:

$$\begin{aligned}
\mathrm{d}\omega(X^*, Y^*) &= X^*(\omega(Y^*)) - Y^*(\omega(X^*)) - \omega([X^*, Y^*]) \\
&= 0 - 0 - \omega([X^*, Y^*]) \ (X \text{ and } Y \text{ are constant, so } X^*Y = Y^*X = 0) \\
&= [X, Y]^* \ (\text{since } [X^*, Y^*] = [X, Y]^*) \\
&= -[\omega(X), \omega(Y)]
\end{aligned}$$

and so both sides of the desired equality vanish. The mixed case is left to the reader. □

11.4 Connections on Associated Vector Bundles

We now explain how a connection on a principal bundle $P \longrightarrow M$ as we just defined leads to the usual notion of a connection on any associated vector bundle. By *associated bundle* we mean the twisted product

$$[W] = P \times_V W,$$

where W is a V-module and the equivalence relation is defined as $(p, w) \sim (pv, v^{-1}w)$, with equivalence class $[p, w]$. We assume that the representation $\rho : V \to \mathrm{GL}(W)$ defining the V-module structure on W is differentiable. In that case the derivative $\dot{\rho} : \mathrm{Lie}(V) \longrightarrow \mathrm{End}(W)$ is defined.

A connection enables us to associate with a smooth curve $\gamma : I \longrightarrow M$ a unique *horizontal lift* $\tilde{\gamma} : I \longrightarrow P$. By this we mean that, first, $\pi \circ \tilde{\gamma} = \gamma$, i.e., $\tilde{\gamma}$ is a lift and second, that the tangent at every point $p \in P$ is horizontal. To paraphrase this, consider the manifold $\pi^{-1}(\gamma(I)) \subset P$. At every point p of this manifold there is a unique horizontal direction mapping to the tangent vector X_p of γ at $\pi(p)$ (always nonzero because γ is supposed to be smooth). Also the vector field X tangent to γ has horizontal lift X^{hor}, obtained by finding a maximal integral curve of $\tilde{\gamma}$ along this vector field. These exist and are unique, once we specify the initial point.

Since sections of $[W]$ are V-equivariant functions:

$$s : P \longrightarrow W, \quad s(p \cdot v) = v^{-1} \cdot s(p), \tag{11.10}$$

we can derive these in the horizontal direction on P given by lifting vector fields on M horizontally. This gives the *derivative of s along X*:

$$\nabla_X s \stackrel{\mathrm{def}}{=} X^{\mathrm{hor}}(s), \tag{11.11}$$

i.e., $\nabla_X s$ is the function $p \mapsto X_p^{\mathrm{hor}}(s)$. And indeed, we obtain in this way a connection on the vector bundle $[W]$.

Proposition 11.4.1 *Let $[W]$ be the bundle associated with a principal V-bundle P and V-representation W. Using the definition* (11.11), *a connection on P defines a connection*

$$\nabla : [W] \longrightarrow A^1[W], \quad \nabla_X s = X^{\mathrm{hor}}(s)$$

on $[W]$.

The proof is a routine verification which we leave to the reader. See Problem 11.4.3.

Remark 11.4.2 As before, let γ be a curve with tangent vector field X. If $\nabla_X(s) = 0$ we say that the section is *parallel* along the curve γ; it means that the function $P \to W$ associated with the section[2] s is constant along $\tilde{\gamma}$, say equal to $w \in W$. Then $\underline{w}(t) \stackrel{\text{def}}{=}$ class of $[\tilde{s}(t), w]$ is said to be obtained from $\underline{w}(0)$ by *parallel transport*. Equivalently, s is parallel along the curve γ if the ordinary differential equation $\nabla_{\dot{\gamma}(t)} s(t) = 0$ holds. The unique solution $s(t) = s(\gamma(t))$, with initial vector $s(0)$, gives the parallel transport $\underline{w}(t)$. Given a curve γ with initial point at $m = \gamma(0) \in M$, parallel transport defines a linear isomorphism $[W]_{\gamma(0)} \xrightarrow{\sim} [W]_{\gamma(t)}$ between fibers of $[W]$ at m and at $\gamma(t)$.

We end this section by showing how to obtain a local expression for the connection using a local section $f : U \to P$ and then, how this leads to an expression for the curvature, which is the main result, Proposition 11.4.5.

First we pull back the form ω defined by the connection to U and get the $\mathfrak{v}$-valued one-form $f^*\omega$. The differential $\dot{\rho}$ of the representation ρ defining W enables one to view the latter as a $\mathfrak{gl}(W)$-valued one-form $f^*\dot{\rho}\omega$. Fix a basis $\underline{w} = \{w_i\}$ for W and use it to identify $\mathfrak{gl}(W)$ with $\mathfrak{gl}(n)$. So we now have a form

$$\theta_f = f^*(\dot{\rho}\omega) \in A^1(U, \mathfrak{gl}(n, \mathbf{R})). \tag{11.12}$$

The frame $\{s_i\}$, with $s_i(u) = [f(u), w_i]$ defined by the basis $\{w_i\}$ of W, defines a trivialization of $[W]|U$, and we identify sections of this bundle over U with column vectors $\vec{s}(u)$. We claim that the form θ_f is the connection matrix for the connection relative to the frame f.

Proposition 11.4.3 *Let $f : U \to P$ be a local section of P and let $\{w_i\}$ be a basis for W. Then the bundle $[W]|U$ is trivialized by means of the sections $s_i(u) = [f(u), w_i]$. Let $\vec{s}$ be the vector-valued function representing a section s of $[W]|U$ in this frame. Then the connection ∇ induced on $[W]$ as prescribed by Proposition 11.4.1 is given by*

$$\nabla\vec{s} = \mathrm{d}\vec{s} + \theta_f(\vec{s}),$$

where θ_f is given by (11.12). *In particular, this form is the connection form for the induced connection.*

To show this we first show what happens if we change frames. So, suppose we have two local sections $f, \tilde{f} : U \to P$. These are related by

$$\tilde{f}(u) = f(u)v(u) \in P, \quad v : U \to V$$

and the two forms θ_f and $\theta_{\tilde{f}}$ are related as follows.

Lemma 11.4.4 (Change of frame formula)

$$\theta_{\tilde{f}} = v^{-1} \circ \theta_f \circ v + v^{-1}\,\mathrm{d}v,$$

[2] See Eq. (11.10).

where we consider $\mathrm{d}v$ *as a* $\mathfrak{gl}(n, \mathbf{R})$*-valued one-form.*

Proof Let γ be a curve through u with tangent ξ. Then

$$\begin{aligned}\tilde{f}_*\xi &= \frac{d}{dt}\left[f(\gamma(t)v(\gamma(t))\right]|_{t=0} \\ &= \frac{\mathrm{d}}{\mathrm{d}t}\left[f(\gamma(t)v(u)\right]|_{t=0} + \frac{\mathrm{d}}{\mathrm{d}t}\left[f(u)v(\gamma(t))\right]|_{t=0} \\ &= \left[R_{v(u)}\right]_*(f_*\xi) + \left[V_{\tilde{f}(u)}\right]_*(v_*\xi)\end{aligned}$$

and hence

$$\begin{aligned}\tilde{f}^*\omega(\xi) &= \omega(\tilde{f}_*\xi) \\ &= v(u)^{-1}f^*\omega(\xi)v(u) + v^*\omega_V(\xi).\end{aligned}$$

Next, one uses the representation $\rho : V \longrightarrow \mathrm{GL}(W)$ which realizes V as a matrix group. Any choice of basis for W makes it then possible to consider $v : U \to V$ as a matrix-valued function which one keeps writing as v. So v is the matrix-function $u \mapsto (x_{ij}v(u))$, where the x_{ij} are the standard coordinates on $\mathrm{End}(W)$ defined by the basis for W.

Since the matrix-valued one-form $v^*\omega_V$ at u is equal to $\left[x_{ij}(v(u)\right]^{-1} \cdot \left[\mathrm{d}(x_{ij}\circ v)\right]_u$, with this convention we can write $v^*\omega_V = v^{-1}\,\mathrm{d}v$. So, finally,

$$\theta_{\tilde{f}} = v^{-1}\theta_f v + v^{-1}\,\mathrm{d}v. \qquad \square$$

Now we are ready for the proof of Proposition 11.4.3.

Proof of Proposition 11.4.3 We may replace $\{e_i\}$ by any frame convenient for calculation. Indeed, both sides transform in the same way under a change $f \mapsto \tilde{f} = f \cdot g$. For the right-hand side, the transformation law is given by Lemma 11.4.4. In particular, choosing a tangent direction $\xi \in T_mM$ and a curve γ through m with that tangent, we may assume that the frame restricted to γ is obtained by parallel translation along γ. Then the left-hand side is $\dot{\vec{s}}(0)$ by construction.

The right-hand side is also equal to $\dot{\vec{s}}(0)$: the second term, $\theta_f(\xi) = \omega(\tilde{\gamma}_*\xi)$ vanishes because $\tilde{\gamma}_*\xi$ is horizontal by construction. $\square$

Let us next consider the curvature. Recall (Def. 11.3.8) that we already introduced the curvature form, a two-form on P with values in $\mathfrak{v}$ and that it satisfies

$$\Omega = \mathrm{d}\omega + \frac{1}{2}[\omega, \omega].$$

Next, consider the expression $\dot{\rho}\circ\Omega$ where $\dot{\rho} : \mathfrak{v} \longrightarrow \mathrm{End}(W)$ is the derivative of the representation $\rho : V \longrightarrow \mathrm{GL}(W)$ defining $[W]$. This is an $\mathrm{End}(W)$-valued two-form on P related to the curvature of ∇.

Proposition 11.4.5 *Let $\rho : V \to \mathrm{GL}(W)$ be a representation, P a V-principal bundle and $[W]$ the associated vector bundle. Suppose that ∇ is the connection on $[W]$ induced by a connection on P and let $f : U \to P$ be a local section. A local expression on U for the curvature of ∇ is given by*

$$F_U = f^*(\dot{\rho}\circ\Omega).$$

Proof Since by Proposition 11.4.3 the one-form $\theta_f = f^*(\dot{\rho}\circ\omega)$ is the connection matrix, the curvature of the connection ∇ is given by

$$\mathrm{d}\theta_f + \theta_f \wedge \theta_f = f^*\left(\dot{\rho}\circ(\mathrm{d}\omega + \frac{1}{2}[\omega, \omega])\right)$$

showing the local expression. □

Problems

11.4.1 Show that a principal V-bundle $P \to M$ is isomorphic to the product bundle $V \times M$ if and only if it admits a cross section $M \to P$.

11.4.2 Verify that the connection on the tautological bundle over $\mathbf{G}(n, d)$ associated with the connection of the Stiefel manifold of unitary $(d+1)$-frames in $\mathbf{C}^{n+1}$ given by the Maurer–Cartan form on $\mathrm{U}(n+1)$ is indeed the one that we considered in the previous section.

11.4.3 Show that the definition of ∇ given above is independent of choices and defines a connection operator on the vector bundle $[W]$.

11.5 Totally Geodesic Submanifolds

Let M be a Riemannian manifold and $S \subset M$ a submanifold. Equip it with the induced metric. Let ∇^M be any metric connection on the tangent bundle T_M and ∇^S the induced connection on T_S. Recall that a *geodesic curve* by definition has the property that its tangent vector field is parallel for the connection. Geodesics at $p \in S$ which are tangent to S at p need not stay on S; if they do, we say that S is geodesically immersed at p and if this holds for all points of S we say that *S is a totally geodesic submanifold.*

In Section 11.2 we explained that the second fundamental form $\sigma : \mathscr{A}^0(T_S) \to \mathscr{A}^1(\nu_{S/M})$ measures the difference $\nabla^M|S - \nabla^S$. It follows that if it vanishes identically, parallel tangent fields along a curve in S for ∇^S are also parallel for ∇^M and so in that case S is a totally geodesic submanifold.

There is also a criterion in terms of the curvature, since, as we have seen

(Lemma 11.2.1), the difference $F(\nabla_S) - F(\nabla^M)|_S$ of the curvatures is given by the negative form ${}^\mathsf{T}\bar{\sigma} \wedge \sigma$. This difference vanishes exactly when $\sigma = 0$. We summarize what has been shown.

Lemma 11.5.1 *A submanifold S of a Riemannian manifold M, equipped with a metric connection ∇^M, is a totally geodesic submanifold of M if one of the following equivalent properties holds:*

(i) *a geodesic curve in M tangent to S at some point stays within S;*
(ii) *the second fundamental form vanishes;*
(iii) *the curvature of ∇^S is the restriction to S of the curvature of ∇^M.*

Examples 11.5.2 (i) Let $D = G/V$ be a reductive homogeneous space equipped with a G-equivariant Riemannian metric. Let $\mathfrak{g} = \mathfrak{v} \oplus \mathfrak{m}$ be a reductive splitting. The adjoint representation of G identifies the tangent bundle as the associated bundle $T_D = G \times_{\mathrm{Ad}} \mathfrak{m}$. Hence there is a Maurer–Cartan induced connection ∇^D on the tangent bundle. It is a metric connection and it can be described as follows. As before, let

$$X^* \overset{\text{def}}{=} L_{g*}X,$$

the vector field on D obtained from $X \in T_{D,o} \simeq \mathfrak{m}$ by left G-translation ($o \in D$ is the coset of $1 \in G$). Then

$$\nabla^D X^* = 0. \tag{11.13}$$

We claim that the orbit S of the base point $o \in D$ under a Lie subgroup $U \subset G$ is a totally geodesic submanifold of D. Indeed, If $X \in T_{S,o} = \mathrm{Lie}(U) \cap \mathfrak{m}$, then X^* restricts to a left-invariant vector field for H on S and so, by (11.13), $\nabla^S X^* = 0$.

Of course, the point $o \in D$ does not play a special role: every other point is of the form $L_g(o)$ and so $L_g S$ is a totally geodesic manifold passing through $L_g(o)$.

Remark Classically, totally geodesic submanifolds are characterized by Lie triple systems contained in $\mathfrak{m}$. Here a *Lie triple system* $\mathfrak{s}$ is a subspace $\mathfrak{s} \subset \mathfrak{g}$ with the property that $[[\mathfrak{s}, \mathfrak{s}], \mathfrak{s}] \subset \mathfrak{s}$ (see Helgason, 1978, Ch IV, Theorem 7.2). The fact that we have a reductive pair $(\mathfrak{v}, \mathfrak{m})$ implies that $\mathfrak{m}$, as well as $\mathfrak{m} \cap \mathrm{Lie}(U)$ is a Lie triple system.

To see that a Lie triple system $\mathfrak{s} \subset \mathfrak{m}$ gives rise to a totally geodesic submanifold S we argue as follows. Making use of $\mathfrak{s}$ being a Lie triple system, we check that $[\mathfrak{s}, \mathfrak{s}]$ is a Lie subalgebra of $\mathfrak{v}$ and hence $\mathfrak{u} = [\mathfrak{s}, \mathfrak{s}] \oplus \mathfrak{s}$ is a Lie subalgebra of $\mathfrak{g}$. Let U be the connected Lie subgroup of G whose Lie algebra is $\mathfrak{u}$ and consider $S = U \cdot o$. We have already seen that S is totally geodesic. By construction, $T_{S,o} = \mathfrak{u} \cap \mathfrak{m} = \mathfrak{s}$.

(ii) In the setting of example (i), note that $\mathrm{Lie}(U)$ gets an induced reductive splitting, i.e., any U-orbit defines a reductive subdomain. If, however, we start with an injective homomorphism j of Lie algebras each of which have a reductive splitting, this splitting need not be preserved under j, and we need to impose this as a condition. Let us explain this in detail. So, suppose that we have two reductive domains $X_1 = G_1/V_1$ and $X_2 = G_2/V_2$, say with reductive splittings

$$\mathrm{Lie}(G_\alpha) = \mathfrak{v}_\alpha \oplus \mathfrak{m}_\alpha, \ \alpha = 1, 2.$$

An injective homomorphism of (real or complex) algebraic groups $\rho : G_1 \to G_2$ such that $\rho(V_1) \subset V_2$ induces an equivariant map $f : X_1 = G_1/V_1 \to X_2 = G_2/V_2$. The base point $o \in G_1$ has V_1 as stabilizer and likewise for $f(o) \in X_2$. Then $\rho_* : \mathrm{Lie}(G_1) \to \mathrm{Lie}(G_2)$ induces the tangent map

$$T_{D_1,o} = f_* : \mathrm{Lie}(G_1)/\mathfrak{v}_1 \to \mathrm{Lie}(G_2)/\mathfrak{v}_2 = T_{D_2,f(o)}.$$

Hence ρ_* preserves the flag defined by the reductive decomposition, but not necessarily the splitting itself. Suppose, however, that in addition we have $\rho_*\mathfrak{m}_1 \subset \mathfrak{m}_2$. We say in that case that f is *compatible with the reductive splittings*. In this case $f(X_1)$ is indeed a reductive subdomain of X_2. Hence as before, it is a totally geodesic submanifold.

(iii) We now treat an example of a complex submanifold S of a Hermitian manifold (M, h) with T_M equipped with the Chern connection ∇_h. For simplicity of notation we denote it ∇. We claim that S is totally geodesic in M if and only if there is a holomorphic and orthogonal direct sum splitting

$$T_M|S = T_S \oplus R, \qquad (\text{hence } R \simeq \nu_{S/M}).$$

To show the direct implication, suppose that the second fundamental form vanishes. We need to show that $R = T_S^\perp$ is a holomorphic subbundle. Since the second fundamental form vanishes, $\nabla R \subset R \otimes \mathscr{A}^1_M$ and because $\nabla^{0,1} = \bar\partial$, this implies $\nabla_\eta R \subset R$ for any tangent vector η of type $(0, 1)$. Hence, the subbundle R is indeed a holomorphic subbundle.

To show the converse, since this is a local question we may test this, using holomorphic vector fields. So let s be a holomorphic vector field on M, tangent to S, and let r be a holomorphic section of R. Then for all tangent vectors ξ of type $(1, 0)$, we have $\nabla_{\bar\xi}(s) = 0$ since s is holomorphic and $\nabla^{0,1} = \bar\partial$. The metric h being Hermitian and ∇ being metric implies

$$0 = h(\nabla_{\bar\xi} s, r) + h(s, \nabla_\xi r) = h(s, \nabla_\xi r) \implies \nabla_\xi r \in R.$$

For vector fields η of type $(0, 1)$, we directly see that $\nabla_\eta r = 0$. Hence, $\nabla R \subset R \otimes \mathscr{A}^1_S$ and so, in view of the decomposition (11.6), the second fundamental form indeed vanishes.

Bibliographical and Historical Remarks

Most of the material in this chapter is classical and can be found for instance in the book by Kobayashi (1987). The term "Chern connection" for the unique metric connection whose (0, 1)-part is the $\overline{\partial}$-operator is quite new; the characterization given in Theorem 11.1.3 is widely used in the literature but mostly without proof or references to Koszul and Malgrange (1958).

The principle of plurisubharmonicity is formulated explicitly for the first time here, but of course it has been used in Hodge theory, as will become clear in later chapters.

The explicit formulas for the curvature of certain homogeneous bundles seem to be new and are used in the next chapter to simplify the calculations in Griffiths and Schmid (1969).

In this edition we added a section on totally geodesic submanifolds. The classical reference is Helgason (1962, Section IV.7).

12

Structure of Period Domains

In Chapter 4 we defined the notion of period domain and showed how such objects arise in nature as targets of period mappings. We shall now study the geometry, especially the differential geometry, of these spaces, following Griffiths and Schmid (1969). We start by giving an efficient, e.g., group-theoretic, description of the tangent bundle and other important bundles on homogeneous spaces and in particular, on period domains. Then we discuss canonical connections on reductive spaces and the implications for period domains. Next comes a section on Higgs bundles. This will become important in Chapter 14. The horizontal and vertical tangent bundles for period domains are discussed in the next section, followed by a final section containing a detailed discussion of the implications of standard real Lie theory on the structure of period domains, in particular on the Hodge metric and the Weil operator.

12.1 Homogeneous Bundles on Homogeneous Spaces

To begin, let D be a space homogeneous for the (left) action of a group G, where the isotropy group of $o \in D$,

$$V = \{g \in G \mid g \cdot o = o\},$$

is compact. Let

$$p : G \to D = G/V$$

be the natural projection.

A bundle E on D is said to be *homogeneous* if it is given with a left G-action that is compatible with the action on D, i.e.,

$$p(g \cdot e) = g \cdot p(e), e \in E.$$

The tangent bundle of a homogeneous space is homogeneous in a natural way, as is any bundle functorially associated with it, e.g., the dual, the exterior powers, etc.

Among the sections of a homogeneous bundle are distinguished the equivariant ones, which satisfy the relation

$$s(g \cdot x) = g \cdot s(x),$$

where $x \in D$. Denote the space of such sections by $\Gamma_G(D, E)$ and observe that the evaluation map is an isomorphism:

$$\Gamma_G(D, E) \xrightarrow{\cong} E_o.$$

Moreover, the fiber E_o is a representation space for the isotropy group V. Thus, we have a natural transformation that assigns to each G-homogeneous vector bundle a V-module. This transformation can be inverted as follows. Given a V-module W, set

$$[W] = G \times_V W = \{\text{equivalence classes } (g, w) \text{ under } \sim\},$$

where $(g, w) \sim (gv, v^{-1}w)$. Observe that the projection $G \times W \longrightarrow G$ descends to a projection $G \times_V W \longrightarrow G/V$ on the level of equivalence classes. The left action of G is given by $g' \cdot [g, w] = [g' \cdot g, w]$, where $[g, w]$ denotes the class of (g, w). This construction is of course precisely the one explained in Section 11.3 starting from the principal bundle $G \to G/V$ and the V-module W.

Example 12.1.1 Consider the tangent bundle T_D. To this end, consider the map

$$e : \mathfrak{g} \longrightarrow T_{D,o}$$

defined by

$$e(\xi) = \left.\frac{\mathrm{d}}{\mathrm{d}t}\right|_{t=0} \exp t\xi \cdot o.$$

Because elements of V fix the base point, the kernel of e is $\mathfrak{v}$, the Lie algebra of V, and we have an induced isomorphism,

$$e : \mathfrak{g}/\mathfrak{v} \xrightarrow{\cong} T_{D,o}.$$

Now V acts on curves $\gamma : (-\epsilon, \epsilon) \longrightarrow D$ such that $\gamma(0) = o$ via the rule $(v \cdot \gamma)(t) = v \cdot \gamma(t)$, $v \in V$. Consequently, it acts on the velocity vectors $\dot{\gamma}(0)$ and hence on $T_{D,o}$. Explicitly, we have

$$v(\dot{\gamma}(0)) = \left.\frac{\mathrm{d}}{\mathrm{d}t}\right|_{t=0} (v \cdot \gamma(t)).$$

Via the isomorphism e, the space $\mathfrak{g}/\mathfrak{v}$ then inherits a natural V-action. Indeed, we have

$$v \cdot e(\xi) = v \cdot \left.\frac{\mathrm{d}}{\mathrm{d}t}\right|_{t=0} \exp t\xi \cdot o = \left.\frac{\mathrm{d}}{\mathrm{d}t}\right|_{t=0} (v \exp t\xi v^{-1}) \cdot v \cdot o$$
$$= \left.\frac{\mathrm{d}}{\mathrm{d}t}\right|_{t=0} \exp tv\xi v^{-1} \cdot o = e(\mathrm{Ad}(v)\xi) \cdot o,$$

where we recall (11.7) that $\mathrm{Ad}(g)\xi = g\xi g^{-1}$ is the adjoint representation of G on $\mathfrak{g}$. This representation, restricted to V, preserves $\mathfrak{v}$, and so acts on the quotient $\mathfrak{g}/\mathfrak{v}$. Consequently, the tangent space at the reference point and $\mathfrak{g}/\mathfrak{v}$ are isomorphic as V-modules, from which we deduce an isomorphism:

$$G \times_V \mathfrak{g}/\mathfrak{v} \xrightarrow{\cong} T_D.$$

12.2 Reductive Domains and Their Tangent Bundle

Recall (Definition 11.3.3) that a homogeneous space G/V is a reductive domain, if there is a an $\mathrm{Ad}(V)$-invariant decomposition

$$\mathfrak{g} = \mathfrak{v} \oplus \mathfrak{m}.$$

The "reductive complement" $\mathfrak{m}$ is a V-module isomorphic to $\mathfrak{g}/\mathfrak{v}$, so, in view of Example 12.1.1, we have an identification

$$[\mathfrak{m}] = G \times_V \mathfrak{m} \xrightarrow{\cong} T_D. \tag{12.1}$$

We apply this to the situation where D is a period domain. In this case, we have a finite-dimensional weight-w Hodge structure (H, F) polarized by a $(-1)^w$-symmetric bilinear form b, $G = \mathrm{GL}(H, b)$, and V is the subgroup preserving F. On the level of complex Lie algebras we have a decomposition

$$\mathfrak{g}_{\mathbb{C}} = \mathrm{End}(V_{\mathbb{C}}, b_{\mathbb{C}}) = \bigoplus\nolimits_p \mathfrak{g}^{-p,p}.$$

Recall that $\mathfrak{g}^{-p,p}$ denotes the endomorphisms sending the Hodge component $H^{r,s}$ to $H^{r-p,s+p}$, $r + s = w$.

Proposition 12.2.1 *A period domain $D = G/V$ is a reductive homogeneous domain. Indeed,*

$$\mathfrak{g} = \mathfrak{v} \oplus \mathfrak{m},$$

where

$$\mathfrak{v} = \mathrm{Lie}(V) = \text{Real points of } \mathfrak{g}^{0,0},$$
$$\mathfrak{m} = \text{Real points of } \sum_{p \neq 0} \mathfrak{g}^{p,-p}.$$

Proof In the notation just introduced,

$$\mathfrak{v}_{\mathbb{C}} = \mathfrak{g}^{0,0}, \quad \mathfrak{v} = \mathfrak{v}_{\mathbb{C}} \cap \mathfrak{g}.$$

Thus, $\mathfrak{v}$ is the underlying set of real points of $\mathfrak{v}_{\mathbb{C}}$. The real points of the complement $\sum_{p\neq 0} \mathfrak{g}^{p-p}$ make up the Lie algebra $\mathfrak{m}$. It is clear that the Lie bracket is a morphism of Hodge structures, i.e., that

$$[\mathfrak{g}^{p,q}, \mathfrak{g}^{r,s}] \subset \mathfrak{g}^{p+r,q+s}.$$

It follows that $\mathfrak{v}$, the real morphisms of type $(0,0)$, leave invariant the set of real morphisms of given type $(p,q)+(q,p)$, i.e., $[\mathfrak{v},\mathfrak{m}] \subset \mathfrak{m}$ so that we obtain our desired V-invariant splitting $\mathfrak{g} = \mathfrak{v} \oplus \mathfrak{m}$. □

We just saw (12.1) that the tangent bundle of D is the bundle obtained from $\mathfrak{m}$. To find the *holomorphic* tangent bundle of D we consider the splitting

$$\mathfrak{m}_{\mathbb{C}} = \mathfrak{m}^{+} \oplus \mathfrak{m}^{-},$$

where

$$\mathfrak{m}^{+} = \bigoplus_{p>0} \mathfrak{g}^{p,-p}, \quad \mathfrak{m}^{-} = \bigoplus_{p<0} \mathfrak{g}^{p,-p}.$$

This is a V-invariant decomposition into spaces satisfying $\overline{\mathfrak{m}^{+}} = \mathfrak{m}^{-}$ and so defines a complex structure.

Lemma 12.2.2 *Define $J : \mathfrak{m}_{\mathbb{C}} \longrightarrow \mathfrak{m}_{\mathbb{C}}$ by letting* i *act by $\pm$*i *on $\mathfrak{m}^{\pm}$. Then J is real, i.e. $J : \mathfrak{m} \longrightarrow \mathfrak{m}$ and $J^2 = -\mathrm{id}$, i.e. J defines a complex structure on $\mathfrak{m}$.*

Proof Suppose $u^{+} + u^{-} \in \mathfrak{m}^{+} \oplus \mathfrak{m}^{-}$ is real. Then u^{+} and u^{-} are each other's complex conjugates. We have $J(u^{+}+u^{-}) = \mathrm{i}u^{+} - \mathrm{i}u^{-}$. The complex conjugate of this element equals $-\mathrm{i}\overline{u^{+}} + \mathrm{i}\overline{u^{-}} = -\mathrm{i}u^{-} + \mathrm{i}u^{+} = \mathrm{i}u^{+} - \mathrm{i}u^{-}$, i.e. $J(u^{+}+u^{-})$ is real. So $J : \mathfrak{m} \longrightarrow \mathfrak{m}$ and obviously $J^2 = -\mathrm{id}$. □

It is this complex structure that determines the holomorphic tangent bundle which we are now going to describe. Recall that at each point $F \in D$ defines a Hodge flag $F^{\bullet}$ on $H_{\mathbb{C}}$ and hence an induced Hodge flag on $\mathfrak{g}_{\mathbb{C}} = \mathrm{End}(H_{\mathbb{C}}, b)$. Then $F^0\mathfrak{g}_{\mathbb{C}}$ is the subalgebra of endomorphisms preserving F. Using this, the holomorphic tangent bundle can be identified as follows.

Lemma–Definition 12.2.3 *Let $D = G/V$ be a period domain. Then*

$$T_D^{\mathrm{hol}} = [\mathfrak{m}^{-}] = [\mathfrak{g}_{\mathbb{C}}/F^0\mathfrak{g}_{\mathbb{C}}],$$

the homogeneous bundle associated with the adjoint action of V on $\mathfrak{m}^{-} = \mathfrak{g}_{\mathbb{C}}/F^0\mathfrak{g}_{\mathbb{C}}$.

Proof Let us give two arguments, one rather formal, the other more geometric in spirit. For the first, note that the isotropy group B in $G_{\mathbb{C}}$ of the Hodge filtration $o \in \check{D} = G_{\mathbb{C}}/B$ has Lie algebra $\mathfrak{v}_{\mathbb{C}} \oplus \mathfrak{m}^+$ and hence $\mathfrak{m}^-$ is a complement which gives the complexified tangent space. To see that the bundle structure comes from the adjoint representation, apply Example 12.1.1.

Alternatively, one can use the characterization of the period domain as an open set in a flag variety of (partial) flags satisfying the first bilinear relation. The flag variety consists of a subvariety in the product of Grassmannians of certain subspaces F^i of a fixed complex vector space $H_{\mathbb{C}}$. At the point F^i the tangent space at the Grassmannian is $T_{F^i} = \mathrm{Hom}(F^i, H_{\mathbb{C}}/F^i)$. The F^i forming a flag translates into compatibilities between the various $X^i \in \mathrm{Hom}(F^i, H_{\mathbb{C}}/F^i)$ which in our case are equivalent to saying that the X_i induces maps $Y_{ij} : H^{i,w-i} \to H^{j,w-j}$ with $i < j$. So the (Y_{ij}) can be assembled into a map $Y : H_{\mathbb{C}} \to H_{\mathbb{C}}$. The first bilinear relation imposes further restrictions on Y stating that $Y \in \mathfrak{g}_{\mathbb{C}}$. The upshot is that $Y \in \mathfrak{g}_{\mathbb{C}}$ corresponds to a holomorphic tangent vector in $T_{D,o}$ if and only if $Y \in \sum_{j<0} \mathfrak{g}^{j,-j}$. □

12.3 Canonical Connections on Reductive Spaces

We turn now to the differential-geometric properties of period domains. These are examples of reductive spaces. A reductive homogeneous space always has a natural connection in the sense defined in the previous chapter. Let us recall how this is done. At the identity element of G, the splitting of the tangent bundle is $\mathfrak{g} = \mathfrak{v} \oplus \mathfrak{m}$. At other points it is gotten by pushing this decomposition forward by left translation, i.e., by the differential of the map $L_g(x) = gx$. We also saw that a connection is given by a $\mathfrak{v}$-valued one-form that transforms under the adjoint action of V. We claim that for the canonical connection this form simply comes from the Maurer–Cartan form

$$\omega = g^{-1}\,\mathrm{d}g.$$

It is a matrix-valued one-form invariant with respect to left translation, because

$$L_h^*\omega = (hg)^{-1}\,\mathrm{d}(hg) = g^{-1}\,\mathrm{d}g = \omega.$$

Note that under right translation it behaves differently:

$$R_h^*\omega = (gh)^{-1}\,\mathrm{d}(gh) = h^{-1} \cdot g^{-1}\,\mathrm{d}g \cdot h = \mathrm{Ad}(h^{-1})\omega.$$

This is the correct transformation behavior, but ω is not $\mathfrak{v}$-valued and so we have to split it as

$$\omega = \phi + \tau, \quad \phi \stackrel{\text{def}}{=} \omega^{\mathfrak{v}}, \quad \tau \stackrel{\text{def}}{=} \omega^{\mathfrak{m}}.$$

If suitable bases are used to represent matrices of G, then these components are defined by setting certain blocks of the Maurer–Cartan matrix to 0. In any case, the component $\omega^{\mathfrak{v}}$ is the fundamental form required to define the principal connection (Lemma 11.3.7). Indeed, by construction, it is now 0 on horizontal vectors and equal to the matrix $A \in \mathfrak{v}$ on any vertical tangent vector of the form A^*_g, $g \in G$.

To compute the curvature of the canonical connection form $\phi = \omega^{\mathfrak{v}}$ for a reductive homogeneous space, we observe that the Maurer–Cartan form is flat, i.e., satisfies

$$d\omega + \frac{1}{2}[\omega, \omega] = 0,$$

where the convention given by Eq. (11.9) has been used. Decomposing this relation into components of type $\mathfrak{v}$ and type $\mathfrak{m}$, we obtain the *structure equations*

$$\begin{array}{lcl} d\phi + \frac{1}{2}[\phi, \phi] & = & -\frac{1}{2}[\tau, \tau]^{\mathfrak{v}}, \\ d\tau & = & -\frac{1}{2}[\tau, \tau]^{\mathfrak{m}}. \end{array} \tag{12.2}$$

Thus the curvature, which is the left-hand side of the first equation in (12.2), equals the right hand side, i.e.

$$\Omega = -\frac{1}{2}[\omega^{\mathfrak{m}}, \omega^{\mathfrak{m}}]^{\mathfrak{v}}. \tag{12.3}$$

Example 12.3.1 In the case of a symmetric space (see Definition 12.5.1), this simplifies to

$$\Omega(\xi, \eta) = -\frac{1}{2}[\xi, \eta].$$

Let us now examine the canonical connection ∇ on associated bundles whose construction is explained in Proposition 11.4.1.

Lemma 12.3.2 *The canonical connection on the principal bundle $G \longrightarrow G/V = D$ defined by a period domain D induces connections ∇ on any associated bundle $[W]$. If W is a complex V-module equipped with a V-invariant Hermitian metric, the connection ∇ is a metric connection with respect to the induced Hermitian metric. In particular, this holds for the Hodge bundles equipped with the Hodge metric h_C.*

Recall that in Proposition 11.4.5 we explained that the curvature F_∇ for these associated connections ∇ can be given in terms of the curvature form Ω, the representation ρ which defines the vector bundle and a local section f for the principal bundle:

$$F_\nabla = f^*(\dot{\rho} \circ \Omega). \tag{12.4}$$

Theorem 12.3.3 *Let W be a complex vector space equipped with a V-invariant Hermitian metric. The bundle $[W]$ admits a unique holomorphic structure for which the canonical connection ∇ is the Chern connection. In particular, its curvature is given by the form (12.3).*

This holds in particular for the bundle $[H_{\mathbb{C}}]$ and the Hodge bundles (equipped with the Hodge metrics).

Proof Since ∇ is a metric connection, by Theorem 11.1.3 it suffices to show that the curvature has pure type (1, 1). But the holomorphic tangent bundle is the homogeneous bundle associated with the representation space

$$\mathfrak{m}^- = \sum_{p<0} \mathfrak{g}^{p,-p}.$$

Clearly $[\mathfrak{m}^-, \mathfrak{m}^-] \subset \mathfrak{m}^-$, so that $[\mathfrak{m}^-, \mathfrak{m}^-]^{\mathfrak{v}} = 0$. By the formula (12.3), we have

$$\Omega(\xi, \eta) = -\frac{1}{2}[\xi, \eta]^{\mathfrak{v}}$$

for $\xi, \eta \in \mathfrak{m}^-$. Consequently, Ω vanishes when applied to a pair of (1, 0) vectors. Thus, it has no component of type (2, 0). By an analogous argument it has no component of type (0, 2). Using (12.4), the claim now follows. □

Remark 12.3.4 The most classical case is the computation of the curvature of the tangent bundle of a Riemannian symmetric space G/K with respect to the canonical G-invariant Killing metric. We find, using Example 12.3.1,

$$F(\xi, \eta) = -\frac{1}{2}[\xi, \eta], \quad \xi, \eta \in \mathfrak{p},$$

where, as before, $\mathfrak{g} = \mathfrak{k} \oplus \mathfrak{p}$ is the splitting of the Lie algebra of G into the Lie algebra of K and its (with respect to the Killing form) orthogonal complement $\mathfrak{p}$. In more traditional notation, the curvature tensor R satisfies

$$R(\xi, \eta)\zeta = -[[\xi, \eta], \zeta].$$

12.4 Higgs Principal Bundles

Our discussion so far is based on a principal V-bundle over D, P, with total space G. For use in Chapter 13 when we treat Higgs bundles, we should consider another natural principal G-bundle, defined as follows.

Definition 12.4.1 Let $D = G/V$ be a homogeneous space. Its associated *Higgs principal bundle* is the G-principal bundle

$$P_{\text{Higgs}} = G \times_V G \cong D \times G,$$

where the isomorphism is a G-equivariant map of G-principal bundles by means of the map $[g, g']$ to $([g], gg')$. The *Higgs connection* ω_{Higgs} on this bundle is obtained by pulling back the Maurer–Cartan form on G to $D \times G$.

It is clear that the Higgs connection is flat. Consider the map $g \mapsto [g, e]$, which imbeds $P = G$ in $P_{\mathrm{Higgs}} = G \times_V G$ equivariantly with respect to the natural right actions by V. One checks that the connection form on $G \times_V G$ pulls back to the Maurer–Cartan form on G under this embedding.

Thus, we have the following setup: on P_{Higgs} there is a flat connection defined by ω_{Higgs} which defines a connection ∇_{Higgs} on any associated G-bundle. When the connection form is restricted to P_{Higgs} it decomposes as

$$\omega_{\mathrm{Higgs}} = \omega^{\mathfrak{v}} + \omega^{\mathfrak{m}} \stackrel{\mathrm{def}}{=} \phi + \tau.$$

Suppose that the vector space W has the structure of a G-module. Then it inherits a V-structure as well and

$$[W] = P_{\mathrm{Higgs}} \times_G W = G \times_V W.$$

Let ∇_V be the V-connection on $[W]$. It is related to the Higgs connection by the formula

$$\nabla_{\mathrm{Higgs}} = \nabla_V + \tau.$$

Decomposing the relation $\nabla^2_{\mathrm{Higgs}} = 0$ into components of type $\mathfrak{v}$ and type $\mathfrak{m}$, we obtain the *Higgs relations*:

$$\begin{aligned} \nabla_V^2 + \tfrac{1}{2}[\tau, \tau]^{\mathfrak{v}} &= 0, \\ \nabla_V(\tau) + \tfrac{1}{2}[\tau, \tau]^{\mathfrak{m}} &= 0. \end{aligned}$$

Such holomorphic vector bundles $[W]$ on D are the prototype of *Higgs bundles*, a concept we introduce later in Section 13.1.

Example 12.4.2 (Hodge bundles and variations of Hodge structure) Note that on D the G-bundle $[H_{\mathbb{C}}]$ has a canonical flat G-connection ∇. As a V-bundle it decomposes into a sum of Hodge bundles $[H^{p,q}]$, each of which has a canonical V-connection ∇_V. The Hodge bundle is the direct sum of these bundles and thus has a V-connection ∇_V, and they are related by $\nabla_{\mathrm{Higgs}} = \nabla_V + \tau$. The Lie algebra-valued one-form τ can be viewed as a canonical map

$$\tau_D : T_D \to \mathfrak{m}.$$

Now consider a holomorphic map $f : M \longrightarrow D$. Then all of the above apparatus – flat bundles, Hodge bundles, connections, Lie algebra-valued one-forms

– pulls back to objects on M to which we assign the same name. The form τ now becomes the canonical map

$$\tau = \mathrm{d}f \circ \tau_D : T_M \longrightarrow \mathfrak{m}.$$

In the rest of this section, we use the convention $\alpha = \alpha' + \alpha''$ for the splitting into types for vector bundle-valued complex one-forms α instead of the more cumbersome notation $\alpha = \alpha^{1,0} + \alpha^{0,1}$ that we used above.

Let us suppose further that f is a variation of Hodge structure. This is equivalent to the assertion that τ' is $\mathfrak{g}^{-1,1}$-valued, where τ' has type $(1,0)$ as a matrix of forms. Consequently, τ'' is $\mathfrak{g}^{1,-1}$-valued. Now

$$[\tau, \tau] = [\tau', \tau'] + 2[\tau', \tau''] + [\tau'', \tau''].$$

Because $\mathfrak{v} = \mathfrak{g}^{0,0}$, the first structure equation implies

$$\nabla_V^2 = -[\tau', \tau'']. \tag{12.5}$$

This formula shows:

Proposition 12.4.3 *The curvature of the Hodge bundle for a variation of Hodge structure as above is given by Eq.* (12.5).

We come back to this in Chapter 13 where we bring in the Chern connection for the Hodge bundle. See in particular Proposition 13.1.1.

Let us continue our investigation of the splitting and note that there is no $(2,0) + (0,2)$ component of $[\tau, \tau]$ which is at the same time $\mathfrak{v}$-valued. Hence we have

$$\nabla_V''^2 = 0 = \nabla_V'^2. \tag{12.6}$$

The second structure equation breaks up into two separate equations according to the $\mathfrak{g}^{p,q}$-type of the matrix-valued forms. The first equation collects the forms with values in $\mathfrak{g}^{-1,1} + \mathfrak{g}^{1,-1}$:

$$\nabla_V(\tau) = 0.$$

This equation can be split up further. For instance, the $(1,1)$-component having values in $\mathfrak{g}^{-1,1}$ reads

$$\nabla_V''(\tau') = 0. \tag{12.7}$$

The other equation gathers the $(\mathfrak{g}^{2,-2} + \mathfrak{g}^{-2,2})$-valued forms and can be further split. For instance, the $(2,0) + \mathfrak{g}^{-2,2}$-component gives

$$[\tau', \tau'] = 0. \tag{12.8}$$

The harmonic equation as well as the three equations (12.6), (12.7), and (12.8), which are called the *pluriharmonic equations*, play a central role in Chapter 14.

Problems

12.4.1 Consider the Riemannian symmetric space $\mathrm{SO}(a+b)/\mathrm{SO}(a)\times\mathrm{SO}(b)$. Calculate the curvature tensor R on basis elements of $\mathfrak{p}$ and show that the form given by $-B(R(\xi,\eta)\eta,\xi)$ is negative semidefinite. Here B is the Killing form. Conclude that the bisectional curvature is ≤ 0. Here we recall that the bisectional curvature s in the direction of the plane spanned by ξ and η with respect to a metric g and curvature tensor R is defined by

$$s(\xi,\eta) = \frac{-g(R(\xi,\eta)\xi,\eta)}{\mathrm{vol}(\xi,\eta)}.$$

12.4.2 Show that if W is a G-module then the map $[g,w]\mapsto(gV,gw)$ defines a trivialization of $[W]$.

12.5 The Horizontal and Vertical Tangent Bundles

Let $D = G/V$ be a period domain classifying b-polarized Hodge structures of given weight and Hodge type. We have seen in Proposition 4.4.4 that G is a simple Lie matrix group. We need some background on simple and semisimple groups, for which we refer to Appendix D.3. In particular we shall use the Killing form on the Lie algebra $\mathfrak{g} \stackrel{\text{def}}{=} \mathrm{Lie}(G)$:

$$B(\xi,\eta) = \mathrm{Tr}\,(\mathrm{ad}(\xi)\circ\mathrm{ad}(\eta)),$$

a nondegenerate symmetric form invariant under $\mathrm{Ad}(G)$. Let K be the maximal compact subgroup of G containing V. The Killing form is negative definite on $\mathfrak{k} = \mathrm{Lie}(K)$ and positive definite on the orthogonal complement $\mathfrak{p}$ of $\mathfrak{k}$. Moreover, the form B induces a G-invariant metric on G/K.

Definition 12.5.1 Let G be a simple Lie matrix group and K a maximal compact subgroup. The domain G/K is called a *symmetric space*. The associated *Cartan decomposition* is the B-orthogonal decomposition

$$\mathfrak{g} = \mathfrak{k}\oplus\mathfrak{p}.$$

The *Cartan involution* $\theta : \mathfrak{g}\to\mathfrak{g}$ by definition is such that $\theta|\mathfrak{k} = \mathrm{id}$ and $\theta|\mathfrak{p} = -\mathrm{id}$. There is a corresponding Cartan involution on G whose fixed point locus is K; it induces an isometry with respect to the canonical K-invariant metric on G/K. The term "symmetric" space comes from the existence of such an involutive symmetry. For an example, see Problem 12.5.1.

The above Cartan decomposition is reflected on the level of the tangent bundle of D as follows. With respect to the canonical projection

$$\omega : D = G/V \to G/K,$$

we can split the tangent space at any point $o \in D$ into the horizontal and vertical tangent spaces

$$\begin{aligned} T^{\text{vert}}_{D,o} &= T_o(\text{fiber of } \omega \text{ through } o), \\ T^{\text{hor}}_{D,o} &= \text{orthogonal complement of } T^{\text{vert}}_{D,o}, \end{aligned}$$

where the orthogonal complement is taken with respect to the Killing form on $\mathfrak{g}$.

Lemma 12.5.2 *The ± 1-eigenspaces of the Cartan involution on $\mathfrak{g}_{\mathbb{C}}$ are given by*

$$\begin{aligned} \mathfrak{k}_{\mathbb{C}} &= \bigoplus\nolimits_{j \text{ even}} \mathfrak{g}^{-j,j}, \\ \mathfrak{p}_{\mathbb{C}} &= \bigoplus\nolimits_{j \text{ odd}} \mathfrak{g}^{-j,j}. \end{aligned}$$

Moreover, we have canonical identifications

$$\begin{aligned} T^{\text{vert}}_{D,o} &= \mathfrak{k}/\mathfrak{v}, \\ T^{\text{hor}}_{D,o} &= \mathfrak{p}. \end{aligned}$$

Proof The fibration $\omega : G/V \to G/K$ can be described concretely as follows. Let $H_{\mathbb{C}} = \bigoplus_{p+q=w} H^{p,q}$ be the Hodge decomposition corresponding to $o \in D$. Extending scalars, the polarizing form b induces the $(-1)^w$-Hermitian form $(x, y) \mapsto h(x, y) = b(x, \bar{y})$ on $H_{\mathbb{C}}$. Introduce the subspaces $H^+_{\mathbb{C}} := \bigoplus_{p \text{ even}} H^{p,q}$ and $H^-_{\mathbb{C}} := \bigoplus_{p \text{ odd}} H^{p,q}$. In the case of even weight $w = 2\nu$, the spaces $H^{\pm}_{\mathbb{C}}$ are complexifications of real vector spaces $H^{\pm}$. The second bilinear relation implies that the form $(-1)^{\nu} h$ is positive on H^+ and negative on H^-. Now G/K is the set of real subspaces of H of maximal dimension $a = \sum_{p \text{ even}} h^{p,q}$ on which $(-1)^{\nu} h$ is positive and we have

$$\omega(o) = [H^+] \in G/K.$$

In the case of odd weight $w = 2\nu - 1$, $\dim H^+_{\mathbb{C}} = \dim H^-_{\mathbb{C}}$ and the form h is anti-Hermitian. Because of the first bilinear relations the spaces $H^{\pm}$ are total isotropic of maximal dimension and the second bilinear relation implies that the *Hermitian* form $(x, y) \mapsto (-1)^{\nu+1} \mathrm{i} b(x, \bar{y}) = (-1)^{\nu+1} \mathrm{i} h(x, y)$ is positive definite on $H^+_{\mathbb{C}}$ and negative on H^-. If we identify G/K with the set of totally b-isotropic complex subspaces $H^+_{\mathbb{C}} \subset H_{\mathbb{C}}$ of maximal dimension on which the latter form is positive definite, we have

$$\omega(o) = [H^+_{\mathbf{C}}] \in G/K.$$

In both cases, the *complexified* tangent space of G/K at the point $\omega(o)$ is isomorphic to the subspace of $\mathrm{Hom}_{\mathbf{C}}(H^+_{\mathbf{C}}, H_{\mathbf{C}}/H^+_{\mathbf{C}})$ consisting of $b_{\mathbf{C}}$-preserving maps $X : H^+_{\mathbf{C}} \to H^-_{\mathbf{C}}$. Such a map belongs to $\mathfrak{p}_{\mathbf{C}}$ and $Y = X + \bar{X} \in \mathfrak{p}$. The converse is not necessarily true since $\mathfrak{p}_{\mathbf{C}}$ also contains maps $X \in \mathrm{Hom}_{\mathbf{C}}(H^-_{\mathbf{C}}, H^+_{\mathbf{C}})$. However, for such maps $\bar{X} \in \mathrm{Hom}_{\mathbf{C}}(H^+_{\mathbf{C}}, H^-_{\mathbf{C}})$. It follows that any $Y \in \mathfrak{p}$ can be written in a unique fashion as a sum $Y = X + \bar{X}$, where $X \in \mathrm{Hom}_{\mathbf{C}}(H^+_{\mathbf{C}}, H^-_{\mathbf{C}})$. This correspondence between X and Y yields the desired identification $[T^{\mathrm{hor}}_o D] = \mathfrak{p}$. Note that $B_0(X_1, X_2) = 0$ when X_1, X_2 are in different components of the Hodge decomposition of $\mathfrak{g}_{\mathbf{C}}$ since $X_1 \circ X_2$ is nilpotent and hence has trace zero. So the B_0-orthogonal complement of $\mathfrak{p}$ in $\mathfrak{g}/\mathfrak{v}$ is $\mathfrak{k}$ taken modulo $\mathfrak{v}$. This is the kernel of the map induced by ω which gives the tangent space to the fiber of ω. □

The vertical tangent spaces fit into a holomorphic bundle T^{vert}_D, because the fibers are complex submanifolds of D, as we saw in Problem 4.4.2. The horizontal spaces, however, just define a smooth vector bundle T^{hor}_D, which in general admits no complex structure. But its complexification $T^{\mathrm{hor}}_D \otimes \mathbf{C}$ does contain an important holomorphic subbundle, namely the one associated with $\mathfrak{g}^{-1,1}$. Summarizing, we have the following.

Proposition 12.5.3 *Let $D = G/V$ be a period domain. Let $K \subset G$ be the maximal compact connected subgroup of G containing V, and let G/K the associated homogeneous domain with projection $\omega : G/V \to G/K$. Recalling the notation T_ω for the bundle of tangents along the fibers of ω, there is a canonical splitting*

$$T_D = T^{\mathrm{vert}}_D \oplus T^{\mathrm{hor}}_D, \quad T^{\mathrm{vert}}_D = T_\omega.$$

These bundles are homogeneous under the adjoint action of V on $\mathfrak{g}$:

$$T^{\mathrm{vert}}_D = [\mathfrak{k}/\mathfrak{v}], \quad T^{\mathrm{hor}}_D = [\mathfrak{p}].$$

The bundle T^{vert}_D is a holomorphic bundle, T^{hor}_D in general not. The subbundle $T^{-1,1}_D$ of $T^{\mathrm{hor}}_D \otimes \mathbf{C}$ has a holomorphic structure as subbundle of the holomorphic tangent bundle of D and is called the holomorphic horizontal tangent bundle.

Additional bundles can be constructed because G acts on H. For example, H itself is obviously V-invariant and so defines a homogeneous bundle. In general, if a representation of V is the restriction of a representation of G, then the resulting bundle is trivial (see Problem 12.4.2). This is the case here: $[H] \cong D \times H$.

Definition 12.5.4 The subspaces $H^{p,q}$ and F^p of $H_{\mathbf{C}}$ are V-invariant; the homogeneous vector bundles they define are the so-called *Hodge bundles.* The flag bundle $[F^\bullet] \subset [H_{\mathbb{C}}]$ is called the *tautological Hodge flag*.

As we have see (Theorem 12.3.3), both $[H_{\mathbb{C}}]$ and the Hodge bundles $[H^{p,q}]$ and $[F^p]$ have natural holomorphic structures.

Remark 12.5.5 (1) Although the bundle $[F^p]$ inherits the holomorphic structure from $[H_{\mathbb{C}}]$, this is not the case for $[H^{p,q}]$.
(2) The tautological flag $[F^\bullet]$ on $D = G/V$ is a holomorphic flag, but it is in general *not* a variation of Hodge structure since the horizontality condition fails in general. Later we shall prove that we do get a variation precisely when D is a bounded symmetric domain. See Section 15.3.

Problems

12.5.1 Consider the homogeneous space $D = \mathrm{SO}^0(a,b)/\mathrm{SO}(a+b) \cap [\mathrm{O}(a) \times \mathrm{O}(b)]$. Put $n = a + b$ and consider the associated Lie-algebra $\mathfrak{so}(a,b)$ and its subalgebra $\mathfrak{k} = \mathfrak{so}(a) \times \mathfrak{so}(b)$. Decompose $n \times n$ matrices correspondingly into blocks and let $\mathfrak{p}$ be the set of matrices which have 0 on the diagonal. Introduce the involution θ on $\mathrm{SO}(a,b)$ defined by the adjoint action of the matrix

$$I_{a,b} = \begin{pmatrix} \mathbf{1}_a & 0 \\ 0 & -\mathbf{1}_b \end{pmatrix}.$$

(a) Prove that θ is the Cartan involution sending $X \in \mathfrak{g}$ to $-{}^{\mathsf{T}}X$. Describe the two eigenspaces as $\mathfrak{k}$ and $\mathfrak{p}$.

(b) Show that θ induces an involution of D.

(c) Show that the trace form $\operatorname{Tr} \xi \circ \eta$ induces a V-invariant metric on the tangent bundle $T_D = [\mathfrak{p}]$. Deduce that θ is an isometry.

12.5.2 (a) Show that the Killing form B is indeed invariant under the adjoint action of G on $\mathfrak{g}$ and that it is symmetric. Deduce that $B([\xi,\eta],\zeta) = B(\xi,[\eta,\zeta])$ for all triples $\xi,\eta,\zeta \in \mathfrak{g}$.

(b) Suppose that G is a compact Lie group. Using a G-invariant inner product, show that the Killing form is negative definite.

(c) Now let $G = \mathrm{SO}(n)$. Compute the Killing form in this case and compare it with the trace form.

(d) Let G be any *complex* simple Lie group and let B' be any bilinear form on its Lie-algebra $\mathfrak{g}$ that is $\mathrm{Ad}(G)$-invariant. Show that for some $A \in \mathrm{End}(\mathfrak{g})$ we have $B'(\xi,\eta) = B'(A\xi,\eta)$ and that

$A[\xi, \eta] = [\xi, A\eta]$. Deduce that any eigenspace $U \subset \mathfrak{g}$ of A is an ideal and hence $U = \mathfrak{g}$. Conclude that $B' = \lambda B$. What does this imply when B' happens to be skew-symmetric?

12.5.3 Let $H = \bigoplus H^{p,q}$ be a Hodge decomposition of weight w corresponding to a reference point $o \in D = G/V$. Let b be the polarization and $h_\mathbf{C}$ the corresponding Hodge metric. A Hodge frame consists of a $h_\mathbf{C}$-unitary frame for each of the Hodge components which, moreover, is invariant under the complex conjugation.

(a) Show that Hodge frames always exist. Determine the matrix of the polarization in such a frame, where we take the ordering such that the frame for $H^{w,0}$ comes first, followed by the one for $H^{w-1,1}$, etc.
(b) A Hodge frame defines a matrix representation of $\mathfrak{g}_\mathbf{C}$. Identify $\mathfrak{v}_\mathbf{C}$ and $\mathfrak{m}^\pm$ in terms of block matrices.
(c) Give explicit bases for $\mathfrak{g}^{-p,p}$.

12.6 On Lie Groups Defining Period Domains

Real Forms

Let (H, b) a polarized weight w Hodge structure and $G = \mathrm{Aut}(H, b)$ its group of automorphisms. Using the terminology of Appendix D.3, the group $G_\mathbf{C} = \mathrm{Aut}(H_\mathbf{C}, b)$ is the complexification of G and G is a real form of $G_\mathbf{C}$. Now G is a semisimple matrix group (it is even a simple Lie group). Hence, according to what has been explained in § D.3, $G_\mathbf{C}$ also has a *compact real form*. We shall make this explicit in this situation.

Let $K \subset G$ be the maximal compact subgroup containing V. As above, we let $\mathfrak{g}$, $\mathfrak{v}$, and $\mathfrak{k}$ be the Lie algebras of G, V, and K, respectively. We have seen that $\mathfrak{v} \otimes \mathbf{C}$ coincides with the Lie subalgebra $\mathfrak{g}^{0,0}$ of $\mathfrak{g}_\mathbf{C}$ of endomorphisms of $H_\mathbf{C}$ preserving the Hodge structure (of a reference point) and that $\mathfrak{v}$ has a V-invariant complement:

$$\mathfrak{g} = \mathfrak{v} \oplus \mathfrak{m}, \quad \mathfrak{m}_\mathbf{C} = \bigoplus_{p \neq 0} \mathfrak{g}^{-p,p}.$$

We say that $\mathfrak{g}$ is a *noncompact real form* of $\mathfrak{g}_\mathbf{C}$ in the sense that $\mathfrak{g}_\mathbf{C}$ is the complexification of $\mathfrak{g}$ and it is the Lie algebra of the noncompact Lie group G. This reflects the fact that on the level of groups, G is a noncompact real form of $G_\mathbf{C}$.

To find the compact real form on the level of Lie algebras we need another real structure on $\mathfrak{g}_\mathbf{C}$ that leads to a compact Lie group. We recall from Appendix D.3 how this can be done using the Cartan decomposition $\mathfrak{g} = \mathfrak{k} \oplus \mathfrak{p}$ we discussed in the previous subsection. Recall that

$$\mathfrak{k}_{\mathbb{C}} = \bigoplus_{p \text{ even}} \mathfrak{g}^{-p,p},$$

and that $\mathfrak{k}$ has a K-invariant complement fitting in the Cartan decomposition (see Appendix D.3)

$$\mathfrak{g} = \mathfrak{k} \oplus \mathfrak{p}, \quad \mathfrak{p}_{\mathbb{C}} = \bigoplus_{p \text{ odd}} \mathfrak{g}^{-p,p}.$$

Hence the corresponding Cartan involution θ on $\mathfrak{g}$ is characterized by its complex linear extension to $\mathfrak{g}_{\mathbb{C}}$ which equals multiplication with $(-1)^p$ on $\mathfrak{g}^{-p,p}$. Note that this extension, also denoted by θ, is *not* the Cartan involution $\theta^{\mathbb{C}}$ on $\mathfrak{g}_{\mathbb{C}}$. Indeed, as we shall see, $\theta^{\mathbb{C}} = \theta \circ \tau$, where τ is complex conjugation with respect to the previously defined complex structure on $\mathfrak{g}_{\mathbb{C}}$. Its fixed point set is the subalgebra

$$\mathfrak{n} \overset{\text{def}}{=} \mathfrak{k} \oplus \mathrm{i} \cdot \mathfrak{p}.$$

It is the Lie algebra of a unique *compact* connected matrix Lie group $N \subset G_{\mathbb{C}}$, the *compact real form* of $G_{\mathbb{C}}$ with Lie algebra $\mathfrak{n}$.

Although the above statements follow directly from the general theory outlined in Appendix D.3, we shall give two independent proofs. The first proof is spelled out in Example 12.6.1, the case of even weight, but we leave the case of odd weight to the reader (see Problems 12.6.1 and 12.6.2). A second and more intrinsic proof using the Hodge metric is given below (Proposition 12.6.2). This proof is in fact equivalent to the classical proof given in the Appendix.

Example 12.6.1 If the weight is even, the Lie algebra is of the form $\mathfrak{g} = \mathfrak{so}(a,b)$. Set $n = a + b$. Consider the corresponding block-decomposition of the n-by-n matrices

$$X = \begin{pmatrix} U & V \\ {}^{\mathsf{T}}V & W \end{pmatrix} \in \mathfrak{so}(a,b).$$

Because $X \in \mathfrak{so}(a,b)$, the matrices U and W are skew-symmetric. The involution θ sends $\begin{pmatrix} U & V \\ {}^{\mathsf{T}}V & W \end{pmatrix}$ to $\begin{pmatrix} U & -V \\ -{}^{\mathsf{T}}V & W \end{pmatrix}$ and so $\mathfrak{k}$ consists of those matrices for which $V = 0$, and the matrices in $\mathfrak{p}$ are the ones with $U = W = 0$. Then $\mathfrak{n} = \mathfrak{k} + \mathrm{i}\mathfrak{p}$ embeds in $\mathfrak{su}(n)$ and one readily verifies that the map

$$\begin{pmatrix} U & \mathrm{i}V \\ \mathrm{i}\,{}^{\mathsf{T}}V & W \end{pmatrix} \mapsto \begin{pmatrix} U & V \\ -{}^{\mathsf{T}}V & W \end{pmatrix}$$

induces an isomorphism of $\mathfrak{n}$ onto $\mathfrak{so}(n)$, the Lie algebra of the compact group $\mathrm{SO}(n)$.

The Role of the Hodge Metric and the Weil Operator

The reference point $o \in D$ corresponds to a Hodge structure $H_{\mathbf{C}} = \bigoplus H^{p,q}$ with Weil operator C, and polarized by b. Hence the form

$$h_C(u,v) = b(Cu,\bar{v}), \quad u,v \in H_{\mathbf{C}}$$

is a Hermitian inner product on $H_{\mathbf{C}}$. If $\xi \in \mathrm{End}(H_{\mathbf{C}})$ we define its Hermitian conjugate ξ^* as usual by the rule

$$h_C(\xi(u),v) = h_C(u,\xi^*(v)).$$

Using h_C-unitary bases we identify $H_{\mathbf{C}} = \mathbf{C}^n$, $n = \dim_{\mathbf{C}} H_{\mathbf{C}}$, so $\mathrm{GL}(H_{\mathbf{C}})$ gets identified with the group of invertible n by n complex matrices. The Cartan involution sends $g \in \mathrm{GL}(H_{\mathbf{C}})$ to $(g^*)^{-1}$ and on the level of the Lie algebra $\mathrm{End}(H_{\mathbf{C}})$ it operates as follows:

$$\theta^{\mathbf{C}}(\xi) = -\xi^*, \quad \xi \in \mathrm{End}(H_{\mathbf{C}}). \tag{12.9}$$

We are going to describe $\theta^{\mathbf{C}}$ in terms of the Weil operator. First recall that we have seen before (Lemma 5.3.1), that the reference Hodge structure induces a weight 0 Hodge structure on $\mathfrak{g}_{\mathbf{C}}$. If C denotes the Weil operator of the reference Hodge structure, the Weil operator of $g_{\mathbf{C}}$ is $\mathrm{Ad}(C)$, i.e. C acting on $\mathfrak{g}_{\mathbf{C}}$ through the adjoint operation and turns out to be related to the Cartan involution.

Proposition 12.6.2 *Let τ be complex conjugation on $\mathfrak{g}_{\mathbf{C}}$.*
(i) *The Weil operator* $\mathrm{Ad}(C)$ *is real and on $\mathfrak{g}$ coincides with the Cartan involution θ.*
(ii) *The Cartan involution $\theta^{\mathbf{C}}$ on $\mathfrak{g}_{\mathbf{C}}$ equals* $\mathrm{Ad}(C)\circ\tau$ *with Cartan decomposition $\mathfrak{g}_{\mathbf{C}} = \mathfrak{u} \oplus \mathrm{i}\mathfrak{u}$, where the subalgebra $\mathfrak{u} = \mathfrak{k} \oplus \mathrm{i}\mathfrak{p}$ consists of the skew-Hermitian maps with respect to h_C; and* $\mathrm{i}\mathfrak{u}$ *is the space of h_C-Hermitian maps. In particular, $\mathfrak{u}$ is a compact real form of $\mathfrak{g}_{\mathbf{C}}$.*

Proof (i) Recall that $\mathfrak{g}_{\mathbf{C}}$ consists of $\xi \in \mathrm{End}(H_{\mathbf{C}})$ that are b-orthogonal. Now for pure type $x,y \in H_{\mathbf{C}}$ we have $b(x,y) = 0$, unless $x \in H^{p,q}, y \in H^{q,p}$, and in this case, using $C|H^{p,q} = \mathrm{i}^{p-q} = C^{-1}|H^{q,p} = \bar{C}|H^{q,p}$, we see that $b(Cx,Cy) = b(x,y)$. So $C \in G_{\mathbf{C}}$ and C commutes with complex conjugation τ on $H_{\mathbf{C}}$ and hence C is real. We claim that its adjoint action on $\mathfrak{g}_{\mathbf{C}}$ is defined by

$$\begin{array}{rccl} \mathrm{Ad}(C): \mathfrak{g}_{\mathbf{C}} & \longrightarrow & \mathfrak{g}_{\mathbf{C}} & \\ \xi & \longmapsto & (-1)^p\xi, & \xi \in \mathfrak{g}^{-p,p}. \end{array}$$

Indeed, if $\xi \in \mathfrak{g}^{-p,p}$ and $u \in H^{r,s}$, we have $C\xi C^{-1}u = C^{-1}\xi Cu = (-1)^p\xi u$.

(ii) Let $\xi \in \mathfrak{g}^{-p,p}$. Then, using that $C\xi C^{-1} = (-1)^p \xi$ and ξ is skew for b we get

$$\begin{aligned} h_C(\xi u, v) = b(\xi u, C^{-1}\bar{v}) &= -b(u, \xi C^{-1}\bar{v}) \\ &= (-1)^{p+1} b(u, C^{-1}\xi\bar{v}) = (-1)^{p+1} b(Cu, \xi\bar{v}) \\ &= (-1)^{p+1} b(Cu, \overline{\bar{\xi} v}) = (-1)^{p+1} h_C(u, \tau(\xi)v). \end{aligned}$$

Since – by definition – this last expression equals $h_C(u, \xi^* v)$, one concludes that $-\xi^* = (-1)^p \bar{\xi} = \mathrm{Ad}(C)\tau(\xi)$. Now use (12.9). □

From now on we revert to the previous notation $\bar{\xi}$ instead of $\tau(\xi)$, We deduce the following simple consequence.

Corollary 12.6.3 (i) *For $\xi \in \mathfrak{p}_{\mathbf{C}}$, respectively $\xi \in \mathfrak{k}_{\mathbf{C}}$, we have $\xi^* = \bar{\xi}$, $\xi^* = -\bar{\xi}$.*

(ii) *For $\xi \in \mathfrak{n}$, respectively $\xi \in i\mathfrak{n}$, we have $C\xi = \bar{\xi}$, respectively $C\xi = -\bar{\xi}$.*

(iii) *Let $\xi \in \mathfrak{p}_{\mathbf{C}}$ with $[\xi, \bar{\xi}] = 0$, then $\xi = 0$.*

Proof (i) On $\mathfrak{p}_{\mathbf{C}}$ the operator C acts as $-\mathrm{id}$ and hence, if $\xi \in \mathfrak{p}_{\mathbf{C}}$ we have $C(\xi) = -\bar{\xi}^* = -\xi$ which shows the result; a similar reasoning holds for $\mathfrak{k}_{\mathbf{C}}$.
(ii) Note that $\mathfrak{n} = \mathfrak{k} \oplus i\mathfrak{p}$ and on $\mathfrak{k}$, respectively $\mathfrak{p}$, the Cartan involution is the identity, respectively minus the identity.
For (iii) observe that $[\xi, \bar{\xi}] = 0$ implies that the imaginary part of ξ, which is skew-Hermitian with respect to h_C, commutes with the real part, an h_C-Hermitian operator. And so, each being diagonalizable, they are simultaneously diagonalizable. But then ξ is also diagonalizable. Because $\xi \in \mathfrak{p}_{\mathbf{C}}$ it is nilpotent as well (see Problem 12.5.3), we must have $\xi = 0$. □

Polarization and Invariant Metric on the Lie Algebra

We can now introduce an invariant metric on $\mathfrak{g}$. Start out with the trace form

$$B_0(\xi, \eta) = \mathrm{Tr}\, \xi \circ \eta, \quad \xi, \eta \in \mathfrak{g}_{\mathbf{C}}.$$

This a $\mathbf{C}$-linear symmetric form, invariant under the adjoint action of $G_{\mathbf{C}}$ on $\mathfrak{g}_{\mathbf{C}}$, and it is real on $\mathfrak{g}$.

Remark 12.6.4 We have seen (Problem 12.5.2 (d)) that the form B_0 is a multiple of the Killing form B. Since in our case the Lie algebras have a real structure this multiple is real. In fact, this is a positive multiple (Helgason, 1978, Ch. II §8) and so in the rest of this subsection we could also have used the Killing form instead of B_0.

Using Proposition 12.6.2 we can now prove the following.

Proposition 12.6.5 *The form $-B_0$ polarizes the Hodge structure $\mathfrak{g}_{\mathbb{C}} = \bigoplus_p \mathfrak{g}^{-p,p}$. In particular, the form*

$$(\xi, \eta) \stackrel{\text{def}}{=} - B_0(\operatorname{Ad} C(\xi), \bar{\eta})$$

is a positive definite Hermitian metric on $\mathfrak{g}_{\mathbb{C}}$, invariant under the adjoint action of V.

Proof First, if $\xi \in \mathfrak{g}^{p,-p}$ and $\eta \in \mathfrak{g}^{q,-q}$ with $p \neq q$, the linear map $\xi\bar{\eta}$ permutes all of the subspaces $H^{p,q}$ nontrivially among themselves and so $\xi\bar{\eta}$ must have zero trace.

Recall that $\mathfrak{g}_{\mathbb{C}} = \mathfrak{n} \oplus \mathrm{i}\mathfrak{n}$ and that any linear map $\xi \in \mathfrak{n}$ is skew-Hermitian with respect to h_C. So ξ is diagonalizable with all the eigenvalues purely imaginary and if $\xi \neq 0$, at least one of these is nonzero. It follows that $-B_0(\xi, \xi) > 0$. On the other hand, on $\mathfrak{n}$ the Weil operator is just complex conjugation (see Corollary 12.6.3 (ii)) so that $(\xi, \xi) = -B_0(C\xi, \bar{\xi}) = -B_0(\bar{\xi}, \bar{\xi})$ is positive by what we just showed. Finally, if $\eta = \mathrm{i}\xi \in \mathrm{i}\mathfrak{n}$ is nonzero, we have $(\eta, \eta) = (\mathrm{i}\xi, \mathrm{i}\xi) = (\xi, \xi) > 0$. □

Problems

12.6.1 Verify that, for odd weight, $\mathfrak{n}$ is the Lie algebra of the compact group $\mathrm{Sp}(n)$ consisting of *complex* $2n$ by $2n$ unitary matrices X with $X J_n {}^{\mathsf{T}}X = J_n$, where, as before, $J_n = \begin{pmatrix} 0 & \mathbf{1}_n \\ -\mathbf{1}_n & 0 \end{pmatrix}$.

12.6.2 Let $H = \bigoplus H^{p,q}$ be a Hodge decomposition of weight w corresponding to a reference point $o \in D = G/V$. Let b be the polarization and h_C the corresponding Hodge metric. Using Hodge frames (see Problem 12.5.3), describe the form of the matrix (X_{ij}) corresponding to $X \in \mathfrak{g}^{-p,p}$ and determine the matrix of the complex conjugate of X. Verify Proposition 12.6.5.

Bibliographical and Historical Remarks

This chapter contains the basic calculations in the fundamental article by Griffiths and Schmid (1969). We avoid the use of roots by working directly with the Cartan involution. In this edition, classical real Lie group theory plays a more dominant role. See also Schmid (1973, § 8).

The notion of "Higgs principal bundle" is ours and it is motivated by its relation with the Higgs vector bundles which come up in Chapters 13 and 14.

13

Curvature Estimates and Applications

We do two types of curvature calculation in this chapter. The first applies to variations of Hodge structures over any complex manifold and is treated in Section 13.1. As a first application, we prove that over a *quasi-projective* manifold, flat (parallel) sections of a variation of Hodge structure have flat (p, q)-components. This has important consequences. First, any flat section of a variation of Hodge structure over a quasi-projective manifold has flat Hodge components (theorem of the fixed part). Second, we have the rigidity theorem, which says that if we have two polarized variations of Hodge structure over a compact manifold and an isomorphism of the underlying local systems which is a morphism of Hodge structures at one point, it must be so everywhere.

More applications are in Section 13.4, first a bound on the degrees of Hodge bundles of a variation of Hodge structure over quasi-projective curves, second an application on hyperbolicity of the base curve, and finally, one on rigidity.

In order to do this, in Section 13.3 we make a digression on Higgs bundles. Among the examples we have variations of Hodge structure as well complex systems of Hodge bundles. The latter generalize variations of Hodge structure in the sense that the underlying local system may be a local system of complex vector spaces without a given real structure. Since we need base manifolds that are not necessarily compact, the technique of logarithmic extensions is needed, which is explained in the preceding Section 13.2.

Then, in Section 13.6, we do curvature calculations on the period domains and show that the sectional curvature is negative and bounded away from zero in flat directions. This implies that period maps are distance decreasing. As an application, we give in Section 13.7 Borel's proof of the monodromy theorem for a one-variable degeneration, i.e., a variation of Hodge structure over the unit disk minus a point. Geometrically, this corresponds to a family of varieties over the unit disk with smooth fibers except over the origin, which is considered to be a "degenerate" fiber, hence the term "degeneration." This theorem gives important topological restrictions on the sort of degeneration that can appear and explains why one often assumes that the local monodromy operators of a variation of Hodge structure are quasi-unipotent or even unipotent.

13.1 Higgs Bundles, Hodge Bundles, and their Curvature

Consider now a polarized variation of Hodge structure. To fix notation, let $\mathbf{H}$ be a local system on a complex manifold S, and let $\mathcal{H}$ be the associated holomorphic vector bundle. The Hodge filtration is denoted as usual by

$$\mathcal{H} = \mathcal{F}^0 \supset \mathcal{F}^1 \supset \cdots \supset \mathcal{F}^w \supset 0,$$

where $\mathcal{F}^i$, $i = 0, \ldots, w$ are holomorphic subbundles. Let us denote by b the polarizing bilinear form and recall that we have put a Hermitian metric on $\mathcal{H}$; the Hodge metric

$$h_C(\cdot, \cdot) \overset{\text{def}}{=} b(C\cdot, \bar{\cdot}),$$

where C is the Weil-operator acting by multiplication with $\mathrm{i}^{p-(w-p)} = (-\mathrm{i})^w \cdot (-1)^p$ on $H^{p,w-p}$. We now consider the graded quotients $\mathcal{F}^p/\mathcal{F}^{p+1}$. These are *holomorphic* bundles which are differentiably isomorphic to the C^∞ Hodge bundles $\mathcal{H}^{p,q}$. We assemble these in the *Hodge bundle*

$$\boxed{\mathcal{H}_{\text{Hdg}} \overset{\text{def}}{=} \text{Gr}_{\mathcal{F}} \mathcal{H} = \bigoplus_{p=0}^{w-1} \mathcal{F}^p/\mathcal{F}^{p+1}, \quad \mathcal{F}^p/\mathcal{F}^{p+1} \simeq \mathcal{H}^{p,w-p}.}$$

Remark The complex structure on the Hodge bundle $\mathcal{H}_{\text{Hdg}}$ is different from the one on $\mathcal{H}$. A holomorphic section for $\mathcal{H}$ induces one for the Hodge bundle $\mathcal{H}_{\text{Hdg}}$, but not conversely. In what follows we let $\bar{\partial}$ be the $\bar{\partial}$-operator for the complex structure on the Hodge bundle.

We have encountered the Hodge bundle before, in the framework of Higgs principal bundles: in Example 12.4.2 we showed that this bundle is associated with the Higgs G-principal bundle $G \times_V G$, where $D = G/V$ is the period domain. By Theorem 12.3.3 the induced connection ∇_V we introduced there is the Chern connection on the Hodge bundle. We already identified the connection ∇_{Higgs} as the flat connection coming from the Gauss–Manin connection and that $\nabla_{\text{Higgs}} = \nabla_V + \tau$, where the matrix-valued one-form τ is split into two parts $\tau = \tau^{1,0} + \tau^{0,1}$. The form $\tau' = \tau^{1,0}$ is related to the second fundamental form associated with the direct sum splitting of the Hodge bundle. The form $\tau'' = \tau^{0,1}$ can be shown to be its adjoint.

To explain this further, we study in more detail how the Gauss–Manin connection ∇ acts on the Hodge flag $\mathcal{F}^\bullet$ The second fundamental form associated with $\mathcal{F}^p \subset \mathcal{H}$ is a $\mathrm{Hom}(\mathcal{F}^p, \mathcal{H}/\mathcal{F}^p)$-valued one-form (see Section 11.2) which by Griffiths' transversality belongs to $\mathcal{A}^1_S(\mathrm{Hom}(\mathcal{F}^p, \mathcal{F}^{p-1}/\mathcal{F}^p))$; by the same argument applied to $\mathcal{F}^{p+1}$ we see that it induces

$$\sigma^p \in \mathcal{A}^{1,0}_S(\mathrm{Hom}(\mathrm{Gr}^p_{\mathcal{F}}\mathcal{H}, Gr^{p-1}_{\mathcal{F}}\mathcal{H})).$$

Collecting these together we put

$$\boxed{\sigma \overset{\text{def}}{=} \bigoplus\nolimits_p \sigma^p \in \mathcal{A}^{1,0}_S(\mathrm{End}^{-1,1}(\mathcal{H}_{\mathrm{Hdg}})).}$$

Here, as usual, we let $\mathrm{End}^{p,-p}$ be the endomorphisms which map the Hodge bundle $\mathrm{Gr}^r_{\mathcal{F}}\mathcal{H}$ to the Hodge bundle $\mathrm{Gr}^{r+p}_{\mathcal{F}}\mathcal{H}$.

Of course, in the old notation we have $\tau' = \sigma$. We make some further change of notation: in place of ∇_V we use ∇_{Hdg}, since this is the natural connection on the Hodge bundle. Lastly, the Gauss–Manin connection transported to $\mathrm{Gr}_{\mathcal{F}}\mathcal{H}$ will be called ∇ instead of ∇_{Higgs} to stress that it is the flat connection on the Hodge bundle. Hence the splitting $\nabla = \sigma + \nabla_{\mathrm{Hdg}} +$ adjoint of σ, where ∇_{Hdg} is the Chern connection on the Hodge bundle. Summarizing, we have shown the following.

Proposition 13.1.1 *The operator*

$$\nabla : \mathcal{H}_{\mathrm{Hdg}} \to \mathcal{H}_{\mathrm{Hdg}} \otimes \mathcal{A}^1_S$$

decomposes as follows:

$$\mathcal{F}^p/\mathcal{F}^{p+1} \xrightarrow{\nabla} \underbrace{\mathcal{A}^{1,0}_S(\mathcal{F}^{p-1}/\mathcal{F}^{p-2})}_{\substack{\downarrow \\ \sigma}} \oplus \underbrace{\mathcal{A}^1_S(\mathcal{F}^p/\mathcal{F}^{p+1})}_{\substack{\downarrow \\ \nabla_{\mathrm{Hdg}}}} \oplus \underbrace{\mathcal{A}^{0,1}_S(\mathcal{F}^{p+1}/\mathcal{F}^p)}_{\substack{\downarrow \\ \sigma^*}}$$

$$\sigma \quad + \quad \nabla_{\mathrm{Hdg}} \quad + \quad \sigma^*.$$

Here ∇_{Hdg} is the Chern connection on $\mathcal{H}_{\mathrm{Hdg}}$, the unique metric connection (for the Hodge metric) on the Hodge bundle whose $(0,1)$-part is the $\bar\partial$-operator on that bundle. The operator σ^* is the adjoint of σ with respect to h_C, i.e. $h_C(\sigma(s), s') = h_C(s, \sigma^*(s'))$ for all local smooth sections s, s' of $\mathcal{H}_{\mathrm{Hdg}}$.

This proposition is central to our calculations. We want to give a short proof.

Direct proof First observe that if s is a holomorphic local section of $\mathcal{F}^p$, then $\nabla s \in \Omega^1_S(\mathcal{F}^{p-1})$ and hence the above decomposition simplifies to $\sigma(s) + \nabla_{\mathrm{Hdg}}(s)$. In particular, since there is no $(0,1)$-part, $\nabla^{0,1}_{\mathrm{Hdg}} s = 0$. By the argument on page 311, this characterizes the holomorphic structure on $\mathcal{H}_{\mathrm{Hdg}}$, i.e., $\nabla^{0,1}_{\mathrm{Hdg}} = \bar\partial_{\mathcal{H}_{\mathrm{Hdg}}}$.

To show that $\nabla_{\rm Hdg}$ is the Chern connection, we first note that the Gauss–Manin connection preserves the indefinite Hermitian metric

$$h(\cdot,\cdot) \stackrel{\rm def}{=} (-{\rm i})^w b(\cdot,\bar{\cdot}),$$

which differs from h_C by the sign $(-1)^p$ on the Hodge bundle ${\rm Gr}^p_{\mathcal{F}} \mathcal{H}$. Using this we shall show that $\nabla_{\rm Hdg}$ is a metric connection with respect to h_C by working out the condition

$$dh(s,s') = h(\nabla s, s') + h(s, \nabla s').$$

Indeed, if s, s' have the same type, say (p,q), then since $h(\sigma s, s') = h(s, \sigma^* s') = 0$, one has

$$\begin{aligned}(-1)^p\, dh_C(s,s') = dh(s,s') &= h(\nabla s, s') + h(s, \nabla s') \\ &= h(\nabla_{\rm Hdg} s, s') + h(s, \nabla_{\rm Hdg} s') + h(\sigma s, s') + h(s, \sigma^* s') \\ &= (-1)^p (h_C(\nabla_{\rm Hdg} s, s') + h_C(s, \nabla_{\rm Hdg} s')).\end{aligned}$$

If they have different type, one only needs to consider "nearby types" , say for s of type (p,q) and s' of type $(p-1, q+1)$, one has

$$\begin{aligned}0 = dh(s,s') &= h(\nabla s, s') + h(s, \nabla s') \\ &= h(\nabla_{\rm Hdg} s, s') + h(s, \nabla_{\rm Hdg} s') + h(\sigma s, s') + h(s, \sigma^* s') \\ &= (-1)^p (h_C(\sigma s, s') - h_C(s, \sigma^* s')) \\ &= 0.\end{aligned}$$

□

Lemma 13.1.2 *The morphism $\sigma : \mathcal{H}_{\rm Hdg} \to \mathcal{H}_{\rm Hdg} \otimes \mathcal{A}^{1,0}_S$ is holomorphic, that is, $\sigma \in \mathcal{E}nd\mathcal{H}_{\rm Hdg} \otimes \Omega^1_S$. Moreover, the composition*

$$\sigma \wedge \sigma : \mathcal{H}_{\rm Hdg} \to \mathcal{H}_{\rm Hdg} \otimes \Omega^2_S$$

is zero.

Proof Note that σ can be considered as an endomorphism of Hodge type $(1,0)$ with values in the one-forms of type $(1,0)$. Now expand the equation $\nabla\circ\nabla = 0$ into Hodge and form-types, making use of the equality

$$\nabla = \sigma + \nabla^{1,0}_{\rm Hdg} + \bar\partial + \sigma^*.$$

Observe that

$$\bar\partial(\sigma) \stackrel{\rm def}{=} \bar\partial\circ\sigma + \sigma\circ\bar\partial$$

is the only part of Hodge type $(1,0)$ and form type $(1,1)$, and so must vanish. It says that σ is a *holomorphic* endomorphism.

Likewise, the second equation,

$$\sigma \wedge \sigma = 0,$$

follows since this is the only part in ∇^2 of Hodge type $(2,0)$ and form type $(2,0)$. □

Remark To tie this up with Example 12.4.2, as remarked above, we have $\tau' = \sigma, \tau'' = \sigma^*$ and $\nabla_V = \nabla_{\mathrm{Hdg}}$. The harmonic equation Eq. (12.7) assert that σ is holomorphic while Eq. (12.8) means that $\sigma \wedge \sigma = 0$.

The above lemma states that $(\mathcal{H}_{\mathrm{Hdg}}, \sigma)$ is a so-called Higgs bundle. Moreover, the Hodge metric h_C is a harmonic metric, explaining the terms "Higgs equations" and "harmonic equation".

Definition 13.1.3 (i) A *Higgs bundle* (E, σ) consists of a holomorphic bundle E equipped with an $\mathrm{End}(E)$-valued holomorphic one-form σ, the *Higgs field*, such that the $\mathrm{End}(E)$-valued holomorphic two-form given by $\sigma \wedge \sigma$ vanishes.
(ii) A Hermitian metric k on a Higgs bundle is called *harmonic* if its Chern connection D_k combines with σ and its k-conjugate σ^* to give a flat connection $\sigma + D_k + \sigma^*$.

Hence, variations of Hodge structure are examples of Higgs bundles equipped with a harmonic metric. But there are certainly more examples of Higgs bundles, among which are the complex systems of Hodge bundles of given weight, defined as follows.

Definition 13.1.4 (i) A *complex system of Hodge bundles* on a complex manifold S consists of a complex local system $\mathbf{H}$ on S such that the corresponding vector bundle $\mathcal{H}$ admits a direct sum decomposition into complex C^∞ subbundles

$$\mathcal{H} = \bigoplus\nolimits_p \mathcal{H}^p$$

with the property that the Gauss–Manin connection ∇ on $\mathbf{H}$ (defined by $\nabla(v \otimes f) = v \otimes \mathrm{d}f, v \in h, f \in \mathcal{O}_S$) satisfies the analogue of the transversality condition of Proposition 13.1.1 above. The bundles $\mathcal{H}^p$ are not necessarily holomorphic, but the transversality condition implies that the Hodge filtration defined by

$$\mathcal{F}^p \stackrel{\text{def}}{=} \bigoplus_{r \geq p} \mathcal{H}^r$$

is holomorphic (while the opposite filtration $\bigoplus_{r \leq p} \mathcal{H}^r$ is anti-holomorphic). Note also that each graded piece $\mathcal{F}^p / \mathcal{F}^{p+1}$ is holomorphic and that it is differentiably isomorphic to $\mathcal{H}^p$.
(ii) Define an operator S on $\mathcal{H}$ by $S|\mathcal{H}^p = (-1)^p$. If there is a sesquilinear form

h on H which is preserved by ∇ such that $h_S(x, y) = h(Sx, y)$ gives a Hermitian metric on H, we say that *h polarizes* the complex system and together these define a *complex polarized variation of Hodge structure.*

Of course, the usual (real) variations of given weight w are examples of complex variations: set $\mathcal{H}^r = \mathcal{H}^{r,w-r}$. If it is polarized by b, the form h in the above definition is just $(-\mathrm{i})^w h$, since $C = (-\mathrm{i})^w S$. A real variation satisfies the additional *reality constraint* $\mathcal{H}^{p,q} = \overline{\mathcal{H}^{q,p}}$. Conversely, specifying a weight, a complex system of Hodge bundles H together with its complex conjugate defines a real variation of Hodge structures on $\mathcal{H} \oplus \overline{\mathcal{H}}$ in the obvious manner. In passing, we observe, however, that we can have complex systems of Hodge bundles of any given pure type. If for instance $\mathcal{H}^p$ has the property that it is preserved by ∇, i.e., if it is a flat subbundle, it is itself a complex system of Hodge bundles.

Let us summarize what we have seen so far.

Lemma 13.1.5 *A complex system of Hodge bundles $\mathcal{H} = \oplus_r \mathcal{H}^r$ on a complex manifold defines a Higgs bundle $\mathrm{Gr}_{\mathcal{F}} \mathcal{H}$, $\mathcal{F}^p = \oplus_{r \geq p} \mathcal{H}^r$ with Higgs field the second fundamental form induced by the Gauss–Manin connection. A complex polarized variation of Hodge structure gives a Higgs bundle with a harmonic metric, namely the Hodge metric.*

There is an obvious operation by $\mathbf{C}^*$ on the set of Higgs bundles: for a Higgs bundle (E, σ) and $t \in \mathbf{C}^*$ the pair $(E, t\sigma)$ is also a Higgs bundle. If $E = \mathcal{H}$ is a complex system of Hodge bundles, one has a commutative diagram

$$\begin{array}{ccc} \mathrm{Gr}^p_{\mathcal{F}} \mathcal{H} & \xrightarrow[\sim]{\cdot t^p} & \mathrm{Gr}^p_{\mathcal{F}} \mathcal{H} \\ \downarrow{\scriptstyle t \cdot \sigma^p} & & \downarrow{\scriptstyle \sigma^p} \\ \mathrm{Gr}^{p-1}_{\mathcal{F}} \mathcal{H} \otimes \Omega^1_S & \xrightarrow[\sim]{\cdot t^{p-1}} & \mathrm{Gr}^{p-1}_{\mathcal{F}} \mathcal{H} \otimes \Omega^1_S. \end{array}$$

This means that $(\mathcal{H}, t \cdot \sigma) \simeq (\mathcal{H}, \sigma)$. In other words, on the moduli space of *isomorphism classes* of Higgs bundles over S the complex systems of Hodge bundles are fixed points for the $\mathbf{C}^*$-action. Simpson (1992) proved the converse and also that this implies that if S is a compact Kähler manifold, these admit a harmonic metric coming from a polarization, i.e., the fixed points of the $\mathbf{C}^*$-action correspond to polarized complex variations of Hodge structure.

Theorem 13.1.6 *A Higgs bundle $(\mathcal{H}, \sigma)$ over a compact base manifold underlies a polarized variation of Hodge structure if and only if its isomorphism class is $\mathbf{C}^*$-invariant. In other words: among the complex representations of the fundamental group of S, those which carry a complex polarized variation* *of Hodge structure are exactly the semisimple ones fixed by the action of* $\mathbf{C}^*$ *as defined above.*

In the remainder of this section and in the next section we will investigate complex polarized variations of Hodge structure. Many results for ordinary polarized variations of Hodge structure remain true for these, such as the curvature calculations (Eq. (12.5) in Chapter 12) which in our case reads

$$F_{\nabla_{\mathrm{Hdg}}} = -[\sigma, \sigma^*].$$

We calculate this in every graded piece and observe that since for any two tangent vectors X, Y the graded pieces of $[\sigma(X), \sigma^*(Y)]$ come from the compositions of two arrows in the diagram

$$\mathrm{Gr}^{p+1}_{\mathcal{F}} \mathcal{H} \underset{(\sigma^*)^p(X)}{\overset{\sigma^{p+1}(Y)}{\rightleftarrows}} \mathrm{Gr}^{p}_{\mathcal{F}} \mathcal{H} \underset{(\sigma^*)^{p-1}(Y)}{\overset{\sigma^{p}(X)}{\rightleftarrows}} \mathrm{Gr}^{p-1}_{\mathcal{F}} \mathcal{H}, \tag{13.1}$$

and since $(\sigma^p)^* = (\sigma^*)^{p-1}$, the contribution from $\mathrm{Gr}^p_{\mathcal{F}} \mathcal{H}$ to the curvature is equal to $-\left((\sigma^p)^* \circ \sigma^p + \sigma^{p+1} \circ (\sigma^{p+1})^*\right)$. Because of the way the metric contraction for h_C is defined (see Eq. (11.1)), this shows

$$\begin{aligned} h_C(-[\sigma, \sigma^*]s, s) &= \sum_p h_C(\sigma^p(s), \sigma^p(s)) + \sum_p h_C((\sigma^*)^{p+1}(s), (\sigma^*)^{p+1}(s)) \\ &= h_C(\sigma(s), \sigma(s)) + h_C(\sigma^*(s), \sigma^*(s)). \end{aligned}$$

Note also, that these same conventions mean that the first term is of the shape $\sum h_{ij}\, \mathrm{d}z_i \wedge \mathrm{d}\bar{z}_j$ with (h_{ij}) a positive definite Hermitian matrix, while the second is of the form $\sum h'_{ij}\, \mathrm{d}\bar{z}_i \wedge \mathrm{d}z_j = -\sum h'_{ij}\, \mathrm{d}z_i \wedge \mathrm{d}\bar{z}_j$ with (h'_{ij}) positive definite. We summarize our results.

Theorem 13.1.7 *Let there be given a complex polarized variation of Hodge bundles of weight w. The Gauss–Manin connection transported to the Hodge bundle decomposes as*

$$\nabla = \sigma + \nabla_{\mathrm{Hdg}} + \sigma^*$$

with ∇_{Hdg} *the Chern connection on the Hodge bundle equipped with the Hodge metric. Its curvature is given by* $F_{\nabla_{\mathrm{Hdg}}} = -[\sigma, \sigma^*]$. *Moreover,*

$$\boxed{h_C(F_{\nabla_{\mathrm{Hdg}}} s, s) = h_C(\sigma(s), \sigma(s)) + h_C(\sigma^*(s), \sigma^*(s)).}$$

The two summand are $(1, 1)$*-forms with the following positivity property:*

$$\begin{aligned} \mathrm{i} h_C(\sigma(s), \sigma(s)) &\geq 0, \\ \mathrm{i} h_C(\sigma^*(s), \sigma^*(s)) &\leq 0. \end{aligned}$$

Let us for clarity write out the preceding computation in a *Hodge frame*, i.e., we choose an h_C-unitary frame for each of the $\mathcal{H}^{p,q}$ in such a way that the basis for $\mathcal{H}^{p,q}$ is the complex conjugate of the basis for $\mathcal{H}^{q,p}$.The connection matrix for the flat metric is then of the form (the blocks are ordered from $(w, 0)$ to $(0, w)$)

$$A = \begin{pmatrix} A_w & T^w & 0 & \cdots & 0 \\ S^w & A_{w-1} & T^{w-1} & 0 & \vdots \\ \vdots & \ddots & \ddots & \ddots & \vdots \\ \vdots & 0 & S^2 & A_1 & T^1 \\ 0 & \cdots & 0 & S^1 & A_0 \end{pmatrix}.$$

If we split this matrix into the diagonal and off-diagonal parts,

$$A = A' + S + T,$$

then A' is a block-diagonal matrix which is the connection matrix of the metric connection on $\mathcal{H}_{\text{Hdg}}$. The matrix S is lower diagonal with blocks just below the diagonal, while T is upper diagonal with blocks just above the diagonal. Their relation is as follows:

$$S = \left(\sum_\alpha s_{ij}^\alpha \, \mathrm{d}z_\alpha\right)_{i,j} \implies T = \left(\sum_\alpha \bar{s}_{ji}^\alpha d\bar{z}_\alpha\right)_{i,j} = S^*$$

so that $(S + T) \wedge (S + T)$ has diagonal blocks of the form

$$\begin{pmatrix} (S^w)^* \wedge S^w & 0 & & \cdots & 0 \\ 0 & \begin{matrix} S^w \wedge (S^w)^* \\ +(S^{w-1})^* \wedge S^{w-1} \end{matrix} & & \ddots & \vdots \\ \vdots & & \ddots & \vdots & \vdots \\ \vdots & & \cdots & \begin{matrix} (S^2) \wedge (S^2)^* \\ +(S^1)^* \wedge S^1 \end{matrix} & 0 \\ 0 & & \cdots & 0 & (S^1) \wedge (S^1)^* \end{pmatrix}.$$

Writing out $0 = A \wedge A + \mathrm{d}A$ and comparing the latter diagonal blocks, we see that the curvature of the metric connection for h_C is given by $A' \wedge A' + \mathrm{d}A' = -(S \wedge S^* + S^* \wedge S)$ which is the matrix expression for $-[\sigma, \sigma^*]$.

Remark The vanishing of the curvature of the Hodge bundle is equivalent to $[\sigma, \sigma^*] = 0$, which in turn is equivalent to $\sigma = 0$. This follows quite directly from the definitions. Compare this also with Corollary 12.6.3. For Hodge bundles the vanishing of the Higgs field, $\sigma = 0$, means that we have a locally constant period map.

Lemma 13.1.8 *Let* $(\mathrm{Gr}_{\mathcal{F}} \mathcal{H}, \sigma)$ *be the Higgs bundle for a variation of Hodge structure. The Higgs field* σ *vanishes if and only if the period map for the variation is constant.*

Proof The homomorphism σ is the derivative of the period map and so it vanishes if and only if the period map is locally constant. □

Remark 13.1.9 Strictly speaking, a period map has been defined for an integral (or rational) variation of Hodge structure, but a local version exists of course also for real and even for complex variations of Hodge structure.

A word of warning is in order here. Starting with a rational variation of Hodge structure, the associated Higgs bundle is in fact the graded bundle for the complexified variation: the Higgs structure does not "see" the rational structure. Below (Corollary 13.3.3) we shall see that a Higgs bundle can be split into indecomposable pieces each of which is associated with a *complex* variation of Hodge structure which inherits a polarization from the polarization on the Hodge bundle, if present. This polarization need not be **Q**-valued, in fact, it generally will be complex-valued. If the Higgs field vanishes, the underlying local system has monodromy in the unitary group for the Hodge metric, but since the polarization need not be defined over the rationals, this monodromy group may very well be infinite.

So, although the period map for this summand is constant, there may be a large monodromy group present. In particular, if we are interested in the period map for the total variation, it is misleading to break up a rational polarized variation of Hodge structure into its indecomposable complex polarized summands.

We give a few applications of these calculations in Problems 13.3.1–13.3.3 below. Let us mention the following.

Application *Let there be given a variation of Hodge structure over a smooth projective variety. Then the image of the period map is an algebraic variety.*

A classic application of the preceding curvature calculations is the theorem of the fixed part.

Theorem 13.1.10 (Theorem of the fixed part) *Let* S *be a quasi-projective complex manifold*[1] *and* $(\mathbf{H}, h)$ *a complex polarized variation of Hodge structure of weight* w. *Any global flat section of* $\mathbf{H}$ *has flat Hodge components.*

[1] The proof applies more generally to any compactifiable complex base manifold S.

Proof Let us first assume that S is compact and write down the type decomposition of a given flat section $s = \sum_{j\geq t} s_{j,w-j}$. We recall the set-up (13.1) which gives the curvature of the Hodge bundles:

$$F(\mathcal{H}^{p,w-p}) = -(\sigma^p)^* \wedge \sigma^p - \sigma^{p+1} \wedge (\sigma^{p+1})^*.$$

For $p = t$ we have $\sigma^t(s_{t,w-t}) = 0$ and so only the second summand remains, which gives a negative contribution to the curvature:

$$ih(F(\mathcal{H}^{t,w-t})s_{t,w-t}, s_{t,w-t}) \leq 0.$$

We therefore can apply the principle of plurisubharmonicity (Proposition 11.1.6) and hence $s_{t,w-t}$ is flat. Repeating the argument with $s - s_{t,w-t}$ shows that $s_{t+1,w-t-1}$ is flat. In this way, we inductively see that all the Hodge components are flat.

If S is quasi-projective, note that the proof proceeds in exactly the same way, once the principle of plurisubharmonicity can be shown to be valid. Because of Remark 11.1.7 (2), this follows as soon as one shows that the Hodge metrics remain bounded near $\Sigma = T - S$, where T is a good compactification of S, i.e., T is smooth and Σ is a normal crossing divisor. This follows from Schmid (1973, Corollary (6.7′)) stating that sections of the bundle $\mathcal{H}$, invariant under local monodromy around Σ, have bounded Hodge norm near Σ. □

Corollary 13.1.11 (Rigidity theorem) *Suppose that we have two polarized variations of Hodge structure over a quasi-projective complex manifold S and an isomorphism between the underlying local systems that respects the Hodge filtration in one fiber. Then it preserves the Hodge structure in all fibers, i.e., the two variations are isomorphic.*

Proof This follows immediately from the theorem of the fixed part. The given isomorphism between the local systems $\mathbf{H}$ and $\mathbf{H}'$ is a flat section j of $\mathrm{Hom}(\mathbf{H}, \mathbf{H}')$. There is a natural variation of Hodge structure of weight 0 on this local system and j is of type $(0, 0)$ at one point of S. So it must be of type $(0, 0)$ everywhere by the preceding theorem. □

We can now deduce an important special case.

Corollary 13.1.12 *Suppose S is a quasi-projective complex manifold. Let there be given a local system $\mathbf{H}$ underlying a polarized variation of $\mathbf{Q}$-Hodge structure. The local subsystem of invariants of $\mathbf{H}$ under the monodromy group inherits from $\mathbf{H}$ the structure of a variation of Hodge structure.*

Proof Look at the local system $\mathbf{H}_{\mathbf{C}}$ and its invariants $\mathbf{H}_{\mathbf{C}}^{\Gamma}$ under the monodromy group Γ at $o \in S$. Consider the orthogonal projection $p \in \mathbf{E} = \mathrm{End}(\mathbf{H}_{\mathbf{C}})$ of $\mathbf{H}_{\mathbf{C}}$

onto $\mathbf{H}^{\Gamma}_{\mathbf{C}}$. At o this flat section of $\mathbf{E}$ has type $(0,0)$. Hence it is of type $(0,0)$ everywhere. This means that $H^{\Gamma}_{\mathbf{C}}$, the maximal constant local subsystem of $\mathbf{H}_{\mathbf{C}}$, is a variation of a sub-Hodge structure. □

13.2 Logarithmic Higgs Bundles

We wish to extend the discussion of the previous section to the following situation. Let S be a Kähler manifold, $\Sigma \subset S$ a simple normal crossing divisor. Set $S_0 = S - \Sigma$. Suppose that we have a Higgs bundle over S_0. When does it extend to S? Usually we need to allow certain singularities along Σ which leads to the following definition.

Definition 13.2.1 Let $\mathcal{H}$ be a holomorphic vector bundle on S and let ∇ be a connection of $\mathcal{H}|S_0$.
(i) ∇ is said to have *logarithmic poles along* Σ if it extends to a morphism

$$\nabla : \mathcal{H} \longrightarrow \Omega^1_S(\log \Sigma) \otimes_{\mathcal{O}_S} \mathcal{H}$$

which satisfies Leibniz' rule

$$\nabla(fs) = f\nabla(s) + \mathrm{d}f \otimes s,$$
$$f \text{ a local section of } \mathcal{O}_S,\ s \text{ a local section of } \mathcal{H}.$$

(ii) Given local coordinates $(x_1, \ldots, x_n)$ on S such that $\Sigma = \{x_1 \cdots x_k = 0\}$, the *residue of* ∇ along $\{x_j = 0\}$ is the endomorphisms $\frac{1}{2\pi\mathrm{i}} \cdot R \in \operatorname{End} \mathcal{H}$, where R_j appears in the following local expression of the covariant derivative:

$$\nabla(s) = \sum_{j=1}^{k} R_j(s)\frac{\mathrm{d}x_j}{x_j} + \text{holomorphic form, } s \text{ a local holomorphic section of } \mathcal{H}.$$

The simplest situation arises when we have a local system $\mathbf{H}$ of complex vector spaces on a quasi-projective *curve* S_0. Then $\mathbf{H}$ carries the standard flat connection ∇. We shall see that the corresponding holomorphic bundle extends over Σ and the connection extends as a logarithmic connection. This is a local question. So, let us assume that S is the unit disk Δ with coordinate t and that $\Sigma = \{0\}$. Let γ be a loop based around 0 oriented in counterclockwise direction and let T be the induced monodromy action on the typical fiber of $\mathbf{H}$. *We shall assume that the monodromy is unipotent*, say $(T - \mathbf{1})^{k+1} = 0$. Then

$$N \overset{\text{def}}{=} \frac{\log T}{2\pi\mathrm{i}} = \frac{-1}{2\pi\mathrm{i}} \sum_{j=1}^{k} \frac{(\mathbf{1} - T)^j}{j}$$

is well defined. The motivation for the unipotency assumption ultimately comes from geometry: when one starts with a projective family over Δ which is smooth away from the origin, it is well known (Peters and Steenbrink, 2008, Theorem 11.11) that one may replace the fiber at 0 by a normal crossing divisor. In general this has multiplicities, but again, by extracting a suitable root, the resulting family still has a normal crossing divisor over 0, but with multiplicities 1. In this setting the monodromy operator T is also known to be unipotent (Peters and Steenbrink, 2008, Cor. 11.19).

We are working in the abstract setting where the local system underlies a variation of Hodge structure. For a *rational polarized* variation of Hodge structure we shall prove in a later section (Theorem 13.7.3) that T is quasi-unipotent so that some power, say T^k becomes unipotent. This can be achieved by extracting a k-th root.

Consider the universal cover

$$\mathfrak{e} : \mathfrak{h} \to \Delta^0, \quad z \mapsto t = \exp(2\pi \mathrm{i} z).$$

Fix a flat frame $\{e_1, \dots, e_n\}$ for $\mathbf{H}$. The sections e_j are multi-valued, but they pull back to univalent sections of $\mathfrak{e}^*\mathbf{H}$. Let us use the same symbol for those; we have

$$e_j(z+1) = Te_j(z).$$

The twisted sections of the holomorphic bundle $\mathfrak{e}^*\mathbf{H} \otimes \mathcal{O}_{\mathfrak{h}}$ given by

$$\widetilde{e_j}(z) \stackrel{\text{def}}{=} \exp(-2\pi \mathrm{i} z N) e_j(z)$$

satisfy

$$\begin{aligned} \widetilde{e_j}(z+1) &= \exp(-2\pi \mathrm{i}(z+1)N)e_j(z+1) \\ &= \exp(-2\pi \mathrm{i} zN) \circ T^{-1} \circ Te_j(z) \\ &= \widetilde{e_j}(z) \end{aligned}$$

and hence they give holomorphic sections of $\mathbf{H} \otimes \mathcal{O}_{\Delta^0}$ and these by definition extend to a holomorphic frame for the bundle $\mathcal{H} = \mathbf{H} \otimes \mathcal{O}_{\Delta}$.

Then ∇ extends to a connection, also denoted

$$\nabla : \mathcal{H} \to \mathcal{H} \otimes \Omega^1_{\Delta}(\log(0)),$$

which is a logarithmic connection with residue $-N$ since

$$\begin{aligned} \nabla \widetilde{e_j}(z) &= \nabla \left(\exp(-2\pi \mathrm{i} N z) e_j(z) \right) \\ &= \mathrm{d}\exp(-2\pi \mathrm{i} N z) e_j(z) \text{ (by flatness of } e_j) \\ &= -2\pi \mathrm{i} N \circ \exp(-2\pi \mathrm{i} N z) e_j(z) \otimes dz \\ &= -N(\widetilde{e_j}(z)) \otimes \frac{\mathrm{d}t}{t}. \end{aligned}$$

In this case, of course $\nabla \circ \nabla = 0$ since this is already the case on Δ^0. In the general normal crossing situation we have commuting local monodromy operators and the same calculations show that we have also have a canonical extension in this setting.

Suppose next that $\mathbf{H}$ underlies a (complex) polarized variation of Hodge structure. So, $\mathcal{H}|S_0$ carries a Hodge filtration $\mathcal{F}^\bullet$ and Griffiths' transversality relation holds. By Schmid (1973, Theorem 4.13) the Hodge bundles can be shown to extend as subbundles of $\mathcal{H}$ well. It is not so straightforward to show this: it uses Schmid's deep work on estimates on the Hodge norms on these bundles. Also, it uses in an essential way that the local monodromy operators are unipotent and not just quasi-unipotent. The bundle $\mathcal{H}$ as well as the extended Hodge bundles are called the *Deligne extension bundles*.

Next, extend the associated graded bundle we had on S_0 in the obvious way:

$$\mathcal{H}_{\text{Hdg}} = \text{Gr}_{\mathcal{F}} \mathcal{H}.$$

As before, one has a splitting $\nabla = \sigma + \nabla_{\text{Hdg}} + \tau$, where this time the Higgs field has become logarithmic:

$$\sigma \in \text{End}\, \mathcal{H} \otimes \Omega^1_S(\log \Sigma).$$

Obviously, $\sigma \circ \sigma = 0$ since this holds on S_0. The new pair $(\mathcal{H}_{\text{Hdg}}, \sigma)$ is called a *logarithmic (complex) variation of Hodge structure*. This gives the primary example of what is called a logarithmic Higgs bundle.

Definition 13.2.2 (Logarithmic Higgs bundles) A *logarithmic Higgs bundle* (E, σ) is a vector bundle E on S together with $\sigma \in H^0(S, \text{End}(E) \otimes \Omega^1_S(\log \Sigma))$ such that $\sigma \wedge \sigma = 0$ considered as an identity in $H^0(S, \text{End}(E) \otimes \Omega^2_S(\log \Sigma))$. The associated de Rham complex is given by

$$E \xrightarrow{\sigma} E \otimes \Omega^1_S(\log \Sigma) \xrightarrow{\wedge\sigma} E \otimes \Omega^2_S(\log \Sigma) \xrightarrow{\wedge\sigma} \cdots.$$

It should be clear what we mean by a *Higgs subbundle* of a (logarithmic) Higgs bundle. Subvariations $\mathcal{G}$ of a variation of Hodge structure $\mathcal{H}$ give examples: $\mathcal{G}_{\text{Hdg}}$ is a Higgs subbundle of $\mathcal{H}_{\text{Hdg}}$. Conversely, Higgs subbundles of $\mathcal{H}_{\text{Hdg}}$ do not necessarily give subvariations of $\mathcal{H}$. We discuss this below. See Simpson's theorem 13.3.5.

Let us now consider logarithmic Higgs bundles obtained from geometric variations. In Section 6.5 we gave the Katz–Oda description of the Gauss–Manin connection (6.11) on the bundle $\mathcal{H} = \mathbf{R}^w f_* \Omega^\bullet_{X/S}(\log \Sigma)$ in the case of a normal crossing degeneration $f : (X, D) \to (S, \Sigma)$. The trivial filtration (Fig. 3.1) defines the extension of the Hodge filtration $\mathcal{F}^\bullet \subset \mathcal{H}$; the Higgs components are

$$E^{p,q} = \text{Gr}^p_{\mathcal{F}} \mathcal{H}$$

with Higgs field components

$$\sigma^p : E^{p,q} \to E^{p-1,q+1} \otimes \Omega^1_{X/S}(\log \Sigma).$$

We will mainly be interested in the following example.

Example 13.2.3 In the weight 1 case the only nontrivial Higgs field is

$$\sigma^1 : f_* \Omega^1_{X/S}(\log D) \to R^1 f_* \mathcal{O}_X \otimes \Omega^1_{X/S}(\log \Sigma).$$

13.3 Polarized Variations Give Polystable Higgs Bundles

Let $\mathcal{H}$ be a logarithmic polarized variation of Hodge structure over (S, Σ). One of the goals of this section is to show that the associated Higgs bundle $\mathcal{H}_{\mathrm{Hdg}}$ is semistable whenever $\Sigma = \emptyset$ or when S is a curve.

We state a characterization of flat sections of $\mathcal{H}$ which we need later on.

Lemma 13.3.1 *Suppose S is a smooth quasi-projective complex manifold. A holomorphic section s of the bundle $\mathcal{H}$ satisfies $\sigma(s) = 0$ if and only if $\nabla(s) = 0$. In other words, the flat sections are precisely the holomorphic sections of the bundle $\mathcal{H}$ killed by the Higgs field.*

Proof Recall the Bochner formula (Proposition 11.1.5)

$$\mathrm{i}\partial\bar{\partial} h_C(s,s) = \mathrm{i}h_C(\nabla_{\mathrm{Hdg}} s, \nabla_{\mathrm{Hodge}} s) - \mathrm{i}h_C(F_{\nabla_{\mathrm{Hdg}}} s, s). \tag{13.2}$$

Since the first term is positive definite, the formula for the curvature of Hodge bundles (Theorem 13.1.7) implies

$$\mathrm{i}\partial\bar{\partial} h_C(s,s) \geq \underbrace{-\mathrm{i}h_C(\sigma(s), \sigma(s))}_{\leq 0} + \underbrace{-\mathrm{i}h_C(\sigma^*(s), \sigma^*(s))}_{\geq 0}.$$

By Schmid's results (Schmid, 1973, Corollary (6.7′)) the Hodge metric remains bounded near Σ and hence, by Remark 11.1.7 this implies that when $\sigma(s) = 0$, the function $h_C(s,s)$ is plurisubharmonic and therefore constant and the left-hand side of (13.2) vanishes. Since in this case also $\sigma^*(s) = 0$, we get $F_{\nabla_{\mathrm{Hdg}}}(s) = [\sigma, \sigma^*](s) = 0$ and the Bochner formula implies that $\nabla_{\mathrm{Hdg}}(s) = 0$. Since $\nabla = \sigma + \nabla_{\mathrm{Hdg}} + \sigma^*$ we deduce that $\nabla(s) = 0$.

Conversely, a holomorphic section of $\mathcal{H}$ is obviously flat. It induces a holomorphic section on $\mathcal{H}_{\mathrm{Hdg}}$ and then we see that $\sigma(s) = 0$ by type considerations. □

We start the study of stability by proving that the first Chern form for Higgs subbundles is always negative semidefinite. Recall that for a vector bundle E the first Chern form is defined in Appendix C.2 as follows:

$$\gamma_1(E, h) = \frac{\mathrm{i}}{2\pi}(\text{trace of the curvature of the Chern connection}). \qquad (13.3)$$

So we need to estimate the curvature (with respect to the Hodge metric) of a graded logarithmic Higgs subbundle $\mathscr{G} = \bigoplus_p \mathscr{G}^{p,w-p}$, i.e., $\mathscr{G}^{p,w-p} \subset \mathscr{H}^{p,w-p}$ and $\sigma\mathscr{G}^{p,w-p} \subset \mathscr{G}^{p-1,w+p+1} \otimes \Omega^1_S(\log \Sigma)$. Recall that by Theorem 13.1.7 the curvature of the Hodge bundle is just

$$F_{\nabla_{\mathrm{Hdg}}} = -[\sigma, \sigma^*].$$

The curvature estimate we are after reads as follows.

Lemma 13.3.2 *For a graded logarithmic Higgs subbundle $\mathscr{G} \subset \mathscr{H}_{\mathrm{Hdg}}$, we have*

$$\mathrm{i}\,\mathrm{Tr}\left(F_{\nabla_{\mathrm{Hdg}}}|\mathscr{G}\right) \le 0$$

with equality everywhere if and only if the orthogonal complement $\mathscr{G}^\perp$ with respect to h_C is preserved by σ so that we have a direct sum decomposition of Higgs bundles

$$\mathscr{G} \oplus \mathscr{G}^\perp.$$

Proof If we split $\mathscr{H} = \mathscr{G} \oplus \mathscr{G}^\perp$ into a C^∞-orthogonal sum with respect to the Hodge metric and write σ into block-form, the fact that σ preserves $\mathscr{G}$ means that this block-form takes the shape

$$\sigma = \begin{pmatrix} s & t \\ 0 & s' \end{pmatrix} \in \mathscr{A}^{1,0}_S(\log(\Sigma))(\mathrm{End}(\mathscr{H})).$$

Then we have

$$\begin{aligned} [\sigma, \sigma^*] &= \sigma \wedge \sigma^* + \sigma^* \wedge \sigma \\ &= \begin{pmatrix} s \wedge s^* + s^* \wedge s + t \wedge t^* & * \\ * & s' \wedge (s')^* + (s')^* \wedge s' + t^* \wedge t \end{pmatrix} \end{aligned}$$

and so $-\mathrm{i}\,\mathrm{Tr}[\sigma, \sigma^*]|\mathscr{G} = -\mathrm{i}\,\mathrm{Tr}\, t \circ t^*$. One checks that this is a negative definite $(1,1)$-form and so its trace is 0 if and only if the matrix-valued one-form t vanishes, which means that σ preserves $\mathscr{G}^\perp$. □

Recall that the curvature decreases on subbundles:

$$\mathrm{i}F_{\nabla_{\mathrm{Hdg}}}(\mathscr{G}) \le \mathrm{i}F_{\nabla_{\mathrm{Hdg}}}|\mathscr{G}$$

and using (13.3) we conclude the following from Lemma 13.3.2.

Corollary 13.3.3 *The first Chern form of a graded Higgs subbundle $\mathcal{G}$ of $\mathcal{H}_{\mathrm{Hdg}}$ (with respect to the Hodge metric) is negative semidefinite:*

$$\gamma_1(\mathcal{G}) \leq 0,$$

and it is 0 everywhere if and only if $\mathcal{G}^\perp$ is a (holomorphic) graded Higgs subbundle as well.

We next want to explain that the above considerations can be interpreted in terms of stability. Recall that semistablity is defined by means of the notion of *slope* for a vector bundle E on a projective manifold S, say of dimension m with hyperplane class H. It is given by

$$\mu(E) \stackrel{\text{def}}{=} \deg(E)/\mathrm{rank}(E), \quad \deg(E) = c_1(E)\cdot H^{m-1}.$$

Since $\deg(E) = \deg(\bigwedge^r E)$, $r = \mathrm{rank}\, E$ this motivates the definition of degree for a torsion-free coherent sheaf E: one has to replace E by the line bundle which is the double dual of $\bigwedge^r E$. Observe that for this definition to make sense, one does not need a metric on E. Moreover, we can also define slopes for vector bundles on Kähler manifolds: replace H by the class ω of a Kähler metric:

$$\deg(E) = \int_S \gamma_1(E)\wedge\omega^{m-1}.$$

This expression depends only on the cohomology class $c_1(E)$ of $\gamma_1(E)$, but metrics can be used to investigate whether the above expression converges in the logarithmic situation. It is a nontrivial fact that this is indeed the case whenever S is a curve and E is a Higgs bundle coming from a variation of Hodge structure. This follows from Schmid's estimates. See e.g. Peters (1984, Sections 2,3) for details. In our setting, a Higgs bundle V with a harmonic metric h is called *stable*, or *semistable*, respectively, if for any *proper* Higgs *subsheaf* $W \subset V$, i.e., a coherent submodule preserved by the Higgs field, one has an inequality of slopes

$$\mu(W) < \mu(V) \text{ and } \mu(W) \leq \mu(V), \text{ respectively.}$$

A semistable Higgs bundle is called *polystable* if it is the direct sum of stable Higgs subbundles of the same slope. For Higgs bundles coming from a polarized variation of Hodge structure the slope is zero and we have seen that a semistable Higgs subbundle splits off precisely when its slope is zero. So, we have shown the following.

Lemma 13.3.4 *The Higgs bundle defined by a polarized complex variation of Hodge structure over a compact Kähler manifold is polystable.*

There is a converse, due to Corlette (1988) and Simpson (1992), valid for a *compact* Kähler manifold. The full verson of Simpson's result uses the notion of "harmonic metric" (see Definition 13.1.3 (ii)) and is a combination of Corollary 1.3 in Simpson (1990) and the results of Simpson (1992).

Theorem 13.3.5 (i) *Let S be a compact Kähler manifold. There is an equivalence of categories between*

- *direct sums of irreducible complex local systems on S,*
- *Higgs bundles on S equipped with a harmonic metric,*
- *polystable Higgs bundles E on S with* $\int_S c_1(E)\wedge\omega^{n-1} = 0 = \int_S c_2(E)\wedge\omega^{n-2}$.

(ii) *For logarithmic Higgs bundles over quasi-projective curves with unipotent local monodromy around the punctures this remains true; the third condition simplifies to* $\deg(E) = 0$.

We should mention at this point that Simpson (1992) allows nonunipotent local monodromy, but the price to pay is a refined notion of degree, that of "parabolic degree". We do not make use of this refinement here.

The proof of these results requires hard analysis and falls beyond the scope of this book. It yields the following refinement of the above lemma.

Corollary 13.3.6 *The logarithmic Higgs bundle* $\mathscr{H}_{\mathrm{Hdg}}$ *associated with a complex polarized variation of Hodge structure* $\mathscr{H}$ *over a quasi-projective curve with unipotent local monodromy decomposes into a direct sum of degree* 0 *stable Higgs subbundles. A Higgs subbundle of* $\mathscr{H}_{\mathrm{Hdg}}$ *of degree* 0 *is a direct sum of such subbundles. Each of the stable summands comes from a complex subvariation of* $\mathscr{H}$.

We end this section in a more discursive way, by commenting on important work of Simpson concerning Higgs bundles. Recall (Theorem 13.1.6) that Simpson has characterized those Higgs bundles $(\mathscr{H}, \sigma)$ that come from complex variations of Hodge structure. This theorem together with the following theorem of Simpson yields striking restrictions on *Kähler groups*, i.e., those groups that may occur as the fundamental group of a compact Kähler manifold.

Theorem (Simpson 1992, Cor. 4.2.) *Let S be a compact Kähler manifold. Any representation of* $\pi_1(S)$ *can be deformed into a representation which is a fixed point for the action of* $\mathbf{C}^*$ *as given in Theorem 13.1.6.*

Let us outline how this results in restrictions on Kähler groups. First, the last theorem can be seen to imply that the Zariski closure G of the monodromy group (in the group $\mathrm{GL}(d, \mathbf{R})$) must be very special, namely of Hodge type.

Definition 13.3.7 A connected real algebraic group G is said to be of *Hodge type* if the rank of G equals the rank of the maximal compact subgroup of G.

For example, the group $\mathrm{SL}(d, \mathbf{R})$ has rank $(d-1)$ while its maximal compact subgroup $\mathrm{SO}(d, \mathbf{R})$ has rank $\frac{1}{2}d$ for d even and $\frac{1}{2}(d-1)$ for d odd, and so for $d \geq 3$ this group is not of Hodge type.

We claim that as a result of these two theorems of Simpson, a lattice (a discrete subgroup with quotient of finite volume) Γ in $\mathrm{SL}(d, \mathbf{R})$ (for example, $\mathrm{SL}(d, \mathbf{Z})$) is not Kähler.

To establish this claim, recall that a representation $\rho : \pi \longrightarrow \mathrm{GL}(d, \mathbf{C})$ is called rigid if the $\mathrm{GL}(d, \mathbf{C})$-orbit of ρ under conjugation on $\mathrm{Hom}(\pi, \mathrm{GL}(d, \mathbf{C}))$ is open. Thus, this orbit is a connected component, because $\mathrm{GL}(d, \mathbf{C})$ is reductive. By Simpson's last theorem this component would contain a complex variation of Hodge bundles and thus the Zariski closure of the monodromy group would be of Hodge type.

On the other hand, a result of Margulis (1991) implies that the natural representation of the lattice Γ is indeed rigid and thus, if Γ were Kähler, the previous reasoning implies that $\mathrm{SL}(d, \mathbf{R})$ would be of Hodge type. However, we noted above that $\mathrm{SL}(d, \mathbf{R})$ is *not* of Hodge type. This contradiction proves the claim.

Problems

13.3.1 Show that the curvature of the Hodge metric on $\mathcal{H}^{w,0}$ is nonnegative whereas the curvature on $\mathcal{H}^{0,w}$ is nonpositive.

13.3.2 (a) Show that the first Chern form of the "ends" in the Hodge decomposition vanishes if and only if the curvature on the ends vanishes.

(b) Deduce that the corresponding partial period maps are locally constant if and only if the Chern forms of the corresponding bundles are 0.

(c) If S is a compact Kähler manifold, show that the same holds for the degrees of the bundles L in question, where the degree is measured using a Kähler form ω on S:

$$\deg L \stackrel{\text{def}}{=} \int_S \{\text{first Chern form of } L\} \wedge \omega^{m-1}, m = \dim S.$$

13.3.3 The canonical bundle associated with a Hodge flag $\mathcal{F}^0 \supset \mathcal{F}^1 \supset \cdots \supset \mathcal{F}^w$, by definition, is the bundle $\det(\mathcal{F}^0) \otimes \det(\mathcal{F}^1) \otimes \cdots \otimes \det(\mathcal{F}^{w-1}) = \det(\mathcal{H}^0)^{w,0} \otimes \det(\mathcal{H}^{w-1,1})^{w-1} \otimes \cdots \otimes \det(\mathcal{H}^{0,w-1})$. Show that the Hodge metric induces a positive semidefinite curvature form on this line bundle. More precisely, if $\xi \in T_m S$ then the curvature of the Hodge metric is ≥ 0 when evaluated in the ξ-direction, and it vanishes if and only if $\sigma^{p,q}(\xi) = 0$ for $p = 0, \ldots, w-1$. Translate this in terms of the period map and deduce, using the results of Grauert (1962), that the image of the period map of a projective variety is algebraic. Hint: look at the fibers of the period map.

13.4 Curvature Bounds over Curves

About Arakelov's Theorem

The inequalities we consider came up for the first time in the proof of Arakelov's finiteness theorem (see Arakelov, 1971).

Theorem (Arakelov's theorem) *Fix a compact Riemann surface S and a finite set Σ of points on S. Suppose that $S - \Sigma$ is hyperbolic, that is, $2g(S) - 2 + \#\Sigma > 0$. Then there are at most finitely many nonisotrivial[2] families of curves of given genus over S that are smooth over $S - \Sigma$.*

The proof consists of two parts. First, one establishes *rigidity* for a nonisotrivial family. It follows upon identifying the deformation space of the family with the H^1 of the inverse of the relative canonical bundle, which is shown to be ample. Kodaira vanishing then completes this step.

Secondly, one proves that there are only finitely many families (this is a *boundedness* statement) by bounding the degree d of the direct image of the relative canonical bundle $\omega_{X/S} = K_X \otimes f^* K_S^{-1}$ in terms of the genus $g(S)$ of S, the genus $g(F)$ of the fiber, and the cardinality of the set Σ:

$$d \leq \frac{1}{2}(2g(S) - 2 + \#\Sigma) \cdot g(F).$$

It is this bound that we are going to generalize.

We make some supplementary remarks in order to translate the geometry into Hodge theoretic terms. Recall from Section 6.5 the notion of the relative canonical sheaf $\omega_{X/S}$ for a fibration $f : X \to S$ having at most normal crossing degenerations over $\Sigma \subset S$ (see Eq. (6.12)). We observed that the direct image sheaf $f_* \omega_{X/S}$ is locally free and that the fiber of the associated vector bundle at a noncritical point $s \in S$ of f is just $H^0(K_{X_s}) = H^{w,0}(X_s)$, giving the stalk in s of the Hodge bundle $\mathcal{F}^w$ with $w = \dim X_s$. This bundle extends as a subbundle of $R^w f_* \mathbb{C}_X \otimes \mathcal{O}_S$ across Σ: it is just the Deligne extension we considered in

[2] Isotrivial means that the family becomes trivial after pulling back under a finite unramified cover.

Section 13.2. This is because the local monodromy around a normal crossing fiber is unipotent and not just quasi-unipotent.

As a side remark on terminology, note that when the base S is a curve, f is a normal crossing degeneration at each point $t \in \Sigma$, if X_t is a reduced normal crossing divisor. In this case, one also speaks of a *semistable fibration*. It is not too difficult to show that after a suitable finite cover of the base curve branched in at most points of Σ the pulled back family becomes semistable. In the situation of the Arakelov theorem we have a fibration in curves which is not necessarily semistable and although the definition of the relative canonical bundle as $K_X \otimes f^*K_S^{-1}$ makes sense and agrees with its definition in the semistable situation, the point is that the degree of its direct image behaves well under such branched covers of the base.

Summarizing, in the semistable situation, the geometrically defined direct image sheaf $f_*\omega_{X/S}$ is the same as the Hodge theoretically defined Deligne extension of the Hodge filtration bundle $\mathcal{F}^w$, $w = \dim X_s$, and we are from now on concentrating on this Hodge theoretic situation.

The Generalized Arakelov Inequality

We start with a (complex) polarized variation of Hodge structure of weight w, say $(\mathcal{H}, \sigma)$ over a *curve* (S, Σ) of genus $g(S)$. We also assume that the local monodromy around a point from Σ is unipotent. Moreover, we want the Hodge bundle to be effective in the sense that $\mathcal{H}^{p,q} = 0$ whenever $p < 0$ or $q < 0$. So the associated (logarithmic) Higgs bundle has $(w+1)$ components $(\mathcal{H}^{p,q}, \sigma^{p,q})$ with $p + q = w$ and $p = 0, \dots, w$. An Arakelov type inequality by definition gives bounds for the degree of these components. The Higgs field σ relates these components, which gives a global restriction.

Naturally, kernels and images from σ come up. Since these are subsheaves of a locally free sheaf over a curve they are locally free themselves and their degrees are well defined. Consider in particular the kernel $\mathcal{K}^{p,q}$ and image of a component $(\mathcal{H}^{p,q}, \sigma^{p,q})$:

$$0 \to \mathcal{K}^{p,q} \to \mathcal{H}^{p,q} \xrightarrow{\sigma^{p,q}} \widetilde{\mathcal{H}}^{p-1,q+1} \otimes \Omega^1_S(\log \Sigma) \to 0,$$

where the right-hand side is the image of the bundle map $\sigma^{p,q}$ and so $\widetilde{\mathcal{H}}^{p-1,q+1} \subset \mathcal{H}^{p-1,q+1}$. Here, a useful remark is in order: since the pair $(\mathcal{K}^{p,q}, 0)$ is obviously a Higgs subbundle of $(\mathcal{H}, \sigma)$, by Corollary 13.3.3, its degree is nonpositive and since

$$\deg\left(\widetilde{\mathcal{H}}^{p-1,q+1} \otimes \Omega^1_S(\log \Sigma)\right) = \operatorname{rank}(\widetilde{\mathcal{H}}^{p-1,q+1}) \cdot (2g(S) - 2 + \#\Sigma),$$

we find:

$$\deg(\mathcal{H}^{p,q}) \leq \deg(\widetilde{\mathcal{H}}^{p-1,q+1}) + \operatorname{rank}(\widetilde{\mathcal{H}}^{p-1,q+1}) \cdot (2g(S) - 2 + \#\Sigma). \quad (13.4)$$

This inequality is especially important if we apply it to the *Higgs bundle saturation* of any subbundle $\mathcal{G}^{w,0} \subset \mathcal{H}^{w,0}$ of the left-end component of $\mathcal{H}$. By definition, this is the smallest Higgs subbundle $(\mathcal{G}, \sigma)$ of $(\mathcal{H}, \sigma)$ with end-component $\mathcal{G}^{w,0}$. In particular, we have surjective morphisms:

$$\mathcal{G}^{p,q} \xrightarrow{\sigma^{p,q}} \mathcal{G}^{p-1,q+1} \otimes \Omega^1_S(\log \Sigma), \quad p = w, \ldots, 0.$$

So there is a trivial inequality for the ranks of these bundles:

$$\operatorname{rank}(\mathcal{G}^{w,0}) \geq \operatorname{rank}(\mathcal{G}^{w-1,1}) \geq \cdots \geq \operatorname{rank}(\mathcal{G}^{0,w}). \quad (13.5)$$

Secondly, inequality (13.4) can be applied in conjunction with this rank inequality:

$$\begin{aligned}\deg(\mathcal{G}^{w,0}) &\leq \deg(\mathcal{G}^{w-1,1}) + \operatorname{rank}(\mathcal{G}^{w-1,1}) \cdot (2g(S) - 2 + \#\Sigma)\\ &\leq \deg(\mathcal{G}^{w-k,k})\\ &\quad +(\operatorname{rank}(\mathcal{G}^{w-1,1}) + \cdots + \operatorname{rank}(\mathcal{G}^{w-k,k})) \cdot (2g(S) - 2 + \#\Sigma)\\ &\leq \deg(\mathcal{G}^{w-k,k}) + k \cdot \operatorname{rank}(\mathcal{G}^{w-1,1}) \cdot (2g(S) - 2 + \#\Sigma). \quad (13.6)\end{aligned}$$

We can now state and prove the generalization we are after.

Theorem 13.4.1 *Let $(\mathcal{H}, \sigma)$ be the logarithmic Higgs bundle associated with a complex polarized variation of Hodge structure of weight w over a curve (S, Σ) of genus $g(S)$. We have the inequalities*

$$0 \leq \mu(\mathcal{H}^{w,0}) = \frac{\deg \mathcal{H}^{w,0}}{\operatorname{rank} \mathcal{H}^{w,0}} \leq \frac{w}{2} \cdot \operatorname{rank} \mathcal{H}^{w,0} \cdot (2g(S) - 2 + \#\Sigma).$$

Proof The first inequality is easy: $(\mathcal{H}^{w,0}, 0)$ is a quotient Higgs bundle of $(\mathcal{H}, \sigma)$ and hence, by stability (Corollary 13.3.3), has degree ≥ 0.

For the second inequality, consider the Higgs bundle saturation $\mathcal{G}$ for $\mathcal{H}^{w,0}$ itself. By stability $\deg \mathcal{G} = \sum_k \deg(\mathcal{G}^{w-k,k}) \leq 0$. Now sum up the inequalities (13.6) for $k = 0, \ldots, w$:

$$\begin{aligned}(w+1)\deg \mathcal{H}^{w,0} &\leq \sum_k \deg(\mathcal{G}^{w-k,k})\\ &\quad + \frac{w(w+1)}{2} \cdot \operatorname{rank}(\mathcal{G}^{w-1,1}) \cdot (2g(S) - 2 + \#\Sigma)\\ &\leq 0 + \frac{w(w+1)}{2} \cdot \operatorname{rank}(\mathcal{G}^{w-1,1}) \cdot (2g(S) - 2 + \#\Sigma)\\ &\leq \frac{w(w+1)}{2} \cdot \operatorname{rank}(\mathcal{H}^{w,0}) \cdot (2g(S) - 2 + \#\Sigma),\end{aligned}$$

where the last line uses the first rank inequality from Eq. (13.5). This establishes the inequality. □

Example 13.4.2 If $w = 1$ and we have a family of abelian varieties of dimension g, then the inequality gives $2 \leq g(2g(S) - 2 + \#\Sigma)$. For $g = 1$ and $S = \mathbf{P}^1$ this implies that $\#\Sigma \geq 4$. The minimal cases with $\#\Sigma = 4$ were classified by Beauville (1982).

Higgs Bundles Attaining the Arakelov Bound

Simplifying earlier notation, we now let $(\mathcal{H}, \sigma)$ be a Higgs bundle associated with a real polarized variation of Hodge structure with unipotent local monodromy and we let $\mathbf{H}$ be the underlying local system. Recall that by Lemma 13.3.1 the flat sections are precisely the holomorphic sections of $\mathbf{H} \otimes \mathcal{O}_S$ killed by the Higgs field. So, if we have a complex subvariation of Hodge structures on which σ is the zero morphism, then the metric connection on it coincides with the flat connection. Hence the Hodge metric restricts to a flat metric on the local system underlying the complex variation of Hodge structure. This gives an example of a Higgs bundle with a flat harmonic metric. Flatness of h means precisely that the Higgs bundle comes from a representation of $\pi_1(S)$ in the unitary group of h and so we call such a bundle *unitary*. This leads to the following definition.

Definition 13.4.3 A *unitary Higgs bundle* over S is a Higgs bundle (E, h) equipped with a harmonic metric h if there exists a local system $(\mathbf{E}, h)$ of Hermitian vector spaces on S such that $(\mathbf{E}, h) \otimes \mathcal{O}_S = (E, h)$.

Note that the definition of a harmonic metric implies that the Higgs field as well as its adjoint field are both 0. But these conditions do not suffice to give a unitary Higgs bundle since there should be an underlying local system. In our situation, we consider Higgs subbundles $\mathcal{G}$ of $\mathcal{H}$ equipped with the Hodge metric and ask when these are unitary. By Corollary 13.3.6, $\mathcal{G}$ has an underlying local system as soon as the degree of $\mathcal{G}$ vanishes. But in our situation $\deg(\mathcal{G}) = 0$ is a consequence of the conditions $\sigma|\mathcal{G} = \sigma^*|\mathcal{G} = 0$. It follows that any Higgs subbundle $\mathcal{G} \subset \mathcal{H}$ for which $\sigma|\mathcal{G} = 0 = \sigma^*|\mathcal{G}$ is unitary, and conversely. The maximal such subbundle is denoted by $(\mathcal{H}^{\text{un}}, 0)$. By Corollary 13.3.3, one has an h-orthogonal direct sum decomposition

$$(\mathcal{H}, \sigma) = (\mathcal{H}^{\text{un}}, 0) \oplus (\mathcal{H}^{\max}, \sigma). \tag{13.7}$$

By maximality, the subbundle $\mathcal{H}^{\max}$ has no unitary Higgs subbundles.

We want to use this splitting to describe what happens if equality holds in the Arakelov inequality from Theorem 13.4.1., i.e.:

$$\deg \mathcal{H}^{w,0} = \frac{w}{2} \cdot \operatorname{rank} \mathcal{H}^{w,0} \cdot (2g(S) - 2 + \#\Sigma). \tag{13.8}$$

The components of $\mathcal{H}^{\max}$ clearly have the same degrees as the corresponding components from $\mathcal{H}$ itself, but since their ranks are smaller or equal (with equality everywhere if and only if $\mathcal{H}^{\mathrm{un}} = 0$), the slope inequality from Theorem 13.4.1 applied to $\mathcal{H}^{\max}$ shows that the above bound can only be attained if the unitary Higgs subbundle $(\mathcal{H}^{\mathrm{un}}, 0)$ is not present. So, let us assume that this is indeed the case. We show that in this case the Higgs field is strictly maximal in the following sense.

Definition 13.4.4 We say that $(\mathcal{H}, \sigma)$ is *strictly maximal* or has *strictly maximal Higgs field*, if $\mathcal{H}^{w,0} \neq 0$ and the Higgs field components are isomorphisms: $\sigma : \mathcal{H}^{p,q} \xrightarrow{\simeq} \mathcal{H}^{p-1,q+1} \otimes \Omega^1_S(\log \Sigma)$ for all $p \geq 1$. For $w = 1$ this means that we have an isomorphism $\sigma : \mathcal{H}^{1,0} \xrightarrow{\simeq} \mathcal{H}^{0,1} \otimes \Omega^1_S(\log \Sigma)$.

We show in fact the following.

Proposition 13.4.5 *Suppose that* $\mathcal{H}^{\mathrm{un}} = 0$. *Then* $\mathcal{H}$ *attains the Arakelov bound* (13.8) *if and only if the Higgs field is strictly maximal.*

Proof The crucial remark is that equality forces $\deg(\mathcal{G}) = 0$ for the bundle $\mathcal{G}$ in the proof of Theorem 13.4.1, where we recall that $\mathcal{G}$ is the Higgs bundle saturation of the component $\mathcal{H}^{w,0}$ of the Higgs bundle $\mathcal{H}$. The assumptions then force all inequalities in the proof to be equalities and this implies that $\mathcal{G} = \mathcal{H}$. Since, moreover, the inequalities in (13.5) must be equalities, all the Higgs bundle components have the same dimension and since the Higgs field is surjective (by the definition of the saturation) it must be an isomorphism. The converse follows directly by inspecting the proof of Theorem 13.4.1. □

Remark As to the left-hand side inequality in Theorem 13.4.1, we have that $\mu(\mathcal{H}^{w,0}) = 0$ precisely if $\sigma^{w,0} = 0$ and since also $\sigma^* = 0$ on $H^{w,0}$, the Higgs subundle $(\mathcal{H}^{w,0}, 0)$ splits off as a summand which comes from a unitary variation of Hodge structure. By duality, the same holds for the other extreme, $(H^{0,w}, 0)$. Now, if $w = 1$ this shows that the variation of Hodge structure is locally constant with a constant period map. If $w \geq 2$, this needs no longer be the case.

13.5 Geometric Applications of Higgs Bundles

Hyperbolicity

Note that the Legendre family (1.1) over $\mathbf{P}^1$ has three singular fibers and $\mathbf{P}^1 - \{0, 1, \infty\}$ is hyperbolic, i.e., its universal covering is the unit disk.

To place this phenomenon in a general context, we consider the combined Arakelov inequalities from Theorem 13.4.1: together these imply that the Euler number of $S - \Sigma$ is nonnegative, which for $S = \mathbf{P}^1$ means at least two singular fibers, and at least three if $\deg \mathcal{F}^w > 0$. In the geometric situation of a semistable fibration $f : Z \to S$ of w-dimensional varieties, we have $\mathcal{F}^w = f_*\omega_{Z/S}$, as we have seen in the discussion following the statement of the original Arakelov theorem. In this case the condition for at least 3 singular fibers becomes $\deg(f_*\omega_{Z/S}) > 0$. This is true if for instance $p_g(Z) \neq 0$, that is, if the canonical sheaf of Z has nontrivial sections. We want to replace this condition by a weaker condition, using the concept of Kodaira dimension (see Appendix C, Definition C.1.4). The result we are going to discuss is as follows.

Theorem 13.5.1 (Viehweg and Zuo (2001)) *Let Z be a complex projective manifold of Kodaira dimension $\kappa(Z) \geq 0$ and let $f : Z \to \mathbf{P}^1$ be a surjective morphism. Then f has at least three singular fibers.*

Since the Kodaira dimension is governed by powers of the canonical sheaf, it is natural to ask for a generalization of Theorem 13.4.1 in which the relative canonical sheaf $\omega_{Z/S}$ has been replaced by a power.

Theorem 13.5.2 (Viehweg and Zuo, 2001, Möller et al., 2006, Prop. 2.1.) *Assume $f : Z \to S$ is a semistable family of n-folds over a curve S and smooth over $S - \Sigma$. Assume that $\#\Sigma \geq 2$ if $S = \mathbf{P}^1$ (otherwise no condition). Then for all $\nu \geq 1$ with $f_*\omega^{\nu}_{Z/S} \neq 0$ we have*

$$\deg f_*\omega^{\nu}_{Z/S} \leq \operatorname{rank} f_*\omega^{\nu}_{Z/S} \cdot n \cdot \frac{\nu}{2} \cdot (2g(S) - 2 + \#\Sigma).$$

We shall not prove this theorem here. Instead, we derive Theorem 13.5.1 from it. There is one other nontrivial result that we need (see Viehweg and Zuo, 2001, §2) based on results of Fujita (1978) and covering tricks of Viehweg (1982):

Lemma 13.5.3 *The sheaf $f_*\omega^{\nu}_{Z/S}$ has the property that all its line bundle quotients have degree ≥ 0.*

Now we can start with the proof of Theorem 13.5.1.

Proof of Theorem 13.5.1 We explain the proof as given in Viehweg and Zuo (2001, p. 797 ff.).

We argue by contradiction. So, assume that the family has ≤ 2 singular fibers. Hence the topological Euler number of $\mathbf{P}^1 - \Sigma$ is nonnegative: $e(\mathbf{P}^1 - \Sigma) \geq 0$. Now apply Stein factorization (Grauert and Remmert, 1977, p. 213) to the morphism f, say

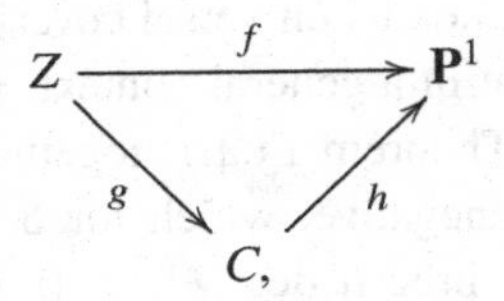

where $h : C \to \mathbf{P}^1$ is finite and g has connected fibers. Then h is finite and unramified outside Σ, hence $e(C - h^{-1}\Sigma) \geq 0$ as well. But this implies that either $C = \mathbf{P}^1$ and $\#\Sigma = 2$ or C is an elliptic curve and $\Sigma = \varnothing$. The latter case is not possible, so we may assume $C = \mathbf{P}^1$, that $\#\Sigma = 2$, and that f has connected fibers. After a finite cover $\mathbf{P}^1 \to \mathbf{P}^1$ branched in the points of Σ we may assume that f is semistable (apply the semistable reduction theorem, see Kempf et al., 1973, p. 53). This procedure does not lower the Kodaira dimension of Z, as one can verify.

Since $\kappa(Z) \geq 0$, for some $\nu \geq 1$ the line bundle $L = \omega_Z^\nu$ has a nonzero section and so has the coherent sheaf f_*L. The coherent sheaf f_*L is of the form $T \oplus M$, with T torsion and M locally free. As a bundle on the projective line it decomposes as a direct sum of line bundles, say $M = Q_1 \oplus \cdots \oplus Q_k$. Now $\omega_{Z/\mathbf{P}^1} = \omega_Z \otimes f^*\omega_{\mathbf{P}^1}^{-1} = \omega_Z \otimes f^*\mathcal{O}_{\mathbf{P}^1}(2)$ and so

$$f_*\omega_{Z/\mathbf{P}^1}^\nu = (T \oplus M) \otimes \mathcal{O}_{\mathbf{P}^1}(2\nu) = T \oplus Q_1(2\nu) \oplus \cdots \oplus Q_k(2\nu).$$

By Lemma 13.5.3 we have $\deg Q_j(2\nu) \geq 0$ for all j, while the nonzero section of M makes appear at least one nontrivial line bundle Q_j with $\deg Q_j \geq 0$, and so $\deg(f_*\omega_{Z/\mathbf{P}^1}^\nu) \geq 1$. The generalized Arakelov inequality from Theorem 13.5.2 then implies $2g(\mathbf{P}^1) - 2 + \#\Sigma \geq 1$, that is, $\#\Sigma \geq 3$, which is a contradiction. □

Rigidity

Let $f : (X, D) \to (S, \Sigma)$ be a family smooth over a quasi-projective curve $S_0 = S - \Sigma$. A *deformation of f keeping source and target fixed* is a morphism $F : (Y, E) \to (S, \Sigma) \times T$ with $T = (T, o)$ a smooth germ of a curve, such that $F|(S, \Sigma) \times \{o\} = f$. It is called *rigid* if every deformation F is induced from f by pullback.

Let $(\mathcal{H}, \sigma)$ be the log-Higgs bundle associated with $(R^w f_* \mathbf{C}_X, b)$ where b is the polarization. Assuming local Torelli holds in this case, the tangent space at f for deformations of f of the above kind is the same as the one for the deformation space of the period map,

$$\mathcal{P}(f) : S_0 \to \Gamma\backslash \{\text{period domain}\}\,.$$

By Peters (2010, Cor. 2.10), the latter is the vector space of those global endomorphisms of $(R^w f_* \mathbf{C}_X, b)$ that have type $(-1, 1)$. Any such endomorphism A of the local system induces an endomorphism of the Higgs bundle. Setting $\mathcal{L} = \Omega^1_S(\log \Sigma)$, we have a commutative diagram

$$\begin{array}{ccccccccc}
 & \mathcal{H}^{w,0} & \xrightarrow[\sigma]{} & \mathcal{H}^{w-1,1} \otimes \mathcal{L} & \xrightarrow[\sigma]{} \cdots \xrightarrow[\sigma]{} & \mathcal{H}^{0,w} \otimes \mathcal{L}^{w} & & \\
 & & \searrow^{A} & & & & \searrow^{A} & \\
\mathcal{H}^{w,0} \otimes \mathcal{L}^{-1} & \xrightarrow[\sigma]{} & & \mathcal{H}^{w-1,1} & \xrightarrow[\sigma]{} \cdots \xrightarrow[\sigma]{} & \mathcal{H}^{0,w} \otimes \mathcal{L}^{w-1} & \longrightarrow & 0.
\end{array}$$

The composition $\sigma \circ \cdots \circ \sigma : \mathcal{H}^{w,0} \to \mathcal{H}^{0,w} \otimes (\Omega^1_S(\log \Sigma))^{\otimes w}$ is called the *Yukawa coupling.*

Theorem 13.5.4 *In the above situation, assume the local monodromy is unipotent and that the local Torelli property holds for the variation of Hodge structure associated with $(R^w f_* \mathbf{C}_X, b)$.*

Assuming moreover that the Yukawa coupling is an isomorphism, [3] *one has rigidity of f. This holds in particular, if* all *components of the Higgs field are isomorphisms, i.e., if the Higgs field is strictly maximal.*

Proof We claim that the assumptions imply that the map A in the preceding diagram is zero, i.e., there are no deformations for the period map and hence, by local Torelli, for the map f itself as well.

To prove the claim, first observe that, since A is an endomorphism of the Higgs bundle, its kernel $\operatorname{Ker} A$ is a Higgs subbundle. Moreover, since A is an endomorphism of the underlying local system, $\operatorname{Ker} A$ comes from a local subsystem of $R^w f_* \mathbf{C}_X$. By Corollary 13.3.6 it defines a complex polarized subvariation of $\mathcal{H}$ and there is a direct sum splitting $\mathcal{H} = \operatorname{Ker}(A) \oplus \mathcal{H}_1$. The endomorphism A restricts to an endomorphism of $\mathcal{H}_1$ and has no kernel on $\mathcal{H}_1$.

The bundle $\operatorname{Ker} A$ contains $\mathcal{H}^{0,w}$ and so, since the Yukawa coupling is an isomorphism, it also contains $\mathcal{H}^{w,0}$. Hence $\mathcal{H}_1$ is possibly nonzero only in degrees $(w-1, 1)$ up to $(1, w-1)$. Since $A = 0$ on $\mathcal{H}_1^{1,w-1}$ and A has no kernel on $\mathcal{H}_1$ we must have $\mathcal{H}_1^{1,w-1} = 0$. Then we see inductively that all components of $\mathcal{H}_1$ must be zero and so $\mathcal{H} = \operatorname{Ker} A$ as claimed. □

Example 13.5.5 By Peters (2010, Cor. 3.5) or Viehweg and Zuo (2005) for a

[3] For $h^{w,0} = 1$, this is equivalent to the nonvanishing of the Yukawa coupling.

family of Calabi–Yau n-folds degenerating at some point with maximally unipotent monodromy (that is, the Jordan matrix of the local monodromy operator consists of one block of maximal size), the Yukawa coupling is an isomorphism. Since for a Calabi–Yau manifold local Torelli holds (see Examples 5.6.2(ii)), such a family is rigid.

13.6 Curvature of Period Domains

We start by carefully constructing a G-invariant metric on D. In fact, we construct a Hermitian metric on the holomorphic tangent bundle T_D^{hol} by giving a G-invariant metric on $[\mathfrak{m}^-]$. To do this, we recall various decompositions of the Lie algebra $\mathfrak{g}$.

We have an $\text{Ad}(V)$-invariant splitting

$$\mathfrak{g} = \mathfrak{v} \oplus \mathfrak{m}.$$

The algebra $\mathfrak{g}$ has a weight 0 Hodge structure with Hodge decomposition

$$\mathfrak{g}_\mathbb{C} = \bigoplus_p \mathfrak{g}^{p,-p}, \quad \mathfrak{g}^{p,-p} = \left\{\xi \in \mathfrak{g} \mid \xi H^{r,s} \subset H^{r+p,s-p}\right\}.$$

Recalling that $\mathfrak{v}_\mathbb{C} = \mathfrak{g}^{0,0}$ and that $\mathfrak{m}_\mathbb{C}$ splits into the two pieces

$$\mathfrak{m}^- \overset{\text{def}}{=} \bigoplus\nolimits_{p<0} \mathfrak{g}^{p,-p},$$
$$\mathfrak{m}^+ \overset{\text{def}}{=} \bigoplus\nolimits_{p>0} \mathfrak{g}^{p,-p},$$

we thus have

$$\mathfrak{v}_\mathbb{C} = \mathfrak{g}^{0,0},$$
$$\mathfrak{m}_\mathbb{C} = \mathfrak{m}^+ \oplus \mathfrak{m}^-.$$

We have $T_D^{\text{hol}} = [\mathfrak{m}^-] = G \times_V \mathfrak{m}^-$ (as V-bundles) and so, to give a G-homogeneous Hermitian metric on D, we need an $\text{Ad}(V)$-invariant metric on $\mathfrak{m}^-$. But we have seen (Proposition 12.6.5) that

$$(\xi, \eta) \overset{\text{def}}{=} -B_0(\text{Ad}\, C(\xi), \bar{\eta}) \tag{13.9}$$

defines a V-invariant Hermitian metric on $\mathfrak{g}_\mathbb{C}$, where B_0 stands for the trace form.

Next, we want to compute the holomorphic sectional curvature of the holomorphic tangent bundle of the period domain in certain directions. First we need the relevant definition.

Definition 13.6.1 Let M be a complex manifold with a Hermitian metric h

on the holomorphic tangent bundle. Let F_h be the curvature $(1,1)$-form of the corresponding Chern connection. For any two $(1,0)$-vector fields ξ, η we put

$$K(\xi,\eta) \stackrel{\text{def}}{=} \frac{h(F_h(\xi,\overline{\eta})\eta,\xi)}{h(\xi)h(\eta)} \quad \text{(the } \textit{holomorphic bisectional curvature} \text{ in the directions } \xi,\eta\text{)},$$

$$K(\xi) = K(\xi,\xi) \quad \text{(the } \textit{holomorphic sectional curvature} \text{ in the direction } \xi\text{)}.$$

Example 13.6.2 (Fundamental example) Suppose $\dim M = 1$. Then $F_h = \overline{\partial}\partial\log(h)$, where $\omega = \frac{\mathrm{i}}{2}h\, dz \wedge d\overline{z}$ is the metric form. We easily find that $K(\partial/\partial z)$ is the Gaussian curvature $K_h = -h^{-1}\cdot \partial^2/\partial z\partial\overline{z}(\log h)$. We can also write this result as follows:

$$-\mathrm{i}F_h = -\text{ Gaussian curvature of } h\cdot\omega.$$

Let us next give special cases of this fundamental example.

(i) Let Δ be the unit disk, with its Poincaré metric

$$h = \frac{1}{(1-|z|^2)^2}\,\mathrm{d}z\otimes\mathrm{d}\overline{z}.$$

An easy computation shows that

$$K(\partial/\partial z) \equiv -1.$$

(ii) The upper half-plane $\mathfrak{h} = \{z\in\mathbf{C} \mid \operatorname{Im} z > 0\}$, with metric

$$h = \frac{1}{|\operatorname{Im} z|^2}\,\mathrm{d}z\otimes\mathrm{d}\overline{z}$$

also has constant Gaussian curvature -1.

(iii) $\Delta^* = \{z\in\mathbf{C} \mid |z|<1,\ z\neq 0\}$ which has $\mathfrak{h}$ as a universal cover. The Poincaré metric on $\mathfrak{h}$ is invariant by translation. So we obtain a metric on Δ^* given by

$$\frac{1}{|\xi|^2(\log|\xi|^2)^2}\,\mathrm{d}\xi\otimes\mathrm{d}\overline{\xi}$$

having constant Gaussian curvature -1.

The fundamental result about the holomorphic sectional curvature for period domains is given by the following theorem.

Theorem 13.6.3 *The holomorphic sectional curvature of a period domain is negative and uniformly bounded away from 0 in nonzero horizontal directions ξ i.e., for some positive constant γ we have*

$$K(\xi) < -\gamma, \quad \xi\neq 0.$$

Proof The Hodge metric on $\mathfrak{m}^- = T_o^{\mathrm{hol}} D$ transports to a G-invariant metric on $T^{\mathrm{hol}} D = [\mathfrak{m}^-]$. As observed before (Lemma-Definition 12.2.3), the horizontal tangent bundle $[\mathfrak{p}^-] = T^{\mathrm{hor}} D$, $\mathfrak{p}^- = \mathfrak{m}^- \cap \mathfrak{p}_{\mathbb{C}}$ is homogeneous under G. To make curvature calculations on this bundle it suffices therefore to do this at o. We note that we may of course also assume that $|\xi| = 1$ and we need to show

$$K(\xi) < 0.$$

The curvature of the canonical connection is given by formula (12.3) in Chapter 12,

$$\Omega = -\frac{1}{2}[\sigma, \sigma]^{\mathfrak{v}},$$

where $\sigma = \omega^{\mathfrak{m}}$ is the $\mathfrak{m}$-part of the Maurer–Cartan form on $\mathfrak{g}$.

The associated bundle T_D^{hol} is homogeneous under the *adjoint action* and we may use Proposition 11.4.5 from Chapter 11. It means that we should consider the above form σ as the canonical $\mathfrak{m}_{\mathbb{C}}$-valued one-form sending each $\xi \in \mathfrak{m}^+$ to itself. Because by Theorem 12.3.3 the canonical connection induces the Chern connection on any associated holomorphic bundle, this is also the curvature of T_D^{hol} equipped with the metric h. Hence, for $\xi, \eta \in \mathfrak{m}^-$, we have

$$F_h(\xi, \overline{\eta}) = -\frac{1}{2} \operatorname{ad} [\xi, \overline{\eta}]^{\mathfrak{v}} \in \operatorname{End}(\mathfrak{m}^-).$$

The left-hand side is possibly nonzero only if we get a component of $[\xi, \overline{\eta}]$ in $\mathfrak{g}^{0,0} = \mathfrak{v}_{\mathbb{C}}$. This only happens if ξ and η belong to the same component $\mathfrak{g}^{-p,p}$. If $\xi, \eta \in \mathfrak{p}^-$ then p is odd, and so, by Corollary 12.6.3 one has $\bar{\xi} = \xi^*$ and similarly for η. In proving the theorem, we may assume that $\xi, \eta \in \mathfrak{g}^{-p,p}$ and by what we just said, in that case

$$F_h(\xi, \eta) = -\frac{1}{2} \operatorname{ad} [\xi, \eta^*].$$

Now we get (recall $|\xi| = 1$)

$$K(\xi) = -\frac{1}{2} h(\operatorname{ad} [\xi, \xi^*]\xi, \xi).$$

Using (13.9) we see that $h(\xi, \eta) = \operatorname{Tr}(\xi \circ \eta^*)$ (on $\mathfrak{p}_{\mathbb{C}}$ the operator C is multiplication with -1, and complex conjugation is the Hermitian conjugate), we find (note the adjoint actions)

$$h([[\xi, \xi^*], \xi], \xi) = \operatorname{Tr}([[\xi, \xi^*], \xi] \circ \xi^*) = \operatorname{Tr}([\xi, \xi^*] \circ [\xi, \xi^*]) = \|[\xi, \xi^*]\|_h^2.$$

The last equation comes from the fact that if $\eta = [\xi, \xi^*]$ we have $\eta^* = \eta$. For the holomorphic sectional curvature we thus find

$$K(\xi) = -\frac{1}{2} \|[\xi, \xi^*]\|_h^2 \leq 0.$$

The implication $[\xi, \xi^*] = 0 \Longrightarrow \xi = 0$ is again Corollary 12.6.3. □

Remark 13.6.4 Similar calculations can be used to show that the canonical bundle K_D is positive in horizontal directions and negative in vertical directions. See Problem 13.6.4.

Problems

13.6.1 (a) Show that the length of the circle $|z| = r$ in the disk Δ^* computed in the Poincaré metric equals $\frac{2\pi}{\log(1/r)}$. In particular, show that this length tends to 0 when r does.

(b) Show that $d_{\mathfrak{h}}(\mathrm{i}y, 1 + \mathrm{i}y) = \frac{1}{y}$.

13.6.2 (a) Show that the holomorphic sectional curvature on $\mathfrak{h}_g$ computed with respect to the trace form satisfies $-1 \leq K(\xi) \leq -\frac{1}{g}$.

(b) Show that the holomorphic sectional curvature on the homogeneous space $\mathrm{SO}_0(2a, b)/\mathrm{U}(a) \times \mathrm{SO}(b)$ computed with respect to the trace form satisfies $-1 \leq K(\xi) \leq -\frac{1}{a}$.

13.6.3 Show that the holomorphic sectional curvature in the direction of vectors in $\mathfrak{k}_{\mathbb{C}}$ is nonnegative.

13.6.4 Consider for $v \in \mathfrak{g}^{-p,p}$ and $w \in \mathfrak{g}^{-q,q}$ the adjoint action of $[v, w]$ on $\mathfrak{g}_{\mathbb{C}}$. Show that the trace vanishes for $p \neq -q$. Show that the trace of $[v, v^*]$ is positive, for instance by using a basis for $\mathfrak{g}_{\mathbb{C}}$ induced by a Hodge frame as in Problem 12.5.3 in Chapter 12. Deduce that the curvature of the canonical bundle K_D is strictly positive in nonzero directions $v \in \mathfrak{p}_{\mathbb{C}}$ and strictly negative in directions $v \in \mathfrak{k}_{\mathbb{C}}$. Hint: use Corollary 12.6.3 in Chapter 12.

13.7 Applications

Recall from the previous section that the Poincaré metric on the unit disk Δ has constant Gaussian curvature (= holomorphic sectional curvature) -1. We let ω_Δ be its associated $(1, 1)$-form. Also, for any metric h with metric form ω_M on a one-dimensional manifold M, the associated Ricci-form is given by $\mathrm{Ric}\,\omega_M = \frac{1}{2}\mathrm{i}\partial\bar{\partial}\log h = -K\omega_M$, where K is the Gaussian curvature.

Lemma 13.7.1 (Schwarz's lemma) *Let $f : \Delta \to M$ be a holomorphic map of the unit disk to a complex manifold equipped with a Hermitian metric ω_M.*

Assume that $f(\Delta)$ is tangent to directions for which the holomorphic sectional curvature is ≤ -1. Then

$$f^*\omega_M \leq \omega_\Delta,$$

i.e., the map f is distance-decreasing.

Proof The assumption of the lemma means that the holomorphic sectional curvature of the metric connection on the tangent bundle in the direction of $f(\Delta)$ is bounded above by -1. By Lemma 11.2.1 we know that curvature decreases on subbundles, so we find that the holomorphic sectional curvature of the tangent bundle of the image $f(\Delta)$ computed with the metric inherited from T_M is bounded above by -1. This is equivalent to the inequality $\text{Ric}\, f^*\omega_M \geq f^*\omega_M$, which is the condition in the following lemma whose proof completes the proof of Schwarz's lemma. It is called Ahlfors' lemma. □

Lemma (Ahlfors' lemma) *Let $f : \Delta_R \to M$ be a holomorphic map of the disk of radius R to a complex manifold equipped with a Hermitian metric ω_M. Suppose that* $\text{Ric}\, f^*\omega_M \geq f^*\omega_M$. *Write*

$$\eta_R = \frac{\mathrm{i} \cdot R^2\, \mathrm{d}z \wedge \mathrm{d}\bar{z}}{(R^2 - |z|^2)^2}$$

for the Poincaré volume form on the disk of radius R. Then $f^*\omega_M \leq \eta_R$.

Proof Put

$$\Psi \stackrel{\text{def}}{=} f^*\omega_M = u\eta_r.$$

Because Ψ remains bounded on any smaller disk, while η_r goes toward infinity when we approach the boundary of the disk with radius r, the function u remains bounded on this disk and so has an interior maximum, say at z_0. At this point we then have

$$0 \geq \mathrm{i}\partial\bar{\partial} \log u = \text{Ric}\, \Psi - \text{Ric}\, \eta_r.$$

The Gaussian curvature of η_r is -1 and so $\text{Ric}\, \eta_r = \eta_r$. Likewise, because the Gaussian curvature of Ψ is bounded above by -1 we have $\Psi \leq \text{Ric}\, \Psi$. Combining these estimates we get

$$\Psi \leq \text{Ric}\, \Psi \leq \text{Ric}\, \eta_r = \eta_r \text{ at } z_0$$

and so $u(z_0) \leq 1$. But u assumes its maximum value at z_0 and so $\Psi \leq \eta_r$. It follows that $f^*\omega_M \leq \eta_r$ for all $r < R$. Now take the limit when $r \to R$. □

We give an application of the preceding distance-decreasing property to period maps over the punctured unit disk Δ^*. Suppose we have a variation of

Hodge structure over it. It gives a period map $p : \Delta^* \to \Gamma\backslash D$, where D is the corresponding period domain $D = G/V$ and Γ is the cyclic group generated by the monodromy operator γ, the image of the generator of $\pi_1(\Delta^*)$ in $G_{\mathbf{Z}}$. The universal cover of Δ^* is the upper half-plane $\mathfrak{h}$ mapping to Δ^* via $\zeta = q(\tau) = e^{2\pi i \tau}$. We have a commutative diagram

$$\begin{array}{ccc} \mathfrak{h} & \xrightarrow{\tilde{p}} & D = G/V \\ {\scriptstyle q}\downarrow & & \downarrow \\ \Delta^* & \xrightarrow{p} & \Gamma\backslash D \end{array}$$

and

$$\tilde{p}(\tau + 1) = \gamma\circ\tilde{p}(\tau).$$

Corollary 13.7.2 *The lifted period map $\tilde{p}$ is distance-decreasing; i.e., if d_D is a (suitably normalized) G-invariant metric on D and $d_{\mathfrak{h}}$ is the Poincaré metric on $\mathfrak{h}$, we have*

$$d_D(\tilde{p}(x), \tilde{p}(y)) \leq d_{\mathfrak{h}}(x, y).$$

Indeed, any locally liftable holomorphic map $p : S \to \Gamma\backslash D$ whose local lifts $\tilde{p}$ are horizontal, in the wider sense that $d\tilde{p}_s : T_s S \to T^{\mathrm{hor}}_{\tilde{p}(s)} D$, is distance-decreasing.

Here we recall that a map $p : S \to \Gamma\backslash D$ is a locally liftable holomorphic map if every point in S has a neighborhood U on which p lifts to a holomorphic map $\tilde{p} : U \to D$.

We use Corollary 13.7.2 to prove a few important results.

Application *Suppose that we have a variation of Hodge structure over $S = \mathbf{C}$. Its period map $p : \mathbf{C} \to D$ then is constant.*

Proof By the Ahlfors lemma, the period map restricted to the disk of radius R satisfies $p^*\omega_D \leq C \cdot \eta_R$ so that, setting $p^*\omega_D = J_p \cdot$ (Euclidean volume form) we get the estimate

$$J_p(0) \leq C \cdot \frac{1}{R^2},$$

and so this tends to 0 when R goes to infinity. This is only possible if p is constant. □

A second, important application is the following.

Theorem 13.7.3 (Monodromy theorem) *The monodromy operator γ is quasi-unipotent, i.e., for some integers m, N we have*

$$(\gamma^m - 1)^N = 0.$$

Proof Because the distance between in and $in + 1$ is exactly $1/n$ we find from the preceding corollary $d_D(\tilde{p}(in), \gamma \circ \tilde{p}(in)) \leq 1/n$. Now write $\tilde{p}(in) = g_n V$. Because the metric on D is G-invariant we have

$$d_D(V, g_n^{-1} \gamma g_n V) = d_D(g_n V, \gamma g_n V) \leq 1/n$$

and so $\lim_{n \to \infty} g_n^{-1} \gamma g_n \in V$. So the conjugacy class of γ has a point of accumulation in the compact subgroup V and so the eigenvalues of γ must have absolute value 1. Because $\gamma \in G_{\mathbb{Z}}$, the eigenvalues are algebraic integers all of whose conjugates occur as eigenvalues as well. A theorem of Kronecker (1857) then implies that any eigenvalue of γ is a root of unity. □

Remark 13.7.4 In the preceding theorem we can take $N = w + 1$, where w is the weight of the Hodge structure. This follows from the SL_2-orbit theorem which is outside the scope of this book. See Schmid (1973) for a proof.

The third application concerns the extension properties of the period map. We continue with the preceding set-up of period maps with source the punctured unit disk Δ^* and γ the monodromy operator generating the cyclic group Γ. We shall see that there is a crucial difference whether or not the order of γ is finite.

Theorem 13.7.5 *Let*

$$p : \Delta^* \to \Gamma \backslash D, \quad \Gamma = \langle \gamma \rangle$$

be a period map.
(i) *If the monodromy operator has finite order, then p extends holomorphically over the origin.*
(ii) *If the monodromy operator has infinite order, then for any sequence of points converging to $0 \in \Delta$, the images in $\Gamma \backslash D$ do not converge.*

Proof (i) The proof is done in three steps. First, by taking a suitable n-th root, we may assume that the monodromy is trivial and we are going to show that any distance-decreasing map of the punctured disk into D extends continuously. The second step is to replace D by a *compact* manifold M. This is a standard technique: by Borel and Harish-Chandra (1962) there exist discrete subgroups $\Gamma \subset G$ that act freely on $D = G/V$, implying that the quotient $M = \Gamma \backslash D$ is smooth and, moreover, is such that M is compact. The G-invariant metric on D induces a metric on M such that the projection $D \to M$ preserves distances. So we may indeed assume that $D = M$. The last step uses the argument principle.

See Problems 13.7.1–13.7.3 for hints for the proof.
(ii) Suppose that there would be a converging sequence $x_n \to 0$ such that $y_n = p(x_n)$ converges to a point $y \in \Gamma\backslash D$. Let $\tilde{y} \in D$ be a lift of y. The period map lifts to a holomorphic map $\tilde{p} : \mathfrak{h} \to D$ and the monodromy operator is reflected on the side of the source by the translation $z \mapsto z + 1$. The fact that the monodromy operator has infinite order means that we can assume that a suitable neighborhood V of $\tilde{y}$ is mapped to an open set $\gamma(V) \subset D$ which is disjoint from V. The distance $d_{\mathfrak{h}}(x+\mathrm{i}y, x+1+\mathrm{i}y)$ equals y^{-1} and so goes to 0 if y goes to ∞. In particular, as a consequence of the distance-decreasing property, this is also true for the distance between their images under $\tilde{p}$. So, if n is large enough, the distance between $\tilde{p}(x_n)$ and $\gamma\tilde{p}(x_n) = \tilde{p}(x_n+1)$ is as small as we want. In particular, $\gamma\tilde{p}(x_n)$ must then remain within V, contradicting the fact that $V \cap \gamma V = \emptyset$. □

We now give an application of the last theorem to a global situation. So we let S be a smooth quasi-projective variety. Moreover, we assume that $S \subset \hat{S}$, where $\hat{S}$ is a projective manifold such that $\hat{S} - S$ consists of a divisor T with normal crossings. This means that locally around each point y on T, holomorphic coordinates $(z_1, \ldots, z_n)$ can be found on $\hat{S}$ so that T is given by the equation $z_1 \ldots z_k = 0$. In this coordinate patch, S is isomorphic to $\Delta^{*k} \times \Delta^{n-k}$ and we have k local monodromy operators γ_i, $i = 1, \ldots, k$ corresponding to loops around the punctures in each of the factors. We may extend the period map to those factors Δ^* for which the local monodromy operator is finite, thereby obtaining an extended period map

$$p' : S' \to \Gamma\backslash D,$$

where S' is the quasi-projective variety obtained by filling in the punctures corresponding to finite local monodromy.

Corollary 13.7.6 *The map p' is proper. The image $p'(S')$ is a closed analytic subvariety of $\Gamma\backslash D$ containing $p(S)$ as the complement of an analytic set.*

Proof It suffices to prove the first statement, because the others follow from the proper mapping theorem (see e.g. Gunning and Rossi, 1965). Assume that p' is not proper. Then there is a divergent sequence $x_n \in S'$ whose images converge in $\Gamma\backslash D$. We may assume that the x_n converges to a point $x \in \hat{S}$ for which at least one local monodromy operator is of infinite order. Restricting the period map to a suitable neighborhood of x we get a period map $p : \Delta^{*k} \times \Delta^{n-k} \to \Gamma\backslash D$, $k \geq 1$. By a slight generalization of the previous theorem (see Problems 13.7.1–13.7.3), we see that $p(x_n)$ diverges, a contradiction. □

Problems

13.7.1 The aim of this problem is to prove that a distance-decreasing map $f : \Delta^* \to M$ with target a compact complex Riemannian manifold M extends continuously over the origin. The original argument is due to Mrs. Kwack (Griffiths, 1970, pp. 163–164). Assume for simplicity first that $\dim M = 1$. Because M is compact, for any sequence $x_n \in \Delta^*$ converging to 0, we may assume that $y_n = f(x_n)$ converges to a certain point $y \in M$. We take a holomorphic coordinate w around y valid in an open neighborhood V of y and we assume that f does not extend continuously. So for some ball B of radius r in V there is a sequence of points x_n^* converging to 0, say with $|x_{n+1}| < |x_n^*| < |x_n|$ such that $|f(x_n^*)| \geq 2r$. In particular, there is a maximal annulus A_n containing the circle through x_n whose image is contained in B. In particular, all these annuli are disjoint and there are points a_n, b_n on the inner and outer boundaries, respectively, whose images are on the boundary ∂B.

(a) Show that passing to subsequences we may assume that $f(a_n) \to a$, $f(b_n) \to b$.

(b) Show that for n sufficiently large the integral of $f(z)/(f(z)-f(x_n))$ over the inner and outer boundaries of A_n must be 0, contradicting the fact that the difference is the (nonzero) integral of $f(z)/(f(z) - f(z_n))$ over the (oriented) boundary of A_n.

(c) Adapt the argument to the case of $\dim M \geq 2$.

13.7.2 Let $\Delta^{*k} \times \Delta^{n-k} \to \Gamma\backslash D$, $k \geq 1$ be a period map and let $x_i = (x_i^1, \ldots, x_i^n)$ be a sequence of points in the source for which $\min(|x_i^1|, \ldots, |x_i^k|)$ tends to 0 when i goes to ∞. Assume that the local monodromy around each puncture is of infinite order. Show that $p(x_i)$ diverges.

13.7.3 (a) Consider a family of projective varieties over $\mathbf{P}^1$ and assume that all the singular fibers are over $x_1, \ldots, x_s \in \mathbf{P}^1$. Consider the complement U and the period map $p : U \to \Gamma\backslash D$ associated with the primitive cohomology (of some rank) of the fibers. Assume that p is nonconstant. Show that $s \geq 3$.

(b) Specifying to the case of curves, show that the hypothesis that the period map be nonconstant means that the smooth part of the family has varying moduli. Equivalently, show that the period map is constant if and only if after a finite base change the family becomes birationally trivial.

(c) Show that the first half of the preceding argument also applies to families of abelian varieties, families of $K3$ surfaces, and families of hypersurfaces in projective space.

Bibliographical and Historical Remarks

We comment on the various sections.

Section 13.1: Higgs bundles on curves were first studied by Hitchin (1987) and play a central role in the work of Simpson (1992, 1994, 1995). The first proofs of the rigidity theorem and the theorem of the fixed part are due to Griffiths (1970). Here we presented the version from Schmid (1973).

Sections 13.2–13.3: The technique of logarithmic extensions is due to Deligne (1970a). Higgs and logarithmic Higgs bundles were first introduced in Simpson (1990, 1992). We have given an elementary proof of the "easy" half of Simpson's stabillity results under the simplifying assumption that the Higgs bundle comes from a variation of Hodge structure.

Section 13.4: In analogy to the case of number fields, Shafarevich conjectured finiteness for non-isotrivial families of curves over a fixed base curve with fixed degeneracy locus. This was proved by Arakelov and Parshin (see Arakelov, 1971). Faltings (1983) obtains the same bounds for weight one variations of Hodge structures over a curve. For higher weight see Peters (2000); Jost and Zuo (2002). For families of higher-dimensional manifolds, see Bedulev and Viehweg (2000); Viehweg and Zuo (2006), and Oguiso and Viehweg (2001).

There are excellent surveys about more recently obtained Arakelov-type (in)equalities, see Viehweg (2008) and Möller et al. (2006). For Arakelov inequalities over surfaces and higher-dimensional varieties see Viehweg and Zuo (2007), and Viehweg and Zuo (2008). Families of $K3$ surfaces have been treated in Sun et al. (2003).

Section 13.5: For further results on hyperbolicity, see Viehweg and Zuo (2003b). For sharper bounds on the number of singular fibers in a family of semistable varieties over $\mathbf{P}^1$ see also the recent preprint Lu et al. (2016).

(Non)rigidity results date back to Faltings (1983) where it is shown that in order for a family of Abelian varieties over curves to be rigid it is generally not enough that the family is not isotrivial. Subsequently, the rigidity statement was generalized in Peters (1990). Using this result Saito and Zucker (1991), respectively Saito (1993), were able to treat completely the case of K3-surfaces, respectively abelian varieties. For an interesting but technically involved application of rigidity see Viehweg and Zuo (2003a).

In relation to Simpson's work, we should note that the boundedness result Simpson (1994, 13.3.4) for slope zero immediately implies that the Chern numbers of Hodge bundles underlying a complex variation of given type over a compact projective variety can only assume finitely many values. It follows that there are *a priori* bounds on these Chern numbers. If the base is a (not necessarily compact) curve, Simpson (1990) shows that the same methods give boundedness for the canonical extensions of the Hodge bundles. Explicit bounds were not given, however. Conversely, knowing that bounds on the degrees of the Hodge bundles exist can be used to simplify the rather technical proof for boundedness; see Peters (2000, Section 4) for details (see also Liu et al., 2011).

Eyssidieux (1997) treats a generalization of the true Arakelov inequality (weight 1 Hodge structures on a compact curve) to higher-dimensional base manifolds. The inequalities concern variations of Hodge structure over a compact Kähler manifold M such that the period map is generically finite onto its image.

Section 13.6: These bounds were given for the first time by Griffiths and Schmid (1969). A much more abstract version of the proof can be found in the Bourbaki report of Deligne (1970b). Here we present a fleshed-out version of Deligne's approach.

Section 13.7: The monodromy theorem in its geometric form is due to Landman (1973). The present proof of the abstract version is due to Borel. It can be found in Schmid (1973). The more refined version on the exponent of nilpotency in the abstract situation can be found there as well. In fact, Landman's geometric approach also gives such a bound.

14

Harmonic Maps and Hodge Theory

As we have seen since the very beginning of this book, much can be gained by studying a distinguished representative of a cohomology class in place of the class itself. The same advantages accrue to the study of homotopy classes of maps. One seeks a distinguished representative which minimizes an appropriate functional (the energy), as a result of which it satisfies a partial differential equation (the harmonic equation). In many cases the Kähler condition forces harmonic maps to satisfy additional equations, and from these one can often deduce useful geometric results. For example, we show that no compact quotient of a period domain of the form

$$D = \mathrm{SO}(2p, q)/\mathrm{U}(p) \times \mathrm{SO}(q),$$

with $p > 1$ and $q \neq 2$, is homotopy equivalent to a compact Kähler manifold (Carlson and Toledo, 1989b). Compare this with the fact that the Hopf manifold is not homotopy-Kähler, a result established via classical Hodge theory (see Examples 2.2.1). See also the historical notes at the end of the chapter.

14.1 The Eells–Sampson Theory

The theory of harmonic maps began with the paper by Eells and Sampson (1964). Let us review the results relevant to us here. To this end, let $f : M \longrightarrow N$ be a smooth map between compact Riemannian manifolds. At any point $x \in M$ for the linear map $\mathrm{d}f_x : T_xM \to T_{f(x)}N$ we define the pointwise trace norm

$$\| \mathrm{d}f \|_x^2 = \mathrm{Tr}\left({}^{\mathsf{T}}(\mathrm{d}f_x) \circ \mathrm{d}f_x \right),$$

where the transpose is in terms of the metrics, i.e., we take a matrix representation with respect to orthonormal bases in source and target tangent spaces. A more geometric way of describing this is that $\| \mathrm{d}f \|_x^2$ is the sum of the squared lengths of the semiaxes of the ellipsoid gotten by applying $\mathrm{d}f_x$ to the unit sphere in $T_{M,x}$. We use this norm to define the *energy density* of f to be the quantity

$$e(f)_x = \frac{1}{2} \| \mathrm{d}f \|_x^2,$$

whose integral is the energy

$$E(f) = \frac{1}{2} \int_M \| \mathrm{d}f \|_x^2 \, \mathrm{dvol}.$$

Then a map is *harmonic* if it is a critical point of the energy, viewed as a function on the space of smooth maps from M to N.

Example 14.1.1 Diffeomorphic self-maps of the circle with the standard metric: these are harmonic if they are geodesic because the energy in this case is half the norm-squared of the length.

We next explain how the usual variational approach applied to $f : M \to N$ leads to a set of linear equations for

$$\mathrm{d}f \in \mathrm{Hom}(T_M, f^*T_N) = A^1(M, f^*T_N).$$

The Levi–Civita connection on T_N induces a connection ∇ on this bundle and it extends as an exterior derivation $\nabla : A^k(M, f^*T_N) \to A^{k+1}(M, f^*T_N)$. The Hodge star-operator (with respect to the Riemannian metrics) is an operator $* : A^k(M, f^*T_N) \to A^{m-k}(M, f^*T_N)$, where $m = \dim M$. Hence $\nabla^* \mathrm{d}f = *\nabla * \mathrm{d}f$ is a section of f^*T_N and it enters in the following Euler–Lagrange type equation for f.

Theorem (Characterization of harmonic maps) *A map $f : M \to N$ between compact Riemannian manifolds is harmonic if and only* $\mathrm{d}f$ *satisfies the* harmonic equation

$$\boxed{\nabla^* \mathrm{d}f = 0.} \tag{14.1}$$

Proof Suppose that we are given a variation of the initial map through a family f_t, which we expand, using local coordinate systems $\{y_j\}$ about $f(p) \in N$:

$$\left(f_t^j\right)_j = (f)_j + t\dot{f} + O(t^2).$$

To make the meaning of $\dot{f}$ more precise, consider what must be done to vary f by a very small amount. At each point $f(p)$ we must assign a vector tangent to N in the direction in which we shall deform the map. Thus, $\dot{f}$ is a section of

f^*T_N, the pullback along f of the tangent bundle of N. The differential of f_t can then be expanded as

$$\mathrm{d}f_t = \mathrm{d}f + t\nabla\dot{f} + O(t^2).$$

Note that $\dot{f}$ must be covariantly differentiated because it is a vector field. Now consider the derivative of the energy,

$$\dot{E}(0) = \int_M \{\mathrm{d}f, \nabla\dot{f}\}\mathrm{dvol}_M = \int_M \{\nabla^* \mathrm{d}f, \dot{f}\}\mathrm{dvol}_M,$$

where ∇^* is the adjoint of ∇ with respect to the metric. We have already seen that $\nabla^* \mathrm{d}f$ is a section of f^*T_N, as is $\dot{f}$, and so the two can be paired by means of the metric. Because $\dot{f}$ is arbitrary, we see that indeed the harmonic equation (14.1) is satisfied.

The converse is obvious. □

With the harmonic equation in hand, we can enlarge somewhat our class of examples.

Example 14.1.2 Recall from §11.5 that a map is totally geodesic if it sends geodesics to geodesics. These are characterized by the differential equation $\nabla \mathrm{d}f = 0$, which implies trace$(\nabla \mathrm{d}f) = 0$. But trace$(\nabla \mathrm{d}f) = \nabla^* \mathrm{d}f$, so totally geodesic maps are harmonic. Totally geodesic maps can be constructed by constructing an embedding of symmetric spaces group-theoretically. Given $X = G/K$, let G' be a closed Lie subgroup such that $X' = G'/K'$ is a symmetric space, where $K' = G' \cap K$. Then $X' \subset X$ is a geodesic inclusion. If Γ and Γ' are compatible discrete groups, then $\Gamma'\backslash X' \longrightarrow \Gamma\backslash X$ is a geodesic immersion. See also Section 17.2 where we discuss special subvarieties.

As for the general existence problem, one may argue informally by thinking of a loop as a piece of string whose energy we attempt to minimize by drawing the string tight. This works well in some cases, e.g., a catenoid of revolution, and establishes the existence of a geodesic in each homotopy class. It works poorly in other cases, such as the sphere. The difference between success in one case and failure in another is partly due to the sign of the curvature, which is negative for the catenoid and positive for the sphere. However, curvature alone is not enough, as our thought experiment shows in the case of the surface obtained by revolving the curve $y = 1/x$. What this limited evidence is meant to suggest is that negative curvature and compactness of the target do guarantee existence.

Theorem 14.1.3 (The Eells–Sampson theorem) *Let $f : M \longrightarrow N$ be a continuous map between Riemannian manifolds, both compact, and where N has nonpositive sectional curvatures. Then f is homotopic to a harmonic map.*

Sketch of the proof as given by Eells and Sampson (1964): Our goal is to deform a continuous map to a harmonic one. This can be done by solving the heat equation,

$$\frac{\partial f_t}{\partial t} + \nabla^* \mathrm{d}f = 0.$$

The first point is that as time increases, the energy of f_t decreases. (The energy is defined because f_t is smooth for all t finite positive.) To see this, substitute the value of $\dot{f} = \partial f_t / \partial t$ given by the preceding equation in the expression for the derivative of the energy to obtain

$$\dot{E} = \int_M \{\nabla^* \mathrm{d}f, \dot{f}\} = - \int_M \|\nabla^* \mathrm{d}f\|^2 \leq 0.$$

The idea now is to take a sequence $\{f_n\}$ for times t_n increasing monotonely to infinity such that $\lim_{n\to\infty} E(f_n) = \inf_{t>0} E(f_t)$ and to set

$$f_\infty = \lim f_n.$$

The problem, of course, is to show that the limit exists and has the right properties. This is the beginning of the difficult analytic part of the argument, which we only outline. The idea is to establish a uniform upper bound on the energy density of the minimizing sequence. Because the energy density measures the size of $\mathrm{d}f$, such a bound implies that the minimizing sequence is equicontinuous and so has a subsequence uniformly convergent to a continuous limit.

Now observe that we do have a bound on the mean energy $E(t)$, since $E(t) \leq E(t_0)$ for all $t \geq t_0$. If the energy were subharmonic ($\Delta E \geq 0$), bounds on the mean would imply pointwise bounds. Although the energy is not in general subharmonic, it does satisfy a useful lower bound in the presence of suitable curvature assumptions. Eells and Sampson compute the Laplacian of the energy density in terms of the gradient of f, the curvature of M, and the curvature of N:

$$\Delta E = |\nabla E|^2 + \mathrm{Ric}^M(\mathrm{d}f) + R^N(\mathrm{d}f).$$

Here $\mathrm{Ric}^M(\mathrm{d}f)$ and R^N are quadratic and quartic, respectively, in $\mathrm{d}f$. If the sectional curvatures of N are nonpositive, then the inequality

$$\Delta E \geq |\nabla E|^2 + CE$$

follows, where the constant C may be negative. This estimate turns out to give the required upper bound. Here we omit the technical details and which can be found in the original article (Eells and Sampson, 1964). □

Problem

14.1.1 Let $f : M \to N$ be a differential map between Riemannian manifolds. There is a natural way to put a connection D on the bundle $\mathrm{Hom}(T_M, f^*T_N)$ using the Levi–Civita connections ∇^M, ∇^N on T_M and T_N, respectively. It is defined by the formula

$$D_\xi(\phi)(\eta) = \nabla^N_{\mathrm{d}f(\xi)}\phi(\eta) - \phi(\nabla^M_\xi \eta),$$

where ξ and η are tangent vectors along M.

(a) Show that this indeed defines a connection.

(b) Explain that the tensor $D(\phi)$ after antisymmetrization can be viewed as a two-form with values in f^*T_N. Show that this two-form is $\nabla(\phi)$, where ϕ is viewed as a one-form with values in f^*T_N, and ∇ is the exterior derivation coming from $\nabla = \nabla^N$ as above.

(c) Show that ∇^2 equals minus the curvature of the Riemannian connection on N.

14.2 Harmonic and Pluriharmonic Maps

Assume that (M, ω) is a Hermitian complex manifold of complex dimension m. There is a particular useful expression for the result of applying the Hodge star-operator to any one-form α with values in f^*TN (see Problem 14.2.1):

$$*\alpha = -\frac{1}{(m-1)!}\omega^{m-1} \wedge C\alpha,$$

where C is the usual Weil-operator. Writing the real operator $\mathrm{d}^c = -\mathrm{i}(\partial - \bar\partial)$ as $\mathrm{d}^c = C^{-1}\,\mathrm{d}C$, we thus have

$$*\,\mathrm{d}f = \frac{1}{(m-1)!}\omega^{m-1} \wedge \mathrm{d}^c f.$$

This implies that the harmonic equation in this case is equivalent to $\nabla(\omega^{m-1} \wedge \mathrm{d}^c f) = 0$. If in addition M is Kähler, this last equation is equivalent to

$$\omega^{m-1} \wedge \nabla(\mathrm{d}^c f) = 0. \tag{14.2}$$

Because for $m = 1$ this equation does not involve ω, the condition that f is harmonic only depends on the metric on N and the complex structure on M, motivating the following definition.

Definition 14.2.1 (i) Let M be a complex manifold and N any Riemannian manifold. A map $f : M \to N$ is *pluriharmonic* if its restriction to any holomorphic curve inside M is harmonic, or equivalently, if its differential $df \in \mathrm{Hom}(T_M, f^*TN)$ satisfies the *pluriharmonic equation*

$$\boxed{\nabla(d^c f) = 0.}$$

(ii) In terms of complex bundles $T_N^{\mathbf{C}} := T_N \otimes \mathbf{C}$, we may consider the $\mathbf{C}$-linear extension of df to $\mathrm{Hom}_{\mathbf{C}}(T_M, f^*T_N^{\mathbf{C}})$ and consider

$$d' f = df \big| \mathrm{Hom}_{\mathbf{C}}\left(T_M^{1,0}, f^*T_N^{\mathbf{C}}\right) = d^c f \big| \mathrm{Hom}_{\mathbf{C}}\left(T_M^{0,1}, f^*T_N^{\mathbf{C}}\right).$$

Now ∇ extends $\mathbf{C}$-linearly and the pluriharmonic equation is equivalent to

$$\boxed{\nabla''(d' f) = 0.}$$

Let us relate the pluriharmonic and the harmonic equation. The latter says that the f^*T_N-valued $(2m)$-form on the right of (14.2) vanishes so that for any local frame of the latter, the coefficients of $\nabla(d^c f)$ are primitive two-forms. Now recall that for forms α and β with values in a complex bundle E equipped with a Hermitian metric h, the notation $h(\alpha, \beta)$ stands for the form obtained by composing the $E \otimes E$-valued form $\alpha \wedge \bar{\beta}$ with the metric $h : E \otimes E \to \mathbf{C}$. We apply this to the Riemannian metric $(,)$ on f^*T_N. So $(\nabla d^c f, \nabla d^c f)$ is a four-form on M. Using an orthonormal frame, if the coefficients of $\nabla(d^c f)$ are the two-forms $\alpha_i, i = 1, \ldots, n$ then, because the α_i are primitive, the $(2m)$-form

$$(\nabla d^c f, \nabla d^c f) \wedge \omega^{m-2} = \sum \alpha_i \wedge \alpha_i \wedge \omega^{m-2}$$

is seen to be seminegative and equality holds if and only if all the α_i vanish. So we have now proven the following crucial observation.

Lemma 14.2.2 *If f is harmonic, then $(\nabla d^c f, \nabla d^c f) \wedge \omega^{m-2}$ is a pointwise negative semidefinite $(2m)$-form which vanishes if and only if $\nabla d^c f = 0$, i.e., if and only if f is pluriharmonic.*

We use this to prove an important result due to Sampson (1986).

Theorem 14.2.3 *Let M be a compact Kähler manifold, and N a Riemannian manifold with curvature F satisfying*

$$(F(\xi, \eta)\bar{\xi}, \bar{\eta}) \le 0, \quad \xi, \eta \in T_N^{\mathbf{C}}.$$

If the map $f : M \to N$ is harmonic map it is pluriharmonic and, moreover,

$$(F(\xi, \eta)\bar{\xi}, \bar{\eta}) = 0, \quad \textit{for all } x \in M \textit{ and } \xi, \eta \in d_x f(T^{1,0}M).$$

Proof The key to the proof is a suitable Bochner formula obtained by calculating the derivative of the three-form $(d^c f, \nabla d^c f)$ using the Leibniz rule and multiplying with ω^{m-2}:

$$d'(d'' f, \nabla d'' f)\wedge\omega^{m-2} = (\nabla d'' f, \nabla d'' f)\wedge\omega^{m-2} - (d'' f, \nabla^2 d'' f)\wedge\omega^{m-2}. \quad (14.3)$$

We have already seen that for harmonic maps f the first term is nonnegative and the second term involves the curvature $F = -\nabla^2$ of the Riemannian metric on f^*T_N. The minus sign is accounted for in Problem 14.1.1 of the previous section. This term can be rewritten as $F \wedge (d'' f, d'' f) \wedge \omega^{m-2}$. In terms of the complex linear extension $f^*T_N^{\mathbf{C}}$ and restricting to $(1,0)$-forms, the condition of the theorem is exactly equivalent to the second term being nonnegative as well. So this second term vanishes if and only if F vanishes on the image of $d' f$.

The final step consists of integrating the formula (14.3) over M. This step uses compactness of M and Stokes' theorem to show that the left-hand side integrates to 0, and so if f is harmonic and the curvature of the Riemannian metric of N is negative in the sense stated, both integrals on the right vanish. We can conclude the proof using Lemma 14.2.2. □

Problems

14.2.1 Let α be any one-form on an m-dimensional Kähler manifold. Verify the expression for $*\alpha$ given in this section.

14.2.2 Using the pluriharmonic equation, show that a holomorphic or anti-holomorphic map from a Kähler manifold to a complex manifold is pluriharmonic (and hence harmonic).

14.3 Applications to Locally Symmetric Spaces

We pass now to the study of a particular but important class of harmonic maps, those for which the source manifold is compact Kähler and for which the target is a *locally symmetric space* of negative curvature. These have the form $N = \Gamma\backslash X$ where $X = G/K$ is the quotient of a semisimple noncompact Lie group and K is a maximal compact subgroup.

Examples 14.3.1 (i) The upper half-plane $\mathfrak{h} \cong \mathrm{SL}(2,\mathbf{R})/\mathrm{U}(1)$.
(ii) The unit ball in $\mathbf{C}^n$, $B^n \cong \mathrm{SU}(1,n)/\mathrm{U}(n)$.
(iii) Siegel's upper half-space, $\mathfrak{h}_g \cong \mathrm{Sp}(g,\mathbf{R})/\mathrm{U}(g)$.
(iv) Real hyperbolic space, $\mathfrak{h}_n(\mathbf{R}) \cong \mathrm{SO}(1,n)/\mathrm{SO}(n)$.
(v) The space $\mathrm{SL}(n,\mathbf{R})/\mathrm{SO}(n)$ of positive definite matrices of determinant 1.

Examples (i)–(iii) are Hermitian symmetric, i.e., are symmetric spaces with a G-invariant complex structure. The symmetric space X is homeomorphic to a ball, so that whenever Γ acts without fixed points, N is an Eilenberg–Maclane space, i.e., the only nonvanishing homotopy group is $\pi_1(N) = \Gamma$.

In this case the Lie-algebra $\mathfrak{g}$ of G splits into

$$\mathfrak{g} = \mathfrak{k} \oplus \mathfrak{p},$$

where $\mathfrak{k}$ is the Lie-algebra of K and where $\mathfrak{p}$ is its orthogonal complement relative to the Killing form, or – equivalently – to the trace form B. We treated this decomposition briefly in Remark 12.3.4. Clearly, the tangent space $T_{X,0}$ can be identified with $\mathfrak{p}$ and hence its complexification with $\mathfrak{p}_{\mathbf{C}} = \mathfrak{p} \otimes \mathbf{C}$. In the same remark we observed that the curvature of the metric is given by

$$R(\xi, \eta)\zeta = -[[\xi, \eta], \zeta], \quad \xi, \eta, \zeta \in \mathfrak{p}_{\mathbf{C}}$$

and so

$$(R(\xi, \eta)\bar{\xi}, \bar{\eta}) = -([\xi, \eta], [\bar{\xi}, \bar{\eta}]) \leq 0.$$

It follows that this vanishes precisely when $[\xi, \eta] = 0$.

Suppose next that we have a harmonic mapping $f : M \to N$ with M a Kähler manifold. Then, for all tangent vectors ξ of M at p, the image tangent vector $d' f(\xi)$, lifted to a tangent to X, and transported back to $0 \in X$, can be viewed as an element of $\mathfrak{p}_{\mathbf{C}}$. So if, moreover, M is compact, Sampson's theorem 14.2.3 implies the following.

Theorem 14.3.2 *If M is a compact Kähler manifold, $N = (G/K)/\Gamma$ as above, and $f : M \to N$ harmonic, then f is pluriharmonic, i.e.,*

$$\nabla d' f = 0$$

and

$$[d' f, d' f] = 0,$$

i.e., for each point $x \in M$, the space $T_x^{1,0} M$ maps to an abelian subspace of $\mathfrak{p}_{\mathbf{C}}$.

We recall from Problem 14.1.1, using the natural connection D on the bundle $\mathrm{Hom}(T_M, f^*T_N)$, that the antisymmetrization of the tensor $D(df) \in A^1_M \otimes A^1_M \otimes f^*T_N$ equals $\nabla(df)$. In the remainder of this section, we use this D. The harmonic equations can thus be restated as

$$D''(d' f) = 0.$$

Lemma 14.3.3 *For the curvature D^2 of this other natural connection we have:*

$$D''^2 = 0 = D'^2.$$

Proof First note that the curvature is a sum of two contributions

$$D^2 = -F^M \otimes 1 + 1 \otimes F^N.$$

To establish the formula just given, we note that D''^2 is the $(0,2)$-component of the curvature. We proved in Section 9.1 that the curvature of the metric connection on any Kähler manifold has type $(1,1)$ as a two-form, so the first summand has type $(1,1)$. The second summand can be written as

$$1 \otimes F^N = 1 \otimes \mathrm{ad}([\mathrm{d}f(\cdot), \mathrm{d}f(\cdot)]).$$

Its $(2,0)$ term vanishes because $[\mathrm{d}' f(\xi), \mathrm{d}' f(\eta)] = 0$ for any two tangent vectors ξ, η of type $(1,0)$. Since the expression is real, its $(0,2)$-component vanishes as well, and so it is of type $(1,1)$. So both terms have type $(1,1)$, thus completing the proof. □

Corollary 14.3.4 *If M is a compact Kähler manifold, $N = \Gamma\backslash(G/K)$ as above. Then $f : M \to N$ is harmonic if and only if it is pluriharmonic. This is the case if and only if the $(0,1)$-part of the natural connection D defines a holomorphic structure on*

$$E := \mathrm{Hom}\,(T^{1,0}M, f^*T_N^{\mathbb{C}})$$

and $\mathrm{d}' f$ is a holomorphic section of it.

Another equivalent statement is that in this setting the pluriharmonic equations:

$$\begin{aligned} D''^2 &= 0, \\ D''\mathrm{d}' f &= 0, \\ [\mathrm{d}' f, \mathrm{d}' f] &= 0 \end{aligned} \tag{14.4}$$

hold.

Proof We showed most of the assertions. We showed that if f is harmonic it is pluriharmonic which means that the middle pluriharmonic equation holds. That the other two hold is the content of the previous proposition.

Conversely, D'' defines an integrable holomorphic structure on E if and only if the integrability condition $D''^2 = 0$ holds (Theorem 11.1.3). Our curvature computation in proving Lemma 14.3.3 in fact shows that this condition is equivalent to $[\mathrm{d}' f, \mathrm{d}' f] = 0$. Now $\mathrm{d}' f$ being a holomorphic section of E is equivalent to f being pluriharmonic, i.e., $D''\mathrm{d}' f = 0$. □

We next want to indicate how these extra conditions lead to the sort of statements found in Siu's rigidity theorem stated below. We start by noting that these equations do indeed give more structure to the map f. The fact that $\mathrm{d}' f$ is a holomorphic section of a holomorphic bundle implies that the set along which

$d' f$ is of less than maximal rank is a complex subvariety. So, roughly speaking, f behaves as if the target were a complex manifold and f itself were holomorphic. Even if N is a complex manifold, we cannot yet conclude that f is either holomorphic or antiholomorphic. Indeed, we need $d' f$ to take values in either $\mathfrak{p}^{1,0}$ or $\mathfrak{p}^{0,1}$. The idea is to investigate when this is true using the third equation. Eventually, one can in this way, for instance, prove Siu's rigidity theorem.

Theorem 14.3.5 (Siu's rigidity theorem (Siu, 1980)) *Let $f : M \longrightarrow N$ be a harmonic map from a compact Kähler manifold to a compact discrete quotient of a Hermitian symmetric space. There is a number $\sigma(N)$ such that* $\mathrm{rank}_x\, \mathrm{d}f > \sigma(N)$ *at a single point implies that f is either holomorphic or antiholomorphic. Here "rank" stands for the* real *rank.*

The invariant $\sigma(N)$ depends only on the Hermitian symmetric domain. In the case of quotients of the ball, it is 2. In general it is the (real) dimension of the largest Hermitian symmetric subspace that contains $\mathfrak{h}_1$ as a factor. Thus, for the Siegel upper half-space of genus g, it is $g(g-1)+2$ because $\mathfrak{h}_g$ contains $\mathfrak{h}_1 \times \mathfrak{h}_{g-1}$.

We return below to the details of the proof of Siu's theorem (our remarks concerning the proof lead up to Remark 14.3.9), but for now we want to show how the third equation does lead to strong constraints.

Theorem 14.3.6 (Sampson, 1986) *Let $f : M \longrightarrow N$ be a harmonic map of a compact Kähler manifold to a real hyperbolic space, i.e., any compact quotient of real hyperbolic space $\mathfrak{h}_n(\mathbf{R}) \cong \mathrm{SO}(1,n)/\mathrm{SO}(n)$. Then $\mathrm{d}f$ has rank at most 2.*

Proof For the proof, set

$$W_x = d' f(T^{1,0}_{M,x})$$

and observe that

$$\begin{aligned} \mathrm{rank}\, \mathrm{d}f_x &= \dim_{\mathbf{R}}(\mathrm{d}f(T_{M,x})) \\ &= \dim_{\mathbf{C}}\left(\mathrm{d}f(T^{\mathbf{C}}_{M,x})\right) \\ &= \dim_{\mathbf{C}}\left(\mathrm{d}f(T^{1,0}_{M,x}) + \mathrm{d}f(T^{0,1}_{M,x})\right) \\ &= \dim_{\mathbf{C}}(W_x + \overline{W}_x). \end{aligned}$$

Thus,

$$\mathrm{rank}\, \mathrm{d}f_x \le 2 \dim_{\mathbf{C}} W_x,$$

and equality holds if and only if $W_x \cap \overline{W}_x = 0$.

To bound the dimension of W_x, we observe that the last equation in (14.4) implies

$$[W_x, W_x] = 0.$$

One naturally expects abelian subspaces to be fairly small and of a quite special nature. The first result of this kind seems to have been proved by Schur (1905), who showed that the dimension of a space of commuting trace zero matrices of size n is bounded by $[n^2/4]$. In the case of even n the bound is attained by the space of (nilpotent) matrices of the form

$$\begin{pmatrix} 0 & * \\ 0 & 0 \end{pmatrix},$$

where this is meant to indicate a block-decomposition into four equal parts. In any case, let us follow Sampson's analysis for abelian subspaces of $\mathfrak{p}$. To begin, we need a concrete description of the Lie algebra in question. It is given by the set of matrices satisfying

$${}^{\mathsf{T}}MQ + QM = 0,$$

where

$$Q = \begin{pmatrix} -1 & 0 \\ 0 & \mathbf{1}_n \end{pmatrix}.$$

Thus,

$$M = \begin{pmatrix} 0 & a \\ {}^{\mathsf{T}}a & b \end{pmatrix},$$

where b is skew-symmetric. This we decompose as

$$M = X(b) + Y(a), \quad X(b) = \begin{pmatrix} 0 & 0 \\ 0 & b \end{pmatrix}, \quad Y(a) = \begin{pmatrix} 0 & a \\ {}^{\mathsf{T}}a & 0 \end{pmatrix}.$$

The Cartan decomposition follows the above decomposition:

$$\mathfrak{k} = \{\text{matrices } Y(a)\}\,,$$
$$\mathfrak{p} = \{\text{matrices } X(b)\}\,.$$

A short calculation shows that the components of $[Y(a), Y(a')] = X(b)$ are $a_i a'_j - a_j a'_i$, i.e., are the components of $a \wedge a'$. Thus, $\dim_{\mathbb{C}} W \leq 1$, as claimed. □

The result just described is really quite remarkable, because the dimension of M may far exceed that of N. It can be interpreted as saying that hyperbolic manifolds of dimension greater than 2 are very, very different from Kähler manifolds. In fact, one can say somewhat more (Carlson and Toledo, 1989b, p. 196).

Theorem 14.3.7 (Factorization theorem) *Let $f : M \longrightarrow N$ be a nonconstant harmonic map from a compact Kähler manifold to a real hyperbolic space. Then f has either rank 1 or rank 2. In the first case, $f(M)$ is a closed geodesic in N. If the rank is 2, f factors as gh, where Y is a smooth Riemann surface, $h : M \to Y$ is holomorphic, and g is a harmonic map.*

Before commenting on the proof, let us see what the previous theorem has to say concerning the relation between Kähler groups and hyperbolic groups, where we recall that a group is called Kähler if it can be realized as the fundamental group of a compact Kähler manifold. The result we want to discuss is as follows.

Theorem 14.3.8 *Let Γ be a torsion-free lattice in* $\mathrm{SO}(1,n)$*, where $n > 2$ such that $\Gamma \backslash \mathrm{SO}(1,n)$ is compact. Then Γ is not Kähler.*

Proof Suppose that M is a compact Kähler manifold with Γ as its fundamental group. Let $N = \Gamma\backslash X$, where X is real hyperbolic n-space. Then N is an Eilenberg–Maclane space for $\pi_1 \cong \Gamma$. This means that N has Γ as its fundamental group and has no higher homotopy groups. It is known from homotopy theory that this characterizes N up to homotopy equivalence.

Construct $f : M \to N = \Gamma\backslash \mathfrak{h}_n(\mathbf{R})$ so that $f_* : \pi_1(M) \to \Gamma$ is the given isomorphism as follows. The manifold M is a cell complex whose 1-skeleton M_1 contracts to a bouquet of circles corresponding to the generators of Γ, defining a continuous map $f_1 : M_1 \to N$ inducing a surjection on fundamental groups. Then consider the 2-skeleton M_2. It is obtained from M_1 by attaching 2-cells D_i for each relation in the fundamental group. Then $(f_1)_*[D_i] = 1$ by construction. This means that f_1 extends to $f_2 : M_2 \to N$ inducing the given isomorphism. We can extend f_2 successively to the k-skeleta because the obstruction to extending lies in $H^k(M, \pi_{k-1}(N))$, and for $k \geq 3$ this group vanishes. Because N is negatively curved, we may, by the Eells–Sampson theorem, assume that f is already harmonic. By the preceding result, it factors through Y. Because f_* is an isomorphism on fundamental groups, the first factor h induces an injection on π_1. Thus Γ operates freely on $\tilde{Y}$, the universal cover of Y. Since $\tilde{Y}$ is contractible, $\Gamma\backslash\tilde{Y}$ is an Eilenberg–Maclane space. Now the cohomology of a group is just the cohomology of its Eilenberg–Maclane space. Thus, the cohomological dimension of Γ, i.e., the largest dimension in which an Eilenberg–Maclane space has nontrivial cohomology, is at most 2. However, N, which is also an Eilenberg–Maclane space, has dimension at least 3, and moreover is compact and orientable, and so has a fundamental class. Thus, the cohomological dimension of Γ is at least 3. The resulting contradiction completes the proof. □

Sketch of the proof of Theorem 14.3.7 We note first that the rank-1 case is an earlier theorem of Sampson (1978).

In the rank-2 case, M being a manifold, the set of regular values of the map f form a two-dimensional manifold Y' over which f is a locally trivial fibration. For simplicity we assume that its fibers C_y are connected. Now $T_{C_y,x} = \mathrm{Ker}(\mathrm{d}f) \subset T_{M,x}$ is a complex subspace of the tangent space because

$\mathrm{d}f(T_{M,x})(\xi) = 0$ if and only if $\mathrm{d}f(T_{M,x})(\xi - \mathrm{i}J\xi) = 0$. This last assertion follows because $\mathrm{d}f(T_{M,x})$ does not contain real vectors. Now look at the image $L_x = \mathrm{d}f(W_x)$, a complex line in the two-dimensional complex vector space $T^{\mathbb{C}}_{Y',y}$ varying holomorphically with x but not containing any real vector. Because C_y is compact and connected, L_x must be constant. This defines a complex structure on Y'.

For the next step, we need to extend the holomorphic map $h' = f \mid M' \to Y'$, where $M' = f^{-1}Y'$. One can use the first two pluriharmonic equations to show that M' is the complement of an analytic subvariety, but this does not suffice. The map h could have essential singularities along this subvariety. To show that it does not, we use the fact that complex submanifolds of a Kähler manifold M have a kind of Hilbert scheme, the *Barlet space* $C(M)$, introduced in Barlet (1975).

For a good introduction explaining the crucial properties, see Lieberman (1978). We need only the facts

- all components of $C(M)$ are compact if M is compact;
- the points of $C(M)$ are in one-to-one correspondence with the analytic cycles of M;
- holomorphic families are induced by universal families over a reduced and normal analytic base space.

In our case, the family $C_y, y \in Y'$ is thus induced by a certain universal cycle $Z \subset Y \times M$ from a certain holomorphic map $i : Y' \to Y$ defined by the fact that $y \in Y'$ is sent to the point in $C(M)$ corresponding to C_y. We let $p : Z \to Y$ and $q : Z \to M$ be the maps induced by the projections. Now $q(Z) = M$ because it contains the open dense set M' and because two distinct fibers, $C_y = Z_y$ and $C_{y'} = Z_{y'}$, of h are disjoint and $\dim Y = 1$; so, because it is normal, Y is a compact Riemann surface. The fundamental cohomology classes of the fibers Z_y of the projection p are all the same, say $[Z_y] = z$. Obviously, $z \cup z = 0$, but because M is Kähler, $z \neq 0$ and so two irreducible fibers do not meet inside M and two reducible fibers can meet inside M only along common components. These facts translate into q being finite-to-one and generically one-to-one. But M is a manifold and so q must be a biholomorphic map which means that up to a biholomorphic isomorphism the projection gives our desired extension $h : M \to Y$ of the map h', or rather of $i \circ h'$, where $i : Y' \to Y$ is the defining map we introduced above. This map is an embedding (because $M' \to M$ is) and it is not hard to see that f is also constant on the fibers of h over the points of $Y - Y'$. A final argument then shows that the resulting map $g : Y \to M$ is harmonic.

The only complication is that the fibers of f might consist of several components and in that case one has to replace h' by its Stein factorization, i.e., h' :

$M' \longrightarrow Y'' \longrightarrow Y'$, where the first map has connected fibers and the second map is finite; it is the first map that gets extended, and the second (as a holomorphic map of Riemann surfaces) then automatically gets extended over the punctures. □

Remark 14.3.9 (Further remarks on Siu's rigidity theorem) The factorization theorem (Theorem 14.3.7) should be viewed as an analogue of Siu's rigidity theorem, which is a holomorphicity statement for harmonic maps of sufficiently large rank. Indeed, as a next step toward the proof of this theorem, let us now study in more detail the abelian subspaces W of $\mathfrak{p}_{\mathbf{C}}$, where $\mathfrak{g} = \mathfrak{k} \oplus \mathfrak{p}$ corresponds to a Cartan decomposition of a symmetric space, each factor of which is assumed to be simple and of noncompact type. By the latter we mean that $X = G/K$ is noncompact, or, equivalently, that it has nonpositive curvature in the natural (invariant) metric. One general result is the following.

Proposition 14.3.10 (Carlson and Toledo, 1989b, pp. 183–186.) *Let $W \subset \mathfrak{p}_{\mathbf{C}}$ be an abelian subspace in the context just defined. Then* $\dim W \leq \dim \mathfrak{p}_{\mathbf{C}}/2$. *Moreover, if equality holds, G/K is a Hermitian symmetric space and for each direct summand of $\mathfrak{g}$, say $\mathfrak{g}_i = \mathfrak{k}_i \oplus \mathfrak{p}_i$ not isomorphic to $\mathfrak{sl}(2, \mathbf{R})$, the corresponding direct summand $W \cap \mathfrak{p}_i$, of W, is equal to either $\mathfrak{p}_i^{1,0}$ or $\mathfrak{p}_i^{0,1}$.*

The last assertion gives a weak form of Siu's rigidity theorem, because a map f is holomorphic if $d' f$ takes vatlues in $\mathfrak{p}^{1,0}$. The strong form follows from a more careful analysis of the abelian subspaces, the point being to ensure one of the two inclusions by a suitable hypothesis on the rank. In any case, note that with the exception of SL(2, **R**), the maximal abelian subspaces of $\mathfrak{p}_{\mathbf{C}}$ in the Hermitian symmetric case are just the obvious ones given by the complex structure. Turning to the non-Hermitian case, the proposition implies the following.

Corollary *Let $f : M \longrightarrow N$ be a continuous map of a compact Kähler manifold to a negatively curved, irreducible locally symmetric space, which is not a quotient of a Hermitian symmetric space. Then f is homotopic to a map whose image is a CW-complex of dimension strictly lower than* $\dim N$.

Proof Apply Eells–Sampson to deform f to a harmonic map, and then apply the preceding bound on the rank. □

Note that the result implies that there are no maps of nonzero degree from a compact Kähler manifold to a space of the kind considered.

With Proposition 14.3.10 in hand, we can finally prove the result mentioned in the introduction to this chapter.

Theorem 14.3.11 *Compact quotients of period domains for Hodge structures* *of even weight that are not Hermitian symmetric are not homotopy equivalent to a Kähler manifold.*

Proof The argument is by contradiction. Suppose that there is a homotopy equivalence $f : M \longrightarrow N$, where M is compact Kähler and N is the quotient of the period domain in question. Let $Y = \Gamma\backslash X$ be the locally symmetric space associated with N and let $\pi : N \longrightarrow Y$ be the canonical projection. By Proposition 14.3.10, $(\pi \circ f)^*[Y] = 0$, where $[Y]$ stands for the fundamental class.

On the other hand, consider a volume form Ω_Y, i.e., a form that represents $[Y]$, and let ω be the first Chern class of the canonical bundle of N. We have seen (Remark 13.6.4) that K_N restricted to fibers of π is positive and so ω pulled back to fibers is also positive. By Fubini's theorem,

$$\int_M \pi^*\Omega_Y \wedge \omega^k = c \int_N \Omega_Y > 0,$$

where $c > 0$ is the volume of the fiber as computed by ω^k, and where k is the fiber dimension. Consequently, the cohomology class of $\pi^*\Omega_N$ is nonzero, i.e., $\pi^*[Y] \neq 0$. Because f is a homotopy equivalence, $f^*\pi^*[Y]$ is also nonzero, and this contradicts what has already been established. The proof is now complete. □

We remark that by an argument of Gromov (oral communication), compact quotients of period domains, except those that are of the Hermitian symmetric type, are non-Kähler. The argument runs as follows: look at the fibers of the associated quotient map $q : \Gamma\backslash D \to \Gamma\backslash X$. Looking carefully at limits of volumes of deformations of the fibers when they become horizontal, one sees that these can become arbitrarily large. So one concludes that the component of the Barlet space containing such a fiber cannot be compact, contrary to what would happen if $\Gamma\backslash D$ were Kähler.

The problem of proving or disproving the stronger homotopy statement for odd weight is open. There are bounds on the rank that are considerably sharper than those given in Carlson and Toledo (1989b), e.g., the following result.

Proposition (Carlson and Toledo, 1993, Theorem 7.1) *Let $f : M \longrightarrow N$ be a harmonic map from a compact Kähler manifold to a compact discrete quotient of the symmetric space associated with* SO$(2p, 2q)$*, where $p, q \geq 3$, $(p, q) \neq (3, 3)$. Then* rank d$f \leq$ dim $N/2$. *If equality holds, then f factors as gh, where $h : M \longrightarrow Y$ is a holomorphic map to a discrete quotient of the symmetric space associated with* SU(p, q). *Moreover, g is a geodesic immersion.*

14.4 Harmonic and Higgs Bundles

We shall now translate the notion of harmonic maps into the language of bundles and connections. This has a number of technical advantages, among which is a one-line proof of the following fact.

Proposition 14.4.1 *Let $f : M \longrightarrow N$ be a period mapping, and let $\pi : N \longrightarrow Y$ be the canonical map to the associated locally symmetric space. Then $\pi \circ f$ is pluriharmonic (and hence harmonic).*

The proposition gives a way of constructing nontrivial harmonic maps, e.g., maps that do not come from geodesic immersions. This type of construction occurred in other contexts much earlier (see Bryant (1985); Calabi (1967), and Lawson (1980). We note that in some cases it is possible to lift harmonic maps to variations of Hodge structure.

Proposition (Carlson and Toledo, 1993, p. 38) *Let $f : M \longrightarrow N$ be a harmonic map from a compact Kähler manifold to a compact discrete quotient of the symmetric space associated with* $\mathrm{SO}(2p, 2q)$, *where $p, q \geq 3$, $(p, q) \neq (3, 3)$. Assume that* $\mathrm{rank}(\mathrm{d}f) > 2(p-1)(q-1)+1$. *Then f lifts to a variation of Hodge structure.*

Observe that the critical number $2(p-1)(q-1)+1$ is the dimension of the largest geodesically embedded, reducible Hermitian symmetric space, namely that associated with $\mathrm{SU}(1,1)\times\mathrm{SU}(p-1, q-1)$. Thus, both the critical dimension and the result follow the pattern pioneered by Siu: harmonicity and sufficiently large rank imply a kind of holomorphicity. Note that the critical dimension is just that necessary to assure that W consists of nilpotent transformations. This is a necessary condition, because if f comes from a period mapping, $W \subset \mathfrak{g}^{-1,1}$. However, $\mathfrak{g}^{-1,1}$ consists of strictly block triangular matrices and so consists of nilpotent transformations.

Let us now turn to our translation work, which uses the language of principal bundles in the setting of Definition 12.4.1. In the present situation, $X = G/K$ is a symmetric space and $\Gamma \subset G$ is a torsion-free discrete group. Let $f : M \longrightarrow N = \Gamma\backslash G/K$ be any map and let $\tilde{f} : \tilde{M} \longrightarrow X$ be the lift to the level of universal covers. We have the Higgs bundle $P_{\mathrm{Higgs}} = X \times G$ and its Maurer–Cartan form ω_{Higgs}. Recall that we have set

$$\sigma := \omega^{\mathfrak{p}}_{\mathrm{Higgs}}.$$

The tangent bundle T_X has a K-structure, and the corresponding connection ∇_K is related to the flat Higgs connection by

$$\nabla_{\mathrm{Higgs}} = \nabla_K + \sigma.$$

Using the canonical trivialization

$$\sigma_X : T_X \xrightarrow{\sim} [\mathfrak{p}]$$

obtained by left-translation, we can interpret σ as

$$\sigma : T_{\tilde{M}} \xrightarrow{\mathrm{d}f} T_X \xrightarrow{\sigma_X} [\mathfrak{p}].$$

The translation can be summarized by the following lemma.

Lemma 14.4.2 (i) *f is harmonic if and only if $\nabla_K^* \sigma = 0$.*
(ii) *f is pluriharmonic if and only if*

$$\begin{aligned} \nabla_K''^2 &= 0, \\ \nabla_K''(\sigma') &= 0, \\ [\sigma', \sigma'] &= 0. \end{aligned}$$

We are now ready for the proof of Proposition 14.4.1.

Proof of Proposition 14.4.1 Consider a variation of Hodge structure $f : M \longrightarrow N$ and let $\pi : N \longrightarrow Y$ be the canonical projection to the associated locally symmetric space. We shall make this explicit, but for simplicity's sake, we assume that the weight is even; it is easy to adjust the definitions to take care of the case of odd weight using the same principles. In any case, from the data of the variation of Hodge structure associated with f we may construct a cruder object, namely the flat bundle E with decomposition

$$E = E^+ \oplus E^-,$$

where

$$E^+ = \bigoplus_{p \text{ even}} E^{p,q},$$

$$E^- = \bigoplus_{p \text{ odd}} E^{p,q},$$

and where the congruence is modulo 2. All the objects just defined are defined over the reals, i.e., $E = \overline{E}$, $\overline{E}^{\pm} = E^{\pm}$, so we may pass to the underlying real bundles (without changing notation). The resulting system defines a map from M to the locally symmetric space $Y = \mathrm{SO}_0(2p, q)/\mathrm{SO}(2p) \times \mathrm{SO}(q)$ associated with $\mathrm{SO}(2p, q)$, where $2p = \dim E^+$ and $q = \dim E^-$.

Now because the map f is holomorphic and hence harmonic, the harmonic equations hold for $\mathrm{d}f$ with respect to the full Hodge decomposition and hence they also hold for the cruder decomposition. We have shown (see Example

12.4.2) that the pluriharmonic equations hold for the full Hodge decomposition. So they continue to hold for the cruder decomposition as well. In other words, the map $\pi \circ f$ is harmonic as well as pluriharmonic. □

We end this chapter by giving the relation with the Higgs bundles as defined in Chapter 12. We start out by noting that additional relations hold, but these are consequences of $\nabla^2_{\text{Higgs}} = 0$ and so are not deep, i.e., do not depend on the harmonic equation. In any case, we have

$$\nabla^2_{\text{Higgs}} = \nabla^2_K + \nabla_K(\sigma) + [\sigma, \sigma] = 0.$$

The first and third terms are of "type $\mathfrak{k}$," i.e., are $\mathfrak{k}$-valued two-forms, whereas the middle term is of "type $\mathfrak{p}$." Thus both terms vanish separately, yielding

$$\begin{aligned} \nabla^2_K + [\sigma, \sigma] &= 0, \\ \nabla_K(\sigma) &= 0. \end{aligned}$$

As a result of our discussion, we see that ∇''_K turns any bundle associated with the K-structure into a holomorphic bundle. Moreover, if we decompose σ according to types as a differential form, say $\sigma' + \sigma''$, then the second pluriharmonic equation asserts that it is holomorphic as a section of $[\mathfrak{p}_{\mathbb{C}}] \otimes \Omega^1_M$. For an application of these ideas, see Problem 14.4.1.

Problem

14.4.1 Show how a harmonic map $f : M \to \Gamma\backslash(G/K)$ can be used to construct a Higgs bundle on M. Hint: write G as a matrix group $G \subset \mathrm{GL}(V)$, where V is a real vector space and the representation of $\pi_1(M)$ on V defining f can be used to construct a local system. Show that its complexification carries a natural holomorphic structure induced by ∇'' and that σ' acts as a holomorphic Ω^1_M-valued endomorphism of this bundle, making it into a Higgs bundle as in Definition 13.1.3.

Bibliographical and Historical Remarks

The foundational results on which this chapter is built are due to Eells and Sampson (1964). The chapter is focussed on explaining a strand of interconnected results by Siu (1980); Sampson (1978, 1986), and Carlson and Toledo (1989b, 1993). See related results by Corlette (1988).
The main result, Theorem 14.3.11, has recently been generalized to include also compact quotients of odd weight period domains which are not bounded

symmetric domains. See Carlson and Toledo (2014).
There is a recent result Griffiths et. al (2015) which complements these results: *any* quotient, compact or not, of a period domain which is not a bounded symmetric domain is not algebraic. Note that the authors don't claim the stronger statement that a discrete quotient of such a domain cannot have a Kähler structure in these cases.

Part FOUR

ADDITIONAL TOPICS

15

Hodge Structures and Algebraic Groups

We recast in Section 15.1 the definition of a (polarized) Hodge structure in terms of algebraic representations. This leads (Section 15.2) to Mumford–Tate groups. As we see there, the Mumford–Tate group of a rational Hodge structure on a real vector space H is the largest algebraic subgroup of $\mathrm{GL}(H)$ that fixes all self-dual tensors built from H. In a variation of Hodge structure, the Mumford–Tate groups vary, but the generic one is constant, say equal to M and by their very definition, period maps factor through a quotient of a subdomain of the period domain on which M acts transitively. Such subdomains are the Mumford–Tate subdomains. The structure of M as a reductive group can be exploited to describe in detail how the period map factors over (quotients of) such Mumford–Tate domains. See Theorem 15.3.14. This structure theorem is one of the main recent achievements in the theory of period maps.

15.1 Hodge Structures Revisited

The Deligne Torus

Recall (Definition 1.2.4(i)) that a real Hodge structure $(H, H^{p,q})$ of weight k consists of a finite dimensional real vector space H such that

$$H_{\mathbf{C}} = H \otimes_{\mathbf{R}} \mathbf{C} = \bigoplus\nolimits_{p+q=k} H^{p,q}, \quad H^{q,p} = \overline{H^{p,q}}.$$

If $R \subset \mathbf{R}$ is a subring of $\mathbf{R}$ and there is an R-module H_R such that $H = H_R \otimes_R \mathbf{R}$, we speak of an R-Hodge structure.

There is a way to phrase this using representation theory as follows. We place ourselves in the context of algebraic groups. Using the notion of Weil restriction as introduced in Appendix D.1, the group that plays a central role is

$$\boxed{\mathbf{S} = \mathrm{Res}_{\mathbf{C}/\mathbf{R}}\, \mathbf{G}_m \qquad \text{(the } \textit{Deligne torus}\text{)}.}$$

This group is the algebraic group $\mathbf{C}^\times$ considered as an $\mathbf{R}$-algebraic group. One may think of the real points of this group as the invertible *real* matrices of the form $\begin{pmatrix} a & -b \\ b & a \end{pmatrix}$. However, there are many isomorphic models. This one is obtained by considering $\mathbf{R}^2$ as an an $\mathbf{R}(J)$-algebra where J operates as complex multiplication by i on $\mathbf{C} = \mathbf{R}^2$. We will use this identification when we write $z = a + ib \in \mathbf{S}(\mathbf{R}) = \mathbf{C}^\times$.

The group $\mathbf{S}$ is a torus of rank 2 since $\mathbf{S}(\mathbf{C}) \simeq \mathbf{C}^\times \times \mathbf{C}^\times$. To see this, note that a point of $\mathbf{S}(\mathbf{C})$ consists of an invertible *complex* matrix $\begin{pmatrix} a & -b \\ b & a \end{pmatrix}$. Such a matrix has exactly two eigenvalues $z = a+ib$ and $w = a-ib$. In this model the complex conjugation acts (nontrivially) by $(z, w) = (a + \mathrm{i}b, a - \mathrm{i}b) \mapsto (\bar{a} - \mathrm{i}\bar{b}, \bar{a} + \mathrm{i}\bar{b}) = (\bar{w}, \bar{z})$. The Deligne torus is not $\mathbf{R}$-split (i.e., $\mathbf{S}(\mathbf{R}) \not\simeq \mathbf{R}^\times \times \mathbf{R}^\times$), the maximal anisotropic subtorus of $\mathbf{S}$ being the circle

$$\boxed{\mathbf{U} \overset{\text{def}}{=} \left\{ \begin{pmatrix} a & -b \\ b & a \end{pmatrix} \middle| \; a^2 + b^2 = 1 \right\}.}$$

Indeed, $\mathbf{U}(\mathbf{R}) = S^1$, the circle group.

The character group of $\mathbf{S}_\mathbf{C}$ is generated by z and w. For the Deligne torus $z = a+ib$ with $a, b \in \mathbf{R}$ and thus on $\mathbf{S}$ the character w is the conjugate of z. The character group of $\mathbf{U}$ is cyclic generated by z (note that on the circle one has $\bar{z} = z^{-1}$). The maximal split torus of $\mathbf{S}$ is the diagonal torus. This is the image of the natural *weight co-character*

$$w : \mathbf{G}_m \to \mathbf{S}, \quad a \mapsto \begin{pmatrix} a & 0 \\ 0 & a \end{pmatrix}. \tag{15.1}$$

So

$$\mathbf{S} = \mathbf{U} \cdot w(\mathbf{G}_m). \tag{15.2}$$

Since the intersection of the above two tori is the finite group $\{\mathrm{id}, -\mathrm{id}\}$ the product map $\mathbf{G}_m \times \mathbf{U} \to \mathbf{S}$ is a degree 2 isogeny. The torus $\mathbf{U}$ also appears as a quotient: there is an exact sequence of algebraic groups

$$1 \to \mathbf{G}_m \xrightarrow{w} \mathbf{S} \xrightarrow{\mathrm{sq}} \mathbf{U} \to 1, \quad \mathrm{sq}_\mathbf{C}(z) = z/\bar{z}.$$

When we restrict the last map to $\mathbf{U} \subset \mathbf{S}$ we get the squaring operation which explains the notation.

Hodge Structures as Representations

Lemma–Definition 15.1.1 *A Hodge structure of weight k (on a real vector space H) is the same as a finite-dimensional* algebraic *representation $h : \mathbf{S} \to \mathrm{GL}(H)$ such that*

$$h \circ w : \mathbf{R}^\times \longrightarrow \mathrm{GL}(H),$$
$$t \longmapsto t^k \,\mathrm{id}_H,$$

where w is the weight co-character (15.1). *If $H = H_R \otimes_R \mathbf{R}$ for some R-module H_R, $R \subset \mathbf{R}$, this gives an R-Hodge structure of weight k.*

Proof of the Lemma We have seen that z and $\bar{z}$ generate the group of characters of the torus $\mathbf{S}$. Hence, since we have an algebraic representation, the complex vector space $H_\mathbf{C} = H \otimes_\mathbf{R} \mathbf{C}$ is a direct sum of weight spaces $H^{p,q}$ on which $\mathbf{S}$ acts as multiplication by $z^p \bar{z}^q$. Then, since $H_\mathbf{C}$ has a real structure, the conjugate of an eigenvector is an eigenvector with conjugate eigenvalue. Hence $\overline{H^{p,q}} = H^{q,p}$. Finally, if we look at the effect of the weight co-character, we must have $h(w(t)) = t^{p+q} = t^k$ which implies that $H^{p,q} = 0$ unless $p + q = k$.

For the converse statement see Problem 15.1.1. □

Examples 15.1.2 (i) The Tate Hodge structure $\mathbf{Q}(1)$ comes from the representation on $\mathbf{R} \cdot (2\pi \mathrm{i})$ sending z to multiplication by $(z\bar{z})^{-1}$. Similarly, $\mathbf{Q}(r)$ comes from $z \mapsto$ (multiplicaton by $(z\bar{z})^{-r}$) on $\mathbf{R} \cdot (2\pi\mathrm{i})^r$.
To direct sums (or tensor products) of Hodge structures correspond direct sums (or tensor products) of the corresponding representations. In particular for rational $\mathbf{Q}$-Hodge structures, we have the Tate twists $H(r) = H \otimes \mathbf{Q}(r)$.
(ii) The dual $H^\vee$ of a Hodge structure H corresponds to the contragredient representation (see Eq. (D.2) in Appendix D).
(iii) Combining the above two examples, we have the tensor spaces

$$\boxed{T^{m,n} H = H^{\otimes m} \otimes (H^\vee)^{\otimes n}.}$$

If H has weight k, these have weight $k(m-n)$. If H has a rational Hodge structure, the *Hodge tensors* $\mathrm{Hdg}(T^{m,n}H)$ by definition are the rational tensors in $T^{m,n}H$ of pure Hodge type. We are especially interested in those of weight zero, denoted $[\mathrm{Hdg}(T^{m,n}H)]^{w=0}$. Depending on the weight of H, these are given by

$$\boxed{\begin{cases} \text{if } H \text{ has weight zero, then} & [\mathrm{Hdg}(T^{m,n}H)]^{w=0} = T^{m,n}H_\mathbf{Q} \cap (T^{m,n}H_\mathbf{C})^{0,0}, \\ \text{otherwise, } m = n \text{ and} & [\mathrm{Hdg}(T^{m,m}H)]^{w=0} = T^{m,m}H_\mathbf{Q} \cap (T^{m,m}H_\mathbf{C})^{0,0}. \end{cases}}$$

We may likewise speak of Hodge tensors in spaces that are finite direct sums of the tensor spaces $T^{m,n}H$.

If we drop the condition on the weights, we still get a $z^p\bar{z}^q$-decomposition of $H_{\mathbf{C}}$ with respect to h and we may assemble those having the same weight. Since h is a real representation we then get a *Hodge structure*, by definition a direct sum of Hodge structures of possibly different weights. However, if now H has a rational structure, this decomposition is not automatically defined over $\mathbf{Q}$ since the different weight spaces, although individually defined over $\mathbf{R}$, need not be defined over $\mathbf{Q}$. However, there is a simple way to ensure this.

Lemma 15.1.3 *Let H be a real vector space and $h : \mathbf{S} \to \mathrm{GL}(H)$ an algebraic representation. It defines a real Hodge structure. If H has a rational structure and $h \circ w : \mathbf{S} \to \mathrm{GL}(H)$ is defined over $\mathbf{Q}$, then h defines a rational Hodge structure.*

What happens if we consider instead algebraic representations $u : \mathbf{U} \to \mathrm{GL}(H)$ of the circle group? Such a representation is not in general the restriction to $\mathbf{U}$ of a morphism $h : \mathbf{S} \to \mathrm{GL}(H)$ giving a weight k Hodge structure; a necessary condition is $u(-1) = (-1)^k \,\mathrm{id}_H$. However, an extension h giving a direct sum of Hodge structures always exists.

Proposition 15.1.4 *Let H be a real vector space and let $u : \mathbf{U} \to \mathrm{GL}(H)$ be an algebraic representation. The following assertions hold.*

- *The morphism u extends to give a Hodge structure. If $u(-1) = 1$ its weight can be any even number; if $u(-1) = -1$ its weight must be odd. In general the Hodge structure has different weights. One gets a rational Hodge structure provided H has a rational structure and if, moreover, $w \circ u$ is defined over $\mathbf{Q}$.*
- *Consider the following commutative diagram of homomorphisms*

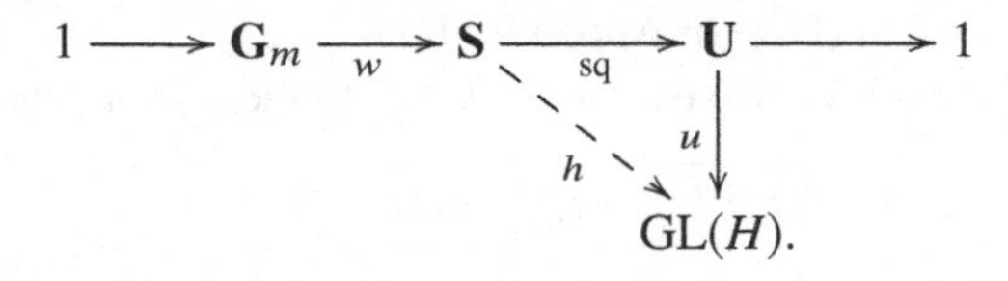

The homomorphism h is a weight zero real Hodge structure. It is a rational Hodge structure if H has a rational structure.

Conversely, if h is a given homomorphism defining a weight zero Hodge structure, a representation u making the diagram commutative exists. If the nonzero character spaces for u occur only among $\{z^{-p}, \ldots, 1, \ldots, z^p\}$, the nonzero Hodge numbers for h occur only among $\{h^{-p,p}, \ldots, h^{0,0}, \ldots, h^{p,-p}\}$.

Proof Let $\chi : \mathbf{U} \to \mathbf{G}_m$ be the character $\chi(z) = z$ and split $H_{\mathbf{C}}$ into character spaces for the action of $\mathbf{U}$, say $H_{\mathbf{C}} = \oplus_r H_{\chi^r}$. Assemble those with even n into $H^+_{\mathbf{C}}$. Since u is real, $H^+_{\mathbf{C}}$ has a real structure, say $H^+_{\mathbf{C}} = H^+ \otimes \mathbf{C}$. Indeed

H^+ is the eigenspace of $u(-1)$ with eigenvalue 1. If H^- is the eigenspace for the eigenvalue -1, we have $H = H^+ \oplus H^-$ and $H^-_{\mathbf{C}}$ is the direct sum of the character spaces for χ^r with r odd. Next, suppose $r = 2m$, fix an even number $k = 2n$, and set $H^{n+m,n-m}_{\mathbf{C}} := H_{\chi^{2m}}$. This gives H^+ an even weight real Hodge structure. Similarly, for $r = 2m+1$ and a given odd number $k = 2n+1$, setting $H^{n+m+1,n-m}_{\mathbf{C}} := H_{\chi^{2m+1}}$ gives H^- the structure of an odd weight real Hodge structure. The claim about rationality follows from Lemma 15.1.3.
The action of $u\circ\mathrm{sq}(z)$ on the character space H_{χ^r} is multiplication by $z^r\bar{z}^{-r}$ and hence we indeed get a weight 0 Hodge structure on H. Note that it is different from the one that extends u! Since it has pure weight zero it is rational as soon as H has a rational structure. Conversely, if h is a given weight 0 Hodge structure, define u by specifying that the character space for χ^r is just $H^{r,-r}$. Indeed $h(z)$, $z \in \mathbf{U}$ acts on it by multiplication with z^{2r} while $u(z)$ acts by definition as multiplication with z^r. The two match under the squaring operation. □

Polarizations

To begin with, observe that if we view a Hodge structure as a representation $h : \mathbf{S} \to \mathrm{GL}(H)$, the Weil operator is just $C = h(\mathrm{i})$.

Definition 15.1.5 Let $h : \mathbf{S} \to \mathrm{GL}(H)$ be a rational (respectively real) Hodge structure of weight k. A *polarization for h* is a morphism of Hodge structures

$$\beta : H \otimes H \to \mathbf{Q}(-k) \text{ (respectively } \mathbf{R}(-k)\text{)};$$

such that

$$(2\pi \mathrm{i})^k \beta(Cx \otimes y) \tag{15.3}$$

is symmetric and positive definite.

Claim 15.1.6 *This notion is the same as the one from Definition 1.2.4(ii).*

Proof If b is a polarization in the "classical sense", let $b_{\mathbf{C}}$ be the $\mathbf{C}$-linear extension of b. Then $b_{\mathbf{C}}(x, y) = 0$, $x \in H^{p,q}$, $y \in H^{r,s}$ unless $(p, q) = (s, r)$. In that case $h(z)x = z^p\bar{z}^q \cdot x$ and $h(z)y = z^q\bar{z}^p \cdot y$ so that $b_{\mathbf{C}}(h(z)x, h(z)y) = (z\bar{z})^k b_{\mathbf{C}}(x, y)$. Hence this equality holds in general. Now set $\beta(x \otimes y) = (2\pi\mathrm{i})^{-k} b(x, y)$. Since $h(z) \cdot b(x, y) = z^{p+q}\bar{z}^{p+q} = (z\bar{z})^k b(x, y)$ the map β is a morphism of Hodge structures. We leave the proof that the form (15.3) is symmetric and positive definite to the reader.

Conversely, if $\beta : H \otimes H \to \mathbf{Q}(-k)$ is a morphism of Hodge structures, for any $x \in H^{p,q}$, $y \in H^{r,s}$ we have $\beta_{\mathbf{C}}(x, y) = 0$ unless $p + r = k = q + s$ which means $(p, q) = (s, r)$. So, for the bilinear form b defined by $b = (2\pi\mathrm{i})^k\beta$, the first bilinear relation holds. The second bilinear relation is equivalent to the stated positivity property. We leave it to the reader to show that the property $b(h(z)x, h(z)y) = (z\bar{z})^k b(x, y)$ for all $z \in \mathbf{S}(\mathbf{R})$ and $x, y \in H$ implies that $b(x, y) = (-1)^k b(y, x)$. □

Remarks 15.1.7 (1) The polarization β can be viewed as a Hodge tensor in $\mathrm{Hom}(H \otimes H, \mathbf{Q}(-k)) = T^{0,2}H(-k)$ of type $(0, 0)$.
(2) Given a polarized Hodge structure (h, β), the morphism $h : \mathbf{S} \to \mathrm{GL}(H)$ is *not* compatible with the polarization since $\beta(h(z)x \otimes h(z)y) = \|z\|^k \beta(x \otimes y) \neq \beta(x \otimes y)$ unless $k = 0$ or $\|z\| = 1$. So what is true is that $h|_{\mathbf{U}}$ preserves β.
(3) As an amusing application of the above, let us give a simple proof of Lemma 5.3.1, namely that the Lie algebra $\mathfrak{g} = \mathrm{Lie}(H, b)$ is a weight 0 Hodge substructure of the one on $\mathrm{End}(H)$. Indeed, consider the adjoint representation

$$\mathrm{Ad} : \mathrm{GL}(H, b) \to \mathrm{GL}(\mathfrak{g}) \subset \mathrm{GL}(\mathrm{End}(H)).$$

Then $\mathrm{Ad} \circ h$ gives a representation of $\mathbf{S}$ in $\mathrm{End}(H)$ which restricts to $\mathfrak{g}$. The representation on $\mathrm{End}(H) = T^{1,1}H$ has weight 0 and so $\mathfrak{g}$ receives an induced weight 0 Hodge structure.

Polarizations and Algebraic Groups

Suppose that we have a Hodge structure $h : \mathbf{S} \to \mathrm{GL}(H)$. When is it polarizable? We need in particular a Hodge tensor in $\mathrm{Hom}(H \otimes H, \mathbf{R}(-k))$ inducing a bilinear form b invariant under the action of the circle $\mathbf{U} \subset \mathbf{S}$. The second condition says that h_C should be symmetric and positive definite. The existence criterion comes from the crucial observation from Proposition 12.6.2: polarizability of b implies that $\mathrm{Ad}(C)$ is a Cartan involution for the Lie algebra of $G = \mathrm{GL}(H, b)$. This is equivalent to saying that $\mathrm{Int}(C)$ =conjugation by C is a Cartan involution for G. This motivates the following definition.

Definition 15.1.8 Let G be an $\mathbf{R}$-algebraic group, $C \in G(\mathbf{R})$ such that $C^2 \in Z(G)$. [1] A C-polarization for a real G-representation H is a G-invariant $\mathbf{R}$-bilinear form b such that

$$b_C(u, v) \stackrel{\text{def}}{=} b(Cu, v)$$

is a metric on H, i.e., b is symmetric and positive definite.

This concept is directly related to Cartan involutions on reductive groups.

Proposition 15.1.9 *Let G be an $\mathbf{R}$-algebraic group. Then we have the following characterization of C-polarizations for G:*

[1] This implies that conjugation by C on G is an involution.

(i) *Suppose that the involution defined by conjugation by C is a Cartan involution.*[2] *Then every finite-dimensional real representation of G carries a C-polarization.*
(ii) *Conversely, if some* faithful *real representation H carries a C-polarization, then* $\mathrm{Int}(C)$ *is a Cartan involution. In particular G is reductive.*

Proof (i) G being reductive, by the results summarized in Appendix D.3, there is a compact real form N of G, i.e., $N(\mathbf{R})$ is compact and $N(\mathbf{C}) = G(\mathbf{C})$. By compactness, there is an $N(\mathbf{R})$-invariant bilinear positive definite form k on H. Define

$$b(x, y) \stackrel{\text{def}}{=} k(C^{-1}x, y), \quad x, y \in H.$$

Then $b_C = k$ is symmetric and positive definite. It remains to show that b is $G(\mathbf{R})$-invariant. Instead we shall prove that the $\mathbf{C}$-bilinear extension $b_{\mathbf{C}}$ of b is invariant under the action of $\mathrm{Lie}(G(\mathbf{C})) = \mathrm{Lie}(N(\mathbf{R})) \oplus \mathrm{i}\,\mathrm{Lie}(N(\mathbf{R}))$:

$$b_{\mathbf{C}}(\xi(u), v) + b_{\mathbf{C}}(u, \xi(v)) = 0, \quad \forall \xi \in \mathrm{Lie}(G(\mathbf{C})), u, v \in H_{\mathbf{C}}.$$

For $\xi \in \mathrm{Lie}(N(\mathbf{R}))$ this follows from the $N(\mathbf{R})$-invariance of b and since $b_{\mathbf{C}}$ is $\mathbf{C}$-bilinear, it also follows for $\mathrm{i}\xi$ with $\xi \in \mathrm{Lie}(N(\mathbf{R}))$.
(ii) If b is any bilinear form on H, and σ the complex conjugation on $G_{\mathbf{C}}$, the form

$$(u, v) \mapsto h(u, v) \stackrel{\text{def}}{=} b_{\mathbf{C}}(C\sigma u, v)$$

is a positive definite hermitian form since b is a C-polarization. Let $g \in K$, the subgroup of G fixed by $\mathrm{Int}(C)\circ\sigma$ Then

$$\begin{aligned} h(gu, gv) &= b_{\mathbf{C}}(C\circ\sigma \cdot g \cdot (\sigma u), gv) \\ &= b_{\mathbf{C}}(g \cdot C\circ(\sigma u), gv) \text{ (since } C\circ\sigma \cdot g = g \cdot C) \\ &= b_{\mathbf{C}}(C\circ(\sigma u), v) \text{ (since } b \text{ is } g\text{-invariant)} \\ &= h(u, v). \end{aligned}$$

But then the subgroup fixed by $\mathrm{Int}(C)\circ\sigma$ is compact and hence $\mathrm{Int}(C)$ is a Cartan involution on the real group G. □

When does a morphism from the Deligne torus to an algebraic group G induce a Hodge structure? This requires some definitions.

Definition 15.1.10 Let G be a reductive $\mathbf{Q}$-algebraic group G and $h : \mathbf{S} \to G_{\mathbf{R}}$ a morphism. With w the weight co-character (15.1), define the *weight* of $h : \mathbf{S} \to G_{\mathbf{R}}$ to be the homomorphism

$$w_h = h\circ w : \mathbf{G}_m \to \mathbf{S} \to G_{\mathbf{R}}. \tag{15.4}$$

[2] In particular, G must be reductive.

Lemma 15.1.3 gives an answer to the above question.

Lemma 15.1.11 *With $h : \mathbf{S} \to G$ as above suppose that w_h is defined over $\mathbf{Q}$. Then any $\mathbf{Q}$-representation $\rho : G \to \mathrm{GL}(H)$ defines by composition a $\mathbf{Q}$-Hodge structure $\rho \circ h : \mathbf{S} \to \mathrm{GL}(H)$.*

We are now ready to translate Proposition 15.1.9 in the framework of polarized real Hodge structures (H, b). We have seen (see Proposition 12.6.2) that $\mathrm{Ad}\, C$, i.e., conjugation with the Weil operator C on the real Lie algebra $\mathrm{Lie}(H, b)$, is a Cartan involution. On the level of Lie groups this is equivalent to $\mathrm{Int}(C)$ being a Cartan involution on $\mathrm{GL}(H, b)$. Suppose that the corresponding Hodge representation $h : \mathbf{S} \to \mathrm{GL}(H, b)$ factors over an algebraic subgroup $G \subset GL(H, b)$. Then $\mathrm{Int}(C)$ is a Cartan involution for G. The preceding proposition, reformulated in this setting, states that being polarized is equivalent to this.

Corollary 15.1.12 (Polarizability criterion) *Let G be an $\mathbf{R}$-algebraic group.*
(i) *Assume that conjugation by the Weil operator $C = h(\mathrm{i})$ is a Cartan involution for G. Then for any real representation $\rho : G \to \mathrm{GL}(H)$, the (real) Hodge structure $\rho \circ h$ is polarized by a G-invariant bilinear form.*
(ii) *Conversely, given a* faithful *real representation $\rho : G \to \mathrm{GL}(H)$ for which $\rho \circ h$ is a polarizable Hodge representation, then G is reductive and the Hodge structure on H is polarizable by a G-invariant form.*

If, moreover, we assume that G is defined over $\mathbf{Q}$, H has a rational structure and the weight w_h of h is defined over $\mathbf{Q}$, then the above assertions hold with "real Hodge structure" replaced by "$\mathbf{Q}$-Hodge structure".

Remark In the preceding corollary it is crucial that the representation be faithful: just take the trivial representation. This is a Hodge representation if and only if H is a direct sum of Tate structures and such a Hodge structure is always polarizable. The group G could be any group, reductive or not.

Problems

15.1.1 Show the converse statement of Lemma-Defiition 15.1.1, namely that a given weight k real Hodge structure on H with Hodge decomposition $H_{\mathbf{C}} = \oplus_{p+q=k} H^{p,q}$ defines an algebraic representation $h : \mathbf{S} \to \mathrm{GL}(H)$ upon setting $h(z) = z^p \bar{z}^q$ on $H^{p,q}$.

15.1.2 Let $h : \mathbf{S}(\mathbf{C}) \to \mathrm{GL}(H)$ be a complex representation space of $\mathbf{S}(\mathbf{C}) = \mathbf{C}^\times \times \mathbf{C}^\times$ such that $h \circ w(z) = z^k$. Decompose H into eigenspaces $H^{p,q}$ for the character $(u, w) \mapsto u^p w^q$. This gives what is sometimes called a *complex Hodge structure of weight k*.

(a) Let u be the restriction of h to the subgroup

$$\mathbf{U}(\mathbf{C}) = \{(z, z^{-1}) \mid z \in \mathbf{C}^{\times}\} \subset \mathbf{S}(\mathbf{C}).$$

Show that $H^{p,q}$ is a character space for this group and show that, conversely, a decomposition into character spaces gives a direct sum of complex Hodge structures of possibly different weights.

(b) Suppose that H is the complexification of a real vector space $H_{\mathbf{R}}$ and h is defined over $\mathbf{R}$. Show that $H_{\mathbf{R}}$ is a direct sum of real Hodge structures.

15.2 Mumford–Tate Groups

Basic Definitions

In this section we assume that the real vector space H has a rational structure, $H = H_{\mathbf{Q}} \otimes \mathbf{R}$ and that $h : \mathbf{S} \to \mathrm{GL}(H)$ is a weight k Hodge structure. The image of h then is an $\mathbf{R}$-algebraic subgroup of $\mathrm{GL}(H)$ but it need not be defined over $\mathbf{Q}$. Indeed, if it were, for instance the Hodge (p, p)-classes in $H_{\mathbf{C}}$ would be the rational points of $H^{p,p}$ which is rarely the case: the Hodge classes span in general a proper subspace $H_{\mathrm{alg}} \subset H^{p,p}$. For $t \in H_{\mathrm{alg}}$ we have $h(z)t = |z|^{2p}t$ and so t can only be invariant if $p = 0$; in general t is only invariant under $h(\mathbf{U})$. However, a (p, p)-class $t \in H_{\mathbf{C}}$ which is invariant under the smallest $\mathbf{Q}$-algebraic subgroup of $\mathrm{GL}(H)$ containing $h(\mathbf{U})$ must belong to a rationally defined subspace; indeed, it must belong to H_{alg}. This motivates a study of this group which – exactly for this reason – was first considered by Mumford and Tate.

Definition 15.2.1 (i) The *Mumford–Tate group* $\mathrm{MT}(h)$ of h is the smallest $\mathbf{Q}$-algebraic subgroup of $\mathrm{GL}(H)$ over which h factors.[3]
(ii) The *special Mumford–Tate group* or *Hodge group* $\mathrm{SMT}(h)$ of h is the smallest $\mathbf{Q}$-algebraic subgroup of $\mathrm{GL}(H)$ over which $h|\mathbf{U}$ factors.[4]

Remark (i) Since $\mathbf{S}$ and $\mathbf{U}$ are connected, both $\mathrm{MT}(h)$ and $\mathrm{SMT}(h)$ are connected algebraic groups.
(ii) The definition and the decomposition (15.2) imply that we have

$$\mathrm{MT}(h) = \mathrm{SMT}(h) \cdot h \circ w(\mathbf{G}_m).$$

If the weight $k = 0$ we have of course $\mathrm{MT}(h) = \mathrm{SMT}(h)$ but otherwise the preceding formula realizes $\mathrm{MT}(h)$ as a group isogenous to $\mathrm{SMT}(h) \times \mathbf{G}_m$.

[3] This is also known as the $\mathbf{Q}$-Zariski closure of $h(\mathbf{S})$ in $\mathrm{GL}(H)$.
[4] We automatically land in the special linear group, since $\det \circ h(\mathbf{U}) \subset \mathbf{G}_m$ is compact and connected.

In the special case of a polarized Hodge structure h we can say more. Let us first recall the following.

Definition 15.2.2 Let (H, b) be a K-vector space equipped with a nondegenerate bilinear form b. The *group of similitudes with respect to* b is defined as

$$\begin{aligned} \mathrm{CGL}(H,b) \stackrel{\text{def}}{=} \{g \in \mathrm{GL}(H) \mid \exists c_g \in K^\times \text{ with} \\ b(gx, gy) = c_g b(x,y), \forall\, x, y \in H\}. \end{aligned} \tag{15.5}$$

Note the inclusions

$$\mathrm{SL}(H,b) \stackrel{\text{def}}{=} \mathrm{Aut}(H,b) \cap \mathrm{SL}(H) \subset \mathrm{Aut}(H,b) \subset \mathrm{CGL}(H,b).$$

Lemma 15.2.3 *For a polarized Hodge structure h we have*

$$\mathrm{SMT}(h) = \mathbf{Q}\text{-}\textit{Zariski closure of } h(\mathbf{U}) \textit{ in } \mathrm{SL}(H,b). \tag{15.6}$$

Thus $\mathrm{MT}(h) \subset \mathrm{CGL}(H,b)$.

Proof By definition $b(h(z)x, h(z)y) = b(x,y)$ whenever $z\bar{z} = 1$, i.e., if $z \in \mathbf{U}(\mathbf{R})$. So $h(\mathbf{U}(\mathbf{R})) \subset \mathrm{Aut}(H,b)$ from which the conclusion follows. □

Examples 15.2.4 (i) The Mumford–Tate group of $\mathbf{Q}(k)$ is $\{1\}$ if $k = 0$ and $\mathbf{G}_m$ for $k \neq 0$. The special Mumford–Tate group in this case is always $\{1\}$.

(ii) Consider a polarized weight one $\mathbf{Q}$-Hodge structure with $h^{1,0} = 1$. Its Hodge group is contained in $\mathrm{Sp}(1) = \mathrm{SL}(2)$. It cannot be trivial since $h(-1) = -1$. Below (Proposition 15.2.6) we show that the Hodge group is reductive. The only nontrivial reductive subgroups of $\mathrm{SL}(2)$ are $\mathrm{SL}(2)$ itself (which is semisimple) or a maximal torus defined over $\mathbf{Q}$. In fact, the Hodge structure comes from an elliptic curve C and if C has no complex multiplication, the endomorphisms of C are just the homotheties and so the center of the Mumford–Tate group is $\mathbf{G}_m$. Hence $\mathrm{MT}(H^1(C)) = \mathrm{GL}(2)$. Recall that an elliptic curve $C = \mathbf{C}/\Lambda$ has complex multiplication by a complex number u if $u\Lambda \subset \Lambda$ (see Example 1.1.4). These complex numbers form a subring of an imaginary quadratic field extension $K = \mathbf{Q}(\sqrt{-d})$ of $\mathbf{Q}$ (with $d \in \mathbf{Q}$ positive). In this case the center of the $\mathbf{Q}$-Mumford–Tate group is isomorphic to $K^\times$ and thus

$$\mathrm{MT}(H^1(C))(\mathbf{Q}) = \mathrm{Res}_{K/\mathbf{Q}}\, K^\times \simeq \left\{ \begin{pmatrix} a & bd \\ -b & a \end{pmatrix} \in \mathrm{GL}(2,\mathbf{Q}) \;\middle|\; a, b \in \mathbf{Q} \right\},$$

the Weil restriction of $K^\times$ to $\mathbf{Q}$. The Hodge group $\mathrm{SMT}(H^1(C))(\mathbf{Q})$ is also abelian and equal to the torus $\{z \in K \mid z\bar{z} = 1\}$, which over $\mathbf{Q}$ is the circle group of matrices $\begin{pmatrix} a & bd \\ -b & a \end{pmatrix}$ of determinant 1, i.e., with $a^2 + b^2 d = 1$.

There is a left action of $\mathrm{SL}(2,\mathbf{Q})$ on the upper half-plane. It sends a point whose Mumford–Tate group is an algebraic torus to an algebraic torus. In other words, this action preserves the set $E_{\rm CM}$ of elliptic curves with complex multiplication. Since $\mathrm{SL}(2,\mathbf{Q})$ is dense in $\mathrm{SL}(2,\mathbf{R})$, the set $E_{\rm CM}$ is thus dense in the upper half plane. The action of $\mathrm{SL}(2,\mathbf{Q})$ induces Hecke correspondences on modular curves. We come back to this in greater generality. See Eq. (16.2).

The last example motivates the following concept.

Definition 15.2.5 Let $h : \mathbf{S} \to \mathrm{GL}(H)$ be a polarizable Hodge structure. We say that (H, h) is a *CM-Hodge structure* if $\mathrm{MT}(h)$ is abelian.

We also observe a further important property for polarized Hodge structures.

Proposition 15.2.6 *The Mumford–Tate group of a polarizable* $\mathbf{Q}$*-Hodge representation is a reductive algebraic group.*

Proof By Lemma 15.2.3 a rational polarized Hodge structure on (H, b) induces a representation $u : \mathbf{U} \to G = \mathrm{GL}(H, b)$. The special Mumford–Tate group $\mathrm{SMT}(h)$ contains $h(\mathrm{i}) = C$, the Weil operator, so is stable under conjugation by C. Since, by Theorem 12.6.2, $\mathrm{Ad}(C)$ coincides with the Cartan involution on the Lie algebra of G, it follows that $\mathrm{SMT}(h)(\mathbf{R})$ is preserved by the Cartan involution. According to Proposition D.3.13 this implies that $\mathrm{SMT}(h)(\mathbf{R})$ is reductive. Since, by Eq. (15.6) in Section 15.2, the Mumford–Tate group either equals this group or is isogeneous to the product $\mathrm{SMT}(h)(\mathbf{R}) \times \mathbf{R}^{\times}$, also the Mumford–Tate group is reductive. □

Relation with Hodge Tensors

If a representation $h : \mathbf{S} \to \mathrm{GL}(H)$ defines a rational Hodge structure, any $\mathbf{Q}$-representation $\rho : \mathrm{GL}(H) \to \mathrm{GL}(H')$ defines a rational Hodge structure on H'. We have the following.

Lemma 15.2.7 *A* $\mathbf{Q}$*-subspace* $S \subset H'$ *is a Hodge substructure if and only if* S *is an* $\mathrm{MT}(h)$*-submodule.*

Proof Consider the $\mathbf{Q}$-Zariski closure $M \subset \mathrm{GL}\, H$ of the group of those $g \in \mathrm{GL}\, H$ for which $\rho_g S \subset S$. Since S is a Hodge substructure of H' it follows that $h(\mathbf{S})$ belongs to M and by the minimality property of the Mumford–Tate group, it follows that $\mathrm{MT}(h) \subset M$ and so S is a $\mathrm{MT}(h)$-module. The converse is obvious. □

We apply this to a finite direct sum T of tensor representations $T^{m,n}H$, as discussed in Example 15.1.2 (iv) and we take for S a rationally defined line $L \subset T$.

By the Lemma, L is left stable by $\mathrm{MT}(h)$ if and only if it is a Hodge substructure. This means that L is spanned by a Hodge tensor. We need a bit more.

Lemma 15.2.8 *Let $H = H_{\mathbf{Q}} \otimes \mathbf{R}$ be a rational Hodge structure and let T be a finite direct sum of tensor spaces $T^{m,n}H$ viewed as a Hodge structure. Then* (i) *a vector $t \in T$ is a weight 0 Hodge tensor if and only if t is a fixed vector for the Mumford–Tate group* $\mathrm{MT}(h)$*;*
(ii) *if T has pure weight, $t \in T$ is a Hodge tensor if and only if t is a* fixed vector *for the special Mumford–Tate group* $\mathrm{SMT}(h)$.

Proof (i) In view of the remarks preceding this lemma, we only need to show that if $\mathrm{MT}(h)$ leaves the line L spanned by t invariant, then it actually fixes t. To show this consider $T^{0,0}$, the one-dimensional tensor representation for which $h = \mathrm{id}$. So $1 \in T^{0,0}$ is a Hodge vector and so the line spanned by $(1, t) \in T^{0,0} \oplus L$ is left invariant by any $g \in \mathrm{MT}(h)$. Since $g(1, t) = (1, g(t)) \in L$ we must have $g(t) = t$.
(ii) Pure weight is used to ensure that all elements of $\mathbf{G}_{m,\mathbf{R}}$ acts as homotheties on T, and so if $t \in T$ is fixed by $\mathrm{SMT}(h)$, the line L spanned by $t \in T_{\mathbf{Q}}$ is left invariant by the action of $\mathrm{MT}(h)$. It then follows from Lemma 15.2.7 that t is a Hodge tensor. □

More importantly, this characterizes the (special) Mumford–Tate group.

Theorem 15.2.9 *Let $h : \mathbf{S} \to \mathrm{GL}(H)$ be a $\mathbf{Q}$-Hodge structure.*
i) *The Mumford–Tate group of h is the largest algebraic subgroup of* $\mathrm{GL}(H)$ *which fixes weight 0 Hodge tensors in any finite direct sum T of tensor representations $T^{m,n}H$. In other words,*

$$\mathrm{MT}(h) = \bigcap_T Z_{\mathrm{GL}(H)}[\mathrm{Hdg}(T)]^{w=0},$$
$$T = T_J = \bigoplus\nolimits_{(m,n)\in J} T^{m,n}H, \ J \text{ runs over finite subsets of } \mathbf{Z}_{\geq 0} \times \mathbf{Z}_{\geq 0}.$$

Moreover, there exists already a finite *set $\mathfrak{T}$ of weight 0 pure Hodge tensors (i.e. in some $T^{m,n}H$) such that*

$$\mathrm{MT}(h) = Z_{\mathrm{GL}(H)}(\mathfrak{T}).$$

ii) *The special Mumford–Tate group of h is the largest algebraic subgroup of* $\mathrm{GL}(H)$ *which fixes Hodge tensors in any finite direct sum T of tensor representations $T^{m,n}H$ of pure weight. In other words,*

$$\mathrm{SMT}(h) = \bigcap_T Z_{\mathrm{GL}(H)}\mathrm{Hdg}(T),$$
$$T = T_J = \bigoplus\nolimits_{(m,n)\in J} T^{m,n}H, \quad (m-n)k = N(J),$$
$$J \text{ runs over finite subsets of } \mathbf{Z}_{\geq 0} \times \mathbf{Z}_{\geq 0}.$$

The proof of this theorem uses some well known facts from representation theory which we review now. First of all since $\mathrm{GL}(H)$ is reductive, we may use the following.

Lemma 15.2.10 (Deligne et al., 1982, Prop. 1.3) *Let K be a field of characteristic zero. Any finite dimensional representation of a K-reductive affine group $G \subset \mathrm{GL}(H)$ is contained in a finite direct sum of tensor representations of H.*

Proof Let G be a reductive group and H' a finite dimensional G-representation. This yields a surjective morphism $G \times H' \twoheadrightarrow H'$ given by $(g, v) \mapsto g(v)$. The resulting injection

$$H' \hookrightarrow K[G] \otimes H' \simeq K[G]^{\oplus n}, \quad n = \dim H',$$

can be viewed as an equivariant morphism where G acts on the right "term by term". So we have embedded the given representation in sums of the regular representation $K[G]$. We want to replace the regular representation by a finite-dimensional one. To do this, note that

$$G \subset \mathrm{GL}(H) \subset \mathrm{End}(H) \simeq K^N \implies K[G] = K[X_1, \ldots, X_N]/I_G,$$

where I_G is the ideal of G. The regular representation is filtered by polynomial degree and since I_G is finitely generated, for some d, H' is contained in the n-fold direct sum of

$$K[G]_{\leq d} = T/I_G \cap T, \quad T = \bigoplus\nolimits_{m \leq d} S^m(\mathrm{End}(H)) \simeq \bigoplus\nolimits_{m \leq d} T^{m,m} H.$$

Since G is reductive, its representations are fully reducible (see e.g. Satake, 1980, I.3) and so the surjection $T \twoheadrightarrow K[G]_{\leq d}$ splits and hence we get an embedding $K[G]_{\leq d} \hookrightarrow T$ and it follows that H' embeds in $T^{\oplus n}$. □

The second fact we now use is that subgroups $G \subset \mathrm{GL}(H)$ can be completely characterized by the lines they leave invariant in such direct sums. More precisely, we have the following.

Theorem 15.2.11 (Theorem of Chevalley (Humphreys, 1981, 11.2)) *Let G' be an affine algebraic group and $G \subset G'$ a closed subgroup. For every algebraic representation $\rho : G' \to \mathrm{GL}(T)$ there exists a line $L \subset T$ such that G is exactly the subgroup of G' that leaves the line L invariant:*

$$G = \left\{ g \in G' \mid \rho_g(L) = L \right\}.$$

We apply this to our G with $G' = \mathrm{GL}(H)$. By what we just said, we may take for T a finite direct sum of tensor representation for H. This indeed characterizes

G: determine the G-invariant lines L in such sums of tensor spaces and then

$$G = \bigcap_L \mathrm{Stab}_{\mathrm{GL}(H)} L.$$

But, in view of the characterization of Hodge tensors as invariants, we need to replace "invariant line" by "invariants". This cannot be done for an arbitrary closed subgroup $G \subset \mathrm{GL}(H)$. We need special properties of the character group of G.

Corollary 15.2.12 *Assume that $G \subset \mathrm{GL}(H)$ is a closed subgroup and that every character of G is the restriction of a character of $\mathrm{GL}(H)$. Then there exists $t \in T$, T a finite direct sum of tensor representations of H, such that $G = Z_{\mathrm{GL}(H)} t$.*

Proof By the preceding remarks, there exists T as in the theorem containing a G-invariant line L such that $G = \mathrm{Stab}_{\mathrm{GL}(H)} L$. If every character of G is the restriction of a character of $\mathrm{GL}(H)$, the G-invariant line $L \subset T$ (which corresponds to a character of G) is in fact a $\mathrm{GL}(H)$-module. Suppose $t \neq 0$ spans the line $L \otimes L^\vee$ within the $\mathrm{GL}(H)$-module $T \otimes L^\vee$. The vector t is then fixed by $g \in \mathrm{GL}(H)$ if and only if $g \in G$. □

Proof of Theorem 15.2.9 i) We first verify that the assumptions of Corollary 15.2.12 are valid for $G = \mathrm{MT}(h)$. So we need to show that characters for the Mumford–Tate group extend to all of $\mathrm{GL}(H)$. Such a character determines a real line L left stable by $\mathbf{S}$. Then L is a simultaneous eigenspace for $h(z)$, $z \in \mathbf{S}$. Since we may assume that $\mathbf{S}$ does not act as the identity on the line L, this is only possible if the weight is nonzero and even, say $k = 2p$, and $L = L_{\mathbf{R}}^{p,p}$. Then $\mathbf{S}$ acts as a homothety on L with corresponding character $\det^k$. This character lifts to $\mathrm{GL}(H)$.

Now we can complete the proof. By Corollary 15.2.12 there exists a rationally defined line $L \subset T$, T a direct sum of tensor representations for H, whose stabilizer is $G = \mathrm{MT}(h)$. Then, by the first step of the proof, we may assume that G leaves L point-wise fixed. But then L is spanned by a weight 0 Hodge tensor t. On the other hand, suppose that t is a weight 0 Hodge tensor in some direct sum of tensor representations for H. Then, by Lemma 15.2.8 it is fixed by $G = \mathrm{SMT}(h)$.

These two remarks imply that

$$G = Z_{\mathrm{GL}(H)} t = \bigcap_T Z_{\mathrm{GL}(H)} [\mathrm{Hdg}(T]^{w=0},$$

where T runs over all finite sums of tensor representations $T^{m,n} H$.

Finally, if $t = \sum t_{m,n}$, $t_{m,n} \in T^{m,n} H$, clearly $G = \bigcap_{m,n} Z_{\mathrm{GL}(H)} t_{m,n}$ and we can take $\mathfrak{T} = \{t_{m,n}\}$.

ii) The proof is the same, except that now one has to use Lemma 15.2.8.2. □

Remark 15.2.13 If $h : \mathbf{S} \to \mathrm{GL}(H)$ defines a $\mathbf{Q}$-Hodge structure, the group $\mathbf{S} \times \mathbf{G}_m$ acts on Tate-twisted tensor spaces $T^{m,n}H \otimes \mathbf{Q}(p)$ in the obvious way. The *extended Mumford–Tate group* of h is the smallest $\mathbf{Q}$-algebraic subgroup of $\mathrm{GL}(H) \times \mathbf{G}_m$ such that $\mathrm{Im}(h(\mathbf{C})) \times \mathbf{G}_m(C) \subset G(\mathbf{C})$. The extended Mumford–Tate group is the largest subgroup of $\mathrm{GL}(H) \times \mathbf{G}_m$ which leaves invariant all weight 0 Hodge tensors in finite direct sums of the form $T^{m,n}H \otimes \mathbf{Q}(p)$. See Problem 15.2.2.

Problems

15.2.1 Let $H = \mathbf{R}e \oplus \mathbf{R}f$ equipped with the standard symplectic form and let $H^{1,0} = (e + \mathrm{i}f) \cdot \mathbf{C} \subset H_{\mathbf{C}}$. Give the associated Hodge representation and calculate its Mumford–Tate group. Do the same with $H^{1,0} = (e + \pi\mathrm{i}f) \cdot \mathbf{C}$.

15.2.2 (a) Calculate the extended Mumford–Tate group for $\mathbf{Q}(p)$.
(b) How does the extended Mumford–Tate group in general compare to the Mumford–Tate group?
(c) Show that the extended Mumford–Tate group is the largest subgroup of $\mathrm{GL}(H) \times \mathbf{G}_m$ which leaves invariant all weight 0 Hodge tensors in finite direct sums of the form $T^{m,n}H \otimes \mathbf{Q}(p)$.

15.2.3 Let $H_1 \neq 0$ and $H_2 \neq 0$ be two Hodge structures of the same weight.
(a) Show that $\mathrm{MT}(H_1 \oplus H_2) \subset \mathbf{G}_m \cdot (\mathrm{SMT}(H_1) \times \mathrm{SMT}(H_2))$ and conclude that one never has $\mathrm{MT}(H_1 \oplus H_2) = \mathrm{MT}(H_1) \times \mathrm{MT}(H_2)$.
(b) Let $H = H_1^{\oplus m_1} \oplus H_2^{\oplus m_2}$ and consider the embedding

$$j : \mathrm{GL}(H_1) \times \mathrm{GL}(H_2) \hookrightarrow \mathrm{GL}(H),\ (g_1, g_2) \mapsto (\underbrace{g_1, \ldots, g_1}_{n_1 \text{ times}}, \underbrace{g_2, \ldots, g_2}_{n_2 \text{ times}}).$$

For the natural Hodge structure on H show that $\mathrm{MT}(H) = \mathrm{MT}(H_1 \oplus H_2)$.
(c) Generalize (b) to an arbitrary number of summands.

15.3 Mumford–Tate Subdomains and Period Maps

Mumford–Tate Subdomains

Consider a period domain which classifies polarized Hodge structures on a (real) vector H of the same type as a given Hodge structure h on H:

$$D = G(\mathbf{R})/V, \quad G = \mathrm{Aut}(H, b),\ V = \mathrm{Stab}_{G(\mathbf{R})}o,\ o \in D \text{ corresponding to } h.$$

We have seen in Section 4.3 that these are complex manifolds which are open subsets in compact complex homogeneous manifolds $\check{D} = G(\mathbf{C})/B$. Moreover, D is the entire orbit $G \cdot o$ of G inside $\check{D}$. In this section we consider orbits of subgroups of G that are also complex manifolds of D.

Definition 15.3.1 The *Mumford–Tate subdomain* associated with $o \in D$ is the orbit $\mathrm{MT}(h) \cdot o \subset D$, where h is a Hodge structure associated with o.

Indeed, for these we show the following.

Proposition 15.3.2 *A Mumford–Tate subdomain of D is a complex submanifold of D.*

Proof By Theorem 15.2.9 the group M is the stabilizer in G of a finite set $\mathfrak{T}$ of weight 0 Hodge tensors. By assumption, for $t \in \mathfrak{T}$ and $z \in \mathbf{S}(\mathbf{R})$ we have $h(z)t = t$. So, if $g(t) = t$ we have $(h(z)^{-1} \circ g \circ h(z))\, t = t$. So any $h(z)$-conjugate of g fixes all tensors from $\mathfrak{T}$ and so belongs also to the Mumford–Tate group M. Now the result follows from Lemma 15.3.3. □

Lemma 15.3.3 *Let $M \subset G = \mathrm{Aut}(H, b)$ be a $\mathbf{Q}$-algebraic subgroup with the property that $h(\mathbf{S})$ normalizes M. Then $\mathrm{Lie}(M(\mathbf{R}))$ is a $\mathbf{Q}$-Hodge substructure of the weight 0 Hodge structure on $\mathfrak{g} = \mathrm{Lie}(G)$ induced by $\mathrm{Ad} \circ h$. In particular, the orbit $M(\mathbf{R}) \cdot o \subset D$ is a complex manifold (homogeneous under $M(\mathbf{R})$).*

Proof The assumption implies

$$\mathrm{Ad} \circ h(\mathbf{S}(\mathbf{R})) \subset \mathrm{GL}(\mathrm{Lie}(M(\mathbf{R}))) \subset \mathrm{GL}(\mathfrak{g})$$

and so $\mathrm{Lie}(M(\mathbf{R}))$ receives a Hodge structure as a Hodge substructure of the one on $\mathfrak{g}$. This is an argument we already employed: see Remark 15.1.7 (3). We now use Lemma 12.2.2 which explains how a Hodge decomposition on $\mathfrak{g}_{\mathbf{C}}$ induces a complex structure on the tangent space $T_{D,o}$. Indeed $T_{D,o} \otimes \mathbf{C} = \mathfrak{g}_{\mathbf{C}}/\mathfrak{v}_{\mathbf{C}} = \mathfrak{m}^+ \oplus \mathfrak{m}^- = \mathfrak{m}_{\mathbf{C}}$ is the decomposition into the $\pm$i-eigenspaces of the complex structure J induced on the complexified tangent space.

The tangent space to $M(\mathbf{R}) \cdot o$ at o equals $\mathrm{Lie}(M(\mathbf{R}))/[\mathfrak{v} \cap \mathrm{Lie}(M(\mathbf{R}))] \simeq \mathrm{Lie}(M(\mathbf{R})) \cap \mathfrak{m}$ and so, if $\mathrm{Lie}(M(\mathbf{R})) \subset \mathfrak{g}$ is a Hodge substructure, on its complexification we have an induced decomposition

$$(\mathrm{Lie}(M(\mathbf{R})) \cap \mathfrak{m}^+) \oplus (\mathrm{Lie}(M(\mathbf{R})) \cap \mathfrak{m}^-) = (\mathrm{Lie}(M(\mathbf{R})) \cap \mathfrak{m})_{\mathbf{C}}$$

into the $\pm$i-eigenspaces of the complex structure J. This means precisely that the tangent space $T_{M(\mathbf{R}) \cdot o, o}$ is a complex subspace of $T_{D,o}$. We leave it to the reader to verify that it then follows that $M(\mathbf{C}) \cdot o$ is a complex manifold containing $M(\mathbf{R}) \cdot o$ as an open subset. □

Noether–Lefschetz Phenomena

We are going to prove a structure theorem for period maps which takes full advantage of Mumford–Tate subdomains, but before we do that, in this subsection we explain how to select a good base point in the period domain.

The situation is as follows. We have a polarized variation of Hodge structure with quasi-projective smooth base S defining a period map

$$\mathcal{P} : S \to \Gamma\backslash D.$$

The Mumford–Tate groups M_s of the Hodge structure over the points $s \in S$ vary; however, outside countably many proper subvarieties of S the Mumford–Tate group is the same. This can be seen as follows. Let H be the vector space underlying the Hodge structures. Recall that M_s is the largest algebraic subgroup of $\mathrm{GL}(H)$ fixing the tensors that are Hodge of weight 0 with respect to $s \in S$. Each of these tensors t remain Hodge over a closed subvariety $Z(t) \subset S$. See Problem 7.5.2. Consider those t for which $Z(t)$ is a proper subvariety and introduce

$$S_{\mathrm{gen}} \stackrel{\mathrm{def}}{=} S - Z, \quad Z \stackrel{\mathrm{def}}{=} \left\{ \bigcup_{t \text{ a Hodge tensor of weight } 0} Z(t) \,\middle|\, Z(t) \neq S \right\}.$$

Remark 15.3.4 The variety $Z(t)$ is also called the *Noether–Lefschetz locus* for the class t. The classical case is when $\mathcal{P}$ is the period map associated with a smooth family of projective manifolds.

Since there are at most countably many t for which $Z(t)$ is a proper subvariety of S, the subset S_{gen} is connected dense subset of S. By definition, for all $s \in S_{\mathrm{gen}}$ the Mumford–Tate group M_s is the same subgroup of $\mathrm{GL}(H)$. This motivates the following definition.

Definition 15.3.5 Let there be given a polarized variation of Hodge structure over a smooth base S with period map $\mathcal{P} : S \to \Gamma\backslash D$. A point $s \in S_{\mathrm{gen}}$ is called *$\mathcal{P}$-Hodge generic* and

$$\mathrm{MT}(\mathcal{P}) \stackrel{\mathrm{def}}{=} \mathrm{MT}(h_s), \quad s \in S_{\mathrm{gen}}$$

is called the *Mumford–Tate group of the variation.*

Observe also by construction, for all points $s \in S$ we have

$$\mathrm{MT}(h_s) \subset \mathrm{MT}(h_{s_{\mathrm{gen}}}), \quad s_{\mathrm{gen}} \in S_{\mathrm{gen}}, \tag{15.7}$$

i.e., the Mumford–Tate group at any given point is always contained in the generic Mumford–Tate group and might only become smaller.

Example 15.3.6 We go back to the Legendre family $y^2 = x(x-1)(x-\lambda)$, $\lambda \in S = \mathbf{P}^1 - \{0, 1, \infty\}$. The period map is a $(6:1)$ morphism $\mathcal{P} : S \to \mathrm{SL}(2, \mathbf{Z})\backslash\mathfrak{h}$. Since the generic fibre is an elliptic curve without complex multiplication, by Example 15.2.4 (ii), $\mathrm{MT}(\mathcal{P}) = \mathrm{SL}(2)$. There is a *dense* set Z of points $s \in S$ corresponding to elliptic curves admitting complex multiplication in $K = \mathbf{Q}(\sqrt{-\delta})$ and so $h(\mathcal{P}(s)) = \mathrm{Res}_{K/\mathbf{Q}}\,\mathbf{G}_m$ which is strictly smaller. In this example $S_{\mathrm{gen}} = S - Z$ which shows that in defining Hodge-genericity, it does not suffice to delete a proper Zariski closed subset from S.

Structure of Period Maps

From the previous discussion together with the containment (15.7), it follows that the period map $\tilde{\mathcal{P}}$ has its image in the corresponding Mumford–Tate domain $D(\mathrm{MT}(\mathcal{P}))$. We now consider the action of Γ on D.

Lemma–Definition 15.3.7 *Let Γ^{Zar} be the smallest $\mathbf{Q}$-algebraic subgroup of* $\mathrm{Aut}(H, b)$ *containing Γ. Its connected component* $\mathrm{Mon}(\mathcal{P})$ *is called the* algebraic monodromy group.

- *There exists a group of finite index of Γ^{Zar} which is contained in* $\mathrm{MT}(\mathcal{P})$. *In particular, we have an inclusion*

$$\mathrm{Mon}(\mathcal{P}) \subset \mathrm{MT}(\mathcal{P}).$$

- *Set*

$$D(\mathcal{P}) \stackrel{\mathrm{def}}{=} \Gamma\text{-orbit of } D(\mathrm{MT}(\mathcal{P})) \text{ in } D.$$

This set has finitely many connected components; we have

$$\mathcal{P} : S \to \Gamma\backslash D(\mathcal{P}).$$

Proof To see finiteness we argue as follows. The monodromy group Γ acts as a finite group on Hodge tensors since the polarization has a fixed sign on these. Hence a finite index subgroup Γ' of Γ fixes Hodge tensors. By Theorem 15.2.9 the Mumford–Tate group is the stabilizer in $G = \mathrm{Aut}(H, b)$ of the weight 0 Hodge tensors at s, and hence Γ' as well as its Zariski closure are contained in the Mumford–Tate group at s. Apply this to $s \in S_{\mathrm{gen}}$. □

The algebraic monodromy group turns out to be a *normal* subgroup of the Mumford–Tate group. To show this, we use André's criterion for normality.

Proposition 15.3.8 (André, 1992, Lemma 1) *Let $k \subset \mathbf{C}$ be some field. Assume that $G' \subset \mathrm{GL}(H)$ is a reductive k-algebraic matrix group and that G is* *a closed algebraic subgroup of G'. Suppose that for all characters χ of G and all tensors $T^{r,s}H$ the subspace*

$$(T^{r,s}H)^{\chi} \stackrel{\text{def}}{=} \{x \in T^{r,s}H \mid g(x) = \chi(v)x, \quad \forall g \in G\}$$

is preserved by G'. Then G is normal in G'.

Proof Chevalley's theorem 15.2.11 states that $G = \mathrm{Stab}_{G'}(L)$ for some line L in some representation of G'. Since G' is reductive, by Lemma 15.2.10, we may take for this representation a finite direct sum T of tensor representations of H. The line L defines a character χ for G and we may form the G'-submodule $S \subset T$ generated by this line. By assumption T^{χ} is preserved by G' and hence, also $S^{\chi} = S \cap T^{\chi}$ is a G'-module. So it must coincide with the irreducible G'-module S: i.e., G acts on *all* of S through the character χ.

Next, consider the induced G'-representation $\mathrm{End}(S)$. We are going to show that the kernel of this representation is precisely $G \subset G'$, from which the result follows.

First of all, as we have just seen, G acts on all of S through the character χ, and so, since G acts on $\mathrm{End}(S)$ by conjugation, it fixes all elements: G is contained in the kernel of this representation. Conversely, using that by construction S is an irreducible G'-module, if $g \in G'$ acts as the identity on $\mathrm{End}(S)$, Schur's lemma implies that $g = c \cdot \mathrm{id}$, $c \in k$. In particular, $gL = L$, i.e., $g \in G$. □

Now recall Corollary 13.1.12. It can be reformulated as follows: given a variation of **Q**-Hodge structures over a quasi-projective S, the largest constant local subsystem (i.e., the one on which the monodromy group Γ acts as the identity) is a constant variation of Hodge structure. We use this as a crucial ingredient in the proof of the following result.

Proposition 15.3.9 (André, 1992, Theorem 1) *Assume that S is a quasi-projective manifold and carries a rational variation of Hodge structure. Then the algebraic monodromy group is a normal subgroup of the Mumford–Tate group of the variation.*

Proof Assume for simplicity that Γ is already contained in $\mathrm{MT}(\mathcal{P})$ (this can be achieved upon replacing Γ by $\Gamma \cap \mathrm{Mon}(\mathcal{P})$, a group of finite index in Γ). So $\mathrm{Mon}(\mathcal{P}) \subset \mathrm{MT}(\mathcal{P})$ and we are going to apply the criterion of Proposition 15.3.8 to $G' = \mathrm{MT}(\mathcal{P})$ which is indeed reductive by Theorem 15.2.6: it is the Mumford–Tate group of a Hodge-generic point. The role of G will be played by $\mathrm{Mon}(\mathcal{P})$. To apply the criterion we note that the G-representation space $T^{r,s}H$ gives a variation of Hodge structure, say $\mathbf{T}$.

Now we use that the Hodge structure is rational and that $\Gamma \subset \mathrm{GL}(H_{\mathbf{Q}})$. It implies that if Γ stabilizes a line, this must be a rational line and hence the corresponding character χ of $\mathrm{Mon}(\mathcal{P})$ has values in ± 1. Since $\mathrm{Mon}(\mathcal{P})$ is connected we have $\chi = 1$. In particular, $(T^{r,s}H)^{\chi}$ is the subspace on which the monodromy acts as the identity. It corresponds to the largest local system $\mathbf{T}^{\mathrm{fix}}$ of $\mathbf{T}$ on which monodromy acts as the identity. So, as we just recalled, because of the Theorem of the fixed part (Corollary 13.1.12) $\mathbf{T}^{\mathrm{fix}}$ is a constant variation of Hodge structure: it is stable under $G = \mathrm{MT}(\mathcal{P})$, and so the condition in Proposition 15.3.8 is verified. □

We can say a bit more. Recall (see Appendix Section D.3) that the derived group G^{der} of a reductive group G is semisimple; it is the largest semisimple subgroup of G. Furthermore, multiplication gives a surjective isogeny

$$G^{\mathrm{der}} \times Z(G) \twoheadrightarrow G.$$

Apply this to $G = \mathrm{MT}(\mathcal{P})$. Since any abelian subgroup of G such as $Z(G)$ acts trivially by conjugation on G, it belongs to the stabilizer V of any lift in $D = G/V$ of $\mathcal{P}(s)$, s a Hodge generic point. But V is compact and so the discrete group Γ meets it in at most a finite group. We conclude the following.

Corollary 15.3.10 *We have*

$$\mathrm{Mon}(\mathcal{P}) \triangleleft \mathrm{MT}(\mathcal{P})^{\mathrm{der}}.$$

There is an important case where we have equality.

Proposition 15.3.11 (André, 1992, Prop. 2) *If S has a CM-point, then* $\mathrm{Mon}(\mathcal{P}) = \mathrm{MT}(\mathcal{P})^{\mathrm{der}}$.

Proof For $s \in S$, we let M_s be its Mumford–Tate group. Consider now a Hodge generic point $s \in S_{\mathrm{gen}}$. By passing to a finite unramified cover, we may assume that the monodromy group at s of the variation is contained in $\mathrm{Mon}(\mathcal{P})$. Consider any Hodge tensor $t \in T$ of weight zero fixed by the action of $\pi = \pi_1(S, s)$, the fundamental group of S based at s. It suffices to show that the M_s-module spanned by t, say W_s, has an abelian action of M_s. Indeed, this shows that $M_s/\mathrm{Mon}(\mathcal{P}) = \mathrm{MT}(\mathcal{P})/\mathrm{Mon}(\mathcal{P})$ is abelian which forces the claimed equality.

The module W_s is π-invariant. This is a consequence of the theorem of the fixed part, Theorem 13.1.10. Indeed, the tensor t defines a global flat section and so its Hodge components $t^{p,-p}$ are flat and so are preserved by π. It follows that for all $g \in M_s$ we have

$$\pi[(gt)^{p,-p}] = \pi[z^p \bar{z}^{-p} t^{p,-p}] = z^p \bar{z}^{-p} \pi[t^{p,-p}] = z^p \bar{z}^{-p} t^{p,-p} = (gt)^{p,-p}.$$

Hence there is a constant subvariation of Hodge structure with underlying local system $\mathbf{W} \subset \mathbf{T}$ on S with W_s as stalk at s. The assumption is that there is a CM-point $o \in S$. Hence the group M_o is abelian and since $\mathbf{W}$ is a constant variation of Hodge structure, this group acts in the same way as M_s on the stalk W_s of $\mathbf{W}$ at s. This is an action through an abelian group. □

We use André's results in connection with the general results on reductive groups recalled in Lemmas D.3.7 and D.3.8. In our case we obtain the following decomposition.

Lemma 15.3.12 *Write* $M = \mathrm{Mon}(\mathscr{P})$ *and introduce the following* $\mathbf{Q}$*-algebraic groups*

$$\begin{aligned} M^{\mathrm{nc}} &\overset{\mathrm{def}}{=} G_1 \cdots G_k \subset M, \quad G_i \text{ a } \mathbf{Q}\text{-simple factor of } M \text{ with } G_i(\mathbf{R}) \text{ not compact,} \\ M^{\mathrm{c}} &\overset{\mathrm{def}}{=} G_{k+1} \cdots G_\ell \subset M, \quad G_i \text{ a } \mathbf{Q}\text{-simple factor of } M \text{ with } G_i(\mathbf{R}) \text{ compact,} \\ R &\overset{\mathrm{def}}{=} G_{\ell+1} \cdots G_m \subset \mathrm{MT}(\mathscr{P})^{\mathrm{der}}, \text{ product of the remaining } \mathbf{Q}\text{-simple factors} \\ &\qquad \text{of } \mathrm{MT}(\mathscr{P})^{\mathrm{der}}. \end{aligned}$$

Then the multiplication map

$$M^{\mathrm{nc}} \times M^{\mathrm{c}} \times R \longrightarrow \mathrm{MT}(\mathscr{P})^{\mathrm{der}} \tag{15.8}$$

is a surjective isogeny.

We now discuss the meaning of the corresponding period domains for these groups. Fix a lift $o \in D(\mathrm{MT}(\mathscr{P}))$ of the period image $\mathscr{P}(s)$ of a Hodge generic point s. For any subgroup $G' \subset \mathrm{MT}(\mathscr{P})$ we shall write

$$D(G') \overset{\mathrm{def}}{=} G'\text{-orbit of } o \text{ in } D(\mathrm{MT}(\mathscr{P})).$$

Proposition 15.3.13 *We have* $D(\mathrm{MT}(\mathscr{P})) = D(\mathrm{MT}(\mathscr{P})^{\mathrm{der}})$*. The domains* $D(G_i)$ *associated with the simple factors* G_i *of* $\mathrm{MT}(\mathscr{P})^{\mathrm{der}}$ *are complex submanifolds of* D*. The homomorphism* (15.8) *induces a surjective holomorphic map*

$$D(M^{\mathrm{nc}}) \times D(M^{\mathrm{c}}) \times D(R) \twoheadrightarrow D(\mathrm{MT}(\mathscr{P}))$$

with finite fibers.

Proof The group $Z(\mathrm{MT}(\mathscr{P}))$ is a connected abelian group and belongs to the stabilizer $V = \mathrm{Stab}_G o$ of o in G. This shows the first assertion. Next, each factor G_i is normal in the Mumford–Tate group and so $h(\mathbf{S})$ normalizes G_i and we can apply Lemma 15.3.3 to see that $D(G_i)$ is a complex submanifold of D. The last assertion follows from the preceding surjective isogeny (15.8). □

Let us now study the period maps. The monodromy group Γ meets M, the connected component of Γ^{Zar}, in a group Γ^0 of finite index. We introduce

$$\begin{aligned} \Gamma^{\mathrm{nc}} &\overset{\mathrm{def}}{=} \Gamma^0 \cap M^{\mathrm{nc}}, \\ \Gamma^{\mathrm{c}} &\overset{\mathrm{def}}{=} \Gamma^0 \cap M^{\mathrm{c}}. \end{aligned}$$

We can now formulate the result we are after.

Theorem 15.3.14 (Structure theorem for the period map) *Let $\mathcal{P} : S \to \Gamma\backslash D$ be a period map. Its image belongs to $\Gamma\backslash D(\mathcal{P})$, where $D(\mathcal{P})$ is the Γ-orbit of the Mumford–Tate domain associated with the Mumford–Tate group $M(\mathcal{P})$ of the generic point of the variation. Possibly after replacing S by a finite unramified cover, the period map factors as*

$$\mathcal{P} : S \xrightarrow{(\mathcal{P}_{M^{\rm nc}},\mathcal{P}_{M^{\rm c}},\mathcal{P}_R)} \Gamma^{\rm nc}\backslash D(M^{\rm nc}) \times \Gamma^{\rm c}\backslash D(M^{\rm c}) \times D(R).$$

Moreover, concerning the factors of $\mathcal{P}$ we have:
(i) *the map $\mathcal{P}_R$ is constant; correspondingly, the variation $\mathcal{P}$ is a direct sum of a variation whose Mumford–Tate group is the full group $M(\mathcal{P})$ and an isotrivial one;*
(ii) *in addition, if the image of $\mathcal{P}_R$ is the point x_3, for all points $x_1 \in \mathcal{P}_{M^{\rm nc}}$ the intersection* $\mathrm{Im}(\mathcal{P}) \cap (x_1 \times \Gamma^{\rm c}\backslash D(M^{\rm c}) \times x_3)$ *is finite.*

Proof Let us first explain why we might need to pass to a cover. The group Γ^0 acts on $D(M)$ and the Γ-orbit of $D(M)$ in D consists of the union $\widetilde{D(M)}$ of $\#(\Gamma/\Gamma^0)$ distinct copies of $D(M)$. If we want to work with the original period map $\mathcal{P} : S \to \Gamma\backslash D$ we have to replace $D(M)$ by $\widetilde{D(M)}$. Alternatively, if we want to stay working with connected domains, we may need to pass to a finite unramified covering of S.

Proof of (i) Let π_3 be the projection onto the third factor of the target. Since S is quasi-projective, by Corollary 13.7.6 the composition $\pi_3 \circ \mathcal{P}$ extends to a holomorphic map

$$p : \hat{S} \to D(R(\mathcal{P})),$$

where $\hat{S}$ is a smooth compactification of S by a normal crossing divisor. We now invoke the following function-theoretic result.

Proposition (Griffiths and Schmid, 1969, Corollary 8.3) *A horizontal holomorphic map $p : \hat{S} \to D$ of a connected compact analytic space $\hat{S}$ to D is constant.*

As a consequence, p is indeed constant and the resulting variation is isotrivial.

Proof of (ii) We first study what happens relative to $M^{\rm c}$, the product of the compact simple factors.

Lemma 15.3.15 *Consider a point $F \in D(M^{\rm c}) \subset D$. Then $T_{D(M^{\rm c}),F} \subset T^{\rm vert}_{D,F}$.*

Proof of the Lemma. Let M' be a compact simple factor of M. The tangent space at F of M' is canonically identified with $\mathrm{Lie}(M')/\mathrm{Lie}(V \cap Z_G M') \subset$

$\mathrm{Lie}(G)/\mathrm{Lie}(V) = \mathfrak{p} \oplus \mathfrak{k}/\mathfrak{v}$. On the other hand, by Lemma 12.5.2 $T^{\mathrm{vert}}_{D,F} = \mathfrak{k}/\mathfrak{v}$. So it suffices to show that $\mathrm{Lie}(M') \subset \mathfrak{k}$. Recall that $\mathfrak{k} \subset \mathfrak{g}$ carries a weight 0 Hodge substructure: $\mathfrak{k}_{\mathbf{C}} = \mathfrak{g}^{\mathrm{even}} = \oplus_a \mathfrak{g}^{2a,-2a}$. There is a natural polarization $-B_0$ on $\mathfrak{g}$ (Proposition 12.6.5) and so, since the Weil operator is the identity on $\mathfrak{k}$, $B_0|\mathfrak{k}$ is negative definite while it is positive definite on $\mathfrak{p} = \mathfrak{g}^{\mathrm{odd}}$. On the other hand, by Remark 12.6.4 the Killing form B is a positive multiple of B_0. Since $\mathrm{Lie}(M')$ comes from a compact simple Lie group, $B|\,\mathrm{Lie}(M')$ is negative definite (see Helgason, 1978, Ch. II, Prop. 6.6). It follows that

$$\mathrm{Lie}(M') = (\mathfrak{m}')^{\mathrm{even}} \subset \mathfrak{k}. \qquad \square$$

Completion of the proof of (ii) By the preceding lemma, the fibers

$$x_1 \times D(M^{\mathrm{nc}})/\Gamma^{\mathrm{nc}} \times x_3$$

of the projection onto the first factor meet the image of $\mathcal{P}_M$ in points. $\square$

Bibliographical and Historical Remarks

About Section 15.1: The idea to formula Hodge theory in terms of representations of $\mathbf{S}(\mathbf{R})$ goes back to Deligne (1971c). Much of our treatment is based on this article and also on Deligne (1979). We also profited very much from Moonen (1999, 2016).

About Section 15.2: The name Mumford–Tate group has been introduced by Deligne to honor a statement in Mumford (1966). Our treatment of the Mumford–Tate group is largely based on Deligne et al. (1982, I.3).

About Section 15.3: The main reference here is Green et al. (2012).

16

Mumford–Tate Domains

In this chapter we treat Mumford–Tate domains in an axiomatic fashion. Historically, the concept of "Shimura domain" arose first. Such domains have a representation as bounded symmetric domains. The tautological variation of Hodge flag over such a domain satisfies Griffiths' transversality, i.e., we have a variation of Hodge structure over it. This characterizes them among Mumford–Tate domains and makes them easier to study directly from a Hodge theoretic perspective, which we do in Section 16.1. To make the transition to the axiomatic treatment, one views a Mumford–Tate domain as an entire conjugacy class of a given Hodge structure. To get a polarizable Hodge structure, the connected group of automorphisms of the domain must be a reductive group of Hodge type and, conversely, these are the groups that act transitively on Mumford–Tate domains. We explain this in Section 16.2. In Section 16.3 we give the promised axiomatic treatment of Mumford–Tate varieties parallel to Deligne's axiomatic treatment of Shimura varieties. The chapter ends with Section 16.4 where we give examples of Hodge structures given by representations of the classical simple groups.

16.1 Shimura Domains

Basic Properties and Classification

Definition 16.1.1 A *Shimura domain* is a Mumford–Tate subdomain D of some period domain such that the tautological Hodge flag restricts to a variation of Hodge structure on D.

Remark It may very well happen that the same Shimura domain can be embedded in different period domains and so the tautological variation can *a priori* be different. However, there is a way to give a more abstract treatment of Shimura domains which starts with the algebraic group under which the domain is homogeneous. See Proposition 16.3.3.

Our main result concerning Shimura domains is Proposition 16.1.9. Before we state it, let us first give some examples.

Examples 16.1.2 (i) The Siegel upper half space $\mathfrak{h}_g \simeq \mathrm{Sp}(g)/\mathrm{U}(g)$ classifies polarizable weight one Hodge structures with $h^{1,0} = g$ and the tautological flag $\mathcal{F}^1 \subset \mathcal{F}^0$ automatically satisfies Griffiths' transversality.

(ii) The type IV domain domain $\mathrm{SO}(2, n)/\mathrm{S}(\mathrm{O}(2) \times \mathrm{O}(n))$ classifies polarizable weight two Hodge structures with $h^{2,0} = 1, h^{1,1} = n$. We have encountered these domains before (see Eq. (4.1) in Chapter 4). That the tautological Hodge flag $\mathcal{F}^2 \subset \mathcal{F}^1 \subset \mathcal{F}^0$ satisfies transversality is not completely obvious. See Problem 16.1.1.

(iii) The domain $I_{p,p} = \mathrm{SU}(p, p)/\mathrm{S}(\mathrm{U}(p) \times \mathrm{U}(p))$ can be embedded in $\mathfrak{h}_{2p} = \mathrm{Sp}(2p)/\mathrm{U}(2p)$ by sending $A + \mathrm{i}B \in \mathrm{SU}(p, p)$ to the matrix $\begin{pmatrix} A & B \\ -B & A \end{pmatrix}$ which then is symplectic with respect to the form $\begin{pmatrix} 0 & \Delta \\ -\Delta & 0 \end{pmatrix}$, $\Delta = \mathrm{diag}(\mathbf{1}_p, -\mathbf{1}_p)$. This shows that $I_{p,p}$ is also a Shimura domain.

The above examples 16.1.2 all have representatives which are bounded symmetric domains. This is no coincidence; below we show that this is always the case.

Let us first recall the relevant definitions.

Definition (i) A bounded open connected subset D of $\mathbf{C}^n$ is called a *bounded symmetric domain* if each point of D is an isolated fixed point of a biholomorphic involution ι_0.

(ii) A *Hermitian symmetric space* is a connected complex Hermitian manifold such that each point $o \in D$ is an isolated fixed point of a unique holomorphic isometric involution.

By Helgason (1978, Ch. VIII, §7) a bounded symmetric domain is holomorphically isomorphic to a Hermitian symmetric space of the noncompact type and conversely. Let us discuss some relevant facts about the structure of those domains.

Proposition 16.1.3 (i) *A bounded symmetric domain decomposes uniquely into a product of irreducible such domains (Helgason, 1978, Ch. VIII, Prop. 4.4).*

(ii) *Let D be an irreducible bounded symmetric domain and $M = \mathrm{Aut}^0(D)$, the connected component of the group of holomorphic automorphisms of D. Then* *(see Helgason, 1978, Ch. VIII, Lemma 4.3)* $M = M^{\mathrm{ad}}$ *is a connected simple* $\mathbf{R}$*-algebraic group and* $D = M/K$*, where* K *is a maximal compact subgroup of* M *whose center* $Z(K)$ *is a circle* S^1 *(in particular, it is connected).*

(iii) *Let* M *be a connected simple* $\mathbf{R}$*-algebraic matrix group such that the center of a maximal compact subgroup* K *is a circle. Then* $M(\mathbf{R})/K$ *is an irreducible bounded symmetric domain.*

Remark 16.1.4 The group M from the above Proposition 16.1.3 is a $\mathbf{Q}$-algebraic group. This is not shown in Helgason (1978). Let us outline a proof. First, note that M being an adjoint group, there is an injective homorphism of groups

$$M \subset G \stackrel{\text{def}}{=} \mathrm{GL}(\mathrm{Lie}(M)).$$

The group G is a $\mathbf{Q}$-algebraic group. Consider the corresponding inclusion of Lie algebras $\mathrm{Lie}(M) \subset \mathrm{Lie}(G)$. Since M is semisimple, $M = [M, M]$, and we can apply Borel (1997, 7.9): $\mathrm{Lie}(M)$ is algebraic in the sense that there is a $\mathbf{Q}$-algebraic subgroup $H \subset G$ with $\mathrm{Lie}(H) = \mathrm{Lie}(M)$. Since M is connected, M is the component of the identity of H and hence is a $\mathbf{Q}$-algebraic group.

Irreducible domains have been classified. Apart from two exceptional domains of dimension 16 and 27, there are four infinite series of irreducible Hermitian symmetric spaces.

Theorem 16.1.5 (Helgason, 1978, Ch. X, §6.4) *The irreducible bounded symmetric domains fall into the following classes:*

- *the type AIII domains* $\mathrm{SU}(p,q)/\mathrm{S}(\mathrm{U}(p)\times\mathrm{U}(q))$ *with realization as a bounded domain*

$$I_{p,q} \stackrel{\text{def}}{=} \left\{Z \in M_{p,q}(\mathbf{C}) \,\middle|\, {}^{\mathsf{T}}Z\bar{Z} < \mathbf{1}_q\right\}, \quad p, q \geq 1;$$

- *the type DIII domains* $\mathrm{SO}^*(2g)/\mathrm{U}(g)$*, where* $\mathrm{SO}^*(2g) = \mathrm{Aut}(\mathbf{C}^{2g}, \tilde{J}_g) \cap \mathrm{SO}(2g, \mathbf{C})$ *and where* $\tilde{J}_g$ *is the Hermitian form* $(z, w) \mapsto {}^{\mathsf{T}}\bar{z}J_g w$*; they are realized as the bounded domain*

$$II_g \stackrel{\text{def}}{=} \left\{Z \in M_g(\mathbf{C}) \,\middle|\, Z = -{}^{\mathsf{T}}Z,\ {}^{\mathsf{T}}Z\bar{Z} < \mathbf{1}_g\right\}, \quad g \geq 3;$$

- *Siegel's upper half space* $\mathfrak{h}_g = \mathrm{Sp}(g)/\mathrm{U}(g)$*, a type CI domain with bounded realization*

$$III_g \stackrel{\text{def}}{=} \left\{Z \in M_g(\mathbf{C}) \,\middle|\, Z = {}^{\mathsf{T}}Z,\ {}^{\mathsf{T}}Z\bar{Z} < \mathbf{1}_g\right\}, \quad g \geq 2;$$

- *the type BDI domains* $\mathrm{SO}(2,n)/\mathrm{S}(\mathrm{O}(2)\times\mathrm{O}(n))$ *with bounded realization*

$$IV_n \stackrel{\text{def}}{=} \left\{X \in M_{2,n}(\mathbf{R}) \,\middle|\, {}^{\mathsf{T}}XX < \mathbf{1}_2\right\}, \quad n \geq 2.$$

The complex structure J sends the first row of X to minus the second row;

- *the exceptional domains EIII of dimension* 16 *associated with the Lie algebra E_6 and EVII of dimension* 27 *associated with E_7.*

In view of their special role, the groups occurring as automorphism groups of bounded symmetric domains deserve a name:

Definition 16.1.6 A reductive $\mathbf{Q}$-algebraic group M is called *of Hermitian type* if $M^{\text{der}}(\mathbf{R})/(\text{center})$ is the group of automorphisms of a bounded symmetric domain. Such groups can be characterized as follows: for each maximal compact subgroup K of $M(\mathbf{R})$, the connected component of its center intersects each simple $\mathbf{R}$-factor in the circle group.

Structure as a Shimura Domain

In this subsection we shall show that a bounded symmetric domain D has the structure of a Shimura domain, in particular, that it carries a polarizable variation of Hodge structure. To start, note that there is a lot more structure around coming from the metric on D and the symmetries ι_x with $x \in D$ as fixed point. These act on the tangent spaces as isometries and in particular on $T_o D$. This space can be identified with $\mathfrak{m}/\mathfrak{k}$ where we set $\mathfrak{m} = \text{Lie}(M(\mathbf{R}))$, $\mathfrak{k} = \text{Lie}(K)$. Since o is an isolated fixed point $d(\iota_o) = -\,\text{id}$. The Cartan involution acts on $\mathfrak{m}$ with $(+1)$-eigenspace $\mathfrak{k}$ and (-1)-eigenspace $\mathfrak{p}$, the orthogonal complement of $\mathfrak{k}$ with respect to the Killing form B. We have the following.

Proposition 16.1.7 (Helgason, 1978, Ch. VIII, §6) *Let $M(\mathbf{R})$ be the connected component of the group of holomorphic automorphism of an irreducible bounded symmetric domain $D = M(\mathbf{R})/K$. There is a homomorphism of real algebraic groups*

$$u : \mathbf{U} \to Z(K) \subset M$$

such that for all $z \in \mathbf{U}(\mathbf{R})$ multiplication by $u(z)$ on D (which fixes o) gives complex multiplication by z on the real tangent space $T_o D \simeq \mathfrak{p}$:

$$\text{Ad}(u(z))|\mathfrak{p} = z \cdot \text{id}\,.$$

Moreover, $\text{Ad}(u(-1))$ *is a Cartan involution of $\mathfrak{g}$ with fixed point set $\mathfrak{k}$.*

Example 16.1.8 The automorphism group of $\mathfrak{h}_g$ is $\text{Sp}(g,\mathbf{R})/\pm\mathbf{1}_{2g}$, a group of adjoint type. Note that

$$\text{Lie}(\text{Sp}(g,\mathbf{R})) = \begin{pmatrix} A & B \\ C & -{}^{\mathsf{T}}\!A \end{pmatrix}, \quad B = {}^{\mathsf{T}}B,\ C = {}^{\mathsf{T}}C$$

and that the tangent space at o is identified with the symmetric matrices therein:

$$\mathfrak{p} = \begin{pmatrix} A & B \\ B & -A \end{pmatrix}, \quad B = {}^{\mathsf{T}}B,\ A = {}^{\mathsf{T}}A.$$

With $o = \mathrm{i}\mathbf{1}_g \in \mathfrak{h}_g$ and $z = \exp(2\pi \mathrm{i}t)$, one puts

$$u(t) = \begin{pmatrix} \cos(t/2)\mathbf{1}_g & -\sin(t/2)\mathbf{1}_g \\ \sin(t/2)\mathbf{1}_g & \cos(t/2)\mathbf{1}_g \end{pmatrix} \quad \mathrm{mod}\ \pm \mathbf{1}_{2g}.$$

Let us show that in this example this gives the desired morphism. Note first that conjugation with $u(t)$ on points of $\mathfrak{p}$ is given by:

$$\begin{pmatrix} A & B \\ B & -A \end{pmatrix} \mapsto \begin{pmatrix} A\cos t - B\sin t & A\sin t + B\cos t \\ A\sin t + B\cos t & -A\cos t + B\sin t \end{pmatrix}.$$

If one makes the identification

$$T_o\mathfrak{h}_g = \mathbf{C}^{\frac{1}{2}g(g+1)} \ni A + B\mathrm{i} \iff \begin{pmatrix} A & B \\ B & -A \end{pmatrix},$$

we see that conjugation with $u(t)$ corresponds to complex multiplication by $\exp(2\pi \mathrm{i}t)$, since $(\cos t + \mathrm{i}\sin t)(A + \mathrm{i}B) = A\cos t - B\sin t + \mathrm{i}(A\sin t + B\cos t)$.

The adjoint operation of $u(-1)$ on $\mathrm{Lie}(\mathrm{Sp}(g, \mathbf{R}))$ is given by

$$\begin{pmatrix} A & B \\ C & -{}^{\mathsf{T}}A \end{pmatrix} \mapsto \begin{pmatrix} -{}^{\mathsf{T}}A & -C \\ -B & A \end{pmatrix}.$$

Since $B = {}^{\mathsf{T}}B$, $C = {}^{\mathsf{T}}C$, this implies that the subalgebra fixed under this action is the algebra of the skew-symmetric matrices, i.e., the Lie algebra of K. So $\mathrm{Ad}\,u(-1)$ is indeed a Cartan involution.

For later use we describe the Cartan decomposition $\mathfrak{m} = \mathfrak{k} \oplus \mathfrak{p}$ in terms of the character spaces of $\mathbf{U}$ under the adjoint action of u. We have $T_o^{\mathrm{hol}} \simeq \mathfrak{p}^- = \mathfrak{g}_{(\chi=z)}$ which is another way of saying that $u(z)$ acts by multiplication by z on the holomorphic tangent space. It follows that

$$\begin{array}{ccccccc} \mathfrak{m}_{\mathbf{C}} & = & \mathfrak{k}_{\mathbf{C}} & \oplus & \mathfrak{p}^+ & \oplus & \mathfrak{p}^-. \\ & & \circlearrowleft & & \circlearrowleft & & \circlearrowleft \\ & & \mathrm{id} & & z^{-1} & & z \end{array}$$

So far we have looked at $o \in D$ but this point is not special. The conjugate morphism gug^{-1} maps the circle into gKg^{-1}, another maximal compact subgroup of M, the stabilizer of $x = g \cdot o$. Since gug^{-1} only depends on x we set

$$u_x := gug^{-1} : \mathbf{U}(\mathbf{R}) \to M(\mathbf{R}), \quad x \in D = M(\mathbf{R})/K.$$

By Lemma 15.1.4 $\mathrm{Ad}\circ u_x(-1)\circ\mathrm{sq}$ gives a weight 0 Hodge structure:

$$h_x = \mathrm{Ad}\circ u_x\circ\mathrm{sq} : \mathbf{S} \xrightarrow{z\mapsto z/\bar{z}} \mathbf{U} \xrightarrow{u_x} M \xrightarrow{\mathrm{Ad}} \mathrm{GL}(\mathfrak{m}),$$

For $u = u_o$, the decomposition of $\mathfrak{m}_{\mathbb{C}}$ according to the characters for the corresponding $\mathbf{S}$-action is then as follows:

$$\begin{array}{ccccccc} & & \mathfrak{k}_{\mathbb{C}} & & \mathfrak{p}^- & & \mathfrak{p}^+ \\ & & \| & & \| & & \| \\ \mathfrak{m}_{\mathbb{C}} & = & \mathfrak{m}^{0,0} & \oplus & \mathfrak{m}^{-1,1} & \oplus & \mathfrak{m}^{1,-1}. \\ & & \circlearrowleft & & \circlearrowleft & & \circlearrowleft \\ & & \mathrm{id} & & \bar{z}/z & & z/\bar{z} \end{array}$$

The corresponding Hodge flag depends on $x \in D$:

$$F_x^\bullet = \left\{F^{-1} \supset F^0 \supset F^1\right\}_x.$$

This dependence is holomorphic, since the $M(\mathbf{R})$-action on D is by holomorphic automorphisms. Griffiths' transversality is automatic: we get a variation of weight 0 Hodge structures on D with underlying vector space the Lie algebra $\mathfrak{m}$ of M. It is also polarizable. We leave this as an exercise (see Problem 16.1.2). Let b be a polarization. Then D embeds in the period domain which parametrizes weight 0 Hodge structures of type (h_x, b) on $\mathfrak{m}$. If we choose $x \in D$ Hodge generic, the Mumford–Tate group of h_x must act transitively on D, since we have a polarized variation of Hodge structure on all of D. This completes the proof that D has the structure of a Shimura domain.

We have shown half of the following characterization of Shimura domains:

Proposition 16.1.9 (i) *Given a period domain D and $o \in D$ with corresponding Hodge structure h. Then $D(h) = \mathrm{MT}(h)\cdot o$, is a Shimura domain if and only if h induces a weight* 0 *Hodge structure on* $\mathrm{Lie}(\mathrm{MT}(h))$ *of level ≤ 2, that is at most Hodge types* $(-1,1), (0,0), (1,-1)$ *occur. This implies that $D(h)$ is a bounded symmetric domain.*
(ii) *Conversely, every bounded symmetric domain has the structure of a Shimura domain.*

Proof We still need to show (i). The crucial condition is the horizontality condition

$$\mathrm{Ad}(\xi)F^pH_{\mathbb{C}} \subset F^{p-1}H_{\mathbb{C}} \quad \forall \xi \in T^{\mathrm{hol}}_{D(h),o}.$$

Recall (see Definition 12.2.3) that for the full period domain $T^{\mathrm{hol}}_{D,o} = \mathfrak{m}^- = \mathfrak{g}_{\mathbb{C}}/F^0\mathfrak{g}_{\mathbb{C}}$, where $\mathfrak{g}_{\mathbb{C}} = \mathrm{Lie}(G_{\mathbb{C}})$. Similarly, we have

$$T^{\mathrm{hol}}_{D(h),o} = \mathrm{Lie}(M_{\mathbb{C}})/F^0\,\mathrm{Lie}(M_{\mathbb{C}}).$$

So horizontality forces $\mathrm{Lie}(M_{\mathbf{C}}) = F^{-1}\mathrm{Lie}(M_{\mathbf{C}}) = (\mathfrak{g}^{-1,1} \oplus \mathfrak{g}^{0,0} \oplus \mathfrak{g}^{1,-1}) \cap \mathrm{Lie}(M_{\mathbf{C}})$ which gives exactly the restriction on the Hodge types.

Note that $D(h) \subset D = M/V$ is a complex manifold homogeneous under $\mathrm{MT}(h)$. So, to see that we have a bounded symmetric domain it suffices to show that $V \cap \mathrm{MT}(h)$ is a maximal compact subgroup in $\mathrm{MT}(h)$. The restrictions on the Hodge types imply that the vertical tangent space at o (which comprises Hodge types $(2p, -2p)$, $p > 0$) is zero. But this means exactly that $\mathfrak{v} \cap \mathfrak{m} = \mathfrak{k} \cap \mathfrak{m}$, which proves the assertion. □

Problems

16.1.1 Show that the tautological Hodge flag on $\mathrm{SO}(2,n)/\mathrm{S}(\mathrm{O}(2) \times \mathrm{O}(n))$ is a polarized variation of Hodge structures of weight two.

16.1.2 Let $D = M(\mathbf{R})/K$ be an irreducible bounded symmetric domain.

(a) Consult Helgason (1962, Chapter IX, Table II) to verify that the center of K is one-dimensional if D is noncompact (and of Hermitian type) and otherwise is a finite group.

(b) Apply Corollary 15.1.12 to conclude polarizability of the weight 0 Hodge structure on $\mathrm{Lie}(M(\mathbf{R}))$.

16.2 Mumford–Tate Domains

Let us rephrase some salient facts we derived in this and the previous chapter as follows.

Proposition 16.2.1 *Let $h : \mathbf{S} \to G$ be a homomorphism corresponding to a point $o \in D = G/V$, a given period domain. Then $Z_G h = V$ and the points of $D = G(\mathbf{R})/V$ are in bijection with the $G(\mathbf{R})$-conjugates of the given homomorphism h.*

Suppose that the image of h lands in an $\mathbf{R}$-algebraic subgroup $M \subset G$; then $Z_M h = M(\mathbf{R}) \cap V$ and the points in the orbit $M(\mathbf{R}) \cdot o$ are in bijection with the $M(\mathbf{R})$-conjugates of h. This holds in particular for the Mumford–Tate group $\mathrm{MT}(h)$.

Proof We claim that if o corresponds to h, then for $g \in G$ the translate $g \cdot o$ corresponds to the conjugate homomorphism ghg^{-1}. This can be seen as follows. Let $\bigoplus H_o^{p,q}$ be the Hodge decomposition corresponding to the base point o. Then the Hodge decomposition for $g.o$ is just $\bigoplus gH_o^{p,q}$. So $gH_o^{p,q}$, the $z^p\bar{z}^q$-eigenspace for this Hodge structure is precisely the $z^p\bar{z}^q$-eigenspace for $gh(z)g^{-1}$. We see furthermore that $gh(z)g^{-1} = h(z)$ for all $z \in \mathbf{S}(\mathbf{R})$ precisely when $gH^{p,q} = H^{p,q}$ for all Hodge summands $H^{p,q}$, i.e., if and only if $g \in V$. □

Remarks 16.2.2 (i) Suppose that we can write $h = u \circ \mathrm{sq}$ for some homomorphism $u : \mathbf{U} \to G$. Then points in D are in bijection with the $G(\mathbf{R})$-conjugates of u. This happens for bounded symmetric domains.
(ii) If for some $\mathbf{R}$-algebraic subgroup $M \subset G$ one has $\mathrm{Im}(h) \subset M$, the $M(\mathbf{R})$-orbit of $h \in D$ is the same as the $M(\mathbf{R})$-conjugacy class of h. If, $h = u \circ \mathrm{sq}$, one can equally use u instead of h.

The appropriate concept which we now need is that of a group of Hodge type (see Definition 13.3.7). Recall that a real algebraic group which is connected is called "of Hodge type" if its rank equals the rank of a maximal compact subgroup. Since the rank of an algebraic group is the dimension of the maximal algebraic torus it contains, we can rephrase this as follows.

Remark *A connected reductive* **R***-algebraic group is of Hodge type if and only if it contains a maximal (real) torus which is anisotropic,* i.e., *compact.*

We note the following consequences of the definition.

Observation (a) A product of groups of Hodge type is of Hodge type if and only if each of its factors are.
(b) If two connected groups are isogeneous, then one is of Hodge type if and only if the other is.
Consequently, *a connected semisimple* **R***-algebraic group is of Hodge type if and only if its simple factors are.*

These preliminary remarks being out of the way, we can now state the main definition:

Definition 16.2.3 A complex manifold homogeneous under a group of Hodge type is called a *Mumford–Tate domain.*

The previous discussion then shows.

Corollary 16.2.4 (i) *The Mumford–Tate group of a polarizable Hodge structure is a group of Hodge type.*
(ii) *A Mumford–Tate subdomain of a period domain is a Mumford–Tate domain.*

Proof (i) Let M be the Mumford–Tate group of a polarizable Hodge structure $\mathbf{S} \xrightarrow{h} \mathrm{GL}(H)$. We show that the maximal torus T of $M(\mathbf{R})$ containing the image

S of $h(\mathbf{R})$ is compact.[1] To prove this, observe that the centralizer of S in $M(\mathbf{R})$ is a compact subgroup in $M(\mathbf{R})$; indeed, by Proposition 16.2.1, $Z_M(S) = Z_M(h) = M(\mathbf{R}) \cap V$. The maximal algebraic subtorus T of $M(\mathbf{R})$ containing S, being abelian, centralizes S. So it is contained in the compact subgroup $Z_M S = M(\mathbf{R}) \cap V$. Hence T is also compact and hence M is of Hodge type.
(ii) Let D be the period domain for the Hodge structure h. The Mumford–Tate subdomain corresponding to h is the $M(\mathbf{R})$-orbit of h in D and since M is of Hodge type, this orbit is a Mumford–Tate domain. □

Examples 16.2.5 (i) Consider a bounded symmetric domain $D = M(\mathbf{R})/K$ where M is a semisimple $\mathbf{R}$-algebraic group. By Proposition 16.1.3 the connected component of the center of a maximal compact subgroup of $G = M^{\mathrm{ad}}(\mathbf{R})$ meets each of the irreducible factors of $M(\mathbf{R})$ in a circle group S. Since $Z_S(G) = K$, the argument used to prove the preceding corollary applies to this situation: the maximal algebraic subtorus of G belongs to the compact group K and so G is of Hodge type. The group M, being isogeneous to G, therefore is also of Hodge type.
(ii) If $D = G(\mathbf{R})/V$ is a period domain, the group G is of Hodge type. This is easily verified by inspecting the groups G that occur. See e.g. Table 16.1 in which these group are placed in boxes. Since there are period domains which are not bounded symmetric, the class of Hodge groups is strictly larger than the collection of groups of Hermitian type.
(iii) Compact groups or, more generally, reductive groups G with G^{der} compact are of Hodge type. The corresponding Mumford–Tate domains are compact and they all belong to a polarizable Hodge structure. Among them figure the groups $\mathrm{SO}(k)$ which give rise to compact period domains (Theorem 4.4.4). For example $\mathrm{SO}(k)$ admits a (trivial) Hodge structure $h : \mathbf{S} \to \mathrm{SO}(k) \subset \mathrm{GL}(\mathbf{R}^k)$ corresponding to the constant Hodge structure of type $(0, 0)$ on $\mathbf{R}^k$. The period domain is a point in this case. One can show (see Sect. ion 16.4) that there are also nontrivial Hodge structures with more interesting compact Mumford–Tate groups.

A group of Hermitian type is automatically of Hodge type and a Shimura domain is a Mumford–Tate domain. Example 16.2.5.(ii) exhibits groups of Hodge type that are not of Hermitian type while Example 16.2.5.(iii) shows some groups of Hodge type that do not act transitively on period domains.

There are also simple groups which are not of Hodge type.

Examples 16.2.6 (i) The group $\mathrm{SL}(n, \mathbf{R})$ for $n \geq 3$ is not of Hodge type. To see this, note that the rank of $\mathrm{SL}(n, \mathbf{R})$ is $n - 1$ while that of $\mathrm{SO}(n, \mathbf{R})$, a maximal compact subgroup, is $[n/2]$.

[1] S is either a circle or $e \in G$.

(ii) A *complex* simple Lie group such as $\mathrm{SL}(n, \mathbf{C})$, $\mathrm{SO}(n, \mathbf{C})$, or $\mathrm{Sp}(n, \mathbf{C})$ has a noncompact maximal torus of *complex* dimension n and so these can never arise as a Mumford–Tate group.
(iii) Another classical $\mathbf{R}$-simple group that is not Hodge is $\mathrm{SO}(p, q)$ with $p = 2a + 1$ and $q = 2b + 1$. Indeed, a maximal torus is $(S^1)^a \times (S^1)^b \times \mathbf{G}_m(\mathbf{R})$ and so the rank is $a + b + 1$ while the rank of $S(\mathrm{O}(2a + 1) \times \mathrm{O}(2b + 1))$ equals $a + b$. This is in agreement with the fact that for a polarizaton of signature (p, q) the sum of the Hodge numbers $h^{s,t}$ either with s odd or with s even has to be even. It follows that no Hodge structure on a vector space of dimension $p + q$ factors over $\mathrm{SO}(p, q)$ under the assumption that p and q are both odd. In fact, one can prove that under these assumptions on p and q, no Hodge structure can factor over a representation of $\mathrm{SO}(p, q)$. See Green et al. (2012, IV.C.3).

Given an algebraic group M defined over $\mathbf{Q}$ which is simple as a $\mathbf{Q}$-group, can we decide whether the group is of Hodge type? The group $M(\mathbf{R})$ need not be simple as a real group and hence we look at its simple factors. For each of those the maximal algebraic subtorus should be compact. This can be decided by consulting the list of real simple Lie groups. By Example 16.2.6.(ii) we can exclude the complex simple Lie groups. The results are shown in Table 16.1.

We have already seen (Remark 16.1.4) that the adjoint group of a real linear algebraic group can be defined over $\mathbf{Q}$ and so if M is a real form, in this table $M^{\mathrm{ad}}(\mathbf{R})$ can be defined over $\mathbf{Q}$. The table indicates which simple real Lie algebras come from groups of Hodge type and if so, when these are of Hermitian type. Those that are boxed are those which can act transitively on period domains.

The next question is of course whether conversely, for any group of Hodge type, can one find a representation and a Hodge structure factoring over it? For the sake of brevity, let us introduce suitable terminology for this.

Definition 16.2.7 Let G be a reductive group. A (polarizable) Hodge structure $h : \mathbf{S} \to \mathrm{GL}(H)$ is called a *(polarizable) G-Hodge structure* if h factors over a representation of G in H.

This question is the same as asking whether a Mumford–Tate domain is a Mumford–Tate subdomain of some period domain. The next result answers this.

Lemma 16.2.8 (i) *Let M be a connected algebraic group defined over $\mathbf{Q}$ and let $h : \mathbf{S}(\mathbf{R}) \to M(\mathbf{R})$ be a homorphism such that its weight* (15.4) *is defined over $\mathbf{Q}$. Then* $\mathrm{Ad} \circ h$ *induces a weight 0 Hodge structure on* $\mathrm{Lie}(M)$. *This Hodge structure can be polarized by minus the Killing form B if and only if the following condition is satisfied:*

Table 16.1 *Which noncomplex real Lie groups are of Hodge type?*

Dynkin diag.	Real group[a]	Hodge type?	Hermitian type?
A_r	$\mathrm{SU}(p,q)$, $p+q=r+1$, $p=0,\ldots,r+1$	Yes	Yes, if $p,q>0$
	$\mathrm{SL}(n,\mathbf{R})$	For $n=2$	For $n=2$
	$\mathrm{SL}(n;\mathbf{H})$	No	No
B_r	$\boxed{\mathrm{SO}(2p,2q+1)}$, $p+q=r$, $p=0,\ldots,r$	Yes	For $p=1$
C_r	$\mathrm{Sp}(p,q)$, $p+q=r$, $p=0,\ldots,r$	Yes	No
	$\boxed{\mathrm{Sp}(g,\mathbf{R})}$	Yes	Yes
D_r	$\mathrm{SO}(2p+1,2q+1)$, $p+q=r$, $p=0,\ldots,r$	No	No
	$\boxed{\mathrm{SO}(2p,2q)}$, $p+q=r$, $p=0,\ldots,r$	Yes	For $p=0,1$
	$\mathrm{SO}^*(2r)$	Yes	Yes
E_6	EI	No	No
	EII	Yes	No
	$EIII$	Yes	Yes
E_7	EIV	No	No
	EV, EVI	Yes	No
	$EVII$	Yes	Yes
E_8	$EVIII, EIX$	Yes	No
F_4	FI, FII	Yes	No
G_2	G	Yes	No

[a] The Lie algebra is uniquely determined, finitely many groups give the same Lie algebra; only one of them figures in the list.

$$\textbf{MT} \qquad Ad\circ h(\mathrm{i}) \text{ is a Cartan involution for } \mathrm{Lie}(M).$$

If this is the case, M is reductive and of Hodge type.
(ii) *Conversely, for a group M of Hodge type, there exist homomorphisms* $h : \mathbf{S}(\mathbf{R}) \to M(\mathbf{R})$ *for which* **MT** *holds and then* $\mathrm{Ad}\circ h$ *is a polarizable Hodge structure on* $\mathrm{Lie}(M)$. *In other words, nontrivial polarizable M-Hodge structures exist.*

Proof The first assertion follows from the polarizability criterion Corollary 15.1.12 together with Lemma 15.3.3. For the second, we may assume that M is a simple group of Hodge type. The classical groups G as enumerated in Table D.1 have G-Hodge structures. This is shown in Section 16.4. We show that this is the case for either the standard representation or the adjoint representation. If the standard representation is an example, by functoriality, also the adjoint representation is Hodge. Since the Hodge structure we construct is polarized, condition **MT** holds.

Since a given simple group G always is an unramified cover of $\mathrm{Ad}(G)$ (see Knapp, 2005, VII.1), this shows that all classical groups G have G-Hodge structures, not just the ones in the list.

As to the exceptional groups, this is more involved. We have seen (Theorem 16.1.5) that some of the exceptional groups are of Hermitian type for which we know the result already. For the other exceptional groups see Green et al. (2012, IV.D). □

Remark 16.2.9 To place even the simplest variations of Hodge structure in this framework, one should allow semisimple groups that are not necessarily of adjoint type. It is even useful to allow reductive groups. The reason is that, although the surjection $M(\mathbf{R}) \twoheadrightarrow M^{\mathrm{ad}}(\mathbf{R})$ allows every representation of $G^{\mathrm{ad}}(\mathbf{R})$ to be lifted to $M(\mathbf{R})$, there might be nonliftable representations. Consider for example the upper half space $\mathfrak{h}_g = M(\mathbf{R})/K$ with $M = \mathrm{Sp}(g)$, $G^{\mathrm{ad}} = \mathrm{Sp}(g)/\pm\mathbf{1} = \mathrm{P\,Sp}(g)$. The tautological representation of G in $\mathbf{R}^{2g}$ is not a lift of representation for the group G^{ad}. To describe the corresponding tautological variation of weight one Hodge structures it suffices to do this for the base point, which corresponds to $\mathrm{i}\cdot\mathbf{1}_g$. The Hodge representation corresponding to this matrix is given by

$$\mathbf{S}(\mathbf{R}) \ni z = a + b\mathrm{i} \xrightarrow{h} \begin{pmatrix} a\mathbf{1}_g & b\mathbf{1}_g \\ -b\mathbf{1}_g & a\mathbf{1}_g \end{pmatrix} \in \mathrm{Sp}(g).$$

It defines a weight one Hodge structure on $H = \mathbf{R}^{2g}$. The circle $S^1 \subset \mathrm{P\,Sp}(g)$ which is the image of $\mathbf{S}(\mathbf{R})$ by means of the map $z \mapsto h(z/\bar{z})$ gives the tautological Hodge representation of weight 0 on $\mathfrak{sp}(g)$. Indeed, it is the subrepresentation of $\mathrm{End}(H)$ given by the skew-symmetric endomorphisms.

16.3 Mumford–Tate Varieties and Shimura Varieties

Classical Point of View

We change our point of view; we no longer focus on period domains but on Mumford–Tate domains. Such a domain is identified with a single conjugacy class X under the action of a group G of Hodge type on homomorphisms $\mathbf{S} \to G$. This leads to the following notion.

Definition 16.3.1 Let G be a $\mathbf{Q}$-algebraic group of Hodge type and let $h : \mathbf{S} \to G_{\mathbf{R}}$ be a homomorphism whose weight $w_h = h \circ w$ is defined over $\mathbf{Q}$. Its *Mumford–Tate group* $\mathrm{MT}(h)$ is the smallest Zariski subgroup of G which is defined over $\mathbf{Q}$ and such that $h(\mathbf{S}(\mathbf{R})) \subset \mathrm{MT}(h(\mathbf{R}))$.

If we require that $\mathrm{Ad} \circ h$ define a polarizable Hodge structure on $\mathrm{Lie}(G)$, by Lemma 16.2.8 this is the case if only if the following condition holds:

MT	$Ad \circ h(\mathrm{i})$ is a Cartan involution for $\mathrm{Lie}(G^{\mathrm{ad}}(\mathbf{R}))$.

This property then holds for any $G(\mathbf{R})$-conjugate of h. Now we are ready to define the concept of a Mumford–Tate datum:

Definition 16.3.2 A *Mumford–Tate datum* is a pair (G, X) with G a reductive $\mathbf{Q}$-algebraic group, and with

$$X = X(G, h) = \{\text{ A } G(\mathbf{R})\text{-conjugacy class of } h : \mathbf{S} \to G \text{ for which } \mathbf{MT} \text{ and } \mathbf{S0} \text{ hold.}\}$$

A *Shimura datum* is a pair (G, X) with G a reductive $\mathbf{Q}$-algebraic group and

$$X = X(G, h) = \{\text{ A } G(\mathbf{R})\text{-conjugacy class of } h : \mathbf{S} \to G \text{ for which } \mathbf{MT}, \mathbf{S0}, \mathbf{S1} \text{ and } \mathbf{S2} \text{ hold}\},$$

where

S0 : the weight $w_h = h \circ w$ is defined over $\mathbf{Q}$;
S1 : at most the characters $\{1, z/\bar{z}, \bar{z}/z\}$ of the group $\mathbf{S}(\mathbf{C})$ occur in the representation $\mathrm{Ad} \circ h : \mathbf{S}(\mathbf{C}) \to \mathrm{GL}(\mathfrak{g}_{\mathbf{C}})$; in other words, h induces a weight 0 Hodge structure on $\mathfrak{g}$ of length 3;
S2 : with $p : G^{\mathrm{ad}} \to H$ the projection onto a $\mathbf{Q}$-simple factor, the composition $p \circ h$ is not constant.

The preceding discussion shows that this terminology is adequate.

Proposition 16.3.3 (i) *A reductive $\mathbf{Q}$-algebraic group is of Hodge type (Definition 13.3.7) if and only if a homomorphism $h : \mathbf{S} \to G_{\mathbf{R}}$ exists such that $(G, X(h))$ is a Mumford–Tate datum. In that case $X(h)$ is a Mumford–Tate domain.*
(ii) *A reductive $\mathbf{Q}$-algebraic group is of Hermitian type (see Definition 13.3.7)*

if and only if $h : \mathbf{S} \to G_{\mathbf{R}}$ *exists such that* $(G, X(h))$ *is a Shimura datum. In that case* $X(h)$ *is a Shimura domain.*

Proof By Lemma 16.2.8 a Mumford–Tate datum (G, X) gives rise to a Mumford–Tate domain X. If (G, X) is a Shimura datum, X is a bounded symmetric domain and hence is a Shimura domain. □

Remarks 16.3.4 (i) If $G(\mathbf{R})$ is not connected, a conjugacy class $X = X(h)$ of a morphism $h : \mathbf{S} \to G$ under the action of $G(\mathbf{R})$ need not be connected in the natural topology. So we allow disconnected Mumford–Tate domains.
(ii) We defined the Mumford–Tate group of a morphism $h : \mathbf{S} \to G_{\mathbf{R}}$ with the property that $h \circ w$ is defined over $\mathbf{Q}$ at the beginning of this section (see Definition 16.3.1). We shall prove later (Proposition 17.1.3) that, although G need not be the Mumford–Tate group of the morphism h, for a $\mathbf{Q}$-*simple* algebraic group this is true of a *generic* conjugate of h.
(iii) In the presence of **S1** and **MT**, the last property **S2** is equivalent to G^{ad} being of noncompact type: for none of its simple factors H, $H(\mathbf{R})$ is compact. Indeed, if $H(\mathbf{R})$ is a connected simple group of adjoint type, the Cartan involution is the identity precisely when $H(\mathbf{R})$ is compact (recall from Definition D.3.12 that a Cartan involution fixes a maximal connected compact subgroup). In this case **MT** implies that $u(-1) = 1$ since $Z(H(\mathbf{R})) = 1$. But since **S1** implies that at most the characters $1, z, z^{-1}$ occur in the representation $\mathrm{Ad} \circ u$, in fact z, z^{-1} cannot occur. So $u(z) = 1$ for all z. Hence, if **MT** and **S1** hold, **S2** implies that $H(\mathbf{R})$ cannot be compact. Conversely, if u is constant, **MT** implies that $H(\mathbf{R})$ is compact. It follows that the conditions **M**, **S1** and **S2** ensure that the image of h in each simple factor separately satisfy **MT** and **S1** and so, by what we just said, each factor gives rise to an irreducible bounded symmetric domain.
(iv) Observe that a $\mathbf{Q}$-simple factor need not be simple over $\mathbf{R}$ and so a $\mathbf{Q}$-simple factor is of noncompact type precisely when *not all* its $\mathbf{R}$-simple factors are compact Lie groups. For nontrivial examples see Section 17.4 where we treat embedded Shimura curves that come from analyzing such groups.

The natural moduli objects are quotients of Mumford–Tate domains by the action of certain groups which act properly and discontinuously on it. The argument given in Section 4.5 for period domains applies and shows the following.

Theorem *Let* Γ *be a discrete subgroup of* $G(\mathbf{R})$ *and let* (G, X) *be a Mumford–Tate datum. Then* $\Gamma\backslash X$ *is a (possibly disconnected) complex analytic variety; it is smooth when* Γ *contains no elements* $\neq e$ *of finite order.*

We cannot expect the quotient to be algebraic, unless we start with Shimura data. For this to be true some restrictions on Γ are needed. To explain these, we choose a fixed representation of G as a matrix subgroup of $\mathrm{GL}(n)$.

Definition 16.3.5 (i) A subgroup Γ of $G(\mathbf{Q})$ is called *arithmetic* if it is commensurable with $G_{\mathbf{Z}} := G(\mathbf{Q}) \cap \mathrm{GL}(n; \mathbf{Z})$ (i.e., $\Gamma \cap G_{\mathbf{Z}}$ has finite index in both Γ and $G_{\mathbf{Z}}$).
(ii) A *congruence subgroup* of $G(\mathbf{Q})$ is a subgroup of $G(\mathbf{Q})$ containing

$$G(\mathbf{Q}) \cap \{g \in \mathrm{GL}(n; \mathbf{Z}) \mid g \equiv \mathbf{1}_n \bmod N\}$$

as a subgroup of finite index. A congruence subgroup is arithmetic.
(iii) A subgroup of $G(\mathbf{Q}) \subset \mathrm{GL}(n; \mathbf{Q})$ is *neat* if for any given element its eigenvalues generate a torsion free subgroup of $\mathbf{C}^\times$ (in particular it cannot have finite order).

Not all arithmetic subgroups are congruence, but by Borel (1969, 17.4) they contain neat congruence subgroups of finite index.
We introduce the following notions.

Definition 16.3.6 Let (G, X) be a Mumford–Tate datum and let Γ be any discrete subgroup of $G(\mathbf{R})$. The associated *Mumford–Tate variety* is the complex variety $\Gamma\backslash X$. If (G, X) is a Shimura datum and $\Gamma \subset G$ is a congruence subgroup, the quotient $\Gamma\backslash X$ is called a *Shimura variety*.

A Shimura variety has a minimal compactification, the so-called *Baily–Borel compactification*, named after the article by Baily and Borel (1966). It is a projective variety.

Theorem (Baily and Borel, 1966) *A Shimura variety $\Gamma\backslash X$ is a quasi-projective variety. If, moreover, Γ is neat, $\Gamma\backslash X$ is smooth.*

Remarks (1) The Baily–Borel compactification typically is very singular. There are other compactifications, the *toroidal compactifications* systematically studied in Ash et al. (1975). The advantage is that for neat Γ such compactification can be chosen to be smooth and such that what has been added to $\Gamma\backslash X$ is a normal crossing divisor.
(2) Since in this case the algebraic variety $\Gamma\backslash X$ is *a priori* only defined over $\mathbf{C}$ there is no reason to believe that it can be defined over $\mathbf{Q}$ or a number field k. Surprisingly, Shimura (1963; 1965; 1966) has shown that there is a model over some number field. The field k over which a model of the Shimura variety is defined, may depend on which congruence subgroup one chooses. See page 445 for more details.

Examples 16.3.7 (i) **CM-points**. Recall (Definition 15.2.5) that a Hodge structure whose Mumford–Tate group is an abelian group (and hence an algebraic torus) is called a *Hodge structure with complex multiplication*, or, more succinctly, a *CM-Hodge structure*.

To give a further motivation for this definition, we first recall that a CM-field E is a totally imaginary extension of $\mathbf{Q}$ containing a totally real subfield F with $[E : F] = 2$. One can show (Green et al., 2012, Prop. V.2.) that if (H, h) is an irreducible Hodge structure which is polarizable, and whose Mumford–Tate group is a nontrivial abelian group, then $E \stackrel{\text{def}}{=} \operatorname{End}(H, h)$ is a CM-field. This implies that H is a one-dimensional vector space over E. Conversely, if E contains an embedded field E' with $[E' : \mathbf{Q}] = \dim H$, then $\operatorname{MT}(H, h)$ is abelian.

The corresponding Mumford–Tate subdomains are points: the torus is entirely contained in the isotropy group. The converse is also true. This is easy: by definition the Mumford–Tate group $\operatorname{MT}(H, h)$ first of all commutes with $\operatorname{End}(H, h)$ and, second, being contained in the isotropy group means that $\operatorname{MT}(H, h) \subset Z_h \operatorname{GL}(H) \subset \operatorname{End}(H, h)$.

Such zero-dimensional Mumford–Tate domains are called *CM points* or *special points*.

(ii) **Elliptic modular curves**. The next simplest examples are *elliptic modular curves* which have an analytic description as $\Gamma\backslash\mathfrak{h}$, where Γ is a modular group, i.e., a congruence subgroup of $\operatorname{SL}(2; \mathbf{Z})$, for instance

$$\Gamma(N) \stackrel{\text{def}}{=} \operatorname{Ker}(\operatorname{SL}(2; \mathbf{Z}) \to \operatorname{SL}(2, \mathbf{Z}/N\mathbf{Z})).$$

The corresponding Shimura datum is $(\operatorname{SL}(2), \operatorname{SL}(2, \mathbf{R})/\operatorname{U}(1))$. It was well known before Shimura proved his result that elliptic modular curves are all defined over $\mathbf{Q}$.

(iii) **Moduli spaces of abelian varieties**. We can define $\Gamma(N) \subset \operatorname{Sp}(g, \mathbf{Z})$ in a way similar to what we did for $g = 1$, and consider $\mathcal{A}_g(N) = \Gamma(N)\backslash\mathfrak{h}_g$ belonging to the Shimura datum $(\operatorname{Sp}(g), \operatorname{Sp}(g)/\operatorname{U}(g))$. For $N \geq 3$ the space $\mathcal{A}_g(N)$ is smooth quasi-projective and it is a fine moduli space for g-dimensional principally polarized abelian varieties with some additional structure, called level-N structure.

((iv) **Hilbert modular surfaces**. Let $K = \mathbf{Q}(\sqrt{\delta})$, $\delta > 0$ be a real quadratic field and $G = \operatorname{Res}_{K/\mathbf{Q}} \operatorname{PGL}(2, K)$. The group of its real points is $G(\mathbf{R}) = \{(A, A') \in \operatorname{PGL}(2, \mathbf{R}) \times \operatorname{PGL}(2, \mathbf{R})\}$, where A' is obtained from A by applying the Galois action $\sqrt{\delta} \mapsto -\sqrt{\delta}$ to each of the matrix entries. Then $G(\mathbf{R})$ acts componentwise on $\mathfrak{h} \times \mathfrak{h}$ and hence $\mathfrak{h} \times \mathfrak{h} = G(\mathbf{R})/K$, where K is a maximal compact subgroup. If $\Gamma \subset G(\mathbf{Q})$ is arithmetic, the surface $\Gamma\backslash\mathfrak{h} \times \mathfrak{h}$ is called a *Hilbert modular surface*.

(v) **Picard modular surfaces**. We have seen in Example 14.3.1 (ii) that the complex two-ball B^2 can be viewed as the homogeneous space $\operatorname{SU}(2, 1)/\operatorname{U}(2)$.

Suppose that h is a Hermitian form of signature $(2, 1)$ with coefficients in the imaginary quadratic field $K = \mathbf{Q}(\sqrt{-\delta})$, $\delta > 0$. Then $G = \mathrm{Res}_{K/\mathbf{Q}} \mathrm{SU}(h)$ is a $\mathbf{Q}$-simple group and hence for any congruence subgroup Γ of $G(\mathbf{Q})$ the quotient $\Gamma \backslash B^2$ is a Shimura variety, a so-called *Picard modular surface*.

The Adelic Point of View

We sketch now a more arithmetic point of view. First we recall that the ring of the finite adèles is given by

$$\mathbf{A}_f = \{(x_p)_p \in \prod_{p \text{ prime}} \mathbf{Q}_p \mid x_p \in \mathbf{Z}_p \text{ for almost all } p\}.$$

This $\mathbf{Q}$-algebra carries a unique smallest topology such that the subring $\prod_p \mathbf{Z}_p \subset \mathbf{A}_f$ and all its translations in $\mathbf{A}_f$ are compact and open subsets.

Let G be a $\mathbf{Q}$-algebraic group. Such a group has points in each $\mathbf{Q}_p$ and hence also in $\mathbf{A}_f$. Congruence subgroups of $G(\mathbf{Q})$ are given by compact and open subgroups K of $G(\mathbf{A}_f)$. For these, the intersection

$$\Gamma = K \cap G(\mathbf{Q})$$

is a congruence subgroup, and conversely, any congruence subgroup can be written in this way. The *adèlic definition of a Mumford–Tate variety* is as follows:

$$\mathsf{MT}_K(G, X) = G(\mathbf{Q}) \backslash X \times G(\mathbf{A}_f) / K.$$

Here $G(\mathbf{Q})$ acts on X and $G(\mathbf{A}_f)$ from the left, and K acts on $G(\mathbf{A}_f)$ from the right:

$$q(x, a)k = (qx, qak), \quad q \in G(\mathbf{Q}),\ x \in X,\ a \in G(\mathbf{A}_f),\ k \in K.$$

To compare this with the previous definition, we need some more notation. Let (G, X) be a Mumford–Tate datum. Since we do not assume G connected, also X can have several connected components. Fix one which we call X^0. For ease of notation, for any congruence subgroup Γ, we shall write

$$X^0(\Gamma) \overset{\text{def}}{=} \Gamma \backslash X^0.$$

For what follows, we need the groups

$$\Gamma_g \overset{\text{def}}{=} (gKg^{-1}) \cap G(\mathbf{Q}) \cap G(\mathbf{R})_0 \quad \text{for } g \in G(\mathbf{A}_f),$$

where $G(\mathbf{R})_0 = \{x \in G(\mathbf{R}) \mid \mathrm{ad}(x) \in (G^{\mathrm{ad}}(\mathbf{R}))^0 \text{ under the adjoint morphism}\}$. Now we can state the comparison result.

Lemma 16.3.8 (Milne, 2004, Lemma 5.12) *There is a finite set $\mathscr{C}_K$ of elements $g \in G(\mathbf{A}_f)$ and an isomorphism*

$$\bigsqcup_{g \in \mathscr{C}_K} X^0(\Gamma_g) \xrightarrow{\simeq} \mathsf{MT}_K(G, X), \quad [x] \mapsto [x, g], \tag{16.1}$$

i.e., $\mathsf{MT}_K(G, X)$ is a disjoint union of connected Mumford–Tate varieties. This isomorphism is a homeomorphism when we use the classical topology on the left and the adèlic topology on the right.

Mumford–Tate varieties for different congruence subgroups of G can be compared with the help of the bijections

$$T_g : \mathsf{MT}_K(G, X) \xrightarrow{\sim} \mathsf{MT}_{K^{(g)}}(G, X), \quad K^{(g)} \stackrel{\text{def}}{=} g^{-1}Kg$$

induced by the right action of elements $g \in G(\mathbf{A}_f)$ on the compact group K. So, if $K' \subset G$ is a second compact subgroup, and since $g^{-1}Kg \cap K'$ is open and compact as well, this action can be used as a "conjugation proof" comparison between the two varieties $\mathsf{MT}_K(G, X)$ and $\mathsf{MT}_{K'}(G, X)$. More generally, one defines the *Hecke correspondence* $T_{g,K,K'}$ associated with K, K' and g as follows:

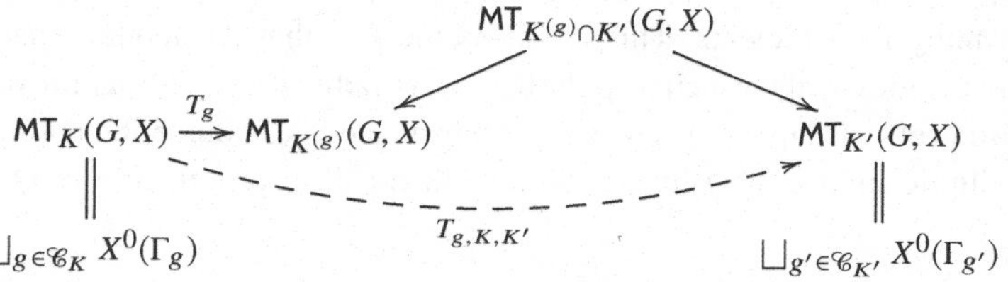

$$\begin{array}{ccccc} & & \mathsf{MT}_{K^{(g)} \cap K'}(G, X) & & \\ & \swarrow & & \searrow & \\ \mathsf{MT}_K(G, X) \xrightarrow{T_g} \mathsf{MT}_{K^{(g)}}(G, X) & & \overset{T_{g,K,K'}}{\dashrightarrow} & & \mathsf{MT}_{K'}(G, X) \\ \| & & & & \| \\ \bigsqcup_{g \in \mathscr{C}_K} X^0(\Gamma_g) & & & & \bigsqcup_{g' \in \mathscr{C}_{K'}} X^0(\Gamma_{g'}) \end{array} \tag{16.2}$$

The two arrows going down are the natural projections coming from the inclusions of $(gKg^{-1}) \cap K'$ in gKg^{-1} and K', or – equivalently – the inclusions of $\Gamma_g \cap \Gamma_{g'}$ into Γ_g and $\Gamma_{g'}$.

As will be explained shortly, with any Mumford–Tate datum (G, X) one associates a number field $E(G, X)$, its *reflex field*. This field plays a central role for any Shimura variety originating from (G, X): these quasi-projective varieties turn out to be all defined over the *same* number field $E(G, X)$! So, in particular, the corresponding finite union (16.1) can be defined over this field. That this is indeed the case was completely unexpected at the time when Shimura (1964; 1965; 1966) showed this. Note that the Galois action of $E(G, X)$ permutes the individual irreducible components of $\mathsf{MT}_K(G, X)$ but it is not clear (and not true in general) that these themselves are definable over the reflex field. This is one of the reasons to pass to the adèlic language. See also Deligne (1971c) and Milne (2004) for pertinent references; the last notes also may serve as an introduction.

We shall now show that the definition of the reflex field as given e.g. in Milne

(2004, Chapter 12) also applies in the situation of a general Mumford–Tate datum. To explain this, pick $h : \mathbf{S} \to G_{\mathbf{R}}$ such that $x = [h] \in X$. Consider the associated cocharacter

$$\mu_x : \mathbf{G}_m(\mathbf{C}) \to G(\mathbf{C}), \quad z \mapsto h\begin{pmatrix} z & -1 \\ 1 & z \end{pmatrix}.$$

The group $G(\mathbf{R})$ acts on X by conjugation and this action induces one on the group of co-characters. It follows that the co-character $\mu_x \in X^*(G(\mathbf{C}))$ defines a unique conjugation class

$$c(X) \in \mathscr{C}(\mathbf{C}) \stackrel{\text{def}}{=} G(\mathbf{C})\backslash X^*(G(\mathbf{C})),$$

completely determined by the given Mumford–Tate datum. One can assume that this class is defined over the algebraic numbers, e.g. $c(X) \in \mathscr{C}(\overline{\mathbf{Q}})$. By definition, $E(G, X)$ is the fixed field inside $\overline{\mathbf{Q}}$ for the subgroup of $\mathrm{Gal}(\overline{\mathbf{Q}}/\mathbf{Q})$ that fixes $c(X)$. For details we refer to Milne (2004, Chapter 12).

Example 16.3.9 The reflex field of $(\mathrm{Sp}(g), \mathfrak{h}_g)$ is $\mathbf{Q}$. This is so because $\mu_{\mathrm{i}}(z^2) = \begin{pmatrix} z & -1 \\ 1 & z \end{pmatrix}$ so that the graph of the morphism μ_{i} is given by equations with integer coefficients. This confirms the fact that the moduli space of abelian varieties with principal polarization is defined over $\mathbf{Q}$, as proven in Mumford (1965, Chapter 7). The simplest instance is of course the moduli space for elliptic curves, the j-line ($\mathbf{P}^1 - \infty$) which clearly is defined over $\mathbf{Q}$.

Problems

16.3.1 Let G be a simple algebraic $\mathbf{R}$-matrix group of adjoint type and suppose that there exists a morphism $h : \mathbf{S}(\mathbf{R}) \to G$ satisfying the two conditions **M** and **S1**. Let K be the connected subgroup of G such that $\mathrm{Lie}(K)$ is the fixed point locus of the Cartan involution on $\mathrm{Lie}(G)$. In particular, K is a maximal compact subgroup of G.

(a) Show that there exists a morphism $u : \mathbf{U}(\mathbf{R}) \to G$ such that $h(z) = u(z/\bar{z})$. Hint: use Proposition 15.1.4.

(b) Show that**S1** implies that G is not compact.

(c) Show that that $\mathrm{Im}(u) \subset Z(K)$ and hence that $\dim Z(K) \geq 1$.

(d) Going through the list of noncompact $\mathbf{R}$-simple groups (see for example Knapp, 2005, AppendixC3), conclude that $\dim Z(K) = 1$ and that G/K has the structure of a bounded symmetric domain.

16.3.2 Recall (15.5) that $\mathrm{CSp}(g, \mathbf{R})$ is the group of symplectic similitudes. Show that this is a reductive group. Show that the subgroup $\mathrm{CU}(g)$ of unitary similitudes is the connected group $\mathbf{R}^+\mathrm{U}(g)$ and thus that $\mathrm{CSp}(g, \mathbf{R})/\mathrm{CU}(g)$ has two connected components. This gives an example of a nonconnected Shimura datum.

16.4 Examples of Mumford–Tate Domains

In this section we investigate the classical $\mathbf{R}$-simple groups in more detail. We use the following matrices.

$$I_{p,q} = \begin{pmatrix} -\mathbf{1}_p & 0 \\ 0 & \mathbf{1}_q \end{pmatrix}, \quad K_{p,q} = \begin{pmatrix} I_{p,q} & 0 \\ 0 & I_{p,q} \end{pmatrix},$$

$$J_g = \begin{pmatrix} 0 & -\mathbf{1}_g \\ \mathbf{1}_g & 0 \end{pmatrix}, \quad R_g(t) = \begin{pmatrix} \cos(t)\mathbf{1}_g & -\sin(t)\mathbf{1}_g \\ \sin(t)\mathbf{1}_g & \cos(t)\mathbf{1}_g \end{pmatrix}, \quad R(t) = R_1(t).$$

We now go down the list of the classical groups.

The group $\mathrm{SU}(p, q)$

There is a Shimura domain D associated with this group: see Theorem 16.1.5. It has the property that the stabilizer of $[h] \in D$ is a maximal compact subgroup of $\mathrm{SU}(p, q)$. Later we give an example (see Example 17.1.8) of a geometric construction of a variation of weight 1 Hodge structures over D.

In this section, however, we consider other Mumford–Tate domains for this group for which the stabilizer is strictly contained in a maximal compact subgroup. So we start by recalling that the Cartan involution on its Lie algebra is given by

$$\theta = \mathrm{Ad}(I_{p,q}).$$

For p even $I_{p,q}$ belongs to $\mathrm{SU}(p, q)$ and for p odd a suitable multiple of $I_{p,q}$ (with the same adjoint action) does. The maximal torus is the (compact) diagonal torus of rank $p + q - 1$. Let us consider the $\mathrm{SU}(p, q)$-Hodge structure of weight 0 given by

$$\begin{aligned} h : \mathbf{U} &\to \mathrm{SU}(p, q) \\ z &\mapsto \mathrm{diag}(z^{\ell_1}, \dots, z^{\ell_p}, 1, \dots, 1), \quad \sum \ell_j = 0. \end{aligned}$$

For simplicity, assume that p is even. One has $\mathrm{Ad}\, h(\mathrm{i}) = \theta$ if and only if all ℓ_j are 2 modulo 4. This makes h polarized by the standard form. The standard representation is $\mathbf{C}^{p+q} = \mathbf{R}^{2(p+q)}$ and the characters $e^{2\pi i t} \mapsto e^{2\pi i \ell_j t}$ act as a rotation over $2\pi i \ell_j t$ in each coordinate plane. Hence it gets a Hodge structure of weight 0 with Hodge numbers those $h^{a,-a}$ according to which ℓ_j are nonzero. For instance, take $\ell_1 = \cdots = \ell_{p-1} = 2$ and $\ell_p = -2(p-1)$. Then $h^{1,-1} = h^{-1,1} = p - 1$ and $h^{p-1,1-p} = h^{1-p,p-1} = 1$, $h^{0,0} = 2q$. The associated Mumford–Tate domain can be found by computing the centralizer of h. This is clearly contains the compact subgroup $\mathrm{U}(q)$. If we write a matrix in obvious block form $\begin{pmatrix} A & B \\ C & D \end{pmatrix}$ then $h(z)$ acts on the entry A_{ij} of A by multiplication with $z^{\ell_i - \ell_j}$. If we assemble the ℓ_j thus in groups of equal numbers we can read off the subfactors of $\mathrm{U}(p)$ in the stabilizer. The above example where all but one ℓ_j equals 2 gives the subgroup $\mathrm{U}(p-1)$. One thus gets $\mathrm{S}(\mathrm{U}(p-1) \times \mathrm{U}(1) \times \mathrm{U}(q))$. Other combinations can give other compact subgroups.

If q is even the same argument works by interchanging p and q and if p and q are both odd, their sum is 2 modulo 4 and the Cartan involution is the adjoint action of $\mathrm{i}I_{p,q} \in \mathrm{SU}(p,q)$. We now need to replace h by a more complicated morphism. We leave this as an exercise.

As a final remark, notice that the group $\mathrm{SU}(p)$ gives rise to compact Mumford–Tate domains, namely flag varieties. These then embed in compact period domains. The above example with $q = 0$ gives an embedding

$$\mathbf{P}^{p-1} \hookrightarrow \mathrm{SO}(2p)/\mathrm{U}(1) \times \mathrm{U}(p-1),$$

where the left-hand side is a compact complex manifold of dimension $\frac{1}{2}p(p-1)$.

Groups Acting Transitively On Period Domains

The groups $\mathrm{SO}(2p, m)$ and $\mathrm{Sp}(g, \mathbf{R})$ act transitively on period domains and, by construction, every point in the period domain defines a polarizable Hodge structure. One can also view this from the actual perspective, starting from exhibiting their Cartan involutions. See Problems 16.4.1 and 16.4.2.

The group $\mathrm{Sp}(p, q)$

Recall that these are the matrices in $\mathrm{Sp}(g, \mathbf{C})$ which leave invariant the Hermitian form $Z^* K_{p,q} Z$, where we put $Z^* = {}^{\mathsf{T}}\bar{Z}$. The Cartan involution is $\mathrm{Ad}\, K_{p,q}$. Consider

$$h|\mathbf{U} \to \mathrm{Sp}(p,q), \quad z \mapsto \mathrm{diag}(\underbrace{z^2, \ldots, z^2}_{p \text{ times}}, \mathbf{1}_q, \underbrace{z^{-2}, \ldots, z^{-2}}_{p \text{ times}}, \mathbf{1}_q).$$

Again, Ad $h(\mathrm{i})$ is the Cartan involution. One has (see Helgason, 1962, p. 341)

$$\mathrm{Lie}(\mathrm{Sp}(p,q)) = \begin{array}{c} \\ p \\ q \\ p \\ q \end{array}\!\!\begin{array}{c} \begin{array}{cccc} p & q & p & q \end{array} \\ \begin{pmatrix} Z_1 & U & V & W \\ U^* & Z_2 & {}^{\mathsf{T}}W & T \\ -V^* & \bar{W} & \bar{Z}_1 & -\bar{U} \\ W^* & -\bar{T} & -{}^{\mathsf{T}}U & \bar{Z}_2 \end{pmatrix} \end{array} \text{ s. t. } \begin{array}{ll} Z_1^* = -Z_1, & Z_2^* = -Z_2, \\ V = {}^{\mathsf{T}}V, & T = {}^{\mathsf{T}}T. \end{array}$$

The adjoint action of $h(z)$, $z \in \mathbf{U}$ of this Lie algebra can be schematically given as

$$\begin{array}{c} \\ p \\ q \\ p \\ q \end{array}\!\!\begin{array}{c} \begin{array}{cccc} p & q & p & q \end{array} \\ \begin{pmatrix} \mathrm{id} & z^{-4} & \mathrm{id} & z^{-4} \\ z^4 & \mathrm{id} & z^{-4} & \mathrm{id} \\ \mathrm{id} & z^4 & \mathrm{id} & z^{-4} \\ z^4 & \mathrm{id} & z^4 & \mathrm{id} \end{pmatrix} \end{array}$$

and we get a polarized Hodge structure on $\mathfrak{sp}(p,q)$ with Hodge numbers $h^{2,-2} = h^{-2,2} = 2pq$, $h^{0,0} = 2p^2 + 2q^2 + 4pq + p + q$. Since this group is not of Hermitian type, there can be no action of $\mathbf{S}$ on the Lie algebra with characters $\{1, z\bar{z}^{-1}, z^{-1}\bar{z}\}$ and the given action is the simplest of non-Hermitian type.

The group $\mathrm{SO}^*(2r)$

Recall that the group $\mathrm{SO}^*(2r)$ is the subgroup of $\mathrm{SO}(2r, \mathbf{C})$ leaving invariant the skew Hermitian form given by the skew matrix J_r. It contains $\mathbf{U}(r)$ embedded through

$$A + \mathrm{i}B \mapsto \begin{pmatrix} A & -B \\ B & A \end{pmatrix}.$$

The maximal torus is given by embedding the diagonal torus $\mathbf{U}^r$ of $\mathbf{U}(r)$ in $\mathrm{SO}^*(2r)$ via this embedding. Then $\mathrm{Ad}(J_r)$ is the Cartan involution on the Lie algebra. In block form this algebra reads

$$\left\{ \begin{pmatrix} Z & W \\ -\bar{W} & \bar{Z} \end{pmatrix} \,\middle|\, Z = -{}^{\mathsf{T}}Z,\ W = W^* \right\}$$

and we find

$$\mathfrak{k} = \left\{ \begin{pmatrix} X & U \\ -U & X \end{pmatrix} \,\middle|\, X + \mathrm{i}U \in \mathfrak{u}(r) \right\},$$

$$\mathfrak{p} = \left\{ \mathrm{i}\begin{pmatrix} Y & V \\ V & -Y \end{pmatrix} \,\middle|\, Y, V \in \mathfrak{so}(r) \right\}.$$

The morphism $h|\mathbf{U} \to \mathbf{U}(r) \to \mathrm{SO}^*(2r)$ which is the composition of the diagonal embedding and the above embedding $\mathbf{U}(r) \to \mathrm{SO}^*(2r)$ sends i to J_r and hence gives a polarized Hodge structure on Lie($\mathrm{SO}^*(2r)$). One sees that $z \in \mathbf{U}$ acts via this representation as a rotation

$$z = e^{it} \mapsto \begin{cases} \begin{pmatrix} X \\ U \end{pmatrix} \to \begin{pmatrix} X \\ U \end{pmatrix} & \text{on } \mathfrak{k}, \\ \begin{pmatrix} Y \\ V \end{pmatrix} \to R_r(2t) \begin{pmatrix} Y \\ V \end{pmatrix} & \text{on } \mathfrak{p}. \end{cases}$$

This shows that the Hodge numbers are $h^{1,-1} = h^{-1,1} = \frac{1}{2}r(r-1)$ and $h^{0,0} = r^2$. Note that this conforms that we have a group of Hermitian type.

Domains for the Group SU(2, 1)

To finish this section, we show that this group defines several distinct Mumford–Tate domains which are all identical as C^∞-manifolds. So the complex structures on them are not isomorphic.

Consider the morphism

$$\begin{aligned} &\mathbf{S} \to \mathrm{SU}(2,1), \\ &z \mapsto \mathrm{diag}(z^{2+4k}, z^{2+4\ell}, z^{-4k-4\ell-4}), \quad k \neq \ell. \end{aligned}$$

Then T, the (two-dimensional) diagonal torus, is the centralizer of h and $\mathrm{SU}(2,1)/T$ receives a complex structure depending on k, ℓ. This can be seen as follows. The adjoint action of $\mathrm{Lie}(T) \otimes \mathbf{C}$ on $\mathrm{Lie}(\mathrm{SU}(2,1)) \otimes \mathbf{C} = \mathfrak{sl}(3, \mathbf{C})$ has eigenvectors E_{ij}, $i \neq j$, where the matrix E_{ij} has zeroes everywhere except entry (i, j) which is 1. Indeed

$$\mathrm{ad}\,(\mathrm{diag}(z_1, z_2, z_3))E_{ij} = (z_i - z_j)E_{ij}.$$

This determines the characters of $\mathrm{ad}(h(z))$. We get

$$\left. \begin{aligned} z \mapsto &\pm 4(k-\ell)z \text{ for } E_{12}, E_{21}, \\ &\pm [4(2k+\ell)-2]z \text{ for } E_{13}, E_{31}, \\ &\pm [4(2\ell+k)-2]z \text{ for } E_{23}, E_{32}. \end{aligned} \right\} \tag{16.3}$$

On the other hand, a complex structure on $\mathrm{Lie}(\mathrm{SU}(2,1))/\mathrm{Lie}(T) = \mathfrak{p}$ is determined by giving an orientation on each of the three two-dimensional real subspaces of $\mathfrak{p} \subset \mathrm{Lie}(\mathrm{SU}(2,1))$ given by $\mathfrak{p}_{12} = \{zE_{12} - \bar{z}E_{21} \mid z \in \mathbf{C}\}$, $\mathfrak{p}_{13} = \{zE_{13} + \bar{z}E_{31} \mid z \in \mathbf{C}\}$, $\mathfrak{p}_{23} = \{zE_{23} + \bar{z}E_{32} \mid z \in \mathbf{C}\}$. In the case at hand, the action of $\mathrm{Ad}(h(z))$ on each of these planes is a rotation and the sense of this rotation determines a complex structure. This is given by the sign of the integers from (16.3). The complex structure can also be described as induced

Table 16.2 *Complex structures on* $\mathfrak{p}$

	sign $(k-\ell)$	sign $(2k-\ell)$	sign $(2\ell-k)$	Flag type
1	+	+	+	(13, 132, 3)
1*	–	–	–	(12, 123, 2)
2	+	+	–	(132, 2, 32)
2*	–	–	+	(1, 132, 13)
3	–	+	–	(1, 12, 123)
3*	+	–	+	(321, 32, 3)

from an embedding in a suitable flagmanifold:

$$\mathrm{SU}(2,1)/T \subset \mathrm{SL}(3,\mathbf{C})/P_{\mathbf{C}}, \quad \mathrm{Lie}(P) = \mathfrak{p};$$

the type of flag is determined by the 6 possible orders of the basis elements of $\mathbf{C}^3$. Using the standard bases $\{e_1, e_2, e_3\}$ for $\mathbf{C}^3$ we have for instance the parabolic

$$(13, 132, 3) = \begin{pmatrix} * & * & 0 \\ 0 & * & 0 \\ * & * & * \end{pmatrix}$$

fixing the flag $\{\mathbf{C}e_3 \subset \mathbf{C}e_1 + \mathbf{C}e_3 \subset \mathbf{C}e_1 + \mathbf{C}e_2 + \mathbf{C}e_3\}$. We get the pairs of conjugate complex structures j and j^* in Table 16.2.

Problems

16.4.1 Consider the group $\mathrm{SO}(2p, m)$ with Cartan involution on the Lie algebra given by

$$\theta = \mathrm{Ad}\, I_{2p,m}.$$

(a) Determine a compact maximal $T \subset \mathrm{SO}(2p, m)$ and a suitable morphism $\mathbf{S} \to T \subset \mathrm{SO}(2p, m)$ inducing a Hodge structure on $\mathrm{Lie}(\mathrm{SO}(2p, m))$, polarized by minus the trace form.

(b) Determine the Hodge numbers of this Hodge structure.

16.4.2 Consider $\mathrm{Sp}(g)$ with Cartan involution on $\mathrm{Lie}(\mathrm{Sp}(g, \mathbf{R}))$ given by $\mathrm{Ad}(J_g)$. Consider the morphism

$$h|\mathbf{U} \to \mathrm{Sp}(g, \mathbf{R}), \quad e^{it} \mapsto R_g(2t)$$

(a) Show that $\operatorname{Ad} h(\mathrm{i})$ is the Cartan involution.
(b) Determine adjoint the action of $h(z)$ on $\operatorname{Lie}(\operatorname{Sp}(g, \mathbf{R}))$.
(c) Show that this gives a weight 0 Hodge structure with Hodge numbers $h^{-1,1} = h^{1,-1} = \frac{1}{2}g(g+1)$, $h^{0,0} = g^2$.

16.4.3 Let G be one of the compact classical groups Hodge type. Exhibit nontrivial Mumford–Tate domains for these.

Bibliographical and Historical Remarks

About Section 16.1: This is classical and based on Cartan's work on bounded symmetric domains as summarized and extended in the books by Helgason (1962, 1978).

About Section 16.2: The name Mumford–Tate domain was coined in Green et al. (2012). Our treatment is largely based on the latter monograph. However, we found it more transparent to exploit Deligne's notion of a C-polarization which we treated in an earlier chapter. Also, some arguments have been adapted since we also wanted to treat Shimura varieties. The notion of Hodge group has been introduced in Simpson (1992, Section 4). He defines it slightly differently, but for real algebraic groups that are connected in the classical topology, the notion coincides with our notion. All of this has been heavily influenced by Deligne's essay (Deligne, 1979) on Shimura's works.

About Section 16.3: The, by now classical, axiomatic treatment of Shimura varieties stems from Deligne (1979). See also Milne (2004). We added as an observation that also Mumford–Tate domains can be treated in a similar axiomatic fashion.

About Section 16.4: We give here examples of Hodge representation associated with the classical groups. The verification of this uses the characterization via C-polarizations and this simplifies the way this is done in Green et al. (2012).

17

Hodge Loci and Special Subvarieties

In this chapter we show that a generic point h in a Mumford–Tate domain $X = G/V$ has Mumford–Tate group G (provided G is a connected $\mathbf{Q}$-simple group). Hence so is the Mumford–Tate group of the Hodge structure $\rho{\circ}h$, where $\rho : G \to \mathrm{GL}(H)$ is a faithful representation of G. For an arbitrary $[h] \in X$, generic or not, the orbit under $\mathrm{MT}(h)$ is a Mumford–Tate subdomain. Such domains are precisely the Hodge loci, i.e., sets consisting of those points $[h] \in X$ for which a given collection of vectors in H (or in some tensor representation of H) is Hodge with respect to the Hodge structure $[h]$. We show this in Section 17.1. We also show that Hodge loci contain a dense set of CM-points (i.e., points whose Mumford–Tate group is a nontrivial algebraic torus).

To place the study of Hodge loci in a functorial context, in Section 17.2 we consider equivariant maps between Mumford–Tate domains. Their images are not always Hodge loci, but they turn out to be totally geodesic subdomains. If they contain a CM-point we get a Hodge locus.

A noteworthy geometric example arises for cubic surfaces. Although their period map is constant, looking at the associated triple cover of $\mathbf{P}^3$ branched in the cubic surface we get a second period map which factors over a Mumford–Tate domain. The latter turns out to be the unit 4-ball. Furthermore, via the period map the moduli space gets identified with a Zariski open subset of the Shimura variety which is an arithmetic quotient of this unit 4-ball. This example is investigated in Section 17.3.

A Hodge locus inside a Shimura variety is also called a "special subvariety". Obvious examples of special subvarieties are the CM-points. In dimension one we have the special curves. In Section 17.4 we consider these inside a Siegel upper half-space. We restrict the discussion to two classical examples: embedded modular curves (via the classical Satake embeddings), and the Mumford-type curves. These give rise to the only Shimura curves rigidly embedded in quotients of the Siegel upper half-space.

In Section 17.5, we consider examples where special subvarieties can be characterized numerically.

17.1 Hodge Loci

Recall from Section 7.5 that the Hodge locus is the locus where certain cohomology classes are algebraic. More generally, we can look at the locus where all classes in some (or all) tensor representations that are Hodge at a given point $o \in S$ remain Hodge. This makes sense at the level of period domains, or, more generally, Mumford–Tate domains.

Definition 17.1.1 Let $X = G/V$ be a Mumford–Tate domain, $h : \mathbf{S} \to G_{\mathbf{R}}$ a homomorphism and $\rho : G \to \operatorname{Aut}(H_{\mathbf{Q}}, b)$ a $\mathbf{Q}$-representation for which $\rho_{\mathbf{R}} \circ h$ defines a rational Hodge structure on H polarized by b. Let $o \in D$ be the class $1 \cdot V$ and let T be a finite direct sum of tensor spaces for H. In this section we denote the vector space of rational tensors in T which are Hodge with respect to $x \in X$ by $\mathrm{Hdg}_x T$. In other words

$$\mathrm{Hdg}_x T \stackrel{\text{def}}{=} T_{\mathbf{Q}} \cap F_x^{\text{even}} T, \quad F_x^\bullet \text{ corresponding to } x.$$

(i) The *Hodge locus* for a rational tensor $t \in T_{\mathbf{Q}}$ is defined as the locus in X where t is a Hodge tensor:

$$\mathrm{HL}(X; t) \stackrel{\text{def}}{=} \left\{ x \in X \mid t \in \mathrm{Hdg}_x T \right\}.$$

(ii) The *Hodge locus* $\mathrm{HL}_o(X)$ at $o \in X$ is defined as the locus in X where all weight zero Hodge tensors at o remain Hodge:

$$\mathrm{HL}_o(X) \stackrel{\text{def}}{=} \left\{ x \in X \mid [\mathrm{Hdg}_o T]^{w=0} \subset [\mathrm{Hdg}_x T]^{w=0} \text{ for all } T \right\},$$

T running over all finite direct sums of tensor spaces for H.

Hodge loci are defined by analytic equations and so are analytic subvarieties of X which may or may not be irreducible. The irreducible component which passes through o is smooth and can be linked to its Mumford–Tate group (as redefined in Definition 16.3.1).

Proposition 17.1.2 *The component X_o of $\mathrm{HL}_o(X)$ passing through o is the orbit of o under the connected component of the Mumford–Tate group $\mathrm{MT}(o)$. In particular, X_o has the structure of a Mumford–Tate domain.*

Proof Since $g \in \mathrm{MT}(o)$ acts as the identity on weight zero Hodge tensors at

o, we have an inclusion $[\mathrm{Hdg}_o T]^{w=0} \subset [\mathrm{Hdg}_{g\cdot o} T]^{w=0}$. Hence, the point $g \cdot o$ belongs to the Hodge locus:

$$Y = \mathrm{MT}(o) \cdot o \subset \mathrm{HL}_o(X).$$

On the other hand, putting

$$\mathfrak{m} \stackrel{\text{def}}{=} \mathrm{Lie}(\mathrm{MT}(o)), \quad \mathfrak{m}^- = \mathfrak{m} \cap \bigoplus_{i>0} \mathfrak{g}^{-i,i},$$

we have an equality

$$T_{Y,o} = \mathfrak{m}^-.$$

It suffices therefore to establish an inclusion

$$T_{\mathrm{HL}_o(X),o} \subset \mathfrak{m}^-.$$

To show this, we remark first that there exist some tensor space T and a tensor $t \in T$ which is Hodge at o for which we have

$$\mathfrak{m}_{\mathbb{C}} = \{\xi \in \mathfrak{g}_{\mathbb{C}} \mid \xi \cdot t = 0\}. \tag{17.1}$$

This is the translation of Theorem 15.2.9 in terms of Lie algebras.

Next, consider the condition that all rational tensors $t \in T_{\mathbb{Q}}$ that are of Hodge type (0, 0) at o, remain of type (0, 0) in the direction of $\xi \in T_{X,o}$. This characterizes vectors ξ which are tangent at o to the Hodge locus $\mathrm{HL}_o(X)$. The necessary and sufficient condition for this is that $\xi \cdot t \in F^0_o T = \oplus_{\alpha \geq 0} T^{\alpha,-\alpha}$. On the other hand, since $\xi \in T_{X,o} = \mathfrak{g}^-$, say $\xi \in \sum_{\beta>0} \mathfrak{g}^{-\beta,\beta}$, we have $\xi \cdot t \in \sum_{\beta>0} T^{-\beta,\beta}$. The two assertions imply $\xi \cdot t = 0$. Hence, by (17.1), $\xi \in \mathfrak{m}_{\mathbb{C}} \cap \mathfrak{g}^- = \mathfrak{m}^-$. □

Recall that we introduced Hodge generic points in relation to period maps (see Lemma-Definition 15.3.5). We now broaden this notion to the abstract setting of Mumford–Tate domains. Consider those tensors t for which $\mathrm{HL}(X;t)$ is a proper closed subvariety of X. This is the case for at most countable many tensors t_α, $\alpha \in I$ and we then say that

$$x \in X \text{ is } \textit{Hodge generic} \iff x \in X - \bigcup_{\alpha \in I} \mathrm{HL}(X; t_\alpha).$$

So, assuming X to be connected (which is the case if G is connected), we have:

$x \in X$ is Hodge generic	$\iff$	X is the component of the Hodge locus passing through x.

We write x_{gen} for such a point. For arbitrary x we then have:

$$X_x \subset X_{x_{\mathrm{gen}}} = X \implies \mathrm{MT}(x) \subset \mathrm{MT}(x_{\mathrm{gen}}). \tag{17.2}$$

In fact, for irreducible domains this "generic" Mumford–Tate group is equal to G; even more is true:

Proposition 17.1.3 *Let $X = G/V$ be a connected Mumford–Tate domain with G a $\mathbf{Q}$-simple group. Then the Mumford–Tate group of a Hodge generic point is G.*

Proof The Mumford–Tate group of x_{gen} is the $\mathbf{Q}$-Zariski closure M of $\text{Im}(h)$ for a generic point $[h] \in X$. Clearly, for all $g \in G(\mathbf{Q})$, the Zariski closure of the image of ghg^{-1} equals gMg^{-1}. This is exactly the Mumford–Tate group of the Hodge structure gx_{gen}. Since x_{gen} is Hodge generic, by (17.2) we have $gMg^{-1} \subset M$. Hence $M(\mathbf{Q})$ is a normal subgroup of $G(\mathbf{Q})$ and so, by the assumption of simplicity, $M = G$. □

The second result (Corollary 17.1.5) concerning Mumford–Tate groups describes how small these can be. First an auxiliary result.

Lemma 17.1.4 *Let G be a reductive algebraic group defined over a number field $E \subset \mathbf{R}$. Let $T \hookrightarrow G$ be an algebraic subtorus. Then some $G(\mathbf{R})$-conjugate of T is contained in a maximal subtorus of G defined over E.*

Proof Consider $H = Z_G(T)$, the centralizer of T in G. The group H is a group defined over $\mathbf{R}$ and by Borel and Springer (1968, Prop. 7.10) there is a a maximal torus T_1 in H containing T.

As a first step in the proof we show that T_1 is even a maximal torus in G. To show this, let $T_2 \subset G$ be any torus containing T_1. Since T_2 also contains T and is abelian, it centralizes T, i.e., $T_2 \subset Z_G T = H$ and so, because of maximality, $T_2 = T_1$.

As our second step we use the result of Baily and Borel (1966, Prop. 2.5). In this setting it states that any two maximal tori defined over $\mathbf{R}$ are conjugate over $\mathbf{R}$. Since G, being reductive and defined over E, contains at least one maximal torus defined over E, the torus T_1 must be conjugate (over $\mathbf{R}$) to such a torus. □

At this point, recall (Example 16.3.7.(i)) that a CM-point or special point $[h] \in X$ corresponds to a morphism $h : \mathbf{S} \to G$ with $\text{MT}(h)$ an algebraic torus.

Corollary 17.1.5 *A given morphism $h : \mathbf{S} \to G$ with $[h] \in X$ is $G(\mathbf{R})$-conjugate to a morphism h' with* MT(h') *contained in a maximal torus defined over $\mathbf{Q}$. In particular, if (G, X) is a (connected) Mumford–Tate datum, X contains special points $[h']$. Special points are dense in X.*

Proof Apply the above lemma to the torus $h(\mathbf{S}) \subset G$ and the field $\mathbf{Q}$. So some conjugate of h, say h', factors over an algebraic subtorus T of G defined over

Q. It follows that $\mathrm{MT}(h')$ is an algebraic subquotient of T and hence is abelian. By definition, $[h']$ is then special.

To show density, let $[h']$ be a special point and observe that conjugation by $g \in G(\mathbf{Q})$ maps the abelian group $\mathrm{MT}(h')$ to $\mathrm{MT}(g \cdot [h'])$ and so $g([h'])$ is also a special point. Since $G(\mathbf{Q})$ is dense in $G(\mathbf{R})$ the assertion follows. □

Given a Mumford–Tate domain $X = G/V$ with G a $\mathbf{Q}$-simple group, we have just shown (see Theorem 17.1.3) that G is the Mumford–Tate group of the generic point. For an arbitrary point $o \in X$ one may have $\mathrm{MT}(o) \subsetneq G$, and the orbit of o under this group is the component $X_0 \subsetneq X$ of the Hodge locus through o.

Definition 17.1.6 Let $\Gamma \subset G$ be a neat congruence subgroup.[1] With the above notation, the image of X_o in $\Gamma\backslash X$ is called a *subvariety of Hodge type*.

Note that a subvariety of Hodge type is irreducible, but that it may be singular: a Γ-orbit may intersect X_o in more than one point.

Remark 17.1.7 Let us come back to period maps $\mathcal{P} : S \longrightarrow \Gamma\backslash D$. Lift this map to the universal cover $\tilde{S}$ of S, say $\tilde{p} : \tilde{S} \longrightarrow D$. Consider a point $o \in D$ which is not Hodge generic and $D_o \subsetneq D$ the component through o of the Hodge locus. Suppose moreover that $\tilde{P}(\tilde{S}) \subset D_o$. In this case not all points of D are $\mathcal{P}$-Hodge generic and $\tilde{\mathcal{P}}$ in fact factors through the Mumford–Tate subdomain D_o. This shows clearly why Mumford–Tate domains give crucial extra information about period maps.

Example 17.1.8 We construct a Hodge locus $M \subset D$, where D is a weight 2 period domain with Hodge numbers $h^{2,0} = p$ and $h^{1,1} = 2q$ and group $\mathrm{SO}(2p, 2q)$. The locus M will be the set of polarized Hodge structures in D which admit a compatible almost complex structure.

To construct the locus, let V be a $\mathbf{Q}$-vector space of dimension $n = 2(p + q)$ on which we have a nondegenerate symmetric form b of signature $(2p, 2q)$. Suppose that there exists an almost complex structure $\mathbf{J}$ for (V, b), i.e., an automorphism of V with $\mathbf{J}^2 = -\,\mathrm{id}$ and $b(\mathbf{J}x, \mathbf{J}y) = b(x, y)$ for all $x, y \in V$. These do exist: take

$$b = \begin{pmatrix} \mathbf{1}_{2p} & 0 \\ 0 & -\mathbf{1}_{2q} \end{pmatrix}, \quad \mathbf{J} = \begin{pmatrix} 0 & -\mathbf{1}_p & 0 & 0 \\ \mathbf{1}_p & 0 & 0 & 0 \\ 0 & 0 & 0 & -\mathbf{1}_q \\ 0 & 0 & \mathbf{1}_q & 0 \end{pmatrix}.$$

We can now be more precise about the domain D: it classifies weight two Hodge structures on V, polarized by b and with Hodge numbers $h^{2,0} = h^{0,2} = p$ and

[1] See Definition 16.3.5.(iii).

$h^{1,1} = 2q$. By definition, points in M correspond to those Hodge structures in D for which $\mathbf{J}$ is a morphism of Hodge structure. These are called $(b, \mathbf{J})$*-polarized Hodge structures*. Such a Hodge structure is thus compatible with the splitting of the complexification $V_{\mathbb{C}}$ of V into the $\pm$i-eigenspaces $V_{\pm}$ for $\mathbf{J}$:

$$V_{\mathbb{C}} = \underbrace{V^{2,0} \oplus V_+^{1,1}}_{V_+} \oplus \underbrace{V_-^{1,1} \oplus V^{0,2}}_{V_-}.$$

With the above choice of $(b, \mathbf{J})$, an example of such a Hodge decomposition can easily be written down. We leave this as an exercise.

The centralizer of $\mathbf{J}$ in $\mathrm{SO}(p, q)$ can be seen to be a $\mathbf{Q}$-simple group G with $G(\mathbf{R}) \simeq \mathrm{U}(p, q)$. The stabilizer of such a Hodge structure turns out to be equal to $\mathrm{S}(\mathrm{U}(p) \times \mathrm{U}(q))$, a maximal compact subgroup K of G. So, indeed, $M = G/K$, a Shimura type domain. It carries the tautological Hodge flag which satisfies the Griffiths' transversality relation. This follows since the tangent space to M can be viewed as the (-1)-eigenspace of the Cartan involution on the Lie algebra of G a group of Hermitian type. So transversality holds automatically. More is true: M is a maximal closed horizontal submanifold of D. By "maximal" one means that if $M \subset M^+$ and M^+ is also a closed horizontal submanifold, then $M = M^+$. This follows from a general bound (Carlson, 1986) for the dimension of such submanifolds, namely $\dim M^+ \le \frac{1}{2} h^{2,0} h^{1,1} = pq = \dim M$.

It remains to show that M is indeed of Hodge type. To see this, remark that a $(\mathbf{J}, b)$-polarized Hodge structure is nothing but a homomorphism h' factoring as

$$h' : \mathbf{S} \xrightarrow{h} G \hookrightarrow \mathrm{SO}(2p, 2q) = \mathrm{Aut}(V, b).$$

The domain M is the set of G-conjugates of h while D is the set of $\mathrm{SO}(2p, 2q)$-conjugates of h'. Replacing h by a generic enough G-conjugate, we may assume that the Mumford–Tate group of h equals G and since M can be viewed as the G-orbit of $[h]$ inside D, by definition M is of Hodge type inside D.

From a somewhat different, but essentially equivalent perspective, this example has been studied by Carlson and Simpson (1987). They also show that there is another representation which realizes M as a family of abelian varieties of dimension $p + q$ with an automorphism $\mathbf{J}'$, where $\mathbf{J}'^2 = -1$. Thus M, which started out as an abstract variation of Hodge structure, is in fact an object coming from geometry.

17.2 Equivariant Maps Between Mumford–Tate Domains

We start with the following result about Hodge loci, a result which is of differential geometric nature.

Proposition 17.2.1 *The component $X_o \subset X$ of the Hodge locus passing through a point o of a Mumford–Tate domain $X = G/V$ is a totally geodesic submanifold of X.*

Proof We have seen that X_o is an orbit of a subgroup $M \subset G$ of Hodge type. The same argument we used in Section 13.6 for period domains shows that a Mumford–Tate domain admits an M-equivariant Hermitian metric. Example 11.5.2 (i) then shows that X_o is a totally geodesic submanifold of X. □

As indicated in Example 11.5.2 (ii), there is a general way to produce a totally geodesic submanifold: take the image $f(X_1) \subset X_2$ under an equivariant map f between Mumford–Tate domains induced by an injective homomorphism of algebraic groups $\varphi : G_1 \to G_2$.

In this setting we want f to be holomorphic and so we recall how the holomorphic structure on a Mumford–Tate domain $X = G/V$ comes about. At the base point o one has a canonical identification of the tangent space with $\mathrm{Lie}(G)/\mathfrak{v} \simeq \mathfrak{m}$, where $\mathrm{Lie}(G) = \mathfrak{v} \oplus \mathfrak{m}$ is the reductive splitting. The complex structure on this tangent space is induced by the morphism $h : \mathbf{S} \to G$ defined by the base point: as in Lemma 12.2.2 there is a complex structure $J : \mathfrak{m} \to \mathfrak{m}$ which is defined as multiplication by i on $\mathfrak{m}^{p,-p}$ for $p > 0$ and by $-$i on $\mathfrak{m}^{p,-p}$ for $p < 0$.

It is now natural to consider equivariant maps f which not only respect the reductive splittings, but also are holomorphic, i.e., $f_* \circ J_1 = J_2 \circ f_*$, where J_1, J_2 are the two complex structures we just described. Such maps are also called *strongly equivariant*. By its very definition f induces a holomorphic injection of reductive domains compatible with the reductive splitting and hence, as in Example 11.5.2 (ii) we deduce the following result.

Lemma 17.2.2 *The image of a strongly equivariant map $f : X_1 \to X_2$ is a totally geodesic submanifold of X_2.*

We now come to a special class of strongly equivariant maps.

To motivate this, recall that points in a Mumford–Tate domain $X = G/V$ are the G-conjugates of some fixed homomorphism $\mathbf{S} \to G$ and that the G-action is by conjugation. So, given a homomorphism $\varphi : G_1 \to G_2$ between groups of Hodge type and a homomorphism $h_1 : \mathbf{S} \to G_1$ representing a point $[h_1] \in X_1$, the composition $h_2 = \varphi \circ h_1$ defines a point $[h_2] \in X_2$. Clearly the assignment $h_1 \mapsto h_2$ induces a strongly equivariant map. Note that we may furthermore alter the map by conjugation in G_2, getting a new strongly equivariant map. The morphisms φ that are defined over $\mathbf{Q}$ give rise to the notation of morphism between Mumford–Tate data.

Definition 17.2.3 A *morphism of Mumford–Tate data* consists of an equivariant pair of maps

$$(G_1, X_1) \xrightarrow{(f,\varphi)} (G_2, X_2)$$

such that φ is a morphism of $\mathbf{Q}$-algebraic groups and f is the morphism induced by φ up to possible $G_2(\mathbf{Q})$-conjugacy. In other words, with $k^{(g)}$ denoting conjugation of $k \in \mathrm{Hom}(\mathbf{S}, G)$ by $g \in G$, we have:

$$f[h] = g \cdot [\varphi{\circ}h] = [\varphi{\circ}h^{(g)}], \quad [h] \in X_1,\ g \in G_2(\mathbf{Q}).$$

Note that a morphism of Mumford–Tate data, being defined over $\mathbf{Q}$, preserves the Mumford–Tate groups.

Lemma 17.2.4 *Let $f : X_1 \to X_2$ be an equivariant holomorphic map coming from a morphism of Mumford–Tate data. Suppose that $\Gamma_\alpha \subset G_\alpha$, $\alpha = 1, 2$ are neat congruence subgroups such that $\varphi(\Gamma_1) = \Gamma_2$ with induced morphism*

$$\bar{f} : \Gamma_1\backslash X_1 \to \Gamma_2\backslash X_2.$$

Each connected component of $f(X_1) \subset X_2$ is a Hodge locus, and each irreducible component of the image of $\bar{f}$ inside $\Gamma_2\backslash X_2$ is then a subvariety of Hodge type.

Proof Let $[h]$ be a Hodge generic point of some connected component of X_1. There is a commutative diagram

$$\begin{array}{ccc} G_1 & \xrightarrow{\ \varphi\ } & G_2 \\ {\scriptstyle\rho_1}\downarrow & & \downarrow{\scriptstyle\rho_2} \\ \mathrm{GL}(H_1) & \longrightarrow & \mathrm{GL}(H_2), \end{array}$$

making $\rho_1{\circ}h$ and $\rho_2{\circ}\varphi{\circ}h$ Hodge structures. For simplicity we identify G_α with its image in $\mathrm{GL}(H_\alpha)$, $\alpha = 1, 2$. Then $G_1 = \mathrm{MT}(h)$, because $[h]$ is Hodge generic. Since φ is a morphism between $\mathbf{Q}$-algebraic groups, its image in G_2 is precisely the Mumford–Tate group of $\varphi{\circ}h$. So the orbit of $f([h])$ under $\varphi(G_1)$ is a Hodge locus inside X_2. □

Remark 17.2.5 Note that X_1 can indeed have several connected components and so the image of $\bar{f}$ can have several irreducible components. This causes a potential problem since, depending on the choice of the congruence subgroups, these might or might not appear in the image of $\bar{f}$. To deal with this we also should allow Hecke correspondences (see Eq. (16.2)). However, we choose to ignore this here.

Remarks 17.2.6 (i) If we replace the morphism f by $f^{(g)}$, where $g \in G_2(\mathbf{R})$, the image of X_1 under the resulting equivariant map is of course still totally geodesic in X_2, but the image need not be a Hodge locus inside X_2.
(ii) Although special points are dense in varieties of Hodge type, they need not be dense on subvarieties which come from period maps since their images have to be horizontal, i.e., verify Griffiths transversality. Indeed, there is no guarantee that a maximal horizontal subvariety of a Hodge locus contains a CM-point. See also Green et al. (2012, VIII.B).

We now consider the special situation where we embed a Mumford–Tate variety holomorphically and equivariantly in a *Shimura variety*. We have seen (Lemma 17.2.2) that the image is totally geodesic. It need not be a variety of Hodge type but if it is, it is customary to call it a "special subvariety".

Definition 17.2.7 A subvariety of Hodge type inside a Shimura variety is called a *special subvariety*.

Note that a special subvariety carries the tautological variation of Hodge structure from the Shimura variety in which it lies and so, since we know that it is a Mumford–Tate variety, it must in fact be a Shimura variety. As we noticed before, a subdomain of a Shimura domain does not always give rise to a special subvariety. If it does, it has at least one CM-point on it. The converse is also true.

Proposition 17.2.8 *Let*

$$\iota : \Gamma_1\backslash X_1 \hookrightarrow \Gamma_2\backslash X_2$$

be an equivariant holomorphic embedding of Shimura varieties. Then the image is special if (and only if) it contains a CM-point.

Proof Let $X_1 \longrightarrow X_2$ be a lift of the given embedding and consider the subgroup Γ_2 which centralizes the image of X_1. We may assume that it equals $\Gamma_1 = \pi_1(Y_1)$, $Y_1 = \iota(\Gamma_1\backslash X_1)$, and also that its Zariski-closure M in G_2 is connected, where (G_2, X_2) is the Shimura datum for X_2. Hence M is the algebraic monodromy group of the tautological variation of Hodge structure on Y_1. By André's result, Proposition 15.3.11, the Mumford–Tate group of this variation is the maximal possible, $\iota(G_1)$. This implies that Y_1 is special indeed. □

One can show that the condition of the preceding Proposition holds, if the embedding is rigid.

Proposition 17.2.9 (Abdulali, 1994) *In the situation of Proposition 17.2.8, assume that ι is a rigid embedding. Then the image is a special subvariety of $\Gamma_2\backslash X_2$.*

17.3 The Moduli Space of Cubic Surfaces is a Shimura Variety

Main Result

In Problem 1.3.3 in Section 1.3 we introduced some basic information about cubic surfaces. The two salient facts are that they depend on four moduli (since S is isomorphic to $\mathbf{P}^2$ blown up in 6 points, 4 of which can be chosen fixed by the transitive action of PGL(3)), whereas their Hodge structure depends on none. Indeed, $H^2(S) = H^{1,1}(S)$, so that as S varies, its Hodge structure remains constant.

Nonetheless, it is possible to associate with S a variable Hodge structure and hence define a second period map. This second period map takes values in the quotient of the unit ball in complex 4-space by the action of a certain arithmetic group, say

$$\mathcal{M}_S \longrightarrow \Gamma\backslash B^4.$$

This "ball quotient" is the period space for a set of weight 3 Hodge structures with a special symmetry of order 3. The source, $\mathcal{M}_S$, is the moduli space of "stable" cubic surfaces. By "stable" in this setting, we mean having at most nodes as singularities. The image of the locus of nodal cubic surfaces under the second period map is the quotient by Γ of a configuration of geodesically embedded complex hyperbolic hyperplanes

$$\tilde{\Delta} = \bigcup \mathcal{H}_\delta.$$

Let us proceed by giving the details of this construction. Let $F(X_0, \ldots, X_3) = 0$ be an equation for a cubic surface S. Define a cubic threefold T by adding the cube of a new variable to the original equation:

$$T = \left\{[X] \in \mathbf{P}^4 \mid F(X_0 \ldots, X_3) + X_4^3 = 0\right\}.$$

Thus T presents itself as a 3-1 branched cover of $\mathbf{P}^3$ with branch locus S. We shall call such cubic threefolds *cyclic*. The period map we shall study is the one which associates with S the Hodge structure $H^3(T)$.

To calculate the period map, we use Theorem 3.2.10. First note that the middle cohomology group, $H^3(T)$, is primitive in this case, or, equivalently, it is the variable cohomology. Hence the Hodge filtration $F^k H^3(T)$ can be found by using rational differential forms on $\mathbf{P}^4$ with poles on T. Those which land in $F^k H^3(T)$ have the form

$$\Omega(A) = \frac{A\omega_4}{G^{4-k}}, \quad \omega_4 = \sum_{\alpha=0}^{4} X_\alpha \cdot dX_0 \wedge \cdots \widehat{dX_\alpha} \cdots \wedge dX_4,$$

and where $G(X) = F(X_0, \ldots X_3) + X_4^3$ is an equation for T. Since $\Omega(A)$ is required to be homogeneous of degree zero, this implies that $\deg(A) =$

$3(4 - k) - 5 = 7 - 3k$. It follows that $0 = F^3H^3(T) = H^{3,0}(T)$ and that $F^2H^3(T) = H^{2,1}(T)$ has dimension 5. Consequently, the Hodge decomposition of the cubic threefold is given by

$$H^3(T) = H^{2,1} \oplus H^{1,2},$$

where the two summands have dimension 5. Via this construction, we associate with a cubic surface S the Hodge structure $H^3(T)$. Griffiths' residue calculus, as made explicit in Theorem 3.2.10, shows that this Hodge structure varies as F varies. Even more is true: according to the celebrated theorem by Clemens and Griffiths (1972) the polarized Hodge structure $H^3(T)$ determines T itself. These facts, after some work, yield the following result.

Theorem 17.3.1 *The period map $S \to H^3(T)$ defines an isomorphism $\mathcal{M}_s \longrightarrow \Gamma\backslash B^4$, where*
(a) $\mathcal{M}_s$ is the moduli space of stable cubic surfaces and
(b) $\Gamma\backslash B^4$ is the quotient of the unit ball in $\mathbf{C}^4$ by the action of an arithmetic group.

The unit ball B^4 is realized as the subset of projective space defined by the inequality

$$h(z) < 0, \quad h(z) = -|Z_0|^2 + |Z_1|^2 + \cdots + |Z_4|^2.$$

The arithmetic group Γ is the group of h-Hermitian matrices with coefficients in the ring of Eisenstein integers

$$\mathcal{E} \stackrel{\text{def}}{=} \mathbf{Z}[\omega], \quad \omega = \frac{-1 + \sqrt{-3}}{2}.$$

In subsequent subsections, we shall prove this result.

Period Domain and Mumford–Tate Subdomain

Let T be a cyclic cubic threefold; that is, a cubic threefold that is a three-sheeted branched cover of $\mathbf{P}^3$. Let $\langle\, | \,\rangle$ be the standard symplectic form on $\mathbf{Z}^{10}$ and fix a marking

$$m : (H^3(T), \text{cup product}) \xrightarrow{\simeq} H_{\mathbf{Z}} \stackrel{\text{def}}{=} (\mathbf{Z}^{10}, \langle\, | \,\rangle).$$

A marking is equivalent to the choice of a symplectic cohomology basis. Thus, if $\{e_i\}$ is the standard basis for $\mathbf{Z}^{10}$, then $\{m^{-1}(e_j)\}$ is a symplectic cohomology basis for $H^3(T)$.

Let D be the period domain for the Hodge structures of a general cubic threefold. It is the space of polarized Hodge structures for a unimodular symplectic form $\langle\, | \,\rangle$ and so is a Siegel upper half-space of genus 5:

$$D = \{m(H^{1,2}) \subset \Lambda_{10} \otimes \mathbf{C}\} = \mathfrak{h}_5 = \mathrm{Sp}(5, \mathbf{R})/\mathrm{U}(5).$$

Note that its dimension is fifteen, whereas the dimension of the moduli space of cubic threefolds is ten.

Let σ be a generator of the group of automorphisms of T over $\mathbf{P}^3$:

$$\sigma(X_0, \ldots, X_3, X_4) = (X_0, \ldots, X_3, \omega X_4). \tag{17.3}$$

It induces a real linear operator on $H_{\mathbf{R}} = H_{\mathbf{Z}} \otimes \mathbf{R}$ and $H_{\mathbf{C}} = H_{\mathbf{R}} \otimes \mathbf{R}$, which we shall denote σ^*. Clearly, for a marked cyclic threefold σ induces a Hodge isometry, i.e., an automorphism of polarized Hodge structures of $H_{\mathbf{Z}}$. Thus we are led to introduce D', the subspace of D for which σ^* is an isometry of Hodge structures:

$$D' \stackrel{\text{def}}{=} \left\{ m(H^{1,2}) \subset \Lambda_{10} \otimes \mathbf{C} \mid \sigma^*(H^{1,2}) = H^{1,2} \right\}.$$

We shall call the corresponding Hodge structures *cyclic cubic Hodge structures.* Thus on those the eigenvalues of σ^* are cube roots of unity. Since a fixed point corresponds to a three-dimensional cohomology class on $\mathbf{P}^3$, σ^* acts without nonzero fixed points on the complexification. Therefore there is a decomposition

$$H_{\mathbf{C}} = H_{\omega} \oplus H_{\bar{\omega}},$$

where the summands are the eigenspaces of σ^*. Since σ^* is a morphism of Hodge structures, the Hodge decomposition is compatible with the eigenspace decomposition, and so we have, for example, that

$$H_{\bar{\omega}} = H_{\bar{\omega}}^{2,1} \oplus H_{\bar{\omega}}^{1,2}.$$

To compute the dimensions of the summands on the right-hand side above, note that by (17.3) σ^* acts on the rational differential forms

$$\Omega(X_i) = \frac{X_i \Omega}{F + X_4^3} \tag{17.4}$$

with eigenvalue ω if $i < 4$ and on $\Omega(X_4)$ with eigenvalue $\bar{\omega}$. Hence

$$\dim H_{\bar{\omega}}^{2,1} = 1, \dim H_{\bar{\omega}}^{1,2} = 4.$$

To describe D' more explicitly, we consider the composition of $\mathbf{R}$-linear maps

$$q : H_{\mathbf{R}} \hookrightarrow H_{\mathbf{C}} \twoheadrightarrow H_{\bar{\omega}}, \tag{17.5}$$

where the first is the natural inclusion and the second the natural projection. The map q is an isomorphism of real vector spaces. Next, we need to bring into play the polarization given by the symplectic form $\langle \mid \rangle$. It is preserved by σ^*. To exploit this fact, observe that the real vector space $H_{\mathbf{R}}$ is a σ^*-module and that this induces a complex structure on $H_{\mathbf{R}}$ given by

$$\mathbf{J} \stackrel{\text{def}}{=} \frac{1}{\sqrt{3}}\theta, \quad \theta \stackrel{\text{def}}{=} \sigma^* - (\sigma^*)^{-1}.$$

Indeed, the minimum polynomial of σ^* equals $x^2 + x + 1$ and so $\theta^2 = -3\,\mathrm{id}$. The **R**-linear complex valued form on $H_{\mathbf{R}}$ defined by

$$\left.\begin{aligned} h(x,y) &\stackrel{\text{def}}{=} \frac{1}{2}\langle \theta x \mid y\rangle + \frac{1}{2}(\omega - \omega^{-1})\langle x \mid y\rangle \\ &= \frac{1}{2}\sqrt{3}\big[\langle \mathbf{J}x \mid y\rangle + \mathrm{i}\langle x \mid y\rangle\big] \end{aligned}\right\} \tag{17.6}$$

is Hermitian with respect to this complex structure. Indeed, this follows directly from our definition of a Hermitian metric (see Definition 2.2.2) with **J** playing the role of ω. We claim that

$$h(x,y) = \mathrm{i}\sqrt{3}\langle q(x) \mid \overline{q(y)}\rangle. \tag{17.7}$$

To show this, since h and the right-hand side are Hermitian, one may suppose $y = x$. Since σ preserves the skew form $\langle\,|\,\rangle$ one has $h(x,x) = \langle \sigma^* x \mid x\rangle$ and so we need to show

$$\langle \sigma^* x \mid x\rangle = \mathrm{i}\sqrt{3}\langle q(x) \mid \overline{q(x)}\rangle.$$

To calculate the latter, for $x \in H_{\mathbf{C}}$, write its decomposition in the two eigenspaces for σ^* as $x = x_\omega + y_\omega$. It gives two equations for x_ω, y_ω:

$$\begin{aligned} \sigma^* x &= \omega x_\omega + \bar{\omega} y_\omega, \\ (\sigma^{-1})^* x &= \bar{\omega} x_\omega + \omega y_\omega. \end{aligned}$$

Now solve for x_ω, y_ω. Since y_ω gets identified with $q(x)$ and x_ω with is conjugate, we may substitute these values in the expression $\langle y_\omega \mid x_\omega\rangle$ to check the desired equality.

In short: h corresponds up to a real multiple to the Hermitian form induced by the polarizing form. To calculate its signature on $H_{\bar{\omega}}$ is then straightforward since by the Riemann relations $\langle Cx \mid \bar{x}\rangle > 0$ for nonzero x and hence $h(x,x) < 0$ on $H^{2,1}$ while $h(x,x) > 0$ on $H^{1,2}$. It follows that the signature of h on $H_{\bar{\omega}}$ is $(1,4)$.

So, giving a polarized cyclic cubic Hodge structure on H is equivalent to giving a decomposition $H_{\bar{\omega}} = H^{2,1}_{\bar{\omega}} \oplus H^{1,2}_{\bar{\omega}}$ such that the Hermitian form $-h$ is positive on the first summand and negative on the second. The group $\mathrm{SU}(1,4)$ acts transitively on the set of flags $H^{2,1}_{\bar{\omega}} \subset H_{\bar{\omega}}$. Indeed, a real linear transformation of $H_{\mathbf{R}}$ that takes one cyclic cubic Hodge structure to another is determined by its restriction to $H_{\bar{\omega}}$ and on that space it preserves the Hermitian form h. The isotropy group of a reference structure is isomorphic to $\mathrm{U}(4)$. But this describes the homogeneous space isomorphic to the unit ball in $\mathbf{C}^4$. Summarizing,

$$D' = \mathrm{SU}(1,4)/\mathrm{U}(4) = B^4.$$

The Arithmetic Subgroup

The lattice $H_{\mathbf{Z}}$ enjoys the fixed-point free action of the automorphism σ. This action endows it with the structure of a free $\mathscr{E}$-module, where we recall that $\mathscr{E} = \mathbf{Z}[\omega]$ is the ring of Eisenstein integers. As such $H_{\mathbf{Z}}$ has rank 5. Moreover, using the comparison (17.7) and the fact that $\langle\,|\,\rangle$ is unimodular implies that the Hermitian structure given by h as defined previously (see (17.6)) is also unimodular, and has signature $(-1, 1, 1, 1, 1)$. Therefore there is a unitary unimodular basis $\{\gamma_i \mid 0 \le i \le 4\}$ for the $\mathscr{E}$-module $H_{\mathbf{Z}}$ such that $\|\gamma_i\||^2 = 1$ for $i > 0$ and $\|\gamma_0\|2 = -1$. Let us call such a basis *standard*. The naturally acting arithmetic subgroup is therefore

$$\begin{aligned}\Gamma &\stackrel{\text{def}}{=} \{\gamma \in \mathrm{GL}(5, \mathscr{E}) \mid h(\gamma x, \gamma y) = h(x, y),\ x, y \in H_{\mathbf{Z}}\} \\ &\simeq \mathrm{U}(1, 4; \mathscr{E}).\end{aligned}$$

We finish this subsection by giving a more classical description of the period map in this case. The Hodge decomposition is completely determined by the complex line spanned by

$$\Omega_T \stackrel{\text{def}}{=} \operatorname{Res} \Omega(X_4) \in H^{2,1}_{\bar{\omega}}$$

inside the five-dimensional complex space $H^3_{\bar{\omega}}(T)$. To describe this via integration we use that the *homology* $H_3(T; \mathbf{Z})$ has the structure of a unimodular $(\mathscr{E}, h)$-module structure via $m^{-1}q^{-1}$, where m is the marking and q the map (17.5). Therefore it has a unitary unimodular basis $\{\gamma_i \mid 0 \le i \le 4\}$ such that $\|\gamma_i\|^2 = 1$ for $i > 0$ and $\|\gamma_0\|^2 = -1$. Let us call such a homology basis *standard*. The period vector

$$Z(T) = (Z_0, Z_1, Z_2, Z_3, Z_4), \quad Z_j = \int_{\gamma_j} \Omega_T \tag{17.8}$$

then gives the period map

$$\mathscr{M} = \{\text{moduli space of cyclic threefolds}\} \longrightarrow \Gamma\backslash D^4.$$

Local Properties of the Period Map

Let us continue our investigation of the period map for cyclic threefolds T. Under the $\mathbf{R}$-linear map from (17.5) the Hodge filtration on $H^3(T)$ projects to a flag

$$H^{2,1}_{\bar{\omega}} \subset H^3_{\bar{\omega}}.$$

This flag gives the period map with target the unit ball $D' = B^4$. To show that this map is an immersion at F, replace F in Eq. (17.4) by $F + tG$, where G is another cubic form in the X_i with $i < 4$. Then

$$\frac{d}{dt}\Omega(X_4)\Big|_{t=0} = -\frac{GX_4\Omega}{(F + X_4^3)^2}.$$

The residue of the rational differential on the right lies in $H^3_{\bar{\omega}}$ and projects to an element of $H^3_{\bar{\omega}}/H^{2,1}_{\bar{\omega}}$. The projection is nonzero if and only if GX_4 is nonzero modulo the Jacobian ideal of F. Let us take $F = X_0^3 + \cdots + X_3^3$. Then the Jacobian ideal for $F + X_4^3$ is generated by the quadratic functions X_i^2. If G is a linear form in $X_0, \ldots X_3$, then GX_4 is not in the Jacobian ideal. Consequently the derivative of the period map for cyclic threefolds is of maximal rank and hence is an open map.

So, at this point we know that the dimensions of source and target space are the same and that the period map is open. As recalled in the introduction to this section, by the result of Clemens and Griffiths (1972) we already know that the period map is injective. So, to prove that the image is *all* of the source, we have to show that there is a *proper* extension of the period map to stable cyclic cubic threefolds. Denote the moduli space of semistable cubics by $\mathcal{M}_{ss}$. It is the one-point compactification of $\mathcal{M}_s$, where the point added corresponds to the semistable cubic surface with one A_2 singularity. We must consider both types of singularity, i.e., A_1 (i.e., at most nodal singularities) and A_2. These properties about stable and semistable cubic surfaces we take for granted, but require of course some computation (see Allcock et al., 2002).

Recall at this point that we discussed the GIT quotients of hypersurfaces in Section 8.4, where we remarked that the set of semistable points gives rise to a projective quotient while the subset of stable points is Zariski open in it. So $\mathcal{M}_{ss}$ is indeed compact. To make this further extension of the period map, we must replace the target by its Baily–Borel compactification, discussed in Section 16.2. Since this extension is clearly a proper map, necessarily of degree one, this will complete the proof of the main theorem.

First Extension of the Period Map

Consider a degeneration of cyclic cubic threefolds where the underlying cubic surface, the branch surface, acquires an A_1 singularity. The local form of the degeneration for the branch surface is $x^2 + y^2 + z^2 = t$ and for the threefolds it is

$$x^2 + y^2 + z^2 + w^3 = t. \tag{17.9}$$

The theorem by Sebastiani and Thom (1955) gives a convenient calculus for understanding the monodromy of such "diagonal" degenerations. To state the result, consider a degeneration

$$f(x_1, \ldots, x_m) + g(y_1, \ldots, y_n) = t.$$

Let V_f be the space of vanishing cycles for $f(x) = t$ and likewise for V_g. Then

$$V_{f+g} = V_f \otimes V_g.$$

The identification is such that if $\alpha \in V_f$ and $\beta \in V_g$, then $\alpha \otimes \beta$ corresponds to their join[2] $\alpha \star \beta$. Moreover, if μ_f is the monodromy transformation on V_f, then

$$\mu_{f+g} = \mu_f \otimes \mu_g.$$

Finally, note that if V_f and V_g carry inner products, then V_{f+g} carries the inner product given by $\langle a \otimes b \mid c \otimes d \rangle = \langle a \mid c \rangle \cdot \langle b \mid d \rangle$.

Consider now the simplest of all degenerations – the case of dimension zero, where d distinct points coalesce. This is the case $x^d = t$. The variety X_t for nonzero t consists of the d points

$$x_k(t) = |t|^{1/d} e^{2\pi i k/d},$$

where by $|t|^{1/d}$, we understand the unique positive real root. The module of vanishing cycles is the module $V(d)$ consisting of cycles of degree zero. It is generated by the differences $\delta_k(t) = x_k(t) - x_{k-1}(t)$. Call this set of generators *standard*.

The monodromy for $V(d)$ is generated by rotations through an angle $2\pi/d$. Let $\mu(d)$ denote the just-described generator of the monodromy. Relative to the standard basis, it acts by $\mu(d)\delta_k = \delta_{k+1}$, where the subscript is taken modulo d. The resulting matrix comes from a permutation, e.g.

$$\mu(3) = \begin{pmatrix} 0 & -1 \\ 1 & -1 \end{pmatrix}.$$

Observe that the minimal polynomial of $\mu(3)$ is $x^2 + x + 1$ and so $V(3)$ can be identified with the lattice of the Eisenstein integers with monodromy multiplication by ω. The module $V(d)$ carries a natural quadratic form defined by declaring the 0-cycles x_i to be an orthogonal norm one basis for $H_0(X_t)$. The matrix in the standard basis for the restriction of this inner product to $V(d)$ is the Gram matrix for the A_{d-1} Dynkin diagram, e.g., for $d = 3$ we have

$$\begin{pmatrix} 2 & -1 \\ -1 & 2 \end{pmatrix}.$$

[2] For two subsets of disjoint subvector-spaces of a given real vector space V, the join consists of the union of real lines in V joining two points in the subsets.

By the Sebastiani–Thom theorem, the space of vanishing cycles for the degeneration

$$x^{d_1} + \cdots + x^{d_n} = t$$

can be identified with

$$V = V(d_1) \otimes \cdots \otimes V(d_n),$$

and the tensor product of cycles $\xi_1 \otimes \cdots \otimes \xi_n$ to the join $\xi_1 \star \cdots \star \xi_n$. The monodromy on V is given by the tensor product

$$\mu = \mu(d_1) \otimes \cdots \otimes \mu(d_d).$$

The inner product of vanishing cycles is given by

$$\langle \xi_1 \otimes \cdots \otimes \xi_d \mid \nu_1 \otimes \cdots \otimes \nu_d \rangle = \langle \xi_1 \mid \nu_1 \rangle \cdots \langle \xi_d \mid \nu_d \rangle.$$

As an exercise, consider the inner product on $V(2) \otimes V(2)$, and consider the element

$$\nu = \delta_1 \otimes \delta_1 - \delta_2 \otimes \delta_2.$$

One finds that ν is isotropic, i.e., $\langle \nu \mid \nu \rangle = 0$.

We now come back to our degeneration (17.9). The space of vanishing cycles is

$$V = V(2) \otimes V(2) \otimes V(2) \otimes V(3).$$

Identify $V(2)$ with $\mathbf{Z}$, $\mu(2)$ with multiplication by -1, and $V(3)$ with the Eisenstein integers $\mathscr{E}$, we find that the monodromy matrix with respect to the standard generators is multiplication with $-\omega$.

In this case the natural intersection form for the tensor product is nondegenerate, and so the homology splits over $\mathbf{Q}$ as $V \oplus V^{\perp}$, with trivial monodromy on $V^{\perp}$. Consequently the monodromy on the full homology is also of order 6. We can, in fact, give a formula for it. Let δ be a generator of V. Since the homology splits orthogonally and since h is unimodular, we conclude that $h(\delta, \delta) = \pm 1$. In fact, $h(\delta, \delta) = 1$. Then the monodromy is given by

$$T = 1_{V^{\perp}} \oplus -\omega 1_V.$$

This identifies T as a *complex reflection* of order 6 in the complex hyperbolic space $V^{\perp}$ of codimension one.

Let us now study the period vector $Z(t)$, introduced in (17.8) for a nodal degeneration T_t of cyclic threefolds, whose local form is given by (17.9). Since $h(\delta, \delta) = 1$, we may take the vanishing cycle δ as part of a standard basis. To be precise, of the five components $Z_j(t)$, $j = 0, \ldots, 4$ of $Z(t)$ consisting of integrals over cycles γ_j, we may assume that $\gamma_4 = \delta \in V$. We now invoke a general result from Malgrange (1974, §4.7). This states that when we have a degeneration with finite monodromy of order d, the periods as functions of t have the following shape:

$$\sum_{k=0}^{d-1} a_k(t) t^{k/d},$$

where the a_k are holomorphic in t. Moreover, if $a_k \neq 0$, $\exp(2\pi i k/d)$ is an eigenvalue of the monodromy. The integral $Z_4(t)$ is of this latter form and since the monodromy has eigenvalues $-\omega$ and $-\bar{\omega}$ on the real 2-dimensional vector space $V \otimes \mathbf{R}$, the integral can be written as

$$Z_4(t) = a(t)t^{1/6} + b(t)t^{5/6}.$$

where $a(t)$ and $b(t)$ are holomorphic in t at $t = 0$. It follows that

$$\lim_{t \to 0} Z_4(t) = 0.$$

Thus the period vector of a nodal cubic surface is defined at $t = 0$ and satisfies $Z(0) \in \delta^{\perp}$. This gives a local extension of the period map from $\mathcal{M}$ to $\mathcal{M}_s$, and shows that the extension maps the local part of the difference $\mathcal{M}_s - \mathcal{M}$ to the geodesic hypersurface $\delta^{\perp}$.

Second Extension of the Period Map

The second type of degeneration has the local structure

$$x^2 + y^2 + z^3 + w^3 = t. \tag{17.10}$$

By the Sebastiani–Thom theorem, the monodromy on

$$V \cong V(2) \otimes V(2) \otimes V(3) \otimes V(3)$$

is of order 3. However, in this case the intersection form on the vanishing cohomology is degenerate, i.e., has a null vector, namely, $v = \delta_1 \otimes \delta_2 - \delta_2 \otimes \delta_1$ in $V(3) \otimes V(3)$. Let T_t be the one-parameter degeneration with local form (17.10). Since the intersection form on the 5-dimensional $\mathscr{E}$-module $m^{-1}H^{\vee}_{\mathbf{Z}} = H_3(T_t, \mathbf{Z})$ is nondegenerate, there is a v' such that $k = v \cdot v'$ is nonzero. The monodromy on the space spanned by v and v' has matrix

$$\begin{pmatrix} 1 & k \\ 0 & 1 \end{pmatrix}$$

as in the earlier discussion on degenerations of genus 2 curves in Section 1.2. Consider the family of cohomology classes

$$\Omega(t) = \operatorname{Res}\Omega(X_4) \in H^3(T_t; \mathbf{C}).$$

The asymptotic study of the period integral gives

$$\int_{\nu'} \Omega(t) = A(t)(\log t^3) + B(t^3)$$

for some $A(t)$ with $A(0) \neq 0$, and where $B(t^3)$ is holomorphic. With respect to a $\mathbf{Q}(\omega)$-basis $\{\nu, \nu', \gamma_2, \gamma_3, \gamma_4\}$ for $H_3(T_t, \mathbf{Z})$, the period of $\Omega(t)$ can be considered as the line

$$\mathbf{C} \cdot \left(\int_{\nu} \Omega(t), \int_{\nu'} \Omega(t), \int_{\gamma_2} \Omega(t), \int_{\gamma_3} \Omega(t), \int_{\gamma_4} \Omega(t) \right)$$

in $\mathbf{C}^5$. Dividing by $\log(t^3)$ we see that as t approaches $t = 0$ along a radius, this line converges to the isotropic line $\mathbf{C}(1, 0, 0, 0, 0))$ corresponding to $\int_{\nu}$.

Isotropic lines correspond to cusps in the Baily–Borel compactification of a ball quotient. This convergence statement implies that the period map extends to the full GIT quotient and maps the semistable-but-not-stable locus to cusps. That is, the period map extends from the stable locus to the GIT moduli space, mapping the latter into to the Baily–Borel compactification.

Having completed this second extension of the period map, our program for the proof of the main theorem 17.3.1 of this subsection has now been achieved.

17.4 Shimura Curves and Their Embeddings

Contrary to some practices, for us a Shimura curve is a Shimura variety of dimension 1. The prototypical example is furnished by modular curves which are noncompact. We review this in the first subsection. These come from the $\mathbf{Q}$-simple group SL(2). The only other types come from a $\mathbf{Q}$-simple group G such that $G(\mathbf{R})$ is a product of factors, one of which is $\mathrm{SL}(2, \mathbf{R})$, but the others are compact. If these are present, this forces the curve to be compact. The second subsection explains their construction. We proceed next to enumerate their embeddings in quotients of the Siegel upper half-spaces and discuss which of these embeddings are rigid. We finish with a section where we make the connection with Higgs bundles.

Classification

Before giving the classification, we start with some elementary examples.

Examples 17.4.1 The description $\mathrm{SL}(2,\mathbf{R})/\mathrm{SO}(2)$ for the upper half-plane $\mathfrak{h}$ has as associated Shimura datum $(\mathrm{SL}(2),\mathfrak{h})$ which leads to *modular curves* $\Gamma\backslash\mathfrak{h}$, where by definition a *modular group* $\Gamma\subset\mathrm{SL}(2,\mathbf{Q})$ is an arithmetic subgroup. If Γ acts faithfully and without fixed points on $\mathfrak{h}$, one has a modular family of elliptic curves

$$E_\Gamma := (\Gamma\times\mathbf{Z}^2)\backslash(\mathbf{C}\times\mathfrak{h}) \longrightarrow C_\Gamma = \Gamma\backslash\mathfrak{h},$$

where $\left(\begin{pmatrix} a & b\\ c & d\end{pmatrix}, m, n\right)\in\Gamma\times(m,n)$ acts as follows on $\mathbf{C}\times\mathfrak{h}$:

$$(z,\tau)\mapsto\left(\frac{z+m\tau+n}{c\tau+d},\frac{a\tau+b}{c\tau+d}\right).$$

Standard examples of modular groups are:

$$\Gamma_0(N)=\left\{\begin{pmatrix} * & *\\ 0 & *\end{pmatrix}\mod N\right\},\quad \Gamma_1(N)=\left\{\begin{pmatrix} 1 & *\\ 0 & 1\end{pmatrix}\mod N\right\},$$
$$\Gamma(N)=\left\{\begin{pmatrix} 1 & 0\\ 0 & 1\end{pmatrix}\mod N\right\},$$

which yield the modular curves

$$X(N)=\Gamma(N)\backslash\mathfrak{h},\quad X_1(N)=\Gamma_1(N)\backslash\mathfrak{h},\quad X_0(N)=\Gamma_0(N)\backslash\mathfrak{h}.$$

These parametrize elliptic curves with additional structure on N-torsion points: for the first, the extra structure is an isomorphism of the N-torsion with $(\mathbf{Z}/N\mathbf{Z})^2$, for the second, it is the choice of one of the N^2 torsion points of order N, while for the last, one singles out any N-th order cyclic subgroup of N-torsion points.

Digression (Quaternion algebras) We start with a field F of characteristic 0. A *quaternion algebra* D over F is a 4-dimensional F-vector space with basis $1,\mathbf{i},\mathbf{j},\mathbf{k}$ made into an algebra by the rules

$$\mathbf{i}^2=\alpha,\mathbf{j}^2=\beta,\mathbf{ij}=-\mathbf{ji}=\mathbf{k},\quad \alpha,\beta\in F.$$

The shorthand notation is $D=(\alpha,\beta)_F$. We call D *split*, if it is isomorphic over F to the matrix algebra $M_2(F)$. Otherwise, D is called *nonsplit*. Any quaternion algebra has an involution, called *standard involution*, given by

$$(a+b\mathbf{i}+c\mathbf{j}+d\mathbf{k})^*=a-b\mathbf{i}-c\mathbf{j}-d\mathbf{k}.$$

The reduced norm and trace are given by

$$\mathrm{Nrd}_D(x)=xx^*,\quad \mathrm{Trd}_D(x)=x+x^*.$$

The first leads to a group we need below:

$$\mathrm{SU}(1,D) = \left\{x \in D^\times \;\middle|\; \mathrm{Nrd}_D(x) = 1\right\}.$$

Next, assume that F is a *number field*. Then $\mathrm{SU}(1,D)$ has canonical discrete subgroups, making use of the ring $\mathfrak{O}_F$ of integers and the associated orders in D: an *order* in D is an $\mathfrak{O}_F$-submodule which is a subring. It is called *maximal order*, if it is maximal among all orders with respect to inclusion. Maximal orders are unique up to conjugation by elements of D and are denoted by $\mathfrak{O}_D$. Then

$$\Gamma_D = \{x \in \mathfrak{O}_D \mid \mathrm{Nrd}_D(x) = 1\}$$

is a discrete subgroup of $\mathrm{SU}(1,D)$.

Now we can state the following theorem.

Theorem (Satake, 1980) *There are two classes of Shimura curves:*
(i) *Noncompact Shimura curves are precisely the modular curves. They come from the Shimura datum* $(\mathrm{SL}(2,\mathbf{R}),\mathfrak{h})$.
(ii) *Compact Shimura curves arise as follows. Let F be a totally real number field and D a quaternion algebra over F which splits at exactly one place. Then* $(\,\mathrm{Res}_{F/\mathbf{Q}}\,\mathrm{SU}(1,D),\mathfrak{h}\,)$ *is a Shimura datum and* $C_{\Gamma_D} := \Gamma_D\backslash\mathfrak{h}$ *is a compact Shimura curve.*

We refer to the introduction of Addington (1987) for an excellent discussion of Satake's classification (Satake, 1965, 1966). We don't give a proof of the above theorem, but we only make some comments.

Consider the different real embeddings $i_\alpha : F \hookrightarrow \mathbf{R}$, $\alpha = 1,\ldots,m$. By assumption, for exactly one embedding, say i_1, one has $D \otimes_{i_1} \mathbf{R} \simeq M_2(\mathbf{R})$, i.e., D splits at the place i_1, while at all other real places D does not split: for $\alpha \geq 2$ we have $D \otimes_{i_\alpha} \mathbf{R} \simeq \mathbf{H}$, the quaternions. Hence

$$D(\mathbf{R}) \simeq M_2(\mathbf{R}) \oplus \underbrace{\mathbf{H} \oplus \cdots \oplus \mathbf{H}}_{m-1}, \qquad m = [F:\mathbf{Q}],$$

$$G(\mathbf{R}) \simeq \mathrm{SL}(2,\mathbf{R}) \times \underbrace{\mathrm{SU}(2) \times \cdots \times \mathrm{SU}(2)}_{m-1}.$$

This shows that, indeed, the upper half-plane is the associated bounded domain.

If $m = 1$ one finds back the modular situation which leads to noncompact Shimura curves. However, if $m \geq 2$ there are compact factors and then, as is well known (Mostow and Tamagawa, 1962; Borel and Harish-Chandra, 1962), for any $\Gamma \in \mathrm{SU}(1,D)$ commensurable with Γ_D, the quotient $\Gamma\backslash\mathfrak{h}$ now is a compact curve.

Rigidly Embedded Special Curves

In this subsection we discuss the classification of embedded Shimura curves in quotients of Siegel's upper half-space $\mathfrak{h}_g$ by arithmetic subgroups of $\mathrm{Sp}(g;\mathbf{Q})$. In other words, we want to classify the morphisms between Shimura data

$$(G,\mathfrak{h}) \longrightarrow (\mathrm{Sp}(g),\mathfrak{h}_g)), \quad G = \mathrm{Res}_{F/\mathbf{Q}}\,\mathrm{SU}(1,D),$$

where $G \longrightarrow \mathrm{Sp}(g)$ is defined over $\mathbf{Q}$ and injective. The congruence groups we are interested in are

$$\Gamma_{g,m} = \{\gamma \in \mathrm{Sp}(g,\mathbf{Z}) \mid \gamma \equiv \mathrm{id} \mod m\},\ m \geq 3.$$

These act freely on $\mathfrak{h}_g$ with quotient

$$\mathcal{A}_{g,m} \stackrel{\text{def}}{=} \Gamma_{g,m}\backslash\mathfrak{h}_g.$$

This quasi-projective smooth variety is a fine moduli space of polarized abelian varieties of dimension g with level m-structure.

There are many embedded Shimura curves in $\mathcal{A}_{g,m}$; this depends on the arithmetic of the corresponding embedding of $\mathbf{Q}$-algebraic groups $G \longrightarrow \mathrm{Sp}(g)$, where G is the $\mathbf{Q}$-simple group of the type introduced above (i.e., $G(\mathbf{R})$ has precisely one noncompact factor, $\mathrm{SL}(2,\mathbf{R})$). Their classification is a special case of the classifcation due to Addington (1987). Here, we shall only consider Shimura curves which are rigidly embedded.

Examples 17.4.2 (i) The classical "Satake embedding" (Satake, 1980) is induced by the following symplectic embedding. Set $V_k = (\mathbf{Q}^{2k}, J_k)$, $J_k = \begin{pmatrix} 0 & \mathbf{1}_k \\ -\mathbf{1}_k & 0 \end{pmatrix}$. The direct sum $\oplus_k V_1$ is isomorphic to the symplectic space V_k. Whence a faithful symplectic representation ρ_k of $\mathrm{SL}(2)$:

$$\begin{pmatrix} a & b \\ c & d \end{pmatrix} \xrightarrow{\rho_k} \begin{pmatrix} a\mathbf{1}_k & b\mathbf{1}_k \\ c\mathbf{1}_k & d\mathbf{1}_k \end{pmatrix} \in \mathrm{Sp}(V_k,\mathbf{Q}).$$

For any $k = 1,\ldots,g$ the direct sum representation $\rho_k \oplus \underbrace{1 \oplus \cdots \oplus 1}_{g-k}$ induces a holomorphic embedding $\mathfrak{h} \hookrightarrow \mathfrak{h}_g$. It gives the noncompact embedded Shimura curves starting from the Shimura datum $(\mathrm{SL}(2),\mathfrak{h})$. Such curves are called *Shimura curves of Satake type*. There is no locally constant factor if and only if $k = g$.

(ii) Symplectic embeddings when compact factors are present. Recall that one starts off with a real number field F and a quaternion algebra over F which splits at exactly one place and the group G is the group $\mathrm{Res}_{F/\mathbf{Q}}\,\mathrm{SU}(1,D)$. For simplicity, assume that $F/\mathbf{Q}$ is Galois. Each of the real embeddings $i_\alpha : F \hookrightarrow \mathbf{R}$,

$\alpha = 1, \ldots, m$, defines

$$G^\alpha = \mathrm{SU}(1, D) \otimes_{i_\alpha} \mathbf{R}, \quad G^\alpha(\mathbf{C}) = \mathrm{SL}(2, \mathbf{C}).$$

So $G = \prod_\alpha G^\alpha$. Do the same for the corresponding algebras. So let D^α be the twisted F-algebra D obtained via the embedding i_α and form the "corestriction": the $\mathbf{Q}$-algebra

$$E = \mathrm{Cor}_{F/\mathbf{Q}} D \stackrel{\text{def}}{=} D^1 \otimes_F \cdots \otimes_F D^m.$$

There are two possibilities for this algebra:

$$E = \begin{cases} M_{2^m}(\mathbf{Q}) & \text{(split case)}, \\ M_{2^{m-1}}(H),\ H \text{ a quaternion algebra}/\mathbf{Q} & \text{(nonsplit case)}. \end{cases}$$

The G-representation space with $G^\alpha(\mathbf{C})$ acting on $V_\alpha = \mathbf{C}^2$ given by

$$W = \begin{cases} V_1 \otimes V_2 \otimes \cdots \otimes V_m & \text{split case}, \\ (V_1 \otimes V_2 \otimes \cdots \otimes V_m)^{\oplus 2} & \text{nonsplit case} \end{cases}$$

turns out to be defined over $\mathbf{Q}$. Moreover, in both cases W carries a G-invariant symplectic form so that we get a symplectic representation

$$\mathrm{Res}_{F/\mathbf{Q}} \mathrm{SU}(1, D) \hookrightarrow \mathrm{Sp}(g, \mathbf{Q}), \quad g = 2^{m-1}, \text{ or } 2^m.$$

See Addington (1987, Sect. 4) or Viehweg and Zuo (2004, Section 5). The corresponding compact curves in $\mathscr{A}_{g,m}$ are called *embedded Shimura curves of Mumford type*. Indeed, for the special case where F is cubic, Mumford (1969a) constructs a family of abelian varieties $\{A_t \mid t \in C\}$ of dimension 4 over a compact curve C (note that $m = 3$ in this case). The representation space is $H^1(A_t; \mathbf{Q})$, a symplectic rational vector space of of dimension $8 = 2^3$, as should be the case.

The above examples indeed give rigid embedding. This follows from some general observations explained in Saito (1993, §6). Suppose we have an embedding $G \hookrightarrow \mathrm{Sp}(W)$ of $\mathbf{Q}$-algebraic groups where G is as in the above examples. In particular D is either the field $\mathbf{Q}$ and $G = \mathrm{SL}(2, \mathbf{Q})$ or a quaternion algebra over F and $G = \mathrm{Res}_{F/\mathbf{Q}} \mathrm{SU}(1, D)$. Set $G' = Z_G \mathrm{Sp}(W)$. For the first example $W = \mathbf{Q}^{2g}$ and $G' \simeq \mathrm{SO}(g)$ is compact. For the second example $W = \mathbf{Q}^{2g}$ with $g = 2^{m-1}$ or $g = 2^m$, and $G'(\mathbf{R})$ is a product of g factors isomorphic to $\mathrm{SO}(2, \mathbf{R}) \simeq S^1$ and hence is also compact. Now, by Saito (1993, Cor. 6.23) this implies that the corresponding embedding is rigid. These are essentially the only possibilities.

Proposition 17.4.3 *Each of the two types of Shimura curve C_Γ, respectively C_{Γ_D}, can be rigidly embedded in $\mathscr{A}_{g,m}$ in essentially one way.*

(i) *The induced family over* C_Γ *is isogeneous to*

$$\underbrace{E_\Gamma \times \cdots \times E_\Gamma}_{g\ copies.}.$$

(ii) *The induced family over* C_{Γ_D} *is isogeneous to the universal family* $A_{\Gamma_D} \to C_{\Gamma_D}$ *(and then* $g = 2^{m-1}$ *or* $g = 2^m$*).*

Sketch of the proof It remains to show that there are no other rigid embeddings. For noncompact Shimura curves, $G = \mathrm{SL}(2, \mathbf{Q})$ and by Addington (1987) it follows easily that the only primary symplectic representations are the classical Shimura representations. For compact Shimura curves we first have to see that a rigid representations in Addington's sense are precisely the ones that give rigidly embedding Shimura curves. This is again because a symplectic embedding $G \hookrightarrow \mathrm{Sp}(W)$ of $\mathbf{Q}$-groups is rigid if and only if the centralizer $G' = Z_G \mathrm{Sp}(W)$ is compact. This happens precisely when $G'(\mathbf{R})$ has only compact factors. It follows from Saito (1993, §6) that for some integer u we have

$$G'(\mathbf{R}) = \mathrm{SO}^*(2u) \times \underbrace{\mathrm{SO}(2u) \times \cdots \times \mathrm{SO}(2u)}_{g-1}.$$

So in our case this group is compact precisely when $u = 1$, which brings us back to the Mumford-type example. □

Remark A different proof of this classification can be found in Viehweg and Zuo (2004).

Relation with Higgs Bundles

Recall the notion of strictly maximal (logarithmic) Higgs field from Definition 13.4.4: all Higgs field components are isomorphisms. We showed just after this definition that a strictly maximal Higgs field has no unitary summand. In particular, there can be no locally constant summand.

In this subsection we want to explain that strictly maximal logarithmic Higgs fields of weight 1 over a curve correspond exactly to rigidly embedded Shimura curves. Let us start with Shimura curves themselves. They are quotients $C_0 = \Gamma\backslash\mathfrak{h}$ with Γ a discrete subgroup of $G(\mathbf{R}) \simeq \mathrm{SL}(2, \mathbf{R}) \times K$, K a, possibly trivial, compact group. On $\mathfrak{h}$ we have the standard variation of Hodge structure of weight 1 which, upon projection $G(\mathbf{R}) \twoheadrightarrow \mathrm{SL}(2, \mathbf{R})$ defines a weight 1 real variation of Hodge structure $\mathscr{L}$ on C_0 with underlying local system, say $\mathbf{L}$, of rank 2. The period map is an isomorphism and the associated logarithmic Higgs field $\tau^{1,0} : \mathscr{L}^{1,0} \to \mathscr{L}^{0,1} \otimes \Omega^1_C(\log \Sigma)$ is an isomorphism. Note that this implies that $\mathscr{M} = \mathscr{L}^{1,0}$ is a root of the line bundle $\Omega^1_C(\log \Sigma)$.

Proposition 17.4.4 *Consider a logarithmic polarized* $\mathbf{Q}$*-variation of Hodge structure of weight* 1 *and rank* $2g$ *over a curve* (C, Σ)*. Put* $C_0 = C - \Sigma$*,* Γ= *the monodromy group, and let* $C_0 \to \Gamma\backslash\mathfrak{h}_g$ *the period map, assumed to be an immersion. The Higgs field associated with the variation of Hodge structure is strictly maximal if and only if* C_0 *is a rigidly embedded Shimura curve and hence comes from a rigid variation of Hodge structure.*

Proof " $\Longrightarrow$ ": Look at Examples 17.4.2. Let us first consider the embedded Satake modular curve $C_0 = C_\Gamma$ with smooth compactification C. Let $\mathbf{V}$ be the local system underlying the variation of Hodge structure $\mathcal{V}$. Since the representation $\mathrm{SL}(2) \to \mathrm{Sp}(g)$ defining the embedding in this case is defined over $\mathbf{Q}$, the resulting splitting $\mathbf{V} = \oplus^g \mathbf{L}$ (where $\mathbf{L}$ has been constructed just above) splits over $\mathbf{Q}$. We may write

$$U \stackrel{\text{def}}{=} \mathrm{Hom}(\mathbf{L}, \mathbf{V}) \Longrightarrow \mathbf{V} = U \otimes \mathbf{L}, \dim U = g.$$

Then $\mathcal{V} = U \otimes \mathcal{L}$ and the vector space U inherits a Hodge structure of type (0, 0). The Higgs field is an isomorphism induced by the polarization and extends as an isomorphism on the corresponding logarithmically extended Higgs bundles

$$\begin{array}{ccc}
\mathcal{V}^{1,0} & \xrightarrow[\simeq]{\sigma} & \mathcal{V}^{0,1} \otimes \Omega^1_C(\log \Sigma) \\
\| & & \| \\
U \otimes_{\mathcal{O}_C} \mathcal{L}^{1,0} & \xrightarrow[\simeq]{\sigma} & U \otimes_{\mathcal{O}_C} \mathcal{L}^{0,1} \otimes_{\mathcal{O}_C} \Omega^1_C(\log \Sigma).
\end{array}$$

For rigid Mumford-type embeddings the local system $\mathbf{V}$ comes from an irreducible symplectic $\mathbf{Q}$-local system. However, over $\mathbf{R}$ it splits as before. In this case $S = S_0$ and we have an ordinary Higgs bundle. In both cases we have a strictly maximal Higgs field, because the Higgs field is an isomorphism.
"$\Longleftarrow$": Conversely, we have to show that a weight 1 variation $\mathcal{V} = \mathbf{V} \otimes \mathcal{O}_{C_0}$ over a curve C_0 with strictly maximal Higgs field or – equivalently – reaching the Arakelov bound, comes from a rigidly embedded Shimura curve. We give a sketch of the proof under the assumption that the number of points in Σ is even. If it is odd, the argument is slightly more involved. For this, see Viehweg and Zuo (2004) on which the following proof is modeled. It is divided into the following steps.

Step 1: Construction of a tensor decomposition of the Higgs bundle.
The simplifying assumption means that we can extract a root $\mathcal{M}$ from the line bundle $\Omega^1_C(\log \Sigma)$. This defines a graded Higgs bundle $\mathcal{L} = (\mathcal{M} \oplus \mathcal{M}^{-1}, \tau)$ with Higgs field given by

$$\tau^{1,0} : \mathcal{M} \xrightarrow{\simeq} \mathcal{M}^{-1} \otimes \Omega^1_C(\log C), \quad \tau^{0,1} = 0.$$

We make now use of Simpson's deep results (Theorem 13.3.5) which imply that this Higgs bundle comes from a polarized variation of Hodge structure, say with underlying local system **L**. It has a natural real structure.

The idea now is to find a local system **U** of rank g underlying a locally constant variation of type $(0,0)$ such that $\mathbf{U}\otimes\mathbf{L}=\mathbf{V}$. So we need a candidate for the bundle $\mathbf{U}\otimes\mathcal{O}_S$. It should be of rank g and degree 0 and since it should underly a variation of Hodge structure, by Proposition 13.3.6, it should be polystable. A natural candidate is the bundle $\mathcal{U}=(\mathcal{V}^{1,0}\otimes\mathcal{M}^{-1},0)$; it has the correct rank and degree. The crucial point now is that it polystable as well. Indeed, let $\mathcal{G}^{1,0}$ be a subbundle of $\mathcal{V}^{1,0}$ The Higgs bundle $(\mathcal{G}^{1,0},0)$ is a subbundle of $(\mathcal{V}^{1,0},0)$ and so, $\mu(\mathcal{G}^{1,0})\le\mu(\mathcal{V}^{1,0})$ and the same inequality holds for subbundles of $\mathcal{U}$. Hence $\mathcal{U}$ is semi-stable. By considering the Higgs bundle saturation $\mathcal{G}$ of $\mathcal{G}^{1,0}$ within $\mathcal{V}$, one infers that $\mathcal{U}$ is polystable. By Remark 13.3.5 it comes from a local system, and hence, by the discussion surrounding Definition 13.4.3, the Hodge metric on this system is flat, i.e., we have a unitary local system **U**.

The isomorphism of Higgs fields

$$\sigma^{1,0}:\mathcal{V}^{1,0}\xrightarrow{\simeq}\mathcal{V}^{0,1}\otimes\Omega^1_C(\log\Sigma)$$

induces an isomorphism

$$\begin{array}{ccc}\mathcal{U}\otimes\mathcal{M}^{-1} & \xrightarrow[\varphi]{\hspace{6em}} & \mathcal{V}^{0,1}\\ \Vert & & \Vert\\ \mathcal{V}^{1,0}\otimes\mathcal{M}^{-2} & \xrightarrow{\simeq} & \mathcal{V}^{0,1}\otimes\mathcal{M}^{-2}\otimes\Omega^1_C(\log C).\end{array}$$

Consequently, the Higgs field can be written as

$$\mathcal{V}^{1,0}=\mathcal{U}\otimes\mathcal{M}\xrightarrow{\varphi(\mathbf{1}\otimes\tau^{1,0})}\mathcal{U}\otimes\mathcal{M}^{-1}\otimes\Omega^1_C(\log\Sigma)=\mathcal{V}^{0,1}\otimes\Omega^1_C(\log\Sigma).$$

It follows that the Higgs bundles $(\mathcal{V},\sigma)$ and $(\mathcal{U}\otimes\mathcal{L},\mathbf{1}\otimes\tau)$ are isomorphic, and $\mathbf{V}\simeq\mathbf{U}\otimes\mathbf{L}$. Comparison of types shows that $\mathcal{U}$ carries a (real) variation of Hodge structure of pure of type $(0,0)$ and hence the flat Hodge metric on $\mathcal{U}$ comes from a "real" metric and the monodromy of the system coming from a discrete group inside a compact group, must be finite. So **U** can be taken to be constant, say with underlying vector space U.

Step 2: Constructing a rigid equivariant embedding of $\mathfrak{h}$.

The real weight 1 variation on $\mathcal{L}=\mathcal{M}\oplus\mathcal{M}^{-1}$ over the curve C_0 comes from a representation $\rho:\pi_1(C_0)\to\mathrm{SL}(2,\mathbf{R})$. Since the variation is not constant (it gives a maximal degree Higgs bundle) this gives a nontrivial period map from the universal cover $\widetilde{C_0}$ of C_0 to $\mathfrak{h}$. In fact, the period map then is a biholomorphism, and we may identify $\widetilde{C_0}=\mathfrak{h}$.

Since the tensor representation of the Higgs bundle decomposition is adapted to the tensor representation $\mathbf{V} = U \otimes \mathbf{L}$, it follows that the symplectic representation associated with the variation of Hodge structure defined by $\mathcal{V}$ factors over ρ. Hence, there is an equivariant holomorphic map $\iota : \mathfrak{h} \to \mathfrak{h}_g$. From Theorem 13.5.4 we know that the embedding is rigid. By Lemma 17.2.2, the embedding ι is necessarily totally geodesic. From Proposition 17.2.9 we conclude that $\iota(C)$ is a special subvariety. □

Remark For maximal Higgs bundles $\mathcal{H}$, i.e., for Higgs bundles for which the summand $\mathcal{H}^{\max}$ of $\mathcal{H}$ (see Eq. (13.7)) is strictly maximal, one has parallel results: first of all the unipotent part comes from a family of abelian varieties which becomes trivial after a finite unramified base change. This means that $\mathcal{H}^{\mathrm{un}}$ is a locally constant Higgs field (with finite monodromy). Secondly, the nonconstant part $\mathcal{H}^{\max}$ is again a rigid family as in the above theorem and hence comes from a rigidly embedded Shimura curve. See Viehweg and Zuo (2004) for details.

17.5 Characterizations of Special Subvarieties

Numerical Criteria for Special Subvarieties

In a previous chapter we considered Hilbert modular surfaces and Picard modular surfaces as examples of Shimura varieties. See Examples 16.3.7. We now consider some suitable smooth projective compactification, say X, a compactification of $X^0 = \Gamma\backslash\mathfrak{h} \times \mathfrak{h}$ or $X^0 = \Gamma\backslash B_2$. Under which conditions does a curve $C \subset X$ give rise to a special curve $C^0 = C \cap X^0$ on X^0? In this section we find numerical criteria in these two cases. The ultimate goal is to generalize this for closures of special subvarieties in arbitrary Shimura varieties.

Theorem 17.5.1 (Hirzebruch, 1976) *Let X be the minimal resolution of singularities of the Baily–Borel compactification* [3] *of a Hilbert modular surface $X^0 = \Gamma\backslash(\mathfrak{h} \times \mathfrak{h})$ and let $C \subset X$ be a reduced curve not contained in the boundary $D = X - X^0$. Set $C^0 = C \cap X^0$. With $\varphi : \tilde{C} \to C$ the normalization of C, introduce*[4]

$$\delta(C) = p_a(C) - g(\tilde{C}),$$
$$\epsilon(C) = D \cdot C - \deg \varphi^{-1}(D \cap C)_{\mathrm{red}}.$$

Assume that Γ is neat and that D is a strict normal crossing divisor. Then one

[3] See Baily and Borel (1966).

[4] $\delta(C)$ measures how singular C is; e.g. if C has δ ordinary double points, $\delta(C) = \delta$. The invariant $\epsilon(C)$ measures how the intersection of C and D fails to be transversal.

has the relative proportionality inequality saying that

$$2C^2 + (K_X + D)\cdot C \geq 4\delta(C) + 2\epsilon(C).$$

If equality holds, C^0 is a special curve.

Proof In the proof we are using some ideas already explained in Example 9.1.8. The standard representation of Γ in $\mathrm{Sp}(2,\mathbf{R})$ gives a rank 4 (symplectic) local system $\mathbf{V}$. It underlies a logarithmic Higgs bundle $\mathscr{V}$ over X. Over X^0 this is the universal variation of Hodge structure of weight 1. The tangent bundle of $\mathscr{A}_g(N)$ equals the bundle of symmetric tensors of $\mathscr{V}^{0,1}$ and hence we have an inclusion

$$T_X(-\log D) \hookrightarrow \mathrm{Sym}^2\mathscr{V}^{0,1}. \tag{17.11}$$

Next, with R the order in the ring of integers of K defining the Hilbert modular surface, there is a splitting of local systems

$$\textstyle\bigwedge_R^2 \mathbf{V} \oplus \mathbf{W} = \bigwedge^2 \mathbf{V}.$$

Let $(\mathscr{W},\sigma)$ be the Higgs bundle associated with $\mathbf{W}$. We claim that we have an isomorphism

$$\sigma^{2,0} : \mathscr{W}^{2,0} \otimes T_X(-\log D) \xrightarrow{\sim} \mathscr{W}^{1,1}. \tag{17.12}$$

Since $\sigma^{2,0}$ is a vector bundle morphism, for rank reasons, it is enough to show that it is injective. Remark that $\bigwedge_R^2$ only contributes to the $(1,1)$ part and so $\mathscr{W}^{2,0} = \bigwedge^2\mathscr{V}^{1,0}$ and $\mathscr{W}^{1,1} \subset \mathscr{V}^{1,0}\otimes\mathscr{V}^{0,1}$. Hence it is enough to show that

$$\textstyle\bigwedge^2\mathscr{V}^{1,0}\otimes T_X(-\log D) \to \mathscr{V}^{1,0}\otimes\mathscr{V}^{0,1}$$

is injective. Since $\bigwedge^2\mathscr{V}^{1,0}$ is a line bundle, we have

$$T_X(-\log D) \to \mathscr{V}^{1,0}\otimes\mathscr{V}^{0,1}\otimes(\textstyle\bigwedge^2\mathscr{V}^{1,0})^\vee = \mathscr{V}^{0,1}\otimes\mathscr{V}^{0,1},$$

which factors over the symmetric tensors and hence is injective by (17.11).

We now consider the saturated Higgs bundle $\mathscr{E}$ on $\tilde{C}$ generated by $\varphi^*\mathscr{W}^{2,0}$. Note that the bundle being saturated implies that $\mathscr{E}^{0,2} = \varphi^*\mathscr{W}^{0,2}$. Setting

$$D_{\tilde{C}} = \varphi^{-1}(D\cap C)_{\mathrm{red}},$$

we then also have an injection

$$\sigma^{2,0} : \mathscr{E}^{2,0} \to \mathscr{E}^{1,1}\otimes\Omega^1_{\tilde{C}}(\log D_{\tilde{C}}).$$

Since $\mathscr{W}^{2,0}$ and $\mathscr{W}^{0,2}$ are dual bundles, their degrees on $\tilde{C}$ add up to zero and we get

$$0 = \deg\varphi^*\mathscr{W} = \underbrace{\deg\mathscr{E}^{2,0} + \deg\mathscr{E}^{0,2}}_{=0} + \deg\varphi^*\mathscr{W}^{1,1} = \deg\varphi^*\mathscr{W}^{1,1}.$$

We then find, using (17.12), that

$$\begin{aligned} 2 \deg \mathscr{E}^{2,0} &= -\deg \varphi^* T_X(-\log(D)) \\ &= (K_X + D) \cdot C. \end{aligned} \tag{17.13}$$

Since $\mu(\mathscr{W}|C) = 0$, by stability $\mu(\mathscr{E}) = \deg \mathscr{E}^{1,1} \le 0$. The Higgs field injects $\mathscr{E}^{2,0}$ into $\mathscr{E}^{1,1} \otimes \Omega^1_{\tilde{C}}(\log(D_{\tilde{C}}))$ and hence

$$\begin{aligned} \deg \mathscr{E}^{2,0} &\le \deg (\mathscr{E}^{1,1} \otimes \Omega^1_{\tilde{C}}(\log(D_{\tilde{C}})) \\ &\le \deg \Omega^1_{\tilde{C}}(\log(D_{\tilde{C}})) = \deg \Omega^1_{\tilde{C}} + \deg D_{\tilde{C}} \\ &= 2g(\tilde{C}) - 2 + \deg D_{\tilde{C}} \\ &= 2p_a(C) - 2 - 2\delta(C) + \deg D_{\tilde{C}} \\ &= K_X C + C^2 - 2\delta(C) + \deg D_{\tilde{C}}. \end{aligned}$$

The last equation uses the adjunction formula (C.2). Together with (17.13) this gives

$$\begin{aligned} (K_X + D) \cdot C &= 2 \deg \mathscr{E}^{2,0} \\ &\le 2(K_X C + C^2 - 2\delta(C) + \deg D_{\tilde{C}}), \end{aligned}$$

and hence

$$K_X \cdot C + 2C^2 + D \cdot C - 2\underbrace{(D \cdot C - \deg D_{\tilde{C}})}_{\epsilon(C)} - 4\delta(C) \ge 0.$$

If equality holds, then all the previous inequalities are equalities. So the Higgs bundle $\mathscr{E}$ is a direct summand of $\mathscr{W}|C$ of degree zero. So by Corollary 13.3.3 it has an orthogonal complement with trivial Higgs field and hence comes from a unitary representation. In particular this holds for the component

$$\mathscr{E}^{1,1} = T_{\tilde{C}}(-\log D_{\tilde{C}}) \otimes \mathscr{E}^{2,0} \subset \varphi^* \mathscr{W}^{1,1}$$

and hence we have an orthogonal direct sum decomposition

$$\varphi^* (T_X(-\log D)) = T_{\tilde{C}}(-\log D_{\tilde{C}}) \oplus (\text{line bundle}).$$

By Example 11.5.2 (iii) this implies that the smooth part of C^0 is a totally geodesic submanifold. Moreover, since we also know that the Yukawa coupling is nonzero, by Theorem 13.5.4 the curve C^0 is rigidly immersed and hence, by Proposition 17.2.9 the curve C^0 is special. □

Remark If we have a Picard modular surface $\Gamma\backslash B_2$, a similar result holds; for example, if C^0 is embedded and C is smooth and meets the boundary transversally, the inequality reads

$$3 \cdot C \cdot C \ge -(K_X + D) \cdot C,$$

with equality if and only if C^0 is special.

Example 17.5.2 (Ball quotient, Picard modular surfaces) Picard (1883) looked at the following family of curves:

$$C_{s,t} : y^3 = x(x-1)(x-s)(x-t), \quad (s,t) \in \mathbf{C}^2 \subset \mathbf{P}^2(\mathbf{C}).$$

We leave it as an exercise to show that these curves have genus 3. They all have an extra $\mathbf{Z}/3\mathbf{Z}$ automorphism. The discriminant locus $\Delta \subset \mathbf{P}^2$ of this family consists of 6 lines and has 4 cusps (3-fold intersections):

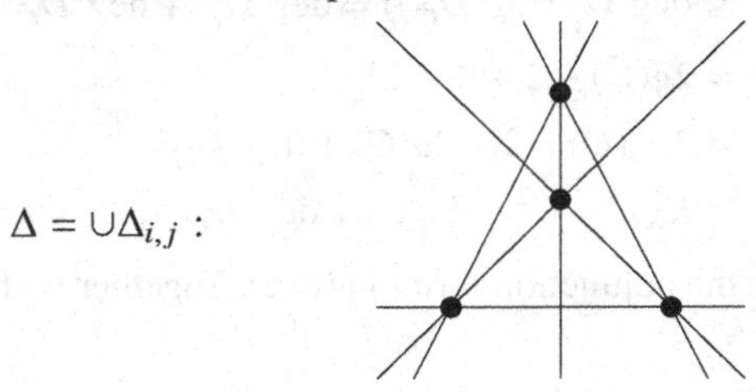

Since the curves have genus 3, the de Rham cohomology has dimension 6 and hence the family has 6 independent periods which are functions of (t, s). The Picard–Fuchs equations for them form a system of partial differential equations of so-called Euler type. See Holzapfel (1986b, p. 152). These define a multivalued map

$$\mathbf{P}^2 - \Delta \longrightarrow B_2 = \{|z_1|^2 + |z_2|^2 < 1\} \subset \mathbf{C}^2 \subset \mathbf{P}^2(\mathbf{C}).$$

The two-ball B_2 is a homogenous space for $U(2,1)$, the group of matrices preserving the Hermitian form with signature $(1,1,-1)$. Picard proved that

$$\mathbf{P}^2 - \text{cusps} \cong \Gamma \backslash B_2,$$

where

$$\Gamma = \{\gamma \in U(2,1)(\mathscr{E}) \mid \gamma \equiv \begin{pmatrix} 1 & 0 & 0 \\ 0 & 1 & 0 \\ 0 & 0 & 1 \end{pmatrix} \mod (1-\omega)\}.$$

Here $\mathscr{E} = \mathbf{Z}[\omega]$ are the Eisenstein numbers with $\omega = \frac{-1+\sqrt{-3}}{2}$, $\omega^3 = 1$. This surface does not carry a nice universal family over it.

Later Holzapfel (1986a) constructed more examples carrying universal families as ramified coverings of Picard's example. Here is one of them.

Example 17.5.3 Let X be the blow-up of $E \times E$ in 3 points (P_i, P_i), where

$$E = \{y^2 z = x^3 - z^3\}$$

is a CM elliptic curve with automorphism $x \mapsto \omega x$ of order 3 and fixed points P_1, P_2, P_3. This is a (birational) covering of $\mathbf{P}^2$ as follows: Blow up the 4 cusps in the preceding configuration of lines $\Delta_{i,j}$ in $\mathbf{P}^2$. Next, blow down all 3 strict transforms of the lines $\Delta_{i,4}$. One ends up with the following configuration of lines in $\mathbf{P}^1 \times \mathbf{P}^1$:

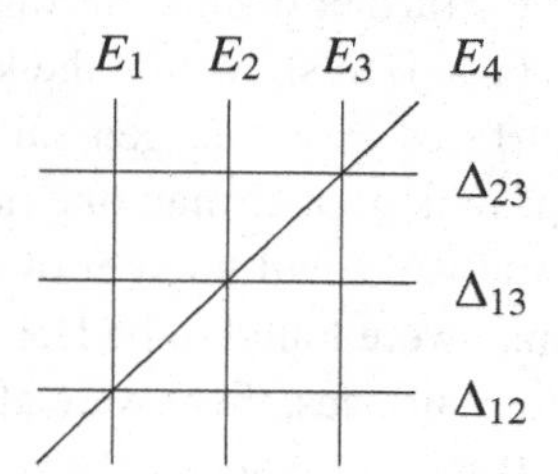

Then take the 3-1 cyclic cover isomorphic to $E \times E$ branched along horizontal and vertical lines. The diagonal curve will then split into 3 elliptic curves corresponding to strict transforms of the 3 diagonals

$$(z, w), \quad (z, \omega w), \quad (z, \omega^2 w) \subset E \times E.$$

Together we get 6 elliptic cusp curves $X_1, \ldots, X_6$:

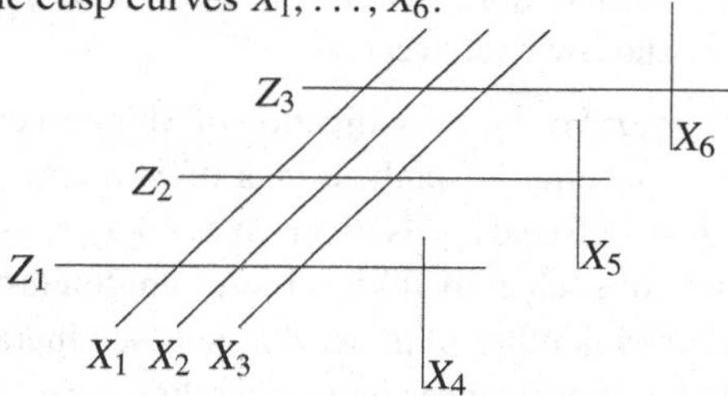

Lemma 17.5.4 *The exceptional curves Z_1, Z_2, Z_3 are rational modular curves with 4 cusps. They carry a special family of Jacobians of type $E \times S^2(E_\lambda)$. There are 3 more modular elliptic curves $\tilde{\Delta}_{ij}$.*

Proof We check the proportionality equality condition from theorem 17.5.1: $K_X = \sum Z_i$, $S = \sum X_j$ and hence for $C = Z_1$ for example we have $(K_X + S \cdot C) = Z_1^2 + (X_1 + X_2 + X_3 + X_6) \cdot Z_1 = 3$, hence $3C^2 + (K_X + S \cdot C) = 0$. The strict transform of the three vertical elliptic curves in the product through the points where we blow up are modular curves with $C^2 = -1$ and $(C \cdot K_X + S) = 3$ as well, since they intersect two of the X_j and one exceptional divisor Z_i. □

Bibliographical and Historical Remarks

About Section 17.1: Apart from some simplifications the material first appeared in the monograph by Green, Griffiths and Kerr (2012).

About Section 17.2: Satake (1965, 1980) noted that strongly equivariant embeddings give rise to totally geodesic submanifolds. The relation with varieties of Hodge type has been investigated in Abdulali (1994), and Moonen (1998).

About Section 17.3: This is a condensed version of Allcock et al. (2002) which the reader may consult for further details. Second period maps were also used by Carlson and Toledo (1999) to show that the kernel of the monodromy representation for hypersurfaces in $\mathbf{P}^n$ is, generally speaking, quite large, e.g., contains a free group of rank greater than one (see Carlson and Toledo, 1999, Theorem 1.2). In that study, a small number of period domains in edge cases for the second period map were found to be Hermitian symmetric.

In addition to the cubic surfaces, there were also the cubic threefolds. Cubic threefolds depend on 10 moduli, and for them, the story unfolds along lines much as in the case of surfaces. The standard period map takes values in a Siegel upper half-space of genus 5 and dimension 15. The second period map takes values in an arithmetic quotient of a ten-dimensional ball. For this map one also has an identification of a suitable moduli space with this ball-quotient, up to some blowing up and blowing down at the boundary. The precise statement is, however, somewhat more technical (Allcock et al., 2011, Theorem 6.1). See also Looijenga and Swierstra (2007).

About Section 17.4: For the classification of Shimura curves see Satake (1980). A classification scheme for embeddings of Shimura varieties in quotients of the Siegel upper half-space was given in Satake (1965, 1967) and carried out for large classes of such embeddings by Addington (1987).

For embeddings in other Shimura domains see Ihara (1967).

For rigidity questions concerning embeddings see Abdulali (1994) and Moonen (1998).

The idea of using Higgs bundle methods to classify certain types of Shimura varieties is due to Viehweg and Zuo (2004).

For other results see Müller-Stach et al. (2011, 2012b).

About Section 17.5: There are generalizations of Hirzebruch's theorem to other situations: for orthogonal or unitary Shimura varieties see Müller-Stach et al. (2009) and Müller-Stach and Zuo (2011), for subvarieties of $\mathscr{A}_g$, and for other Mumford–Tate varieties see Mohajer et al. (2016).

Moduli spaces of Calabi–Yau manifolds of dimension $n \geq 3$ tend to be non-modular, i.e., they are not isomorphic to Zariski open subsets of any Shimura subvariety of the period domain. For example, the moduli space of double solids ramified along 8 planes in general position is not modular by a result of Gerkmann, Sheng and Zuo (2007).

It is interesting to find maximal modular subvarieties. Viehweg and Zuo

(2005) have for example constructed a 4-dimensional Shimura family of quintic 3-folds.

We finish with some remarks concerning the André–Oort problem and the Coleman–Oort conjecture. The first (see André, 1989; Oort, 1997) asks whether the Zariski closure of any subset of CM-points in a given Shimura variety is a finite union of special subvarieties. This is known under the validity of the generalized Riemann hypothesis (see Klingler and Yafaev, 2014; Ullmo et al., 2014) and, unconditionally, for $\mathscr{A}_g(N)$ by Tsimerman (2015) building on ground breaking work by Yuan and Zhang (2015). See also Edixhoven and Yafaev (2003).

Concerning the second conjecture, we make the following remarks. The Torelli locus by definition is the closure of the image of the period map $\mathscr{M}_g \to \mathscr{A}_g$. The conjecture by Oort (1997) states that for $g \gg 0$ there are no special subvarieties of positive dimension in the Torelli locus which intersect the image of the Torelli map in a dense open subset, while that of Coleman (1987) claims that for $g \gg 0$ there are only finitely many smooth, projective curves C with a Jacobian of CM-type. Assuming the André–Oort conjecture, this is equivalent to the Oort conjecture, since the Zariski closure of infinitely many CM-points would be a special subvariety of positive dimension in the Torelli locus. The conjecture is known for rigidly embedded Shimura curves (see Lu and Zuo, 2014, 2015; Peters, 2016) and for certain higher-dimensional embedded Shimura varieties (see Chen et al., 2015).

Appendix A

Projective Varieties and Complex Manifolds

Let us start with the first and most basic example.

Definition A.1 (Complex projective space) $\mathbf{P}^n$ is the set of complex lines through the origin in $\mathbf{C}^{n+1}$. The standard coordinates induce homogeneous coordinates on $\mathbf{P}^n$ which we denote by $(z_0 : \cdots : z_n)$.

Note that there is a natural map $\mathbf{C}^{n+1} - \{0\} \to \mathbf{P}^n$ sending a nonzero vector to the line it defines. We equip $\mathbf{P}^n$ with the quotient topology and as such it becomes a compact Hausdorff space (see Problem A.1). Projective space is covered by the standard open sets $U_j = \{z_j \neq 0\}$ homeomorphic to $\mathbf{C}^n$. On U_j we have the affine coordinates $z_i^j = z_i/z_j, i = 0, \ldots, j-1, j+1, \ldots, n$. Suppose that $F(z_0, \ldots, z_n)$ is a homogeneous polynomial. Inside $\mathbf{P}^n$ its zerolocus $V(F)$ is well defined and is called a hypersurface. If F is irreducible we say that the hypersurface is *irreducible* as well. This is an instance of a projective variety of dimension $(n-1)$.

Definition A.2 A *projective algebraic set* is the zero locus inside $\mathbf{P}^n$ of finitely many homogeneous polynomials. If the ideal they generate is a prime ideal, we speak of a *projective variety*.

It should be clear what is meant by a (closed) subvariety of a given projective variety. These generate a topology, the *Zariski topology*. Any Zariski-open subset of a projective variety by definition gives a *quasi-projective variety*.

If we look at that part of a projective algebraic set which lives in the standard open set U_j we get what is called an affine set, and similarly for an *affine variety*. Intrinsically, an affine set is the zero locus in $\mathbf{C}^n$ of a finite number of polynomials and it is an affine variety if these generate a prime ideal.

The next step is to define a suitable notion of maps between projective or affine varieties.

Definition A.3 A collection of $(m+1)$ homogeneous polynomial expressions $\{G_0, \dots, G_m\}$ in $(z_0, \dots, z_n)$ of the same degree and without common factors defines the rational map

$$\psi : \mathbf{P}^n \dashrightarrow \mathbf{P}^m$$
$$p \mapsto (G_0(p) : \cdots : G_m(p)).$$

If $X \subset \mathbf{P}^n$ is a projective variety, a *rational map* $X \dashrightarrow \mathbf{P}^m$ by definition is the restriction to X of a rational map of the ambient projective space.

Note that a rational map as in the definition is well defined outside the locus of common zeroes of the G_j, the indeterminacy locus of ψ, and so if our X avoids the indeterminacy locus, the restriction $\phi = \psi|X : X \to \mathbf{P}^m$ is everywhere defined and is the prototype of a regular map. To have a better understanding of rational maps, we need a version that works locally in the affine setting.

Definition A.4 A quotient $f = P/Q$ of two polynomials P and Q (not necessarily homogeneous or of the same degree) in the variables $(z_1, \dots, z_n)$ is a *rational function*. We say that $f = P/Q$ is defined on the affine variety $V \subset \mathbf{C}^n$ if Q does not vanish identically on V. The rational functions P/Q and P'/Q' by definition give the same function on V if $PQ' - QP'$ vanishes identically on V. It is called regular at a given point $v \in V$, if there is a representative $h = P/Q$ for h with $Q(v) \neq 0$. A *regular function* is a rational function that is everywhere regular. Any m-tuple of rational functions $(h_1, \dots, h_m)$ defined on V gives a rational map $H : V \dashrightarrow \mathbf{C}^m$. If the h_j are all regular, we say that H is a *regular map*.

We now return to the projective situation. Let $V \subset \mathbf{P}^n$ be a projective variety and let $\phi : V \dashrightarrow \mathbf{P}^m$ be a rational map defined by the homogeneous polynomials $G_0, \dots, G_m$. The map ϕ can very well be regular at v, even if $G_j(v) = 0, j = 0, \dots, m$. Indeed, if for instance $G_j/G_0 = H_j/H_0$ on V and $H_0(v) \neq 0$, the function G_j/G_0 is regular at v. Indeed, the function G_j/G_0 can be viewed as a function in the affine coordinates and as such defines a rational function of the sort considered above. In general, we say that ϕ is regular at $v \in V$ if for some $i = 0, \dots, m$ the rational functions $G_j/G_i, j \neq i$ are all regular at v; and if ϕ is everywhere regular, we say that ϕ is *regular* or a *morphism*.

It is a nontrivial theorem that the image of a projective variety under a morphism is again projective. This result is usually shown by means of elimination theory (see Mumford, 1976, Section 2C, and Harris, 1992, Lecture 3).

Ultimately, we shall be interested in projective manifolds. We have already defined the notion of a projective variety. We now introduce the notion of a complex manifold.

Definition A.5 A Hausdorff topological space M with countable basis for the topology is an n-dimensional *complex manifold* if it has a covering $\{U_i\}$, $i \in I$ by open sets that admit homeomorphisms $\varphi_i : U_i \longrightarrow V_i$ with V_i an open subset of $\mathbf{C}^n$ and such that for all $i \in I$ and $j \in I$ the map $\varphi_i \circ \varphi_j^{-1}$ is a holomorphic map on the open set $\varphi_j(U_i \cap U_j) \subset \mathbf{C}^n$ where it is defined.

A function f on an open set $U \subset M$ is called *holomorphic* if for all $i \in I$ the function $f \circ \varphi_i^{-1}$ is holomorphic on the open set $\varphi_i(U \cap U_i) \subset \mathbf{C}^n$. Also, a collection of functions $z = (z_1, \ldots, z_n)$ on an open subset U of M is called a *holomorphic coordinate system* if $z \circ \varphi_i^{-1}$ is a holomorphic bijection from $\varphi_i(U \cap U_i)$ to $z(U \cap U_i)$ with holomorphic inverse. The open set on which a coordinate system can be given is then called a chart. Finally, a map $f : M \longrightarrow N$ between complex manifolds is called *holomorphic* if it is given in terms of local holomorphic coordinates on N by holomorphic functions.

Examples A.6 (i) Complex projective space $\mathbf{P}^n$ is a compact n-dimensional manifold.

(ii) A hypersurface of $\mathbf{P}^n$ given by the equation $F = 0$ usually is not a manifold. But if the gradient of F does vanish identically, it is a manifold (Problem A.3). Examples include the Fermat hypersurfaces $z_0^d + z_1^d + \cdots + z_n^d = 0$.

(iii) A complex manifold of dimension 1 is also called a *Riemann surface*. Compact Riemann surfaces are topologically classified by their genus, the "number of holes." As an example, consider the Riemann surface of the function $\sqrt{x(x-1)(x-\lambda)}$, $\lambda \neq \{0, 1, \infty\}$, which we encountered in Chapter 1, Section 1.1. We saw there that it is topologically a torus. On the other hand, it can also be given by the equation $y^2 = x(x-1)(x-\lambda)$ in the (x, y)-plane and hence by homogenizing by the cubic equation $Y^2Z = X(X-Z)(X-\lambda Z)$. This defines a smooth hypersurface in $\mathbf{P}^2$, i.e., a nonsingular projective curve.

(iv) A *complex n-torus* is the quotient of $\mathbf{C}^n$ by a lattice of maximal rank. It is an n-dimensional manifold. Any affine-linear map $\mathbf{C}^n \longrightarrow \mathbf{C}^m$ sending a lattice Γ_1 to a lattice Γ_2 induces a holomorphic map from the n-torus $\mathbf{C}^n/\Gamma_1$ to the m-torus $\mathbf{C}^m/\Gamma_2$. The converse is also true. See Problem A.6.

(v) The set of k-planes in a complex vector space V of dimension n carries the structure of a projective manifold, the *Grassmann manifold* $G(k, V)$ of dimension $k(n-k)$. Indeed, it becomes a subset of $\mathbf{P}(\bigwedge^k V)$ by sending the k-plane $W \subset V$ to the line $\bigwedge^k V$ considered to be an element of $\mathbf{P}(\bigwedge^k V)$. The equations are determined by the fact that $[w] \in \mathbf{P}(\bigwedge^k V)$ is of the form $[w] = [\wedge^k v]$ if and only if w is fully decomposable. Now $w = w' \wedge v$, $w' \in \bigwedge^{k-1} V$, $v \in V$ if and only if $w \wedge v = 0$. If this is the case, we say that w is divisible by v; w is fully decomposable if and only if the space of vectors dividing w has dimension k. This in turn is the case if the map $\varphi_w : V \longrightarrow \wedge^{k+1} V$ sending v to $w \wedge v$ has rank exactly $(n-k)$ and hence rank $\leq n-k$. The map $\bigwedge^k V \to \mathrm{Hom}(V, \bigwedge^{k+1} V)$ which sends w to φ_w is linear and so in choosing a basis for V, the map φ_w can be represented by a matrix whose entries are homogeneous coordinates on $\mathbf{P}(\bigwedge^k V)$. Then w is fully decomposable if and only if the $(n-k+1)\times(n-k+1)$ minors of this matrix vanish. This indeed gives polynomial equations for $G(k, V)$. Finally, to see that it is a complex manifold, note that the general linear group $\mathrm{GL}(V)$ acts transitively on $G(k, V)$. Looking at the stabilizer, we can easily see that the dimension equals $k(n-k)$ as claimed.

(vi) For any increasing set of integers $a_1 < a_2 < \cdots < a_k < n$ we have the *variety of flags*

$$\begin{aligned} F(a_1, \ldots, a_k; V) &= \{W_1, \ldots, W_k \mid W_1 \subset \cdots \subset W_k\} \\ &\subset G(a_1, V) \times \cdots \times G(a_k, V). \end{aligned}$$

That this is a smooth subvariety of the products of the Grassmannians on the right is left as an exercise (Problem A.7).

(vii) A *Hopf manifold* is a quotient of $\mathbf{C}^n - \{0\}$ by the infinite cyclic group generated by a dilatation $z \mapsto 2z$. It is easy to see that this is an n-dimensional manifold homeomorphic to $S^1 \times S^{2n-1}$.

The preceding examples show that a complex projective variety need not be a complex manifold. Indeed, it may have singular points, i.e., points around which no holomorphic chart can be found. But (and this is *a priori* not obvious) the singular points themselves form a proper subvariety and so "most" points are manifold points. If there are no singular points, we say that the variety is nonsingular. We also speak of projective manifolds. So a *projective manifold* is a submanifold of projective space. The converse is also true.

Theorem A.7 (Chow's theorem) *Any complex submanifold of* $\mathbf{P}^n$ *is a projective variety.*

It is also true that a holomorphic map between submanifolds of projective spaces is a regular map. See Problem A.4. This important fact allows us to shift between algebraic methods on the one hand and topological and complex-analytic methods on the other.

It is a profound fact that any compact Riemann surface admits a realization as a submanifold of some projective space and so, by Chow's theorem, it is projective. See Bibliographical Remarks.

Problems

A.1 Show that $\mathbf{P}^n$ is a quotient of S^{2n+1} viewed as the unit sphere in $\mathbf{C}^{n+1}$ by the equivalence relation $(z_0, \ldots, z_n) \sim (w_0, \ldots, w_n)$ if and only if $z_k = \alpha w_k$, $k = 0, \ldots, n$, $|\alpha| = 1$. Deduce that $\mathbf{P}^n$ is Hausdorff and compact.

A.2 Prove that the product of two projective varieties is projective. Hint: use the Segre embedding $\mathbf{P}^n \times \mathbf{P}^m \longrightarrow \mathbf{P}^{nm+m+n}$. One may consult Mumford (1976, Section 2B) for details.

A.3 Let F be a homogeneous polynomial of degree d. Consider the open set U_j. Dividing through by z_j^d we arrive at the corresponding inhomogeneous polynomial $f^j(z_1, \ldots, z_n)$. Compare the singular set of $V(f^j)$ and $V(F)$ and show that $V(F)$ is singular at those points where all the partial derivatives of F vanish.

A.4 Prove that holomorphic maps between projective manifolds are morphisms (consider the graph of the holomorphic map).

A.5 Show that any constant holomorphic function on a connected compact complex manifold must be constant. Hint: use the maximum principle for holomorphic functions on open subsets in $\mathbf{C}^n$.

A.6 Prove that holomorphic maps $u : T_1 \longrightarrow T_2$ between tori are realized by affine linear maps between their universal covers. Hint: any lift $\tilde{u} : \mathbf{C}^n \longrightarrow \mathbf{C}^m$ has the property that the function $x \mapsto \tilde{u}(x+\gamma) - \tilde{u}(x)$, $\gamma \in \Gamma_1$, does not depend on x. What does this imply for the partial derivatives of $\tilde{u}$?

A.7 Prove the assertion about flag varieties.

Bibliographical and Historical Remarks

The monograph Mumford (1976) is a good introduction to complex projective geometry. The book Harris (1992) gives many motivating examples and serves as complementary reading.

A rather self-contained proof of Chow's theorem can be found in Mumford (1976). For a considerably shorter proof, see Griffiths and Harris (1978, p. 167). This proof, however, uses the so-called proper mapping theorem, a partial proof of which is supplied in Griffiths and Harris (1978, pp. 395–400). The fact that every compact Riemann surface is projective is a special case of Kodaira's embedding theorem, which is proved, for instance, in Griffiths and Harris (1978, p. 181).

Appendix B

Homology and Cohomology

B.1 Simplicial Theory

Of the various homology theories existing today, historically, simplicial theory came first. This theory has an intuitive appeal and leads to invariants that can be computed in a purely mechanical fashion. However, it has the disadvantage that it works only for triangulable spaces, i.e., spaces homeomorphic to the support of a simplicial complex. We explain this concept in some detail below. The spaces we are working with turn out to be triangulable and so simplicial theory can be applied.

The basic building block of a simplicial complex is the *affine-linear p-simplex*. By definition this is a set σ, affine-linearly isomorphic to the standard p-simplex $\Delta_p \subset \mathbf{R}^{p+1}$, the convex hull in $\mathbf{R}^{p+1}$ of the $p+1$ standard unit vectors

$$\Delta_p = \left\{ (x_0, \ldots, x_p) \mid x_i \geq 0, \sum_{i=0}^{p} x_i = 1 \right\}.$$

The faces of σ correspond to the lower-dimensional simplices contained in the boundary $\bigcup_{q=0}^{p} \Delta_p \cap \{x_q = 0\}$. In particular, we have the vertices of σ, i.e., the zero-dimensional faces. An orientation of σ is an ordering of its vertices up to an even permutation. The standard simplex comes with a natural orientation such that all its faces are coherently oriented. See Fig. B.1 for an oriented two-simplex.

A *simplicial complex* in some $\mathbf{R}^N$ is a set K of simplices with the properties that:

(i) with every simplex in K all of its faces belong to K; and
(ii) every two simplices in K either have empty intersection or they intersect in a simplex from K.

The *support* $|K|$ is the union of all the simplices of K with the topology

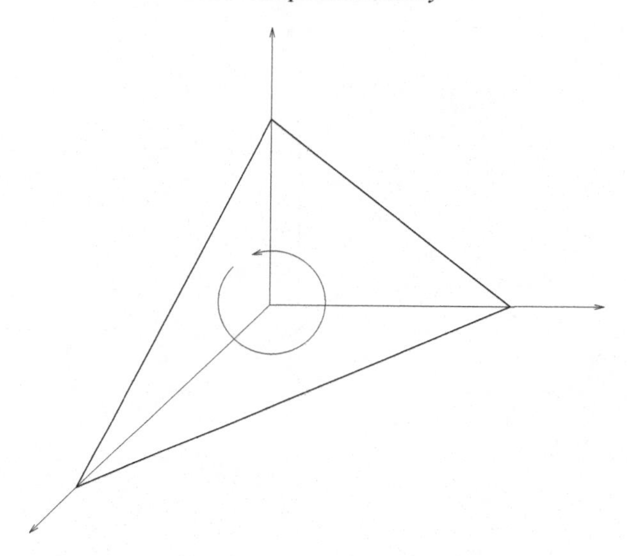

Figure B.1 Standard two-simplex with its orientation.

induced from the surrounding $\mathbf{R}^N$. The complex K is allowed to have infinitely many simplices, but we restrict ourselves to locally finite complexes K: given $x \in |K|$ there should be a neighborhood $U \subset \mathbf{R}^N$ of x meeting at most finitely many simplices of K.

It should be clear what is meant by a *subcomplex* of a given simplicial complex and what we mean by *simplicial maps*.

As mentioned above, we say that a topological space is *triangulable* if it is homeomorphic to the support of such a simplicial complex.

To define the homology groups $H_q(K)$ of a simplicial complex K, we first choose orientations for all simplices in K inducing order-preserving identifications of any p-simplex with the standard p-simplex. This is always possible, but not in a unique fashion. Fix some ring R. A p-chain is a finite R-linear combination of p-simplices. Two such chains can be added and multiplied by an element from R making the set of p-chains into an R-module $C_p(K)$. The boundary map

$$\delta_p : C_p(K) \to C_{p-1}(K)$$

is defined to be the R-linear extension that on simplices σ is induced by the map given on the standard p-simplex by

$$\delta_p(\Delta_p) = \sum_{q=0}^{p} (-1)^q \Delta_p^q, \quad \Delta_p^q = \Delta_p \cap \{x_q = 0\}.$$

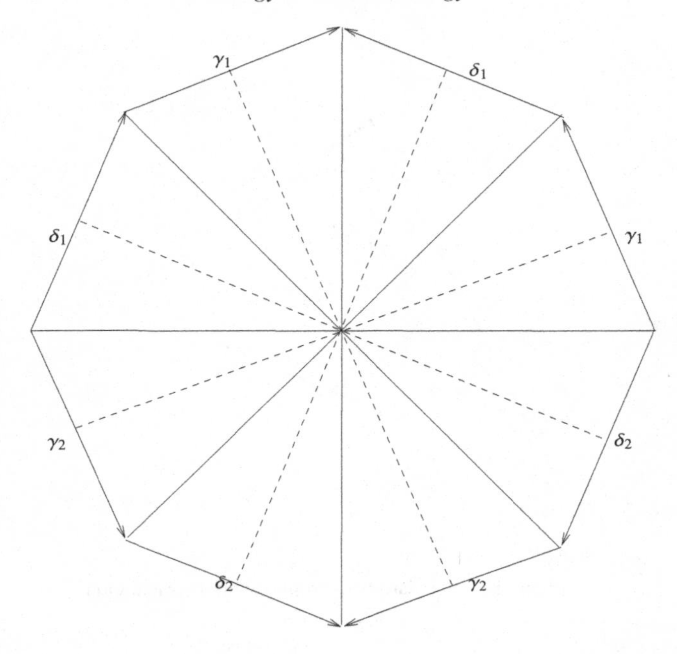

Figure B.2 Triangulated Riemann surface C_g for $g = 2$.

One verifies that $\delta \circ \delta = 0$ so that $C_\bullet(K)$ forms a complex. Its homology by definition is the *simplicial homology* group

$$H_p(K;R) \stackrel{\text{def}}{=} \frac{\operatorname{Ker}\left(C_p(K) \to C_{p-1}(K)\right)}{\operatorname{Im}\left(C_{p+1}(K) \to C_p(K)\right)}.$$

Simplicial maps $f : K \to L$ induce homomorphisms $H_p(f) : H_p(K;R) \to H_p((L;R)$.

For a triangulable space X, say homeomorphic to $|K|$, we then set

$$H_p(X;R) = H_p(K;R),$$

and it is a nontrivial fact that this group is independent of the chosen triangulation and the chosen orientation of the simplices.

The assignment $\{\text{Top. spaces } X\} \mapsto H_p(X;R)$ follows certain standard rules known as functoriality. See Eqs. (B.1), (B.2).

Examples B.1.1 (i) The n-sphere is homeomorphic to the boundary of the standard $(n+1)$-simplex. We leave it as an exercise to show that the only nonzero homology groups are H_0 and H_n, both infinite cyclic.

(ii) A compact genus g Riemann surface is an oriented closed topological surface C_g and it admits a triangulation (see Fig. B.2). From this, we can easily calculate that

$$H_0(C_g; R) = H_2(C_g; R) = R,$$
$$H_1(C_g; R) = R^{2g},$$

where the cycles $\gamma_1, \ldots, \gamma_g, \delta_1, \ldots, \delta_g$ give a basis for H_1.

(iii) It is a nontrivial theorem that any differentiable manifold admits a triangulation. Moreover, given a submanifold, we can choose the triangulation in such a way that the submanifold becomes a subcomplex (see Bibliographical Remarks at the end of this appendix). One can further show that any compact orientable differentiable manifold M of dimension m admits a triangulation whose m-simplices can be oriented in such a way that the sum over all of these is an m-cycle, the *orientation cycle* or fundamental class of M. It defines a generator $[M]$ of $H_m(M; R)$.

(iv) Let $N \subset M$ be an oriented compact n-dimensional differentiable submanifold of a given oriented differentiable manifold M. Choose a triangulation of M for which N becomes a subcomplex. Choosing appropriate orientations for the simplices, the fundamental class for N defines also an n-cycle for the homology of M and the inclusion $i : M \to N$ induces a map $H_n(i) : H_n(N; R) = R \cdot [N] \to H_n(M; R)$ whose image, the fundamental class of N in M, may or may not be zero.

(v) The complex projective space $\mathbf{P}^n$ only has homology in even degrees between 0 and $2n$. These groups are infinite cyclic. The generator in rank $2k$ is the fundamental class of $\mathbf{P}^k \subset \mathbf{P}^n$ viewed as the subspace where the last $(n-k)$ coordinates vanish.

For oriented differentiable manifolds M there is an intersection product

$$H_p(M) \otimes H_q(M) \to H_{p+q-m}(M), \quad m = \dim M$$

defined as follows. Given a p-cycle z and a q-cycle w, there are homologous cycles $z' \sim z$, $w' \sim w$ such that z' and w' are in general position in the sense that both are smooth submanifolds that intersect each other transversally in the correct dimension. Now choose triangulations such that $z' \cup w'$, z', and w' become subcomplexes. Then choose orientations for the simplices such that $[M]$ is exactly the given orientation class. Now the orientation of z' followed by that of w' is either compatible with the orientation of M or not. In the first case, we put $[z'] \cdot [w'] = [z' \cap w']$ and in the second case, we set $[z'] \cdot [w'] = -[z' \cap w']$.

This product has two obvious disadvantages. First, it requires sufficient room to deform cycles in general position and so *a priori* only works for manifolds.

Second, the product is not compatible with degrees in that the degrees do not add when we intersect. Both disadvantages disappear if we pass to cohomology, which is the theory dual to homology. We do this in a more general context, that of singular theory, which works on all topological spaces. It is a topological invariant by definition and so none of the above problems is present. For triangulable spaces it coincides with simplicial homology so we can freely make use of our above calculations.

B.2 Singular Theory

To define singular homology for a topological space X, we use singular p-simplices instead of ordinary simplices. A *singular p-simplex* is a continuous *map*[1] from the standard p-simplex to X:

$$\sigma_p : \Delta_p \to X.$$

As before, a p-chain is a linear combination of p-simplices, and two p-chains can be added and multiplied by an element from R, thereby giving the R-module $S_p(X)$. The embeddings

$$\begin{aligned} \iota^q : \Delta_{p-1} &\to \Delta_p, \\ (x_0, \ldots, x_p) &\mapsto (x_0, \ldots, x_{q-1}, 0, x_q, \ldots, x_p) \end{aligned}$$

yield boundary maps

$$\begin{aligned} \delta_p : S_p(X) &\to S_{p-1}(X), \\ \delta_p(\sigma) &\mapsto \sum_{q=0}^{p} (-1)^{q+1} \sigma \circ \iota^q . \end{aligned}$$

Again $S_\bullet(X)$ forms a *chain complex*, and its homology by definition is the *singular homology group*

$$H_p(X; R) \stackrel{\text{def}}{=} \frac{\operatorname{Ker}(S_p(X) \to S_{p-1}(X))}{\operatorname{Im}(S_{p+1}(X) \to S_p(X))}.$$

We often write $H_q(X)$ in place of $H_q(X; \mathbf{Z})$. Dually, a singular p-cochain is an R-module homomorphism $S_p(X) \to R$. The singular p-cochains form the R-module $S^p(X)$. The coboundary $\partial^{p-1} : S^{p-1}(X) \to S^p(X)$ is the R-transpose of the boundary map, $\partial^{p-1} = \delta_p^*$, yielding a *cochain complex* $\{S^\bullet, \partial^\bullet\}$ and

$$H^p(X; R) \stackrel{\text{def}}{=} \frac{\operatorname{Ker}(S^p(X) \to S^{p+1}(X))}{\operatorname{Im}(S^{p-1}(X) \to S^p(X))}$$

[1] Note that one does not identify this with its image.

is the p-th *singular cohomology* with coefficients in R, denoted $H^p(X; R)$. We write $H^p(X)$ instead of $H^p(X; \mathbf{Z})$.

We occasionally use *cohomology with compact support*, obtained by calculating the cohomology of the singular complex with compact support. In other words, we let $S_c^\bullet(X)$ be the subcomplex of the singular complex $S^\bullet(X)$ of cochains suppported on a compact subset of X, and then

$$H_c^p(X; R) = H^p(S_c^\bullet(X)).$$

If $A \subset X$ is any subset, the complex of singular chains on A forms a subcomplex of those on X, and the quotient complex by definition is $S_\bullet(X, A)$. Its homology is $H_\bullet(X, A; R)$. Taking duals, one defines $H^\bullet(X, A; R)$, the relative cohomology for the pair (X, A).

If $f : X \to Y$ is any continuous map, there are induced maps on every level of the preceding constructions. For homology we thus get $H_q(f) : H_q(X; R) \to H_q(Y; R)$, but for cohomology the arrows must be reversed:

$$H^p(f) : H^p(Y; R) \to H^p(X; R).$$

The rule that assigns to X the homology group $H_q(X; R)$ and to $f : X \to Y$ an induced morphism $H_q(f)$ of R-modules is called a (covariant) functor, which means

$$H_q(f \circ g) = H_q(f) \circ H_q(g), \tag{B.1}$$

$$H_q(\mathrm{id}) = \mathrm{id}\,. \tag{B.2}$$

More precisely, one says that homology is a covariant functor from topological spaces to R-modules. As for cohomology, we have a contravariant functor between the same categories. This means that the first rule (B.1) has been replaced by the rule $H^q(f \circ g) = H^q(g) \circ H^q(f)$.

Of course, there are also similar functors for pairs of topological spaces (X, A) with $A \subset X$. In all of these cases, if no confusion is likely, we denote the induced maps in cohomology by f^* in place of $H^p(f)$.

Tools for calculating cohomology

1. The long exact sequence associated with a short exact sequence of complexes yields the *long exact sequence for pairs:*

$$\cdots \to H^q(X, A; R) \xrightarrow{j^*} H^q(X; R) \xrightarrow{i^*} H^q(A; R) \xrightarrow{\delta} H^{q+1}(X, A; R) \to \cdots.$$

Here i^* is induced by the inclusion $i : A \hookrightarrow X$, and the homomorphism j^* comes from the inclusion of pairs $(X, \varnothing) \hookrightarrow (X, A)$. The definition of the connecting

homomorphism δ is somewhat involved. For this as well as the functoriality of the long exact sequence, see the Bibliographical Remarks at the end of this appendix.
2. The *homotopy invariance property*: homotopic maps induce the same maps in (co)homology. In particular, if $A \subset X$ is a deformation retract of X, the cohomology of X is that of A. Special cases are contractible spaces, such as a convex subset of $\mathbf{R}^n$: these have the cohomology of a point. The second tool is very powerful. It says that the relative cohomology of (X, A) does not change if we cut out suitable subsets U from A. This *excision* property thus says that inclusion induces isomorphisms

$$H^\bullet(X, A; R) \xrightarrow{\sim} H^\bullet(X - U, A - U; R).$$

Among the sets U that have this property are the open sets with closure contained in the interior of A and deformation retracts of those.
3. The exact *Mayer–Vietoris sequence,* a "patching tool," which enables one to compute cohomology of unions $X = X_1 \cup X_2$ of two open sets from the cohomology of X_1, X_2 and the intersection $X_1 \cap X_2$. With $i_k : X_k \to X$ and $j_k : X_1 \cap X_2 \to X_k$ $(k = 1, 2)$ the inclusion maps, the Mayer–Vietoris sequence reads as follows:

$$\cdots \longrightarrow H^q(X; R) \xrightarrow{(i_1^* + i_2^*)} H^q(X_1; R) \oplus H^q(X_2; R) \xrightarrow{j_1^* - j_2^*} H^q(X_1 \cap X_2; R) \longrightarrow H^{q+1}(X; R) \longrightarrow \cdots .$$

The reader is encouraged to calculate the cohomology groups of the sphere, a compact Riemann surface of genus g, and of the complex projective space using these tools. The results should be the same as we found in the examples in Section B.2.

The ring structure in cohomology

In what follows we omit reference to the coefficient ring. One first introduces a product on $\oplus_p S^p(X)$ by declaring how $f \cup g$ for $f \in S^p(X)$ and $g \in S^q(X)$ evaluates on a singular $(p + q)$-simplex $\sigma : \Delta \to X$: let σ_p be the singular p-simplex obtained by restricting σ on the standard-simplex spanned by the first $p+1$ unit vectors and let σ_q be obtained by restricting σ to the "complementary face" spanned by the last $q + 1$ unit vectors; then define

$$f \cup g(\sigma) = f(\sigma_p) \cdot g(\sigma_q).$$

The relation

$$\partial(f \cup g) = \partial f \cup g + (-1)^p f \cup \partial g$$

shows that this *cup product* induces cup products

$$H^p(X) \otimes H^q(X) \xrightarrow{\cup} H^{p+q}(X)$$

and it makes the direct sum $H^\bullet(X) = \oplus_q H^q(X)$ into a ring, the *cohomology ring* of X. If R has a unit 1, the cohomology ring also has a unit $1 \in H^0(X)$ given by the constant cochain $x \mapsto 1$, $x \in X$. One can show that this makes $H^\bullet(X)$ into a graded commutative ring, where "graded" means the commutation rule:

$$a \cup b = (-1)^{pq} b \cup a, a \in H^p(X), \ b \in H^q(X).$$

If $f : X \to Y$ is continuous, the induced homomorphism $f^* : H^\bullet(Y) \to H^\bullet(X)$ preserves cup products. Dual to cup products, we have *cap products*

$$S_{p+q}(X) \times S^p(X) \xrightarrow{\cap} S_q(X).$$

These are defined as follows: for a $(p+q)$-simplex σ and a p-cochain f we put

$$\sigma \cap f = f(\sigma_p)\sigma_q,$$

where as before σ_p is the "front face" and σ_q is the "back face" of σ.

These cap products can be shown to induce cap products for pairs (X, A):

$$\bigcap : H_{p+q}(X, A) \times H^p(X, A) \to H_q(X),$$
$$\bigcap : H_{p+q}(X, A) \times H^p(X) \to H_q(X, A).$$

Examples The coefficient ring is taken to be $\mathbf{Z}$ so the intersection ring has a unit as well. We come back to these examples after we have made the connection with our previously defined intersection products on manifolds.

(i) The ring structure on $H^\bullet(S^n)$ is very simple. It is generated as a ring (with 1) by the generator x of $H^n(S^n)$ and there is one obvious relation $x \cup x = 0$.

(ii) The ring structure on $H^\bullet(\mathbf{P}^n)$ is likewise rather simple. It is generated as a ring by a generator $h \in H^2(\mathbf{P}^n)$ with the (obvious) relation $h^{n+1} = 0$. What has to be verified is that $h^k \in H^{2k}(\mathbf{P}^n)$ generates the infinite cyclic group at hand (see Greenberg, 1967).

(iii) The compact oriented Riemann surface C_g. Using the dual of the ordered basis $\{\gamma_1, \ldots, \gamma_g, \delta_1, \ldots, \delta_g\}$, as in Fig. B.2, and identifying $H^2(C_g, \mathbf{Z})$ with $\mathbf{Z}$ (through the orientation), the cup product becomes the standard symplectic form

$$J = \begin{pmatrix} 0 & \mathbf{1}_g \\ -\mathbf{1}_g & 0 \end{pmatrix}.$$

There is no need to compute cohomology and homology separately, nor does one need to compute these for arbitrary coefficients. This follows from the "universal coefficient theorem." We shall not use the full scope of this theorem for which we refer the reader to the Bibliographic Remarks at the end of this appendix. In fact, the following suffices.

Facts B.2.1 (i) *The "evaluation" pairing between chains and cochains induces an isomorphism (here k is any field)*

$$H^q(X, A; k) \xrightarrow{\sim} \mathrm{Hom}_k(H_q(X, A); k),$$
$$[f] \mapsto \{[c] \mapsto f(c)\}.$$

(ii) *For any field k, the natural k-linear map*

$$H^q(X; \mathbf{Z}) \otimes k \to H^q(X; k)$$

is an isomorphism. In particular, the Betti numbers

$$b_q(X) = \mathrm{rank}\, H_q(X; \mathbf{Z})$$

are equal to the dimension of the vector spaces $H_q(X; k)$ or of $H^q(X; k)$.

B.3 Manifolds

For topological manifolds, the cohomology has extra structure. Because we consider only complex manifolds below, we treat here only the case of an oriented manifold. If M is an oriented n-dimensional compact connected manifold, the only nonzero homology is in the range $0, \ldots, n$. The class of a point generates $H_0(M; R) \cong R$, and the highest-degree homology group $H_n(M)$ is generated by the fundamental homology class $[M]$, so $H_q(M; R) \cong R$.

More generally, if $N \subset M$ is a submanifold of dimension m, one can speak of its *fundamental homology class* $i_*[N] \in H_m(M; R)$, the image of its own fundamental class under the homomorphism induced by the inclusion $i : N \to M$.

Theorem B.3.1 (Poincaré duality) *For a connected oriented n-dimensional manifold M there are isomorphisms*

$$D_M : H^q_c(M; R) \xrightarrow{\sim} H_{n-q}(M; R).$$

In particular, $H^q_c(M) = 0$ when $q > n$, and similarly $H_q(M) = 0$ when $q > n$. If M is compact, D_M is induced by the cap product with the fundamental homology class. Moreover, in this case, the cup product pairing on cohomology translates into the intersection pairing on homology.

This theorem implies that for compact manifolds M, one needs to know only half of the cohomology. Also, we now understand why the intersection pairing behaves better after passing to cohomology: it now becomes graded commutative.

Examples Let us now return to some of the examples in order to explain the ring structure in cohomology.

(i) Complex projective space $\mathbf{P}^n$. In homology, the generator of the intersection ring is the fundamental class $[H] \in H_{2n-2}(\mathbf{P}^n)$ of a hyperplane. Then $[H]^k$ is the fundamental class of the intersection of k hyperplanes in general position, i.e., a codimension k-plane. So the cohomology is generated by h, the Poincaré dual of $[H]$.
(ii) The standard cycles $\{\gamma_1, \ldots, \gamma_g, \delta_1, \ldots, \delta_g\}$ on the compact Riemann surface C_g intersect according to the standard symplectic matrix J and so the ring structure is as claimed above.

There is one more sophisticated tool to compute cohomology that is especially useful in the context of algebraic manifolds, although it is of a purely topological nature. It is the famous theorem of Lefschetz on hyperplane sections. This result follows from an equally important result about the topology of affine algebraic varieties.

An affine manifold, indeed any manifold, has no (co)homology beyond its real dimension. For affine manifolds a significantly stronger result is true: there is no cohomology beyond the *complex* dimension. This can be proved using Morse theory (see the Bibliographical Remarks at the end of this appendix).

Theorem B.3.2 *Let V be a smooth affine variety of dimension n. Then $H_k(V) = 0$ for $k > n$. Similarly, $H^k(V) = 0$ for $k > n$.*

A corollary is the fundamental Lefschetz theorem.

Theorem B.3.3 (Lefschetz hyperplane theorem) *Let M be projective variety of dimension $n+1$ and let $i : N \hookrightarrow M$ be any hyperplane section. Suppose that $M - N$ is smooth. Then*

$$i^* : H^k(M) \to H^k(N) \quad \begin{cases} \text{is an isomorphism} & \text{if } k \leq n-1, \\ \text{is injective} & \text{for } k = n. \end{cases}$$

Dually,

$$i_* : H_k(N) \to H_k(M) \quad \begin{cases} \text{is an isomorphism} & \text{if } k \leq n-1, \\ \text{is surjective} & \text{for } k = n. \end{cases}$$

Proof We need a relation between the homology of the affine variety $U := M - N$ and the cohomology of the pair (M, N) known as Lefschetz duality.

This is a sophisticated generalization of Poincaré duality, discussed in Spanier (1966, Chapter 6.2). In our setting this gives

$$H_k(U) = H^{2n+2-k}(M, N).$$

The previous result shows that $H_k(U) = 0$ whenever $k > n + 1$ and so $H^k(M, N) = 0$ whenever $k \leq n$, and the exact sequence for the pair (M, N) then gives the result in cohomology. The dual statement follows indirectly from this by using the "universal coefficient theorem" (Spanier, 1966, Chapter 5.5). □

This theorem implies that the cohomology of a hypersurface N in projective space is rather simple. Suppose that $\dim N = n$ so that there is only cohomology in the range $0, \ldots, 2n$. Then the cohomology is generated by linear sections except in the middle dimension n. In particular, $H^q(N) = 0$ for q odd, $q \neq n$, and $H^q(N) \cong \mathbf{Z}$ for q even, $q = 0, \ldots 2n$, $q \neq n$.

It is possible to compute the middle Betti number by computing the *Euler characteristic* $e(N) = \sum_q (-1)^q b_q(N)$. There are various ways to do this. The most elementary one is outlined in Problem B.5. A more sophisticated computation, using Chern classes, is explained in Appendix C.2.

Problems

B.1 Let $f : M \to N$ be a continuous map between compact oriented differentiable manifolds. Applying Poincaré duality to both sides of the map induced in homology, we obtain the *Gysin homomorphism* in cohomology: $f_* : H^\bullet(M) \to H^{\bullet+e}(N)$. Determine the shift e.

B.2 Show that the product of two simplices is triangulable, and derive from this that the product of two simplicial complexes can be triangulated.

B.3 Let X be a topological space admitting a triangulation and let N_q be the number of q-simplices in the triangulation. Prove that the Euler characteristic of X is equal to $\sum_q (-1)^q N_q$. As a hint, if we have a finite complex of finite-dimensional vector spaces

$$0 \to C_n \to C_{n-1} \to \cdots \to C_0 \to 0,$$

show that $\sum_q (-1)^q \dim C_q = \sum_q (-1)^q \dim H_q(C_\bullet)$.

B.4 Let X be a degree-d smooth plane curve and let $\pi : X \to \mathbf{P}^1$ be the projection of X from a point outside of X to a line. Using equations, determine where the map π is ramified. Using the dual curve, show that for generic choices of the projection center, only two points come together at a ramification point. Using a suitable triangulation, show that $e(X) = -d^2 + 3d$.

B.5 Consider the Fermat hypersurface $F_{d,n}$ in $\mathbf{P}^{n+1}$ with equation $x_0^d + \cdots + x_n^d = 0$. Show that the projection onto one of the coordinate hyperplanes exhibits $F_{d,n}$ as a d-fold cover of $\mathbf{P}^n$ with branch locus $F_{d,n-1}$. Deduce an inductive formula for the Euler characteristic of $F_{d,n}$. Show that

$$e(X) = d\left[\sum_{k=0}^{n}(-1)^k\binom{n+2}{k+2}d^k\right].$$

This in fact gives the Euler number for any smooth degree-d hypersurface in $\mathbf{P}^{n+1}$, because such surfaces are mutually diffeomorphic. Indeed, they appear as fibers of the tautological family and any family is locally differentiably trivial (see § 4.1 and 4.2).

B.6 The *Künneth decomposition* tells us how to express the cohomology of $X \times Y$ into the cohomology of the factors X and Y (see Spanier, 1966, p. 247). For fields k, this reads

$$H^q(X \times Y; k) \cong \bigoplus_{i=0}^{q} H^i(X; k) \otimes H^{q-i}(Y; k).$$

Use this to calculate the cohomology of a torus and of a product of two spheres.

Bibliographical and Historical Remarks

The book Greenberg (1967) is a solid introduction to (co)homology theory. A classical reference book is Spanier (1966). One finds a beautiful introduction to Morse theory in Milnor (1963), where also a proof of Theorem B.3.2 can be found. Among more recent books we mention Hatcher (2002).

Appendix C

Vector Bundles and Chern Classes

C.1 Vector Bundles

We recall the notion of a complex rank r vector bundle E over a differentiable manifold M.

Definition C.1.1 Let M be a differentiable manifold. A rank r *vector bundle* E over M consists of a differentiable manifold E, the total space of the bundle, and a differentiable map $p : E \longrightarrow M$, the projection, whose fibers are r-dimensional complex vector spaces. Moreover, one demands that there exists a trivializing open cover $\{U_i\}_{i \in I}$ of M, i.e., $p^{-1}U_i$ is diffeomorphic to a product $\varphi_i : p^{-1}U_i \xrightarrow{\sim} U_i \times \mathbf{C}^r$ in such a way that p corresponds to the projection onto the first factor. In addition, the map φ_i should send the fiber E_m over $m \in U_i$ linearly and isomorphically onto $\{m\} \times \mathbf{C}^r$. If $r = 1$ we speak of a *line bundle*.

Trivializations over nonempty intersections $U_i \cap U_j$ can be compared:

$$\varphi_{U_j}^{-1}(m, t) = \varphi_{U_i}^{-1}(m, \varphi_{ij}(m)t),$$

where $\varphi_{ij} : U_i \cap U_j \longrightarrow \mathrm{GL}(r, \mathbf{C})$ is differentiable and is called the transition function. These transition functions satisfy a certain compatibility rule,

$$\varphi_{ij} \circ \varphi_{jk} \circ \varphi_{ki} = \mathbf{1} \text{ (cocycle relation).}$$

Conversely, given some covering of M by open sets $U_i, i \in I$ and a collection of transition functions φ_{ij} for subsets U_i and U_j having a nonempty intersection, take the disjoint union of the $U_i \times \mathbf{C}^r$ and identify (m, t) and $(m, \varphi_{ij}(m)t)$ whenever $m \in U_i \cap U_j$. The above compatibility rule ensures that this yields an equivalence relation so that we may form the quotient E.

A *vector bundle homomorphism* between two vector bundles $p : E \longrightarrow M$ and $p' : F \longrightarrow M$ consists of a differentiable map $f : E \longrightarrow F$ such that

(i) $p = p' \circ f$ so that fibers go to fibers, and

(ii) $f|E_m$ is linear.

If f is a diffeomorphism we have a *vector bundle isomorphism.*

For any vector bundle homomorphism $f : E \longrightarrow F$, you can form the kernel $\mathrm{Ker}(f)$, which consists of the union of the kernels of $f|E_m$. If these have constant rank, it can easily be seen that $\mathrm{Ker}(f)$ is a vector bundle. Similarly, $\mathrm{Im}(f) = \bigcup_{m\in M} \mathrm{Im}(f|E_m)$ is a bundle under the same hypothesis, the image bundle.

We shall see that *exact sequences of vector bundles* arise naturally; a two-step sequence of vector bundle homomorphisms

$$E' \xrightarrow{f} E \xrightarrow{g} E''$$

is called exact at E if $\mathrm{Ker}(g) = \mathrm{Im}(f)$ and a sequence of vector bundles

$$\cdots \rightarrow E_{i-1} \xrightarrow{f_{i-1}} E_i \xrightarrow{f_i} E_{i+1} \rightarrow \cdots$$

of arbitrary length is called exact if it is exact at all E_i. Especially, a sequence

$$0 \rightarrow E' \xrightarrow{f} E \xrightarrow{g} E'' \rightarrow 0$$

is exact if and only if f is injective, g is surjective, and $\mathrm{Ker}(g) = \mathrm{Im}(f)$.

A *section* s of a vector bundle $p : E \longrightarrow M$ is a differentiable map $s : M \longrightarrow E$ such that $p \circ s = \mathrm{id}_M$. Sections of a vector bundle E form a vector space denoted by $\Gamma(E)$ or $H^0(M, E)$.

Examples C.1.2 (i) The trivial bundle $M \times \mathbf{C}^r$.

(ii) The complexified tangent bundle $T_M^{\mathbf{C}}$ of a differentiable manifold M.

(iii) If E is a bundle, a subbundle consists of a subset $F \subset E$ such that the projection and trivialization of E gives F the structure of a bundle. For a subbundle $F \subset E$, the fibers F_m are subspaces of E_m and one can form $\bigcup_{m\in M} E_m/F_m$ which inherits the structure of a bundle E/F, the quotient bundle. This can be seen by looking at the transition functions for E which can be written as

$$\left(\begin{array}{c|c} \phi' & * \\ \hline 0 & \phi'' \end{array}\right),$$

where ϕ' is the transition matrix for F and ϕ'' is the one for E/F. This also shows that for any exact sequence

$$0 \rightarrow E' \xrightarrow{f} E \xrightarrow{g} E'' \rightarrow 0,$$

the map f identifies E' with a subbundle of E, and g induces an isomorphism of E/E' with E''.

(iv) If E is a vector bundle, any linear algebra construction on the fibers yields a vector bundle. We already saw the examples of subbundles and quotient bundles. One can also form $E^\vee$, the dual bundle, by taking the transpose inverse transition functions. Similarly, we have the tensor powers $\bigotimes^k E$ and their subbundles, the exterior powers $\bigwedge^k E$ and the symmetric powers $S^k E$. The highest wedge with $k = \operatorname{rank}(E)$ is also called the determinant line bundle

$$\det(E) = \bigwedge^{\operatorname{rank} E} E.$$

Combining these operations and applying them to the previous example, you get the cotangent bundle or bundle of one-forms and its k-fold exterior power, the bundle of k-forms:

$$\mathscr{E}^k_M = \bigwedge^k T^\vee_M.$$

Sections in the bundle of k-forms are precisely the k-forms.

(v) Likewise, if E and F are two bundles, one can form their direct sum $E \oplus F$ and their tensor product $E \otimes F$.

(vi) The tangent bundle T_N of a submanifold N of a manifold M is a subbundle of the restriction $T_M|N$ of the tangent bundle of M to N. The quotient $(T_M|N)/T_N$ is called the *normal bundle* and denoted by $\nu_{N/M}$.

(vii) If $\varphi : M \to N$ is a differentiable map and $p' : F \to N$ a vector bundle, the *pullback bundle* $\varphi^* F$ is defined as follows. Its total space consists of the pairs $(m, f) \in M \times F$ with $\varphi(m) = p'(f)$. Projection comes from projection onto the first factor. One may verify that the trivialization of N induces one for $\varphi^* F$.

(viii) Consider the subbundle of the trivial bundle with fiber $\mathbf{C}^{n+1}$ on the projective space $\mathbf{P}^n$ consisting of pairs $([w], z) \in \mathbf{P}^n \times \mathbf{C}^{n+1}$ with z belonging to the line defined by $[w]$. This is a line bundle, the tautological line bundle $\mathcal{O}(-1)$. See Example C.1.6 for the explanation of the minus sign which is used here.

(ix) Given an exact sequence

$$0 \to E' \xrightarrow{f} E \xrightarrow{g} E'' \to 0,$$

there is an isomorphism

$$\det E' \otimes \det E'' \xrightarrow{\sim} \det E.$$

If M is a complex manifold, a complex bundle E is called a holomorphic vector bundle if in the definition of a complex vector bundle one demands that E be a complex manifold and that the differentiable maps involved be holomorphic. All of the constructions of the previous examples do produce holomorphic bundles out of holomorphic bundles. This holds, for example, for the holomorphic tangent bundle T_M for a complex manifold M.[1] So its dual, the cotangent bundle, has the structure of a holomorphic vector bundle as well as the exterior wedges thereof, which now are denoted as follows:

$$\Omega^k_M = \bigwedge^k T^\vee_M.$$

The line bundle $\det \Omega^1_M$ is called the *canonical line bundle*and is sometimes denoted by K_M. If N is a complex submanifold of M, the normal bundle $\nu_{N/M}$ of N in M is a holomorphic bundle. Applying Example C.1.2 (ix), one arrives at an important formula:

$$K_N \cong K_M \mid N \otimes \det \nu_{N/M} \quad (\textit{canonical bundle formula}). \tag{C.1}$$

Example C.1.3 The collection of holomorphic line bundles on a complex manifold M modulo isomorphisms forms a group under the tensor product. It is called the *Picard group* and denoted by $\operatorname{Pic} M$.

Related to any codimension-1 subvariety D of a complex manifold M we have the line bundle $\mathcal{O}(D)$ on M defined by means of transition functions as follows. Choose a coordinate covering $U_i, i \in I$ of M in which D is given by the equation $f_i = 0$. In $U_j \cap U_j$ the relation $f_i =$ (a nonzero function φ_{ij}) $\cdot f_j$ enables one to form the line bundle given by the transition functions $\varphi_{ij} = f_i/f_j$. (Note that the functions φ_{ij} obviously satisfy the cocycle relation.) Observe that the bundle $\mathcal{O}(D)$ always has a section s_D canonically defined by D. Indeed, over U_i the bundle is trivial and the function f_i defines a section over it. These patch to a section s_D of $\mathcal{O}(D)$ because $f_i = (f_i/f_j)f_j$ in $U_i \cap U_j$. Restricting the bundle $\mathcal{O}(D)$ to D itself, in the case when D is a submanifold, one gets back the normal bundle $\nu_{D/M}$ (see Problem C.1.4). The canonical bundle formula in this case therefore reads

$$K_D \cong (K_M \otimes \mathcal{O}(D))|D \quad (\textit{adjunction formula}). \tag{C.2}$$

By definition a *divisor* is a formal linear combination $\sum_{i=1}^m n_i D_i$ with $n_i \in \mathbf{Z}$ and D_i a codimension-1 subvariety. If the numbers n_i are nonnegative the divisor is called effective. Divisors on M form an abelian group $\operatorname{Div} M$. The line bundle $\mathcal{O}(D)$ is defined by setting $\mathcal{O}(D) = \mathcal{O}(D_1)^{\otimes n_1} \otimes \cdots \otimes \mathcal{O}(D_m)^{\otimes n_m}$ so that it yields a homomorphism $\operatorname{Div} M \to \operatorname{Pic} M$. The trivial bundle on M is therefore denoted $\mathcal{O}_M$ or simply $\mathcal{O}$ if no confusion is likely.

Let us next describe how divisors behave under holomorphic maps $f : M' \to M$. Let g be a local defining equation for D in M. If the image of f avoids the support of D, the function $g \circ f$ is nowhere 0, but otherwise, $g \circ f$ defines a divisor on M' that is independent of the choice of the local defining equation

[1] We identify T_M with the subbundle T'_M of the complexified tangent bundle $T^{\mathbf{C}}_M$.

for D. It is called the *pullback* f^*D. It is related to the pullback of the line bundle $\mathcal{O}(D)$ by means of the relation $\mathcal{O}(f^*D) = f^*(\mathcal{O}(D))$.

In the framework of holomorphic bundles $E \longrightarrow M$, the group of holomorphic sections is also denoted by $\Gamma(E)$ or $H^0(M, E)$. It is true, but by no means trivial, that for compact complex manifolds the space of sections is finite dimensional (see the Bibliographical Remarks at the end of this appendix).

We now turn to vector bundles on projective manifolds. Note that one could have defined algebraic vector bundles using morphisms instead of holomorphic maps. Algebraic vector bundles are holomorphic. The converse is true over a projective manifold. This is the GAGA principle (the terminology is explained in the Bibliographical Remarks at the end of this appendix).

Since a projective manifold is compact, the space of sections of any algebraic bundle on it is finite-dimensional, as we have seen before. If L is a line bundle on a projective manifold M, and its space of sections is not 0, say $(n+1)$-dimensional with basis $x_0, \ldots, x_n$, one can define a rational map

$$\varphi_L : M \dashrightarrow \mathbf{P}^n$$

by associating with $m \in M$ the point $(x_0(m) : \cdots : x_n(m)) \in \mathbf{P}^n$. This map is not defined on the locus where all sections of L vanish. This locus is called the *base locus* and any point in it is called a *base point*. If φ_L is an embedding, the line bundle L is called very ample. If for some integer $k \geq 1$ the k-th tensor power $L^{\otimes k}$ is very ample, L is said to be *ample*.

Definition C.1.4 Let X be a complex compact manifold.
(i) For $m \geq 1$ the m-th plurigenus is defined by $P_m(X) = H^0(X, K_X^{\otimes m})$. If all plurigenera vanish, the Kodaira dimension $\kappa(X)$ is said to be $-\infty$.
(ii) If not all plurigenera vanish, consider the pluricanonical map

$$\varphi_m \stackrel{\text{def}}{=} \varphi_{K_{X^{\otimes m}}} : X \dashrightarrow \mathbf{P}^N, \quad N = P_m(X) - 1.$$

The Kodaira dimension of X is defined as

$$\kappa(X) = \limsup_{m\to\infty} \dim \operatorname{Im}(\varphi_m).$$

As a consequence, $\kappa(X) \in \{\infty, 0, 1, \ldots, \dim(X)\}$.

Examples C.1.5 (i) $\kappa(\mathbf{P}^n) = -\infty$; $\kappa(X \times \mathbf{P}^n) = -\infty$,
(ii) $\kappa(X \times Y) = \kappa(X) + \kappa(Y)$.
(iii) note that $\kappa(X) = 0$ means that $P_m(X) \leq 1$ but not all plurigenera vanish. In particular, if K_X is trivial $\kappa(X) = 0$. Examples: X a complex torus or X a $K3$ surface.
(iv) if K_X is ample, for some m, the map φ_m gives an embedding and so $\kappa(X) = \dim(X)$.

The definition of divisors and of the Picard group for projective manifolds can be modified in the obvious way by using projective codimension-1 subvarieties instead.

Examples C.1.6 (i) The hyperplane in $\mathbf{P}^n$ defines an algebraic line bundle, the hyperplane bundle $\mathcal{O}(1)$. The tautological bundle is the dual of this bundle, which explains the notation $\mathcal{O}(-1)$ for the tautological bundle. The line bundle $\mathcal{O}(d)$ is defined as $\mathcal{O}(1)^{\otimes d}$ for $d > 0$ and as $\mathcal{O}(-1)^{\otimes -d}$ if $d < 0$. The line bundle associated with a hypersurface of degree d is (isomorphic to) $\mathcal{O}(d)$ (see Problem C.1.8).

(ii) The canonical line bundle of projective space $\mathbf{P}^n$ is isomorphic to $\mathcal{O}(-n-1)$ (see Problem C.1.6). Using the canonical bundle formula one finds that the canonical bundle for a smooth degree-d hypersurface D in $\mathbf{P}^n$ is the restriction to D of $\mathcal{O}(d-n-1)$.

Any polynomial P which does not vanish identically on M defines a divisor (P) on M. Any rational function $f = P/Q$ defines the divisor $(f) = (P) - (Q)$. Because one can represent rational functions on M in different ways, it is not *a priori* clear that this definition makes sense. We shall explain how this follows from the fact that the ring of holomorphic functions near the origin in $\mathbf{C}^n$ is a unique factorization domain (Griffiths and Harris, 1978, p. 10). So, let $f \in \mathbf{C}(M)$ be a rational function on M and D be an irreducible hypersurface. Let $p \in M$ and let $f_D = 0$ be a local equation for D at p. The ring $\mathcal{O}_{M,p}$ of germs of holomorphic functions at p being a unique factorization domain, we can write

$$f = f_D^m \cdot (u/v)$$

with u and v not identically 0 along D. It is easily verified that m does not depend on f_D and the chosen point $p \in M$ so that one can now unambiguously define m to be the order of vanishing of f along D, denoted $\mathrm{ord}_D(f)$, and one introduces the divisor of the rational function f by

$$(f) = \sum_{D \text{ an irreducible hypersurface}} \mathrm{ord}_D(f) D.$$

This makes the preceding definition precise.

Divisors of rational functions form the subgroup of *principal divisors* inside $\operatorname{Div} M$. Two divisors D and D' are said to be *linearly equivalent*, $D \equiv D'$, if their difference is the divisor of a rational function. Equivalent divisors define isomorphic line bundles and hence there is a well-defined map

$$\operatorname{Div} M/\text{principal divisors} \longrightarrow \operatorname{Pic} M.$$

This is in fact an isomorphism. That it is injective is not so difficult to show (see

Problem C.1.5). The surjectivity is not entirely trivial. It is a consequence of the fact that for a line bundle L on a *projective* manifold M, the twisted bundle $L \otimes \mathcal{O}(k)$ becomes ample for large enough k. This in turn is a consequence of Kodaira's embedding theorem (Griffiths and Harris, 1978, p. 181).

Rational functions f with the property that $(f) + D$ is effective form a vector space traditionally denoted by $\mathcal{L}(D)$. The resulting effective divisors $(f) + D$ linearly equivalent to D form a projective space $|D|$, which is nothing but $\mathbf{P}\mathcal{L}(D)$. Any projective subspace of $|D|$ is called a *linear system of divisors*, whereas $|D|$ itself is called a *complete linear system.*

Let us now return to the rational map defined by the line bundle $\mathcal{O}(D)$ associated with D. This can be generalized by taking a basis $\{s_0, \ldots, s_n\}$ for any linear subspace W of $\Gamma(\mathcal{O}(D))$, and the rational map $p \mapsto (s_0(p), \ldots, s_n(p))$ then is said to be given by the linear system $\mathbf{P}(W)$. A *fixed component* of the linear system $\mathbf{P}(W)$ is any divisor F which occurs as a component of all divisors in $\mathbf{P}(W)$. The map defined by taking away this fixed part then is the same. The resulting divisors form the *moving part* of $\mathbf{P}(W)$ and now there still can be *fixed points*, although they form at most a codimension-2 subspace.

The notion of ampleness has been introduced in connection with line bundles. A divisor D is called ample if the corresponding line bundle $\mathcal{O}(D)$ is ample.

If D is a hypersurface, the line bundle $\mathcal{O}(D)$ has a section s vanishing along D, and if $f \in \mathcal{L}(D)$, the product $f \cdot s$ is in a natural way a section of $\mathcal{O}(D)$, and every section can be obtained in this way (see Problem C.1.5). So

$$\mathcal{L}(D) \xrightarrow{\cdot s} H^0(\mathcal{O}(D)) \text{ is an isomorphism.}$$

We finish this chapter with an important theorem.

Theorem C.1.7 (Bertini's theorem) *A generic hyperplane section of a smooth projective variety is smooth.*

Proof Let $X \subset \mathbf{P}^n$ be a smooth projective variety and let $(\mathbf{P}^n)^\vee$ be the dual projective space of hyperplanes of $\mathbf{P}^n$. Inside $X \times (\mathbf{P}^n)^\vee$, consider the set B consisting of pairs (x, H) such that the embedded projective tangent space $\mathrm{Tan}_{X,x}$ to X at x and H are *not* transversal, i.e., such that $\mathrm{Tan}_{X,x} \subset H$. If $\dim X = k$, the possible hyperplanes with this "bad" behavior form a projective space parametrized by some $\mathbf{P}^{n-k-1}$ disjoint from $\mathrm{Tan}_{X,x}$. So the projection $B \to X$ realizes B as a projective bundle over X and hence is a variety of dimension $k+n-1-k = n-1$. Consequently, the projection of B into $(\mathbf{P}^n)^\vee$ is not surjective. The complement of this variety parametrizes the "good" hyperplanes. □

Problems

C.1.1 Let

$$0 \to E' \xrightarrow{f} E \xrightarrow{g} E'' \to 0$$

be an exact sequence of vector bundles over a manifold M. Introduce the subbundle F^r of $\bigwedge^k E$ whose fiber over $m \in M$ is the subspace generated by the wedges of the form $e_1 \wedge e_2 \cdots \wedge e_k$ with r of the e_j in $f(E')_m$. Prove that F^{r+1} is a subbundle of F^r and that g induces an isomorphism

$$F^r/F^{r+1} \xrightarrow{\sim} \bigwedge^r E' \otimes \bigwedge^{k-r} E''.$$

In particular, one has an isomorphism

$$\det E' \otimes \det E'' \xrightarrow{\sim} \det E.$$

C.1.2 Let $E \to M$ be a complex vector bundle over a differentiable manifold M. A Hermitian metric h on E consists of a differentiably varying Hermitian metric in the fibers of E. Prove that every complex bundle on a (paracompact) differentiable manifold admits a Hermitian metric.

C.1.3 Let

$$0 \to S \to E \to Q \to 0$$

be an exact sequence of vector bundles on a (paracompact) differentiable manifold. Prove that this sequence splits. Hint: use Hermitian metrics to identify Q with the orthogonal complement of S in E.

C.1.4 Prove that the normal bundle for a smooth complex hypersurface D of M is isomorphic to the restriction of $\mathcal{O}(D)$ to D.

C.1.5 Prove that two divisors D and D' on a projective manifold give isomorphic line bundles if and only if the divisors are linearly equivalent.

C.1.6 Prove that the canonical bundle of $\mathbf{P}^n$ is isomorphic to $\mathcal{O}(-n-1)$.

C.1.7 Prove that any hypersurface in $\mathbf{P}^n$ is linearly equivalent to dH, where d is the degree of the hypersurface and H is a hyperplane. Deduce that $\operatorname{Pic} \mathbf{P}^n \cong \mathbf{Z}$.

C.1.8 Prove that for a smooth hypersurface D of degree d in $\mathbf{P}^n$, the normal bundle is given by $\nu_{D/\mathbf{P}^n} = \mathcal{O}(d)|D$. Calculate the canonical bundle of D and $\dim H^0(K_D)$.

C.1.9 Prove that the tangent space of the Grassmannian $G(k, V)$ at a point $[W]$ is canonically isomorphic to $\operatorname{Hom}(W, V/W)$ and hence $T_{G(k,V)} \cong \operatorname{Hom}(S, Q)$, where S and Q are the tautological subbundle and quotient bundle, respectively.

C.1.10 Let $M = M_1 \times M_2$ and let $p_1 : M \longrightarrow M_1$ and $p_2 : M \longrightarrow M_2$ be the projections onto the factors.

(a) Let $\mathcal{V}_1$ be a vector bundle on M_1 and $\mathcal{V}_2$ a vector bundle on M_2. There is a natural homomorphism

$$H^0(M_1, \mathcal{V}_1) \otimes H^0(M_2, \mathcal{V}_2) \to H^0(M_1 \times M_2, p_1^*\mathcal{V}_1 \otimes p_2^*\mathcal{V}_2).$$

Show that this is an isomorphism. Hint: restrict a section s of $p_1^*\mathcal{V}_1 \otimes p_2^*\mathcal{V}_2$ to the fiber $p_1^{-1}(x)$. This yields a section $s(x) \in H^0(M_2, \mathcal{V}_2) \otimes (\mathcal{V}_1)_x$ depending holomorphically on x.

(b) Prove that $\Omega^1_M = p_1^*\Omega^1_{M_1} \oplus p_2^*\Omega^1_{M_2}$ and that $K_M = p_1^*K_{M_1} \otimes p_2^*K_{M_2}$.

C.1.11 Let D be a projective hypersurface of the projective manifold M and let s_D be a regular section of $\mathcal{O}(D)$ vanishing along D. Let $f \in \mathcal{L}(D)$. Prove that $f \cdot s$ is a regular section of $\mathcal{O}(D)$ and that any regular section of $\mathcal{O}(D)$ is obtained in this way.

C.1.12 The tangent bundle of $\mathbf{P}^n$ can be described by means of the *Euler sequence*

$$0 \to \mathcal{O} \to \bigoplus^{n+1} \mathcal{O}(1) \to T_{\mathbf{P}^n} \to 0.$$

The first map is defined by sending 1 to $(z_0, \ldots, z_n)$, and the second map associates with $(L_0, \ldots, L_n)$, the vector field $\sum_{j=0}^n L_j(\partial/\partial z_j)$. Show that this last map is well defined and prove that the Euler sequence is indeed exact.

C.2 Axiomatic Introduction of Chern Classes

With any complex bundle E on a manifold M one associates Chern classes $c_i(E) \in H^{2i}(M)$, assembled in a total Chern class

$$c(E) = 1 + c_1(E) + \cdots + c_m(E) \in H^\bullet(M), m = \dim M.$$

These obey the following axioms.

Naturality The Chern classes of the pullback bundle f^*E under a smooth map $f : N \to M$ are precisely the Chern classes $f^*c_i(E)$ of E pulled back to N.

Whitney sum formula $c(E \oplus F) = c(E) \cup c(F)$.

Calibration $c_1(\mathcal{O}_{\mathbf{P}^1}(1)) \in H^2(\mathbf{P}^1)$ is the generator h we have seen before.

Because every short exact sequence of vector bundles splits differentiably, the Whitney sum formula implies that for any subbundle S of E with quotient Q one has

$$c(E) = c(S) \cup c(Q).$$

For a complex manifold X the tangent bundle T_X is a complex bundle, and so it makes sense to speak of its Chern classes:

$$c_i(X) = c_i(T_X) \in H^{2i}(X).$$

The highest Chern class $c_n(X)$, $n = \dim X$, has a special interpretation.

Lemma C.2.1 *Let X be an n-dimensional (connected) complex manifold; then*

$$\begin{aligned} c_n(X) &= e(X) \text{ (Euler characteristic of } X) \\ &= 1 - b_1(X) + b_2(X) + \cdots - b_{2n-1}(X) + 1. \end{aligned}$$

For a proof, see Milnor and Stasheff (1974, pp. 130 and 158).

From this axiomatic setup we can calculate the Chern classes of several basic examples from projective geometry.

Examples C.2.2 (i) The Euler sequence (see Problem C.1.1) shows that $c(\mathbf{P}^n) = (1 + h)^{n+1}$, from which we find the Chern classes of $\mathbf{P}^n$ easily.

(ii) For a curve C of genus g we have $c_1(C) = (2 - 2g)[C]$.

(iii) For a product $X \times Y$ we have $c(X \times Y) = p^*c(X) \cup q^*c(Y)$, where p, q are the obvious projections.

(iv) If T_X is trivial as for a torus, we have $c(X) = 1$.

To calculate the Chern classes of bundles produced from E or E, F by means of a linear algebra construction, there is the following useful principle.

Theorem (Splitting principle) *Let E be a complex bundle over a manifold M. There exists a manifold N and a differentiable map $f : N \to M$ such that*

(i) *f^*E splits as a direct sum of line bundles, and*

(ii) *$f^\bullet : H^\bullet(M) \to H^\bullet(N)$ is injective.*

See Hirzebruch (1966, Lemma 4.4.1) for a proof. This lemma together with naturality imply that we formally may write any vector bundle as a direct sum of line bundles and use the Whitney sum formula to find the Chern class of any new bundle obtained by a linear algebra construction. See Problems C.2.2, and C.2.4 for applications.

So far we have said nothing about the existence of Chern classes. In fact, there are at least three approaches, each with its own merits. The first is through obstruction theory and the central tool is the Euler class in the top cohomology whose degree calculates the Euler characteristic. This approach can be found in Milnor and Stasheff (1974). This leads, for instance, naturally to the interpretation of the Euler number as the highest Chern class of the tangent bundle, as mentioned above.

Then we have Grothendieck's approach, which works well in algebraic geometry. We treat this approach here only briefly, but see Husemoller (1966) for the details. The third approach using connections on bundles is treated in more detail in the next section.

Grothendieck's approach to the Chern class of a complex rank r vector bundle E on a manifold M is based on properties of the cohomology of the projective bundle

$$\pi : P = \mathbf{P}(E) \to M,$$

associated with E. This is the bundle obtained upon replacing the fiber E_x above x by the projective space $\mathbf{P}(E_x)$ of lines in E_x. Since $\pi^* : H^\bullet(M) \to H^\bullet(P)$ is injective, we may identify $H^\bullet(M)$ with $\pi^* H^\bullet(M)$. The fundamental observation is as follows: with u the first Chern class of the tautological line bundle over P, there is a degree r polynomial relation for u in $H^\bullet(P)$ whose coefficients are (up to sign) the Chern classes of E:

$$u^r - c_1(E)u^{r-1} + \cdots + (-1)^r c_r(E) = 0.$$

Grothendieck turned this around by using this as the definition of the Chern classes and showed that these satisfy the axioms. This is a consequence of the following description of the ring:

$$\begin{aligned} H^\bullet(P) &= H^\bullet(M)[u]/(u^r - c_1(E)u^{r-1} + \cdots + (-1)^r c_r(E)) \\ &\cong \pi^* H^\bullet(M) \oplus \pi^* H^{\bullet-2}(M)u \oplus \cdots \oplus \pi^* H^{\bullet-2r+2}(M)u^{r-1}, \quad \text{(C.3)} \end{aligned}$$

where the last isomorphism is only additive.

Let us finish this section by giving two examples of how Chern classes can be used.

Examples C.2.3 (i) We first explain how we can calculate the Betti numbers of smooth hypersurfaces $X \subset \mathbf{P}^{n+1}$. The first remark is that we only need to compute the middle Betti number $b_n(X)$, since by Poincaré duality we have $b_{n+k}(X) = b_{n-k}(X)$, which in turn equals $b_{n-k}(\mathbf{P}^{n+1})$ by the Lefschetz hyperplane theorem B.3.3. Equivalently, we may compute the Euler number $c_n(X) = e(X) = -b_n(X)+n+1$ if n is odd and $= b_n(X)+n$ if n is even. Because the normal bundle of X equals $\mathcal{O}(d)|X$, where $d = \deg X$, the Whitney sum formula yields

$$c(X) = (1 + h)^{n+2}/(1 + dh)$$

and we need the coefficient of h^n. This can be done by the usual residue calculus. We find (see Problem C.2.1)

$$c_n(X) = \left[\sum_{k=0}^{n} (-1)^k \binom{n+2}{k+2} d^k\right] h^n.$$

Compare this with Problem B.5.

(ii) Consider a compact complex manifold M with a closed submanifold $i : N \hookrightarrow M$ of codimension r and let

$$\sigma : \mathrm{Bl}_N M \to M$$

be the blowup with center N. Finally, let $j : E \hookrightarrow \mathrm{Bl}_N M$ be the exceptional locus. Since $E = \mathbf{P}(\nu_{N/M})$, we can apply (C.3) to get the cohomology $H^\bullet(E)$. We want to express it in cohomology of N. We first observe that the map σ induces a surjection on the level of homology because we can move any cycle z on M into general position so as to meet N transversally (if at all). Then consider the closure $\bar{z}$ inside $\mathrm{Bl}_N M$ of $z - z \cap N$ considered on $\mathrm{Bl}_N M$. Obviously, the class of $\bar{z}$ maps to the class of z. Now the kernel of $\sigma_* : H_q(\mathrm{Bl}_N M) \to H_q(M)$ consists of cycles entirely supported on the exceptional locus: $\mathrm{Ker}(\sigma)_* = \mathrm{Im}\, j_* H_q(E)$. Dually, σ induces thus an injection on the level of cohomology and we have

$$H^\bullet(\mathrm{Bl}_N M) = \sigma^* H^\bullet(M) + j_* H^{\bullet-2}(E), \tag{C.4}$$

where j_* is the (injective) Gysin homomorphism. Identifying $H^\bullet(N)$ with $\sigma^* H^\bullet(N)$, we now write

$$H^{\bullet-2}(E) = V(M,N) \oplus H^{\bullet-2r}(N)u^{r-1} \tag{C.5}$$

with

$$V(M,N) = H^{\bullet-2}(N) \oplus H^{\bullet-4}(N)u \oplus \cdots \oplus H^{\bullet-2r+2}(N)u^{r-2}.$$

This last space corresponds under Poincaré duality to cycles mapping to cycles of strictly lower dimension in N and so $\sigma_* V(M,N) = 0$. The remaining part in the decomposition (C.5) maps isomorphically to $H^{\bullet-2r}(N)$ under the Gysin map σ_*, and composing with the Gysin map i_* we find that $j_*(H^{\bullet-2r+2}(N)u^{r-2}) = i_* H^{\bullet-2r}(N) \subset H^\bullet(M)$ is exactly the intersection of the two summands in the decomposition (C.4). So we find

$$H^\bullet(\mathrm{Bl}_N M) = \sigma^* H^\bullet(M) \oplus V(M,N).$$

Problems

C.2.1 Complete the residue calculation giving the Euler number of a hypersurface in projective space. Hint: calculate the residue at $z = 0$ of the rational function

$$\frac{(1+z)^{n+2}}{z^{n+1}(1+d \cdot z)}.$$

C.2.2 Express the Chern classes of the dual bundle in the Chern classes of the original bundle.

C.2.3 Show that $c_1(\det E) = c_1(E)$.

C.2.4 Express $c(\mathrm{Hom}(E, E))$ in terms of the Chern classes of E.

C.3 Connections, Curvature, and Chern Classes

A connection ∇ on a (real or complex) vector bundle E is a rule to differentiate sections. If ξ is a vector field on the underlying manifold and s a section of E we denote the derivative in the direction of ξ by $\nabla_\xi(s)$. As any ordinary derivative, the resulting map D_ξ on sections is linear and if we multiply with a smooth function, the Leibniz rule $\nabla_\xi(f \cdot s) = \xi(f) \cdot s + f \cdot \nabla_\xi s$ holds. Because these notions are purely local, we can even consider germs of sections and vector fields, which motivates introducing the bundles $\mathscr{A}^p(E)$ of smooth p-forms with values in E. Then the preceding discussion leads to the following definition.

Definition C.3.1 A *connection* on a vector bundle E is an operator

$$\nabla : \mathscr{A}^0(E) \to \mathscr{A}^1(E),$$

which is linear in the sense that $\nabla(s + s') = \nabla(s) + \nabla(s')$ for any two (germs of) sections s, s' of E and which satisfies the Leibniz rule:

$$\nabla(f \cdot s) = \mathrm{d}f \cdot s + f \cdot \nabla s$$

for any germ of a smooth function f on the underlying manifold and any germ of a section s of E.

The directional derivative of s with respect to a tangent vector ξ, or the *covariant derivative* in the direction ξ, then is

$$\nabla_\xi s \stackrel{\text{def}}{=} \xi(\nabla s).$$

Thus, once a connection operator is defined, it is possible to do calculus on (germs of) sections of vector bundles. Sections satisfying $\nabla s = 0$ are said to be *parallel* and are regarded as "constant."

Example C.3.2 (i) $M = \mathbf{R}^m$, $E = \mathbf{R}^m \times \mathbf{R}^r$. We take for our connection simply differentiation on vector-valued functions. Then ∇_ξ is just the usual directional derivative in the direction of ξ. More generally, we can take $\nabla = d + A$ with A any $r \times r$ matrix of one-forms.

(ii) Take any m-dimensional manifold M and let E be the trivial bundle with fiber $\mathbf{R}^r$. Let $s_1, \ldots, s_r$ be a frame for this bundle, i.e., sections which give a basis in each fiber, and write

$$\nabla s_j = \sum_i A_{ij} s_i$$

so that $A = (A_{ij})$ is some $r \times r$ matrix of one-forms. In the given frame, ∇ is then completely determined by A. In fact, identifying a section s with the column vector $\vec{s}$ of its coordinates, we get

$$\nabla \vec{s} = \mathrm{d}\vec{s} + A\vec{s}.$$

Every vector bundle E being trivial over suitably small open sets U, one can write

$$\nabla|_U = \mathrm{d} + A_U,$$

where A_U is called the (local) *connection matrix*.

(iii) The general situation. One has to patch the connection matrices A_U and A_V over two open sets U and V over which the bundle E is trivial. The transition matrix g_{UV} relating the frames $\{s_1^U, \ldots s_r^U\}$ and $\{s_1^V, \ldots s_r^V\}$ is defined as follows. If $\vec{s}_U$ and $\vec{s}_V$ are the column vectors for the section s in these two frames, then

$$\vec{s}_U = g_{UV}\vec{s}_V, g_{UV} \in \mathscr{A}^0(U \cap V, \mathrm{GL}(r)).$$

Because $(d + A_U)\vec{s}_U = g_{UV}(d + A_V)\vec{s}_V$, we find

$$A_V = g_{UV}^{-1} \circ A_U \circ g_{UV} + g_{UV}^{-1} \circ \mathrm{d}g_{UV}.$$

(iv) If the transition functions are locally constant, the bundle is also called a *local system* of real or complex vector spaces. We can take $\nabla = \mathrm{d}$ i.e., all of the connection matrices are taken to be 0. This is the *canonical flat connection*.

Let us now examine the question of whether connections actually exist, noting that if one does, then many do. For trivial bundles, we have already seen the answer. Because every bundle admits a local trivialization, connections exist locally and global connections can then be constructed by means of partitions of unity.

We repeatedly use the fact that although connections do not behave linearly over C^∞-functions, the difference of two connections does, as follows.

Lemma C.3.3 *Let ∇ and ∇' be two connections of the same bundle E. Then $\nabla - \nabla' \in A^1(\mathrm{End}(E))$, i.e., for all local sections s of E and all C^∞-functions f we have $(\nabla - \nabla')(f \cdot s) = f(\nabla - \nabla')(s)$.*

Proof This follows from the Leibniz rule. Alternatively, if locally $\nabla = d + A$ and $\nabla' = d + A'$, then $\nabla - \nabla' = A - A'$. □

We now define the concept of curvature. Note that using the Leibniz rule, we can extend the connection to operators on E-valued k-forms,

$$\nabla : \mathcal{A}^k(E) \longrightarrow \mathcal{A}^{k+1}(E),$$

where ∇ acts as a (graded) derivation, i.e., by

$$\nabla(\alpha \otimes s) = d\alpha \otimes s + (-1)^k \alpha \wedge \nabla s.$$

We see that our definition generalizes the rules of calculus satisfied by the exterior derivative.

Definition C.3.4 The *curvature* of a connection ∇ on the vector bundle E is the operator

$$F_\nabla \stackrel{\text{def}}{=} \nabla^2 : \mathcal{A}^0(E) \longrightarrow \mathcal{A}^2(E).$$

We say that ∇ is *flat* if its curvature is identically zero.

We claim that this object, which is *a priori* a second-order differential operator, is, by a miracle of cancellation, a tensor, i.e., is linear with respect to the C^∞-functions. In fact, the curvature can be viewed as an $\mathrm{End}(E)$-valued two-form, as the following calculation shows. If $A = A_U$ is a local connection matrix relative to a local frame $\{s_1, \ldots, s_r\}$ defined in an open set U, the curvature matrix is computed from

$$\begin{aligned} \nabla(\nabla s_j) &= \sum_i \nabla(A_{ij} \otimes s_i) = \sum_i \big((\mathrm{d}A_{ij})s_i - A_{ij} \wedge \nabla s_i\big) \\ &= \sum_i (\mathrm{d}A_{ij})s_i - \sum_{i,k} (A_{ij} \wedge A_{ki})s_k \\ &= \sum_k \big((dA)_{kj} + (A \wedge A)_{kj}\big)s_k, \end{aligned}$$

so F_∇ is locally given by the *curvature matrix*

$$F_U \stackrel{\text{def}}{=} \mathrm{d}A_U + A_U \wedge A_U.$$

In the overlap $U \cap V$ we have $F_U = \mathrm{d}A_U + A_U \wedge A_U = g_{UV}^{-1} F_V g_{UV}$ using the relation between A_U and A_V found above. This shows that indeed the curvature is an $\mathrm{End}(E)$-valued form.

Example C.3.5 If E is a local system, its canonical connection (Example C.3.2 (iv)) has been defined by setting it to zero on locally constant sections of E. This is an example of a flat connection.

In order to define Chern classes, from now on we consider complex vector bundles E of rank r. We have already seen that the curvature matrix F_∇ of a connection ∇ on E is an End(E)-valued two-form. Taking its trace defines a global two-form. More invariants can be found using other invariant polynomial functions. Recall that a polynomial function P on $r \times r$ matrices A is called *invariant* if $P(g \cdot A \cdot g^{-1}) = P(A)$ for all invertible g. Clearly, the determinant and the trace are such invariant polynomials. In between are traces of the k-th wedge products

$$\mathrm{Tr}\left(\bigwedge^k A\right) = \frac{1}{k!} \sum_{i_1,\ldots,i_k=1}^{r} \sum_{\sigma \in \mathfrak{S}_k} \mathrm{sgn}(\sigma) A_{i_1\sigma(i_1)} A_{i_2\sigma(i_2)} \cdots A_{i_k\sigma(i_k)}$$

coming up in the expansion

$$\det(I + tA) = \sum_k t^k \mathrm{Tr}\left(\bigwedge^k A\right).$$

Note that $\mathrm{Tr}\left(\bigwedge^k A\right)$ is the k-th elementary symmetric function in the eigenvalues of A. By the fundamental theorem on symmetric polynomial functions, any symmetric polynomial in the eigenvalues is a polynomial expression in these elementary symmetric functions. On the other hand, if P is an invariant polynomial P on $\mathfrak{gl}(r, \mathbf{C})$ it must be symmetric in the eigenvalues. This can be seen as follows. The diagonalizable matrices are Zariski-dense in $\mathfrak{gl}(r, \mathbf{C})$, and P is determined on this set. But the value of P on a diagonalizable matrix equals its value on the corresponding diagonal matrix. The eigenvalues can be permuted by elementary matrices and so P must indeed be symmetric in these eigenvalues. It follows that P is a polynomial expression in the traces of the k-th wedge products of A. We normalize these as follows:

$$\gamma_k(A) = \left(\frac{\mathrm{i}}{2\pi}\right)^k \mathrm{Tr}\left(\bigwedge^k A\right), k = 0, \ldots, r.$$

The constant is put in front of this expression because we will see that it then leads to the desired Chern classes.

Let E now be any rank-r complex vector bundle on a manifold M with a connection whose associated curvature is F. For any invariant polynomial $P(X)$ of degree k, we locally define the expression $P(F_U)$ as the form obtained after substituting $X = F_U$, the local curvature matrix. Because F_U is a matrix of two-forms, and because any two two-forms commute, this makes sense. The transition formula relating F_U and F_V shows that the expressions $P(F_U)$ even define a global $2k$-form, which we denote by $P(F)$. One can show that this form is closed and any other connection leads to a form that differs from $P(F)$ by an exact form (see, e.g., Kobayashi, 1987, Chapter 2). So the cohomology class in $H^{2k}(M; \mathbf{C})$ is independent of the choice of the connection. In particular, this holds for the k-th Chern class

$$c_k(E) = [\gamma_k] \in H^{2k}(M; \mathbf{C}).$$

It can be shown that these classes satisfy the axioms that we stated in the first section of this appendix. See Problem C.3.3 for the verification of the axioms.

Problems

C.3.1 Let $i_v\alpha$ denote contraction of the form α against the vector field v. Using the expression

$$\mathrm{d}A(\xi, \eta) = i_\xi \left(i_\eta(\mathrm{d}A)\right) - i_\eta \left(i_\xi(\mathrm{d}A)\right) - i_{[\xi,\eta]}A$$

for a vector-valued one-form A and two pairs of C^∞ vector fields ξ and η, show that the curvature of a connection ∇ can be expressed as

$$F_\nabla(\xi, \eta) = \nabla_\xi \nabla_\eta - \nabla_\eta \nabla_\xi - \nabla_{[\xi,\eta]}.$$

C.3.2 Let E be a vector bundle on a differentiable manifold M equipped with a connection ∇. Show that the dual bundle $E^\vee$ then admits a connection $\nabla^\vee$ by the formula

$$\mathrm{d}\langle\sigma, s\rangle = \langle\nabla^\vee\sigma, s\rangle + \langle\sigma, \nabla s\rangle.$$

Here $\langle -, - \rangle$ is the evaluation pairing. Calculate its curvature.

C.3.3 Prove the axioms for the Chern classes.

C.3.4 Let E and F be two vector bundles equipped with connections ∇_E and ∇_F. Then $E \oplus F$ carries the connection $\nabla_E \oplus \nabla_F$ and $E \otimes F$ has the connection given by $\nabla_E \otimes \mathbf{1}_F + \mathbf{1}_E \otimes \nabla_F$. Calculate their curvatures.

C.4 Flat Connections

First consider *any* connection on a vector bundle E, flat or not, and its parallel transport along a smooth path $\gamma : I \to X$, say with $\gamma(0) = p$ and $\gamma(1) = q$. Assume that the path is contained in a trivializing chart and that we have chosen a frame for a bundle E of rank r on X. The pullback along γ of a section of E can then be given by a vector-valued function $\vec{x}(t)$ on I. The connection form pulls back to a matrix valued smooth one-form $-A(t)\,\mathrm{d}t$ on I, where the minus sign is convenient in view of the calculations that follow. The section is parallel along γ if $\mathrm{d}\vec{x} = A(t)\vec{x}$, i.e., we have a differential equation

$$\frac{\mathrm{d}\vec{x}}{\mathrm{d}t} - A(t)\vec{x} = 0.$$

By the theory of ordinary differential equations (Ince, 1956), this equation always has a unique solution if the initial value

$$\vec{x}(0) = \vec{v}$$

is fixed.

If we want the result $\vec{\omega} = \vec{x}(1)$ of parallel transport from p to q not to change under homotopies fixing the end points, an integrability condition is needed, which follows from the curvature being zero.

Lemma C.4.1 For a flat connection (i.e., its curvature is zero) the result of parallel transport between two points p and q depends only on the homotopy class of the path (relative to p and q).

Proof Let us suppose that we have a homotopy $\Gamma : I \times I \to X$. Using coordinates (s, t), we have $\Gamma(s, 0) = p$ and $\Gamma(s, 1) = q$. Pulling back the connection matrix under Γ gives

$$\alpha \overset{\text{def}}{=} \Gamma^* A = \vec{a}\,dt + \vec{b}\,ds$$

and we now want to solve the system

$$\frac{\partial \vec{x}}{\partial t} = \vec{a},$$
$$\frac{\partial \vec{x}}{\partial s} = \vec{b}$$

on the unit square with initial conditions $\vec{x}(s, 0) = \vec{v}$ and final conditions $\vec{x}(s, 1) = \vec{w}$. Once we have a solution that obeys the initial condition, the final condition is automatic, since $\Gamma(s, 1)$ is constant, implying that $\vec{b}(s, 1) = 0$, and hence the last equation implies $\vec{x}(s, 1)$ does not depend on s. As to existence, a necessary condition comes from the equality of mixed derivatives which is a translation of $\Gamma^*(\mathrm{d}A - A \wedge A) = 0$. To prove that this indeed guarantees the existence of a solution, we think of s as being fixed. Then the first equation is an ordinary differential equation, and we write down its unique solution $\vec{x}(s, t)$ with initial value $\vec{x}(s, 0) = \vec{v}$. This solution depends smoothly on s. The equality of mixed derivatives then implies that

$$\frac{\partial}{\partial t}\left(\frac{\partial \vec{x}}{\partial s} - \vec{b}\right) = 0,$$

so that the second equation is satisfied up to a vector $\vec{c}(s)$. Now $\vec{x}(s, 0)$ does not depend on s, so $(\partial\vec{x}/\partial s)(s, 0) = 0$. Because $\vec{b}(s, 0) = 0$, we see that $\vec{c}(s) = 0$, completing the proof. □

Corollary C.4.2 *Given a flat connection in a differentiable vector bundle E over a differentiable manifold X, there is a local trivialization of E by a parallel frame. Hence, the existence of a flat connection in E implies that E is a locally constant vector bundle.*

As a result we can now prove the following theorem.

Theorem C.4.3 *There is a one-to-one correspondence between the following objects on M:*

(i) *local systems of* **R***-vector spaces,*

(ii) *flat connections on real vector bundles, and*

(iii) *real representations of the fundamental group.*

Proof Clearly, for a local system the canonical connection as given above is flat. The preceding corollary asserts that the converse is true.

A local system leads to a representation of the fundamental group: cover any closed path γ with open sets over which the system is trivial. This yields a linear isomorphism of the fiber over $\gamma(0) = \gamma(1)$. One verifies that this isomorphism ρ_γ only depends on the homotopy class of γ and that $\rho_{\gamma * \tau} = \rho_\gamma \circ \rho_\tau$. This therefore leads to a representation of the fundamental group.

Conversely, if $\rho : \pi_1(M) \to \mathrm{GL}(r, \mathbf{R})$ is a representation of the fundamental group, we take the quotient of the trivial $\mathbf{R}^r$-bundle on the universal cover $\tilde{M}$ of M by the equivalence relation $(\tilde{m}, x) \sim (\tilde{m}\gamma, \rho_{\gamma^{-1}} x)$. Here the fundamental group acts from the right on $\tilde{M}$ as a group of covering transformations. □

Problems

C.4.1 Prove that the Chern classes of a flat bundle all vanish.

C.4.2 Let E be a complex vector bundle equipped with a Hermitian metric h. Prove that this is equivalent to having a local trivialization of E such that the transition functions have values in the unitary group. Show that these can be chosen to be locally constant if and only if (E, h) admits a flat metric connection.

C.4.3 Prove that a flat line bundle is differentially trivial.

Bibliographical and Historical Remarks

Section C.1: A general reference for this section is Griffiths and Harris (1978). The GAGA principle saying that holomorphic vector bundles on a projective manifold are algebraic is considerably harder to prove than Chow's theorem and uses a lot of sheaf theory and the Kodaira embedding theorem; see Griffiths and Harris (1978, Chapter 1, Section 5), for a proof. The word "GAGA" is an acronym for the title of Serre's article (Serre, 1956).

Section C.2: An excellent introduction to the theory of characteristic classes from the topological viewpoint is Milnor and Stasheff (1974). One should also consult the more scholarly book Husemoller (1966).

Section C.3: The book Milnor and Stasheff (1974) contains a discussion on the differential geometric approach to Chern classes. Also see Kobayashi (1987), although it is far less elementary.

Section C.4: See Donaldson and Kronheimer (1990) as well as Deligne (1970a).

Appendix D

Lie Groups and Algebraic Groups

In this appendix a field K is always a *field of characteristic zero* although some of the results remain valid in more generality.

D.1 First Properties

We collect some basic facts about Lie groups and Lie algebras (see Warner, 1983). We also give a comparison with the analogous results in the algebraic setting for which we refer to Borel (1997) and Humphreys (1981).

Definitions and Examples

Definition • A *Lie group* is a group G that has a compatible manifold structure, meaning that G is a smooth manifold such that the product map $G \times G \longrightarrow G$ given by $(g_1, g_2) \mapsto g_1 g_2$ as well as the inverse map $G \to G$, $g \mapsto g^{-1}$ is smooth. The dimension of G is its dimension as a manifold.

- Homomorphisms of Lie groups are group homomorphisms that are at the same time smooth with respect to the manifold structure. Invertible such homomorphisms are Lie group isomorphisms. A Lie subgroup of a Lie group G by definition is a smooth submanifold of G which inherits a Lie group structure from G.

Consider next $T_{G,e}$, the tangent space of G at the identity. Since the left-translation maps $L_g(x) = gx$ are smooth, every tangent vector $\xi_e \in T_{G,e}$ yields a smooth left-invariant vector field ξ whose value at g is $\xi_g = (L_g)_* \xi_e$. Conversely, any left-invariant vector field on G is completely determined by the value at e and so $T_{G,e}$ can be identified with the vector space of such vector fields.

There is more structure around: the bracket operation between vector fields preserves left-invariant vector fields and induces a bracket. Explicitly, if ξ and η are two such vector fields, and f a germ of a smooth function, the bracket is given by

$$[\xi, \eta]f = \xi \cdot (\eta f) - \eta \cdot (\xi f).$$

This product is a skew-symmetric, nonassociative, and satisfies the *Jacobi identity*:

$$[\xi, [\eta, \zeta]] + [\eta, [\zeta, \xi]] + [\zeta, [\xi, \eta]] = 0. \tag{D.1}$$

So $T_{G,e}$ together with the bracket is an example of a Lie algebra, denoted

$$\mathrm{Lie}(G) = \mathfrak{g} = \left\{T_{G,e}, [,]\right\}.$$

The formal definition is as follows.

Definition A *Lie algebra* $\mathfrak{g}$ over a field K is a K-vector space equipped with a skew-symmetric bilinear bracket operation

$$\mathfrak{g} \times \mathfrak{g} \to \mathfrak{g}$$

which satisfies the Jacobi identity (D.1).

With any $\xi_e \in T_{G,e}$ one can associate a unique Lie group homomorphism $X : \mathbf{R} \to G$, or *one-parameter group* such that $X_*(0)(\mathrm{d}/\mathrm{d}t) = \xi_e$. Indeed, consider the flow defined by the field ξ. It is locally defined around e but left translations define the flow everywhere and unicity of the flow can be used to show that the flow through e is a group. Now we can define the exponential map by

$$\exp : \mathrm{Lie}(G) \longrightarrow G,$$
$$\xi \longmapsto X(1).$$

If $\phi : \mathbf{R} \times G \to G$ is the flow, we have $\phi(t, e) = X(t)$. More generally, we have $\phi_t(g) = \phi_t(e \cdot g) = \phi_t(e) \cdot g = X(t) \cdot g$ and hence

$$\phi_t(g) = L_g(\exp(t\xi)).$$

It is a basic fact that the exponential is a local diffeomorphism and that $\mathrm{Lie}(G_1) \simeq \mathrm{Lie}(G_2)$ if and only if G_1 and G_2 are locally isomorphic in the sense that there exists a connected neighborhood $U \subset G_1$ of 1 which can be diffeomorphically mapped onto a neighborhood of $1 \in G_2$ in a way that is equivariant under the group operations.

Examples D.1.1 (i) The prototypical example of a Lie group is $\mathrm{GL}(n, \mathbf{R})$, the group of all $n \times n$ invertible matrices. It has dimension n^2. Its Lie algebra $\mathfrak{gl}(n, \mathbf{R})$ can be identified with the space $M_n(\mathbf{R})$ of all $n \times n$ real matrices, and the Lie bracket can be identified with the bracket $[A, B] = AB - BA$ of the matrices A and B. The exponential map is simply the matrix exponential: $\exp(tA) = e^{tA}$. The subgroup $\mathrm{SL}(n, \mathbf{R})$ of matrices of determinant 1 has Lie algebra $\mathfrak{sl}(n, \mathbf{R})$, the trace 0 matrices in $M_n(\mathbf{R})$.

(ii) A *matrix group* is a *closed* Lie subgroup of $\mathrm{GL}(n, \mathbf{R})$. Consider, for example, the group of real matrices that preserve a bilinear form b:

$$\left\{ g \in \mathrm{GL}(n, \mathbf{R}) \,\middle|\, {}^{\mathsf{T}}g b g = b \right\}.$$

Substitute $g = e^{tA}$ in the previous relation, differentiate, and evaluate at $t = 0$ to conclude that an element $A \in \mathfrak{g}$, viewed as a matrix, satisfies

$${}^{\mathsf{T}}Ab + bA = 0.$$

Special cases:

- the *orthogonal group* $\mathrm{O}(n, \mathbf{R})$ where we take the form $b = \mathbf{1}_n$; its Lie algebra is the set of skew-symmetric $n \times n$ matrices $\mathfrak{o}(n, \mathbf{R})$ and similarly the *special orthogonal group* $\mathrm{SO}(n, \mathbf{R}) = \mathrm{O}(n, \mathbf{R}) \cap \mathrm{SL}(n, \mathbf{R})$ has Lie algebra $\mathfrak{so}(n, \mathbf{R})$, the skew-symmetric $n \times n$ matrices with trace zero;
- more generally, if we take the form $b = \mathrm{diag}(\mathbf{1}_p, -\mathbf{1}_q)$ we obtain the orthogonal groups $\mathrm{O}(p, q; \mathbf{R})$ with Lie algebra $\mathfrak{o}(p, q; \mathbf{R})$;
- for $b = \begin{pmatrix} \mathbf{0}_n & \mathbf{1}_n \\ -\mathbf{1}_n & \mathbf{0}_n \end{pmatrix}$, the standard symplectic form we get the *symplectic group* $\mathrm{Sp}(n, \mathbf{R})$ with Lie algebra $\mathfrak{sp}(n, \mathbf{R})$.

(iii) Of course, also $\mathrm{GL}(n, \mathbf{C})$, the group of all $n \times n$ invertible matrices with complex entries is a Lie group (of real dimension $2n^2$) with Lie algebra $\mathfrak{gl}(n, \mathbf{C})$ the space $M_n(\mathbf{C})$ of all $n \times n$ complex matrices. It is an example of a *complex Lie group*: a complex manifold with holomorphic group operations. Other examples are the groups $\mathrm{SL}(n, \mathbf{C})$, $\mathrm{O}(n, \mathbf{C})$, $\mathrm{SO}(n, \mathbf{C})$ and $\mathrm{Sp}(n, \mathbf{C})$.

A complex Lie group can have interesting Lie subgroups that are *not* complex Lie groups, for example, if h is a hermitian form, the subgroup $\mathrm{U}(h)$ of matrices preserving this form. The standard examples are the *unitary groups* $\mathrm{U}(n)$ (where $h = \mathrm{diag}(\mathbf{1}_n)$) and $\mathrm{SU}(n) = \mathrm{U}(n) \cap \mathrm{SL}(n, \mathbf{C})$. If we take the diagonal form $\mathrm{diag}(\mathbf{1}_p, -\mathbf{1}_q)$ we get other unitary groups $\mathrm{U}(h)$ which are denoted $\mathrm{U}(p, q)$.

A list of the real forms of the so called *classical Lie groups* is given in Table D.1. Starting from the adjoint group of the classical Lie group (whose center is $\{1\}$), all other Lie groups with the same Lie algebra can be found by taking unramified coverings. If one stays in the algebraic setting this is governed by the so-called fundamental group of the Dynkin diagram which is finite and classifies all finite unramified coverings of the adjoint group which are algebraic.

Furthermore, some overlaps are possible since some Lie algebras are isomorphic, e.g. $\mathfrak{su}(2) \simeq \mathfrak{so}(3)$ (for details, see Fulton and Harris, 1991, Lecture 26).

Table D.1 *List of classical Lie groups*

$$\text{Set } I_{p,q} = \begin{pmatrix} -\mathbf{1}_p & 0 \\ 0 & \mathbf{1}_q \end{pmatrix}, \quad K_{p,q} = \begin{pmatrix} I_{p,q} & 0 \\ 0 & I_{p,q} \end{pmatrix}, \quad J_n = \begin{pmatrix} 0 & \mathbf{1}_n \\ -\mathbf{1}_n & 0 \end{pmatrix}.$$

Group Adjoint grp.	Definition Dynkin diagr., fund. grp.	Max. comp. subgr.
$\mathrm{SL}(n;\mathbf{k})$, $\mathbf{k} = \mathbf{R}, \mathbf{C}, \mathbf{H}$ $\mathrm{PSL}(n;\mathbf{k})$	$\{g \in \mathrm{GL}(n;k) \mid \det g = 1\}$ $\mathbf{k} = \mathbf{R}, \mathbf{C} \implies A_{n-1}, A_{n-1}, \mathbf{Z}/n\mathbf{Z}$ $\mathbf{k} = \mathbf{H} \implies A_{2n-1}, \mathbf{Z}_{2n}$	$\mathrm{SU}(n)$
$\mathrm{SU}(n)$ $\mathrm{PSU}(n)$	$\{g \in \mathrm{SL}(n,\mathbf{C}) \mid g^*g = \mathbf{1}_n\}$ $A_{n-1}, \mathbf{Z}/n\mathbf{Z}$	$\mathrm{SU}(n)$
$\mathrm{SU}(p,q)$, $p+q=n$ $\mathrm{PSU}(p,q)$	$\{g \in \mathrm{SL}(n,\mathbf{C}) \mid g^*I_{p,q}g = I_{p,q}\}$ $A_{n-1}\ \mathbf{Z}/n\mathbf{Z}$	$\mathrm{S}(\mathrm{U}(p) \times \mathrm{U}(q))$
$\mathrm{SO}(n,\mathbf{k}))$, $\mathbf{k} = \mathbf{R}, \mathbf{C}$ $\mathrm{PSO}(n,\mathbf{k}))$	$\left\{g \in \mathrm{SL}(n,\mathbf{k}) \,\middle|\, {}^{\mathsf{T}}gg = \mathbf{1}_n\right\}$ $n = 2m+1 \implies B_m, \mathbf{Z}/2\mathbf{Z}$ $n = 2m \implies D_m, \mathbf{Z}/4\mathbf{Z}$	$\mathrm{SO}(n,\mathbf{R})$
$\mathrm{SO}(p,q)$, $p+q=n$ $p,q>0$ $\mathrm{PSO}(p,q)$	$\left\{g \in \mathrm{SL}(n,\mathbf{R}) \,\middle|\, {}^{\mathsf{T}}gI_{p,q}g = I_{p,q}\right\}$ $n = 2m+1 \implies B_m, \mathbf{Z}/2\mathbf{Z}$ $n = 2m \implies D_m, \mathbf{Z}/4\mathbf{Z}$	$\mathrm{S}(\mathrm{O}(p) \times \mathrm{O}(q))$
$\mathrm{Sp}(n;\mathbf{C})$ $\mathrm{PSp}(n;\mathbf{C})$	$\left\{g \in \mathrm{SL}(n,\mathbf{C}) \,\middle|\, {}^{\mathsf{T}}gJ_ng = J_n\right\}$ $C_n, \mathbf{Z}/2\mathbf{Z}$	$\mathrm{Sp}(n)$
$\mathrm{Sp}(n,\mathbf{R})$ $\mathrm{PSp}(n,\mathbf{R})$	$\mathrm{Sp}(n,\mathbf{C}) \cap \mathrm{SL}(n,\mathbf{R})$ $C_n, \mathbf{Z}/2\mathbf{Z}$	$U(n)$
$\mathrm{Sp}(n)$ $\simeq \mathrm{Sp}(n,\mathbf{C}) \cap U(2n)$ $\mathrm{PSp}(n)$	$\{g \in \mathrm{GL}(n,\mathbf{H}) \mid g^*g = \mathbf{1}_n\}$ $C_n, \mathbf{Z}/2\mathbf{Z}$	$\mathrm{Sp}(n)$
$\mathrm{Sp}(p,q)$, $p+q=n$ $\mathrm{PSp}(p,q)$	$\{g \in \mathrm{GL}(n,\mathbf{H} \mid g^*I_{p,q}g = I_{p,q}\}$ $C_n, \mathbf{Z}/2\mathbf{Z}$	$\mathrm{Sp}(p) \times \mathrm{Sp}(q)$
$\mathrm{SO}^*(2n)$ $\mathrm{PSO}^*(2n)$	$\left\{g \in \mathrm{SU}(n,n) \,\middle|\, {}^{\mathsf{T}}gI_{n,n}J_ng = I_{n,n}J_n\right\}$ $D_n, \mathbf{Z}/4\mathbf{Z}.$	$U(n)$

Algebraic Groups

Recall that the set of zeroes $\subset K^N$ of the polynomials belonging to a prime ideal $I \subset K[X_1, \ldots, X_N]$ is called an *affine variety*. For any extension field L of K, the set $V(L)$ of common zeros in L^N forms the L-valued points of V. Suppose that the generators of I can be chosen to have coefficients in a smaller field k. Then one says that V is *defined over* k.

If K/k is a finite field extension, out of V, defined over K, we can produce a variety, $\mathrm{Res}_{K/k} V$, *the Weil restriction* of V, which is defined over the smaller field k. This is done as follows: write $K = ke_1 \oplus \cdots \oplus ke_s$ and set $X_j = \sum_{\alpha=1}^{s} X_j^\alpha e_\alpha$, $j = 1, \ldots, N$. Then one writes a generator for the ideal of V as

$$f = \sum f_\alpha e_\alpha,$$

where f_α is a polynomial in the new variables X_j^α. The ideal of $\mathrm{Res}_{K/k} V$ is generated by the f_α, $\alpha = 1, \ldots, s$, $f \in I(V)$.

Definition A *K-affine algebraic group* G is a group that is a K-affine variety for which the multiplication map $(g_1, g_2) \mapsto g_1 g_2$ as well as the inverse map $g \mapsto g^{-1}$ are morphisms defined over K.

Remarks (1) One can show Humphreys (1981, Section 8.6) that any K-affine algebraic group can be realized as a closed subgroup of $\mathrm{GL}(n; K)$. In other words, affine algebraic groups are algebraic *matrix* groups.
(2) Note that for any extension field L of K, a K-group G is automatically an L-group. Sometimes one stresses this by writing G_L and one denotes its points by $G(L)$.
(3) If K is a finite extension of k, the Weil restriction $\mathrm{Res}_{K/k} G$ is a k-affine group. This can be seen as follows. As before, suppose that $G \subset \mathrm{GL}(N, K)$ and that $s = [K : k]$. Consider the operation $\mu : K^{N\times N} \to k^{Ns\times Ns}$ which replaces any entry X_{ij} of a matrix by the matrix which represents multiplication with this number in the field K, identified with the k-vector space k^s (after a choice of a basis). Applied to the matrices defining G, this yields a group $\mu(G)$ whose underlying variety is indeed the Weil restriction.
(4) For $K = \mathbf{R}, \mathbf{C}$ a K-affine algebraic group gives an example of a Lie group.

Examples D.1.2 (i) The additive group $\mathbf{G}_a$ which is defined over any field K as the additive group K^+ of K.

(ii) The multiplicative group $\mathbf{G}_m$, also defined over any field K, is the multiplicative group $K^\times$ of the field. As an algebraic set it is the affine variety $\{(x, y) \in K^2 \mid xy - 1 = 0\}$ and as such it can also be viewed as a matrix group inside $M_2(K)$ consisting of diagonal matrices $\{\mathrm{diag}(x, y) \mid xy = 1\}$.

(iii) The group $\mathrm{GL}(n; K)$ of invertible $n \times n$ matrices with entries in K is the

Zariski open subset of $M_n(K)$, the space of all $n \times n$ matrices with entries in K whose determinant is nonzero. So it is an affine subset of K^{n^2+1} given by the polynomial identity $Y \cdot \det(X_{ij}) = 1$. It can be seen as the matrix subgroup of $M_{n+1}(K)$ consisting of matrices of the form $\begin{pmatrix} A & 0 \\ 0 & b \end{pmatrix}$ obeying the equation $b \cdot \det A = 1$. The multiplication of matrices A and B involves a polynomial operation in the entries of A and B and hence is a morphism. Similarly, the entries of B are polynomial expressions in the entries of A and the determinant of A. The classical groups from Table D.1 provide examples of affine algebraic groups over $\mathbf{R}$. For the orthogonal and symplectic groups there are variants over any field.

D.2 Further Notions and Constructions

Connectedness Properties

A Lie group need not be connected; indeed it may have infinitely many components, for example the discrete group $\mathbf{Z}$. For any Lie group G the connected component G^0 containing e is a closed Lie subgroup and G/G^0 is a discrete Lie group. If G is a K-affine algebraic group this does not make sense; instead we put

$$G^0 \stackrel{\text{def}}{=} \{\text{irreducible component of } G \text{ containing } e\}.$$

In contrast to Lie groups, an algebraic group G can at most have a finite number of components. We keep saying that G^0 is the connected component of G and that G is connected if $G = G^0$. If $K = \mathbf{C}$ this is equivalent to the previous notion. However, already for $\mathbf{R}$ this can be different: look at Example D.3.4. The torus $\mathbf{R}^\times$ is irreducible since it is given by the irreducible equation $XY = 1$. However, it has two connected components $\mathbf{R}^\times_{>0}$ and $-\mathbf{R}^\times_{>0}$ which cannot be distinguished algebraically.

The universal cover of a connected Lie group is a Lie group. The quotient of a simply connected Lie group by a discrete subgroup is a Lie group. On the algebraic side this need not be true: the universal cover of an algebraic group need not be algebraic: the universal cover of $\mathrm{SL}(2;\mathbf{R})$ is not an algebraic group. However, there is an algebraic notion of fundamental group Humphreys (1981, 31.1). This group classifies all the algebraic groups which are finite unramified covers of a given group. This is always a finite group. See e.g. Table D.1 for the classical groups. For instance, in the algebraic sense $\mathrm{SL}(2,\mathbf{R})$ is simply connected: it is the universal cover of $\mathrm{PSL}(2,\mathbf{R})$ since from the table we see that its fundamental group is $\mathbf{Z}/2\mathbf{Z}$.

Functorial Properties

Any closed subgroup of a Lie group is naturally a Lie group. Likewise, any Zariski-closed subgroup of an affine algebraic group is an (affine) algebraic group. If G is an algebraic group over the field $K = \mathbf{R}$, or $\mathbf{C}$, we may consider it as a Lie group and any closed subgroup H (in the ordinary or "classical" topology) is a Lie group, but only its Zariski closure is an algebraic subgroup.

For morphisms $h : G \to G'$ of Lie groups the kernel $\mathrm{Ker}(h)$ is a closed Lie group; the image $\mathrm{Im}(h)$ is a Lie subgroup but need not be closed. Think of the one-parameter subgroup of the torus $T = \mathbf{R}^2/\mathbf{Z}^2$ given by $X(t) = (t, \alpha t)$ with α irrational; the closure of $X(\mathbf{R})$ is the entire torus. For morphisms $h : G \to G'$ between algebraic groups these "pathologies" don't occur: not only is $\mathrm{Ker}(h)$ a Zariski closed subgroup, and so is $\mathrm{Im}(h)$; in addition, the dimensions of the two add up to the dimension of G.

Any morphism $h : G \to G'$ induces a morphism $\mathrm{Lie}(h) : \mathrm{Lie}(G) \to \mathrm{Lie}(G')$ between the respective Lie algebras, and the assignment $h \mapsto \mathrm{Lie}(h)$ is clearly functorial. This remains true in the algebraic setting.

For a given Lie group G there is a one-to-one correspondence between *connected* Lie subgroups of G and Lie subalgebras of $\mathrm{Lie}(G)$. Normal subgroups $N \subset G$ correspond to ideals of the Lie algebra $\mathrm{Lie}(G)$. The quotient G/N carries the structure of a Lie group such that the natural map $G \to G/N$ is a morphism of Lie groups. Conversely, if we have a surjective morphism $G \to G'$ with kernel $\mathrm{Ker}(h) = N$, then $G/N \simeq G'$.

In the algebraic setting closed subgroups do correspond to Lie subalgebras (and normal closed subgroups give ideals) but the converse need not be true: there are Lie subalgebras that are not Lie algebras of Zariski closed subgroups. See Borel (1997, Chapter II.7).

Normalizers and Centralizers

Consider a group of G acting on a set X. Let $Y \subset X$. Recall

$$\mathrm{Stab}_G Y \stackrel{\text{def}}{=} \{g \in G \mid gY \subset Y\}, \text{ the \textit{stabilizer} of } Y,$$
$$Z_G Y \stackrel{\text{def}}{=} \{g \in G \mid gy = y \text{ for all } y \in Y\}, \text{ the \textit{centralizer} of } Y.$$

If X is a smooth manifold with G a Lie group acting by smooth morphisms and $Y \subset X$ a closed submanifold, then $\mathrm{Stab}_G Y$ and $Z_G Y$ are Lie subgroups. Similarly, if G is an algebraic group over K, G acts on X by morphisms, and $Y \subset X$ is a closed subvariety, $\mathrm{Stab}_G Y$ and $Z_G Y$ are algebraic subgroups of G. This applies in particular to the adjoint action of G on $X = G$ given by $(g, x) \mapsto gxg^{-1}$. In this case the stabilizer of a Lie subgroup (respectively a closed algebraic subgroup) H is called the *normalizer* of H in G and denoted

$$N_G H \overset{\text{def}}{=} \mathrm{Stab}_G H = \{g \in G \mid gHg^{-1} \subset H\}.$$

In particular $N_G H$ as well as the centralizer subgroup

$$Z_G H = \{g \in G \mid gxg^{-1} = x \text{ for all } x \in H\}$$

are (closed) Lie subgroups (respectively closed algebraic subgroups) of H. We usually put

$$Z_G = Z_G G, \text{ the } \textit{center} \text{ of } G.$$

Representations

Fix a field L. Recall that a *representation* of a group G in an L-vector space V, possibly of infinite dimension, is a homomorphism

$$\begin{aligned} \rho : G &\longrightarrow \mathrm{GL}(V), \\ g &\longmapsto \rho_g. \end{aligned}$$

Associated with ρ we have (co-gredient) tensor representations

$$\begin{aligned} \rho^{\otimes r} : G &\longrightarrow \mathrm{GL}(V^{\otimes r}), \\ g &\longmapsto \{v_1 \otimes \cdots \otimes v_r \mapsto \rho_g(v_1) \otimes \cdots \otimes \rho_g(v_r)\}, \end{aligned}$$

the dual or *contragredient representation*

$$\begin{aligned} \rho^\vee : G &\longrightarrow \mathrm{GL}(V^\vee), \\ g &\longmapsto \left\{f \mapsto f^{(g)} = \{x \to f(g^{-1}x)\}\right\}, \end{aligned} \tag{D.2}$$

and combinations of the two, the (generalized) *tensor representations*

$$\begin{aligned} \rho^{\otimes r} \otimes (\rho^\vee)^{\otimes s} : G &\longrightarrow \mathrm{GL}(V^{\otimes r} \otimes (V^\vee)^{\otimes s}), \\ g &\longmapsto \rho_g^{\otimes r} \otimes (\rho_g^\vee)^{\otimes s}. \end{aligned}$$

We say that $W \subset V$ is a *subrepresentation* if $\rho_g(W) \subset W$ for all $g \in G$. If no subrepresentations exist except the trivial ones $\{0\}$ and V, we say that ρ is *irreducible*. We say that ρ is *fully reducible* or *semisimple* if W is a direct sum of irreducible subrepresentations. It is well known that all representations of a finite group are semisimple.

If G is a Lie group and $L = \mathbf{R}$ or $L = \mathbf{C}$ and V is finite-dimensional, we demand in addition that ρ be a Lie group homomorphism, i. e., the action morphism $G \times V \to V$ should be a *smooth* map and we speak of a smooth representation. Similarly, if G is a K-affine algebraic group and L a field extension of K, we want this to be an L-morphism. In this case we also say that ρ is a *rational* or *algebraic* representation.

Example D.2.1 (Regular representation) Let $G \subset \mathrm{GL}(V)$ be a K-algebraic group. It is a closed subset of the affine space of dimension n^2 with $n = \dim V$. Let $\{X_{ij}\}$, $1 \leq i \leq j \leq n$ be the coordinates on this space (giving the n^2 entries of the corresponding matrices). The coordinate ring $K[G]$ is the quotient of the polynomial ring $K[X_{11}, \ldots, X_{nn}]$ by the ideal of G. An element $f \in K[G]$ can be viewed as a K-valued function on G, and G hence acts naturally on $K[G]$ just like the contragredient representation: $\lambda_g(f)(x) = f(g^{-1}x)$. In fact, this makes $K[G]$ into an algebraic representation, the *regular representation.*

A linear map $\varphi : V \longrightarrow V'$ between G-representation spaces (V, ρ), (V', ρ') is called *equivariant* if $\varphi \circ \rho_g = \rho'_g \circ \varphi$ for all $g \in G$. Two G-representations are called *isomorphic* or *equivalent* if there is a G-equivariant linear automorphism between them.

A representation of a Lie group $\rho : G \to \mathrm{GL}(V)$ induces a representation of the corresponding Lie algebra through the tangent map at the identity:

$$\dot{\rho} \stackrel{\text{def}}{=} (\rho_*)_e : \mathrm{Lie}(G) \longrightarrow \mathrm{End}\, V.$$

Example D.2.2 The *adjoint representation* of a Lie group G on its Lie algebra $\mathfrak{g}$ is defined as follows. For a fixed $g \in G$ conjugation by g defines the isomorphism $G \xrightarrow{\sim} G$ given by $x \mapsto gxg^{-1}$. Its derivative at $e \in G$ is the adjoint $\mathrm{Ad}(g)$ of g; it is a Lie algebra isomorphism $\mathrm{Ad}(g) : \mathfrak{g} \to \mathfrak{g}$ and for varying g we obtain the adjoint representation

$$\begin{array}{ccc} G & \xrightarrow{\mathrm{Ad}} & \mathrm{GL}(\mathfrak{g}), \\ g & \mapsto & \mathrm{Ad}(g). \end{array}$$

The image of the above map turns out to be a Lie group, the *adjoint group* G^{ad} of G. Suppose that G is connected. Then the kernel of the resulting morphism $G \to G^{\mathrm{ad}}$ is the center of G so that

$$\mathrm{Ad} : G/Z(G) \xrightarrow{\simeq} G^{\mathrm{ad}}.$$

This can also be done for algebraic groups defined over a field[1] and the adjoint representation is clearly algebraic Borel (1997, Chapter I.3).

[1] Here one really needs characteristic zero.

D.3 Classification of Lie Groups and Algebraic Groups

Isogenies

As a first step in classifying algebraic groups, one first looks at the corresponding Lie algebras. Since such a Lie algebra corresponds to finitely many possible algebraic groups, the following concept turns out to be useful.

Definition An *isogeny* $G_1 \to G_2$ between Lie groups (algebraic groups) is a surjective morphism with finite kernel. Two groups G_1 and G_2 are isogeneous if there is a third Lie group (algebraic group G) and two isogenies $G \to G_1$ and $G \to G_2$. Isogeneous groups have isomorphic Lie algebras.

Abelian Lie Groups

A Lie group G is abelian if and only the corresponding Lie algebra is, i.e. $[X, Y] = 0$ for all $X, Y \in \text{Lie}(G)$. A connected abelian Lie group is isomorphic to $\mathbf{R}^r \times (S^1)^s$, i. e., it is a product of a real vector space and a (topological) torus. Hence a simply connected abelian Lie group is isomorphic to the additive group of a real vector space.

Algebraic Tori

An important class of commutative algebraic groups for which the field of definition intervenes is the class of the algebraic tori.

Definition D.3.1 (i) A *split* algebraic torus of rank r is an algebraic group isomorphic to a product of r copies of the multiplicative group $\mathbf{G}_m$.
(ii) A K-algebraic torus is an algebraic group which over some extension field of K is a split algebraic torus.

Tori can be studied via their characters.

Definition D.3.2 A *character* of an (affine) K-algebraic group G is a morphism $\chi : G \to \mathbf{G}_m$; a *co-character* or *rational one-parameter group* is a morphism $\mu : \mathbf{G}_m \to G$. The characters and co-characters of G form an abelian group $X^*(G)$, respectively $X_*(G)$ and there is a natural pairing as follows. For a character μ and co-character λ the composition $\chi \circ \mu$ is a group homomorphism of $K^\times$, hence of the form $t \mapsto t^n$. This integer n by definition is $\langle \chi, \mu \rangle$:

$$\begin{aligned} X^*(G) \times X_*(G) &\longrightarrow \mathbf{Z}; \\ (\chi, \mu) &\mapsto \langle \chi, \mu \rangle, \quad \text{where } \chi \circ \mu : t \mapsto t^{\langle \chi, \mu \rangle}. \end{aligned}$$

Let us now consider characters for algebraic tori. Clearly, for the group $\mathbf{G}_m$ the group of characters is $\mathbf{Z}$ with $n \in \mathbf{Z}$ identified with $x \mapsto x^n$ and for a split torus T of rank r we have $X^*(T) \simeq \mathbf{Z}^r$. We say that a torus T is *anisotropic* if $X^*(T) = \{0\}$. A torus T is split over some finite field extension L of K. Let Γ be the Galois group of $L|K$. Then Γ acts on the torus $T(L)$ and hence on $X^*(T(L))$. One can assemble the characters on which this action is trivial. This gives the following result Borel (1997, Section 8).

Lemma D.3.3 *There is a unique maximal split subtorus $T = T_{\text{split}}$ of T as well as a maximal anistropic subtorus T_{an} of T and the product map $T_{\text{split}} \times T_{\text{an}} \to T$ is an isogeny.*

Example D.3.4 Let $K = \mathbf{R}$. A split torus is $\mathbf{R}^\times$. The Galois group of $\mathbf{C}|\mathbf{R}$ is generated by complex conjugation and acts trivially on the real points of $\mathbf{C}^\times$. Note that $\mathbf{G}_m(\mathbf{C})$ is given by the equation $xy = 1$. Take new coordinates in $\mathbf{C}$ given by $x = u+\mathrm{i}v$, $y = u-\mathrm{i}v$ so that $u^2+v^2 = 1$. This gives an isomorphic model

$$S^1 = \left\{ A = \begin{pmatrix} u & v \\ -v & u \end{pmatrix} \middle| \, u, v \in \mathbf{C},\ \det(A) = 1 \right\}.$$

Clearly, over $\mathbf{R}$ this is the circle group, explaining the notation. Complex conjugation now acts nontrivially on this model. The torus is not split but anisotropic (the image in $\mathbf{R}^\times$ of any character is compact and connected and hence equal to 1). In general, real algebraic tori which are anisotropic are compact and hence such algebraic tori are topological tori, and conversely.

An algebraic torus T over $\mathbf{R}$ is isogeneous to a product $(\mathbf{R}^\times)^r \times (S^1)^s$. For such a torus complex conjugation acts on characters as follows:

$$X^*(T_{\mathbf{C}}) = \mathbf{Z}^{\oplus r} \oplus \mathbf{Z}^{\oplus s},$$

$$(n_1, \ldots, n_r, n_{r+1}, \ldots, n_s) \mapsto (n_1, \ldots, n_r, -n_{r+1}, \ldots, -n_s).$$

Consider for example a one-dimensional anisotropic torus $T = S^1$. One dimensional complex representations are isomorphic to $\mathbf{C}$ with $z \in S^1$ acting by multiplication by z^n. Complex conjugation sends this representation, say $V^{(n)}$, to $V^{(-n)}$ as it should. As a real representation it is $\mathbf{R}^2$ with $\mathrm{e}^{2\pi i t} \in S^1$ acting as a rotation through the angle nt; conjugation sends (x, y) to $(x, -y)$.

Solvable and Nilpotent Groups

Several of the constructions in abstract group theory remain valid in the Lie world. For instance if G is a Lie group, the *derived group* or commutator subgroup

$$G^{\text{der}} = DG = [G, G] = \left\{ \text{group generated by } ghg^{-1}h^{-1} \mid g, h \in G \right\}$$

is a normal Lie subgroup and its Lie algebra can be identified with the ideal generated by the commutators $[X, Y]$, with $X, Y \in \mathrm{Lie}(G)$. If G is an affine algebraic group, the group $[G, G]$ need not be Zariski-closed. However, if G is connected, this is true and in characteristic zero one has $\mathrm{Lie}([G, G]) = [\mathrm{Lie}(G), \mathrm{Lie}(G)]$.

A related construction starts from subgroups H_1, H_2. The commutators $h_1 h_2 h_1^{-1} h_2^{-1}$ with $h_1 \in H_1, h_2 \in H$ generate a group denoted $[H_1, H_2]$ (or sometimes (H_1, H_2)) and if H_1, H_2 are Lie subgroups of G then so is $[H_1, H_2]$. If H_1 and H_2 are connected algebraic groups, so is $[H_1, H_2]$.

A group is called *solvable* if the *derived series* eventually stops:

$$G \supset DG \supset D^2G = D(DG) \supset \cdots \supset D^kG = \{e\}\,.$$

This makes sense for Lie groups and for connected algebraic groups.

Examples D.3.5 (i) Abelian groups.

(ii) The group $\mathsf{T}(n; K)$ of upper triangular matrices in $\mathrm{GL}(n; K)$ or any closed subgroup thereof. One can show that any simply connected solvable Lie group is such a group.

A *nilpotent group* is one whose *lower central series* eventually stops:

$$G \supset G^1 \stackrel{\text{def}}{=} [G, G] \supset G^2 \stackrel{\text{def}}{=} [G, DG] \cdots \supset G^k = \{e\}\,.$$

Again, this makes sense for Lie groups and for connected algebraic groups.

Examples D.3.6 (i) Clearly, a nilpotent group is solvable.

(ii) There are solvable groups that are not nilpotent. For instance $\mathsf{T}(n; K)$ is solvable but not nilpotent. However, its commutator subgroup, the subgroup $\mathsf{U}(n; K)$ of upper triangular matrices with 1 on the diagonal, is nilpotent.

(iii) Unipotent algebraic groups: those whose elements x are unipotent (i.e. x has only 1 as eigenvalue). This is because one can prove (Humphreys 1981, section 17.5; Borel 1997, section 4.8) that they are conjugate to some algebraic subgroup of $\mathsf{U}(n; K)$.

Semisimple and Reductive Groups

Let us first recall some linear algebra. An endomorphism $X : V \longrightarrow V$ of a K-vector space V is *semisimple* if X is diagonalizable over an algebraic closure of K. At the other extreme, X is *nilpotent* if 0 is the sole eigenvalue of X, or, equivalently, if $X^n = 0$ for some n. We call X *unipotent* if X has 1 as the sole eigenvalue, or, equivalently, if $(X - \mathrm{id})$ is nilpotent. Every endomorphism X has a unique decomposition, the *additive Jordan decomposition*

$$X = X_s + X_n, \quad X_s \text{ semisimple, } X_n \text{ nilpotent,}$$

and every *invertible* X can be written uniquely as

$$X = X_s \circ X_u = X_u \circ X_s, \quad X_s \text{ semisimple}, X_u \text{ unipotent}.$$

The latter is called the *multiplicative Jordan decomposition*. If $X \in G$, a matrix group, one can show that $X_s, X_u \in G$.

Next, consider a *connected* Lie group (or algebraic group) G and let $R(G)$ be the largest connected normal Lie (or algebraic) subgroup which is solvable. It is called the *radical* of G. The unipotent elements in $R(G)$ constitute a subgroup $R(G)_u$ which is itself a unipotent group. Moreover, it is normal in G. It is called the *unipotent radical* of G.

Definition If $R(G) = e$ we say that G is *semisimple* and if $R(G)_u = e$ we say that G is *reductive*.

For a reductive G one knows Humphreys (1981, Section 19.5) that $R(G) = Z(G)^0$ is an algebraic torus.[2] In other words, we have an exact sequence

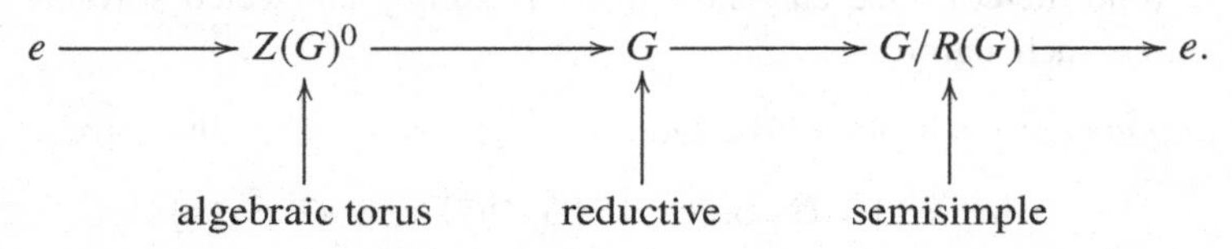

From a different angle, we note that since for any group G the quotient G/DG is the largest abelian quotient of G, in the special case when G is reductive, this implies that DG is semisimple and hence we have the following.

Lemma D.3.7 *The derived subgroup $G^{\text{der}} = DG$ of a reductive group G is a semisimple algebraic group. The multiplication map*

$$G^{\text{der}} \times Z(G) \longrightarrow G$$

is a surjective isogeny.

A semisimple group is called *simple* if it does not contain nontrivial connected normal algebraic subgroups (defined over the same field). If it does, such a subgroup is itself semisimple. The simple subgroups of a semisimple group are called *simple factors*. A similar concept is available for semisimple Lie groups. In both cases we have:

Lemma D.3.8 (Humphreys, 1981, §14.2, Knapp, 2005, Theorem 1.54) *Let G be a semisimple algebraic group or Lie group with simple factors $G_1, \ldots, G_k$. Then multiplication*

$$G_1 \times \cdots \times G_k \longrightarrow G_1 \cdots G_k = G$$

[2] The proof as given in Humphreys (1981) does not use that the field is algebraically closed.

defines a surjective isogeny. This is the semidirect product decomposition *of G into its simple factors.*

For Lie groups one has a useful criterion for semisimplicity:

Proposition D.3.9 (Helgason, 1962, Prop II, 6.1) *Let G be a Lie group. The* Killing form *is the bilinear form on* Lie(G) *given by*

$$\mathrm{Lie}(G) \times \mathrm{Lie}(G) \xrightarrow{B} \mathbf{R}, \quad B(X,Y) \stackrel{\text{def}}{=} \mathrm{Tr}(\mathrm{ad}\,X \circ \mathrm{ad}\,Y).$$

This form is nondegenerate on Lie(G) *if and only if the group G is semisimple.*

Examples D.3.10
- Examples of reductive groups that are not semisimple: algebraic tori, $\mathrm{GL}(n;K)$ and $\mathrm{U}(p,q)$. They are not semisimple since they all have a nondiscrete (and thus at least one-dimensional) center.
- Compact Lie groups are reductive. We see this as follows. Let $\mathfrak{g}$ be the Lie algebra of a compact Lie group. Then by Helgason (1962, Ch. II.6) the Killing form is negative definite on $\mathfrak{g}$ modulo its center. So $G/Z(G)$ is semisimple which implies G itself reductive. Note that G need not be semisimple, for example compact tori are not semisimple. Conversely, if the Killing form of $\mathfrak{g}$ is negative definite, the group G is compact.
- The classical groups from Table D.1 are semisimple. This is not obvious from the definition. However, the Killing form can easily be calculated and in each case turns out to be nondegenerate. Then Proposition D.3.9 tells us that these groups are semisimple (they are even simple).

Real Reductive Groups

Real Forms

Let $G_0 \subset \mathrm{GL}(n,\mathbf{R})$ be a connected $\mathbf{R}$-algebraic matrix group with Lie algebra $\mathrm{Lie}(G_0) = \mathfrak{g}_0$. Note that for the Lie algebra $\mathfrak{g}_0$ of G_0 we have

$$\mathfrak{g}_0 \hookrightarrow \mathfrak{gl}(n,\mathbf{R}); \quad \mathfrak{g}_0 \otimes_{\mathbf{R}} \mathbf{C} \hookrightarrow \mathfrak{gl}(n,\mathbf{C}).$$

This leads to the following concepts.

Definition
- Let $\mathfrak{g}_0$ be any real Lie algebra. Its *complexification* is the complex Lie algebra $\mathfrak{g} \stackrel{\text{def}}{=} \mathfrak{g}_0 \otimes_{\mathbf{R}} \mathbf{C}$.
- If $\mathfrak{g}$ is a complex Lie algebra such that it can be written as $\mathfrak{g} = \mathfrak{g}_0 + i\mathfrak{g}_0$ for some real Lie algebra $\mathfrak{g}_0$, we say that $\mathfrak{g}_0$ is a *real form* of $\mathfrak{g}$.
- If $G_0 \subset \mathrm{GL}(n,\mathbf{R})$, the *complexification* $G_0(\mathbf{C})$ is the (unique) connected matrix group in $\mathrm{GL}(n,\mathbf{C})$ whose Lie algebra is $\mathrm{Lie}(G_0) \otimes \mathbf{C}$.[3]

[3] This is of course the group of complex points of the $\mathbf{R}$-algebraic group G_0; there is no conflict of notation.

- If $G \subset \mathrm{GL}(n, \mathbf{C})$ and $\mathfrak{g}_0$ is a real form of $\mathrm{Lie}(G) = \mathfrak{g}$, the (unique) connected matrix subgroup $G_0 \subset G$ with Lie algebra $\mathfrak{g}_0$ is called a *real form* of G. If G_0 is compact, we say that G_0, respectively $\mathfrak{g}_0$, is a *compact real form* of G, $\mathfrak{g}$, respectively.

Examples D.3.11 (i) Of course $\mathrm{GL}(n, \mathbf{R})$ is a real form of $\mathrm{GL}(n, \mathbf{C})$. There is also a compact real form, as we now show. The Lie algebra $\mathfrak{gl}(n, \mathbf{C})$ consists of arbitrary $n \times n$ matrices $Z \in M_n(\mathbf{C})$. We may write

$$Z = \frac{1}{2}(Z + Z^*) + \frac{1}{2}(Z - Z^*) = \frac{1}{2}(Z + Z^*) + \mathrm{i}\frac{1}{2}(-\mathrm{i}Z + (-\mathrm{i}Z)^*).$$

Hence $\mathfrak{gl}(n, \mathbf{C}) = \mathfrak{u}(n) \oplus \mathrm{i}\mathfrak{u}(n)$. This shows that $\mathfrak{u}(n)$ is a real form of $\mathfrak{gl}(n, \mathbf{C})$ and $\mathrm{U}(n)$ is a (compact) real form of $\mathrm{GL}(n, \mathbf{C})$.

(ii) The complexification of $\mathrm{O}(p, q; \mathbf{R})$ is $\mathrm{O}(p, q; \mathbf{C})$. Let $J_{p,q} = \mathrm{diag}(\mathbf{1}_p, \mathrm{i}\mathbf{1}_q)$. Then $A \mapsto J_{p,q}^{-1} A J_{p,q}$ conjugates $\mathrm{O}(p, q; \mathbf{C})$ to $\mathrm{O}(p + q; \mathbf{C})$ and so both $\mathrm{O}(p, q; \mathbf{R})$ and the compact real group $\mathrm{O}(p + q; \mathbf{R})$ are real forms of $\mathrm{O}(p + q; \mathbf{C})$. This shows that a complex Lie group may have essentially different real forms, some compact, some noncompact.

(iii) The group $\mathrm{Sp}(n, \mathbf{R})$ has $\mathrm{Sp}(n, \mathbf{C})$ as its complexification. Its maximal compact subgroup

$$\mathrm{Sp}(n) = \mathrm{Sp}(n, \mathbf{C}) \cap \mathrm{U}(2n)$$

is a compact real form of $\mathrm{Sp}(n, \mathbf{C})$ in the same way as $\mathrm{U}(n)$ is a compact real form of $\mathrm{GL}(n, \mathbf{C})$.

Reductive Matrix Groups

We have just given examples (see Examples D.3.11) of complex Lie groups G that have several real forms. Any choice G_0 of a real form determines a complex conjugation $\sigma : G \to G$. An important class of groups have a second type of involution with the special property that its fixed point locus is a maximal compact subgroup. These play a major role in this book. The basic example from which all others are derived is the involution

$$\Theta : \mathrm{GL}(n, \mathbf{C}) \longrightarrow \mathrm{GL}(n, \mathbf{C}),$$
$$x \longmapsto (x^*)^{-1}.$$

Indeed, the fixed point locus is $\mathrm{U}(n)$, a maximal compact connected subgroup of $\mathrm{GL}(n, \mathbf{C})$. The real group involution $\Theta_0 \overset{\mathrm{def}}{=} \Theta|_{\mathrm{GL}(n,\mathbf{R})}$ is an example of a so-called Cartan involution.

Definition D.3.12 An involutive group isomorphism Θ_0 on a connected $\mathbf{R}$-algebraic group G_0 with the property that the fixed point set of $\Theta_0 \circ \sigma : G_0(\mathbf{C}) \to$

$G_0(\mathbf{C})$ is compact, is called a *Cartan involution* for G_0. The induced Cartan involution on $\mathrm{Lie}(G_0)$ is denoted θ. It induces the *Cartan decomposition*

$$\begin{array}{ccccc} \mathrm{Lie}(G_0) & = & \mathfrak{k} & \oplus & \mathfrak{p} \\ & & \circlearrowleft & & \circlearrowleft \\ & & \theta = \mathrm{id} & & \theta = -\,\mathrm{id}\,. \end{array}$$

If K is the fixed point set of Θ, the group $K_0 = K \cap G_0$ is a maximal compact subgroup of G_0 with Lie algebra $\mathfrak{k}$. Then $\mathfrak{k} \oplus i\mathfrak{p}$ is the Lie algebra of K and K is a compact real form of G_0. The subclass of matrix groups stable by the involution Θ is familiar.

Proposition D.3.13 (Satake, 1980, Chap I.4) *A connected* $\mathbf{R}$*-algebraic group* G_0 *is reductive if and only if one of the following equivalent conditions hold:*

(i) G_0 *can be realized as a matrix group left stable under* Θ*;*

(ii) G_0 *admits a Cartan involution. If this is the case, any two Cartan involutions are conjugate under the action of* G_0 *and the fixed locus of any Cartan involution is a maximal compact subgroup of* G_0*. Moreover, with* B *the Killing form and* θ *the induced involution on* $\mathrm{Lie}(G_0)$*, the bilinear form given by*

$$B_\theta(u, v) \stackrel{\mathrm{def}}{=} -\, B(u, \theta v)$$

is symmetric and positive definite on $\mathrm{Lie}(G_0)$ *modulo its center.*

Bibliographical and Historical Remarks

This material is standard and can be found for instance in Borel (1997); Humphreys (1981) and Knapp (2005).

References

Abdulali, S. 1994. Conjugates of strongly equivariant maps. *Pac. J. Math.*, **165**, 207–216.

Accola, R. 1979. On Castelnuovo's inequality for algebraic curves, I. *Trans. Am. Math. Soc.*, **251**, 357–373.

Addington, S. 1987. Equivariant holomorphic maps of symmetric domains. *Duke Math. Journal*, **55**, 65–88.

Allcock, D., J. Carlson, and D. Toledo. 2002. The complex hyperbolic geometry of the moduli space of cubic surfaces. *J. Algebraic Geom.*, **11**, 659–724.

Allcock, D., J. Carlson, and D. Toledo. 2011. *The Moduli Space of Cubic Threefolds as a Ball Quotient*. Mem. Amer. Math. Soc., **209**, A.M.S., Providence R.I.

Amerik, E. 1998. On a problem of Noether–Lefschetz type. *Comp. Math.*, **112**, 255–271.

del Angel, P. L. and S. Müller-Stach. 2002. The transcendental part of the regulator map for K_1 on a mirror family of K3 surfaces. *Duke Math. J.*, **112**, 581–598.

André, Y. 1989. *G-functions and Geometry*. Vieweg Verlag, Braunschweig.

André, Y. 1992. Mumford-Tate groups of mixed Hodge structures and the theorem of the fixed part. *Compositio Math.*, **82**, 1–24.

Arakelov, A. 1971. Families of algebraic curves with fixed degeneracies. *Izv. Akad. Nauk. SSSR, Ser. Math*, **35**, 1277–1302.

Asakura, M. 1999. Arithmetic Hodge structures and higher Abel–Jacobi maps. ArXiv;math/990819[math.AG].

Asakura, M. and S. Saito. 1998. Filtrations on Chow groups and higher Abel–Jacobi maps. Preprint.

Ash, A., D. Mumford, M. Rapoport, and Y. Tai. 1975. *Smooth Compactification of Locally Symmetric Varieties.* Math. Sci. Press, Boston. Second edition (2010) Cambridge University Press, Cambridge.

Atiyah, M. F. 1957. Complex analytic connections in fibre bundles. *Trans. Am. Math. Soc.*, **85**, 715–726.

Atiyah, M. F. and F. Hirzebruch. 1962. Analytic cycles on complex manifolds. *Topology*, **1**, 25–45.

Atiyah, M. F. and W. V. D. Hodge. 1955. Integrals of the second kind on an algebraic variety. *Ann. Math.*, **62**, 56–91.

Baily, W. and A. Borel. 1966. Compactification of arithmetic quotients of bounded symmetric domains. *Ann. Math,*, **84**, 442–528.

Bardelli, F. 1989. Curves of genus three on a general abelian threefold and the nonfinite generation of the Griffiths group. *Arithmetic of Complex Manifolds, Erlangen 1988*, Lect. Notes in Maths, **1399**, Springer-Verlag, Berlin, 10–26.

Bardelli, F. and S. Müller-Stach. 1994. Algebraic cycles on certain Calabi–Yau 3-folds. *Math. Z.*, **215**, 569–582.

Barlet, D. 1975. Espace analytique réduit des cycles analytiques compacts d'un espace analytique complexe de dimension finie. *Sém. Francois Norguet*, Springer Lect. Notes Math., **482**, Springer-Verlag, Berlin, 1–158.

Barth, W., K. Hulek, C. Peters, and A. Van de Ven. 1993. *Compact Complex Surfaces* (Second enlarged edition). Ergebnisse der Math., **3**, Springer-Verlag, New York, Berlin.

Beauville, A. 1982. Les familles stables de courbes elliptiques sur $\mathbf{P}^1$ admettant quatre fibres singulières. *C. R. Acad. Sci. Paris, Ser I Math.*, **294**, 657–660.

Beauville, A. 1983. Variétés kähleriennes dont la première classe de Chern est nulle. *J. Diff. Geom.*, **18**, 755–782.

Beauville, A. 1986a. Sur l'anneau de Chow d'une variété abélienne. *Math. Ann.*, **273**, 647–651.

Beauville, A. 1986b. Le groupe de monodromie d'hypersurfaces et d'intersections complètes. Springer Lect. Notes Math., **1194**, Springer-Verlag, Berlin, 1–18.

Bedulev, E. and E. Viehweg, 2000. On the Shafarevich conjecture for surfaces of general type over function fields. *Invent. Math.*, **139**, 603–615.

Bertin, J. and C. Peters. 2002. Variations of Hodge structure, Calabi–Yau manifolds and mirror symmetry. *Introduction to Hodge Theory* (translation of *Introduction à la Théorie de Hodge*), Am. Math. Soc., SMF/AMS Texts and Monographs, **8**.

Bingener, J. 1987. *Lokale Modulräume in der analytischen Geometrie I, II*. Aspects of Mathematics, **1-2**, Vieweg Verlag, Braunsweig.

Bloch, S. 1979. Some elementary theorems about algebraic cycles on Abelian varieties. *Invent. Math.*, **37**, 215–228.

Bloch, S. 1980. *Lectures on Algebraic Cycles*. Duke University Press. Second edition (2010), New Mathematical Monographs, **16**, Cambridge University Press, Cambridge.

Bloch, S. 1986. Algebraic cycles and higher K-theory. *Adv. Math.*, **61**, 267– 304.

Borel, A. 1969. *Introduction aux Groupes Arithmétiques*. Pub. Inst. Math. University Strasbourg, **XV**, Act. Scient. et Industr. **1341**, Hermann, Paris.

Borel, A. 1997. *Linear Algebraic Groups*. Graduate Texts in Mathematics, **126**, 2nd enlarged edition, Springer-Verlag, New York, Berlin, Heidelberg.

Borel, A. and Harish-Chandra. 1962. Arithmetic Subgroups of Algebraic Groups. *Ann. Math.*, **75**, 485–535.

Borel, A. and T. Springer. 1968. Rationality properties of linear algebraic groups II. *Tôhoku Math. J.*, **20**, 443–497.

Bott, R. 1957. Homogeneous vector bundles. *Ann. Math.*, **66**, 203–248.

Bott, R. and L. W. Tu. 1982. *Differential Forms in Algebraic Topology*. Springer-Verlag, New York, Berlin.

Braun, R. and S. Müller-Stach. 1996. Effective bounds for Nori's connectivity theorem. *Higher Dimensional Varieties, Proc. Int. Conf. in Trento*, de Gruyter, 83–88.

Bryant, R. L. 1985. Lie groups and twistor spaces. *Duke Math. J.*, **52**, 223–261.

Burns, D. and M. Rapoport. 1975. On the Torelli problem for Kählerian K-3 surfaces. *Ann. Sci. École Norm. Sup.* (4), **8**, 235–273.

Calabi, E. 1967. Minimal immersions of surfaces in Euclidean spheres. *J. Diff. Geom.*, **1**, 11–125.

Carlson, J. 1980. Extensions of mixed Hodge structures. *Journées de Géométrie Algébriques d'Angers 1979*, Sijthoff-Noordhoff, Alphen a/d Rijn, 107–127.

Carlson, J. 1986. Bounds on the dimension of a variation of Hodge structure. *Trans. AMS.*, **294**, 45–64.

Carlson, J., M. Green, P. Griffiths, and J. Harris. 1983. Infinitesimal variations of Hodge structure I. *Comp. Math.*, **50**, 109–205.

Carlson, J., A. Kasparian, and D. Toledo. 1989. Variation of Hodge structure of maximal dimension. *Duke Math. J.*, **58**, 669–694.

Carlson, J. and P. Griffiths. 1980. Infinitesimal variations of Hodge structure and the global Torelli problem. *Journées de Géométrie Algébriques d'Angers 1979*, Sijthoff-Noordhoff, Alphen a/d Rijn, 51–76.

Carlson, J. and C. Simpson. 1987. Shimura varieties of weight two Hodge structures. *Hodge theory, Proceedings, Sant Cugat, Spain 1985*, Springer Lect. Notes in Math., **1246**, 1–15.

Carlson, J. and D. Toledo. 1989a. Variations of Hodge structure, Legendre submanifolds and accessibility. *Trans. Am. Math. Soc.*, **312**, 319–412.

Carlson, J. and D. Toledo. 1989b. Harmonic mappings of Kähler manifolds to locally symmetric spaces. *Publ. Math. IHES*, **69**, 173–201.

Carlson, J. and D. Toledo. 1993. Rigidity of harmonic maps of maximum rank. *J. Geom. Anal.*, **3**, 99–140.

Carlson, J. and D. Toledo. 1999. Discriminant complements and kernels of monodromy representations. *Duke J. Math.*, **97**, 621–648.

Carlson, J. and D. Toledo. 2014. Compact quotients of non-classical domains are not Kähler. *Hodge Theory, Complex Geometry, and Representation Theory*, Contemp. Math., **608**, Amer. Math. Soc., Providence, RI, 1–57.

Cartan, H. 1957. Quotients d'un espace analytique par un groupe d'automorphismes. *Algebraic Geometry and Topology*, Princeton University Press, Princeton, NJ. 165–180.

Catanese, F. 1979. Surfaces with $K^2 = p_g = 1$ and their period mapping. *Algebraic Geometry, Proc. Summer Meeting Copenhagen 1978*, Springer Lect. Notes Math., **732**, 1–29.

Catanese, F. 1980. The moduli and the global period mapping of surfaces with $K^2 = p_g = 1$: A counterexample to the global Torelli problem. *Comp. Math.*, **41**, 401–414.

Cattani, E., P. Deligne, and A. Kaplan. 1995. On the locus of Hodge classes. *J. Am. Math. Soc.*, **8**, 483–506.

Ceresa, G. 1983. *C* is not equivalent to C^- in its Jacobian. *Ann. of Math*, **117**, 285–291.

Chakiris, K. 1980. Counterexamples to global Torelli for certain simply connected surfaces. *Bull. Am. Math. Soc.*, **2**, 297–299.

Chen, K., X. Lu, and K. Zuo. 2015. On the Oort conjecture for Shimura varieties of unitary and orthogonal types. `arxiv.org/abs/1410.5739`.

Chern, S. 1967. *Complex Manifolds Without Potential Theory*. Van Nostrand Math. Studies, **15**, Princeton.

Chowla, S. and A. Selberg. 1949. On Epstein's zeta function (I). *Proc. N.A.S*, **35**, 371–374.

Ciliberto, C. and A. Lopez. 1991. On the existence of components of the Noether–Lefschetz locus with given codimension. *Manuscr. Math.*, **73**, 341–357.

Ciliberto, C., J. Harris, and R. Miranda. 1988. General components of the Noether–Lefschetz locus and their density in the space of all surfaces. *Math. Ann.*, **282**, 667–680.

Clemens, C. H. 1977. Degenerations of Kaehler manifolds. *Duke Math. J.*, **44**, 215–290.

Clemens, C. H. 1980. *A Scrapbook of Complex Curve Theory*. Plenum University Press, New York, London.

Clemens, C. H. and P. Griffiths. 1972. The intermediate jacobian of the cubic threefold. *Ann. Math.*, **95**, 218–356.

Coleman, R. 1987. Torsion points on curves. *Galois Representations and Arithmetic Algebraic Geometry (Kyoto 1985/Tokyo 1986)*, Adv. Math. Stud. Pure Math., **12**, North-Holland, Amsterdam, 235–247.

Collino, A. 1997. Griffiths' infinitesimal invariant and higher K-theory on hyperelliptic Jacobians. *J. Alg. Geom.*, **6**, 393–415.

Collino, A. and G. P. Pirola. 1995. Griffiths infinitesimal invariant for a curve in its Jacobian. *Duke Math. J.*, **78**, 59–88.

Corlette, K. 1988. Flat G-bundles with canonical metrics. *J. Differential Geom.*, **28**, 361–382.

Cox, D. 1990. The Noether–Lefschetz locus of regular elliptic surfaces with section and $p_g \geq 2$. *Am. J. Math.*, **112**, 289–329.

Cox, D. and R. Donagi. 1986. On the failure of variational Torelli for regular elliptic surfaces with a section. *Math. Ann.*, **273**, 673–683.

Cox, D., R. Donagi, and L. W. Tu. 1987. Variational Torelli implies generic Torelli. *Inv. Math.*, **88**, 439–446.

Debarre, O. and Y. Laszlo. 1990. Le lieu de Noether–Lefschetz pour les variétés abéliennes. *C. R. Acad. Sci. Paris, Ser. I*, **311**, 337–340.

Deligne, P. 1970a. *Équations Différentielles à Points Singuliers Réguliers*. Springer Lect. Notes Math., **163** , Springer-Verlag, Berlin.

Deligne, P. 1970b. *Travaux de Griffiths*. Exp. 376, Sém., Bourbaki (Juin 1970), Springer Lecture Notes Math., **180**, Springer-Verlag, Berlin.

Deligne, P. 1971a. Théorie de Hodge I. *Actes du Congrès International des Mathématiciens, (Nice, 1970), Tome 1*, 425–430. Gauthier-Villars, Paris.

Deligne, P. 1971b. Théorie de Hodge II. *Publ. Math. IHES*, **40**, 5–57.

Deligne, P. 1971c. *Travaux de Shimura*. Séminaire Bourbaki (1970/71), Exp. 389 123–165, Springer Lecture Notes in Math., **244**, Springer, Berlin.

Deligne, P. 1974. Théorie de Hodge III. *Publ. Math. IHES*, **44**, 5–77.

Deligne, P. 1979. Variétés de Shimura: interprétation modulaire, et techniques de construction de modèles canoniques. *Automorphic Forms, Representations and L-functions (Oregon State University, Corvallis, Oregon, 1977), Part 2* , Proc. Sympos. Pure Math., **XXXIII**, Amer. Math. Soc., Providence, R.I., 247–289.

Deligne, P. and A. Dimca. 1990. Filtrations de Hodge et par l'ordre du pôle pour les hypersurfaces singulières. *Ann. Sci. E.N.S.*, **23** , 645–656.

Deligne, P. and L. Illusie. 1987. Relèvements modulo p^2 et décomposition du complexe de De Rham. *Inv. Math.*, **89**, 247–270.

Deligne, P., J. S. Milne, A. Ogus, and K. Shih. 1982. *Hodge Cycles, Motives and Shimura Varieties*. Springer Lecture Notes in Math. **900**, Springer-Verlag, Berlin.

Deninger, Ch. and J. Murre. 1991. Motivic decomposition of abelian schemes and the Fourier transform. *J. reine u. angew. Math.*, **422**, 201–219.

Dieudonné, J. A. and J. B. Carrell.1971. *Invariant Theory, Old and New*. Academic Press, New York.

Donagi, R. 1983. Generic Torelli for projective hypersurfaces. *Comp. Math.*, **50**, 325–353.

Donagi, R. and M. Green. 1984. A new proof of the symmetrizer lemma and a stronger weak Torelli theorem for projective hypersurfaces. *J. Diff. Geom*, **20**, 459–461.

Donagi, R. and L. W. Tu. 1987. Generic Torelli for weighted hypersurfaces. *Math. Ann.*, **276**, 399–413.

Donaldson, S. K. 2011. *Riemann Surfaces*. Oxford Graduate Texts in Mathematics **12**, Oxford University Press, Oxford.

Donaldson, S. K. and P. B. Kronheimer. 1990. *The Geometry of Four-Manifolds*. Oxford Mathematical Monographs, Oxford University Press, Oxford.

Donin, I. F. 1978. On deformations of non-compact complex spaces. *Russ. Math. Surv.*, **33**, 181–182.

Douady, A. 1984/85. Le problème de modules pour les variétés analytiques complexes. *Sém. Bourbaki, Exp.* **277**.

Edixhoven, B. and A. Yafaev. 2003. Subvarieties of Shimura varieties. *Ann. of Math.*, **157**, 621–645.

Eells, J. and J. H. Sampson. 1964. Harmonic mappings of Riemannian manifolds. *Am. J. Math.*, **86**, 109–160.

Ehresmann, C. 1947. Sur les espaces fibrés différentiables. *C. R. Acad. Sci. Paris*, **224**, 1611–1612.

Ein, L. 1985. An analogue of Max Noether's theorem. *Duke Math. J.*, **52**, 689–706.

Eisenbud, D. 1994. *Commutative Algebra with a View Toward Algebraic Geometry*. Springer-Verlag, New York, Berlin.

Enriques, F. 1914. Sulla classificazione delle superficie algebriche e particolarmente sulle superficie di genere $p^1 = 1$ (2 notes). *Atti Acc. Lincei V Ser.*, **23**^{1}.

Esnault, H. and E. Viehweg. 1988. Deligne-Beilinson cohomology. *Special Values of L-Functions, Arbeitsgemeinschaft Oberwolfach 1986*, Perspectives in Math., **4**, Academic Press, Boston, MA, 43–91.

Eyssidieux, Ph. 1997. La caractéristique d'Euler du complexe de Gauss-Manin. *J. reine u. angew. Math.*, **490**, 155–212.

Fakhruddin, N. 1996. Algebraic cycles on generic abelian varieties. *Comp. Math.*, **100**, 101–119.

Faltings, G. 1983. Arakelov's theorem for abelian varieties. *Inv. Math.*, **73**, 337–348.

Flenner, H. 1986. The infinitesimal Torelli problem for zero sections of vector bundles. *Math. Z.*, **193**, 307–322.

Forster, O. 1981. *Lectures on Riemann Surfaces*. Springer-Verlag, New York, Berlin.

Fujita T. 1978. On Kähler fiber spaces over curves. *J. Math. Soc. Japan*, **30**, 779–794.

Fulton, W. and J. Harris. 1991. *Representation Theory, a First Course*. Graduate Texts in Math., **129**, Springer-Verlag, Berlin.

Gerkmann, R., M. Sheng, and K. Zuo. 2007. Disproof of modularity of the moduli space of CY 3-folds of double covers of $\mathbf{P}^3$ ramified along eight planes in general position. arXiv AG/07091051.

Gillet, H. 1984. Deligne homology and Abel–Jacobi maps. *Bull. Am. Math. Soc.*, **10**, 285–288.

Godement, R. 1964. *Théorie des Faisceaux*. Hermann, Paris.

Gordon, B. and J. Lewis. 1998. Indecomposable higher Chow cycles on products of elliptic curves. *J. Alg. Geom.*, **8**, 543–567.

Grauert, H. 1962. Über Modifikationen und exzeptionelle analytische Mengen. *Math. Ann.*, **146**, 331–368.

Grauert, H. 1972. Über Deformationen isolierter Singularitäten analytischer Mengen. *Inv. Math.*, **15**, 171–198.

Grauert, H. 1974. Der Satz von Kuranishi für kompakte komplexe Raüme. *Inv. Math.*, **25**, 107–142.

Grauert, H. and R. Remmert. 1977. *Theorie der Steinschen Räume*. Grundl. Math., **227**, Springer-Verlag, Berlin, Heidelberg.

Green, M. 1984a. Koszul cohomology and the geometry of projective varieties. Appendix: The nonvanishing of certain Koszul cohomology groups (by Mark Green and Robert Lazarsfeld). *J. Diff. Geom.*, **19**, 125–167; 168–171.

Green, M. 1984b. The period map for hypersurface sections of high degree of an arbitrary variety. *Comp. Math.*, **55**, 135–171.

Green, M. 1988. A new proof of the explicit Noether–Lefschetz theorem. *J. Diff. Geom.*, **27**, 155–159.

Green, M. 1989a. Components of maximal dimension in the Noether–Lefschetz locus. *J. Diff. Geom.*, **29**, 295–302.

Green, M. 1989b. Griffiths' infinitesimal invariant and the Abel-Jacobi map. *J. Diff. Geom.*, **29**, 545–555.

Green, M. 1996. What comes after the Abel–Jacobi map? Unpublished manuscript.

Green, M. 1998. Higher Abel–Jacobi maps, Invited Lecture at the International Congress Berlin 1998. *Documenta Math*, **II**, 267–276.

Green, M. and S. Müller-Stach. 1996. Algebraic cycles on a general complete intersection of high multidegree of a smooth projective variety. *Comp. Math.*, **100**, 305–309.

Green, M., P. A. Griffiths, and M. Kerr. 2012. *Mumford-Tate Groups and Domains: Their Geometry and Arithmetic*. Princeton University Press, Princeton, NJ.

Green, M., J. Murre, and C. Voisin. 1994. *Algebraic Cycles and Hodge theory* (ed. A. Albano and F. Bardelli), Springer Lect. Notes Math., **1594**.

Greenberg, M. J. 1967. *Lectures on Algebraic Topology*. W. A. Benjamin, Reading, MA.

Griffiths, P. 1968. Periods of integrals on algebraic manifolds, I, II. *Am. J. Math.*, **90**, 568–626; 805–865.

Griffiths, P. 1969. On the periods of certain rational integrals I, II. *Ann. Math.*, **90**, 460–495; 498–541.

Griffiths, P. 1970. Periods of integrals on algebraic manifolds, III. *Publ. Math. IHÉS*, **38**, 125–180.

Griffiths, P. 1983. Infinitesimal variations of Hodge structure (III): determinantal varieties and the infinitesimal invariant of normal functions. *Comp. Math.*, **50**, 267–324.

Griffiths, P. and J. Harris. 1978. *Principles of Algebraic Geometry*. Wiley, New York.

Griffiths, P. and J. Harris. 1983. Infinitesimal variations of Hodge structure (II): an infinitesimal invariant of Hodge classes. *Comp. Math.*, **50**, 207–266.

Griffiths, P. and J. Harris. 1985. On the Noether–Lefschetz theorem and some remarks on codimension-two cycles. *Math. Ann.*, **271**, 31–51.

Griffiths, P. and W. Schmid. 1969. Locally homogenous complex manifolds. *Acta Math.*, **123**, 145–166.

Griffiths, P., C. Robles, and D. Toledo. 2015. Quotients of non-classical flag domains are not algebraic. `http://arxiv.org/abs/1303.0252`.

Gross, M., D. Huybrechts, and D. Joyce. 2003. *Calabi–Yau Manifolds and Related Geometries (Lectures from the Summer School held in Nordfjordeid, June 2001.)*. Springer-Verlag, Berlin, Heidelberg.

Grothendieck, A. 1958. La théorie des classes de Chern. *Bull. Soc. Math., France*, **86**, 137–154.

Guillemin, V. and A. Pollack. 1974. *Differential Topology*. Prentice-Hall, Englewood Cliffs, NJ.

Gunning, R. C. and H. Rossi. 1965. *Analytic Functions of Several Complex Variables*. Prentice-Hall, Englewood Cliffs, NJ.

Harris, J. 1992. *Algebraic Geometry, a First Course*. Grad. Texts in Math, **133**, Springer-Verlag, Berlin, Heidelberg.

Hartshorne, R. 1975. Equivalence relations on algebraic cycles and subvarieties of small codimension. *Proc. Symp. Pure Math.*, **29**, Am. Math. Soc., Providence, RI. 129–164.

Hartshorne, R. 1987. *Algebraic Geometry*. Springer-Verlag, Berlin, Heidelberg.

Hatcher, A. 2002. *Algebraic Topology*. Cambridge University Press, Cambridge.

Helgason, S. 1962. *Differential Geometry and Symmetric Spaces*. Academic Press, New York, London.

Helgason, S. 1978. *Differential Geometry, Lie Groups and Symmetric Spaces*. Academic Press, New York, London.

Hironaka, H. 1964. Resolution of singularities of an algebraic variety of characteristic zero. *Ann. Math.*, **79**, 109–326.

Hirzebruch, F. 1966. *Topological Methods in Algebraic Geometry.* Grundl. Math. Wiss., **131** (dritte Ausgabe), Springer-Verlag, New York, Berlin.

Hirzebruch, F. 1976. Hilbert's modular group of the field $\mathbb{Q}(\sqrt{5})$ and the cubic diagonal surface of Clebsch and Klein. *Russ. Math. Surv.*, **31** , 96–110.

Hitchin, N. 1987. The self-duality equations on a Riemann surface. *Proc. London Math. Soc.*, **55**, 59–126.

Hodge, W. V. D. 1952. The topological invariants of algebraic varieties. *Proc. Int. Congr. Math. (Cambridge 1950).*, Am. Math. Soc., Providence, RI, 182–192.

Hodge, W. V. D. 1959. *The Theory and Applications of Harmonic Integrals*. Cambridge University Press, Cambridge, UK.

Holzapfel, R.-P. 1986a. Chern numbers of algebraic surfaces – Hirzebruch's examples are Picard modular surfaces. *Math. Nachrichten*, **126**, 255–273.

Holzapfel, R.-P. 1986b. *Geometry and Arithmetic Around Euler Partial Differential Equations*. Reidel Publishing Company, Kluwer, Dordrecht.

Horikawa, E. 1976. Algebraic surfaces of general type with small c_1^2, I. *Ann. Math.*, **104**, 357–387.

Huber, A. and S. Müller-Stach. 2017. *Periods and Nori Motives*. Ergebnisse Series 3rd series **65**, Springer-Verlag, Berlin.

Humphreys, J. 1981. *Linear Algebraic Groups*. Graduate Texts in Mathematics, **21**, Springer-Verlag, Berlin, Heidelberg, New York.

Husemoller, D. 1966. *Fibre Bundles*. MacGraw-Hill, New York.

Huybrechts, D. 1999. Compact hyperkähler manifolds: basic results. *Invent. Math.*, **135**, 63–113.

Huybrechts, D. 2012. A global Torelli theorem for hyperkähler manifolds (after Verbitsky). Séminaire Bourbaki: Vol. 2010/2011. Exposés 1027–1042, *Astérisque*, **348**, 375-403. See also `arxiv:1106.5573`.

Huybrechts, D. 2016. *Lectures on K3 Surfaces*. Cambridge University Press.

Ihara, S-i. 1967. Holomorphic imbeddings of symmetric domains. *J. Math. Soc. Japan*, **19**, 261–302.

Ince, E. L. 1956. *Ordinary Differential Equations*. Dover Publ., New York.

Ivinskis, K. 1993. A variational Torelli theorem for cyclic coverings of high degree. *Comp. Math.*, **85**, 201–228.

Jannsen, U. 2000. Equivalence relations on algebraic cycles. *The Arithmetic and Geometry of Algebraic Cycles Proc. Banff 1998*, Kluwer Acad. Publ., Dordrecht, Ser. C **548**, 225–260.

Joshi, K. 1995. A Noether–Lefschetz theorem and applications. *J. Alg. Geom.*, **4**, 105–135.

Jost, J. and K. Zuo. 2002. Arakelov type inequalities for Hodge bundles over algebraic varieties, Part I: Hodge bundles over algebraic curves with unipotent monodromies around singularities. *J. Alg. Geom.*, **11**, 535–546.

van Kampen, E. R. 1933. On the fundamental group of an algebraic curve. *Am. J. Math.*, **55**, 255–260.

Katz, N. and T. Oda. 1968. On the differentiation of de Rham cohomology classes with respect to parameters. *J. Math. Kyoto University*, **8**, 199–213.

Kempf, G., F. Knudsen, D. Mumford, and B. Saint-Donat. 1973. *Toroidal Embeddings I*. Springer Lect. Notes, **336**, Springer-Verlag, Berlin.

Kiĭ, K. 1973. A local Torelli theorem for cyclic coverings of $\mathbf{P}^n$ with positive class. *Math. Sb.*, **92**, 142–151.

Kiĭ, K. 1978. The local Torelli theorem for varieties with divisible canonical class. *Math. USSR Izv.*, **12**, 53–67.

Kim, Sung-Ock. 1991. Noether–Lefschetz locus for surfaces. *Trans. Am. Math. Soc.*, **324**, 369–384.

Klingler, B. and A. Yafaev. 2014. The André–Oort conjecture. *Ann. Math.*, **180**, 867–925.

Knapp, A. 2005. *Lie Groups Beyond an Introduction* (second edition). Progress in Maths., **140**, Birkhäuser, Boston, Basel, Berlin.

Kobayashi, S. 1987. *Differential Geometry of Complex Vector Bundles*. Princeton University Press, Princeton, NJ.

Kodaira, K. and D. Spencer. 1953. Divisor class groups on algebraic varieties. *Proc. Natl. Acad. Sci. USA*, **39**, 872–877.

Kodaira, K. and D. Spencer. 1962. A theorem of completeness for complex analytic fibre spaces. *Acta Math.*, **100**, 43–76.

Kodaira, K., L. Nirenberg, and D. Spencer. 1958. On the existence of deformations of complex structures. *Ann. Math.*, **90**, 450–459.

Konno, K. 1985. On deformations and the local Torelli problem for cyclic branched coverings. *Math. Ann.*, **271**, 601–618.

Konno, K. 1986. Infinitesimal Torelli theorems for complete intersections in certain homogeneous Kähler manifolds. *Tôhoku Math. J.*, **38**, 609–624.

Konno, K. 1991. On the variational Torelli problem for complete intersections. *Comp. Math.*, **78**, 271–296.

Koszul, J.-L. and B. Malgrange. 1958. Sur certaines structures fibrées complexes. *Arch. Math.*, **9**, 102–109.

Kronecker, L. 1857. Zwei Sätze über Gleichungen mit ganzzahligen Coeffizienten. *J. reine u. angew. Math.*, **53**, 173–175.

Künnemann, K. 1994. On the Chow motive of an abelian scheme. *Motives Seattle WA 1991*, Proc. Symp. Pure Math., **55** (1), Amer. Math. Soc., Providence RI, 189–205.

Kuranishi, M. 1965. New proof for the existence of locally complete families of complex analytic structures. *Proc. Conf. Complex Analysis, Minneapolis, 1964*, Springer-Verlag New York, Berlin, 142–154.

Kynev, V. 1977. An example of a simply connected surface for which the local Torelli theorem does not hold (Russian). *C.R.Ac. Bulg. Sc.*, **30**, 323–325.

Lamotke, K. 1981. The topology of complex projective varieties after S. Lefschetz. *Topology*, **20**, 15–52.

Landman, A. 1973. On the Picard–Lefschetz transformation for algebraic manifolds acquiring general singularities. *Trans. Am. Math. Soc.*, **181**, 89–126.

Lang, S. 1985. *Introduction to Differentiable Manifolds* (second edition 2002). Springer-Verlag, New York, Berlin.

Lawson, H. B., Jr. 1980. *Lectures on Minimal Submanifolds*, Volume I. Mathematics Lecture Series, **9**, Publish or Perish, Inc.

Lefschetz, S. 1924. *L'Analysis Situs et la Géométrie Algébrique*. Gauthiers-Villars Paris.

Lelong, P. 1957. Intégration sur un ensemble analytique complexe. *Bull. Soc. Math. France*, **85**, 239–262.

Leray J. 1959. Le calcul différentiel et intégral sur une variété analytique complexe (Problème de Cauchy III). *Bull. Soc. Math. France*, **87**, 81–180.

Levine, M. 1998. *Mixed Motives*. Mathematical Surveys and Monographs, **57**, AMS, Providence, RI.

Lewis J. D. 1991. *A Survey of the Hodge Conjecture*. Publications C.R.M., Montreal.

Libgober, A. and I. Dolgachev. 1981. On the fundamental group of the complement to a discriminant variety. *Algebraic Geometry,* Springer Lect. Notes Math., **862**, 1–25.

Lieberman, D. 1978. Compactness of the Chow scheme: applications to automorphisms and deformations of Kähler manifolds. *Sém. Norguet 1976*, Springer Lect. Notes Math., **670**, (1978), 140–186.

Lieberman, D., C. Peters, and R. Wilsker. 1977. A theorem of Torelli type. *Math. Ann.*, **231**, 39–45.

Liu, K., A. Todorov, S.-T. Yau, and K. Zuo. 2011. Finiteness of subfamilies of Calabi-Yau N-folds over curves with maximal length of Yukawa-coupling. *Pure and Applied Mathematics Quarterly*, **7** (Special Issue: In honor of Eckart Viehweg), 1585–1598.

Looijenga, E. and R. Swierstra. 2007. The period map for cubic threefolds. *Compos. Math.*, **143**, 1037–1049.

Lopez, A. 1991. *Noether–Lefschetz Theory and the Picard group of Projective Surfaces*. Mem. Am. Math. Soc. monograph, **438**.

Lu, X. and K. Zuo. 2014. On Shimura curves in the Torelli locus of curves. arxiv.org/abs/1311.5858.

Lu, X. and K. Zuo. 2015. The Oort conjecture on Shimura curves in the Torelli locus of curves. arxiv.org/abs/1405.4751.

Lu, X., S.-L. Tan, and K. Zuo. 2016. Singular fibers and Kodaira dimensions. arxiv.org/abs/arXiv:1610.07756.

Macaulay, F. S. 1916. *The Algebraic Theory of Modular Systems.* Revised 1994 reprint of the 1916 original, with an introduction by Paul Roberts, Cambridge Mathematical Library, Cambridge University Press, Cambridge.

Malgrange, B. 1974. Intégrales asymptotiques et monodromie. *Ann. Sci. École Norm. Sup., ser. 4,* **7**, 405–430.

Margulis, G. 1991. *Discrete Subgroups of Semisimple Lie Groups.* Ergebnisse der Math. und ihrer Grenzgebiete, **17**, Springer-Verlag, Berlin.

Matsumura, H. 1989. *Commutative Ring Theory.* Cambridge Studies in Advanced Mathematics, **8**, Cambridge University Press, Cambridge.

Mazza, C, V. Voevodsky, and C. Weibel. 2006. *Lecture Notes on Motivic Cohomology.* CMI/AMS, Claymath Publ., **7**.

Miller, A., S. Müller-Stach, S. Wortmann, Y.-H. Yang, and K. Zuo. 2007. Chow-Künneth decomposition for universal families over Picard modular surfaces. *Algebraic Cycles and Motives Part II* (eds. J. Nagel and Ch. Peters), London Math. Soc. Lecture Notes, **344**, Cambridge University Press.

Milne, J. 2004. *Introduction to Shimura Varieties.* http://www.jmilne.org/math/xnotes/svi.pdf

Milnor, J. 1963. *Morse Theory.* Princeton University Press, Princeton, NJ.

Milnor, J. and J. Stasheff. 1974. *Characteristic Classes.* Ann. Math. Studies, **76**, Princeton University Press, Princeton, NJ.

Mohajer, A., S. Müller-Stach, and K. Zuo. 2016. Special subvarieties in Mumford-Tate varieties. arXiv:1410.4654 [math.AG].

Möller, M., E. Viehweg, and K. Zuo. 2006. Special families of curves, of Abelian varieties and of certain minimal manifolds over curves, *Global Aspects of Complex Geometry*, Springer-Verlag, Berlin, 417–450.

Moonen, B. 1998. Linearity properties of Shimura varieties, I. *J. Alg. Geom*, **7**, 539–567.

Moonen, B. 1999. Notes on Mumford–Tate groups www.math.ru.nl/~bmoonen/Lecturenotes/CEBnotesMT.pdf.

Moonen, B. 2016. Notes on Mumford-Tate groups and Galois representations. Forthcoming.

Morrison, D. 1988. On the moduli of Todorov surfaces. *Algebraic Geometry and Commutative Algebra in Honor of Masayoshi Nagata 1987*, Kinokuniya, Tokyo, 313–355.

Mostow, G. and T. Tamagawa. 1962. On the compactness of arithmetically defined homogeneous spaces. *Ann. of Math.*, **76**, 440–463.

Müller-Stach, S. 1992. On the non-triviality of the Griffiths group. *J. reine u. angew. Math.*, **427**, 209–218.

Müller-Stach, S. 1994. Syzygies and the Abel–Jacobi map for cyclic coverings. *Manuscr. Math.*, **82**, 433–443.

Müller-Stach, S. 1997. Constructing indecomposable motivic cohomology classes on algebraic surfaces. *J. Alg. Geom.*, **6**, 513–543.

Müller-Stach, S. and K. Zuo. 2011. A characterization of special subvarieties in orthogonal Shimura varieties, *Pure and Applied Mathematics Quarterly*, **7** (Special Issue: In honor of Eckart Viehweg), 1599–1630.

Müller-Stach, S., C. Peters, and V. Srinivas. 2012a. Abelian varieties and theta functions as invariants for compact Riemannian manifolds; constructions inspired by superstring theory. *J. Math. Pures et Appliquées*, **97**.

Müller-Stach, S., M. Sheng, X. Ye, and K. Zuo. 2011. On the cohomology groups of local systems over Hilbert modular varieties via Higgs bundles. `math.AG 1009.2011`.

Müller-Stach, S., E. Viehweg, and K. Zuo. 2009. Relative proportionality for subvarieties of moduli spaces of K3 and abelian surfaces. *Pure and Applied Mathematics Quarterly*, **5** (Special Issue: In honor of Friedrich Hirzebruch), 1161–1199.

Müller-Stach, S., X. Ye, and K. Zuo. 2012b. Mixed Hodge complexes and L^2-cohomology for local systems on ball quotients. To appear in *Documenta Mathematica*.

Mumford, D. 1963. *Curves on Algebraic Surfaces*. Princeton University Press, Princeton, NJ.

Mumford, D. 1965. *Geometric Invariant Theory*. Springer-Verlag, Berlin, New York.

Mumford, D. 1966. Families of abelian varieties. *Algebraic Groups and Discontinuous Subgroups (Boulder, Colorado, 1965)*. Proc. Sympos. Pure Math., Amer. Math. Soc., Providence, R.I., 347–351.

Mumford, D. 1969a. A note on Shimura's paper "Discontinuous groups and abelian varieties". *Math. Annalen*, **181**, 345–351.

Mumford, D. 1969b. Rational equivalence of zero-cycles on surfaces. *J. Math. Kyoto University*, **9**, 195–204.

Mumford, D. 1970. Varieties defined by quadratic equations. *Questions on Algebraic Varieties (C.I.M.E., III Ciclo, Varenna, 1969)*, Edizioni Cremonese, Rome, 29–100.

Mumford, D. 1976. *Algebraic Geometry I, Complex Projective Varieties*. Grundl. Math. Wissensch., **221**, Springer-Verlag, Berlin, New York.

Murre, J. 1993. On a conjectural filtration on the Chow groups of an algebraic variety. *Indag. Math.*, **4**, 177–188.

Murre, J., J. Nagel, and C. Peters. 2013. *Lectures on the Theory of Pure Motives*. University Lecture Series, **61**, AMS, Providence RI.

Nagel, J. 1997. *The Image of the Abel-Jacobi Map for Complete Intersections*. Proefschrift (Thesis), Rijksuniversiteit Leiden.

Nagel, J. 2002. Effective bounds for Hodge theoretic connectivity. *J. Alg. Geom.*, **11**, 1–32.

Newstead, P. E. 1978. *Introduction to Moduli Problems and Orbit Spaces*. Springer-Verlag, Berlin, New York.

Nori, M. 1989. Cycles on the generic abelian threefold. *Proc. Indian Acad. Sci. Math. Sci*, **99**, 191–196.

Nori, M. 1993. Algebraic cycles and Hodge theoretic connectivity. *Inv. Math.*, **111**, 349–373.

Oguiso, K. and E. Viehweg. 2001. On the isotriviality of families of elliptic surfaces. *J. Alg. Geom.*, **10** , 569–598.

Oort, F. 1997. Canonical liftings and dense sets of CM points. *Arithmetic Geometry (Cortona 1994)*, Symp. Math., **XXXVII**, Cambridge University Press, Cambridge, 228–234.

Palamodov, V. 1976. Deformations of complex spaces. *Russ. Math. Surv.*, **31**, 129–197.

Paranjape, K. 1991. Curves on threefolds with trivial canonical bundle. *Proc. Ind. Acad. Sci. Math. Sci*, **101**, 199–213.

Paranjape, K. 1994. Cohomological and cycle-theoretic connectivity. *Ann. of Math.*, **140**, 349–373.

Peters, C. 1975. The local Torelli theorem: complete intersections. *Math. Ann.*, **217**, 1–16.

Peters, C. 1976a. Erratum to "The local Torelli theorem. I: complete intersections". *Math. Ann.*, **223**, 191–192.

Peters, C. 1976b. The local Torelli theorem II: cyclic branched coverings. *Ann. Sc. Norm. Sup. Pisa Ser. IV*, **3**, 321–339.

Peters, C. 1984. A criterion for flatness of Hodge bundles over curves and geometric applications. *Math. Ann.*, **268**, 1–19.

Peters, C. 1988. Some remarks about Reider's article 'On the infinitesimal Torelli theorem for certain irregular surfaces of general type.' *Math. Ann.*, **282**, 315–324.

Peters, C. 1990. Rigidity for variations of Hodge structure and Arakelov-type finiteness theorems. *Comp. Math.*, **75**, 113–126.

Peters, C. 2000. Arakelov inequalities for Hodge bundles. *Prépub. de l'Inst. Fourier*, **511**.

Peters, C. 2010. Rigidity, past and present. *Proceedings of Teichmueller Theory and Moduli Problems (HRI, India, January 2006)*. Ramanujan Mathematical Society's Lecture Notes Series, **10**, 529–548.

Peters, C. 2016. Inequalities for semi-stable fibrations on surfaces, and their relation to the Coleman-Oort conjecture. Pure and Applied Math. Q. **12**, 1–30. See also `arxiv.org/abs/1405.4531v2`.

Peters, C. and J. Steenbrink. 1983. Infinitesimal variation of Hodge Structure and the generic Torelli problem for projective hypersurfaces. *Classification of Algebraic and Analytic Varieties*, Birkhäuser Verlag, Basel, 399–464.

Peters, C. and J. Steenbrink, 2008. *Mixed Hodge structures*. Ergebnisse der Math., **52**, Springer-Verlag, Berlin.

Piateckii-Shapiro, I. and I. Shafarevich. 1971. A Torelli theorem for algebraic surfaces of type K3. *Izv. Akad. Nauk. SSSR. Ser.Math.*, **35**, 530–572.

Picard, E. 1883. Sur des fonctions de deux variables indépendentes analogues aux fonctions modulaires. *Acta. Math.*, **2**, 114–135.

Pourcin, G. 1974. Déformations de singularités isolées. *Astérique*, **16**, 161–173.

Quillen, D. 1973. Higher algebraic K-theory, I (*Proc. Conf. Battle Inst. 1972*). Lect. Notes Math., **341**, Springer-Verlag, Berlin., 85–147.

Ravi, M. 1993. An effective version of Nori's theorem. *Math. Z.*, **214**, 1–7.

Reider, I. 1988. On the infinitesimal Torelli theorem for certain types of irregular surfaces. *Math. Ann.*, **279**, 285–302.

Roberts, J. 1972. Chow's moving lemma. *Algebraic Geometry Oslo Conf. 1970*, Wolters–Noordhoff, Groningen, 89–96.

Roĭtman, A. A. 1974. Rational equivalence of zero-dimensional cycles. *Math. USSR Sb.*, **18**, 571–588.

Saint-Donat, B. 1975. Variétés de translation et théorème de Torelli. *C.R.Ac. Sci. Paris, Sér. A.*, **280**, 1611–1612.

Saito, Ma. 1983. On the infinitesimal Torelli problem of elliptic surfaces. *J. Math. Kyoto University*, **23**, 441–460.

Saito, Ma. 1986. Weak global Torelli theorem for certain weighted projective hypersurfaces. *Duke Math. J.*, **53**, 67–111.

Saito, Ma. 1993. Classification of non-rigid families of abelian varieties. *Tôhoku Math. J.*, **45**, 159–189.

Saito, Ma. and S. Zucker. 1991. Classification of non-rigid families of K3-surfaces and a finiteness theorem of Arakelov type. *Math. Ann.*, **289**, 1–31.

Saito, Mo. 2001. Arithmetic mixed Hodge structures. *Inv. Math.*, **144**, 533–569.

Saito, S. 1996. Motives and filtrations on Chow groups. *Inv. Math.*, **125**, 149–196.

Sampson, J. H. 1978. Some properties and applications of harmonic mappings. *Ann. Sci. ÉNS*, **11**, 211–228.

Sampson, J. H. 1986. Applications of harmonic maps to Kähler geometry. *Contemp. Math.*, **49**, 125–133.

Satake, I. 1965. Holomorphic embeddings of symmetric domains into a Siegel space. *Amer. J. Math.*, **87**, 425–461.

Satake, I. 1966. Symplectic representations of algebraic groups. *Algebraic Groups and Discontinous Subgroups*, Proc. of Symposia, **IX**, 352–360.

Satake, I. 1967. Symplectic representations of algebraic groups satisfying a certain analyticity condition. *Acta Math.*, **117**, 215–279.

Satake, I. 1980. *Algebraic Structures of Symmetric Domains*. Princeton University Press, Princeton.

Schmid, W. 1973. Variations of Hodge structure: the singularities of the period mapping. *Inv. Math.*, **22**, 211–320.

Schur, I. 1905. Zur Theorie der vertauschbaren Matrizen. *J. reine u. angew. Math.*, **130**, 66–78.

Sebastiani M. and R. Thom. 1955. Un résultat sur la monodromie. *Invent. Math.*, **13**, 90–96.

Serre, J.-P. 1955. Faiscaux algébriques cohérents. *Ann. Math.*, **61**, 197–278.

Serre, J.-P. 1956. Géométrie algébrique et géométrie analytique. *Ann. Inst. Fourier*, **6**, 1–42.

Serre, J.-P. 1979. *A Course in Arithmetic*. Springer-Verlag, Berlin.

Shimura, G. 1963. On analytic families of polarized abelian varieties and automorphic functions. *Ann. of Math.*, **78**, 149–192.

Shimura, G. 1964. On the field of definition for a field of automorphic functions. I, *Ann. of Math.*, **80**, 160–189.

Shimura, G. 1965. On the field of definition for a field of automorphic functions. II. *Ann. of Math.*, **81**, 124–165.

Shimura, G. 1966. On the field of definition for a field of automorphic functions. III. *Ann. of Math.*, **83**, 377–385.

Shioda, T. 1981. On the Picard number of a complex-projective variety. *Ann. Sc. ENS*, **14**, 303–321.

Shioda, T. 1985. A note on a theorem of Griffiths on the Abel–Jacobi map. *Inv. Math.*, **82**, 461–466.

Silverman, J. 1992. *The Arithmetic of Elliptic Curves*. Graduate Texts in Math., **106**, Springer Verlag, Berlin.

Simpson, C. 1990. Harmonic bundles on non-compact curves. *J. Am. Math. Soc.*, **3**, 713–770.

Simpson, C. 1992. Higgs bundles and local systems. *Publ. Math. IHES*, **75**, 5–95.

Simpson, C. 1994. Moduli of representations of the fundamental group of a smooth variety. *Publ. Math. IHES*, **79**, 47–129.

Simpson, C. 1995. Moduli of representations of the fundamental group of a smooth projective variety, II. *Publ. Math. IHES*, **80**, 5–79.

Siu, Y.-T. 1980. Complex analyticity of harmonic maps and strong rigidity of complex Kähler manifolds. *Ann. Math.*, **112**, 73–111.

Siu, Y.-T. 1983. Every K3-surface is Kähler. *Invent. Math.*, **73**, 139–150.

Spandaw, J. 1992. A Noether–Lefschetz theorem for linked surfaces in $\mathbf{P}^4$. *Indag. Math. New Ser.*, **3**, 91–112.

Spandaw, J. 1994. Noether–Lefschetz problems for vector bundles. *Math. Nachr.*, **169**, 287–308.

Spandaw, J. 1996. A Noether–Lefschetz theorem for vector bundles. *Manuscri. Math.*, **89**, 319–323.

Spandaw, J. 2002. *Noether–Lefschetz Theorems for Degeneracy Loci.* Habilitationsschrift, Hannover (2000), *Mem. Am. Math. Soc.,* **161**.

Spanier, E. 1966. *Algebraic Topology*. Springer-Verlag, Berlin, Heidelberg.

Sun, X., S.-L. Tan, and K. Zuo. 2003. Families of K3-surfaces reaching the Arakelov–Yau upper bounds and modularity. *Math. Res. Letters*, **10**, 323–342.

Szabo, E. 1996. Complete intersection subvarieties of general hypersurfaces. *Pac. J. Math.*, **175**, 271–294.

Tian, G. 1986. Smoothness of the universal deformation space of compact Calabi–Yau manifolds and its Petersson–Weil metric. *Mathematical Methods of Theoretical Physics* (S.-T. Yau, ed.), World Scientific, Hong Kong.

Todorov, A. N. 1980. Surfaces of general type with $p_g = 1$ and $(K, K) = 1$. I. *Ann. Sci. École Norm. Sup. (4)*, **13**, (1980), 1–21.

Todorov, A. N. 1981. A construction of surfaces with $p_g = 1, q = 0$ and $2 \leq K^2 \leq 8$. Counterexamples of the global Torelli theorem. *Inv. Math.*, **63** , 287–304.

Torelli, R. 1913. Sulle varietà di Jacobi. *Rend. Accad. Lincei, Cl. Fis. Mat. Nat.*, **22** (5), 98–103.

Tsimerman, J. 2015. A proof of the André–Oort conjecture. `arXiv:1506.01466`.

Tu, L. W. 1983. *Hodge Theory and the Local Torelli Problem*. Mem. Am. Math. Soc., **279**.

Usui, S. 1976. The local Torelli theorem for non-singular complete intersections. *Jap. J. Math.*, **2**, 411–418.

Usui, S. 1982. Torelli theorem for surfaces with $p_g = c_1^2 = 1$ and K ample and with certain type of automorphism. *Comp. Math.*, **45**, 293–314.

Ullmo, E. and A. Yafaev 2014. Galois orbits and equidistribution of special subvarieties: towards the André–Oort conjecture. *Ann. Math.*, **180**, 823–865.

Verbitsky, M. 2013. Mapping class group and global Torelli theorem for hyperkähler manifolds. *Duke Math. J.*, **162**, 2929–2986.

Viehweg, E. 1982. Die Additivität der Kodaira Dimension für projektive Faserraüme über Varietäten des allgemeinen Typs. *J. reine u. angewandte Math*, **330**, 132–142.

Viehweg, E. 1995. *Quasi-Projective Moduli for Polarized Manifolds*. Springer-Verlag, Berlin, New York.

Viehweg, E. 2008. Arakelov (in)equalities. `arXiv:0812.3350v2`.

Viehweg, E. and K. Zuo. 2001. On the isotriviality of families of projective manifolds over curves. *J. Alg. Geom.*, **10**, 781–799.

Viehweg, E. and K. Zuo. 2003a. Discreteness of minimal models of Kodaira dimension zero and subvarieties of moduli stacks. *Surveys in Differential Geometry*, Vol. **VIII**, Int. Press, Somerville, MA, 337–356.

Viehweg, E. and K. Zuo. 2003b. On the Brody hyperbolicity of moduli spaces for canonically polarized manifolds. *Duke Math. J.*, **118**, 103–150.

Viehweg, E. and K. Zuo. 2004. A characterization of certain Shimura curves in the moduli stack of abelian varieties. *J. Differential Geom.*, **66**, 233–287.

Viehweg, E. and K. Zuo. 2005. Complex multiplication, Griffiths–Yukawa couplings and rigidity for families of hypersurfaces. *J. Alg. Geom.*, **14**, 481–528.

Viehweg, E. and K. Zuo. 2006. Numerical bounds for semistable families of curves or of certain higher dimensional manifolds. *J. Alg. Geom.*, **15**, 771–791.

Viehweg, E. and K. Zuo. 2007. Arakelov inequalities and the uniformization of certain rigid Shimura varieties. *J. Diff. Geom.*, **77**, 291–352.

Viehweg, E. and K. Zuo. 2008. Special subvarieties of A_g. *Third International Congress of Chinese Mathematicians. Part 1, 2*, AMS/IP Stud. Adv. Math., **42**, Amer. Math. Soc., Providence, RI, 111–124.

Voevodsky, V., A. Suslin, and E. Friedlander. 2000. *Cycles, Transfers and Motivic Homology Theories*. Ann. Math. Studies, **143**, Princeton University Press, Princeton, NJ.

Voisin, C. 1988a. Une précision concernant le théorème de Noether. *Math. Ann.*, **280**, 605–611.

Voisin, C. 1988b. Une remarque sur l'invariant infinitésimal des fonctions normales. *C. R. Acad. Sci. Paris*, **307**, 157–160.

Voisin, C. 1989a. Sur une conjecture de Griffiths et Harris. *Algebraic Curves and Projective Geometry (Proc. Conf., Trento/Italy 1988)*, Springer Lect. Notes Math., **1389**, 270–275.

Voisin, C. 1989b. Composantes de petite codimension du lieu de Noether–Lefschetz. *Comm. Math. Helv.*, **64**, 515–526.

Voisin, C. 1990. Sur le lieu de Noether–Lefschetz en degrés 6 et 7. *Comp. Math.*, **74**, 47–68.

Voisin, C. 1991. Contrexemple à une conjecture de J. Harris. *C. R. Acad. Sci., Paris, Ser. I*, **313**, 685–687.

Voisin, C. 1992. Une approche infinitésimale du théorème de H. Clemens sur les cycles d'une quintique générale de $\mathbf{P}^4$. *J. Alg. Geom.*, **1**, 157–174.

Voisin, C. 1994a. Sur l'application d'Abel–Jacobi des variétés de Calabi–Yau de dimension trois. *Ann. Sci. ENS*, **27**, 209–226.

Voisin, C. 1994b. Variations de structure de Hodge et zéro-cycles sur les surfaces générales. *Math. Ann.*, **299**, 77–103.

Voisin, C. 1995. Variations of Hodge structures and algebraic cycles. *Proc. Int. Congr. Zürich 1994*, 706–715.

Voisin, C. 1999a. A generic Torelli theorem for the quintic threefold. *New Trends in Algebraic Geometry (Warwick, 1996)*, London Math. Soc. Lecture Note Ser., **264**, 425–463.

Voisin, C. 1999b. Some results on Green's higher Abel–Jacobi map. *Ann. Math.*, **149**, 451–473.

Voisin, C. 2000. The Griffiths group of a general Calabi–Yau threefold is not finitely generated. *Duke Math. J.*, **102**, 151–186.

Warner, F. W. 1983. *Foundations of Differentiable Manifolds and Lie Groups*. Springer-Verlag, Berlin, Heidelberg.

Wehler, J. 1986. Cyclic coverings: deformations and Torelli theorems. *Math. Ann.*, **274**, 443–472.

Wehler, J. 1988. Hypersurfaces of the flag variety: deformation theory and the theorems of Kodaira–Spencer, Torelli, Lefschetz, M. Noether and Serre. *Math. Z.*, **198**, 21–38.

Weil, A. 1958. *Variétés Kähleriennes*. Hermann, Paris.

Wells, R. 1980. *Differential Analysis on Complex Manifolds*. Springer-Verlag, Berlin, Heidelberg.

Wu, Xian. 1990a. On a conjecture of Griffiths and Harris generalizing the Noether–Lefschetz theorem. *Duke Math. J.*, **60**, 465–472.

Wu, Xian. 1990b. On an infinitesimal invariant of normal functions. *Math. Ann.*, **288**, 121–132.

Xu, Geng. 1994. Subvarieties of general hypersurfaces in projective space. *J. Differ. Geom.*, **39**, 139–172.

Yuan, X. and S.-W. Zhang. 2015. On the averaged Colmez conjecture. arXiv:1507.06903.

Zariski, O. 1971. *Algebraic Surfaces* (second supplemented edition). Springer-Verlag, Berlin, Heidelberg.

Index